Deutsche
Forschungsgemeinschaft

# Basic Research and Technologies for Two-Stage-to-Orbit Vehicles

Deutsche
Forschungsgemeinschaft

# Basic Research and Technologies for Two-Stage-to-Orbit Vehicles

## Final Report of the Collaborative Research Centres 253, 255 and 259

Edited by
Dieter Jacob, Gottfried Sachs
and Siegfried Wagner

WILEY-VCH Verlag GmbH & Co. KGaA

Deutsche Forschungsgemeinschaft
Kennedyallee 40, D-53175 Bonn, Federal Republic of Germany
Postal address: D-53170 Bonn
Phone: ++49/228/885-1
Telefax: ++49/228/885-2777
E-mail: postmaster@dfg.de
Internet: www.dfg.de



Library of Congress Card No.: applied for
A catalogue record for this book is available from the British Library.

Bibliographic information published by Die Deutsche Bibliothek.
Die Deutsche Bibliothek lists this publication in the Deutsche Nationalbibliografie; detailed bibliographic data is available in the Internet at http://dnb.ddb.de

**ISBN-13:** 978-3-527-27735-3
**ISBN-10:** 3-527-27735-8


Printed on acid-free paper



Cover Design and Typography: Dieter Hüsken
Composition: K+V Fotosatz, 64743 Beerfelden
Printing: Strauss GmbH, Moerlenbach
Bookbinding: J. Schäffer GmbH, Grünstadt

Printed in the Federal Republic of Germany

# Contents

1 **Introduction** ... 1

2 **Network Organization of Collaborative Research Centres for Scientific Efficiency Enhancement** ... 3
*Dieter Jacob, Gottfried Sachs, and Siegfried Wagner*

2.1 Introduction ... 3

2.2 Organization of Collaboration ... 4

2.3 Efficiency Enhancement in Research ... 4

2.4 Efficiency Enhancement in Teaching and Education ... 5

2.5 Internationalization ... 6

2.6 Final Remarks ... 7

3 **Overall Design Aspects** ... 9

3.1 Conceptual Design of Winged Reusable Two-Stage-to-Orbit Space Transport Systems ... 9
*Stefan Lentz, Mirko Hornung, and Werner Staudacher*
3.1.1 Background and Introduction ... 9
3.1.2 Concepts for Reusable Space Transports ... 11
3.1.2.1 Single-Stage-to-Orbit SSTO ... 11
3.1.2.2 Two-Stage-to-Orbit TSTO ... 12
3.1.3 Design Procedure ... 13
3.1.3.1 Design Tools and Methods ... 14
3.1.3.2 Baseline Concept ... 15
3.1.3.3 Boundary Conditions and Requirements ... 16

*Basic Research and Technologies for Two-Stage-to-Orbit Vehicles*
DFG, Deutsche Forschungsgemeinschaft

ISBN: 3-527-27735-8

3.1.3.4 Variation of Mission and Staging Mach Number ............. 16
3.1.3.5 Trade Studies ............................................ 17
3.1.3.6 Evaluation and Comparison of the Concepts ................. 17
3.1.4 Variation of Mission and Mach Numbers ................... 18
3.1.4.1 Mission Comparison ...................................... 20
3.1.4.2 Comparison of Mach Number Variation ..................... 21
3.1.4.3 Accelerator Vehicle Concepts ............................ 25
3.1.5 Trade Studies ............................................ 25
3.1.5.1 Airbreathing Second Stage ................................ 26
3.1.5.2 LOX-Collection ........................................... 29
3.1.6 Comparison and Evaluation ................................ 34
3.1.7 Conclusion and Outlook .................................. 35

3.2 Evaluation and Multidisciplinary Optimization of Two-Stage-to-Orbit Space Planes with Different Lower-Stage Concepts ........................................ 38
*Thorsten Raible and Dieter Jacob*
3.2.1 Introduction ............................................. 38
3.2.2 Reference Configurations ................................. 40
3.2.2.1 Concept Design and Mission Requirements ................. 40
3.2.2.2 Space Plane Configuration with Lifting Body Lower Stage ..... 40
3.2.2.3 Space Plane Configuration with Waverider Lower Stage ....... 42
3.2.2.4 Design and Optimization Parameters ...................... 44
3.2.3 Analysis Methods ......................................... 44
3.2.3.1 Quality Criteria ......................................... 44
3.2.3.2 Simulation and Optimization Software .................... 46
3.2.4 Performance of Reference Space Planes ................... 46
3.2.4.1 Mass Breakdown ........................................... 46
3.2.4.2 Design Sensitivities ..................................... 48
3.2.5 Optimization Results ..................................... 50
3.2.5.1 Nominal Optimizations .................................... 50
3.2.5.2 Sensitivity-Based Optimizations .......................... 53
3.2.6 Summary and Conclusions .................................. 54

**4 Aerodynamics and Thermodynamics** ...................... 57

4.1 Low-Speed Tests with an ELAC-Model at High Reynolds Numbers ............................ 57
*Günther Neuwerth, Udo Peiter, and Dieter Jacob*
4.1.1 Introduction ............................................. 58
4.1.2 Wind Tunnel Models ....................................... 59
4.1.3 Pressure Distributions Influenced by Reynolds Number ........ 61
4.1.4 Flow Field Influenced by Reynolds Number ................. 67
4.1.5 Force Coefficients Influenced by Reynolds Number ........... 71
4.1.6 Conclusion ............................................... 75

4.2 Experimental and Numerical Analysis of Supersonic Flow over the ELAC-Configuration ... 77
*Anatoly Michailovich Kharitonov, Mark Davidovich Brodetsky, Andreas Henze, Wolfgang Schröder, Matthias Heller, Gottfried Sachs, Christian Breitsamter, and Boris Laschka*
4.2.1 Introduction ... 77
4.2.2 Experimental Setup ... 78
4.2.3 Numerical Method ... 87
4.2.4 Results ... 88
4.2.4.1 Flow Over the Orbital Stage and the EOS/Flat Plate Configuration ... 88
4.2.4.2 Separation of ELAC1C and EOS ... 96
4.2.5 Conclusions ... 100

4.3 Stage Separation – Aerodynamics and Flow Physics ... 101
*Christian Breitsamter, Lei Jiang, and Mochammad Agoes Moelyadi*
4.3.1 Introduction ... 102
4.3.2 Methodology and Vehicle Geometries ... 102
4.3.3 Numerical Simulation ... 105
4.3.3.1 Flow Solver ... 105
4.3.3.2 Grid Generation ... 106
4.3.4 Experimental Simulation ... 107
4.3.4.1 Models and Facility ... 107
4.3.4.2 Measurement Technique and Test Programme ... 108
4.3.5 Steady State Flow ... 109
4.3.5.1 Dominant Flow Phenomena ... 109
4.3.5.1.1 Inviscid Case – 2D and 3D Simulations ... 109
4.3.5.1.2 Viscous Effects – Laminar and Turbulent Flow ... 112
4.3.5.2 Comparison of Experimental and Numerical Results ... 113
4.3.6 Unsteady Aerodynamics ... 115
4.3.6.1 Longitudinal Motion – Dynamic Separation ... 115
4.3.6.2 Lateral Motion – Disturbance Effects ... 117
4.3.7 Detailed Two-Stage-to-Orbit Configuration ... 119
4.3.8 Conclusions and Outlook ... 122

4.4 DNS of Laminar-Turbulent Transition in the Low Hypersonic Regime ... 124
*Axel Fezer, Markus Kloker, Alessandro Pagella, Ulrich Rist, and Siegfried Wagner*
4.4.1 Introduction ... 124
4.4.2 Numerical Approach ... 125
4.4.2.1 Governing Equations ... 126
4.4.2.2 Spatial and Time Discretization ... 127
4.4.2.3 Initial and Boundary Conditions ... 127
4.4.3 Transition on Flat Plate and Sharp Cone ... 128

4.4.3.1 Application-Specific Details of the Numerical Method . . . . . . . . . 128
4.4.3.2 Results: Simulation of a Controlled Experiment . . . . . . . . . . . . . . 130
4.4.3.3 Results: Flat Plate and Cone at $M$=6.8 . . . . . . . . . . . . . . . . . . . . . 131
4.4.4 Transitional Shock-Wave/Boundary-Layer Interaction at $Ma$=4.8 135
4.4.4.1 Application-Specific Details of the Numerical Method . . . . . . . . . 137
4.4.4.2 Results: Impinging Shock on a Flat Plate vs. Compression Ramp at $Ma$=4.8 . . . . . . . . . . . . . . . . . . . . . . . . . . . . . . . . . . . . . . . . 139
4.4.4.3 Conclusions . . . . . . . . . . . . . . . . . . . . . . . . . . . . . . . . . . . . . . . . 146

4.5 Numerical Simulation of High-Enthalpy Nonequilibrium Air Flows . . . . . . . . . . . . . . . . . . . . . . . . . . . . . . . . . . . . . . . . . . . . . 148
*Farid Infed, Markus Fertig, Ferdinand Olawsky, Panagiotis Adamidis, Monika Auweter-Kurtz, Michael Resch, and Ernst W. Messerschmid*
4.5.1 Aerothermodynamic Aspects of Re-Entry Flows . . . . . . . . . . . . . . 148
4.5.1.1 Inviscid Fluxes . . . . . . . . . . . . . . . . . . . . . . . . . . . . . . . . . . . . . 149
4.5.1.2 Thermal Relaxation . . . . . . . . . . . . . . . . . . . . . . . . . . . . . . . . . . 150
4.5.1.3 Electronic Excitation . . . . . . . . . . . . . . . . . . . . . . . . . . . . . . . . . 150
4.5.1.4 Thermochemical Relaxation . . . . . . . . . . . . . . . . . . . . . . . . . . . . 151
4.5.1.5 Transport Coefficients . . . . . . . . . . . . . . . . . . . . . . . . . . . . . . . . 152
4.5.1.6 Turbulence . . . . . . . . . . . . . . . . . . . . . . . . . . . . . . . . . . . . . . . . 152
4.5.1.7 Electrical Discharge . . . . . . . . . . . . . . . . . . . . . . . . . . . . . . . . . . 152
4.5.1.8 Gas-Surface Interaction Modelling . . . . . . . . . . . . . . . . . . . . . . . 152
4.5.1.9 Radiative Exchange at the Surface . . . . . . . . . . . . . . . . . . . . . . . 153
4.5.1.10 Heat Conduction within TPS Materials . . . . . . . . . . . . . . . . . . . . 154
4.5.2 Numerics and Parallelization . . . . . . . . . . . . . . . . . . . . . . . . . . . . 155
4.5.2.1 Conservation Equations . . . . . . . . . . . . . . . . . . . . . . . . . . . . . . . 155
4.5.2.2 Solver . . . . . . . . . . . . . . . . . . . . . . . . . . . . . . . . . . . . . . . . . . . 156
4.5.2.3 Multiblock . . . . . . . . . . . . . . . . . . . . . . . . . . . . . . . . . . . . . . . . 157
4.5.2.4 Metacomputing . . . . . . . . . . . . . . . . . . . . . . . . . . . . . . . . . . . . . 158
4.5.2.5 Adaptive Grids . . . . . . . . . . . . . . . . . . . . . . . . . . . . . . . . . . . . . 158
4.5.3 Results . . . . . . . . . . . . . . . . . . . . . . . . . . . . . . . . . . . . . . . . . . . 159
4.5.3.1 Simulation of the Re-Entry of the X-38 . . . . . . . . . . . . . . . . . . . . 159
4.5.3.2 Simulation of the Plasma Source RD5 . . . . . . . . . . . . . . . . . . . . . 162

4.6 Flow Simulation and Problems in Ground Test Facilities . . . . . . . 165
*Uwe Gaisbauer, Helmut Knauss, Siegfried Wagner, Georg Herdrich, Markus Fertig, Michael Winter, and Monika Auweter-Kurtz*
4.6.1 Introduction . . . . . . . . . . . . . . . . . . . . . . . . . . . . . . . . . . . . . . . 165
4.6.2 Validation of a Short Duration Supersonic Wind Tunnel for Natural Laminar Turbulent Transition Studies . . . . . . . . . . . . . 170
4.6.2.1 Introduction to the Problem . . . . . . . . . . . . . . . . . . . . . . . . . . . . 170
4.6.2.2 The Shock Wind Tunnel at Stuttgart University . . . . . . . . . . . . . 171
4.6.2.3 Detection Techniques for Flow Disturbance Fields . . . . . . . . . . . 175

4.6.2.4 Free Stream Disturbance Measurements in the Shock Wind Tunnel ... 178
4.6.2.5 Transition Experiments in the Test Section Flow ... 184
4.6.2.6 Conclusion and Aspects ... 189
4.6.3 Plasma Wind Tunnels ... 191
4.6.3.1 Plasma Generators ... 192
4.6.3.1.1 Arc-Driven Plasma Generators (TPG and MPG) ... 192
4.6.3.1.2 Inductively Heated Plasma Generators (IPGs) ... 195
4.6.3.2 Heat Flux Simulation for X-38 Using PWK1 as Example for PWK Investigation ... 197

4.7 Characterization of High-Enthalpy Flows ... 199
*Monika Auweter-Kurtz, Markus Fertig, Georg Herdrich, Kurt Hirsch, Stefan Löhle, Sergej Pidan, Uwe Schumacher, and Michael Winter*
4.7.1 Intrusive Measurement Methods ... 200
4.7.1.1 Material Sample Support System ... 202
4.7.1.2 Heat Flux Measurements ... 203
4.7.2 Non-Intrusive Techniques ... 206
4.7.2.1 Emission Spectroscopy ... 207
4.7.2.2 Laser-Induced Fluorescence ... 209
4.7.2.3 Thomson Scattering for Electron Temperature and Density Determination ... 211
4.7.2.4 High-Resolution Spectroscopy and Fabry Perot Interferometry ... 212
4.7.3 Flight Instrumentation (PYREX, RESPECT, PHLUX, COMPARE) ... 213
4.7.3.1 Description of PYREX-KAT38 (Pyrometric Entry Experiment) ... 214
4.7.3.2 RESPECT (Re-Entry SPECTrometer) ... 216
4.7.3.3 COMPARE ... 217

4.8 Numerical Simulation of Flow Fields Past Space Transportation Systems ... 220
*Andreas Henze, Wolfgang Schröder, and Matthias Meinke*
4.8.1 Introduction ... 221
4.8.2 Numerical Scheme ... 221
4.8.2.1 Basic Equations ... 221
4.8.2.2 Initial and Boundary Conditions ... 222
4.8.2.3 Spatial Discretization in Structured Grids ... 223
4.8.2.4 Spatial Discretization in Unstructured Grids ... 224
4.8.2.5 Structured/Unstructured Coupling ... 225
4.8.2.6 Temporal Integration ... 225
4.8.3 Results ... 226
4.8.3.1 Geometry of the Two-Stage System ... 226
4.8.3.2 Flow Past ELAC ... 228
4.8.3.3 Flow Past ELAC-1c ... 233
4.8.3.4 Simplified Stage Separation ... 240
4.8.4 Conclusions ... 241

4.9 High-Speed Aerodynamics of the Two-Stage ELAC/EOS-Configuration for Ascend and Re-entry . . . 242
*Martin Bleilebens, Christoph Glößner, and Herbert Olivier*
4.9.1 Introduction and Experimental Conditions . . . 242
4.9.2 Measurement Equipment . . . 244
4.9.2.1 Pressure Measurement . . . 244
4.9.2.2 Temperature and Heatflux Measurement . . . 244
4.9.2.3 Force and Moment Measurement . . . 244
4.9.2.4 Flow Visualization . . . 245
4.9.3 Measurements on the ELAC- and EOS-Configurations . . . 246
4.9.3.1 Pressure and Heat Flux Measurements on the ELAC-Configuration . . . 246
4.9.3.2 Force and Moment Measurements on the ELAC-Configuration . . 247
4.9.3.3 Pressure and Heat Flux Measurements on the EOS-Configuration . . . 249
4.9.3.4 Force and Moment Measurements on the EOS-Configuration . . . 251
4.9.4 Detailed Measurements on Ramp-Configurations . . . 254
4.9.4.1 Laminar and Turbulent Shock-Wave/Boundary-Layer Interactions 254
4.9.4.2 Theoretical Considerations . . . 255
4.9.4.3 Ramp Flows with Variation of Surface Temperature . . . 256
4.9.4.4 Description of Ramp Model . . . 258
4.9.4.5 Schlieren Pictures and Position of Separation . . . 260
4.9.4.6 Determination of Pressure Coefficients . . . 262
4.9.4.7 Determination of Stanton Numbers . . . 264
4.9.5 Conclusions . . . 267

**5 Propulsion** . . . 269

5.1 PDF/FDF-Methods for the Prediction of Supersonic Turbulent Combustion . . . 269
*Stefan Heinz and Rainer Friedrich*
5.1.1 Introduction . . . 269
5.1.2 Methods for Turbulent Reacting Flow Calculations . . . 270
5.1.2.1 Basic Methods . . . 271
5.1.2.2 Hybrid PDF/FDF-Methods . . . 272
5.1.3 Some Deficiencies of Existing Hybrid PDF-Methods . . . 272
5.1.3.1 The Transport Problem . . . 273
5.1.3.2 The Mixing Problem . . . 273
5.1.3.3 The Energy Problem . . . 274
5.1.4 New Theoretical Concepts . . . 275
5.1.4.1 The Transport Problem . . . 276
5.1.4.2 The Mixing Problem . . . 276
5.1.4.3 The Energy Problem . . . 276
5.1.5 The Use of PDF Combustion Codes . . . 277
5.1.5.1 The Current Use of PDF/FDF-Methods . . . 277

5.1.5.2 New Developments . . . . . 279
5.1.5.3 Common Activities to Develop a New Combustion Code . . . . . 279
5.1.6 Prospects for Further Developments . . . . . 280
5.1.6.1 The Current and Future Use of Computational Methods . . . . . 280
5.1.6.2 Some Challenges . . . . . 281

5.2 Design and Testing of Gasdynamically Optimized Fuel Injectors for the Piloting of Supersonic Flames with Low Losses . . . . . 284
*Anatoliy Lyubar, Tobias Sander, and Thomas Sattelmayer*
5.2.1 Introduction . . . . . 284
5.2.2 Experimental Setup . . . . . 285
5.2.2.1 Model SCRamjet Combustor . . . . . 285
5.2.2.2 Preheater . . . . . 285
5.2.2.3 Combustion Chamber . . . . . 286
5.2.2.4 Injectors . . . . . 288
5.2.3 Investigation Tools . . . . . 289
5.2.3.1 Shadowgraph Method . . . . . 289
5.2.3.2 Rayleigh Scattering . . . . . 289
5.2.3.3 Raman Scattering . . . . . 289
5.2.3.4 OH-LIF Measurements . . . . . 290
5.2.3.5 Self-Fluorescence Measurements (Chemiluminescence) . . . . . 290
5.2.4 Numerical Modelling . . . . . 291
5.2.4.1 Numerical Simulation with the CFD-Code Fluent 5.5 . . . . . 291
5.2.4.2 Special Features of the Modelling of the Supersonic Combustion 291
5.2.4.3 Reducing the Number of Species . . . . . 292
5.2.4.4 Reaction Mapping by Using of the Polynomials . . . . . 294
5.2.4.5 Validation of the Modelling Approach with Polynomials . . . . . 295
5.2.5 Two Stage Injector . . . . . 297
5.2.5.1 Theoretical Considerations . . . . . 297
5.2.5.2 Shock Stabilization . . . . . 300
5.2.5.3 Combustion . . . . . 303
5.2.6 Conclusions . . . . . 305

5.3 Hypersonic Propulsion Systems: Design, Dual-Mode Combustion and Systems Off-Design Simulation . . . . . 308
5.3.1 Combustion Stability of a Dual-Mode Scramjet – Configuration with Strut Injector . . . . . 308
*Sara Rocci-Denis, Armin Brandstetter, Dieter Rist, and Hans-Peter Kau*
5.3.1.1 Introduction . . . . . 308
5.3.1.2 Experimental Setup . . . . . 310
5.3.1.3 Results and Discussion . . . . . 315
5.3.1.4 Conclusions . . . . . 324

5.3.2 Hypersonic Highly Integrated Propulsion Systems – Design and Off-Design Simulation . . . 327
*Hans Rick, Andreas Bauer, Thomas Esch, Sebastian Hollmeier, Hans-Peter Kau, Sven Kopp, and Andreas Kreiner*
5.3.2.1 Introduction . . . 327
5.3.2.2 Reference Propulsion System for the TSTO Concept . . . 330
5.3.2.3 Engine Integration . . . 330
5.3.2.4 Core Engine . . . 336
5.3.2.5 Numerical Engine Simulation . . . 337
5.3.2.6 Thrust Vectoring . . . 338
5.3.2.7 Real Time Flight Simulation . . . 344
5.3.2.8 Conclusion . . . 345

5.4 Experimental Investigation about External Compression of Highly Integrated Airbreathing Propulsion Systems . . . 347
*Uwe Gaisbauer, Helmut Knauss, and Siegfried Wagner*
5.4.1 Introduction . . . 347
5.4.1.1 Focus on the Problem . . . 348
5.4.1.2 Preliminary Measurements . . . 349
5.4.2 Experimental Facility . . . 350
5.4.3 Wind Tunnel Models and Instrumentation . . . 351
5.4.3.1 Model 1 . . . 351
5.4.3.2 Model 2 . . . 352
5.4.3.3 Model 3 . . . 353
5.4.4 Numerical Model . . . 354
5.4.5 Measurements and Results . . . 354
5.4.5.1 Determination of the Boundary-Conditions . . . 355
5.4.5.2 Measurements in the Field of Shock Boundary Layer Interaction 358
5.4.6 Conclusion and Outlook . . . 362

5.5 Experimental and Numerical Investigation of Lobed Strut Injectors for Supersonic Combustion . . . 365
*Peter Gerlinger, Peter Kasal, Fernando Schneider, Jens von Wolfersdorf, Bernhard Weigang, and Manfred Aigner*
5.5.1 Introduction . . . 365
5.5.2 Experimental Setup and Measurement Techniques . . . 366
5.5.3 Governing Equations and Numerical Simulation . . . 369
5.5.3.1 Multigrid Convergence Acceleration . . . 370
5.5.4 Strut Design and Performance Parameters . . . 371
5.5.5 Supersonic Mixing . . . 373
5.5.6 Supersonic Combustion . . . 374
5.5.6.1 Investigation of Different Lobed Strut Injectors . . . 375
5.5.7 Conclusions . . . 380

5.6 Experimental Studies of Viscous Interaction Effects in Hypersonic Inlets and Nozzle Flow Fields ... 383
*Andreas Henckels and Patrick Gruhn*
5.6.1 Introduction ... 383
5.6.2 Experimental Techniques ... 385
5.6.2.1 Facility and Flow Diagnostics ... 385
5.6.2.2 Wind Tunnel Models ... 386
5.6.3 Inlet Studies ... 388
5.6.4 Nozzle Studies ... 395
5.6.5 Conclusion ... 400

5.7 Intake Flows in Airbreathing Engines for Supersonic and Hypersonic Transport ... 403
*Birgit Ursula Reinartz, Joern van Keuk, Josef Ballmann, Carsten Herrmann, and Wolfgang Koschel*
5.7.1 Introduction ... 404
5.7.2 Physical Model ... 405
5.7.3 Numerical Method ... 406
5.7.4 Results ... 408
5.7.4.1 Turbulent 2D Supersonic Intake Flows with Internal Compression ... 408
5.7.4.2 Laminar 3D Hypersonic Corner Flows ... 411
5.7.4.3 Turbulent 3D Hypersonic Flows through Symmetric/Asymmetric Double-Fin Configurations ... 414
5.7.4.4 Laminar 2D Shock Interactions in Hypersonic Flows with Chemical Non-Equilibrium ... 415
5.7.5 Conclusions ... 418

**6 Flight Mechanics and Control ... 421**

6.1 Safety Improvement for Two-Stage-to-Orbit Vehicles by Appropriate Mission Abort Strategies ... 421
*Michael Mayrhofer, Otto Wagner, and Gottfried Sachs*
6.1.1 Introduction ... 422
6.1.2 Dynamics Model of Two-Stage-to-Orbit Vehicle ... 423
6.1.3 Optimization Problem ... 427
6.1.4 Safety Improved Nominal Trajectory ... 428
6.1.5 Mission Aborts of Carrier Stage ... 430
6.1.6 Mission Aborts of Orbital Stage ... 432
6.1.7 Mission Abort Plan ... 435
6.1.8 Conclusions ... 436

6.2 Optimal Trajectories for Hypersonic Vehicles with Predefined Levels of Inherent Safety ... 438
*Rainer Callies*
6.2.1 Introduction ... 439
6.2.2 Theoretical Background ... 440
6.2.2.1 Classical Problem ... 440
6.2.2.2 Related Boundary Value Problem ... 440
6.2.2.3 Extended Problem (A) ... 441
6.2.2.4 Extended Problem (B) ... 444
6.2.3 Numerical Method ... 446
6.2.4 Model System ... 449
6.2.4.1 Overview ... 449
6.2.4.2 Thrust Model ... 449
6.2.4.3 Atmospheric and Aerodynamic Model ... 450
6.2.4.4 Equations of Motion ... 451
6.2.4.5 Primary Problem ... 452
6.2.4.6 Secondary Problem ... 453
6.2.4.7 Extended Problem (B) ... 454
6.2.4.8 Numerical Results ... 455
6.2.5 Conclusion ... 456

6.3 Hypersonic Trajectory Optimization for Thermal Load Reduction ... 458
*Michael Dinkelmann, Markus Wächter, and Gottfried Sachs*
6.3.1 Introduction ... 459
6.3.2 Modelling of Vehicle Dynamics ... 460
6.3.3 Modelling of Heat Input ... 464
6.3.4 Optimization Problem ... 467
6.3.5 Results ... 469
6.3.5.1 Range Cruise ... 469
6.3.5.2 Return-to-Base Cruise ... 471
6.3.6 Conclusions ... 473

6.4 Flight Dynamics and Control Problems of Two-Stage-to-Orbit Vehicles ... 476
6.4.1 Flight Tests and Simulation Experiments for Hypersonic Long-Term Dynamics Flying Qualities ... 476
*Robert Stich, Timothy H. Cox, and Gottfried Sachs*
6.4.1.1 Introduction ... 477
6.4.1.2 Hypersonic Flight Dynamics ... 478
6.4.1.3 Research Aircraft and Flight Simulator ... 480
6.4.1.4 Results ... 482
6.4.1.5 Conclusions ... 487

6.4.2 Wind Tunnel Tests for Modelling the Separation Dynamics of a Two-Stage-to-Orbit Vehicle ........................ 489
*Christian Zähringer and Gottfried Sachs*
6.4.2.1 Introduction ........................ 489
6.4.2.2 Test Facility and Wind Tunnel Models ........................ 490
6.4.2.3 Results ........................ 492
6.4.2.4 Conclusions ........................ 497

**7 High-Temperature Materials and Hot Structures** ........................ 499

7.1 Ceramic Matrix Composites – the Key Materials for Re-Entry from Space to Earth ........................ 499
*Martin Frieß, Walter Krenkel, Richard Kochendörfer, Rüdiger Brandt, Günther Neuer, and Hans-Peter Maier*
7.1.1 Introduction and Overview ........................ 499
7.1.2 Liquid Silicon Infiltration: Process Development ........................ 500
7.1.3 Microstructural Design of C/C-SiC Composites ........................ 502
7.1.3.1 C/C-SiC Composites Derived from As-Received Carbon Fibres .. 502
7.1.3.2 C/C-SiC Composites Derived from Thermally Pre-Treated Carbon Fibres ........................ 503
7.1.3.3 Graded C/C-SiC Composites ........................ 504
7.1.3.4 C/C-SiC Composites Derived from Graphitized C/C ........................ 508
7.1.4 Macroscopic Design Aspects ........................ 509
7.1.4.1 Dimensional Stability ........................ 509
7.1.4.2 Modular Construction by *In-Situ* Joining ........................ 511
7.1.5 Thermophysical Characterization of C/C-SiC ........................ 512
7.1.5.1 Methods to Measure Thermophysical Properties ........................ 512
7.1.5.2 Materials and Specimen Preparation ........................ 512
7.1.5.3 Specific Heat Capacity ........................ 514
7.1.5.4 Thermal Conductivity ........................ 515
7.1.5.5 Spectral and Total Emissivity ........................ 518
7.1.6 Thermomechanical Characterization of C/C-SiC ........................ 520
7.1.6.1 Failure Mechanism of C/C-SiC Materials ........................ 520
7.1.6.2 Influence of the Temperature on the Stress-Strain Behaviour .... 520

7.2 Behaviour of Reusable Heat Shield Materials under Re-Entry Conditions ........................ 527
*Fritz Aldinger, Monika Auweter-Kurtz, Markus Fertig, Georg Herdrich, Kurt Hirsch, Peter Lindner, Dirk Matusch, Günther Neuer, Uwe Schumacher, and Michael Winter*
7.2.1 Principles and Modelling of Heterogeneous Reactions ........................ 528
7.2.1.1 Heterogeneous Catalysis ........................ 528
7.2.1.2 Redox Reactions Including Active and Passive Oxidation ........................ 531
7.2.1.3 Surface Reaction Model Applied to MIRKA Re-Entry Flow ..... 533

7.2.2 Characterization of High-Temperature Oxidation and Catalytic Behaviour of TPS Materials . . . 535
7.2.2.1 Experimentally Observed Influence of Catalytic Efficiency . . . 535
7.2.2.2 Oxidation Behaviour . . . 537
7.2.3 Developments and Investigations of Protection Layers for Reusable Heat Shield Materials . . . 541
7.2.3.1 Production and Characteristics of Protection Layers . . . 541
7.2.3.2 Diagnostics for the Tests of the Protection Layers in the Plasma Wind Tunnel . . . 542
7.2.3.3 Protection Material Tests and Results . . . 543

7.3 Design and Evaluation of Fibre Ceramic Structures . . . 549
*Bernd-Helmut Kröplin, Richard Kochendörfer, Thomas Reimer, Thomas Ullmann, Ralf Kornmann, Roger Schäfer, and Thomas Wallmersperger*
7.3.1 Introduction . . . 549
7.3.1.1 Concept Design and Manufacturing Studies . . . 551
7.3.1.2 Manufacturing . . . 553
7.3.1.3 Test . . . 554
7.3.1.4 Plasma Sprayed Yttrium Silicates for Oxidation Protection of C/C-SiC Panels . . . 555
7.3.1.5 Flight Experiment . . . 557
7.3.2 Measuring Model Deflections by Thermo-Mechanical Loads in a Plasma Wind Tunnel . . . 559
7.3.2.1 Overview . . . 559
7.3.2.2 Model Design . . . 561
7.3.2.3 Adaptation of the HTGM to the L3K Facility . . . 562
7.3.2.4 Results . . . 565
7.3.3 Material Description of Fibre Ceramics . . . 569
7.3.3.1 Phenomena in C/C-SiC Materials . . . 569
7.3.3.2 Phenomenological Model . . . 571
7.3.3.3 Micromechanically Based Phenomenological Model . . . 573
7.3.3.4 Functionally Graded Materials . . . 575
7.3.4 Conclusions . . . 578

**8 Cooperation with Industry and Research Establishments, Participation in National and International Research Programmes** . . . 581
*Dieter Jacob, Gottfried Sachs, and Siegfried Wagner*

**9 Conclusions and Perspectives** . . . 585
*Dieter Jacob, Gottfried Sachs, and Siegfried Wagner*

**10 Appendix** . . . 587

10.1 Publications . . . 587

10.2 Dissertations . . . 639

10.3 Habilitations . . . 648

10.4 Patents . . . 649

10.5 Number of Diploma Theses . . . 649

10.6 Visiting Researchers . . . 649

10.7 Organization and Projects . . . 656

# 1 Introduction

In 1989 the Deutsche Forschungsgemeinschaft established three Collaborative Research Centres concerned with hypersonic vehicles at the Rheinisch-Westfälische Technische Hochschule Aachen, the Technische Universität München and the Universität Stuttgart. The final report presents a selection of recent research results and an overview of the activities and the organization of the network which evolved during the past fifteen years.

The research was focused on basic aspects of future reusable space transportation systems and covered the areas of overall design, aerodynamics, thermodynamics, flight dynamics, propulsion, materials, and structures. The underlying configuration which served as a guideline for detailed research consisted of a two-stage-to-orbit vehicle with the ability to start horizontally. The first stage had an airbreathing propulsion, the second stage a rocket propulsion. Both stages were designed to return to earth and land horizontally on adequate airports.

A major part of the research dealt with experimental and numerical aerodynamic topics ranging from low-speed to hypersonic flow past the external configuration and through inlet and nozzle. The low-speed flow past the lower stage was investigated for a large range of Reynolds numbers in different wind tunnels including a test period at high Reynolds numbers with a large model in the German-Dutch Wind Tunnel (DNW). The studies at high Mach numbers included the very complex interference between the lower stage and the upper stage during the initial flight and during stage separation and the aero-thermodynamic heating. In all cases experimental and numerical approaches were employed.

Another major part of the research was concerned with flight mechanics. One aspect was trajectory optimization which was dealt with in cooperation of mathematicians and engineers. A further aspect relates to stability, control and flying qualities, the treatment of which includes a collaboration with the NASA Dryden Flight Research Center using their unique simulation and flight test facilities. Moreover, the flight dynamics of the separation manoeuvre was subject of the research activities, employing also wind-tunnel tests at the Institute of Theoretical and Applied Mechanics of the Russian Academy of Sciences in Novosibirsk.

*Basic Research and Technologies for Two-Stage-to-Orbit Vehicles*
DFG, Deutsche Forschungsgemeinschaft

ISBN: 3-527-27735-8

The re-entry phase was investigated both experimentally and numerically. Plasma wind tunnels were built to generate high-enthalpy plasma flows and to investigate the interactions with heat shield materials. The experimental investigation was accompanied by numerical simulation of the flow field inside the ground test facility and around a space vehicle re-entering the Earth's atmosphere. New aero-thermodynamic models enabled a successful post-flight analysis of the MIRKA re-entry. Re-entry experiments for in-flight investigation of plasma flow and material response were successfully flown on capsules such as EXPRESS and MIRKA; others are about to be flown on missions such as EXPERT.

For the overall design investigations a propulsion simulation model including the jet and ramjet modes was developed. The efficiency of supersonic and hypersonic airbreathing propulsion depends strongly on the efficiency of inlets and nozzles. Therefore, several numerical and experimental projects dealt with these components of future space planes. In other projects methods to reach stable supersonic combustion were investigated.

Structural research and development was predominantly coupled to the needs for high-temperature resistant structures for space vehicles. During the re-entry phase from orbit to earth temperatures of more than 1600 °C are reached. For the application in a thermal protection system (TPS) and also as a material for the use in hot structures, like control surfaces, a new type of ceramic matrix composite was developed on the basis of carbon fibres that is called C/C-SiC. The technology of thermal protection systems reached a maturity that allowed a flight experiment with a representative TPS structure on the surface of a Russian FOTON research capsule that was scheduled for a micro-gravity mission in orbit with subsequent re-entry to earth.

This final report presents some of the most recent results obtained in the disciplines required for the design of future space planes. In additional chapters the unique model established for the cooperation of three cooperative research centres at different universities is described and analyzed.

December 2003

Dieter Jacob
Gottfried Sachs
Siegfried Wagner

# 2 Network Organization of Collaborative Research Centres for Scientific Efficiency Enhancement

Dieter Jacob, Gottfried Sachs, and Siegfried Wagner

## 2.1 Introduction

Three initiatives for Collaborative Research Centres of the Deutsche Forschungsgemeinschaft evolved in the late eighties, at the Rheinisch-Westfälische Technische Hochschule Aachen, the Technische Universität München and the Universität Stuttgart. After exploratory and advisory talks with the Deutsche Forschungsgemeinschaft, principles for research planning and cooperation were established, resulting in a concept for a framework of research and organization of the initiatives. This concept is based on the following elements:

- Each initiative for a Collaborative Research Centre proposes its own, independent research programme which can be realized even if another initiative fails. Each research programme has an own concentration on points of emphasis as part of the overall theme.
- The research programmes of the initiatives should be complementary.
- The complementing of the research programmes must not confine the decision such that approval of all Collaborative Research Centres becomes imperative.

After successful passing the review procedure, the following three Collaborative Research Centres have been established by the Deutsche Forschungsgemeinschaft in 1989:

- Collaborative Research Centre 253 "Fundamentals of Space Plane Design" at the Rheinisch-Westfälische Technische Hochschule Aachen,
- Collaborative Research Centre 255 "Transatmospheric Flight Systems" at the Technische Universität München,
- Collaborative Research Centre 259 "High-Temperature Problems of Reusable Space Transportation Systems" at the Universität Stuttgart.

*Basic Research and Technologies for Two-Stage-to-Orbit Vehicles*
DFG, Deutsche Forschungsgemeinschaft

ISBN: 3-527-27735-8

## 2.2 Organization of Collaboration

For organizing the collaboration of the three Collaborative Research Centres, a Compound Network was established. It has the following structure:

1. Steering Committee
A Steering Committee consisting of the Speakers of the three Collaborative Research Centres was established. The Steering Committee meets several times a year and is responsible for the following topics:
- strategic planning for future research programmes;
- coordination of main research activities between the Collaborative Research Centres;
- laying down of principles and goals of the collaboration among the Research Centres as well as with external partners from research institutions and industry;
- planning of joint activities for the presentation of research results;
- planning of joint education activities for students as well as engineers and scientists.

2. The Collaborative Research Centres exchange their research results and inform each other about ongoing and planned work.
3. Data banks concerning the air and combustion gases will be jointly generated.
4. The Collaborative Research Centres will inform each other about test configurations and models and use the same or similar models.
5. The Collaborative Research Centres present research results in a joint manner at national and international scientific meetings, conferences, etc.

## 2.3 Efficiency Enhancement in Research

The Compound Network of the Collaborative Research Centres enabled an enhancement of efficiency in research. Basically, resources and competences of the three Collaborative Research Centres could be brought together to yield synergy effects and improvement of research efforts.

At the working level, advantages resulted from mutual contacts and visits as well as from the exchange of scientists. Furthermore, working groups could be established to address specific subjects (numerical methods, measurement techniques, etc.). Other joint activities relate to the development of computer codes and software or verification of numerical results.

Contacts and cooperations with external scientist groups were also promoted by the Compound Network. This concerns university institutes and research establishments as well as industry companies.

A further enhancement of the research efficiency relates to large research facilities which may not be accessible for a single Collaborative Research Centre. As an example, the three Collaborative Research Centres jointly conducted a large experimental project at the German-Dutch Wind Tunnel.

The establishment of the Compound Network led to an increase in competence. Thus, the position of the three Collaborative Research Centres as cooperation partners of research and industry was strengthened. This resulted in a further advantage since industry expressed their willingness for a continuous support concerning computational techniques, data, and experimental facilities. Another result was the participation of the Collaborative Research Centres in national and international research programmes, like the German Programmes TETRA "Technologien für zukünftige Raumtransportsysteme" and ASTRA "Ausgewählte Systeme und Technologien für zukünftige Raumtransportsystem-Anwendungen" as well as the European Programmes FESTIP "Future European Space Transportation Investigations Programme" and FLPP "Future Launcher Preparatory Programme" (planned). Moreover, working groups with representatives from research and industry were established. In addition, scientists conducted flight experiments related to aero-thermodynamics and materials on the re-entry missions EXPRESS, MIRKA and IRDT. Surface protection layer development for thermal protection system materials of future reusable vehicles is funded within a programme by the State of Baden-Württemberg.

## 2.4 Efficiency Enhancement in Teaching and Education

The establishment of the Compound Network also enabled an enhancement of the efficiency in teaching and education. This is of particular importance for hypersonics because of the backlog demand in this field.

A very effective means were the Space Courses which were jointly held by the three Collaborative Research Centres at the Rheinisch-Westfälische Technische Hochschule Aachen, the Technische Universität München, and the Universität Stuttgart. The Space Courses which had a duration of two or three weeks were offered to graduate students as well as to participants from research establishments, industry companies, and administration agencies. There was great interest in the Space Courses not only from Germany but also from other countries.

There are manifold other activities of the Compound Network supporting and enhancing the efficiency in teaching and education. Joint seminars and workshops were conducted, yielding an exchange of experiences and results between members of the Collaborative Research Centres. Moreover, working groups supported the education of students and young scientists. A further possibility concerns the participation of members of a Collaborative Research Centre in doctoral theses of another one.

Further activities enhancing the efficiency in education relate to joint research programmes, yielding unique experience for the involved young scientists. This concerns the already mentioned experimental programme at the German-Dutch Wind Tunnel. Another activity was a joint wind tunnel test programme at the Institute of Theoretical and Applied Mechanics of the Russian Academy of Sciences, Siberian Branch in Novosibirsk, Russia, offering experience on cooperation with scientists from abroad.

## 2.5 Internationalization

The competence which the Collaborative Research Centres have attained led to a greater visibility, both nationally and internationally. The Compound Network has gained recognition in various countries. Multiple invitations came from Europe, the USA, and Japan to give an overview of the German university research on hypersonics. Furthermore, the joint arrangement of workshops with participants from various countries contributed to the international recognition of the Compound Network. This is also true for the joint organization of sessions in international scientific congresses in Germany as well as in other countries.

Many research activities developed on an international basis. There was a very successful collaboration with the NASA Dryden Flight Research Center in Edwards, California, over many years, leading to the utilization of flight test and simulation facilities which are unique in the world. Another international cooperation effort was the already mentioned wind tunnel test programme of the Institute of Theoretical and Applied Mechanics in Novosibirsk. The participation in the European programmes FESTIP and FLPP is another example. There are many other research activities with scientists from other countries, contributing to the international visibility of the Compound Network.

The international visibility of the Research Network also holds for the teaching and education activities. There were students from other countries, purposefully approaching the Collaborative Research Centres for diploma theses. Other successful activities concern research stays of young scientists at the Collaborative Research Centres abroad and vice versa. The great interest of people from other countries in the Space Courses is also evidence of the international visibility.

A most remarkable activity which gained both national and international recognition is the exhibition "The New Way into Space – Space Transporters of the Next Generation" of the Deutsche Forschungsgemeinschaft. This exhibition which is concerned with the research of the Compound Network was displayed with great success in various cities in Germany, like Bonn, Stuttgart, München, Aachen, Berlin, and others. An international version of the exhibition was shown in several countries within the scope of the Concerted Action "Joint Initiative for the Promotion of Study, Research, and Training in Germany" of the German Federal Ministry of Education and Research, the States of the Federal Republic of Germany, and other institutions. The fact that the Deutsche Forschungsgemeinschaft selected the subject of the Compound Network for these exhibitions is evidence of its successful research work.

## 2.6 Final Remarks

The experience which the involved scientists gained with the Compound Network is very positive. It strengthened their activities in research and teaching. This also holds for their relation and cooperation with external partners from research institutions and industry, both nationally and internationally. To sum up, it can be said that the Compound Network of the three Collaborative Research Centres 253, 255, and 259 turned out as an appropriate means to efficiently organize the research work for a subject which is sufficiently broad.

Evidence of a Compound Network as an efficient possibility of organizing research in a greater framework is also due to the statement of the German Wissenschaftsrat on the development of the programme for the Collaborative Research Centres from 23 January 1998. Here, the Compound Network of the three Collaborative Research Centres 253, 255, and 259 is recognized and a stronger networking of thematically related Collaborative Research Centres is also recommended for the future.

# 3 Overall Design Aspects

## 3.1 Conceptual Design of Winged Reusable Two-Stage-to-Orbit Space Transport Systems

Stefan Lentz, Mirko Hornung *, and Werner Staudacher

### 3.1.1 Background and Introduction

During the Space Race – from the fifties to the eighties – money was almost irrelevant to bring anything alive, which went to orbit and beyond. Those days could be characterized as paradise for rocket scientists, engineers, conceptualists and lots of men with brilliant or weird ideas on both sides of the iron curtain. Some concepts and ideas came to the drawing board and entered life (Apollo and Space Shuttle, Saljut, Buran, Mir etc.) and some – or most of them – went into the drawer or just became paper planes. Eventually, the curtain dropped and space flight slithered into a crisis; money became a factor which could not be disregarded, economical aspects gained in importance and as a consequence lots of concepts died. The cold war and national prestige driven high tech aerospace machinery began to stutter.

The American partly reusable Space Transportation System STS or "Space Shuttle" was still suffering from the Challenger catastrophe and was just too expensive to place satellites into orbit with human assistance. The two major expendable rocket systems Delta and Titan – which were derivatives from intercontinental ballistic missiles concepts – were unreliable and ineffective for a rising demand in commercial payloads. Russian launchers still were not or only hardly accessible. It was the age of the European Ariane rocket which was especially designed to place (commercial) payloads into orbit and, in addition, due to the lack of other competitors. Although Ariane IV was very versatile and

* EADS European Aeronautic Defence & Space Company, Division Military Aircraft, MS61 – A400M Program Management, 81663 München; mirko.hornung@eads.com

*Basic Research and Technologies for Two-Stage-to-Orbit Vehicles*
DFG, Deutsche Forschungsgemeinschaft

ISBN: 3-527-27735-8

successful, it was still an expendable system and thus expensive. In the fading of 20$^{th}$ century it was realized, that exorbitant costs for access to space need to be reduced drastically to give further rise of attractiveness for space transportation and exploitation. Auspicious and ambitious fully reusable concepts like the American National Aerospace Plane, the German Sänger or the British HOTOL emerged from this proposition. National prestige and still a kind of cold-war-thinking led to excessive requirements resulting in insuperable technical and financial obstacles. However these concepts showed the right trend for long-term future launchers. Cost reduction strategies for present expendable rockets led to increasing and multiple payload capabilities and heavier and bulkier systems such as Ariane V, Delta IV and Atlas V, which share a hard-fought market, especially since Russian low cost carriers are available [1].

Today the expense factor is a major and global approved criterion for the development of future space transportation systems. A cost-efficient and reliable launcher is vital to win the leading market position. After initial success in the commercial space market, Europeans need to find alternatives for the cost-intensive expendable systems to maintain or regain their position in the global market. The logical and in the long term only reasonable consequence is the design of reusable space transportation systems, which keep down non recurring costs (design and development) as well as recurring costs (production and operations) and finally allow "aircraft-like" operations. To achieve these requirements several strategies and philosophies exist [2–4].

At the present time, design and development of reusable space transportation systems is still in a conceptual phase. The state-of-the-art is rather seen as prospect and can be classified as highly evolutionary. The continuous quest for solutions and the exigency of an evolutionary process are reflected in a multitude of more ore less favourable design alternatives and concepts, preferences depending on strategies and experiences from the past. An example for a wide-spread investigation is the FESTIP (Future European Space Transportation Investigations Programme) system study of ESA, in which 7 system concept families with 19 variants of space launchers were analyzed. This system study can be seen as a paradigmatic characterization from an European view, since American concepts were also incorporated. On the other side of the Atlantic several American studies exist such as the SLI (Space Launch Initiative) of NASA [3–6].

A major problem is the justification of Reusable Launch Systems (RLV). Current number of world wide launches is below 100 per year but only few of them are performed by at least partially reusable Space Shuttle. This market is shared among the major competitors as the USA, Europe, Russia, Japan and (soon) China. Even the United States' share of the market (including commercial and governmental launches) is too small to overcome the break even point for the development of a fully reusable launch vehicle [7].

### 3.1.2 Concepts for Reusable Space Transports

Basically, reusable space transportation concepts can be divided into two major groups, Single-Stage-to-Orbit (SSTO) and Multi or Two-Stage-to-Orbit (TSTO) concepts. From these, hybrid versions and variations emerge and a ramification of other conceptual particularities arises. In the following, both design concepts are shortly reviewed with respect to their advantages and disadvantages.

#### 3.1.2.1 Single-Stage-to-Orbit SSTO

From operational experiences of the Space Shuttle most US space transportation studies preferred the SSTO concept to reduce costs for access to space. Going orbital with just a single stage means: only a single kind of vehicle has to be developed and finally operated. Hence expenses for development, fabrication and operations (infrastructure for refurbishment and maintenance) could be kept down and so does the payload price. However, these hypothetical advantages come along with a significant increase in dry mass (which also means orbital mass), complexity, a higher technology level (e.g. achieving a dry mass fraction less than 10%) and, in consequence, tremendous design and development risks [1, 3, 4].

SSTO concepts can be divided into rocket-based vertical-take-off-horizontal-landing (VTHL) vehicles (VentureStar, FSSC-1, FSSC-5) or combined-cycle-engine (airbreather/rocket) based vehicles (NASP, HOTOL) with horizontal-take-off-and-landing (HTHL) capability [3, 4]. The fundamental difficulty of SSTO concepts is a high sensitivity towards critical design parameters, such as dry mass and propulsion performance, since all system components need to be carried into orbit (and also back to earth). This affects payload capability and overall performance in an unfavourable manner when trying to balance the required complexity and efforts against overall effectivity and benefit. Concerning non validated statements about technology readiness and availability, design risks arise, which endanger the entire programme even in the late stage of development. This has been observed in the USA during the development of the VentureStar: main issue was the feasibility of achieving SSTO with rocket-based technology. The eventual failure of the required huge composite integral fuel tanks (as shown by its downscaled experimental vehicle X-33) made the dry mass excessive, when replaced with heavier metallic tanks. Coupled with an inevitable mass growth it finally became apparent, that there was no chance for VentureStar to reach orbit as a SSTO. The programme was cancelled in early 2002 after nearly 2 billion US$ were spent. A rocket-based SSTO would require about an 18% decrease in empty weight or a 7% increase in specific impulse referring to current technology [1, 6, 7].

The outcome of this is to use combined cycle engines, such as rocket/ram/scramjet/rocket, to increase overall specific impulse and to reduce the amount of on-board oxidizer. Investigations of ram/scramjet/rocket-propelled SSTO have been carried out in the beginning of the nineties. An American study was the National Aerospace Plane NASP. However, all experiences made with any SSTO concept proved its system complexity required high technology level, unacceptable sensitivities and development and operational risks [3, 4, 6, 8].

#### 3.1.2.2 Two-Stage-to-Orbit TSTO

After realizing that a SSTO is at least for the next generation of STS unrealistic and too risky for access to space – referring to present and near term technology – a reorientation towards fully or at least partially reusable Two-Stage-to-Orbit (TSTO) launch vehicles becomes apparent. In comparison to the former, a TSTO concept requires two autonomous vehicles, which have to be designed, produced and finally operated. This results in higher development and operational costs. Since the required kinetic energy amount is shared by two vehicles, design difficulties ease, although key technologies (such as propulsive, structural and thermal protection concepts) remain design critical. However the "orbital dry mass" of the second stage to deliver the same payload is significantly reduced compared with SSTO. Concerning aerodynamical, flight mechanical and dynamical aspects, stage integration and separation introduce an additional problem. However, by means of suitable staging configurations, a lower technology level can be utilized, which can work out favourably to reduce costs and risks [2–4, 9].

TSTO concepts can be divided in winged rocket-based VTHL vehicles and in aircraft-like HTHL vehicles mostly with airbreathing engines in the first stage and rocket engines in the second stage. Examples for the first group are the biamese FSSC-16 and in particular current US concepts emerging from the Space Launch Initiative (SLI). Vertical take-off TSTO rocket systems might be achievable with current available technology and are certainly the cheapest and least risky near term solution for access to space [3, 4, 6, 7].

Horizontal take-off TSTO systems are more complex and rather mid to long term solutions but offer aircraft-like operations (theoretically) from every airport, with comparably low level of infrastructure requirements. They are featured by a high degree of reusability and operability. Additionally they provide excellent potential of high tech transferability and spin-off for the development of hypersonic transports. Examples are the German Sänger concept, FSSC-12 and more recently Falcon in the US. Due to horizontal take-off, thrust to weight ratio can be kept low in the range of 0.4 to 0.6. Additionally, they can benefit from a high specific impulse of the airbreathing engines (turbojet, ramjet or scramjet) and the relinquishment of the heavy oxidizer in the first stage. This results in an exchange of oxygen mass and volume into mass and volume of the propulsion system, which finally increases dry mass. The orbiter (in a first step must not necessarily be reusable) can be designed more conventionally with rocket engines and does not need to carry unnecessary components into orbit and back. Thus sensitivities and design risks can be reduced [3, 4, 9, 10].

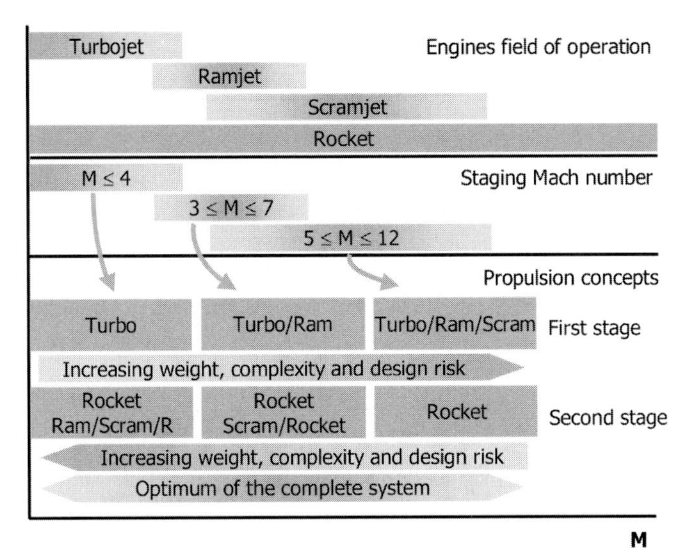


Figure 3.1.1: Engine concepts and effects on complexity and design risk over staging Mach number.

Concerning conceptual aspects, a TSTO provides separation of system components between first and second stage. Though both stages cannot be designed independently, they still can be designed relatively specifically for their field of operation. Here, velocity at stage separation defines the system characteristics of both vehicles, such as propulsive and structural concepts, materials, vehicle size, weights, stage integration and hence design risks and costs (see Fig. 3.1.1). Consequently, the major conceptual task is the determination of an optimum staging velocity or Mach number.

Depending on the mission requirement (payload mass in specified orbit) the "technical design-optimum" is achieving the delivery of the payload at minimum GTOM (Gross-Take-Off-Mass). However, considering in addition development and operational cost aspects a different of "optimum" might be more reasonable and overrule purely energetic stage optimization [3, 4, 11].

### 3.1.3 Design Procedure

Along with the German Hypersonic Technology Programme, the two Collaborative Research Centres 253 (Aachen) and 255 (Munich) scrutinize fully reusable TSTO space transportation systems, ELAC and HTSM (Hypersonic Transport System Munich) respectively. Both systems incorporate airbreathing horizontal take-off and landing concepts, as explained above. Hence, the first stage is featured by all aspects of a hypersonic aircraft and the second stage resembles a rocket-propelled orbital space plane [2, 12].

The conceptual design task is carried out at the Institute for Aeronautics at the University of the German Armed Forces Munich. Cadehyp (Computer Aided Design Hypersonics) is used for design and scaling of the HTSM. This programme resembles a numerical code which emerged from the Sänger system study and was successively upgraded and adapted to current requirements. Verification and validation have been carried out using datasets of the original MBB/DASA Sänger configuration. The creation respectively quality of input data decisively affects the design procedure itself and hence, the process is shortly outlined in the following [2, 9, 12].

### 3.1.3.1 Design Tools and Methods

The Cadehyp software is scaling a so-called basic configuration, described by datasets from the domains of aerodynamics, propulsion, weights, systems and geometry along a dedicated mission profile (see Fig. 3.1.2). Design optimization is performed by means of parameter variations (e.g. overall-mass and drag as well as fuel consumption and thrust). After a design point is found, sensitivity analyses are carried out in which the vehicle-behaviour is examined on possible changes of primary parameters. By doing so, technology requirements, risks and cost drivers can be identified. Additionally, the results are suitable for later evaluation and comparison of the concepts. This procedure is most notably reasonable especially in the early design phase, when necessary key system components like engines are not well defined [13].

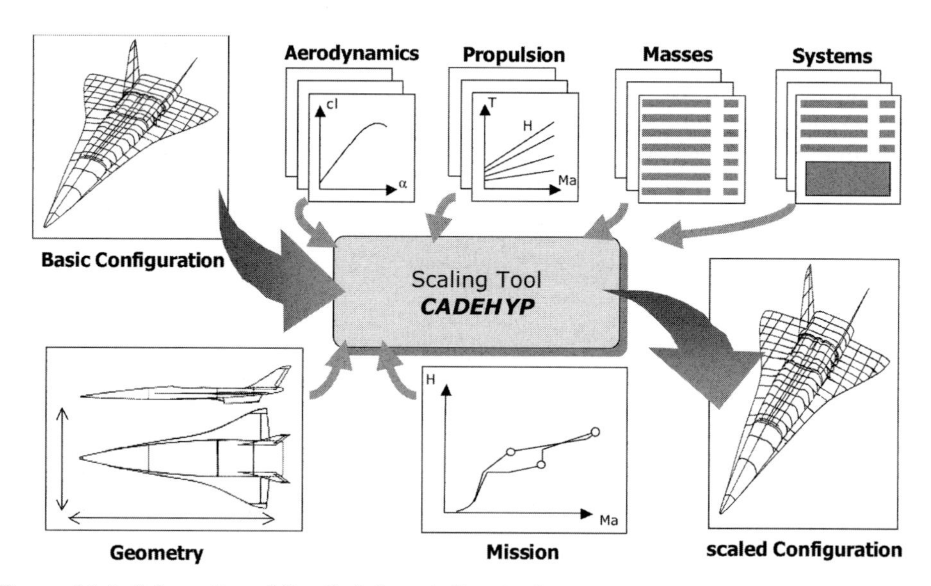


Figure 3.1.2: Schematics of the Cadehyp scaling tool.

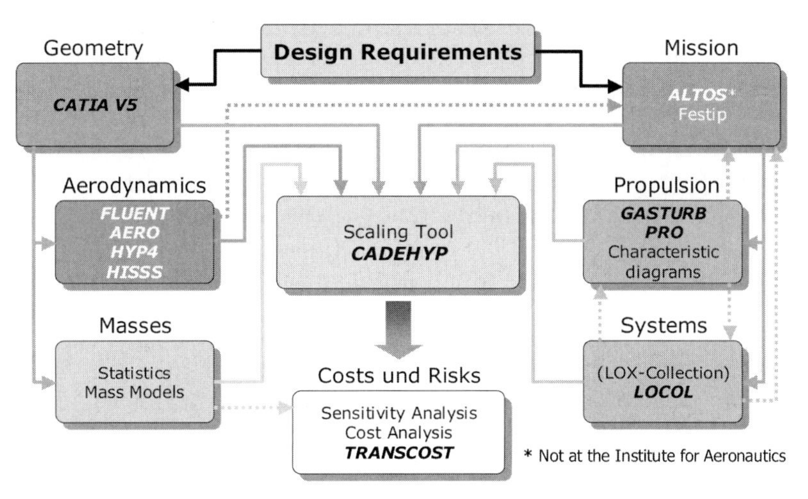


Figure 3.1.3: Methods and tools for data set generation of Cadehyp and within the conceptual design process.

Because Cadehyp does not generate, but scales the basic datasets, explanatory power is strongly dependent on the quality of input. By application of high-grade methods for dataset generation quality of output can be arbitrarily improved. Methods and tools for the design process and their connections are displayed in Fig. 3.1.3.

#### 3.1.3.2 Baseline Concept

Baseline concept for the Collaborative Research Centre 255 is the HTSM (Hypersonic Transport System Munich), which largely corresponds to the 1990 proposed, fully reusable Sänger concept. This TSTO was projected for independent access to space from Europe. Take-off and landing should be possible from southern sites in France or Spain. Stage separation should take place over northern to central Africa, ensuring near equatorial orbits. Hence a cruise-out segment was necessary to reach the designated staging latitude. Optimum separation and cruising mach numbers were found at $M_{\mathrm{SEP}}=6.8$ and $M_{\mathrm{CR}}=4.5$, respectively, using turbo/ramjet combined cycle engines. Concerning weight aspects, the first stage was favourably designed in complex hot structure with appropriate consequences for materials and architecture. Second stage was propelled by a single high-pressure rocket engine. The orbiter was designed in cold structure covered with TPS to withstand heat loads during ascent and reentry. Payload capability was 7.0 tons in a $28.5^\circ$ inclined 250 km circular orbit [9].

After validation and evaluation of the baseline configuration, a first step was to adapt and define current boundary conditions and requirements. In the following, mission and staging Mach number variations are carried out to find an optimum, viable TSTO RLV with minimum weight, complexity and hence reduced risks and comparably low costs.

#### 3.1.3.3 Boundary Conditions and Requirements

The comparison of different concepts implies similar boundary conditions and technical requirements. The most essential requirements and – in particular – design margins for a future European reusable launcher have been adopted from the FESTIP system study, displayed in Tab. 3.1.1. This is to ensure later comparability to concepts of this paradigmatic design study [3, 4].

#### 3.1.3.4 Variation of Mission and Staging Mach Number

Size, complexity, performance and hence risk and costs of a TSTO reusable space transport are strongly depending on mission and flight strategy. For identification of the optimum vehicle configuration, two mission types and different staging velocities have to be investigated. This is primarily performed using the original Sänger configuration and the dedicated cruise-out mission profile from Europe. Secondly, effects on the vehicle are examined for the more realistic equatorial accelerator mission with start and landing site in Kourou. Variations of separation Mach numbers are carried out between $3.5 < M_{SEP} < 6.8$ for both mission types. Mission type and velocity depending technical aspects and changes, such as propulsive and structural concepts, are considered. The out-

Table 3.1.1: Design requirements according to FESTIP for a future European reusable launcher System.

| | |
|---|---|
| Payload mass | 7.0 to (250 km LEO, 5° inclination)<br>2.0 to (250 km LEO, 98° inclination) |
| Payload volume | ∅ 4.7 m×10 m |
| Payload price | <2000 $/kg |
| Reusability | Full |
| Commissioning | 2020 |
| Life cycle | 30 a |
| Probability of loss | $<10^{-3}$/Mission |
| Vehicle turn around time | <20 d |
| Structural mass margin | 14% |
| TPS mass margin | 14% |
| Propulsion and system mass margin | 12% |
| Equipment mass margin | 10% |

come is an optimum vehicle which serves as new baseline configuration for further adaptations or trade studies [5].

#### 3.1.3.5 Trade Studies

The achievement of a broad basis of comparison is mandatory for evaluation of different TSTO concepts and, in particular, of new technologies. Starting from a previously found optimum configuration, trade-off studies are accomplished. In this connection, two system concept are scrutinized, which are of special interest: the application of an airbreathing second stage using scramjet engines and utilization of an in-flight Air-Collection-and-Enrichment-System (ACES). Both concepts provide the potential to reduce overall system mass.

#### 3.1.3.6 Evaluation and Comparison of the Concepts

For final system analysis and evaluation of different concepts or key technologies, comparability of the competitors has to be warranted. For a wide range of different concepts during the conceptual phase some Figures Of Merit (FOM) are helpful to obtain indications or trends for performance, robustness, flexibility, risk, safety and system respectively mission costs. The following figures of merit are used for analysis and evaluation [7]:

*Gross-Take-Off-Mass*

GTOM is particularly not a significant or strong figure of merit because it does not differentiate between propellants and hardware. Hardware is extremely expensive to develop, acquire and maintain. However, some statements about operations and infrastructure related costs will be received.

*Dry Mass*

Dry mass is a good indicator for development and acquisition costs, since it is directly connected to hardware. It is widely used at this stage of analysis, and frequently is input for cost analysis tools, such as e.g. the bottom down cost model Transcost [5].

*Growth Factor and Sensitivities*

These figures directly measure margins to the technical limit and indicate risks and requirements. Growth factors and sensitivities can be determined by variation of primary design parameters as weight, drag, thrust, fuel consumption, etc. [13].

*Complexity*

Overall system complexity indicates development and acquisition costs and risks. Complexity can be as simple as the number of major subsystem types.

*Wetted Area*

Wetted area is a good indicator for structures and thermal protection system related expenses. Especially for the second stage, TPS is a major contributor to maintenance and turn around time as well as to risk and safety.

Above introduced figures of merit of the considered vehicles can be finally incorporated in an evaluation matrix or any other pattern for comparison and identification of potentials and risks, which will lead to overall statements and recommendations on possible actions concerning further investigation on RLV concepts and applied technologies.

### 3.1.4 Variation of Mission and Mach Numbers

For $M_{SEP}$-variation some basic conceptual considerations need to be discussed in advance. Basically the investigated Mach number range spans from 3.5 to 6.8, since within this bandwidth the basic concepts – super-/hypersonic aircraft for the first stage and orbital space plane for the second stage – do not change dramatically. Regarding the lower limit of $M_{SEP}=3.5$, the orbiter becomes too bulky for proper integration in the first stage. Hence, first stage would rather be characterized as a small winged booster where application of airbreathing engines could hardly be justified. The separation Mach number of $M_{SEP}=6.8$ stands for the upper limit of the hot structure and the used heat protection system of the first stage. Exceeding this value a different structural concept is required with separate TPS components resulting in more weight and complexity. Depending on type of orbit, the second stage can become too small for payload integration especially in regard to the required volume of the payload bay for LEO (Low Earth Orbit) missions. Within the observed Mach number range the propulsion concept of the first stage requires a change at $M_{SEP}=4.0$ (the upper limit for application of turbojet engines). Exceeding this value the first stage requires a turbo/ramjet combined cycle engine. Changing the engine concept from turbojet to turbo/ramjet to increase separation Mach number resembles a technology step for the first stage and hence an increase in vehicle complexity and risk. However for the second stage the technology level can be found almost constant regarding to different staging Mach numbers. This very fact could be sufficient to define the staging Mach number at $M_{SEP}=4$ for both, cruising and accelerator mission [9].

Prior to performing mission and Mach number variations, the original Sänger 4/92 configuration (internally marked as HTSM-6811) had to be adapted to current requirements. This was generally done by increasing size of the orbiter payload bay and by adjusting the configuration to the FESTIP derived design margins. This resulted in an heavier complete system, since original mass margins were kept at only 5%. Hence orbiter mass raised from 115 tons to 127 tons and the GTOM of the HTSM-6812 configuration increased to 418 tons instead of 390 tons. Primary parameters and figures of merit of HTSM-6812 are presented in Tab. 3.1.2. The meaning of the *GF* (Growth Factor) and the SF (Sensitivity Factor) can be explained in the following: 1 ton more in dry mass yields to an increase of 2.14 tons in GTOM or 1% more drag results in 0.9 tons more GTOM. The complete *GF* at the end of the table summarizes the effects of payload mass changes on GTOM of the system (1 ton more payload results in 22.2 tons more GTOM) [9].

In a first step, the orbiter has been scaled for different Mach numbers. The mass range was found between 209 tons for $M_{SEP}=3.5$ and 127 tons for $M_{SEP}=6.8$. For both missions (cruise and accelerator mission as introduced below) the size of the second stage has been assumed constant for respective Mach numbers. Staging velocity depending orbiter size and hence drag changes for the complete system are incorporated into the first stage data sets. Then first stage Mach number variation is performed by scaling HTSM-6812 photographically, since this procedure allows to assume most of the aerody-

Table 3.1.2: Comparison of the adapted Sänger cruise configuration HTSM-6812 for a staging Mach number of 6.8 and the accelerator configuration HTSM-4021 for a staging Mach number of 4.0.

| Figure of Merit | HTSM-6812 1$^{st}$/2$^{nd}$ | HTSM-4021 1$^{st}$/2$^{nd}$ |
|---|---|---|
| GTOM [to] | 418.6/127.1 | 362.0/196.0 |
| Dry Mass [to] | 188.2/26.6 | 135.7/34.7 |
| Payload Mass [to] | 127.1/7.0 | 209.1/7.0 |
| Payload Mass Ratio [%] | 1.6/5.5 | 1.9/3.4 |
| GF [to/to] | 2.14/9.0 | 1.27/11.7 |
| SF drag [to/%] | 0.9/0.02 | 0.13/0.03 |
| SF thrust [to/%] | –2.9/–2.8 | –0.76/–5.3 |
| SF consumption [to/%] | 2.6/2.1 | 0.41/4.7 |
| Complexity | very high/medium | high/medium |
| Wetted Area [m$^2$] | 3403/787 | 2400/960 |
| Complete GF [to/to] | 22.2 | 15.4 |

namics as remaining nearly constant. For each Mach number and mission, thrust-to-weight ratio and wing loading was then optimized for minimum vehicle weight. For comparison with the cruise mission baseline concept HTSM-6812 the scaled accelerator concept for a separation Mach number of 4.0 HTSM-4021 is also presented in Tab. 3.1.2. Although orbiter mass increases due to the lower staging velocity the complete system GTOM is reduced by about 56 tons. *GF* and SF of the first stage are significantly lower due to the smaller vehicle but they increase for the orbiter. However, the first stage of HTSM-4021 has lower complexity since it is propelled by turbojet engines only.

### 3.1.4.1 Mission Comparison

*Cruise Mission*

As explained above, the original Sänger concept was designated to start and land in southern continental Europe. Stage separation occurred over northern to central Africa. Hence a cruise-out segment was required to reach the lateral distance of 2280 km. Prior to stage separation an eastward turn was performed. After staging the first stage reached back to the starting point. The cruising Mach number was $M_{CR}=4.5$ in ramjet mode at comparatively low dynamic pressure to minimize drag and fuel consumption. For concepts with lower Mach numbers cruise velocity is adjusted to ensure optimum conditions for turbojet and ramjet operation respectively. GTOM and dry mass for different configurations are displayed over separation Mach number (see Fig. 3.1.4) [9].

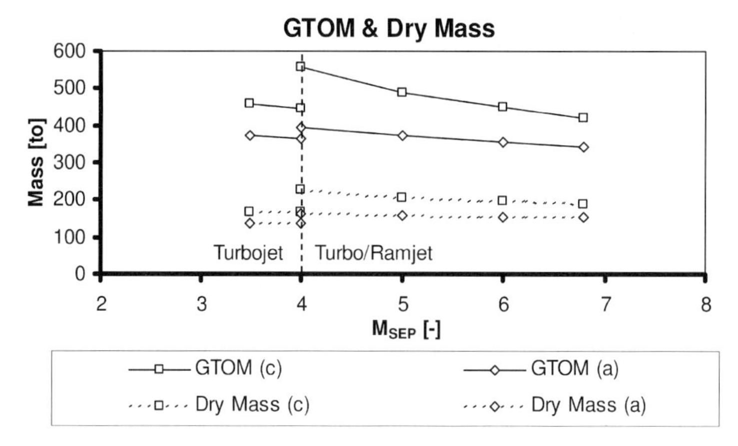


Figure 3.1.4: GTOM of the TSTO and dry mass of the first stage over Mach number for cruise and accelerator mission.

*Accelerator Mission*

During FESTIP system-study the take-off/landing site was shifted to Kourou in French Guayana. Due to equator proximity a cruise-out segment was no longer necessary to reach separation latitude. Hence it became a pure accelerator mission. The TSTO vehicle accelerates along a 50 kPa dynamic pressure trajectory in eastward direction (depending on orbit inclination) up to separation velocity and altitude. After separation of the orbiter, the first stage performs a 180° turn and reaches back to the starting point at Kourou [3, 4].

To avoid the use of complex combined cycle engines – as a FESTIP requirement – the staging Mach number was shifted down to $M_{SEP}=4$. Two configurations emerged: the FSSC-12D delta wing concept with higher aerodynamic quality and the FSSC-12T trapezoidal wing concept with less stage integration but thrust augmentation during atmospheric ascent by cross-fed rocket motors of the second stage [3, 4].

#### 3.1.4.2 Comparison of Mach Number Variation

Figure 3.1.4 presents take-off gross weight and empty weight of the vehicles for both mission types (cruise mission (c), accelerator mission (a)) over a variation of the staging Mach numbers. The black line with squares shows GTOM of the cruise configuration and the grey line with diamonds that of the accelerator configuration. The same applies for the dashed lines, which refer to the dry mass. The dotted black vertical line at M=4 stands for the maximum operation velocity of turbojet engines. If separation Mach number exceeds $M_{SEP}=4$, vehicles with combined cycle turbo/ramjet engines are considered. In this case transition from turbo to ramjet takes place at $M_{TJ/RJ}=3.5$. For a separation Mach number of 4 two vehicles with different engine concepts exist: one with turbojet only and the other one with turbo/ramjet combined cycle engine [9].

Concerning the cruise-mission (c) GTOM decreases with increasing separation Mach number. The change in engine concept at M=4 is responsible for the observed discontinuity. Engine mass rises due to more complex and heavier engines for higher Mach numbers and so does the dry mass. Additionally ramjet performance at comparatively low velocities and high altitudes (required for staging) is rather poor compared to the turbojets. This results in higher vehicle fuel demand and increased GTOM. As we have learned above, the decisive figure of merit is dry mass since it is directly related to hardware and costs. GTOM optimum can still be found at $M_{SEP}=6.8$ as for the baseline Sänger-configuration but for dry mass it is located at $M_{SEP}=4.0$.

Observing the accelerator mission – due to absence of a cruise-out segment and hence the shorter overall flight distance – GTOM and dry mass is much lower compared to the Europe mission. Anyhow, analogous trends compared to the cruise-mission can be found, though mass changes are less significant. The higher the separation Mach number, the longer the distance for acceleration from home base to separation point. This results in a higher fuel demand and in more mass for the first stage. Hence, velocity dependent mass reductions of the second stage does not lead to strong influences on GTOM of the complete system. With respect to dry mass, the same findings as for the cruise-mission also apply for the accelerator mission.

The following diagrams show results of the sensitivity analysis for both mission types. In the first diagram, the volume uncritical (dead weight) and the volume critical weight growth factors (*GF*) are displayed as functions of separation Mach number (Fig. 3.1.5). The second diagram deals with sensitivity factors (SF) concerning variations in drag, thrust and fuel consumption (Fig. 3.1.6).

Dead weight variations are such of e.g. structural mass and cause indirect volume changes of the vehicle. In contrast to that, volume critical variations are mass changes, which directly influence volume. The latter can be described as an increase in fuel mass that does not contribute to thrust generation (cooling, fuel reserve etc.) [10, 13].

In Fig. 3.1.5 the following basic trend is apparent for both mission types and growth factors: at higher Mach numbers sensitivities towards mass changes increase. The cruise mission vehicles react more distinctively compared to those from the accelerator mission. A change of the engine concept from turbojet to turbo/ramjet combined cycle engines results in a discontinuity and a sudden rise of the *GF*, especially for the cruise concepts. This is due to less thrust and specific impulse of the ramjet and increased dry mass from a more complex and heavier engine. The difference between dead weight *GF* and volume critical *GF* is also more significant for cruise vehicles. This higher sensitivity results from the cruise segment where fuel is spent which does not contribute for acceleration or to increase kinetic energy of the vehicle.

Observing the sensitivity factors in Fig. 3.1.6 of both mission types, similar attitudes appear: the cruise vehicles react more sensitive towards changes in

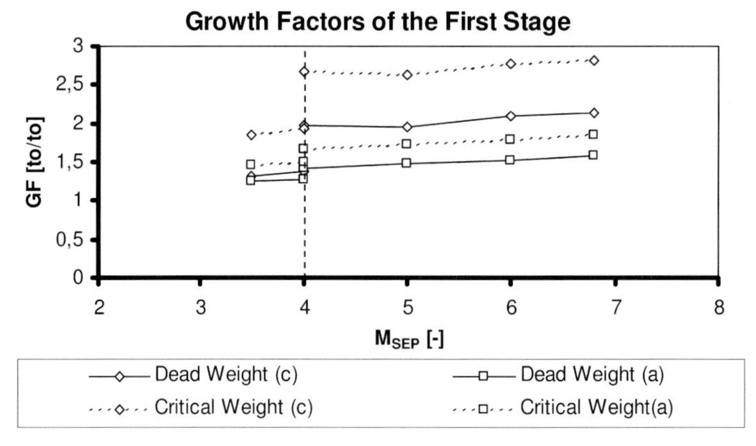


Figure 3.1.5: *GF* of the first stage over staging Mach number for cruise and accelerator mission.

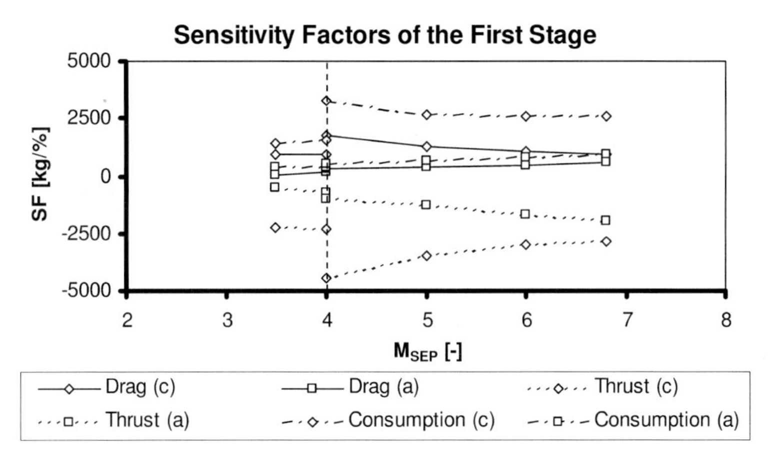


Figure 3.1.6: SF of the first stage over staging Mach number for cruise and accelerator mission.

drag, thrust and fuel consumption. Additionally, a more distinctive discontinuity is found when engine concept is changed. Cruise vehicles with turbo/ramjet combined cycle engine show a reverse SF trend for increasing Mach numbers. Raising the velocity at stage separation – leaving lateral distance constant at 2280 km – means that a larger part of the route is required for acceleration. Therefore, the cruise segment minimizes and a higher fuel fraction is used for acceleration, which finally is responsible for the reverse trend.

The concepts for the Kourou mission become more sensitive with raising $M_{SEP}$. Additional time is required to accelerate to the separation Mach number of 6.8 compared to lower values. Consequently the distance from take-off/landing site increases. The $M_{SEP}=6.8$ vehicle takes nearly the same distance from the home base as its counterpart of the cruise mission. Hence, both SF nearly converge since they originate from the same baseline configuration.

The most sensitive parameter (of both mission types) is thrust followed by fuel consumption – both parameters are engine-related. Consequently, to provide lower risks and costs, profound design knowledge and high demands are required on the propulsion system. Since drag changes have a comparatively low influence on the vehicle, aerodynamic quality is less mandatory, especially for the accelerator mission and low staging Mach numbers.

In Fig. 3.1.7 the orbiter and the complete system growth factors are presented. Concerning the second stage, increasing Mach numbers lead to decreasing orbiter size – since less kinetic energy is required to reach orbit – and so does the *GF*.

The complete growth factor describes the effects on the TSTO system by changing payload mass of the second stage. Resulting mass, size and drag changes of the orbiter are incorporated into scaling of the first stage. This is only valid for slight changes around the design point, so that entire parameters, such as aerodynamics, stability etc. do not change substantially. For preliminary

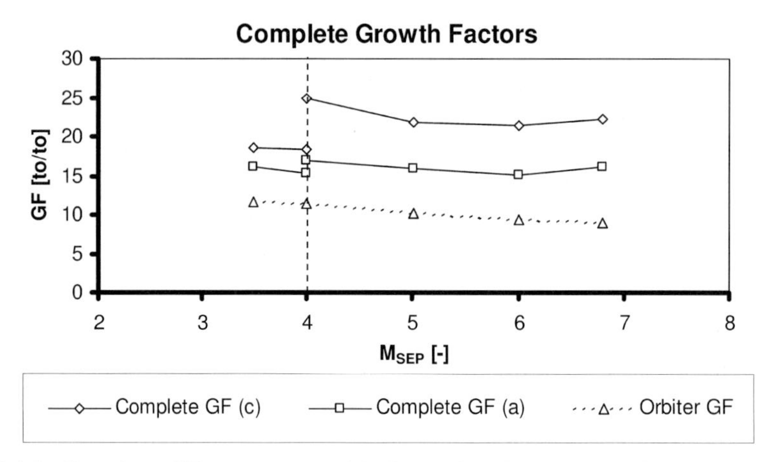


Figure 3.1.7: Complete *GF* over staging Mach number for cruise and accelerator mission and in comparison *GF* of the second stage.

estimations, the *GF* of both stages can be multiplied. The overall system *GF* allows comparison even of SSTO and TSTO concepts.

Similar trends as for the lower stage *GF* and SF are apparent. The growth factors of the cruise mission are more distinctive than of the accelerator mission. The cruise segment – where fuel does not contribute to kinetic energy increase, but has to be stored in the vehicle and accelerated – is responsible for high system growth factors. The values of the accelerator mission sum up at approximately 16 and they do not change much for different Mach numbers. There are two optima apparent: one is located at $M_{SEP}=4$ and the other one is found around $M_{SEP}=6$. However the vehicle with lower staging Mach number has a less complex engine concept. But depending on mission requirements such as orbit altitude for e.g. a GTO mission – which would result in a heavier second stage – a higher $M_{SEP}$ could yield to a lighter complete system. Pertaining to the above defined design requirements the optimum staging Mach number is located at 4.0. This also applies for the cruise mission where the concepts with combined cycle engines have significant growth factors ($\approx 23$ compared to $\approx 18$ with turbojet only). In comparison to that, the FESTIP FSSC-1 SSTO concept has an overall growth factor of $GF > 25$.

Accounting for weight, sensitivity and complexity aspects, the optimum staging Mach number can be found at $M_{SEP}=4$ for both missions. However, the lightest (GTOM and dry mass) and less sensitive vehicle is found for the accelerator mission with a GTOM of 362 tons, a dry mass of 136 tons and a *GF* of 15.4. In terms of minimum weight and cost effectiveness, this kind of mission-type should be the choice for future European TSTO launch vehicles.

#### 3.1.4.3 Accelerator Vehicle Concepts

The findings from FSSC-12 concepts and investigations from above lead to the design of two additional configurations: HTSM-3522 for the lower limit ($M_{SEP}$=3.5) and HTSM-6822 for the upper limit ($M_{SEP}$=6.8). This is to scrutinize accelerator-mission influences on TSTO concepts. Both first stage vehicles are featured by hybrid wings, simpler structural concepts, less stage integration and hence lower aerodynamic quality but also less complexity. The second stage and other primary systems especially the engine concepts remain unchanged referring to separation Mach number. In Tab. 3.1.3 the accelerator concepts HTSM-3522 and HTSM-6822 are compared.

It is apparent that *GF* and SF are significantly higher for the HTSM-6822 first stage compared to the other two concepts. This is based in the utilization of a ramjet engine with lower thrust and specific impulse and hence acceleration potential compared to turbojet engines. It results in additional acceleration time to staging condition and causes a large radius of action (as already mentioned above).

### 3.1.5 Trade Studies

Based on the findings of mission and Mach number variation, two trade studies are investigated. This is primarily to examine and evaluate different concepts and key technologies within the preliminary design process. It also helps for optimization of the overall concept and decision finding and increases design knowledge. Trade-off studies may include minor changes, such as changing wing geometry or major changes as changes in the propulsion concept for both stages [2, 12].

Table 3.1.3: Comparison of the two accelerator TSTO concepts HTSM-3522 with staging Mach number at 3.5 and HTSM-6822 with staging Mach number of 6.8.

| Figure of Merit | HTSM-3522 $1^{st}/2^{nd}$ | HTSM-6822 $1^{st}/2^{nd}$ |
|---|---|---|
| TOGW [to] | 349.3/209.1 | 359.2/127.1 |
| Dry Mass [to] | 109.5/34.7 | 150.1/26.6 |
| Payload Mass [to] | 209.1/7.0 | 127.1/7.0 |
| Payload Mass Ratio [%] | 2.0/3.34 | 1.94/5.5 |
| GF [to/to] | 1.2/11.7 | 2.25/9.0 |
| SF drag [to/%] | 0.28/0.034 | 2.1/0.02 |
| SF thrust [to/%] | –1.0/–5.9 | –5.9/–2.8 |
| SF consumption [to/%] | 0.38/5.1 | 2.8/2.0 |
| Complexity | medium/medium | high/medium |
| Wetted Area [$m^2$] | 2150/990 | 2790/787 |
| Complete GF [to/to] | 16.7 | 25.3 |

A main weight contributor of a TSTO space transport is the required oxidizer mass for the second stage rocket engines. Most technology enhancements aim at the reduction of the amount of liquid oxygen carried aboard. This can be done by consistently expanding application of airbreathing engines or by technologies, which provide atmospheric oxygen for later use in the rocket engines. In regard to the former, influences of a scramjet equipped airbreathing second stage on the overall system efficiency have been examined. And – concerning the latter – a device for collection and enrichment of aerial oxygen during flight has been applied to a TSTO. The potentials of the aforementioned two concepts are explained in the following.

#### 3.1.5.1 Airbreathing Second Stage

Changing rocket propulsion to airbreathing turbo/ramjet engines in the first stage of an TSTO delivered most of the benefits for aircraft-like operations of RLVs. Expanding this type of engine operations to the second stage (while accelerating in the atmosphere) might reduce separation mass of the orbiter due to a higher specific impulse of the airbreather. Potentials of an airbreathing second stage are scrutinized with the concept HTSM-4025.

Integration of the hypersonic airbreathing engine (dual mode ram-/scramjet) in the second stage has one major advantage. Limited thrust-to-weight ratio at high Mach numbers in conjunction with less system mass of the orbiter results in a (at least) sufficient acceleration capability. Scramjet application in the first stage or even in SSTO would reduce acceleration significantly due to the higher dry mass, caused by the complex combined cycle propulsion system (turbojet/ram-/scramjet and rocket).

Airbreathing engines in hypersonic flight regimes require a fully integrated design of the vehicle. The complete bottom side of the fuselage is part of the scramjet engine (Fig. 3.1.8). Precompression on the forebody and expansion along a large semi-expansion-ramp nozzle is necessary to increase efficiency and thrust at high Mach numbers. Although an optimized engine design is integrated, the airbreather lacks the high thrust potential of the rockets (and thrust is crucial for an accelerator vehicle).

For optimum engine performance the trajectory of the orbiter has to be adapted to the airbreathing engine characteristics. Precompression of the ambient air requires high dynamic pressures during ascent. This leads to increased aerodynamic drag and heat loads on the structure. Heat loads on the orbiter might even be higher during hypersonic ascent than during re-entry into the atmosphere.

Consequently the vehicle has to be covered with a high-temperature thermal-protection-system (TPS) over most of the vehicle's surface, thus increasing dry mass. Internal layout (tank configuration, payload bay) and the rocket engine integration in the vehicle's base area had to account for the low drag configuration. This has only been possible with a reduction in payload volume and

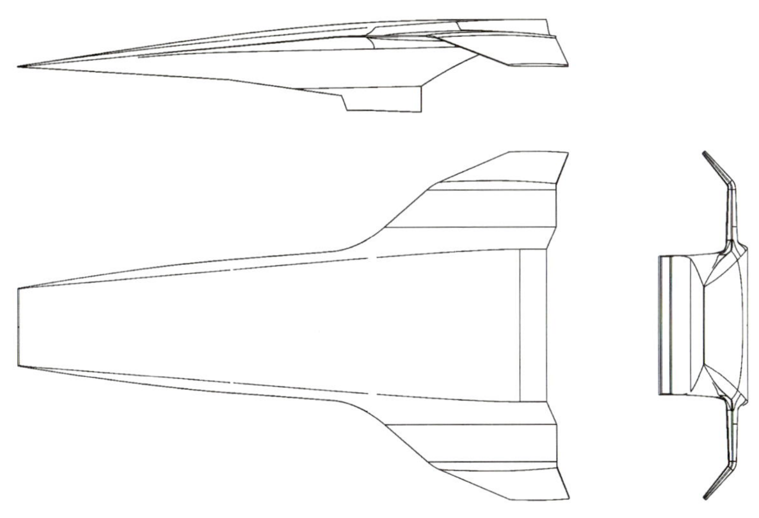

Figure 3.1.8: 3-view of the airbreathing orbiter.

mass (from 7 to down to 5 to). Strong coupling of the engine's performance (thrust vectoring and pressure forces on the forebody) with the aerodynamics requires wing design be part of the engine optimization, to keep the orbiter controllable throughout the mission (Fig. 3.1.9).

Comparison of the airbreathing second stage with a "conventional" rocket orbiter of HTSM-4023, adapted to the same mission requirements (e.g. 5 to payload to LEO), showed no gain in separation mass for the airbreather. The benefits of the increased specific impulse are overcompensated by the large increase in empty mass caused by the dual propulsion system, high-temperature

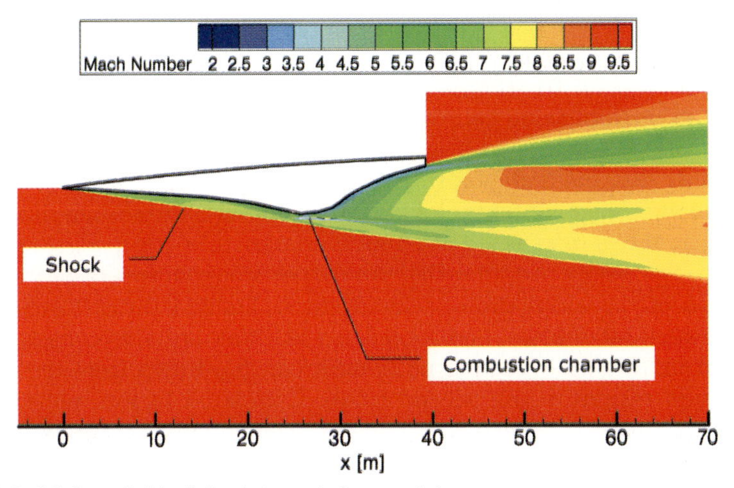


Figure 3.1.9: 2d-flow field of the integrated scramjet.

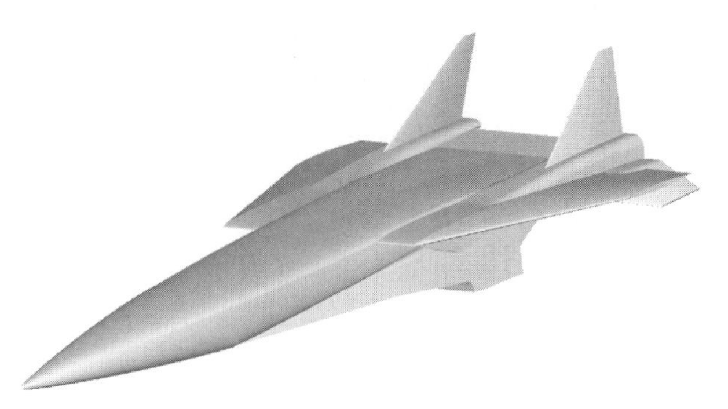

Figure 3.1.10: TSTO with integrated airbreathing orbiter.

TPS and the complex configuration. Scramjet performance is the main design driver. High uncertainties in thrust prediction at the present time may quickly lead to design divergence (separation mass "explodes").

Evaluation of the complete concept has to include the first or carrier stage. According to optimum staging conditions, a simple design, in means of the propulsion system, yields to a separation mach number of $M_{SEP}=4.0$. As with the scramjet engine in the hypersonic flight regime it is also crucial for engine performance – at high supersonic flight – to increase propulsion efficiency by precompression in front of the inlet.

A rocket-propelled orbiter allows for great flexibility in regard to stage integration. System-checks prior to stage separation are necessary, especially for the main propulsion system. In airbreathing mode this means sufficient dynamic pressure with as low distortion as possible. To comply with these boundary conditions an integration of the orbiter is only feasible in the bottom side of the carrier stage (see Fig. 3.1.10).

In this configuration the turbojet engines cannot be integrated in the fuselage, to utilize forebody precompression as in Sänger. Engine Nacelles have to be integrated in the wings for two reasons. Firstly, precompression of the inboard wing panel can be used, and secondly, the engines can be installed far backwards, to prevent large centre-of-gravity (CoG) shifts during stage separation.

Although this method helps to reduce CoG shift, it is not sufficient to maintain controllability. To keep longitudinal stability within margins, a partial integration of the orbiter's aerodynamic surfaces into the first stage aerodynamic hull is necessary. Additionally, the carrier stage suffers from the complex configuration of the airbreathing second stage. Limited thrust-to-weight ratio, increased structural masses and hence sensitivities are the consequences as summarized in Tab. 3.1.4. Again, no benefit can be drawn from the airbreathing propulsion system in the second stage. Compared to the reference TSTO this leads to increased TOGW, empty weight, complexity and design risks.

For an accelerator vehicle the use of airbreathing engines can only result in advantageous designs, if the high specific impulse is accompanied by a high

Table 3.1.4: Comparison of the conventional TSTO with rocket-propelled orbiter HTSM-4023 and the concept with scramjet-propelled second stage HTSM-4025.

| Figure of Merit | HTSM-4023 1st/2nd | HTSM-4025 1st/2nd |
|---|---|---|
| GTOM [to] | 285.3/169.5 | 307.7/171.8 |
| Dry Mass [to] | 87.7/28.0 | 94.4/47.5 |
| Payload Mass [to] | 169.5/5.0 | 171.8/5.0 |
| Payload Mass Ratio [%] | 1.75/2.94 | 1.62/2.91 |
| GF [to/to] | 1.4/10.8 | 1.66/5.2 |
| SF drag [to/%] | 0.436/0.09 | 0.94/0.48 |
| SF thrust [to/%] | –1.4/–5.5 | –2.0/–2.5 *, –1.3 ** |
| SF consumption [to/%] | 0.48/4.8 | 0.62/2.3 *, 0.8 ** |
| Complexity | medium/medium | high/very high |
| Wetted Area [m$^2$] | 2270/849 | 2508/1883 |
| Complete GF [to/to] | 15.5 | 8.6 |

* Rocket propulsion
** Scramjet propulsion

thrust-to-weight ratio. This is not the case for scramjet engines at the present state. Therefore, rocket engines are the favourable choice for this kind of application [14].

#### 3.1.5.2 LOX-Collection

Operation of an Air Collection and Enrichment System (ACES) provides – as primary effect – significant reduction in GTOM of the space transport due to empty LOX-tanks of the orbital stage. This may result in structural and empty mass savings particularly for the first stage. The required oxygen for the upper stage does not need to be accelerated at the beginning of the mission. It can be collected at comparatively high velocities. Another benefit is the possibility to confine the use of airbreathing propulsion to Mach numbers of 4 or even less with pure turbojets instead of combined cycle engines. Using liquid hydrogen for cooling and liquefaction of oxygen prior to thrust generation could yield to additional synergy-effects. However, the aforementioned potentials cause an increase in complexity of the first stage due to an additional system besides and in accordance with the engines. Further, ACES is critical to overall success – failure means mission abort [15, 16].

The integration of the ACES is only viable in the airbreathing first stage. The vehicle needs to be propelled by cryogenic liquid hydrogen to provide cooling capacity for the chemical engineering process of oxygen extraction. Pertaining to the complete system mass at take-off, an exchange of LOX-weight of the second stage into ACES-mass (dry mass) and additional fuel for the first stage will take place. Another important aspect is the purity grade of the extracted oxygen, which is related to the specific impulse of the second stage rocket engines. A 1% dilution of oxygen purity diminishes specific impulse of $LH_2$/LOX rocket engines by 1.5 sec. Compared to ground-based oxygen extraction facilities, ACES can reach a purity of 90–98%. If specific impulse of the rocket engines decreases due to impure oxygen, more LOX and $LH_2$ are required to meet the kinetic energy demand of the second stage. Thus orbiter size will increase, which results in additional mass and drag for the complete system [17, 18].

For aircraft-integration two types of air-separation devices are possible: the rotary separator and the vortex-tube separator. The former was invented and developed up to a prototype plant in the sixties of the last century [19]. Although the rotary separator's functionality was proven, a vortex-tube device is considered for the following investigation due to its higher efficiency, LOX-purity and comparatively low complexity [20]. An arrangement of different heat exchangers and turbo components is necessary to provide the working conditions and high pressures for the separation device. In trade studies it was found, that LOX-collection at subsonic cruise was only viable with strong turbojet interference to provide the high working pressure. At supersonic acceleration most ACES-components, especially precooler, compressor and turbine, need to be adjustable and hence complexity and risks are increased drastically [17, 18].

According to Fig. 3.1.11 air enters the vehicle via a combined variable supersonic intake. During ACES operation upstream of the turbojet engine, a baffle is opened which feeds a fraction of the air stream to the LOX-collection plant. To account for the increased air flux, the intake size has to be designed for more capacity. Air for the oxygen-extraction process does not contribute to the propulsion process. Momentum is generated opposing flight direction, thus resulting in additional drag. This can be partly compensated by ejecting nitrogen enriched air for thrust generation [17, 18].

The ACES-architecture displayed in Fig. 3.1.11 is the most practical for supersonic cruise application. Its thermal and mechanical energy balances can be compensated without additional or auxiliary components (APUs etc). Connections to the engines are only via combined intake and hydrogen flux. This enables a minimum of interference with the turbojets and helps to reduce complexity and risks. Optimum working conditions have been found at a cruising Mach number of $M_{CR}=2.75$ and a comparatively high dynamic pressure of $q=50$ kPa [17, 18].

The potential of LOX-collection strongly depends on mission and flight strategy. To identify the optimum conditions for ACES-application in a TSTO, different mission types (cruise- and accelerator mission) and staging conditions – similar to the analysis explained in the previous chapter – have been investigated using the Sänger as baseline configuration.

Since a cruise segment is required for operating the collection device, the accelerator mission is found to be impracticable, even when using a supersonic loiter at high Mach numbers (to maintain distance from take-off/landing site) instead of a cruise-out segment. Due to longer flight duration and increased

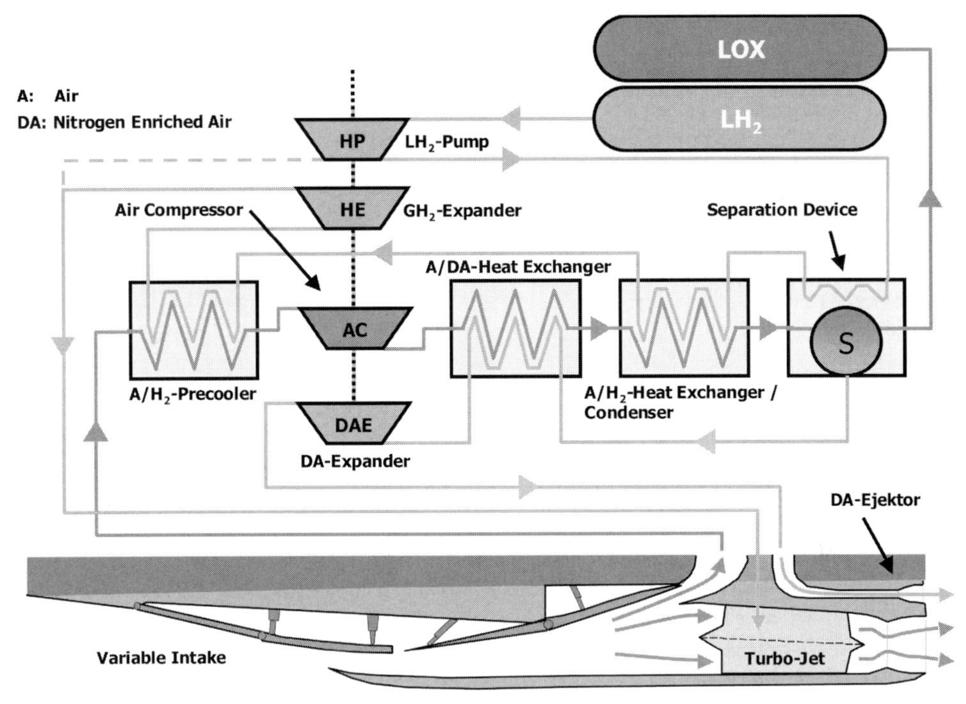


Figure 3.1.11: Schematics of the ACES in connection with propulsion system of the first stage.

fuel demand compared to conventional accelerator vehicles, LOX-collection provides only slight GTOM reductions and non of dry and fuel mass. This hardly justifies its application into TSTO accelerator concepts [17, 18, 21].

Consequences look different for the original Sänger cruise mission from Europe. In Fig. 3.1.12 GTOM and dry mass are displayed over staging Mach number. It can be observed that significant reductions are achievable. This results mainly from mass savings in the wing or structural group and the propulsion group, which will be explained below. Accounting for mass and complexity aspects, the optimum staging Mach number for an ACES-vehicle can also be found at maximum operating velocity of the turbojet engines [21].

To evaluate the effects on the complete launch system a detailed comparison of the cruise vehicles (HTSM-4014 with ACES and HTSM-4012 without ACES) becomes necessary. Maximum mass reductions for the ACES-vehicle were found at lower wing loading and thrust-to-weight ratio (based on GTOM). Due to a lighter vehicle in early mission segments, acceleration (even through the transonic regime) can be performed with less thrust. This results in mass savings for structure and engines. The ACES was scaled for a mass of 3.3 tons plus 1.7 tons of piping for the cross feeding system, which corresponds to an air flux of 280 kg/s. In spite of the ACES as an additional system, dry mass savings of 15 tons are achieved. LOX-collection takes place at $M_{CR}=2.75$ and requires a cruising time of 40 minutes.

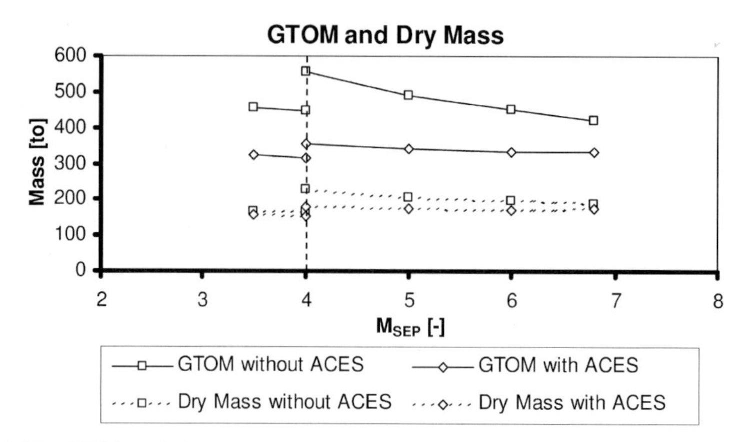


Figure 3.1.12: GTOM of the TSTO and dry mass of the first stage over staging Mach number for the cruise mission and for concepts with and without ACES.

Cruising of the conventional Mach 4 vehicle can be performed at $M_{CR}=3$ and lower dynamic pressure (minimum fuel consumption of turbojet engines.) During LOX-collection the ACES generates additional drag. Furthermore, the weight increase of the vehicle causes a rise in lift demand. This affects a higher flight angle of attack and hence, produces more induced drag. Both aspects are responsible for an increased fuel demand of 20 tons of the ACES vehicle.

Due to impure LOX and hence, less specific impulse of the rocket engines, the second stage of the ACES-vehicle requires more fuel/oxidizer to meet the $\Delta v$-demand. This leads to a mass increase of 9 tons for the orbiter. Eventually, in comparison to the conventional vehicle, GTOM savings of 130 tons are obtained [15, 21].

Concerning sensitivities, the growth factors of the first stage nearly commensurate. But a significant difference is apparent for the growth factors of the

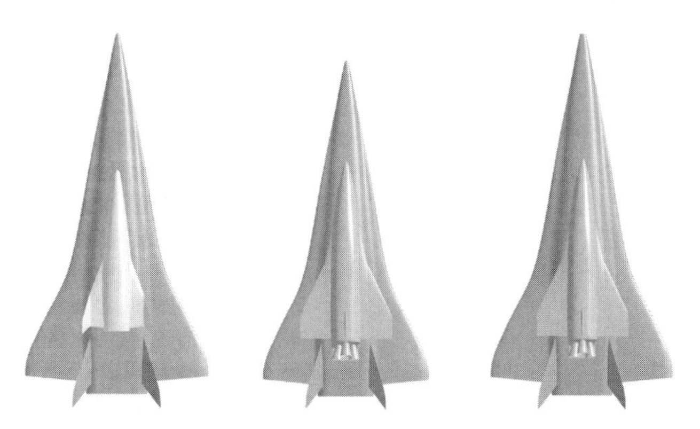

Figure 3.1.13: Top view of the three cruise-vehicles (from left to right) HTSM-6812, HTSM-4012 and HTSM-4014.

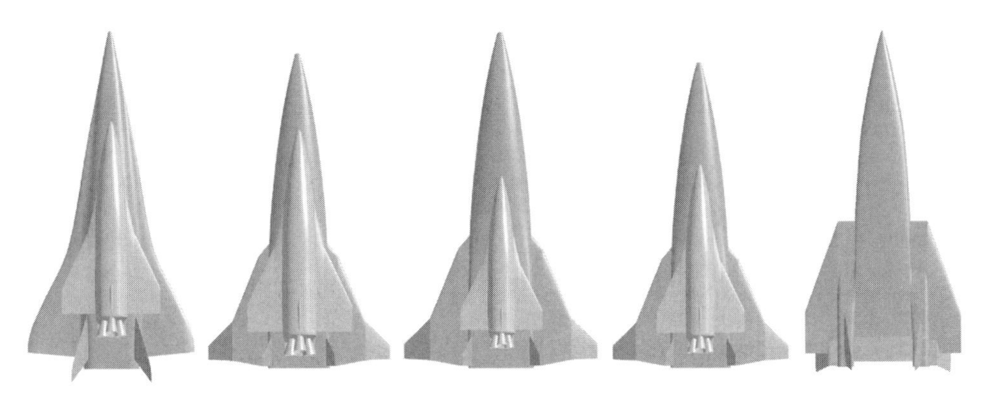

Figure 3.1.14: Top view of the five accelerator vehicles (from left to right) HTSM-4021, HTSM-3522 and HTSM-6822, HTSM-4023 and HTSM-4025.

complete system as apparent in Tab. 3.1.5. Reduced sensitivity of the ACES TSTO is primarily founded in the lower GTOM and dry mass. Due to empty LOX-tanks of the second stage at take-off, its weight changes have less influence on the complete system. The same applies for the drag, thrust and consumption sensitivities, which are lower for the lighter vehicle. Due to ACES as an additional system besides the engines, the complexity of the first stage is higher. A cross feeding system for the LOX supply is required for the orbiter, which would increase its complexity with respect to stage separation [21].

It shows, that the ACES potential to reduce GTOM and dry mass only applies for a turbojet engine propelled TSTO vehicle, which is designed for a cruise mission and a staging Mach number of $M_{SEP}=4$. If there are requirements for direct access to space from continental Europe, an ACES TSTO would provide a viable mid to long term alternative.

Table 3.1.5: Comparison of the HTSM-4012 vehicle without ACES and the HTSM-4014 vehicle with ACES.

| Figure of Merit | HTSM-4012 $1^{st}/2^{nd}$ | HTSM-4014 $1^{st}/2^{nd}$ |
|---|---|---|
| GTOM [to] | 446.4/196.0 | 316.4/205.5 |
| Dry Mass [to] | 166.9/31.6 | 151.9/32.0 |
| Payload Mass [to] | 196.0/7.0 | 205.5/7.0 |
| Payload Mass Ratio [%] | 1.57/3.57 | 2.21/3.41 |
| GF [to/to] | 1.38/11.4 | 1.38/11.5 |
| SF drag [to/%] | 0.95/0.03 | 0.71/0.037 |
| SF thrust [to/%] | –2.3/–5.3 | –1.8/–5.6 |
| SF consumption [to/%] | 1.5/4.7 | 1.0/4.9 |
| Complexity | high/medium | very high/medium |
| Wetted Area [$m^2$] | 2789/960 | 3097/965 |
| Complete GF [to/to] | 18.3 | 11.5 |

Table 3.1.6: Evaluation matrix of the three cruise-vehicles.

| HTSM | 6812 | | 4012 | | 4014 | |
|---|---|---|---|---|---|---|
| Stage | 1st | 2nd | 1st | 2nd | 1st | 2nd |
| GTOM | 4 | 2 | 5 | 3 | 1 | 3 |
| Dry Mass | 5 | 1 | 4 | 2 | 3 | 2 |
| Growth Factor | 5 | 3 | 2 | 4 | 2 | 4 |
| Complexity | 4 | 2 | 3 | 2 | 4 | 3 |
| Wetted Area | 5 | 1 | 3 | 2 | 4 | 2 |
| Stage Average | 4.6 | 1.8 | 3.4 | 2.6 | 2.8 | 2.8 |
| Average | 3.2 | | 3.0 | | 2.8 | |

Table 3.1.7: Evaluation matrix of the five accelerator vehicles. The two concepts marked with * have a payload capacity of 5 tons instead of 7 tons for the other concepts.

| HTSM | 4021 | | 26822 | | 3522 | | 4023 * | | 4025 * | |
|---|---|---|---|---|---|---|---|---|---|---|
| Stage | 1st | 2nd | 1st | 2nd | 1st | 2nd | 1st | 2nd | 1st | 2nd |
| GTOM | 2 | 3 | 2 | 3 | 2 | 5 | 2 | 3 | 3 | 3 |
| Dry Mass | 2 | 2 | 3 | 1 | 1 | 3 | 2 | 2 | 2 | 5 |
| Growth Factor | 1 | 4 | 5 | 3 | 1 | 4 | 1 | 4 | 3 | 3 |
| Complexity | 2 | 2 | 4 | 2 | 2 | 2 | 2 | 2 | 3 | 5 |
| Wetted Area | 2 | 2 | 3 | 1 | 1 | 3 | 2 | 2 | 3 | 5 |
| Stage Average | 1.8 | 2.6 | 3.4 | 1.8 | 1.4 | 3.4 | 1.8 | 2.6 | 2.8 | 4.2 |
| Average | **2.2** | | 2.6 | | 2.4 | | 2.2 | | 3.5 | |

### 3.1.6 Comparison and Evaluation

For comparison and evaluation all examined HTSM concepts are incorporated in a simple evaluation matrix. It is assumed, that none of the five introduced figures of merit has a special emphasis. The pattern consists of the FOM for each first and second stage. It is ranging from 5 which is poor to 1 which is good for the respective design concept. An average value is determined for carrier and orbiter and eventually both values are again averaged to acquire the final result. At first the three cruise concepts are addressed and then the five accelerator concepts. Additionally, the top view of each concept is presented.

The evaluation matrix approves the results from the mission and Mach number variation and the trade studies performed. All cruise mission concepts suffer from the high amount of fuel they have to store and carry. Hence, the carrier vehicles become bulky and heavy. It can work out advantageously to use the cooling capacity of liquid hydrogen to collect aerial oxygen for the second stage rocket engine during cruise. Thus, at least a part of the fuel spent for cruise-out – which does not contribute to increase kinetic energy of the vehicle – can be used to increase energetic efficiency of the overall system. If there are demands for a cruise mission – to gain access to space from continental Europe – the ACES TSTO could be a viable alternative. However, the mission of choice is that of a pure accelerator system.

For accelerator missions a turbojet-propelled first stage at a separation Mach number of 4.0 offers the best results concerning weight, size, sensitivity and provides the concept with the lowest complexity, risk and hence costs. The expansion of airbreathing engine operation to the orbiter – to reduce overall oxidizer mass – yields to an increase in dry mass, complexity and risks primarily for the second stage but also for the first stage. As long as a scramjet engine suffers from a low thrust density it cannot compete with rocket engines especially for accelerator vehicles.

### 3.1.7 Conclusion and Outlook

Different TSTO reusable space transports have been investigated which are featured by an airbreathing super-/hypersonic aircraft as first stage and by an orbital space plane as second stage. Baseline configuration was the original Sänger concept where requirements have been adapted to current values for a first analysis. Thereafter, the cruise mission from Europe and the accelerator mission with Kourou as launch site have been examined for a variation of staging conditions. Sensitivity analyses have been performed to identify possible risks and margins to technical limits. According to low complexity, risks and costs, for both missions the respective optimum vehicle has been found at separation mach numbers of 4.0 – the maximum operation velocity for hydrogen-fuelled turbojet engines. Comparing both mission types, the accelerator mission provides the most viable concept for a TSTO space transport system. Here, efficiency is mainly depending on the accelerating potential where thrust is crucial, which can be observed in high sensitivities. Hence, a major design driver will be hypersonic turbojet engines in contrast to aerodynamic quality which can be abated to reduce structural complexity.

Based on these findings, trade-off studies have been carried out to evaluate other concepts and technologies. For the reduction of the oxidizer amount, a scramjet-propelled second stage and a TSTO equipped with an air collection and enrichment system have been of special interest. Since the scramjet lacks the high thrust potential no mass savings could be obtained. The uncertainties concerning its performance and the high sensitivities of this propulsion system can lead to design divergence and hence to tremendous risks. Thus, application of a scramjet rather increases complexity risks and costs primarily for the second stage but also for the first stage. Consequently, according to today's technology knowledge it is inappropriate for utilization in an accelerator vehicle.

An air collection and enrichment system aboard a TSTO launch vehicle is only suited for a cruise mission. Here energetic efficiency of the complete system can be slightly increased since the fuel, which does not contribute to the acceleration process, is used for oxygen extraction for the second stage. However, complexity and empty weight of the first stage are rising and the device is mission critical. In comparison to accelerator vehicles it cannot compete due to its size and dry mass. Hence, it provides only an alternative for access to space from continental Europe.

Near term space transportation will certainly be performed with rocket derived launchers which successively increase reusability. An example is the German effort for a rocket-based suborbital Hopper. Here, the first stage accelerates to suborbital velocity releases the second stage and returns to earth. The second stage, which is rather a kick stage, is supposed to place the payload into the favoured orbit and is lost afterwards. Current US concepts prefer vertical take-off and horizontal landing TSTO reusable rocket systems. Achievable with current US technology, they are certainly the cheapest and least risky near term solution for access to space.

For mid to long term fully reusable space transports a TSTO with a turbojet-propelled Mach 4 accelerator vehicle as first stage and rocket-propelled space plane as second stage offers a feasible system which can serve as a baseline configuration for new investigations and trade studies. To identify potentials and risks, more detailed investigations are required and have to be carried out. However, requirements should preferably aim at low complexity, risks and costs so that no insuperable obstacles are discovered during design, development and operation.

## References


1. Hammond, W. E.: Design Methodologies for Space Transportation Systems, AIAA Education Series, 2001.
2. Lentz, S., Hornung, M., Staudacher, W.: Integration und Auswirkungen der Ergebnisse der Teilprojekte des SFB 255 auf den Gesamtentwurf eines Transatmosphärischen Raumtransportsystems (HTSM), Finanzierungsantrag des Sonderforschungsbereiches 255, p. 311–340, 2001.
3. FESTIP-reports, ESA, 1994–1998.
4. FESTIP, Final Meeting handout, ESA, 1998.
5. Koelle, D. E.: TRANSCOST 7.0, TCS-TR-168/00, November 2000.
6. Whitmore, S. A., Dunbar, B. J.: Orbital Space Plane: Past, Present and Future, AIAA 2003–2718, 2003.
7. Livingston, J.: Access to Space System Design Options and Technology Availability, AIAA-Oral Presentation, International Air & Space Symposium, The Next 100 Years, 14–17 July 2003, Dayton/Ohio, 2003.
8. Heppenheimer, T. A.: Hypersonic Technologies and the National Aerospace Plane, Pasha Publications Inc. (ISBN 0-935453-33-4), 1990.
9. Leitkonzept Sänger Referenz-Daten-Buch, interner Bericht, MBB, 1993.
10. Staudacher, W.: Entwurfsproblematik luftatmender Raumtransportsysteme, 3. Space Course, S. 63–98, Stuttgart, 1995.

11. Sacher, P.W.: Personal Conversation.
12. Hornung, M., Lentz, S., Staudacher W.: Integration und Auswirkungen der Ergebnisse der Teilprojekte des SFB 255 auf den Gesamtentwurf eines Transatmosphärischen Raumtransportsystems (HTSM), Arbeits- und Ergebnisbericht des Sonderforschungsbereiches 255, p. 629–668, 2001.
13. Staudacher, W., Wimbauer, J.: Design Sensitivities of Airbreathing Hypersonic Vehicles, AIAA 93–5099, AIAA/DGLR fifth International Aerospace Planes and Hypersonics Technologies Conference, Munich, 1993.
14. Hornung, M.: Entwurf einer luftatmenden Oberstufe und Gesamtoptimierung eines transatmosphärischen Raumtransportsystems, Dissertation, University of the German Armed Forces Munich, Neubiberg, 2003.
15. Leingang, J.L., Maurice, J.Q., Carreiro, L.R.: In Flight Oxidizer Collection System for Airbreathing Space Boosters, in Murthy, S.N.B, Curran, E.T.: Developments in High-Speed-Vehicle Propulsion Systems, Progress in Astronautics and Aeronautics, Volume 165, AIAA, 1996, pp. 333–384.
16. Balepin, V.V.: Air Collection Systems, in Murthy, S.N.B, Curran, E.T.: Developments in High-Speed-Vehicle Propulsion Systems, Progress in Astronautics and Aeronautics, Volume 165, AIAA, 1996, pp. 385–420.
17. Lentz, S., Staudacher, W.: Entwurf eines wiederverwendbaren zweistufigen Raumtransportsystems mit einer Anlage zur Sauerstoffgewinnung während des Fluges, DGLR Jahrestagung 23.09.–26. 09. 2002, DGLR-JT2002-180, Stuttgart, 2002.
18. Lentz, S., Staudacher, W.: Application of an Air Collection and Enrichment System aboard a TSTO Space Transport, Notes on Numerical Fluid Mechanics, Vol. 87, Springer-Verlag, 2003.
19. Perry, J.H., Chemical Engineer's Handbook, 5$^{th}$ ed., Section 13, McGraw-Hill, New York, 1973.
20. Nau, R.A., Cambell, S.A.: Rotary Separator, United States Patent 3,779,452, Dec. 1973.
21. Lentz, S., Staudacher, W.: Reusable TSTO Launch vehicles using in-flight LOX-Collection, Poster Presentation, AIAA 2003-PP8095, 2003.

## 3.2 Evaluation and Multidisciplinary Optimization of Two-Stage-to-Orbit Space Planes with Different Lower-Stage Concepts

Thorsten Raible* and Dieter Jacob

### Abstract

Two fully reusable, two-stage-to-orbit space plane configurations with different lower-stage concepts are examined. They both consist of airbreathing lower stages and rocket-propelled upper stages. The first configuration is based on a lifting body vehicle and the second configuration on a waverider vehicle. By means of a specially tailored simulation and optimization software the space planes are studied and evaluated for a reference ascent mission into low earth orbit. The method allows the optimization of configurations with low sensitivity with regard to inaccuracies in the physical computation models. For evaluation, the growth factor is taken as the fundamental quality criterion referring the gross take-off mass to the orbital payload mass.

The reference configurations of the lifting body and the waverider space planes are compared with each other in terms of mass breakdown and design sensitivities. Separate optimizations are carried out for the first and second stages of each space plane configuration as well as for the overall systems. Moreover, optimizations are made to obtain configurations with lowest sensitivity.

### 3.2.1 Introduction

Worldwide, there is a strong interest in future space transportation systems to maintain and improve the access to space for the next decades. The main goals are significantly reduced launching costs per kilogram payload mass as well as more operational reliability and flexibility. One promising, though long-term concept is based on two-stage-to-orbit (TSTO) space planes with horizontal take-off and landing capability [1, 2]. Such systems could consist of airbreathing lower stages and rocket-powered upper stages, and they fly aerodynamically lifted in the lower regions of the atmosphere. However, the design and development process is highly complex, because it invokes many single disciplines such as aerothermodynamics, structures and weight, propulsion technology, trajectories, and flight mechanics. Additionally, the technological fundamentals have not been fully established yet, and there is no experience with existing two-stage-to-orbit space planes. For all these reasons, the development is extremely risky and special analysis tools are required from the preliminary design level on. The project

* Institute of Aerospace Engineering (ILR), Wuellnerstr. 7, 52056 Aachen

"Concepts and Design Analysis of Hypersonic Configurations" (A1) of the Collaborative Research Centre "Fundamentals of Space Plane Design" has developed a simulation and optimization tool that integrates performance risk into the overall design and evaluation process of two-stage-to-orbit space planes. In the following, a brief review containing the most essential aspects of this project is given.

The research started with concept analyses of airbreathing single-stage and two-stage space planes separating at different Mach numbers ($M \approx 3$, $M \approx 7$, $M \approx 12$) [2]. For this purpose, a numerical Fortran code has been written with the objective of simulating the TSTO missions computer-assisted. To quantify overall performance, the growth factor ($GF$) referring the gross take-off mass to the orbital payload mass has been established as the fundamental quality criterion. Initially, the computational models were very simple, based on handbook data. The comparative analysis of six configurations showed that the two-stage-to-orbit space plane with a stage separation Mach number of about 7 is the most favourable concept. Therefore, it has been chosen as the reference concept of the Collaborative Research Centre. This space transportation system consists of the lifting body lower-stage ELAC (Elliptical Aerodynamic Configuration) [3] and the orbital upper-stage EOS (ELAC Orbital Stage).

In the following years, this configuration was examined in more detail throughout all projects of the Collaborative Research Centre [4]. Corresponding the integrative character of project A1, the exchange of data with other projects played a decisive role. On the one hand, many numerical and experimental results obtained in the other projects were continuously integrated into the simulation programme. The aero-thermodynamic results of the projects A2, A3, A4, and B1 as well as the experiments at ITAM (Institute of Theoretical and Applied Mechanics) [5] and in the DNW (Deutsch-Niederländischer Windkanal) wind tunnel [6–8] significantly increased the physical model accuracy of the code. In the area of propulsion technology, the projects C1, C5, and C8 contributed essential results, as project A6 did for the mass and structural models. On the other hand, the project A1 could provide essential information for other projects from the overall system's view. For example, the maximum angle of attack for the force and heat flux experiments of the orbital stage EOS (project A2) was fixed cooperatively and in accordance with its reference trajectory. To identify the most sensible design drivers and to improve the overall system performance, the methodology of the space plane evaluation programme was gradually extended by embedding design sensitivities and an optimization algorithm [9]. Generally, design sensitivities point out the performance of a technical system when inaccuracies in the physical models are applied and hence indicate performance risks. Because of the complexity and multidisciplinarity, a genetic algorithm was chosen as optimization strategy. Moreover, a new approach, called sensitivity-based optimization, was established [9] and successfully used for space plane analysis [9–14]. It combines sensitivity analysis and optimization and even allows to find space plane configurations with best performance when inaccuracies in the physical computation models are assumed.

During the final phases of the Collaborative Research Centre the project A1 was mainly concerned with the design and analysis of an alternative TSTO space plane, where the lower-stage ELAC is replaced by a waverider vehicle. The alternative concept is denoted as WALS/EOS (Waverider Lower Stage and ELAC Orbital Stage). After designing the alternative space plane, the aforementioned sensitivity-based analysis tools were applied to it and comparative studies to the lifting body configuration were carried out [14]. During the model generation phase great emphasis was laid on comparably high model accuracies of both space transportation systems.

### 3.2.2 Reference Configurations

#### 3.2.2.1 Concept Design and Mission Requirements

A Collaborative Research Centre is inherently concerned with fundamental research to study basic phenomena. For this purpose, simplifying assumptions as well as reference systems are required. The configurations were defined such that on the one hand all major physical problems could be investigated. On the other hand, the geometry was kept as simple as possible to reduce the costs of the wind tunnel models and the time required for the grid generation for numerical calculations.

The objective of the project "Concepts and Design Analysis of Hypersonic Configurations" (A1) was to design two-stage-to-orbit space planes for equatorial acceleration missions carrying an orbital net payload mass of 7000 kg into a circular low earth orbit (LEO) of 250 km at an inclination angle up to 5° (Tab. 3.2.1).

The lower stages' propulsion systems consist of six airbreathing engines combining turbojet and ramjet modes and use liquid hydrogen (LH2) as fuel. The orbital stages are propelled by a single rocket motor using LH2, as well, and liquid oxygen (LOX) as oxidizer. Following the recommendation from the initial phase of the Collaborative Research Centre, the stage separation manoeuvre should start at a Mach number of about 7. The reference ascent trajectories for both vehicles including the operating regimes of the propulsion units can be seen from Fig. 3.2.1 [15, 16].

#### 3.2.2.2 Space Plane Configuration with Lifting Body Lower Stage

A slender delta wing with elliptical cross sections called ELAC (Elliptical Aerodynamic Configuration) was chosen for the first stage of the basic space trans-

Table 3.2.1: Basic mission requirements.

| | |
|---|---|
| Orbital net payload mass | 7000 kg |
| Orbit altitude (LEO) | 250 km |
| Inclination angle | 0–5° |

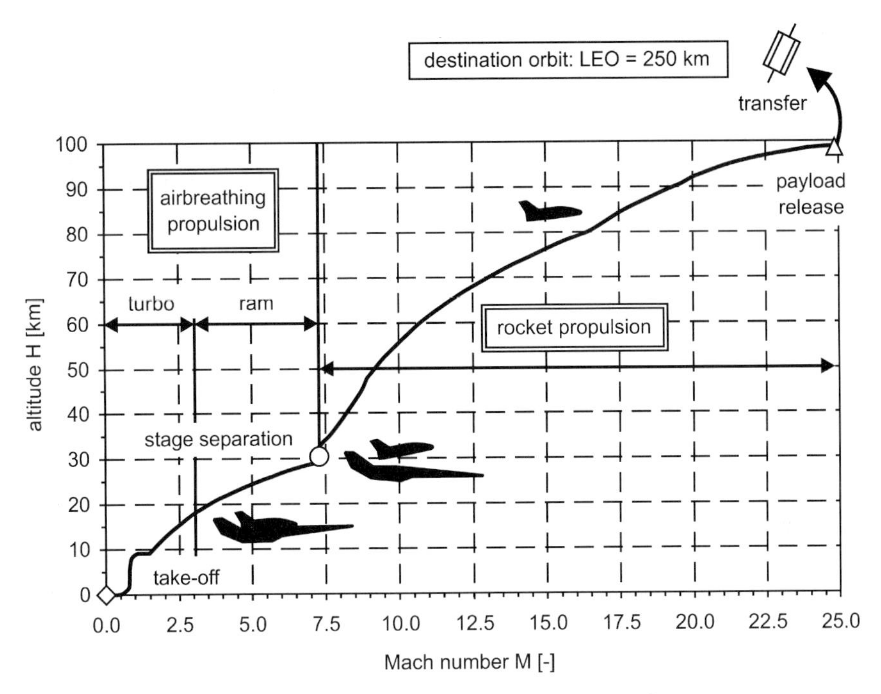


Figure 3.2.1: Reference ascent mission and engine operation modes.

portation system (Fig. 3.2.2) [3]. The cross-sections consist of two half-ellipses with an axis ratio of 1.4 at the upper and 1.6 at the lower surface. The leading edge is swept back by 75° and the reference body length in the symmetry plane is 72 m. These dimensions give a span of 38.6 m, a planform area of 1389 $m^2$, and an aspect ratio of 1.1. Two stabilizer fins with a height of 7.5 m, a dihedral angle of 65°, and symmetrical NACA 0010 profiles are used to control the vehicle. Maximum body thickness of 5.36 m is reached at two-third of the reference body length. To ease airbreathing engine integration, the tail of the lifting body behind the maximum thickness is inclined upwards by 5.6°.

The orbital stage EOS (ELAC Orbital Stage) is a wing-body vehicle whose straked delta-wing has a sweep angle of 65° [17]. The fuselage has a length of 28.8 m, a maximum diameter of 4.96 m, and a maximum height of 4.86 m. The vertical fin is composed of NACA 0010 profiles and placed centrally at the rear upper side of the fuselage. With a span of 16.2 m, the planform area amounts to 239 $m^2$ and the aspect ratio to 1.1. The cross-sections are composed of semi-circles at the top and rectangles at the bottom part. The orbital payload is positioned between 50 and 75% of the fuselage length. The tanks of liquid hydrogen and oxygen are placed in front of as well as behind the payload bay. The rocket engine is integrated in the aft fuselage part with its nozzle standing slightly out.

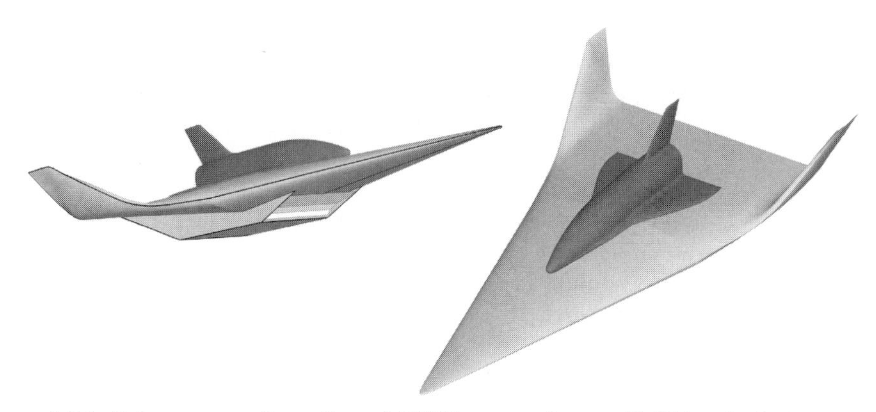

Figure 3.2.2: Reference configuration of TSTO space plane with lifting body lower stage (ELAC/EOS).

### 3.2.2.3 Space Plane Configuration with Waverider Lower Stage

The alternative space plane concept is composed of a waverider lower stage (WALS) carrying the same orbital stage EOS (Fig. 3.2.3). As opposed to the conventional design process in which the flowfield for a given geometry is calculated, a waverider results from inverse design. In practice, the geometry is derived from a known flowfield given in terms of a three-dimensional shock surface at a hypersonic design Mach number. An arbitrary curve on this shock surface serves as the leading edge. To generate the lower surface geometry of the vehicle, streamlines are traced downstream from the leading edge. The upper surface is formed by a freestream surface that intersects with the leading edge.

The fundamental geometry design of the waverider was carried out at the German Aerospace Centre (DLR) Braunschweig according to the method of *osculating cones*. This method was established by Center and Sobieczky [18, 19] and uses several cones for the production of the shock surface. The flowfield

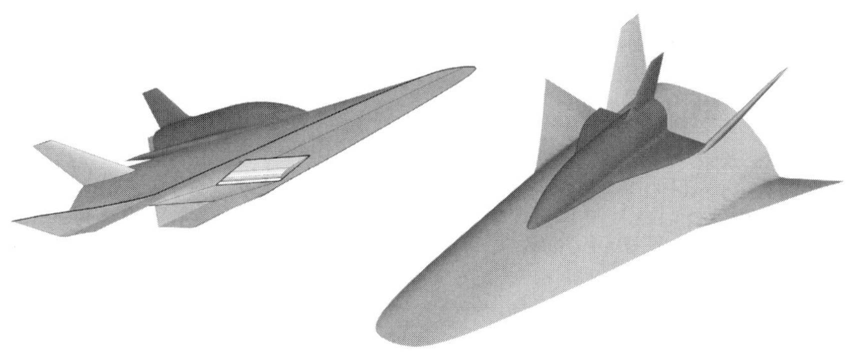

Figure 3.2.3: Reference configuration of TSTO space plane with waverider lower stage (WALS/EOS).

around these cones is treated as locally conical flow regions whose conically axisymmetric flow is known in terms of the Taylor-Maccoll solution. However, for practical application as a lower-stage vehicle, some modifications had to be applied to the waverider derived from this theory [20–23]. The changes comprise an engine bay at the vehicle's lower side, straight hinge lines at the outboard wings for a flap integration, and two medium sized fins.

Finally, the reference waverider resulted in a planform area with a gothic forebody and a delta wing with a 60° sweep angle. In order to facilitate the integration of the EOS upper stage, the same reference body length of 72 m as for ELAC was chosen leading to a span of 45.3 m. With a planform area of 1534 $m^2$ it is about 10% larger than the lifting body's. The body angle in the symmetry plane is 6.3°. Since the upper surface of the waverider is much more curved than the lifting body's is, the upper stage is less integrated into the lower stage.

Table 3.2.2: Reference values of design parameters of space planes ELAC/EOS and WALS/EOS.

| Parameter | ELAC/EOS | WALS/EOS |
|---|---|---|
| Lower stage | | |
| length | 72 m | 72 m |
| sweep angle of delta-wing | 75° | – |
| outboard sweep angle | – | 60° |
| maximum thickness | 5.36 m | 7.73 m |
| design shock angle | – | 8° |
| thrust-to-weight-ratio of turbojet | 0.90 | 0.98 |
| thrust-to-weight-ratio of ramjet | 0.32 | 0.30 |
| switching Mach number: turbojet > ramjet | 3.0 | 3.0 |
| stoichiometric ratio | 1.5 | 1.5 |
| switching Mach number to overstoichiometry | 6.0 | 6.0 |
| flight path angle at subsonic flight | 15° | 15° |
| final height at subsonic flight | 6800 m | 6800 m |
| final Mach number at subsonic climb | 0.8 | 0.8 |
| flight path angle at transonic regime | 0° | 0° |
| maximum dynamic pressure | 50 kPa | 50 kPa |
| initialization Mach number for stage separation | 7.0 | 7.0 |
| Upper stage | | |
| length | 28.8 m | 28.8 m |
| sweep angle | 65° | 65° |
| fuselage diameter | 4.90 m | 4.90 m |
| flight path angle at stage separation | 3° | 3° |
| maximum flight path angle | 6° | 6° |
| thrust-to-weight-ratio of rocket | 1.25 | 1.25 |
| mixture ratio of rocket | 6.0 | 6.0 |
| relative longitudinal displacement of upper stage referred to lower stage length | 0.44 | 0.45 |

#### 3.2.2.4 Design and Optimization Parameters

In total, there are 22 design quantities for each space plane concept covering geometries, propulsion systems, and ascent trajectories of both lower and upper stages (Tab. 3.2.2). Since the upper stages and the ascent trajectories of both configurations are almost identical, the differences exclusively arise in lower-stage geometries, as discussed before, and in propulsion systems. The air-breathing engines of the different space plane concepts vary respectively in their thrust-to-weight-ratios of turbojet and ramjet. Because of the higher aerodynamic efficiency of the waverider configuration in the supersonic and hypersonic regime, lower ramjet thrust is required. In contrast, the lifting body lower stage has a lower thrust-to-weight-ratio of the turbojet engines, because its body is more slender and produces less drag at transonic flight velocities.

### 3.2.3 Analysis Methods

#### 3.2.3.1 Quality Criteria

For performance evaluation of different space plane configurations, the growth factor (*GF*) has been selected as the fundamental quality criterion. It refers the gross take-off mass $m_0$ to the orbital net payload mass $m_{PL}$ (Eq. (1)).

$$GF = \frac{\mathrm{m}_0}{\mathrm{m}_{PL}} \tag{1}$$

Since the payload mass to be carried into low earth orbit is fixed to 7000 kg, the improvement of mission performance aims at decreasing the gross take-off mass and the growth factor *GF*, respectively. As in all numerical calculations, however, the performance is derived from physical models inherently tending to simplify reality. Hence, there might be discrepancies between the computational models and reality. To allow for this, the performance index *GF* can be evaluated in two different ways. In the first approach all physical models are expected to describe reality exactly and the quality criterion was named the *nominal* growth factor. If $\Delta_i$ denotes an inaccuracy in the i-th model and m the total number of physical models, the nominal growth factor can be expressed in terms of Eq. (2).

$$GF = GF(\Delta_i = 0 \quad \forall \quad i = 1, \ldots, \mathrm{n}, \mathrm{n} = \mathrm{m}) \tag{2}$$

The second approach assumes that one or more models, e.g. the propulsion mass, contain inaccuracies. These failures are supposed to worsen the overall performance. The growth factor calculated with certain inaccuracies is called the sensitivity-based growth factor and is marked by an asterisk ($GF^*$). Its mathematical definition is given by Eq. (3).

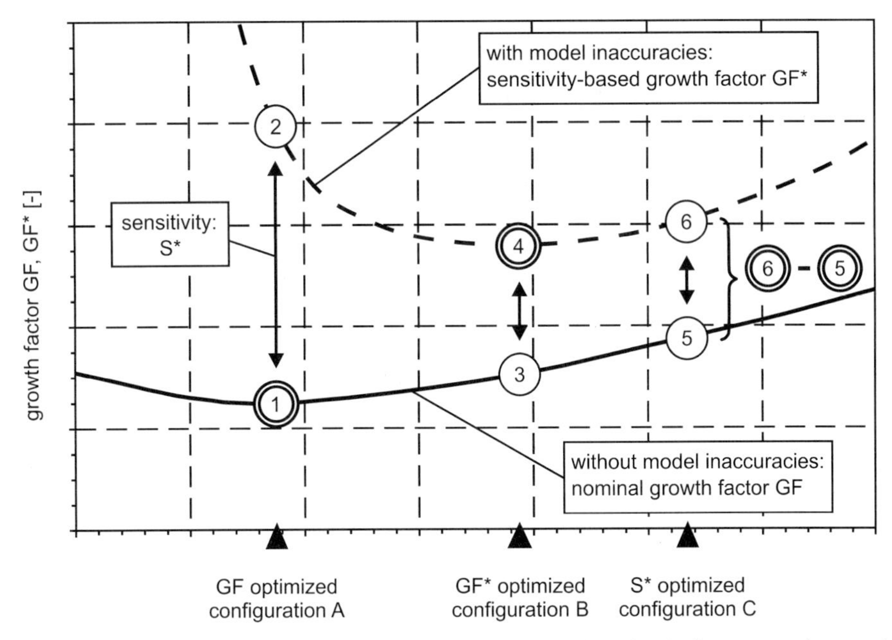


Figure 3.2.4: Quality criteria considering inaccuracies in the physical computation models (principle for one-dimensional design variation).

$$GF^* = GF(\Delta_i \neq 0 \quad \forall \quad i = 1, \ldots, \mathrm{n}, \mathrm{n} \leq \mathrm{m}) \tag{3}$$

The differences between nominal and sensitivity-based growth factors are called sensitivities ($S^*$) (Eq. (4)). They indicate the degradation of a nominally evaluated configuration subjected to model inaccuracies.

$$S^* = GF^* - GF \tag{4}$$

The three different quality criteria are sketched in Fig. 3.2.4 for a one-dimensional design variation. The nominal growth factors are represented by the solid line, the sensitivity-based ones by the dashed line. Assuming that all models specify reality exactly, configuration A would perform best, because its nominal growth factor (point '1') is minimal. The corresponding design is named *GF* optimized. However, in case of model inaccuracies the growth factor would increase to the value of point '2'. Integrating performance risk into the optimization process would lead to configuration B, because its sensitivity-based growth factor is minimal (point '4'). Such a configuration is called $GF^*$ optimized. Furthermore, the nominal performance of configuration B (point '3') in this example would be only slightly worse than that of configuration A (point '1'). A third, though more theoretical aim, might be to reduce a configuration's sensitivity as far as possible. Hence, one would find configuration C for which the sensitivity is minimum (point '6'–'5'); this configuration is called $S^*$ optimized.

#### 3.2.3.2 Simulation and Optimization Software

At the Institute of Aerospace Engineering a numerical tool was established to evaluate and optimize two-stage-to-orbit space planes according to the quality criteria mentioned before (Section 3.2.3.1). The programme ASTOR (Advanced Space Transportation Optimization Routine) is a Fortran 90 code especially tailored to comparative analyses of airbreathing space planes with lifting body and waverider lower-stage configurations. To allow a quick and robust adaptation of the physical models to the results found in the other projects, the programme makes intense use of characteristic diagrams provided in terms of ASCII data sets and files. In total, three different types of calculations that can be influenced by several input files are implemented. When a single simulation is executed, one space plane configuration with a certain parameter set according to Tab. 3.2.2 is evaluated in terms of the growth factor. In the case of a parameter variation study, one single design parameter is varied during several simulations carried out successively. The optimization simulates and evaluates different configurations until a terminating condition, i.e. a maximum configuration number, is reached. Each of these calculation types can be carried out either under nominal conditions or under the assumption of simultaneous inaccuracies in up to 15 models.

A genetic algorithm was selected as an optimization strategy, because it is well suited for such a complex and multidisciplinary task as space plane design [24, 25]. In general, genetic algorithms are heuristic search techniques that try to find best performing configurations or systems by modelling the strategies of evolutionary biological processes according to Darwin. The natural selection processes are imitated and assisted by the evolutionary principles of recombination and mutation. In the present case, the public domain optimizer Pikaia 1.2 was taken as a basis [26, 27] and slightly modified with regard to its generational convergence scheme. For rapid convergence, a steady-state-replace-worst reproduction plan including elitism was chosen. Steady state reproduction means that not all individuals of a generation are replaced, but only its worst. Additionally, the principle of elitism assures that the fittest individual of a generation cannot be selected for random deletion. For the optimization calculations, 500 valid generations were evaluated comprising 100 elements each.

### 3.2.4 Performance of Reference Space Planes

#### 3.2.4.1 Mass Breakdown

In this chapter, the reference configurations of both space planes according to the design quantities of Tab. 3.2.2 are evaluated. Under nominal conditions, i.e. when no model inaccuracies are applied, the lifting body configuration ELAC/EOS has a gross take-off mass of 381.9 t for a zero inclination low earth orbit of

Table 3.2.3: Nominal performance of reference space plane configurations.

| | ELAC/EOS | WALS/EOS |
|---|---|---|
| Growth factor GF | 54.6 | 53.7 |
| Overall take-off mass | 381.9 t | 375.9 t |
| Lower stage mass | 274.4 t | 268.5 t |
| Upper stage mass | 107.4 t | 107.5 t |

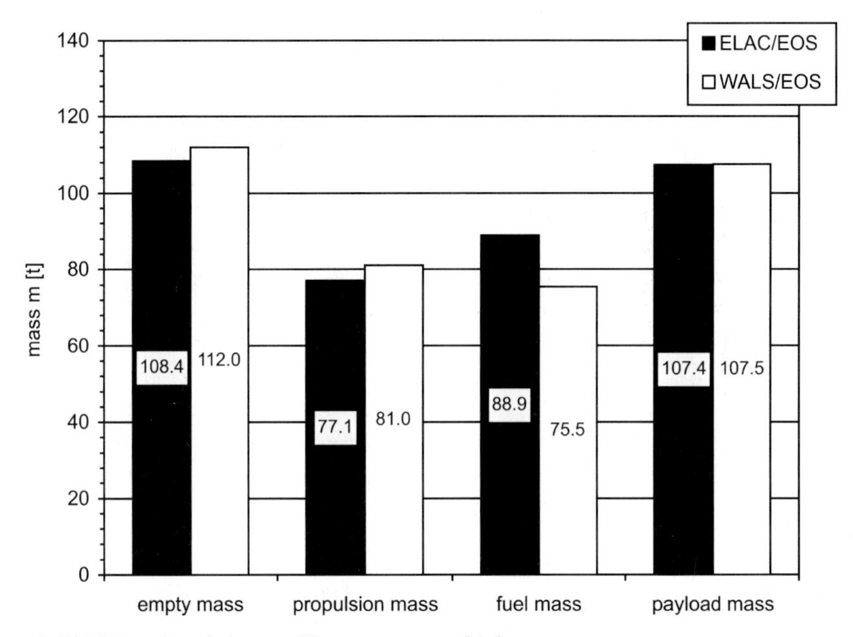


Figure 3.2.5: Mass breakdown of lower stage vehicles.

250 km. Considering the required net orbital payload mass of 7 t gives a growth factor of 54.6. The waverider configuration weighs 375.9 t in total resulting in a nominal growth factor of 53.7. Since the upper stage is largely identical for both configurations, the difference of gross take-off weight is caused by different lower-stage weights.

A more detailed mass breakdown of the lower-stage vehicles is given in Fig. 3.2.5 in terms of mass components and of single masses in Tab. 3.2.4, respectively. Because the upper-stage EOS marks the payload of the lower stage, these masses are identical.

Both empty and propulsion mass of the waverider vehicle lie about 4 t each above the values of the lifting body vehicle. The waverider's higher empty mass primarily results from its different structural concept. While the lifting body is composed of frames with elliptical cross-sections connected by strin-

Table 3.2.4: Single mass breakdown of lower stage vehicles.

| Mass component | ELAC/EOS | WALS/EOS |
|---|---|---|
| Total empty mass | 108.4 t | 112.0 t |
| basic structure | 61.2 t | 67.8 t |
| structure for stage separation | 9.0 t | 9.0 t |
| tail fins | 3.9 t | 2.8 t |
| thermal protection system | 7.0 t | 6.7 t |
| tanks | 9.0 t | 6.8 t |
| landing gear | 11.5 t | 10.9 t |
| electrics | 1.7 t | 2.0 t |
| pneumatics/hydraulics | 1.5 t | 1.6 t |
| avionics | 3.3 t | 4.2 t |
| crew | 0.4 t | 0.4 t |

gers, the waverider is based on a frame construction supported by tensile bars to enhance stiffness. Thus, the waverider's basic structure is about 10% heavier than the lifting body's. The higher propulsion mass of the waverider is caused by a higher thrust-to-drag-ratio of its turbojet engines. Since this configuration is not as slender in the streamwise direction as the lifting body configuration, the transonic drag is higher and more thrust is needed. These drawbacks, however, are overcompensated by the the waverider's lower fuel mass which is about 15% smaller (13.4 t), due to the lower drag at supersonic and hypersonic velocities. Apart from the already discussed structure mass, the other masses of both lower stages lie quite close together even though they result from different mass models. More or less significant differences only occur in the weights of the tail fins and the tanks. The fitting of the waverider's tail fins to the base structure saves mass compared to the integration into the lifting body's outboard wing where higher stiffness is necessary. Since the effective body volume of the waverider is lower and less fuel is required for the mission, tanks can be smaller and their mass is reduced.

#### 3.2.4.2 Design Sensitivities

To trace the most essential design drivers, each space plane configuration has been subjected to sensitivity studies of 15 models. They affect the calculation of masses, aerodynamic coefficients, centres of gravity, and propulsion quantities of both lower and upper stages (Tab. 3.2.5). The model inaccuracies are chosen such that they worsen design performance, which is expressed in terms of higher growth factors and overall take-off masses.

Some single design sensitivities are exemplarily sketched in Fig. 3.2.6 in terms of overall mass changes. The limiting bars indicate that a further increase in model inaccuracy leads to a termination of the simulation programme, due to

Table 3.2.5: Models for sensitivity studies.

| Lower stage | Upper stage |
|---|---|
| structure mass | structure mass |
| propulsion mass | propulsion mass |
| fuel mass flow | fuel mass flow |
| drag coefficient | drag coefficient |
| pitching moment coefficient | pitching moment coefficient |
| center of gravity | center of gravity |
| turbojet thrust | rocket thrust |
| ramjet thrust | |

restricting conditions as tank volume, for instance. The comparison between space planes with lifting body and waverider lower stages points out that the performance loss of the waverider configuration is less intense than in the case of ELAC/EOS. In general, upper-stage sensitivities are lower than the corresponding lower-stage sensitivities and nearly identical for most models, as in the cases of structure mass and rocket thrust, for instance. The higher drag sensitivity of the lifting body's upper stage, however, is due to the fact that the

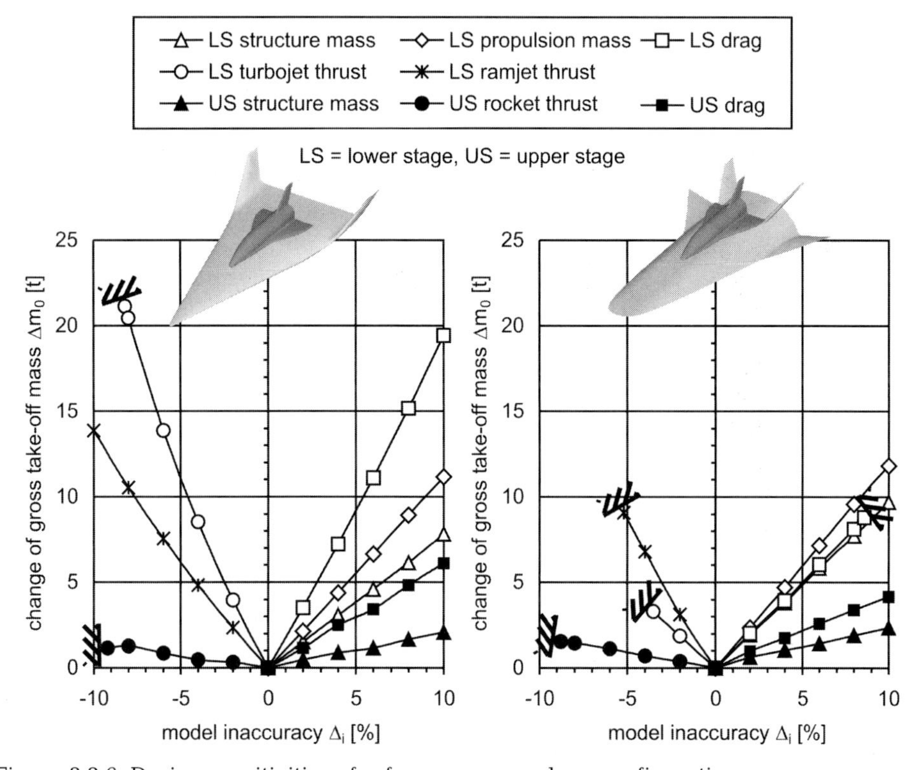


Figure 3.2.6: Design sensitivities of reference space plane configurations.

aerodynamic models of the orbital stages do not only affect their missions after stage separation, but also the ascent flights of the overall systems.

In total, the reference configuration of the waverider replies less dramatically to slight and moderate model inaccuracies in the physical models. However, since the restrictions of model inaccuracies occur more often and at lower absolute values for the waverider configuration, the mission performance is put at risk, as well.

## 3.2.5 Optimization Results

### 3.2.5.1 Nominal Optimizations

The nominal optimizations are based on the assumption that all physical computation models are correct and therefore reproduce reality exactly. Considering the principle sketch of Fig. 3.2.4, point '1' is searched at which the nominal growth factor *GF* is minimum. In a first step the first and second stages of both the lifting body and the waverider configuration were optimized separately (Fig. 3.2.7). In addition, the overall configurations were nominally optimized. All results were related to the space planes' reference configurations which were defined initially and which were not optimized at all.

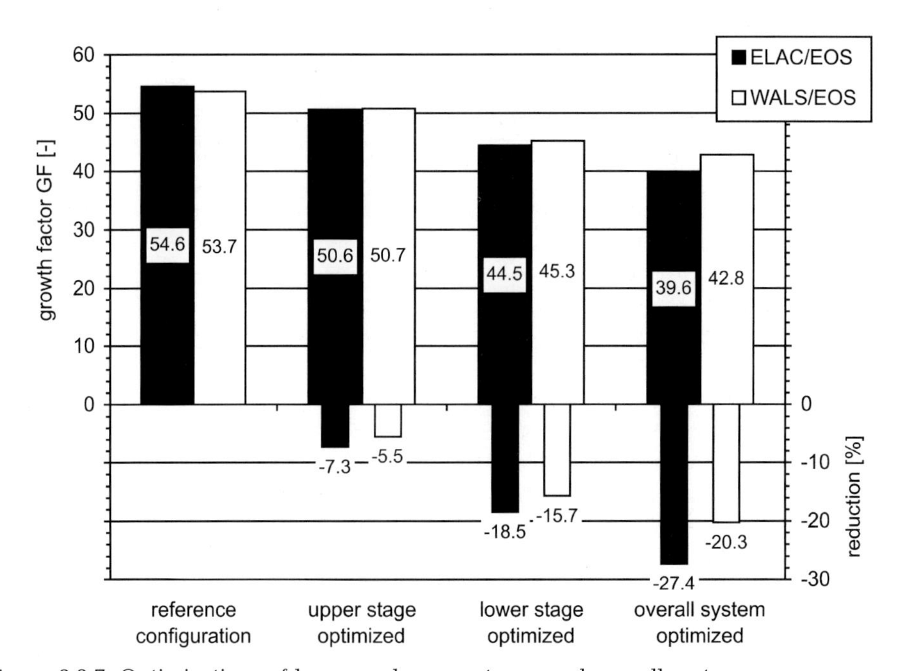


Figure 3.2.7: Optimizations of lower and upper stages and overall systems.

The upper-stage optimizations lead to growth factors of 50.6 (ELAC/EOS) and 50.7 (WALS/EOS) and to a very similar upper-stage design. Since the lifting body reference configuration has a slightly higher growth factor ($GF = 54.6$) than the waverider reference configuration ($GF = 53.7$), its upper-stage optimization yields a nearly 2% higher growth factor reduction. The fuselages of both optimized upper-stage vehicles are slightly shorter and wider, while their sweep angle is increased up to the upper boundary of $71.5°$. Altogether, the effective body and tank volumes remain nearly constant, but the wetted vehicle surface, which directly influences skin friction, is lessened by about 10%. The mixture ratios of the rocket engines are reduced from 6 to 5 which lowers the average fuel density of liquid hydrogen and oxygen from 362.2 $kg/m^3$ to 325.2 $kg/m^3$ and hence saves fuel mass. By choosing a thrust-to-weight-ratio of the rocket motor of 1.0, the fuel mass flow is decreased resulting in longer flight times of 290 s (+10%) at lower thrust. The payload release takes place at a lower altitude of 77 km instead of 92 km for the reference configurations. Of course, more fuel is required for the orbital transfer, but the staging principle is enhanced in so far, as in the final flight phase less mass has to be accelerated. All in all, these design changes lead to a reduction of the total upper-stage masses from 107 t to 97 t. Hereby, 9 t are saved by the lower fuel mass and about 1 t by a correspondingly decreased empty weight.

The exclusive optimizations of the lower-stage vehicles yield significant growth factor reductions for both the lifting body configuration (–18.5%) and the waverider configuration (–15.7%). The direct comparison between them shows, however, that the lower-stage optimized lifting body configuration has 5.6 t less overall take-off mass, because the geometric optimization of the lifting body's lower stage is more effective. By shortening ELAC to 67.6 m, reducing its maximum body thickness to 4.54 m and increasing its sweep angle to $77°$, the planform area as well as the wetted surface can be reduced by 24%. Thus, the structural mass is diminished by 17 t to 44.2 t. Additionally, the smaller wetted surface reduces skin friction and hence, fuel mass. For the same reasons, the optimized lower stage of the waverider space plane is shortened to 65 m. However, to provide enough tank volume, the design shock angle increases from $8°$ to $10°$ thickening the lower stage. In total, both the planform area and the body surface are less reduced than in the case of the lifting body configuration. Consequently, the decrease in structure weight is just about 18% leading to a higher empty mass of 152.1 t (WALS) compared to 144.4 t (ELAC). Even though the optimized waverider lower stage is thicker than its reference configuration, its thrust-to-weight-ratio is scaled down to 0.86 to save fuel mass and in particular, fuel volume being more critical for the waverider. To compensate for the missing thrust at transonic speeds, the flight path angle in this regime is inclined downward ($-5°$). For both space plane configurations the switching from turbojet to ramjet mode occurs at $M \approx 3.4$ instead of $M = 3.0$. Furthermore, stage separation is initiated at $M \approx 6.5$ instead of $M = 7.0$ leading to a more effecting staging effect.

Naturally, the overall system optimizations give the best results in terms of the growth factors. Both space planes show great optimization potential, even

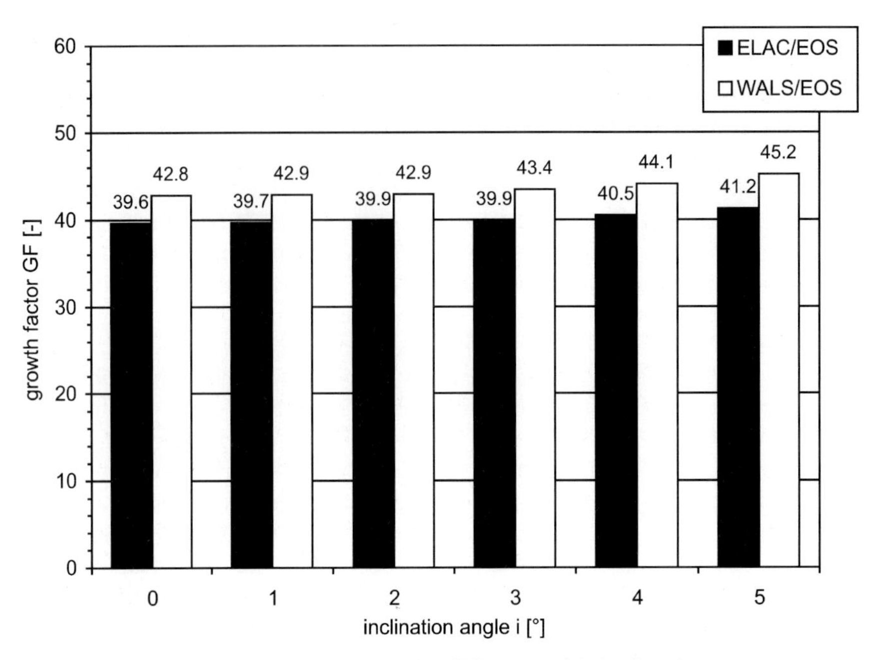


Figure 3.2.8: Overall system optimizations for different orbit inclinations.

though the performance improvement is higher in the case of the lifting body configuration. Its growth factor of 39.6 corresponds to a gross take-off mass of 277.4 t, whereas the completely optimized waverider vehicle weighs 299.7 t ($GF$=42.8). The lower and upper stages are largely identical with those obtained from the separate optimizations of the first and second stages of the lifting body and the waverider configuration, except that the lifting body lower stage is further shortened to a length of 65.0 m decreasing its empty weight even more. Compared with the upper-stage optimized configurations, the flight path angles at stage separations increased to 3.2°, and the maximum flight path angles of the upper stages decreased to about 3°.

With regard to the destination orbit, inclination angles $i$ between 0° and 5° are desired for a net payload mass of 7 t at a height of 250 km (Tab. 3.2.1). Therefore, overall system optimizations for both the lifting body and the waverider space plane have been carried out varying these angles gradually by 1° up to 5° (Fig. 3.2.8).

As can be seen from the figure above, the overall optimizations show significant growth factor reductions for all inclination angles compared to the reference space plane configurations. Again, the growth factors of the optimized lifting body configurations are lower than the corresponding values of the waverider ones. The growth factor increases by 1.6 in the case of ELAC/EOS when varying the inclination angles from 0° to 5°, while it grows more for the waverider configuration (2.4). Since the fuel mass for the orbital transfer is restricted to 3 t, but more fuel is needed the higher the orbital inclination angle is, the payload release height increases from about 82 km ($i$=0°) to 96 km ($i$=5°). To reach the greater height, the upper stages separate at higher Mach numbers ($M \approx 7.1$) from the lower stages whose ramjet propulsions have more thrust for this purpose.

### 3.2.5.2 Sensitivity-Based Optimizations

In the following studies model inaccuracies were integrated into the optimization process and optimizations with regard to the nominal growth factor *GF*, the sensitivity-based growth factor $GF^*$, and the sensitivity $S^*$ were carried out. Because the models concerning the centre of gravity of lower and upper stages massively compromise overall mission performance, they were not included into the sensitivity-based optimizations. The models with inaccuracies rather comprise the remaining 13 models of Tab. 3.2.5. Each of them was supposed to worsen performance by 3%. The optimization results for both the lifting body and the waverider space plane are depicted in Fig. 3.2.9 which categorizes them according to the particular optimization objective (*GF*, $GF^*$, $S^*$). On the left hand side the lifting body results are shown (black and black crossed) and on the right hand side the waverider configuration results (white and white crossed). For clearness, the numbers according to the principle optimization sketch of Fig. 3.2.4 are attached to the vertical bars. All data are obtained for an inclination angle of i=0°.

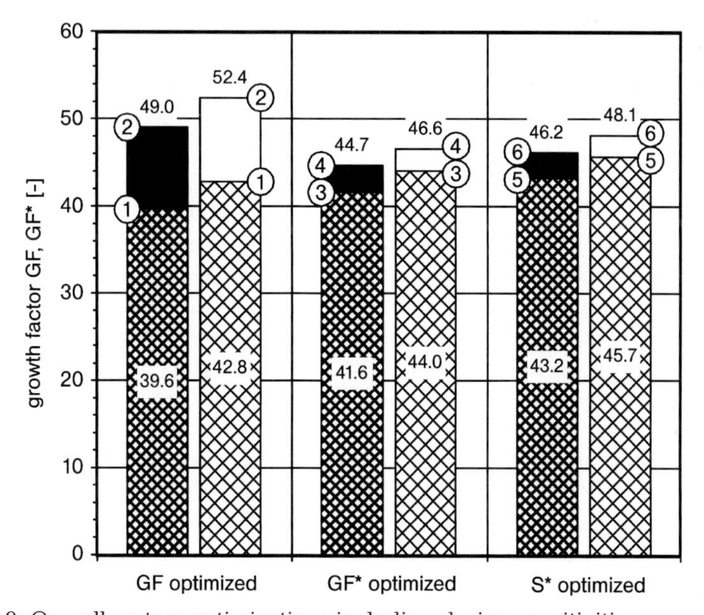


Figure 3.2.9: Overall system optimizations including design sensitivities.

The nominal growth factors at points '1' of ELAC/EOS ($GF$=39.6) and WALS/EOS ($GF$=42.8) correspond to the already discussed results presented in Fig. 3.2.7 (overall system optimized) and Fig. 3.2.8 (inclination angle $i$=0°). If these nominally optimized configurations are simultaneously affected by 13 model inaccuracies, their sensitivity-based growth factors increase by approximately 10 to the values of points '2'.

However, as can be seen from Fig. 3.2.9, it is possible to limit the sensitivity-based growth factors to 44.7 (ELAC/EOS) and 46.6 (WALS/EOS), if the optimizations are performed for models including the degradation corresponding to the inaccuracies mentioned above. The corresponding points in Fig. 3.2.4 are marked with a '4'. Moreover, the nominal growth factors in this case grow only slightly from points '1' to points '3'.

In a third, more academic approach a configuration is optimized for lowest sensitivity $S^*$. These configurations lead, of course, to the lowest sensitivities, but improvements in sensitivity compared with the $GF^*$ optimized configurations are very small ($\Delta S^* \approx 0.15$) and the nominal growth factors increase from points '3' to points '5' by $\Delta GF \approx 1.6$.

Therefore, for practical application, the $GF^*$ optimized configurations are of much greater importance than the $S^*$ optimized configurations. Geometrically, the stages of the $GF^*$ optimized configurations are nearly identical with the nominally optimized ones, i.e. their planform area and body surface is diminished as far as possible for the reasons already discussed. The other design parameters are quite similar, as well. Solely, the ramjets' thrust-to-weight-ratios of the sensitivity-based configurations are increased to 0.34 at a constant separation Mach number of $M \approx 6.5$. In addition, the upper-stage engines are amplified by about 9%. Altogether, this leads to a payload release altitude of about 91 km.

### 3.2.6 Summary and Conclusions

In this paper two fully reusable, two-stage-to-orbit space planes with horizontal take-off and landing capability were examined for a given reference mission to carry a 7 t payload mass into a low earth orbit of 250 km. The space plane concepts differ in the airbreathing lower-stage vehicles, each carrying a rocket powered orbital upper stage. While the first concept is based on a geometrically simple lifting body with elliptical cross-sections, the second concept makes use of a waverider geometry derived from the osculating cones principle and modified for practical application. Both configurations were examined by means of the specially developed numerical simulation and optimization tool ASTOR. This programme established an approach to preliminary design and analysis integrating performance risks into the simulation and optimization studies. Risks are simulated by inaccuracies in certain physical computation models which worsen mission performance. Thus, the design engineer is enabled to find best performing configurations under the assumption that the numerical models do not reproduce reality exactly. In total, three different quality criteria were used. They are all based on the growth factor *GF* which is defined as the ratio of gross take-off mass to orbital payload mass. The nominal growth factor *GF* evaluates configurations assuming that all computational models are correct. In contrast, the sensitivity-based growth factor $GF^*$ assesses space plane configurations subjected to model inaccuracies during their mission. The performance degradation from the nominal growth factor to the sensitivity-based growth factor of a configuration is called its sensitivity. In principle, sensitivities can either refer to a single numerical model or to several models simultaneously.

The comparison of the optimized configurations with the initially defined reference configurations showed that the heavier empty weight of the waverider lower stage is overcompensated by its favourable aerodynamic characteristics and thus leads to a 6 t lower gross take-off mass. The sensitivity studies showed that its design sensitivities were generally lower than the lifting body's.

The nominal optimization of the single stages and the overall systems indicate that the lifting body configuration has more optimization potential, due to its higher geometric efficiency. The sensitivity-based optimizations that were successfully carried out for model inaccuracies of 3% in 13 models clearly confirmed this tendency.

## References


1. M. Bayer, A. Cwielong, H. Grallert, H. Sacher, H. Uebelhack: System Concepts Survey and Preselection, FESTIP System Study FSS-DRI-SC-2300–2301, 1995.
2. R. Janovsky: Analyse und Bewertung horizontal startender Raumtransportsysteme, Dissertation, Institut für Luft- und Raumfahrt RWTH Aachen, VDI-Verlag, ISBN 3-18-324212-5, 1995.
3. G. Neuwerth, R. Staufenbiel: Geometrie der Konfiguration ELAC-1, Interner Bericht, No. 1 TP A4, Institut für Luft- und Raumfahrt, RWTH Aachen, 1990.
4. D. Jacob: Research on the Hypersonic Space Plane Configuration ELAC at the RWTH Aachen, 12th European Aerospace Conference on Space Transportation Systems, Paris 29.11.–1. 12. 1999.
5. A. Kharitonov, M. Brodetsky, N. Adamov: Investigation of Aerodynamic Characteristics of the Models of a Two-Stage Aerospace System, Final Report, Institute of Theoretical and Applied Mechanics, Novosibirsk, Russia, June 2000.
6. U. Peiter, G. Neuwerth, D. Jacob: Aerodynamik der Hyperschallkonfiguration ELAC 1 im Langsamflugbereich, DGLR-Jahrbuch 2000, Band I, DGLR-JT2000-020.
7. G. Neuwerth, U. Peiter, F. Decker, D. Jacob: Reynolds Number Effects on Low-Speed Aerodynamics of the Hypersonic Configuration ELAC 1, Journal of Spacecraft and Rockets, Vol. 36, No. 2, 1999.
8. D. Jacob, G. Neuwerth, U. Peiter: High Reynolds Number Wind Tunnel Test with an ELAC-Model in the DNW, Tagungsheft der ZAMM, GAMM-Jahrestagung, Bremen, April 1998.
9. V. Engler: Bewertung und Optimierung von Raumflugzeugen unter Berücksichtigung der Sensitivität gegenüber Entwurfsunsicherheiten, Dissertation, Institut für Luft- und Raumfahrt, RWTH Aachen, 1999, Shaker Verlag, ISBN 3-8265-4780-2.

10. V. Engler, D. Coors, D. Jacob: Optimization of a Space Transportation System including Design Sensitivities, Journal of Spacecraft and Rockets , Vol. 35, No. 6, November–December 1998.
11. V. Engler, D. Coors, D. Jacob: Optimization of a Space Transportation System including Design Sensitivities, AIAA-98-1553, $8^{th}$ International Space Planes and Hypersonic Systems and Technologies Conference, Norfolk, Virginia, USA, April 1998.
12. V. Engler, D. Coors, D. Jacob: Sensitivity Based Design Optimization of a Space Transportation System, Third European Symposium on Aerothermodynamics for Space Vehicles, ESTEC, Nordwijk, 24–26 November 1998.
13. T. Raible, D. Jacob: Multidisciplinary Spaceplane Design and Optimization using a Genetic Algorithm, CEAS Conference on Multidisciplinary Aircraft Design and Optimization, 25–26 June 2001, Köln, Germany.
14. T. Raible, D. Jacob: Sensitivity-Based Optimization of Two-Stage-To-Orbit Space Planes with Lifting Body and Waverider Lower Stages, AIAA-2003-6955, 12th International Space Planes and Hypersonic Systems and Technologies Conference, Norfolk, Virginia, USA, 15-19 December 2003.
15. R. Janovsky, R. Staufenbiel: Referenzflugbahn der Konfiguration ELAC-I, Interner Bericht, No. 5 TP A1, Institut für Luft- und Raumfahrt, RWTH Aachen, 1991.
16. R. Staufenbiel: Second Stage Trajectories of Airbreathing Space Planes, Journal of Spacecraft and Rockets, Vol. 27, No. 6, November–December 1990.
17. V. Engler: Geometrie der Oberstufenkonfiguration EOS und Integration in die Unterstufe ELAC-1, Interner Bericht, No. 21 TP A1, Institut für Luft- und Raumfahrt, RWTH Aachen, 1997.
18. K.B. Center, H. Sobieczky, F.C. Dougherty: Interactive Design of Waverider Geometries, AIAA-91-1697, 1991.
19. K.B. Center, K.D. Jones, F.C. Dougherty, A.R. Sebass, H. Sobieczky: Interactive Hypersonic Waverider Design and Optimization, Proceedings $18^{th}$ ICAS Congress, Beijing, 20–25 September 1992.
20. T. Eggers: Aerodynamischer Entwurf von Wellenreiter-Konfigurationen für Hyperschallflugzeuge, Institut für Entwurfsaerodynamik, DLR Braunschweig, Forschungsbericht 1999-10, ISSN 1434-8454, März 1999.
21. A. Bardenhagen: Massenabschätzung und Gesamtauslegung der Unterstufe von Hyperschall-Raumtransportern, Institut für Flugzeugbau und Leichtbau, TU Braunschweig, ZLR-Forschungsbericht 98-04, 1998.
22. D. Strohmeyer, T. Eggers, M. Haupt: Waverider Aerodynamics and Preliminary Design for Two-Stage-to-Orbit Missions, Part 1, Journal of Spacecraft and Rockets, Vol. 35, No. 4, July-August 1998.
23. W. Heinze, A. Bardenhagen: Waverider Aerodynamics and Preliminary Design for Two-Stage-to-Orbit Missions, Part 2, Journal of Spacecraft and Rockets, Vol. 35, No. 4, July-August 1998.
24. L.F. Rowell, R.D. Braun, J.R. Olds, R. Unal: Multidisciplinary Conceptual Design Optimization of Space Transportation Systems, Journal of Aircraft, Vol. 36, No. 1, January-February 1999.
25. T. Mosher: Conceptual Spacecraft Design Using a Genetic Algorithm Trade Selection Process, Journal of Aircraft, Vol. 36, No. 1, January-February 1999.
26. P. Charbonneau, B. Knapp: A User's guide to PIKAIA 1.0, NCAR Technical Note 418+IA, Boulder: National Center for Atmospheric Research, 1995.
27. P. Charbonneau: Release Notes for PIKAIA 1.2, NCAR Technical Note 451+STR Boulder: National Center for Atmospheric Research, 2002.

# 4 Aerodynamics and Thermodynamics

## 4.1 Low-Speed Tests with an ELAC-Model at High Reynolds Numbers

Günther Neuwerth*, Udo Peiter, and Dieter Jacob

### Abstract

The hypersonic research configuration ELAC consists of a lifting body with a delta planform of aspect ratio AR = 1.1 and rounded leading edges. In this paper the low-speed aerodynamics of ELAC especially during take-off and landing is described. The flow field of such a configuration at higher angles of attack is dominated by a vortex system at the suction side. For a rounded leading edge the location of the separation line of the primary vortex is not known. It will be shown in this paper that the position of this line and the strenght and location of the vortex system are strongly influenced by the Reynolds number.

Investigations were conducted in several subsonic wind tunnels with ELAC models of different scale. Therefore, an extensive aerodynamic data base at different Reynolds numbers is now available which can be used for code validation. The opportunity to investigate the ELAC flow field in one of the largest low-speed wind tunnels in Europe, the DNW, made it possible to reach Reynolds numbers up to $Re \approx 40 \times 10^6$ (based on the centre chord length $l_i$). Pressure distribution measurements with a high spatial resolution especially at the leading edges, surface oil flow patterns, laser-light-sheet technique and Particle-Image-Velocimetry (PIV) measurements contributed to a detailed interpretation of the flow phenomena.

* RWTH Aachen, Institut für Luft- und Raumfahrt, 52062 Aachen; neuwerth@ilr.rwth-aachen.de

*Basic Research and Technologies for Two-Stage-to-Orbit Vehicles*
DFG, Deutsche Forschungsgemeinschaft

ISBN: 3-527-27735-8

### 4.1.1 Introduction

The low-speed aerodynamic characteristic of the hypersonic research configuration ELAC (Elliptical Aerodynamic Configuration) was investigated in several wind tunnels. But only the measurements in the German-Dutch Wind Tunnel (DNW) made it possible to get flow field data with high resolution especially with respect to pressure distributions at high Reynolds numbers. Therefore, mainly the results of these measurements are desribed below. The tests were conducted 1997. The investigations were a joint task of the three Collaborative Research Centres SFB (Sonderforschungsbereich) 253 (RWTH Aachen), SFB 255 (TU München), SFB 259 (Univ. Stuttgart). In addition, the German Aerospace Centre DLR participated and supplied the PIV measurement technique. The main work and the coordination of the project were conducted by the RWTH Aachen.

The configuration ELAC represents the lower stage of a two-stage space transportation system which takes off and lands horizontally. It is a lifting body with a delta planform of AR=1.1 and rounded leading edges. Figure 4.1.1 shows the geometry and the overall dimensions of this configuration. The configuration was defined by the SFB 253 such that on the one hand all main phenomena of space transporters were represented and on the other hand the geometry was sufficiently simple to facilitate basic experimental and numerical studies (s. Ref. [1]). The cross sections consist of two half ellipses which lead to

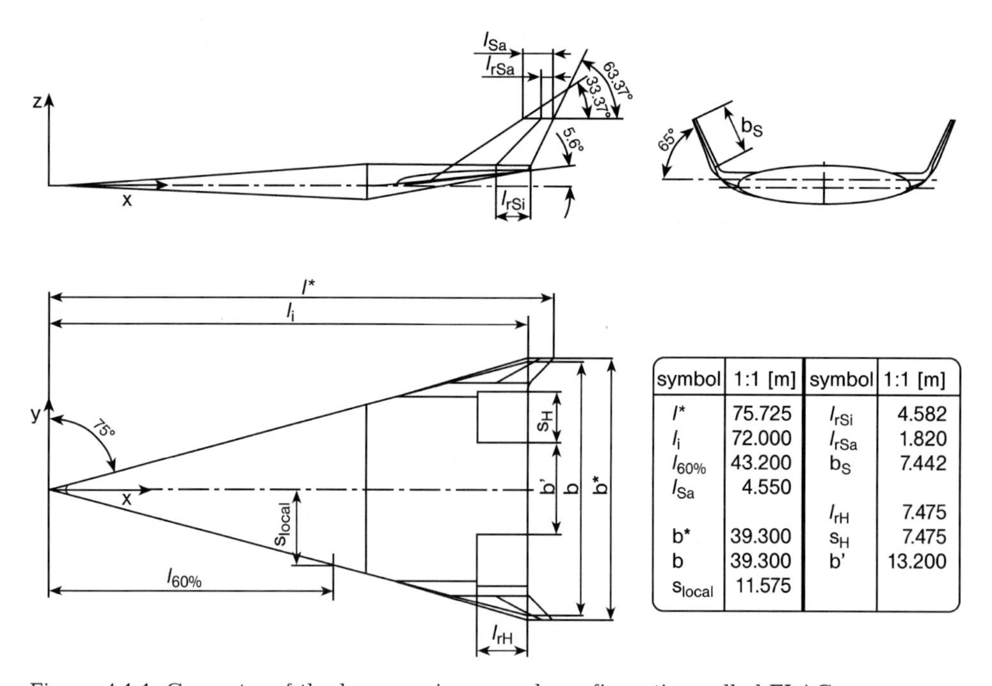


| symbol | 1:1 [m] | symbol | 1:1 [m] |
|---|---|---|---|
| $l^*$ | 75.725 | $l_{rSi}$ | 4.582 |
| $l_i$ | 72.000 | $l_{rSa}$ | 1.820 |
| $l_{60\%}$ | 43.200 | $b_S$ | 7.442 |
| $l_{Sa}$ | 4.550 | | |
| | | $l_{rH}$ | 7.475 |
| $b^*$ | 39.300 | $s_H$ | 7.475 |
| b | 39.300 | b' | 13.200 |
| $s_{local}$ | 11.575 | | |

Figure 4.1.1: Geometry of the hypersonic research configuration called ELAC.

large radii at the leading edges of the delta-wing. These rounded leading edges are important because they are reducing the thermal loads. The air-breathing propulsion system will be integrated at the underside of ELAC. Two stabilizer fins profiled with NACA 0010 are used to control the vehicle.

At higher angles of attack the flowfield around such a configuration is dominated by a vortex system at the suction side. The formation of the vortex system is closely related to flow separation at the rounded leading edges. The flow around a sharp-nosed delta-wing separates directly at the leading edge so that the location of the separation line is known in advance. For a rounded leading edge the exact position of the primary separation line is not known. It will be shown in this paper that its position and the strenght of the vortex system are strongly influenced by the Reynolds number.

There are numerous papers dealing with the low-speed aerodynamics of sharp-edged delta-wings e.g. [2–9] which report force-, moment-, pressure-, and flow field measurements. One result is that the overall forces and moments for these wings are nearly independent of the Reynolds number due to the fixed primary separation lines. Only a few papers address the influence of leading edge geometry although this is of considerable importance, because reusable space planes have rounded leading edges. Refs. [10–12] show that the non-linear vortex lift for rounded leading edges is smaller than for sharp ones, and that the formation of the vortex system starts at higher angles of attack. These effects depend on local and global Reynolds numbers. This paper contributes new results to explain these effects up to Reynolds numbers $Re \approx 40 \times 10^6$ (based on the centre chord length).

## 4.1.2 Wind Tunnel Models

To investigate the low-speed aerodynamics of ELAC, 17 models of this configuration with different scales were built and tested in 6 different low-speed wind tunnels. An extensive aerodynamic data basis is available now for the analyses of the Reynolds number influences and for the validation of aerodynamic codes. Lower Reynolds numbers were investigated in the wind tunnels of the Universities in Aachen and Darmstadt and the NWB of the German Aerospace Centre (DLR) Braunschweig. High Reynolds numbers were reached with small models in the wind tunnel HDG of the DLR Göttingen (due to high pressure) and in the KKK of the DLR Cologne (due to low temperatures). High Reynolds number measurements with a large wind tunnel model were conducted in the $6 \times 8\ \mathrm{m}^2$ test section of the German-Dutch Wind Tunnel (DNW).

The 1:12-scale model for the DNW has a centre chord length $l_i = 6\ \mathrm{m}$, a reference wing area of nearly $10\ \mathrm{m}^2$, and a mass of about 1600 kg. It was designed by the Institute of Aeronautics and Astronautics (ILR) in Aachen. The strength and the stiffness of the load-bearing structure were computed by the Institute of Aerospace Structures in Aachen, TP A6 SFB 253 (s. Fig. 4.1.2).

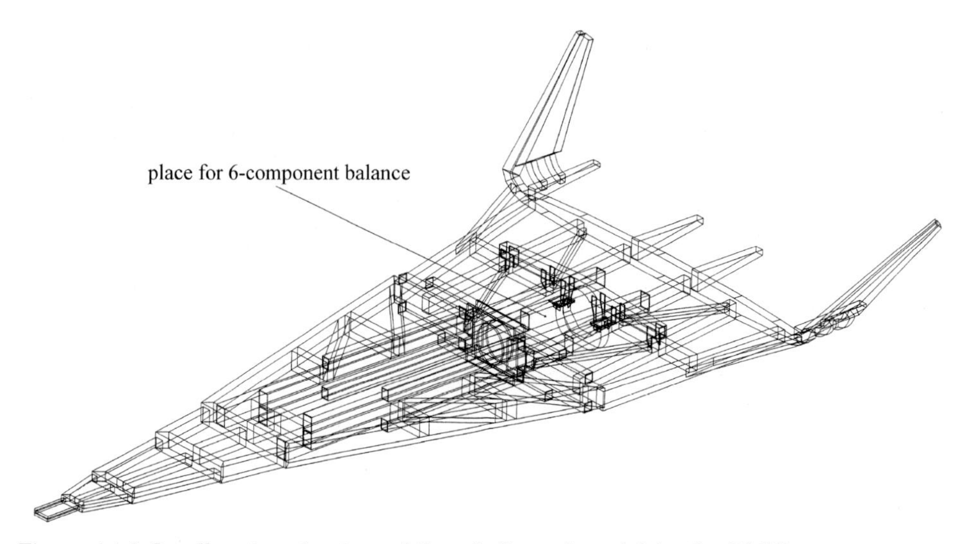


Figure 4.1.2: Loadbearing structure of the windtunnel model for the DNW.

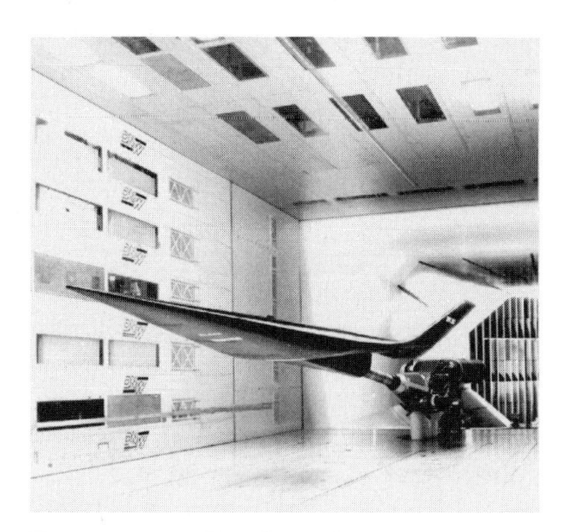

Figure 4.1.3: ELAC-model in the DNW $8 \times 6\ \mathrm{m}^2$ test section.

Figure 4.1.4: Looking from behind at the ELAC-model in the DNW.

Figures 4.1.3 and 4.1.4 show this model in the closed $8 \times 6\ \mathrm{m}^2$ test section of the DNW. The model has an internal six-component strain gage balance and is mounted on a rear sting as can be seen in Fig. 4.1.5. Flaps and rudders are movable. Some of the tests were performed with a flow-through-representation of the engine.

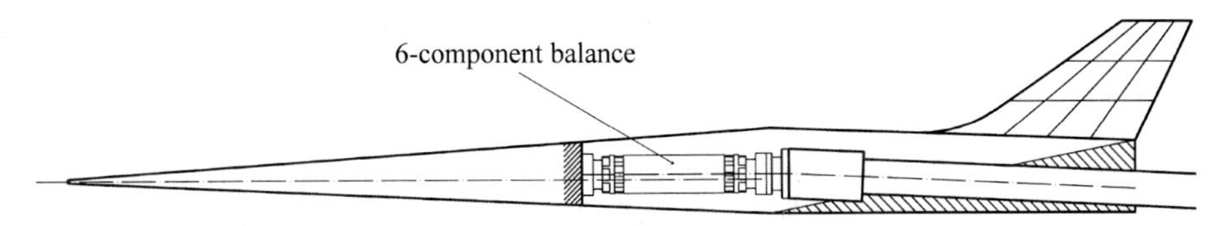


Figure 4.1.5: Side view of the model mounted on a rear sting and the internal six-component strain gage balance.

## 4.1.3 Pressure Distributions Influenced by Reynolds Number

Steady pressures were measured at the surface of ELAC in different wind tunnels using different models. The pressure holes were positioned always in the same 4 sections $y/s_{\text{local}}$ which are shown in Fig. 4.1.6. In comparison to the smaller models the model for the DNW allowed a smaller spacing of the pressure holes such that in total 150 pressure tubes were installed as can be seen in Fig. 4.1.6.

At first the pressure distributions for a smaller ELAC-model (length scale 1:65), the so-called modular model for a Reynolds number of $Re = 3.7 \times 10^6$ shall be discussed. Fig. 4.1.7 shows this model (centre chord length $l_i = 1.1$ m) and the pressure tubes coming out of the rear part.

The second stage EOS which is mounted on ELAC in this figure was not present during the pressure distribution measurements shown in Fig. 4.1.8. For the left half cross section at $x/l_i = 0.6$ the pressure coefficients $c_p$ are plotted. The mea-

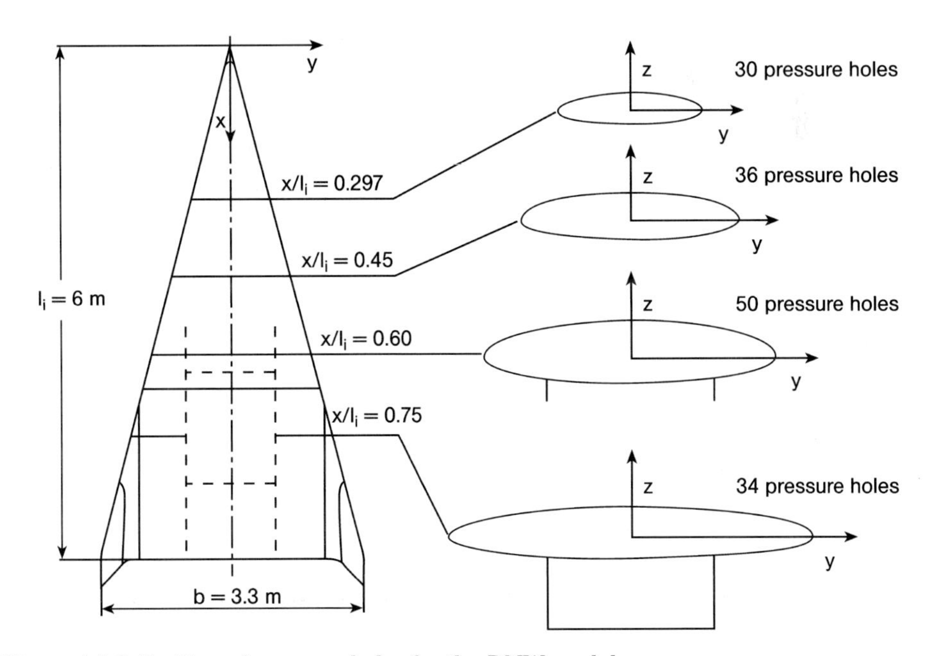


Figure 4.1.6: Position of pressure holes for the DNW model.

Figure 4.1.7: Modular ELAC wind tunnel model, here with second stage EOS.

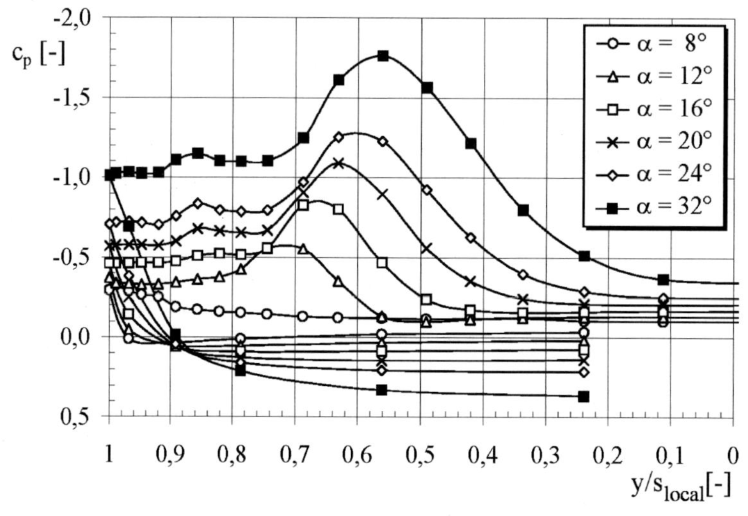


Figure 4.1.8: Pressure distribution of the left half cross section at $x/l_i = 0.6$, $Re = 3.7 \times 10^6$.

surements are described in [13]. At the suction side the $c_p$-values are negative and show the typical behaviour of a delta-wing. For angles of attack higher than $\alpha = 8^\circ$ a primary separation takes place at the leading edge ($y/s_{local} = -1$). The free shear layer rolls up to the primary vortex which induces additional velocities. These are directed outwards and become maximal just below the vortex axis. At that position on the surface a strong pressure minimum results. The oil flow pattern in Fig. 4.1.9 clearly shows this outwards (to the leading edge) directed flow. The oil flow pattern corresponds to the conditions in Fig. 4.10.8 for $\alpha = 20^\circ$. The position with the highest induced outwards directed velocity component and the pressure minimum in Fig. 4.1.8 appear at the same $y/s_{local} = -0.64$. Because the pressure is increasing along these streamlines against the leading edge, a secondary separation is caused. This separation line is marked with a "2" in the oil flow pattern. The secondary vortex can also be seen in the pressure distribution of Fig. 4.1.8 in form of a second and weaker minimum at $y/s_{local} \approx 0.85$. The topology of the

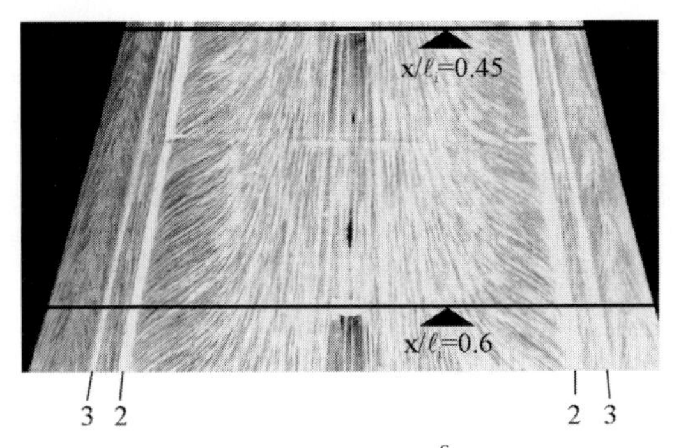


Figure 4.1.9: Oil flow pattern for $\alpha=20^\circ$ , $Re=3.7\times10^6$.

vortex system is outlined in Fig. 4.1.10; the secondary separation line is marked with $S'_{s2}$. The secondary vortex is formed and induces a flow away from the leading edge. In combination with the pressure distribution in this region a tertiary vortex is developed in this case. The corresponding tertiary separation line is marked with a "3" in Fig. 4.1.9 and with $S'_{s3}$ in Fig. 4.1.10. In the pressure distribution of Fig. 4.1.9 this tertiary vortex cannot be detected because this vortex is too weak.

With increasing angles of attack $\alpha$ the circulation in the vortices increases and the low pressures at the surface become more negative as Fig. 4.1.9 shows. Further on the primary vortex is shifted towards the plane of symmetry of the wing.

The condition of the boundary layer and the location of the transition from laminar to turbulent flow are very important for the strength and the position of the vortex system on the suction side. These conditions are strongly influenced by the Reynolds number as shown below.

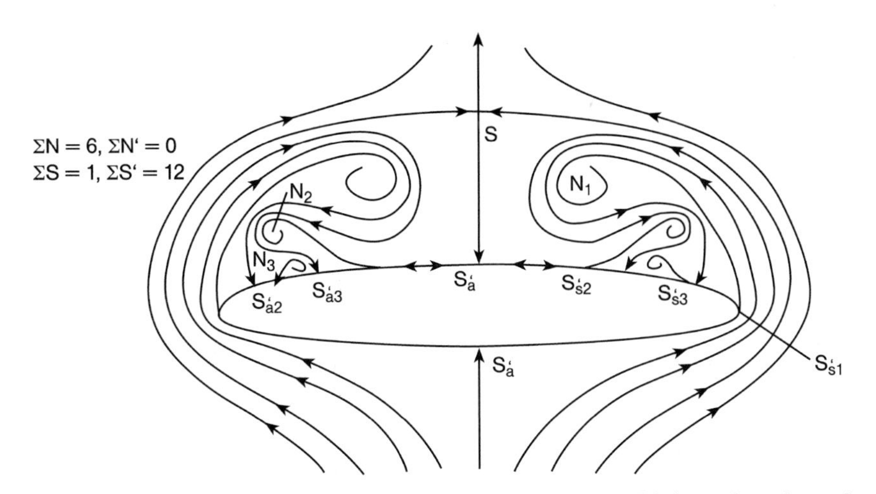


Figure 4.1.10: Topology of the vortex system for a delta-wing at high angles of attack $\alpha$.

In [13] investigations using liquid crystals (LC) are described to get an overall information of the boundary conditions for the modular ELAC-model with a length scale of 1 : 65. The LC's, which are sensitive to temperature differences, make it possible to detect the regions with laminar or turbulent boundaries. The LC-investigations have shown that the entire underside has laminar boundary layers up to the maximum Reynolds number $Re = 4.9 \times 10^6$ which was possible with the modular ELAC-model in the subsonic wind tunnel in Aachen. The transition line moved closer to the leading edges with increasing angle of attack and it can be concluded that the boundary layer at the primary separation line near the leading edge is laminar for Reynolds numbers up to $Re = 4.9 \times 10^6$. But the secondary separation takes place in a turbulent region for a Reynolds number $Re = 3.7 \times 10^6$ in Fig. 4.1.8. Due to turbulent secondary separation the pressure peaks for the secondary vortex in this figure are weak compared to the primary vortex. The turbulent boundary layer in this case is able to resist longer the increasing pressure in the direction to the leading edge. Therefore the secondary separation takes place relatively late and a weak secondary vortex originates. But it could be shown in [13] that for smaller Reynolds number of e.g. $Re = 1.1 \times 10^6$ the secondary separation takes place in a region with laminar boundary layers. Then the separation occurs earlier causing a bigger secondary vortex at a larger distance from the leading edge.

In the Reynolds number range up to $Re = 4.9 \times 10^6$ obtained with the smaller modular ELAC-model the primary separation line was always located at the leading edge. The tests with the large model (centre chord length $l_i = 6$ m) in the German-Dutch Wind Tunnel (DNW) allowed Reynolds numbers up to $Re \approx 40 \times 10^6$ combined with pressure distribution measurements with a sufficient spacial resolution.

In Fig. 4.1.11 the pressure distribution in the cross section $x/l_i = 0.3$ and $\alpha = 12^\circ$ is plotted for four different Reynolds numbers. Interesting is the following feature: For the two Reynolds numbers $Re = 7.9 \times 10^6$ and $Re = 11.9 \times 10^6$ the primary separation occurs directly at the leading edge ($y/s_{local} = -1$) and leads to the well known strong primary vortex at $y/s_{local} = -0.74$. However for the higher Reynolds numbers $Re = 27.7 \times 10^6$ and $Re = 39.0 \times 10^6$ a new phenomenon occurs; the flow does not separate at the leading edge. Therefore now a strong suction peak arises at $y/s_{local} = -1$. The primary separation line has moved to the upper side. This causes a weaker primary vortex as described in detail in [14]. The movement of the primary separation line can be seen in the surface oil-flow pattern (Fig. 4.1.12). Shown is the left leading edge in the upper front part of the model with a flow direction from the left to the right. The white curved line markes the cross section $x/l_i = 0.3$. The picture is plotted for $= 12^\circ$ and $Re = 19.8 \times 10^6$. The primary separation line at $x/l_i = 0.3$ is located on the suction side roughly 6 cm away from the leading edge. Clearly visible are the curved skin friction lines at the rounded leading edges showing the flow direction from the lower to the suction side.

Also the secondary separation line is clearly visible in Fig. 4.1.12. But all pressure distributions for that relatively high Reynolds number range show a weak suction peak for the secondary vortex. This was expected because the

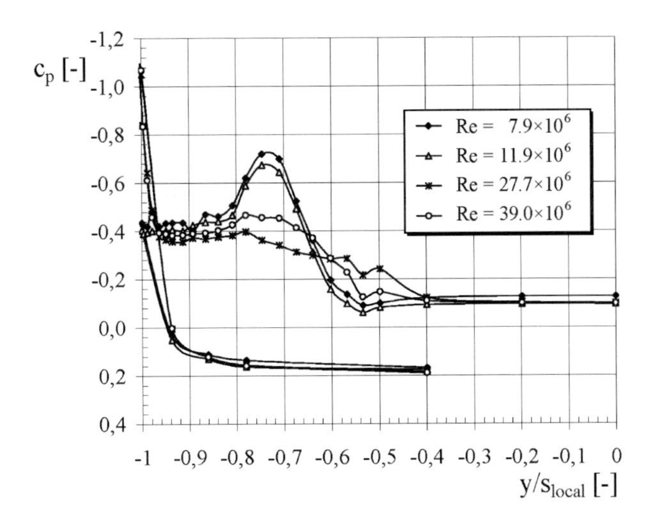


Figure 4.1.11: Influence of high Reynolds numbers on the pressure distribution for $\alpha=12^{\circ}$, $x/l_i=0.3$.

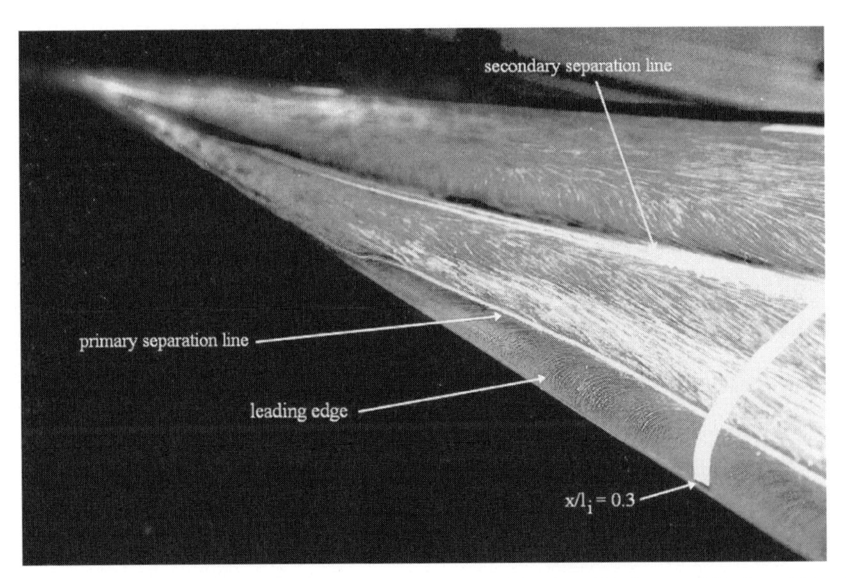


Figure 4.1.12: Oil flow pattern of the left front part of the model for $\alpha=12^{\circ}$, $Re=19.8\times10^{6}$.

boundary layer on the upper side of the ELAC-model in the regions of secondary separations is highly turbulent and leads to a vortex with less circulation.

The boundary layer transition on the suction side of the DNW-model was investigated by the "Aerodynamisches Institut" AIA of the RWTH Aachen using multisensor hot film foils which were sticked on the model at three positions $x/l_i$. In these foils with a thickness of 50 m 192 hot films in total are integrated as descibed in [15]. Their time dependent signals gave informations about the boundary layer conditions. From the pressure distribution shown in Fig. 4.1.11 it can be followed that the boundary layer at the leading edge is laminar for the two smaller Reynolds numbers $Re=7.9\times10^6$ and $Re=11.9\times10^6$. The boundary layers at the lower side of the model stay laminar near the leading edge because the flow is accelerated there due to the strong negative pressure gradients. The signals of the multisensor hot film foils show that the suction side has turbulent boundary layers. For Reynolds numbers of approximately $Re=19.8\times10^6$ and higher the boundary layer is turbulent near the leading edge. The flow can now follow the rounded leading edge for a while in spite of the strong increasing pressures on the leeward side and the primary separation occurs on this side now at about 98% of the local semispan.

Now pressure distribution measurements in the cross section $x/l_i=0.6$ shall be discussed. Fig. 4.1.13 shows for $\alpha=21^\circ$ that the flow separates just on the leading edge for the lower Reynolds number $Re=5.9\times10^6$. This pressure distribution corresponds to Fig. 4.1.8 ($Re=3.7\times10^6$) obtained with a smaller model. But for $Re=7.8\times10^6$ the primary separation has moved to the leeward side leading to a pressure peak at the leading edge ($c_p=-1.4$).

Simultaneously the primary vortex shifts inboard and is weakened. At $x/l_i=0.6$ the relatively small Reynolds number of $Re=7.8\times10^6$ is sufficient now for the movement of the primary separation line to the upper side. This can be explained by the doubling of the leading edge radius from the cross section at $x/l_i=0.3$ to that at $x/l_i=0.6$. The leading edge radius for the ELAC-configuration is 2.8% of the local half span. That leads to the statement that the pressure distributions in different cross sections are similar if the local Reynolds numbers $Re_{local}=x\times v_\infty/\nu$ are the same.

Figure 4.1.13: Influence of increasing Reynolds numbers on the pressure distribution for $\alpha=21^\circ$, $x/l_i=0.6$.

### 4.1.4 Flow Field Influenced by Reynolds Number

To supplement the pressure distribution measurements the Reynolds number influence was also investigated using flow field measurements. The flow field of a vortex system on the upper side of the delta-wing of ELAC is more or less unsteady. In such a case the Particle-Image-Velocimetry (PIV) makes it possible to get good informations because it allows the capture of the flow velocity of large flowfields instantaneously. The PIV measurements in the DNW were conducted by the Institute of Fluid Mechanics, DLR Göttingen. Four pulsed lasers were generating a light sheet on the suction side of ELAC-1 in the cross section at $x/l_i$=0.6. A drawing of the PIV test setup is given in Fig. 4.1.14. Two light pulses (pulse duration 6 ns) were generated, each with an energy of 640 mJ. The separation time between these two pulses had to be chosen between 15 μs and 100 μs as a function of $v_\infty$. Tiny special tracer particles in the flow are lit with the pulses and recorded with CCD cameras (resolution 1008×1018 pixel).

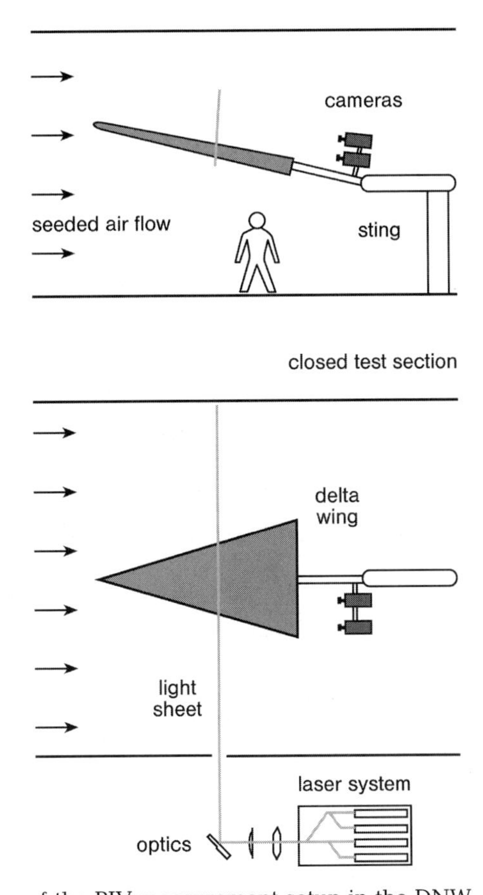


Figure 4.1.14: Drawing of the PIV measurement setup in the DNW.

Analyzing the dislocation of the particles between the two light pulses the velocity distribution can be computed. In this case a 2D-PIV was used. This means that the radial and circumferential components of the vortex flow fields can be measured and analyzed simultaneously. Two CCD cameras had to be installed to cover a sufficiently large part of the cross section.

Figure 4.1.15 shows a 2D-vector plot of the velocity distribution computed from the PIV data with the parameters $x/l_i=0.6$, $\alpha=21^\circ$ and $Re=7.8\times10^6$. This corresponds to the second case in Fig. 4.1.13. The contour of the left suction side of the model's cross-section in the object plane is also drawn. The separate areas observed by the two cameras can be seen. One camera was pointed towards the position of the primary vortex at about 61% of the local semispan and $z/s_{local}=0.47$. The second camera observed the shear layer region emerging from the leading edge and the secondary vortex. In Fig. 4.1.15 average values of the velocities are plotted which are computed from 40 instantaneous flow field PIV measurements in quick succession ($f=5$ Hz). Fig. 4.1.15 shows that the secondary vortex is weak because of the turbulent boundary layer and because it is situated at about 80% of the local semispan. It was not possible to observe the region at the leading edge from the position of the CCD cameras mounted downstream of the model because the cameras had to be mounted high above the model with a disadvantageous oblique orientation relative to the model. However, an extrapolation of the curved shear layer position in Fig. 4.1.15 leads to a primary separation point away from the leading edge on the suction side at about $z/s_{local}=0.04$. This behaviour corresponds to the pressure distribution in Fig. 4.1.13. The pressure minimum in Fig. 4.1.13 due to the primary vortex is located at 60% of the local semispan. At this position the highest

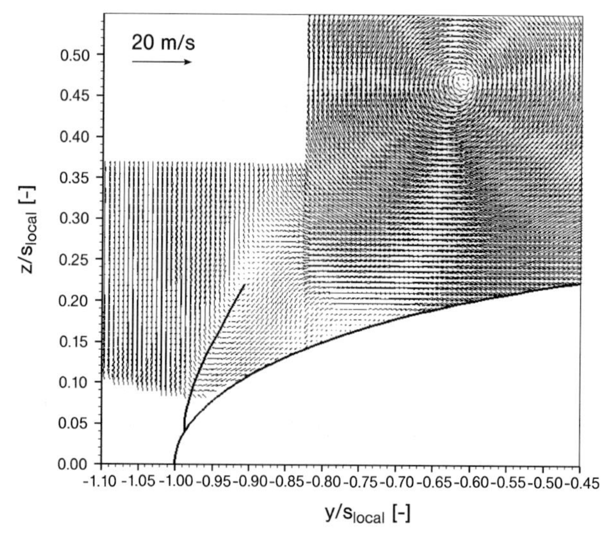


Figure 4.1.15: Time-averaged velocity field computed from PIV data. Shown is the left suction side at $x/l_i=0.6$, $\alpha=21^\circ$ and $Re=7.8\times10^6$.

values of the velocities induced by the primary vortex are measured on the surface.

Figure 4.1.16 shows a time-averaged velocity field, measured with PIV, superimposed on a time-averaged image of the seeded flow, visualized by laser-light-sheet as described in [16]. The parameters are identical with Fig. 4.1.15, but the Reynolds number is increased to $Re=19.8\times10^6$. Looking from behind in upstream direction it can be seen that the shear layer separates near the left leading edge and rolles up to a vortex. The particle images taken by the CCD cameras are of the same quality for the PIV measurements as for the light-sheet visualization, because one and the same seeding source has been utilized in both cases. The positions of the primary vortex centres agree very well in both measurement techniques.

Figure 4.1.17 shows the instantaneous flowfield without time averaging. The possibility to measure such instantaneous data is one of the important features of PIV. The parameters are identical with Fig. 4.1.16. Comparing some of the 40 instantaneous PIV flow field measurements shows that the location of the primary vortex axis is fluctuating with an amplitude of about $\pm0.5\%$ of the local semispan (here $\pm50$ mm). Figure 4.1.17 also shows that the flow near the vortex centre at that moment is not symmetrical. The counter rotating secondary vortex can clearly be seen at $y/s_{local}=0.78$. An interesting feature is that along the shear layer a series of local instantaneous small vortices can be observed. All these vortices are collected and rolled up into the primary vortex.

In Fig. 4.1.18 the vorticity distribution ($\omega_x$-component) is plotted which is computed from the average values of the velocities from 40 instantaneous PIV flowfield measurements. The parameters are the same as in Fig. 4.1.17. Clearly the highly concentrated primary vortex is visible with a high vorticity concentration ($\omega_x=-3110$ 1/s). A further high vorticity (also with negative sign) occurs at the free shear layer starting at the primary separation line. A third region

Figure 4.1.16: Time-averaged velocity field, measured with PIV, superimposed on a time-averaged image of the seeded flow, visualized by laser-light-sheet, $x/l_i=0.6$, $\alpha=21^\circ$ and $Re=19.8\times10^6$.

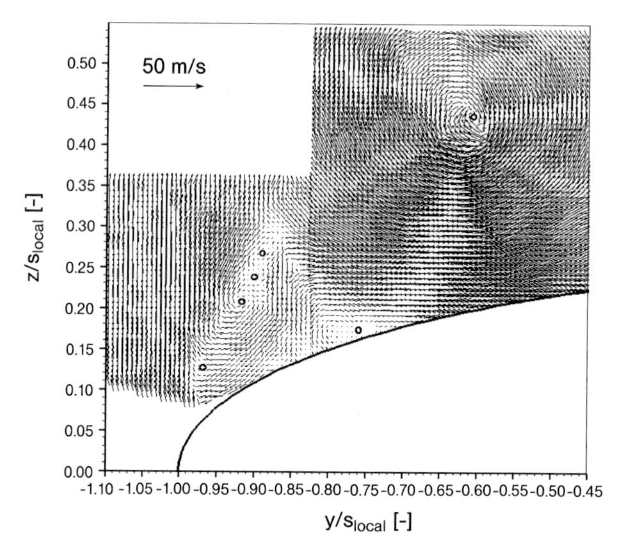


Figure 4.1.17: Instantaneous flow field, measured with PIV, $x/l_i = 0.6$, $\alpha = 21°$, $Re = 19.8\times10^6$.

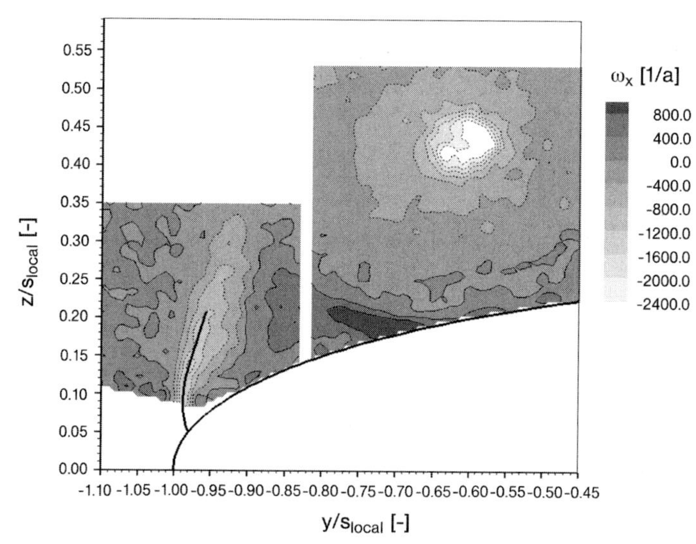


Figure 4.1.18: Vorticity component $\omega_x$ computed from PIV data, $x/l_i = 0.6$, $\alpha = 21°$, $Re = 19.8\times10^6$.

with increased vorticity (now with positive sign) is near the secondary separation line around $y/s_{local} = 0.7$.

Figure 4.1.19 shows the shifting of the shear layer and of the centres of the vortices due to an increase of the Reynolds number from $Re = 5.9\times10^6$ to $Re = 19.8\times10^6$. Using the data of the PIV measurements it can be seen again

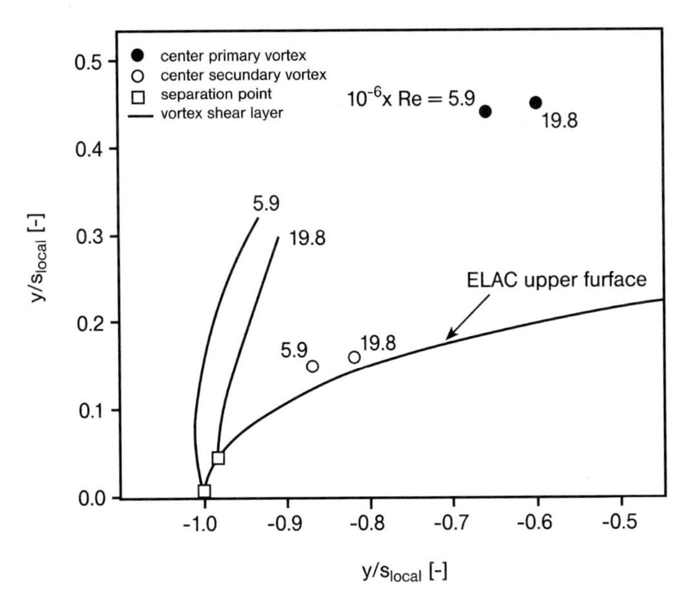


Figure 4.1.19: Shifting of the shear layer and the vortex centres due to *Re*-number.

that the separation point moves from the leading edge to the suction side of ELAC with increasing Reynolds number. The centres of the primary and secondary vortices are both moving inboard which corresponds to the pressure distribution measurements.

### 4.1.5 Force Coefficients Influenced by Reynolds Number

It was shown in the preceeding chapters that the vortex system on the leeward side of ELAC is strongly influenced by the Reynolds number. Now the influence of the Reynolds number on aerodynamic forces shall be discussed.

Six-component force and moment measurements were carried out using internal strain gage balances and different scaled ELAC-models in 7 different wind tunnels. The geometry of the rear sting and its fairing was the same in all cases. A side view was shown in Fig. 4.1.5.

Before the results in the different wind tunnels can be compared it is at first necessary to apply a wind tunnel correction for the different test sections. For the flow field of the configuration ELAC which is dominated by concentrated vortices a special correction code had to be developed based on [17]. The special correction code is described in [18]. Both the interference due to lift and to the displacement are considered. Here only the results for symmetrical flow conditions (without side slip) and without deflected flaps are presented.

Figure 4.1.20 shows the lift coefficient $C_L$ as a function of the angle of attack. The presented results were achieved in 4 different wind tunnels. The angle of attack is related to the plain between the two half ellipses forming the cross sections of ELAC.

It can be seen that the $C_L$-values (with wind tunnel corrections) measured with the different test facilities agree very well. A linear relation between $C_L$ and $\alpha$ exists for about $0° \leq \alpha \leq 9°$. Above $\alpha = 9°$ an additional nonlinear lift force occurs. This lift force is caused by the vortex system on the suction side. Due to the rounded leading edges the formation of these concentrated vortices starts at relatively high angles of attack. In Ref. [11] it is shown that this characteristic has the following differences compared to a delta-wing with sharp leading edges: for a sharp-edged delta-wing the nonlinear lift force due to the vortices begins already near $C_L = 0$ and the nonlinear lift force is much higher. At $\alpha = 20°$ e.g. the additional lift force would be increased in the order of $\Delta C_L = 0.2$ for a delta-wing with sharp leading edges and the same AR. Also the drag coefficients $C_D$ agree very well for the same four different test cases as can be seen in Fig. 4.1.21. For low angles of attack the drag is very small. An increasing $\alpha$ leads to an increase of $C_D$ mainly due to the induced drag.

In Fig. 4.1.22 the function $C_L = f(\alpha)$ is plotted for four different Reynolds numbers from $Re = 8.1 \times 10^6$ up to $Re = 38.4 \times 10^6$ (DNW investigations). The interesting result is that the Reynolds number is not influencing the lift forces. That means that the integral of the pressures over the surface of ELAC is constant although the vortex system and the pressure distributions depend on the Reynolds number as described above. The "Aerodynamisches Institut" AIA of the RWTH Aachen developed a code to compute the subsonic aerodynamics of ELAC (SFB 253, TP A3). The numerical results described in [19] are also plotted in Fig. 4.1.22. The agreement with the measurements is good and they also show a very small Reynolds number dependence.

Figure 4.1.23 shows the drag coefficients for the same parameters as in Fig. 4.1.22. In the range around $\alpha = 10°$ a Reynolds number dependence can be

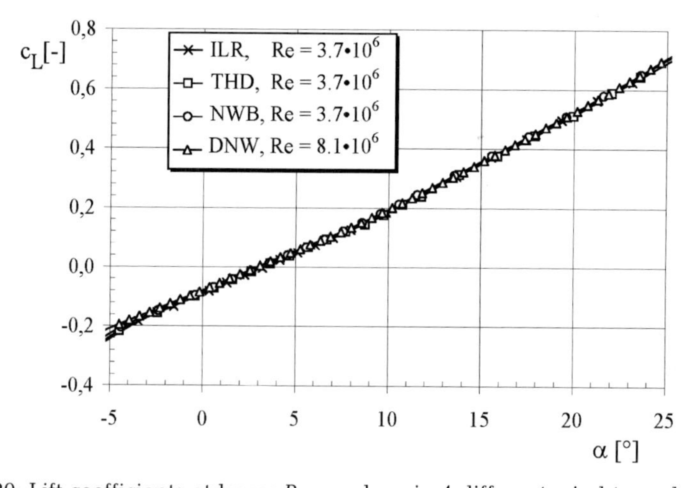


Figure 4.1.20: Lift coefficients at lower *Re*-numbers in 4 different wind tunnels.

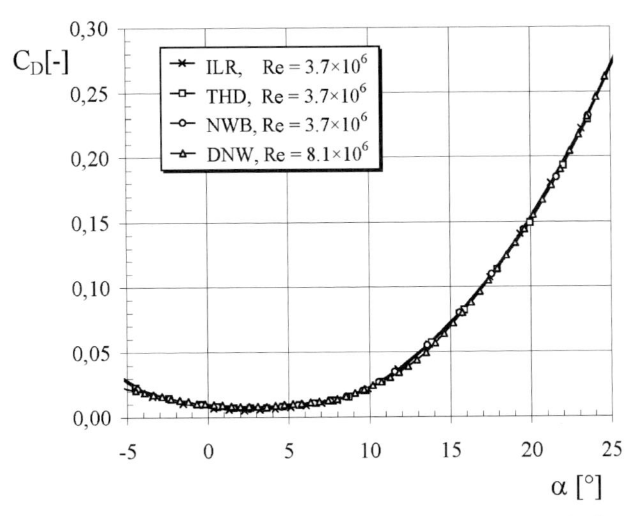


Figure 4.1.21: Drag coefficients at lower $Re$-numbers in 4 different wind tunnels.

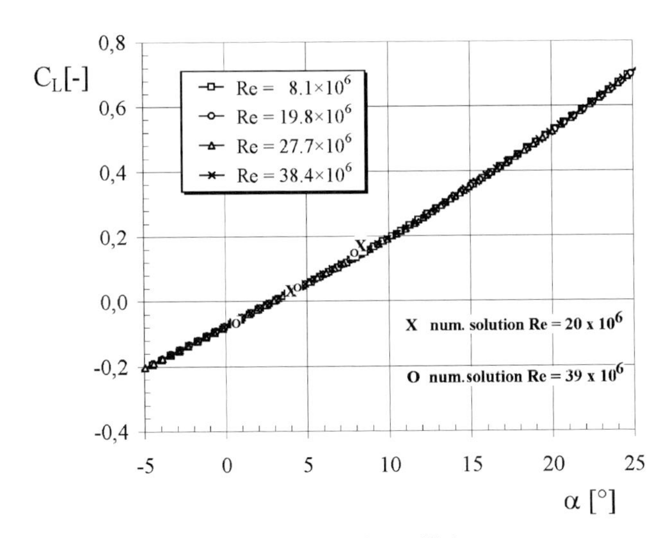


Figure 4.1.22: Influence of $Re$-number on the lift coefficients.

seen. Here the primary separation line moves for increasing $Re$ from the leading edge to the leeward side and an underpressure at the nose of ELAC generates a suction force and reduces the drag. Figure 4.1.24 gives the $C_L/C_D$-values computed with the data from Fig. 4.1.22 and 23. For $7^\circ \leq \alpha \leq 8^\circ$ the best positive $C_L/C_D$-values occur. An increase of $Re$ leads to an increase of $C_L/C_D$ up to about $C_L/C_D{=}11$ for $Re{=}38.4{\times}10^6$.

It should be mentioned that the coefficient of the pitching moment $C_m$ changes noticeably if the Reynolds number is increased. For angles of attack

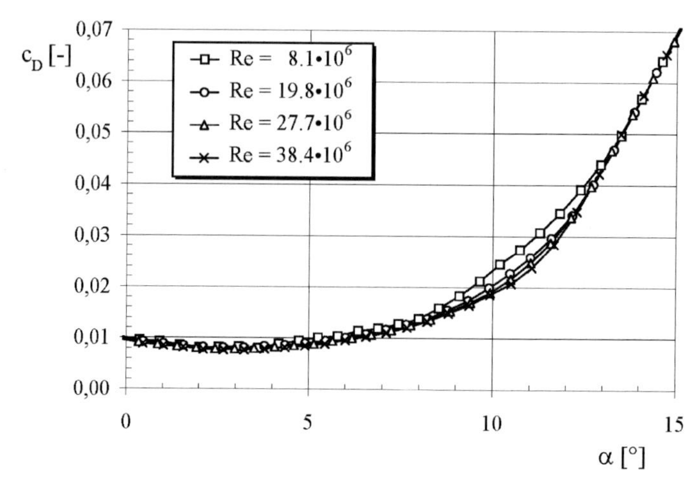


Figure 4.1.23: Influence of *Re*-number on the drag coefficients.

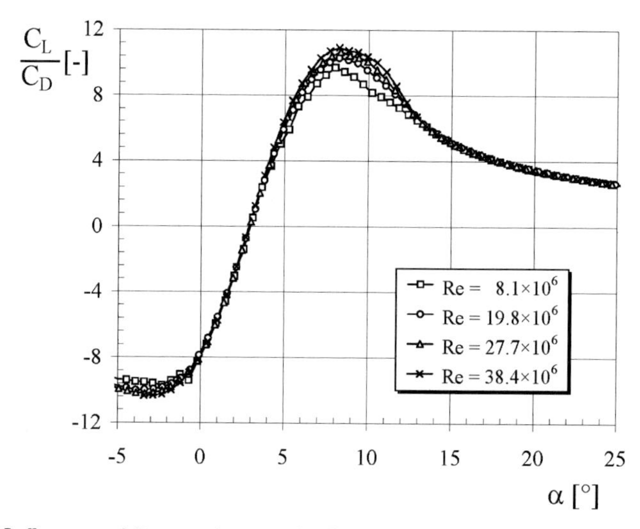


Figure 4.1.24: Influence of *Re*-number on the lift coefficients on lift to drag ratio.

$a > 8°$ $C_m$ becomes more negative due to the fact that the shifting of the primary separation line to the leeward side (generation of suction force) starts in the rearward part of ELAC as is explained in Fig. 4.1.12.

### 4.1.6 Conclusion

The hypersonic research configuration ELAC (Elliptical Aerodynamic Configuration) consists of a lifting body with a delta planform of aspect ratio AR=1.1 and rounded leading edges. In this paper the low-speed aerodynamics of ELAC is described. Investigations were conducted in several low-speed wind tunnels with ELAC-models of different scale. Tests in one of the largest low-speed wind tunnels in Europe, the DNW, made it possible to measure the flow field and pressure distributions with high resolution at high Reynolds numbers. The German Aerospace Centre DLR also participated and provided the PIV measurement technique. Now an extensive aerodynamic data base is available for a Reynolds numbers range up to $Re \approx 40 \times 10^6$ (based on the centre chord length $l_i$), which can be used for validation of aerodynamic codes.

The flowfield of those delta-wings is dominated by a vortex system on the suction side. The investigations have shown that for a delta-wing with rounded leading edges the location of the primary separation line and also the strength and location of the vortex system depend strongly on the Reynolds number. For Reynolds numbers up to $Re=4.9 \times 10^6$ the primary separation line is located always just at the leading edge. For higher Reynolds numbers however the primary separation line moves to the upper side of the delta-wing. A strong suction peak at the leading edge arises and the strength of the primary vortex is decreased. The secondary vortices are weak at high Reynolds numbers because the secondary separation occurs now in a region with turbulent boundary layers. The pressure distributions in different cross sections are similar if the local Reynolds number $Re_{local}$ is the same. The analysis of these flow phenomena was supported by surface oil-flow patterns, laser-light-sheet technique, and particle-image-velocimetry (PIV).

Six-component force and moment measurements, using internal strain gage balances, show a linear relation between $C_L$ and $\alpha$ in the range $0° \leq \alpha \leq 9°$. Above $\alpha=9°$ an additional nonlinear lift force occurs caused by the vortex system. Due to the rounded leading edges the formation of these concentrated vortices starts at higher angles of attack compared with a sharp-edged delta-wing and the generated nonlinear lift force is much lower. An important result is that the Reynolds number is not influencing the lift forces although the vortex system and the pressure distributions depend on the Reynolds number as described above. But in the range around $\alpha=10°$ a Reynolds number dependence of the drag can be seen. Here the primary separation line moves for increasing $Re$ from the leading edge to the leeward side and so an underpressure at the nose of ELAC generates a suction force and reduces the drag. In this $\alpha$-range the maximum $C_L/C_D$-values increase with increasing $Re$ to about $C_L/C_D=11$ for $Re=38.4 \times 10^6$.

## References


1. G. Neuwerth, R. Staufenbiel, Geometrie der Konfiguration ELAC I, Internal Paper A-90-1, SFB 253, Institut für Luft- und Raumfahrt, RWTH Aachen, 1990.
2. P.T. Fink, Wind Tunnel Tests on a Slender Delta Wing at High Incidence, Z. Flugwiss. Weltraumforschung 4, Heft 7, 1956.
3. D.J. Marsden, R.W. Simpson, W.J. Rainbird, An Investigation into the Flow over Delta Wings at Low Speeds with Leading-Edge Separation, Cranfield College of Aeronautics, Rep. No. 114, 1959.
4. D. Hummel, Experimentelle Untersuchung der Strömung auf der Saugseite eines schlanken Deltaflügels, Z. Flugwiss. Weltraumforsch. 13, Heft 7, 1965.
5. D. Hummel, Zur Umströmung scharfkantiger schlanker Deltaflügel bei großen Anstellwinkeln, Z. Flugwiss. Weltraumforschung 15, Heft 10, 1967.
6. N.C. Lambourne, D.W. Bryer, The Bursting of Leading-Edge Vortices – Some Observations and Discussion of the Phenomenon, Aeronautical Research Council, Reports and Memoranda No. 3282, 1962.
7. D. Hummel, Untersuchungen über das Aufplatzen der Wirbel an schlanken Deltaflügeln, Z. Flugwiss. Weltraumforschung 13, Heft 5, 1965.
8. D. Hummel, G. Redeker, Über den Einfluß des Aufplatzens der Wirbel auf die aerodynamischen Beiwerte von Deltaflügeln mit kleinem Seitenverhältnis beim Schiebeflug, Jahrbuch der WGLR, 1967.
9. D. Hummel, On the Vortex Formation over a slender Wing at Large Angles of Incidence, AGARD-CP-247, High Angles of Attack Aerodynamics, Paper 15, 1978.
10. G.E. Bartlett, R.J. Vidal, Experimental Investigation of Influence of Edge Shape on the Aerodynamic Characteristics of Low Aspect Ratio Wings at Low Speeds, Journal of the Aeronautical Sciences, Vol. 22, No. 8, 1955.
11. K. Gersten, D. Hummel, Untersuchungen über den Einfluß der Vorderkantenform auf die aerodynamischen Beiwerte schiebender Pfeil- und Deltaflügel von kleinem Seitenverhältnis, Deutsche Luft- und Raumfahrt, Forschungsbericht 66-86, 1966.
12. W.P. Hendersen, Effects of Wing Leading-Edge Radius and Reynolds Number on Longitudinal Aerodynamic Characteristics of High Swept Wing-Body Configurations at Subsonic Speeds, NASA TN D-8361, 1976.
13. F. Decker, Experimentelle und theoretische Untersuchungen zur Aerodynamik der Hyperschallkonfiguration ELAC-I im Niedergeschwindigkeitsbereich, Diss. RWTH Aachen, Cuvillier Verlag, Göttingen, 1997.
14. G. Neuwerth, U. Peiter, F. Decker, D. Jacob, Reynolds Number Effects on the Low-Speed Aerodynamics of a Hypersonic Configuration, Journal of Spacecraft and Rockets, Vol. 36, No. 2, March-April 1999, pp. 266–272.
15. E. Krause, R. Abstiens, S. Fühling, V.N. Vetlutsky, Boundary Layer Investigations on a Model of the ELAC 1 Configuration at High Reynolds Numbers in the DNW, European Journal of Mechanics-B/Fluids, 19 (2000) 745–764.
16. L. Dieterle, J. Kompenhans, U. Peiter, K. Prengel, Flow investigations on a large delta wing using LSI and PIV, 8th International Symposium on Flow Visualization, Sorrento (NA), Italy, Sept. 1–4, 1998, Conference Proceedings pp. 204.1–204.11.
17. H.C. Garner, E.W.E. Rogers, W.E.A. Acum, E.C. Maskell, Subsonic Wind Tunnel Wall Corrections, AGARDograph 109, 1966.
18. R. Arning, Untersuchung des Einflusses der Windkanalinterferenz auf die aerodynamischen Beiwerte der Hyperschallkonfiguration ELAC-I, Interner Bericht, Institut für Luft- und Raumfahrt, RWTH Aachen, 1994.
19. F.W. Zhou, Numerische Simulation des Strömungsfeldes des zweistufigen Raumflugzeugs ELAC, Dissertation, RWTH Aachen, 2003.

## 4.2 Experimental and Numerical Analysis of Supersonic Flow over the ELAC-Configuration

Anatoly Michailovich Kharitonov, Mark Davidovich Brodetsky, Andreas Henze*, Wolfgang Schröder, Matthias Heller, Gottfried Sachs, Christian Breitsamter, and Boris Laschka

### Abstract

Experimental and numerical findings for the supersonic flow field of the two-stage vehicle ELAC/EOS are presented. The experiments have been performed in the T-313 wind tunnel of the Institute of Theoretical and Applied Mechanics, SB, Russian Academy of Sciences, in Novosibirsk. They include pressure, force, and moment measurements, and schlieren flow visualization. The numerical simulations provide the fully three dimensional data and are used to analyze the flow field by visualizing wall streamlines and the inner structure of the flow field. The first configuration of the investigation is the orbital stage EOS in close vicinity to a flat plate, while in the second part the separation of ELAC and EOS is modelled. The flow fields of these bodies strongly interfere and the influence of the interaction will be discussed in detail.

### 4.2.1 Introduction

One of the most realistic concepts of future reusable space transportation systems is a two-stage configuration whose second stage is an orbital vehicle. If the first stage is equipped by an airbreathing engine, the stage separation occurs within the Mach number range of 6–12 at altitudes of about 30 km and, hence, at high dynamic pressures. Under these conditions, the aerodynamic interference between the stages will have a significant effect on the safety of the separation manoeuvre. In the general case, the supersonic flow around separating winged stages is a complex three-dimensional unsteady gas-dynamic problem. The separation of stages of aerospace systems is accompanied by intricate interactions of the incident and reflected shock waves and expansion waves and with boundary layers. The aerodynamic characteristics of two-stage-to-orbit (TSTO) models are analyzed in [1] for Mach numbers 3 and 6 to determine the separation process. The measured results for aerodynamic characteristics of the orbital Shuttle separating from the fuel tank are given in [2] for $M_\infty = 10$. Pressure fields of two bodies of revolution and their interference with a flat surface are analyzed in detail in [3, 4]. The results of a detailed numerical simula-

* RWTH Aachen, Aerodynamisches Institut, Wüllnerstr. 5–7, 52062 Aachen; andreas@aia.rwth-aachen.de

tion of the flow around aerospace systems in the course of separation of the stages are described in [5]. Certainly, the numerical methods require careful verification with experimental data to confirm their applicability. The study for the hypersonic research configuration ELAC has been initiated in Germany [6–8]. On the agreement with the Aerodynamisches Institut, RWTH Aachen, within the framework of the Collaborative Research Centre SFB253, ITAM has manufactured the ELAC (lower stage) and EOS (upper stage) models and performed experimental studies of the aerodynamic characteristics of these models within the Mach number range $M_\infty$=0.6–6.0. On the agreement between the TU Munich, Collaborative Research Centre SFB255, and ITAM SB RAS, an extensive series of experimental tests on separation of the models of the two-stage ELAC-EOS system was performed. Several experimental and numerical investigations for the supersonic flow field around the space configuration consisting of ELAC and EOS are presented. The experimental results are used for the validation and the verification of the numerical methods obtained in the numerical work of project A1 in SFB 255 in Munich and the project A3 in SFB253 in Aachen. The aerodynamic coefficients are necessary for the analysis of project A1 in SFB 253 as well as for the investigation in the dynamical behaviour of the two-stage system in project C4 in SFB 255. The numerical methods have been verified for various three-dimensional flow fields and the data can easily be exchanged since high compatibility has been reached using similar data structures and grid topologies. The results include wall pressure, force, and moment distributions as well as flow visualizations.

### 4.2.2 Experimental Setup

The present tests were performed in the supersonic wind tunnel T-313 of ITAM SB RAS whose layout is shown in Fig. 4.2.1. This facility is a blowdown wind tunnel and operates on dried air, which is compressed in gas holders of volume $V$=4400 m$^3$.

The basic modules of the wind tunnel are the input pipeline 1 with pressure adapter 2, the plenum chamber 3, the technological insert 4, the nozzle block 5, the pressure chamber with the test section 6, the supersonic diffuser 7, the two ejectors 8, the output pipeline 9, from which the air is fed into the noise-damping chamber 10 and then released into the atmosphere. This tunnel configuration ensures an air flow up to $M_\infty \leq 4$. A further increase in flow velocity requires heating of air to prevent condensation; therefore, in wind tunnel operation with flow velocities at $M_\infty \geq 5$, another input pipeline 11 is used; in this case, compressed air is directed to the nozzle block through the electric heater 12 to avoid the plenum chamber.

A supersonic air flow is formed by a plane contoured nozzle consisting of two parts (inserts) 13. A set of inserts with different throat sizes ensures a discrete variation of the air flow velocity in the test section within $M_\infty$=2–6.

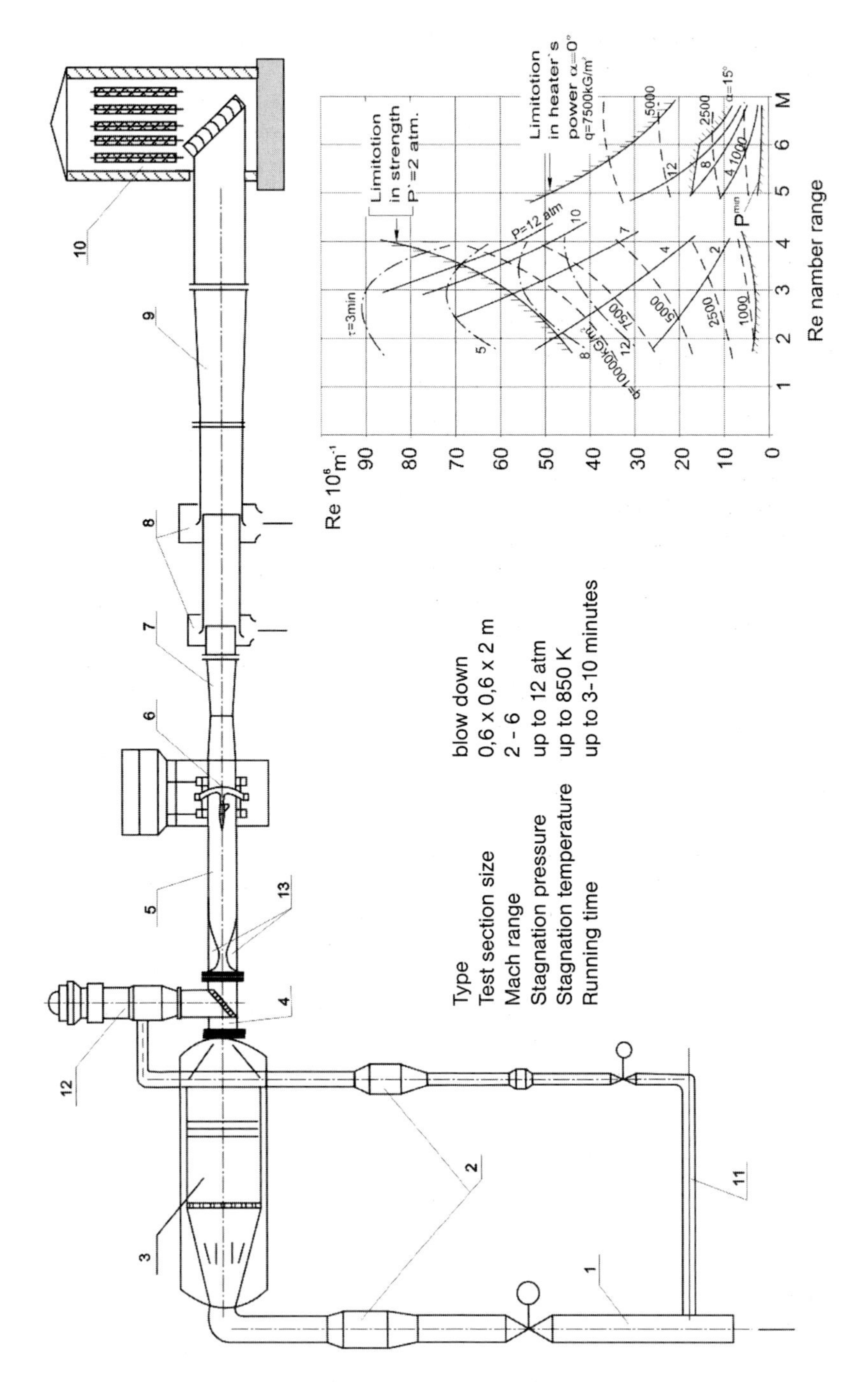


Figure 4.2.1: Sketch of ITAM wind tunnel T-313.

The test section with a cross-section of 0.6 m×0.6 m and a length of 2 m is placed in the pressure chamber, which contains the main elements of the model suspension and the aerodynamic balance including the mechanism of changing the angle of attack of the model. The latter enables an automatic variation of the angle of attack in the range of $a_M$=–4°–22°.

Two ejectors and the maximum pressure in the plenum chamber of ~1.18 MPa enable to run a rather wide range of Reynolds numbers and dynamic pressure in the test section of the wind tunnel (see Fig. 4.2.1).

The running time of the wind tunnel T-313 is determined by the compressed air stored in gas holders. In the most economical regime (ejectors are out of operation and the pressure in the plenum chamber has the minimum possible value for a given $M_\infty$), the tunnel runs for 10 minutes. For low Reynolds numbers, when it is necessary to switch on the injectors, or for Reynolds numbers that require a higher pressure in the plenum chamber, the flow rate of air per second increases and the time of operation of the wind tunnel at the maximum flow rate of air is no more than 3 minutes.

Periodic checking of the degree of nonuniformity of the velocity field in a number of sections along the test section showed that the relative root-mean-square deviations of the Mach number in the zone where models are located are less than 0.5% for $M_\infty$=2–4 and 1% for $M_\infty$=5–6.

The main configurations in the present work were the models of the two-stage space transportation system consisting of the launcher ELAC1C (first stage) and the orbital (second) stage EOS separated from the launcher. In contrast to the initial configuration ELAC1, the configuration ELAC1C has a trough in its upper surface to hold the model of the orbital stage EOS. For tests of Part II (EOS+Plate), a specially extended tail sting with a shield (Fig. 4.2.2) providing a more forward position of the EOS model in the test section was manufactured for mounting the EOS model on the mechanical balance. Hence, the model could be located almost completely in the working field of the shadowgraph and the difference between the coordinates of the prescribed and the actual centre of revolution of EOS with respect to the angle of attack could be reduced. Plate 1 in Fig. 4.2.3 is mounted on the floor of the test section of the wind tunnel by means of guide 2, vertical strut 3, and base plate 4. The plate surface is parallel to the horizontal plane of symmetry, and the mid-plane of strut 3 coincides with the OX axis of the EOS model mounted in the central unit of the arc-shaped suspension 5 of the mechanical balance. The necessary distance between the model and the plate is ensured by the height of strut 3. Therefore, the basic strut and an additional strut-insert were manufactured for defined distances between the lower point of EOS and the plate surface $\bar{h}$=0.150 and 0.225.

Four longitudinal notches are located on the upper surface of plate 1 with a spanwise step of 77 mm. These notches contain dummy inserts 6 (without pressure taps). Guide 2 has a special electric pneumomechanism, which moves plate 4 during the run across the flow (along the OY axis) with a constant step of 3.5 mm at a distance of ±38.5 mm. A sequential replacement of each insert by insert 7, which has 100 pressure taps with a stepsize of 3.5 mm in the central

Figure 4.2.2: Sting and shield for mounting the EOS model on the mechanical balance.

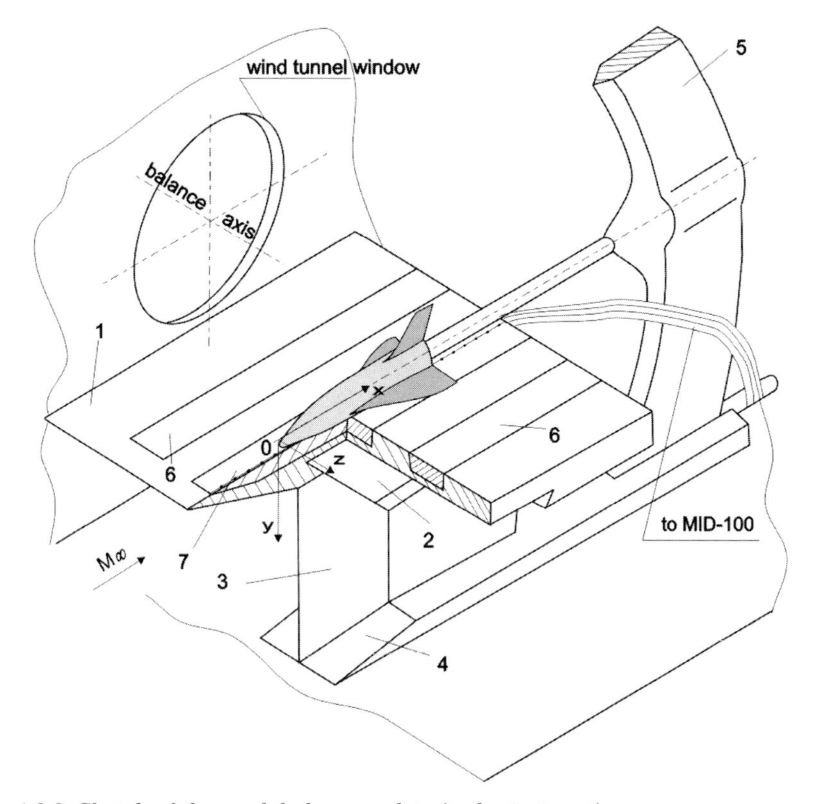


Figure 4.2.3: Sketch of the model above a plate in the test section.

longitudinal section, allows pressure measurements in the domain 346.5 mm× 308.5 mm. The first number indicates the distance between the first (located at a distance of 20 mm from the leading edge of the plate) and the last pressure tap along the OX axis, and the second number is the doubled OY distance between the plane of symmetry XOZ. The pressure was measured by ten original scanivalves of the MID-100 pressure meter [9], which ensure pressure measurements at 100 points within the range of 0–105 [Pa] with an error less than 0.3% of the upper measurement limit.

To eliminate the effect of air overflow from the lower surface of the plate to the upper one, cut-off plates (cheeks) were placed on the side walls, and special pillars-chassis were placed to increase the stability of the plate to starting loads. The EOS/plate model mounted in the test section of the wind tunnel is shown in Fig. 4.2.4.

For Part III (separation of ELAC1C + EOS), the "Separation" test bench of the wind tunnel T-313 of ITAM SB RAS was modified. A sketch of the test bench with orbital stage launch model is shown in Fig. 4.2.5, and its photograph is shown in Fig. 4.2.6.

The "Separation" test bench, as depicted in Fig. 4.2.5, is located in the pressure chamber under the lower wall of the test section of the wind tunnel. The basic element of the test bench is platform 5, which is attached by hinges to base 6; the latter is rigidly attached to the suspension shield beam 4 by four grips 7. The platform 5 has guiding grooves along which carriage 8 moves. The motion of carriage 8 is ensured by the electric driver 9 also mounted on platform 5. The cheeks 10 attached rigidly to suspension shield 4 and adjustment unit 11 ensure the guiding grooves of platform 5 to be parallel to the streamwise centreline of the ELAC1C model. Carriage 8 contains pylon 12, which can be moved in the vertical direction by the rotating screw 13. The upper part of pylon 12 has a groove where the base of the other pylons 14 is located; these pylons are designed for locating the orbital stage model in the test section of the wind tunnel for given angles of attack ($\Delta\alpha$), yaw angles ($\Delta\beta$), and rolling positions ($\Delta\phi$).

The EOS model is fixed in pylon 14 by means of an internal six-component strain-gauge balance; the sketch of the balance and some fragments of the unit for the model attachment are shown in Fig. 4.2.7.

The angular displacement of the EOS model relative to the ELAC1C model in the plane of the angle of attack may be $\Delta\alpha=0$, $2^\circ$, and $5^\circ$; it is established discretely by rotating pylon 14 in the groove of pylon 12 (Fig. 4.2.5) and subsequent bolt fixation using one of the three pairs of high-accuracy orifices. The nominal deviation of the EOS model relative to ELAC1C in terms of the rolling angle is $\Delta\phi=0$, $2^\circ$, and $4^\circ$; it is also established discretely by rotating the EOS model together with the strain-gage balance along the longitudinal axis with

Figure 4.2.4: Model above a plate in a test section (side view).

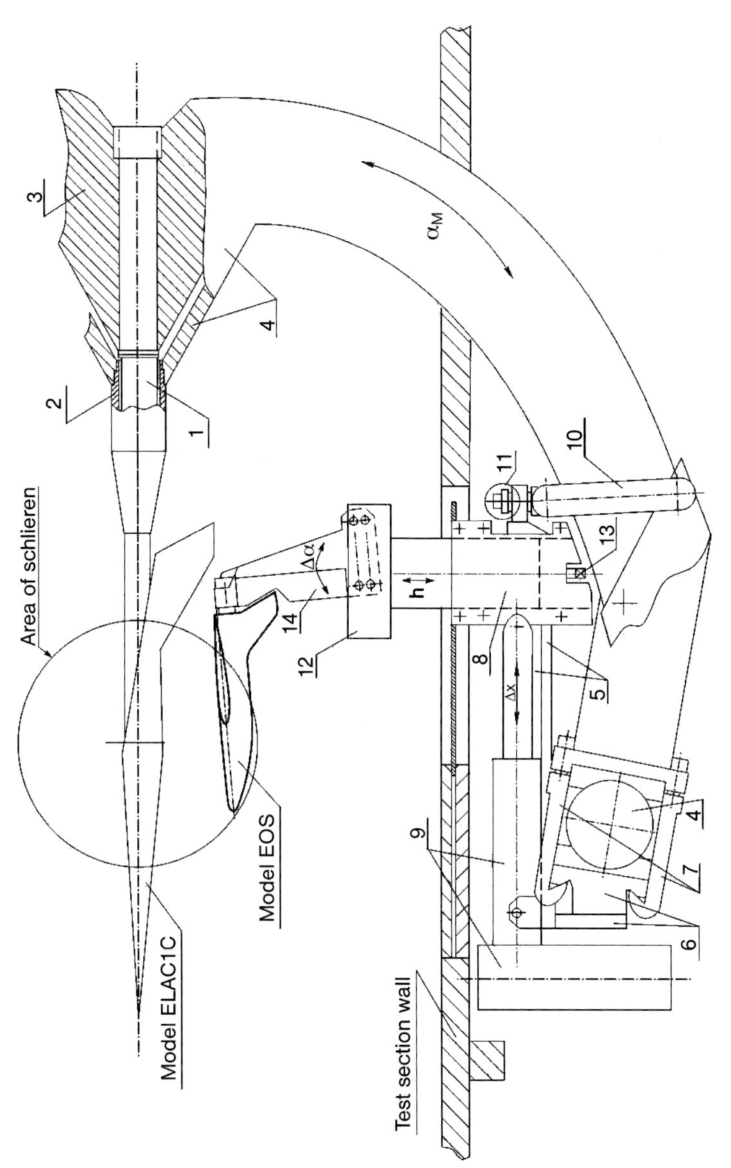


Figure 4.2.5: Sketch of the separation test.

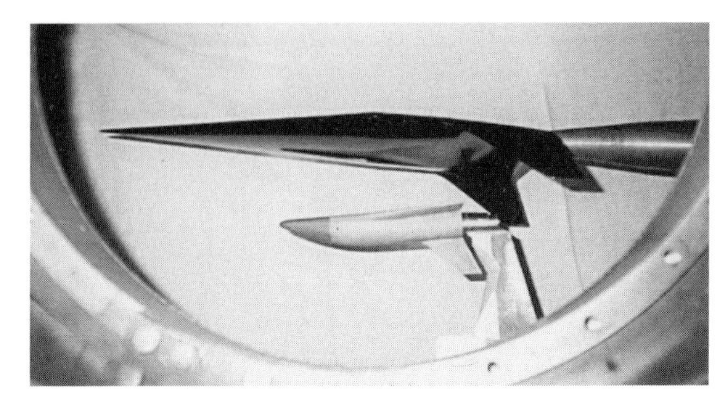

Figure 4.2.6: Orbital stage mounted for the separation test.

subsequent pin fixation using one of the three orifices designed for that. The orbital stage EOS is placed at an angle of yawing angle $\Delta\beta$ relative to the ELAC1C model by using various variants of pylon 14, in which the longitudinal axis of the conical orifice is oriented at a given yawing angle $\Delta\beta$.

Two types of aerodynamic balances were used to measure the aerodynamic loads acting on the models of separating stages.

1. An automatic four-component mechanical balance of type AB-313M. The mechanical balance was used to measure the loads acting on the first stage model ELAC1C. This balance has an arm system to decompose the total aerodynamic force and total aerodynamic moment in the balance coordinate system into the following components:
- lift force $A$,
- drag force $(-W)$,
- pitching moment $M$,
- rolling moment $L$.

2. An internal six-component strain-gage balance I-210-348 which was developed and manufactured in TsNIIMash. The construction of this balance is made in the form of a tail sting. It was used in the experiments to measure all components of the total aerodynamic force and total aerodynamic moment acting on the EOS model in the model-fixed coordinate system.

The principal scheme of the aerodynamic balances is depicted in Fig. 4.2.8, which shows the meters of the components of the mechanical balance (1–4), the meters for the pressure in the base chamber of the ELAC1C model and behind its sting (5, 6), the strain-gage balance (7), the power source (8), the electric joint (9), the amplifiers of the output signal of the strain-gage balance components (10–15), the meter for the pressure in the base chamber of the EOS model (16), the digital voltmeter (17), and the pillar of the stationary automated system acquire and process the measured information IVK-313 (18).

The flow pattern around the separating models was visualized using the standard interference shadowgraph IAB-458, which allows to observe and reg-

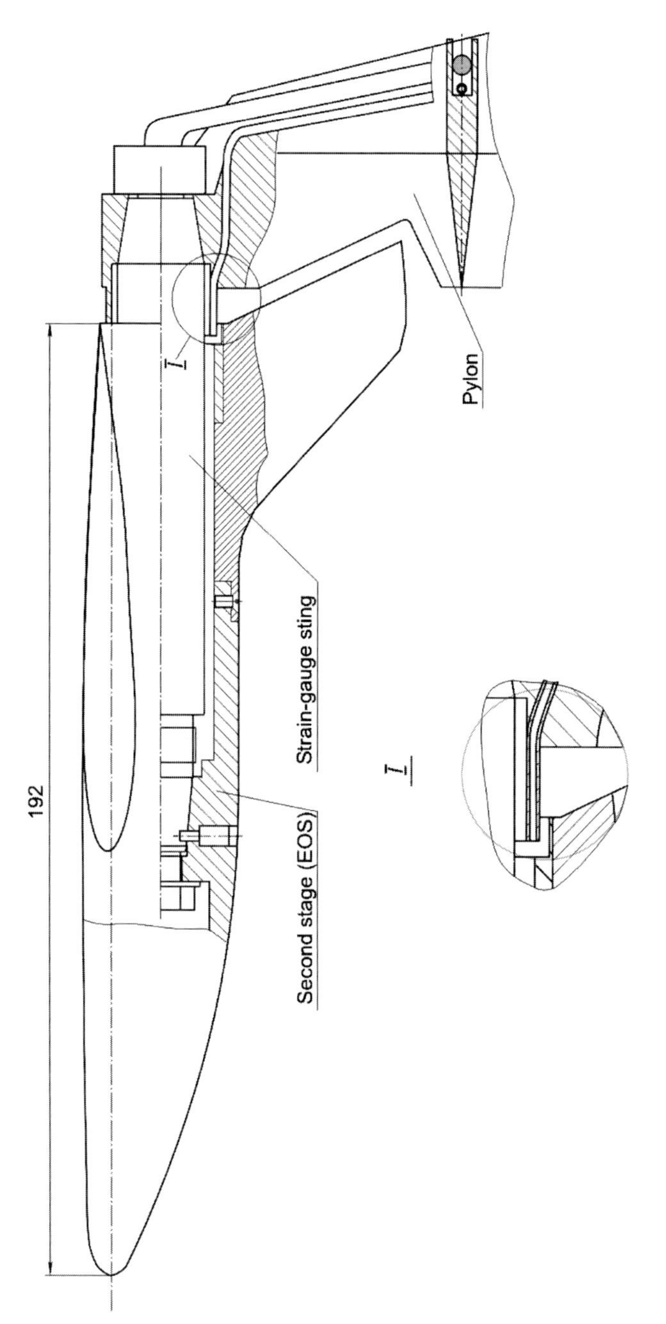


Figure 4.2.7: EOS and the six-component strain-gage balance.

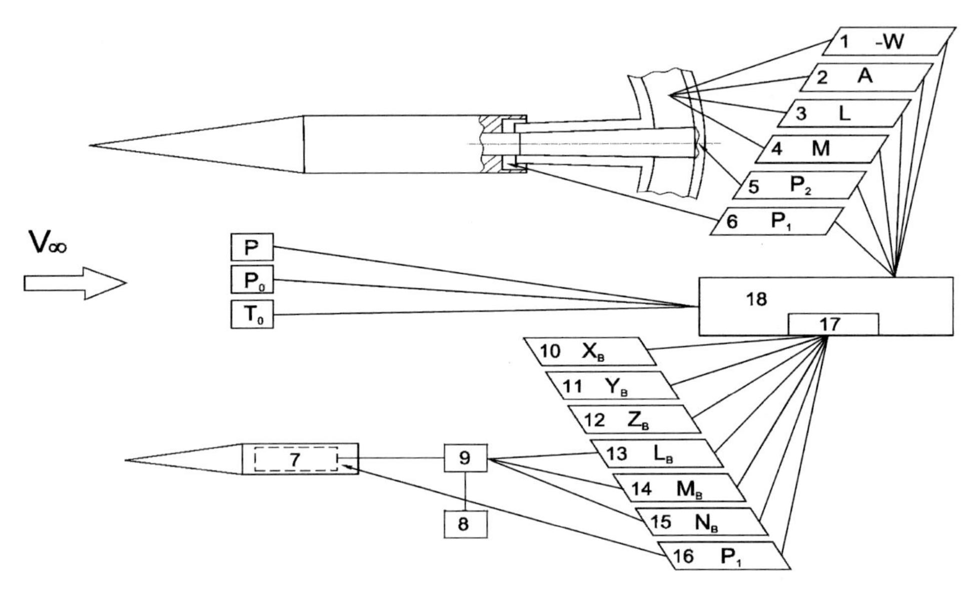


Figure 4.2.8: Principal scheme of the aerodynamic balances.

ister the density gradients in the air flow. The field of vision of the shadowgraph is a circle with a diameter of 230 mm. The wind tunnel has optically transparent glasses 400 mm in diameter in the side walls of the test section and pressure chamber, and the centre of the glass circle (field of vision) coincides with the centre of rotation of the arc-shaped strut of the "$\alpha$"-mechanism of the mechanical balance. Hence, the first stage model ELAC1C is located in the middle of the field of vision of the shadowgraph (see Fig. 4.2.5). The second stage model is shifted to the lower part of the field of vision during separation. Only the lower part of the EOS contour is visible in the case of the maximum distance between the models.

The IAB-458 instrument has a photoadapter for a narrow-film camera. In the experiments, a CCD camera JAI-CVMIORS was used, which was located on the instrument using a specially developed attachment unit. The videosignal from the CCD camera was transferred to a PC-EYE1 frame grabber mounted in a PC.

During the tests of the EOS and ELAC1C models the flow pattern around the models was registered in each run of the wind tunnel for the following nominal values of the angle of attack of the ELAC1C model: $\alpha$=0, 3°, and 6°.

### 4.2.3 Numerical Method

The Navier–Stokes equations are discretized in general curvilinear coordinates with a finite volume method on a node-centred hybrid grid, consisting of structured and unstructured blocks (Fig. 4.2.9).

The equations read in tensor notation for a Cartesian coordinate system

$$\frac{\partial Q}{\partial t} + (F^C_\beta - F^D_\beta)_\beta = 0; \quad Q = [\rho, \rho u_a, \rho E]^T$$

with the vector of conservative variables Q and the vector of convective $(F^C_\beta)$ and diffusive $(F^D_\beta)$ fluxes.

$$F^C_\beta - F^D_\beta = \begin{pmatrix} \rho u_\beta \\ \rho u_a u_\beta \\ u_\beta(\rho E + p) \end{pmatrix} + \frac{1}{Re} \begin{pmatrix} 0 \\ \sigma_{a\beta} \\ u_a \sigma_{a\beta} + q_\beta \end{pmatrix}$$

$$\sigma_{a\beta} = -2\mu \left( S_{a\beta} - \frac{1}{3} S_{\gamma\gamma} \delta_{a\beta} \right) \quad \text{with} \quad S_{a\beta} = \frac{1}{2}(u_{a,\beta} + u_{\beta,a}) \, .$$

The equations are closed using the thermal and calorical equations of state for a perfect gas as well as Fourier's law of heat conduction for the heat flux $q_\beta$. Assuming a constant Prandtl number the dimensionless heat conductivity equals the dynamic viscosity which is obtained using an exponential law

$$q_\beta = -\frac{k}{Pr(\gamma - 1)} T_\beta, \; p = \frac{1}{\gamma} \rho T, \; p = (\gamma - 1)\left(\rho E - \frac{\rho}{2} u_\beta u_\beta\right), \; \mu = T^{0.72}$$

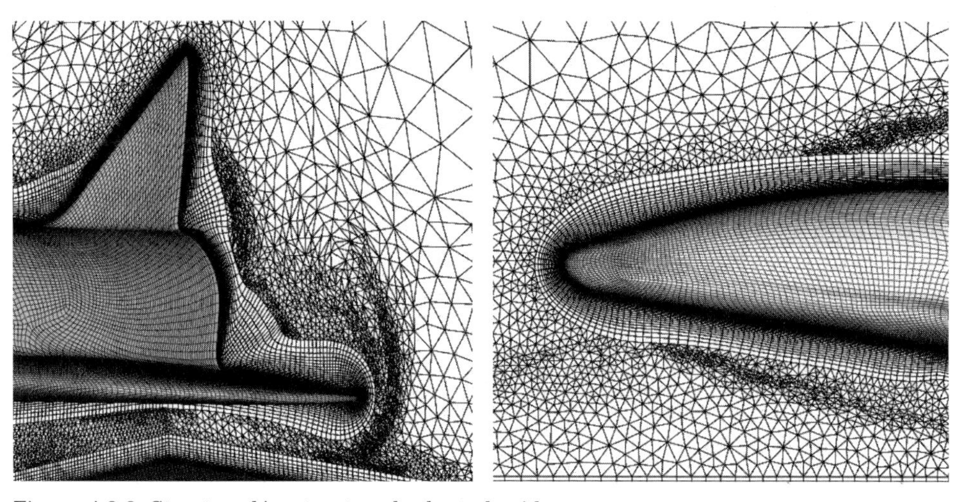

Figure 4.2.9: Structured/unstructured adapted grid.

where $\gamma$ is the ratio of specific heats, $T$ is the temperature and $Pr$ represents the Prandtl number.

The convective terms are computed using a second-order upwind AUSM-scheme [10]. In the structured part of the grid the MUSCL interpolation [11] is used to achieve a second-order scheme. To avoid overshoots near extrema the flux-limiter of van Albada has been implemented in the approach [12]. The viscous stresses are discretized using central differences of second-order accuracy with a modified "Cell-Vertex" scheme [13].

The convective flux across a triangular surface of a control volume reads

$$\vec{F}_C = |\vec{S}_f| \cdot \left(\frac{1}{2}[Ma[\Phi_L + \Phi_R) - |Ma|(\Phi_R - \Phi_L)] + p_f\right) ,$$

with the vectors, $\Phi = (\rho a, \rho au, \rho av, \rho aw, \rho ah)^T$, $p_f = (0, S_x p, S_y p, S_z p, 0)$, and $\vec{S}_f = (S_x, S_y, S_z)$ representing the dimensionless normal vector of the surface. The splitting of the Mach number and the pressure agrees with that of the structured form. For a linear reconstruction of the left and right values the average vector of gradients is computed using Greens formula

$$\int_V \nabla u dV = \int_\Omega u\vec{n} d\Omega \rightarrow \nabla u = \frac{1}{V}\int_\Omega u\vec{n} d\Omega \rightarrow \nabla u_i = \frac{1}{V_i}\sum_{f=1}^{n} \overline{u}_f \vec{S}_f .$$

$V_i$ is the control volume, $n$ is the number of surface elements enclosing the control volume, and $\overline{u}_f$ represents the average values of the corresponding quantity (density, pressure, velocity component). In order to avoid oscillations in the vicinity of discontinuities the formulation of Barth and Jespersen [14] is used.

### 4.2.4 Results

#### 4.2.4.1 Flow Over the Orbital Stage and the EOS/Flat Plate Configuration

A schematic of the models mounted in the test section of the wind tunnel and the coordinate systems is shown in Fig. 4.2.10.

The distance between the leading edge of the plate and the EOS nose model was 0.1 $l_{EOS}$. The tests were performed in the following sequence. First, the pressure was measured on the plate without the EOS model. The plate was in the position of the $\overline{h}=0.150$ case. Then, the EOS model was mounted, and a test was performed with measurements of the pressure distribution over the plate and aerodynamic loads on EOS for $\Delta\alpha=0$ and $3^\circ$. Next, a balance test of the EOS model without the plate was performed within the range $\Delta\alpha=0$–$3^\circ$ with an increment of $1^\circ$. Subsequently, the plate was mounted for the second value $\overline{h}=0.225$, and the joint measurements of pressure and aerodynamic loads were performed for $\Delta\alpha=0^\circ$ and $3^\circ$. Finally, the pressure was measured on the

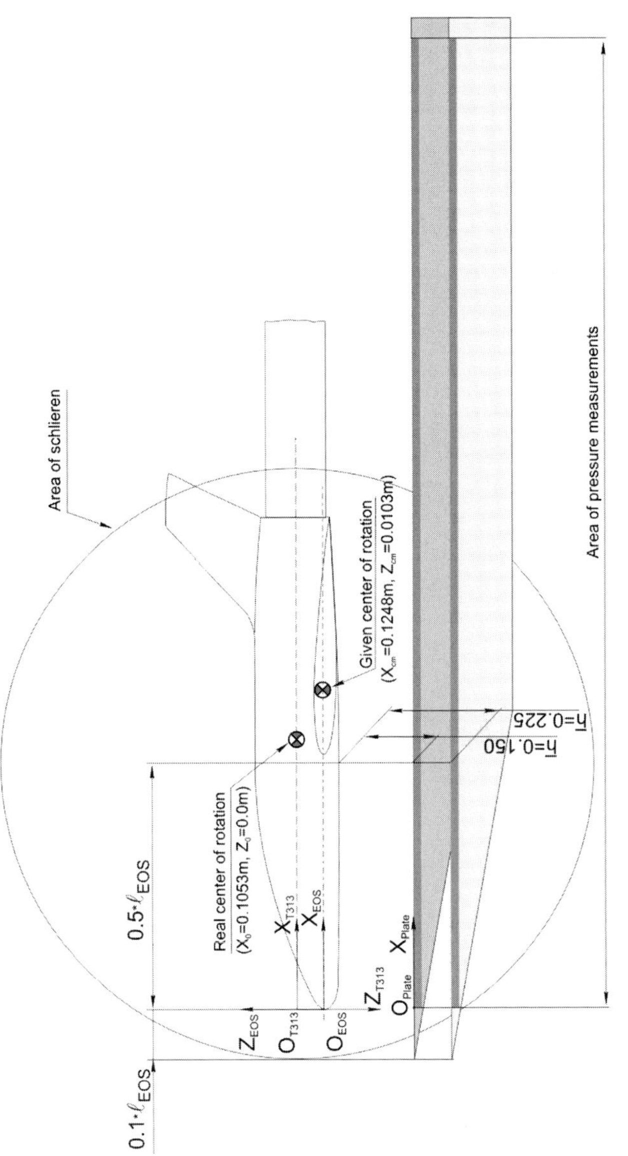


Figure 4.2.10: Coordinate system for the EOS/plate system.

Table 4.2.1: Mean flow parameters.

| $M$ | $P$ [kPa] | $T_0$ [°C] | $q$ [kPa] | $Re_{l=1m}$ |
|---|---|---|---|---|
| 4.04 | 6.4 | 20 | 73.5 | $50\times10^6$ |

plate for the second value of $\bar{h}$ without the EOS model. In joint tests of both models, schlieren pictures were taken for each value of $\Delta\alpha$. The mean flow parameters for all tests are listed in Tab. 4.2.1.

Fig. 4.2.11 shows the distribution of the pressure coefficient $C_P$ over the plate along the x-axis corresponding to the aforementioned test sequence. Thus, Fig. 4.2.11 a shows the values of $C_P$ obtained on a plate mounted for $\bar{h}=0.150$, without the EOS model ($\bar{h}=\infty$) and with EOS for $\Delta\alpha=0°$ and 3°. Figure 4.2.11 b shows the difference $\Delta C_P$ obtained after subtracting the background distribution for $\bar{h}=\infty$, i.e., $\Delta C_{P\Delta\alpha i}-C_{Ph=\infty}$. Similarly, Figs. 4.2.11 c and 4.2.11 d show the distributions of $C_P$ and $\Delta C_P$ for a distance $\bar{h}=0.225$. The schlieren pictures for the flow over the EOS/plate configuration at various $\bar{h}$ and $\Delta\alpha$ values are shown in Fig. 4.2.12. It is evident from the photographs that the bow shock wave of EOS always impinges upon the plate. Immediately downstream, there is a small expansion fan and then a united oblique shock wave from the hinge points of the leading edge of the wing. These features are also observed in the $C_P(\bar{X})$ distributions.

Figure 4.2.11 b evidences that at $\Delta\alpha=0°$ the pressure in the vicinity of the impingement of the bow shock wave from EOS onto the plate ($\bar{X}\approx0.4$) increases rapidly to $\Delta C_P\approx0.08$. Then, due to the expansion fan from the model tip, the pressure on a small section up to $\bar{X}\approx0.55$ decreases intensively to $\Delta C_P\approx0.03$. After that, there is a section with an almost constant pressure distribution with an insignificant decrease to $\Delta C_P\approx0.02$ in the vicinity of $\bar{X}\approx1.0$. Subsequently, the pressure increases slightly due to the impingement of oblique shock waves from the hinge points of the leading edges of the wing. Afterwards, the pressure coefficient starts to decrease monotonically and becomes negative at $\bar{X}\approx1.3$. For $\Delta\alpha=3°$, the nose part of the model moves upwards away from the plate. As a result, the impingement of the bow shock wave is shifted downstream by $\sim0.1\,l_{EOS}$ and the peak value of $\Delta C_{Pmax}$ induced by this shock wave is lower by $\sim0.01$ than that in the case $\Delta\alpha=0$. With increasing distance between the models to h=0.225, the influence region of EOS on the plate is shifted downstream by $\sim0.2\,l_{EOS}$. However, the qualitative and to a large extent the quantitative effect of EOS on the pressure distribution along the plate remains the same as for $\bar{h}=0.150$. Thus, the distances between the impingement points of the bow and wing-induced shock waves for identical angles of attack and their displacements with increasing $\Delta\alpha$ are almost alike.

The effect of the plate on EOS is observed only for $\bar{h}=0.150$ and $\Delta\alpha=0°$ (Fig. 4.2.12 (left)), when the reflected bow shock wave of the EOS model impinges upon the tail part of the model near the base plane. In all other cases, this shock wave shows no interaction with the EOS model, e.g. $\bar{h}=0.150$ and

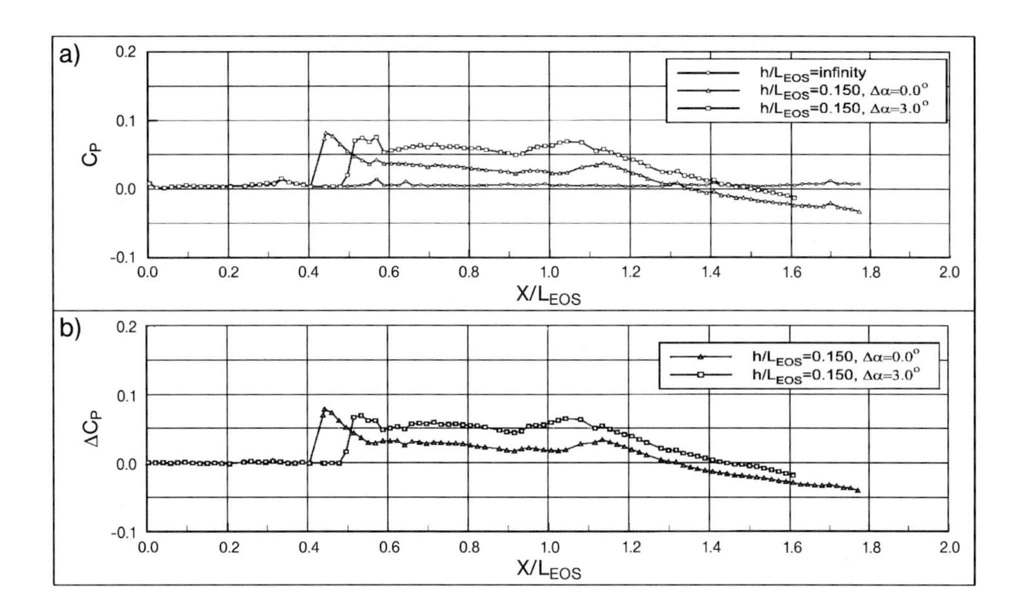


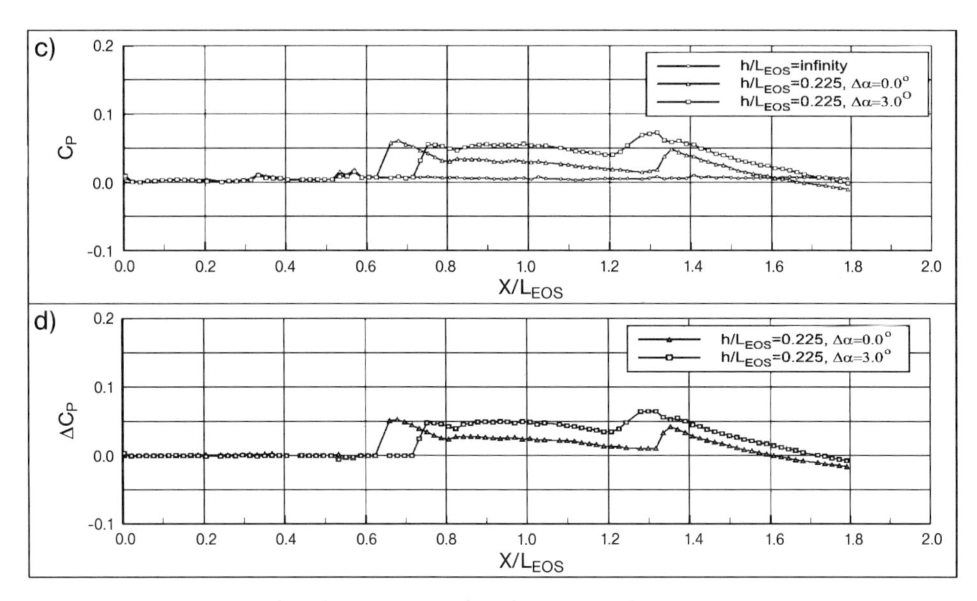


Figure 4.2.11: Pressure distributions over the plate along the x-axis.

$\Delta\alpha=3^\circ$ (Fig. 4.2.12 (right)). The line emanating from the leading edge of the plate, which is visible in all photographs, is a weak characteristic (the thickness of the leading edge of the plate did not exceed 0.05 mm), which did not exert any noticeable effect. Taking into account the rather weak bow shock wave impingement onto the plate (and the strength of the reflected shock is even

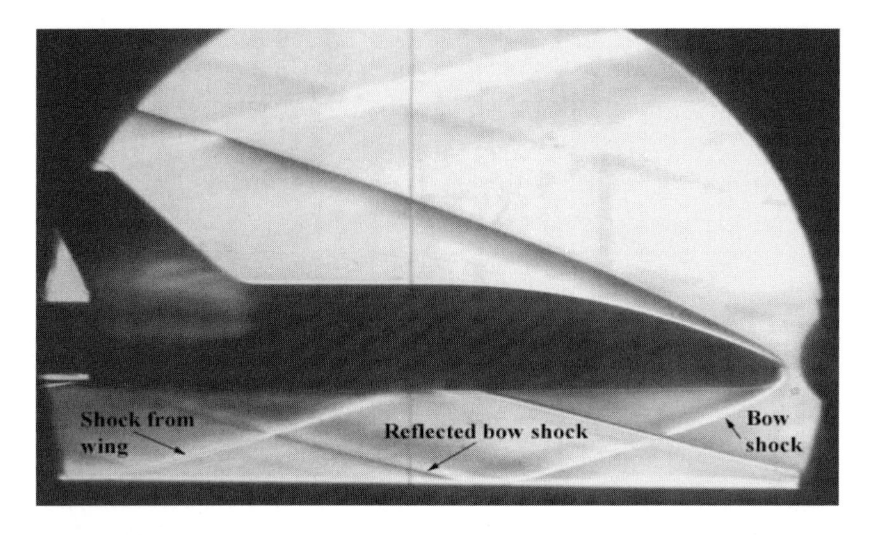


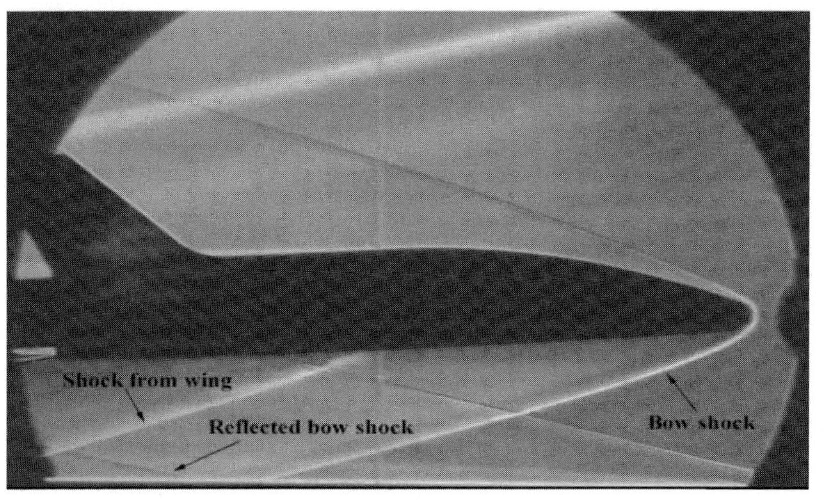


Figure 4.2.12: Schlieren picture for $h/l_{\mathrm{EOS}}=0.150$ (left) and $\Delta a=0$, and $h/l_{\mathrm{EOS}}=0.225$ and $\Delta a=3$ (right).

weaker), we cannot expect any noticeable changes in the aerodynamic characteristics of the EOS model even ($\bar{h}=0.150$ and $\Delta a=0^{\circ}$) when the reflected shock interacts with the tail part of the model. This is also confirmed by the results in Fig. 4.2.13 showing dependences $C_{ZB}(a)$, $C_{XB}(a)$, and $C_M(a)$.

In all the cases considered the aerodynamic coefficients of the EOS model almost coincide with the values obtained without the plate. Thus we can state that the EOS effect on the plate within the parameter range considered leads to a pressure increase between the impingement points of the bow shock wave and the oblique shocks from the hinge points of the leading edge of the wing. Further downstream the pressure decreases and $\Delta C_P$ takes on negative values.

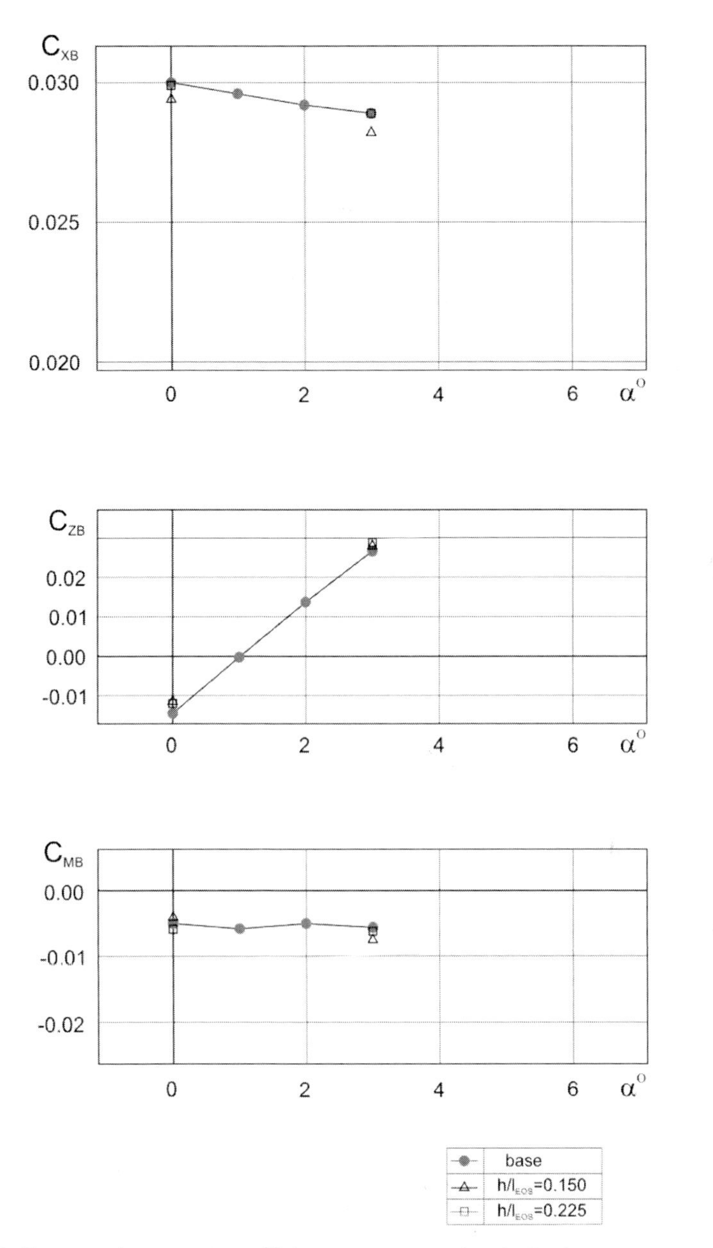


Figure 4.2.13: Force and moment coefficients versus angle of attack.

For both $\overline{h}$ values, the distance between the impingement points of these shocks decreases at increasing $\Delta\alpha$ and the pressure level grows. With increasing $\overline{h}$ and identical $\Delta\alpha$ the pressure level decreases noticeably only near the impinging point of the bow shock wave. Downstream of it, the pressure hardly changes.

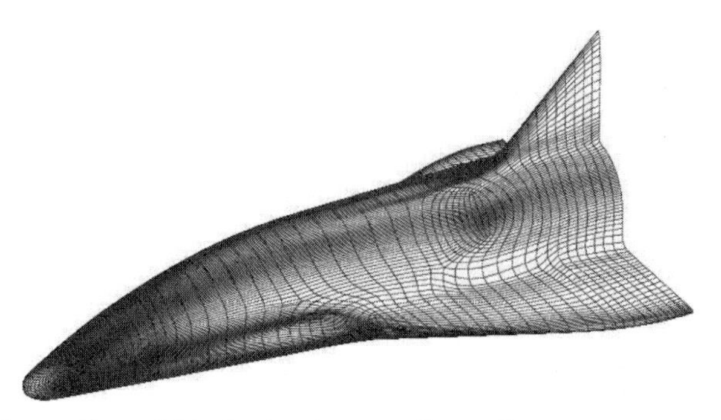

Figure 4.2.14: Geometry and block structure on the wall.

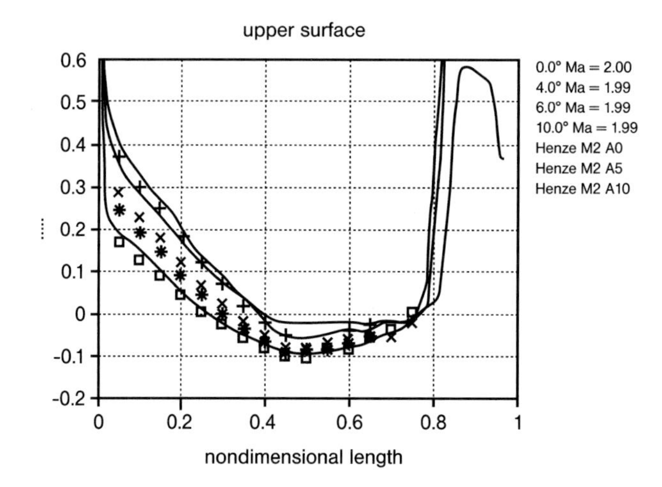


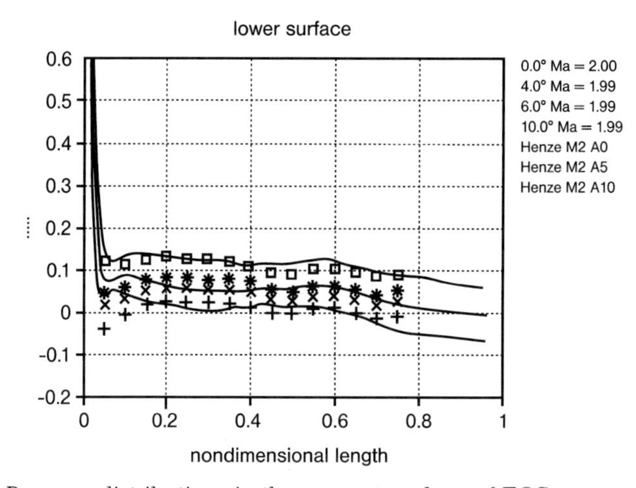


Figure 4.2.15: Pressure distributions in the symmetry plane of EOS.

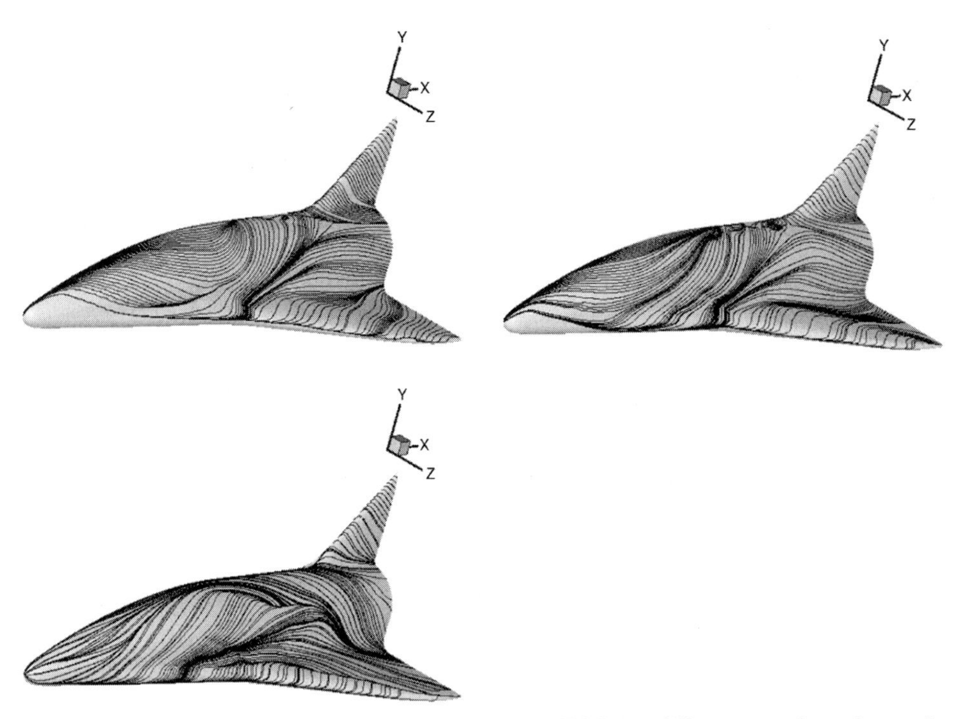


Figure 4.2.16: Computed wallstream pattern on EOS at different angles of attack, $Ma_\infty = 2.0$, $Re = 2 \times 10^6$ (0° top left, 5° top right, 10° bottom).

The numerical solution of the supersonic flow around the orbital stage comprises the comparison between computed and measured $C_P$ values. The experiments have been performed at the trisonic wind tunnel of the Aerodynamisches Institut of RWTH Aachen. The flow conditions are $Ma_\infty = 2.0$, $Re_\infty = 2 \times 10^6$. The grid has approx. 450 000 nodes in 6 blocks. The surface structure of the grid is depicted in Fig. 4.2.14.

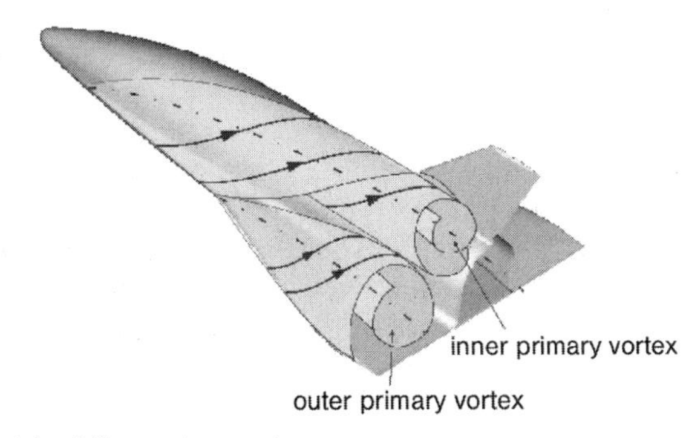


Figure 4.2.17: Sketch of the vortex system.

It was generated using the commercial tool GridPro. The comparison of the experimental and numerical pressure distributions shows satisfactory agreement at $\alpha=0°$ and $\alpha=10°$ (Fig. 4.2.15). In Fig. 4.2.16 the computed wallstream patterns at angles of attack $\alpha=0°$, $5°$ and $10°$ are plotted. At $\alpha=0°$ the attachment of the streamlines at the leading edge of the wing is clearly visible. This accumulation of streamlines moves onto the lower side at $\alpha=5°$ and at $\alpha=10°$ a flow separation occurs on the upper side of the wing. The spatial development of the flow field is sketched in Fig. 4.2.17. A more detailed discussion can be found in [15, 16].

#### 4.2.4.2 Separation of ELAC1C and EOS

The results presented here are only exemplary. The full discussion is in [17]. The staging manoeuvre of EOS and ELAC1C was investigated using the following parameters:

1. The distance $\bar{h}$ between the orbital stage EOS and the launcher ELAC1C is investigated for $\bar{h}=0.225$; 0.325; 0.450.
2. The orientation parameter $\Delta\alpha$ of EOS relative to ELAC1C is varied $\Delta\alpha=0°$, $2°$, and $5°$.
3. The orientation parameter $\Delta\beta$ of EOS relative to ELAC1C is altered $\Delta\beta=0°$, $2°$, and $4°$.
4. The orientation parameter $\Delta\Phi$ of EOS relative to ELAC1C is changed $\Delta\phi=0$, $2°$, and $4°$.

The characteristic axes and points on the models used to change the relative orientation of the configurations are shown in Fig. 4.2.18, where $\bar{h}$ is the distance in the vertical plane of symmetry of the models between the lowest point of the EOS model at 50% of its length and its corresponding point in the trough of the ELAC1C model. The following test conditions were used. The freestream Mach number was $M_\infty=4.04$, the Reynolds number was $Re \cong 48\times10^6\ \text{m}^{-1}$, and the nominal values of the angles of attack for which the aerodynamic loads on the models were measured $\alpha_M=0$, $1°$, $2°$, $3°$, $4°$, $5°$, and $5.7°$.

The schlieren picture of the interaction between ELAC1C and EOS in Fig. 4.2.19 shows that the supersonic flow around the ELAC1C configuration is accompanied by the formation of a bow shock wave. The photograph evidences a boundary layer with increasing thickness and an expansion fan downstream of the hinge of the model contour. The configuration of the trough for holding the second stage on the surface of the first stage gives rise to an expansion fan and two shock waves, that are indicated by the thin white lines on the photograph in Fig. 4.2.19. The flow around each fin is also accompanied by the formation of a shock wave. In detail the picture shows:

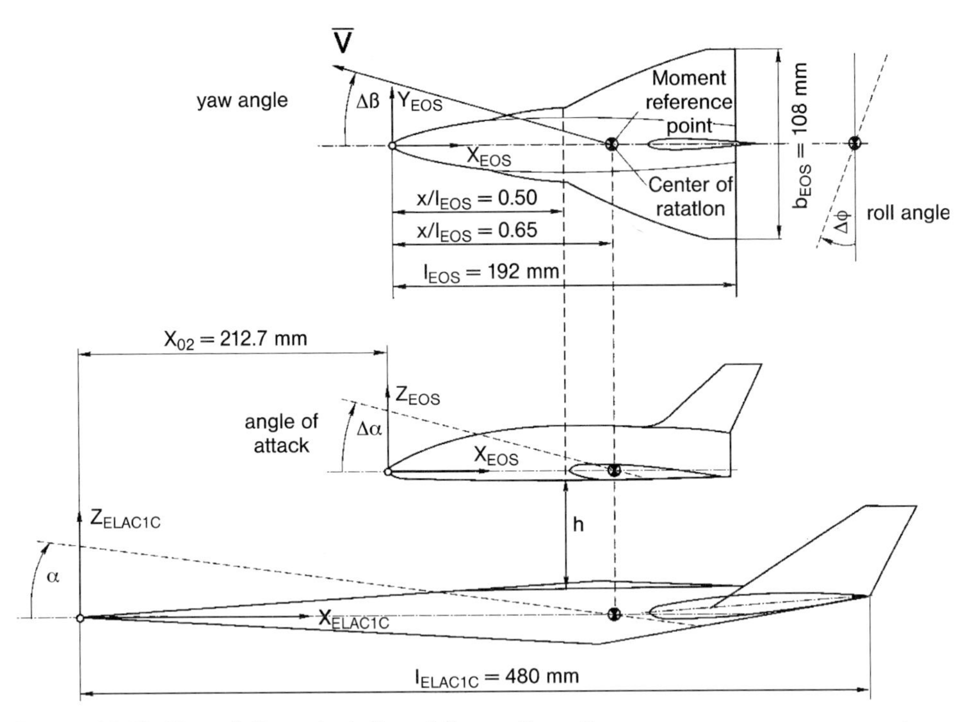


Figure 4.2.18: The relative orientation of the configurations.

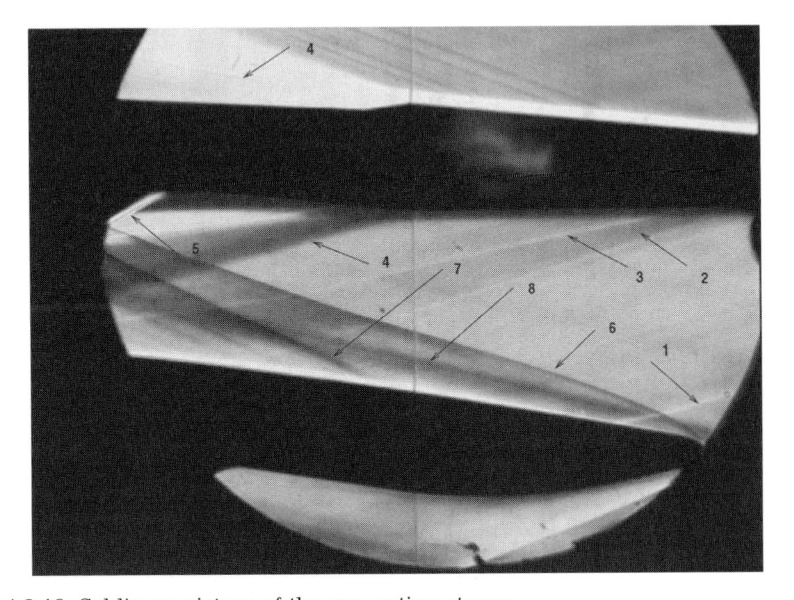


Figure 4.2.19: Schlieren picture of the separating stages.

1. The bow shock wave of ELAC1C.
2. The expansion waves caused by the edge of the trough.
3. The shock wave caused by the reflected flow within the trough.
4. The expansion fan generated downstream of the maximum thickness cross section of ELAC1C.
5. The shock wave from the fin of ELAC1C.
6. The bow shock wave of EOS.
7. The shock wave from the leading edges of EOS wings.
8. The weak characteristic from the nozzle/test section.

Figure 4.2.20 shows examples of the schlieren pictures of the flow around the ELAC1C/EOS models at several gap sizes *h/l*. Figure 4.2.21 illustrates the aero-

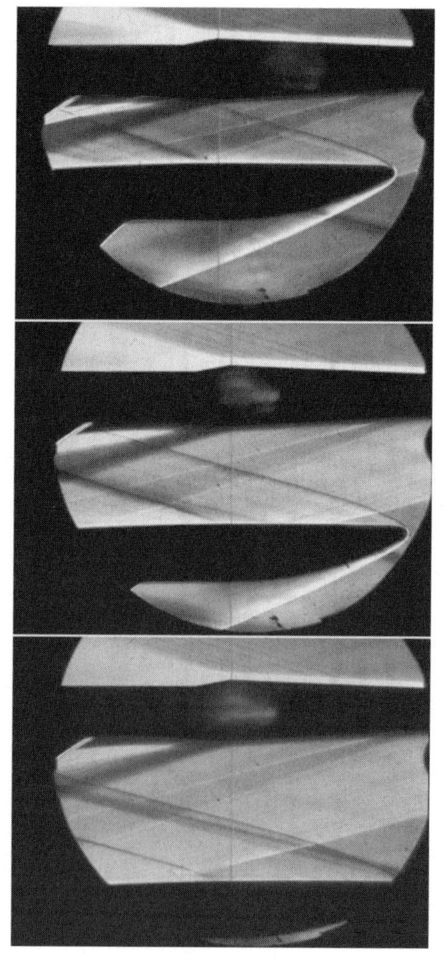

Figure 4.2.20: Schlieren pictures of the separating stages for different distances, $h/l_{EOS}$ = 0.225 (top), 0.325 (center), and 0.45 (bottom).

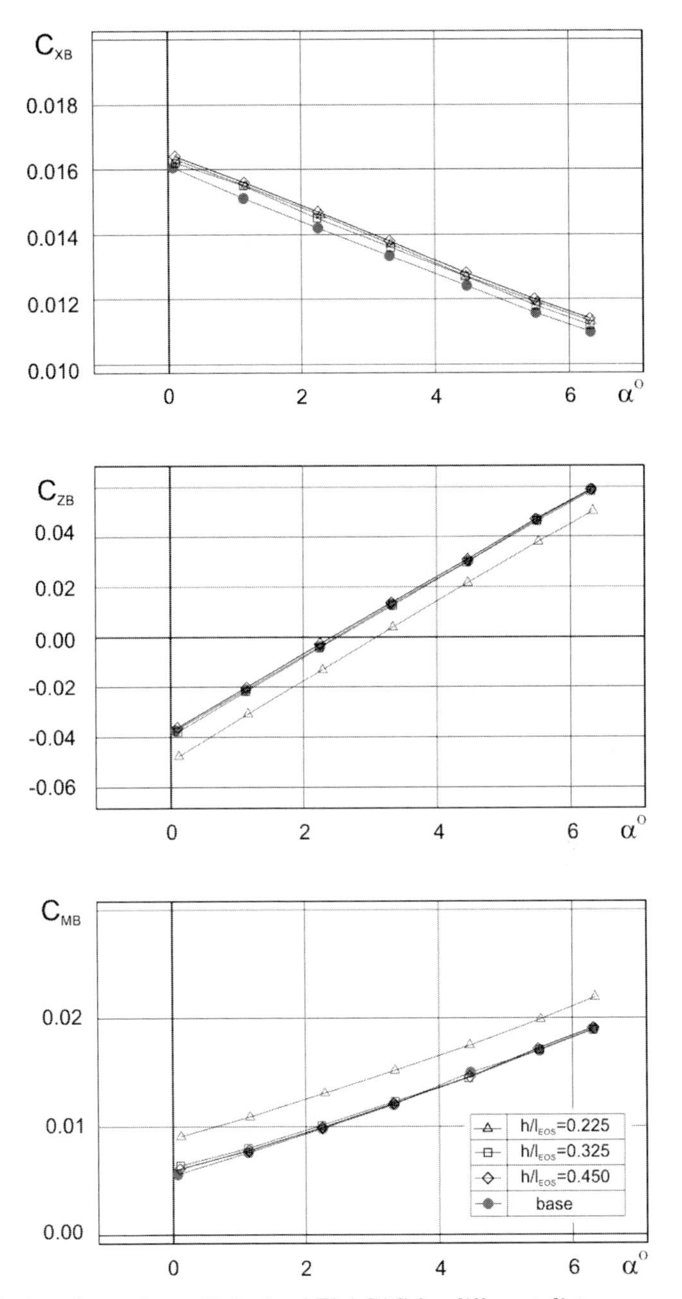


Figure 4.2.21: Aerodynamic coefficients of ELAC1C for different distances.

dynamic coefficients $C_{XB}(a)$, $C_{ZB}(a)$, and $C_{MB}(a)$ of ELAC1C located at distances $\bar{h}$=0.225, 0.325, 0.450, and $\infty$ from the EOS model. The analysis and reconstruction of the schlieren pictures of the flow around the separating stages evidence that the disturbances induced by EOS do not influence the aerodynamics of ELAC1C at a distance larger than $\bar{h}$=0.450. This is true for all the limiting values of the parameters $a$, $\Delta a$, $\Delta\phi$, and $\Delta\beta$.

### 4.2.5 Conclusions

Surface pressure, forces, moments, and flow visualizations were obtained by experimental and numerical investigation. The experiments were performed in the T-313 windtunnel at $Ma_\infty = 4$ and various angles of attack, yaw angles and rolling positions. The numerical solution was achieved by solving the three-dimensional Navier–Stokes equations for laminar flows. The flow around EOS, the interaction of EOS with a flat plate was investigated and the separation of ELAC1C and EOS. The intricate flow field was shown in the schlieren pictures. The impact of the flat plate on the aerodynamics of EOS and the vice versa was presented by schlieren pictures and force measurements, which were performed for various plate/EOS configurations. The separation of ELAC1C and EOS was modeled by varying the distance between the two models, the angle of attack, the yaw angle and the rolling position. At gap sizes larger than $\bar{h}$=0.45 no immediate impact of the EOS flow field on the ELAC1C flow was measured.

### References

1. J.P. Decker, J. Gera, An exploratory study of parallel-stage separation of reusable launch vehicles. NASA TND-4765, 1968.
2. P.T. Bernot, Abort separation study of a shuttle orbiter and external tank at hypersonic speeds. NASA TMX-3212, 1975.
3. M.D. Brodetsky, E.K. Derynov, Aerodynamic interference of two bodies of revolution at supersonic speeds. In: Interference of Complex Spatial Flows, Novosibirsk, 1987, pp. 39–53 (in Russian).
4. M.D. Brodetsky, E.K. Derynov, A.M. Kharitonov, A.E. Lutsky, A.V. Zabrodin, Interference in a supersonic flow around a combination of bodies. Proceedings of the First Europe-US High Speed flow field Database Workshop, Part II, Naples, Italy, November 12–14, 1997.
5. W. Schröder, G. Hartmann, Detailed numerical analysis of hypersonic flows over a two-stage spacecraft, Computational Fluid Dynamics Journal, Vol. 1, No. 4, pp. 375–404, 1993.
6. W. Schröder, G. Hartmann, Analysis of inviscid and viscous hypersonic flows past a two-stage spacecraft, Journal of Spacecrafts and Rockets, Vol. 30, No. 1, pp. 8–13, 1993.
7. E. Krause, German universities research in hypersonics, Fourth International Aerospace Planes Conference, Orlando, Florida, 1992.

8. E. Krause, W. Limberg, A.M. Kharitonov, M.D. Brodetsky, A. Henze, An experimental investigation of the ELAC1 configuration at supersonic speeds. J. Experiments in Fluids 26, p. 4423–4436, 1999.
9. M.D. Brodetsky, S.N. Bruk, A.M. Makhnin, The study of the errors in pressure coefficients at supersonic speeds, A collection of scientific papers, Inst. Theor. Appl. Mech. USSR Acad. Sci., 1977, p. 94–113.
10. Y. Wada, M.S. Liou, A flux splitting scheme with high-resolution and robustness for discontinuities, 32nd Aerospace Sciences Meeting and Exhibit, 1994.
11. B. van Leer, Towards the ultimate conservative difference Schemes, V.A. second-order sequel to Godunov's method, Journal of Computational Physics, 32: 101–136, 1979.
12. G. van Albada, B. van Leer, J. Roberts, A comparative study of computational methods in cosmic gas dynamics, Astron. Astrophysics, Vol. 108, 76–86, 1982.
13. M. Meinke, Numerische Lösung der Navier–Stokes Gleichungen für instationäre Strömungen mit Hilfe der Mehrgittermethode, Dissertation, Aerodynamisches Institut, RWTH Aachen, 1993.
14. T.J. Barth, D.C. Jespersen, The design and application of upwind schemes on unstructured meshes, AIAA-89-0366, 1989.
15. W. Limberg, A. Stromberg, Pressure measurements at supersonic speeds on the research configuration ELAC 1, ZfW April 1993, Vol. 17,2: 82–89.
16. C.-C. Ting, Strömungs- und Wärmeübergangsmessung für das zweistufige Raumtransportsystem ELAC, Dissertation, Aerodynamisches Institut, RWTH Aachen, 2003.
17. A. Kharitonov, M. Brodetsky, L. Vasenyov, N. Adamov, C. Breitsamter, M. Heller, Investigation of aerodynamic characteristics of the models of a two-stage aerospace system during separation, Final report, SFB 255, 2000.

## 4.3 Stage Separation – Aerodynamics and Flow Physics

Christian Breitsamter *, Lei Jiang, and Mochammad Agoes Moelyadi

### Abstract

This article gives an overview about the extensive numerical and experimental investigations conducted on the dominant flow phenomena and aerodynamic characteristics associated with the stage separation problem of two-stage-to-orbit space transport systems. Numerical studies are performed on generic two- and three-dimensional hypersonic bodies as well as on detailed vehicle geometries to critically assess the quality of the developed Euler/Navier-Stokes code. Flow field patterns demonstrating strong interference effects due to incident and reflected shock waves and expansion fans are analyzed in detail as well as corresponding surface pressure distributions and aerodynamic forces and moments of carrier and orbital stages. The influence of viscous effects including laminar and turbulent boundary layers is addressed as well. Very good agree-

* Technische Universität München, Lehrstuhl für Fluidmechanik, Abteilung Aerodynamik, Boltzmannstr. 15, 85748 Garching; chris@flm.mw.tum.de

ment is found comparing numerical and experimental results substantiating the accuracy of the high fidelity flow solver data.

### 4.3.1 Introduction

In the field of space transportation systems research activities were recently intensified by the development of demonstrators, e.g. within the NASA Hyper-X Programme and the X-33 Reusable Launch Vehicle Demonstrator Programme [1, 2]. European and national research projects concentrate on system studies, for example, the Programme on Technologies for Future Space Transportation Systems (TETRA) [3] and the Programme on Selected Systems and Technologies for Future Space Transportation Applications (ASTRA) [4]. Among the various concepts of reusable systems, Two-Stage-To-Orbit (TSTO) systems are strongly investigated showing specific advantages [5, 6].

The here considered TSTO system consists of a carrier stage equipped with an airbreathing engine and an orbital stage with a rocket motor. The carrier stage takes off horizontally carrying the orbital stage on its back. The stage separation takes place within the Mach number range of 4–7 at altitudes of about 30–35 km and, hence, under high dynamic pressures [7]. Optimum aerodynamic and trimming conditions are to be provided for the entire configuration to prepare save separation [8]. The critical phase of the separation manoeuvre begins with releasing the orbital stage, where the distance between the two stages is small. In general, the hypersonic flow around separating winged stages is a complex three-dimensional unsteady gas dynamic problem. Extensive numerical and experimental simulations have demonstrated that during the separation phase complex interactions of the incident and reflected shock waves and expansion fans with each other and with boundary layers occur [9–11]. The investigations presented herein are therefore dedicated to accurately predict and analyze the dominant flow field effects and the aerodynamic characteristics of the carrier and orbital stage associated with the separation manoeuvre.

### 4.3.2 Methodology and Vehicle Geometries

During the five phases (1989–2003) of the Colloborative Research Centre 255 the main objective of the research work reported here was the development of a high fidelity flow solver resulting in a validated simulation tool for stage separation aerodynamics. The road map of Fig. 4.3.1 highlights the stepwise improvement of the flow solver physical modelling and its application capabilities. Consequently, the complexity of vehicle geometry and associated test parameters increases as follows: First, a generic orbital stage consisting of a two- and three-dimensional hypersonic body, respectively, was studied introducing a flat

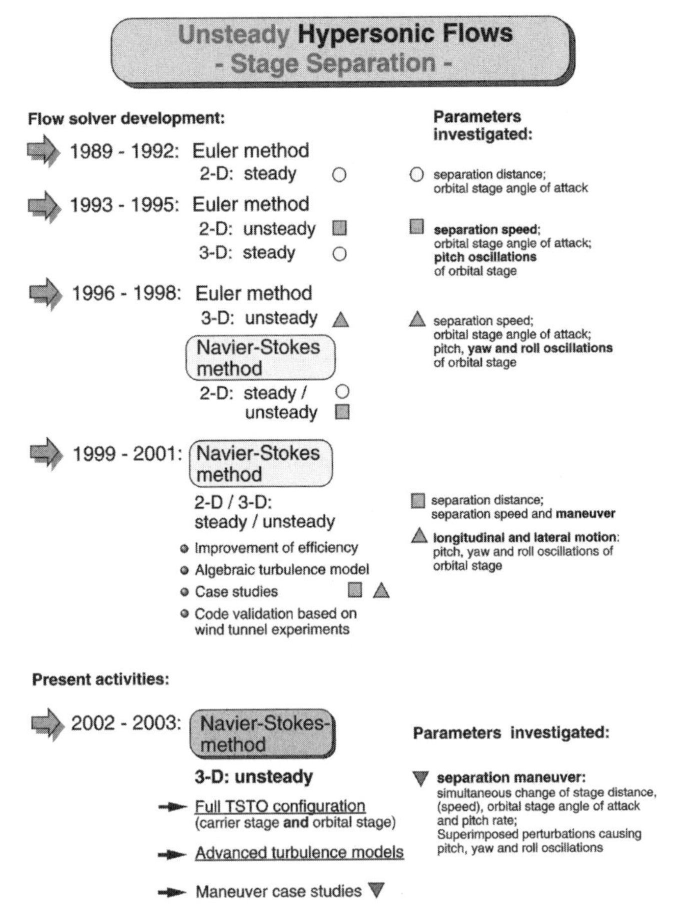


Figure 4.3.1: Historical overview of flow solver development and applications.

plate to simulate carrier stage interference (Fig. 4.3.2). On the one hand the generic configuration was employed to test the flow solver on the stage separation problem requiring reasonable computational costs. On the other hand the results of the generic geometries are ideally suited for assessing numerical accuracy and enhancing the performance of the code. Further, the influence of steady and unsteady separation parameters, the latter with respect to basic harmonic motions, were analyzed. In the next phase the generic orbital stage was replaced by the ELAC Orbital Stage EOS representing a detailed wing-body configuration [12] (Fig. 4.3.3). In the final working phase the flat plate is exchanged for the ELliptical Aerodynamic Configuration ELAC thus establishing a fully realistic TSTO configuration [12] (Fig. 4.3.4).

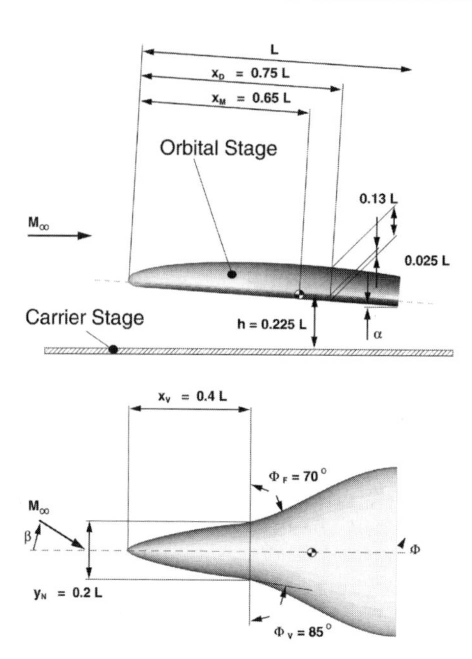


Figure 4.3.2: Geometry of the generic orbital stage.

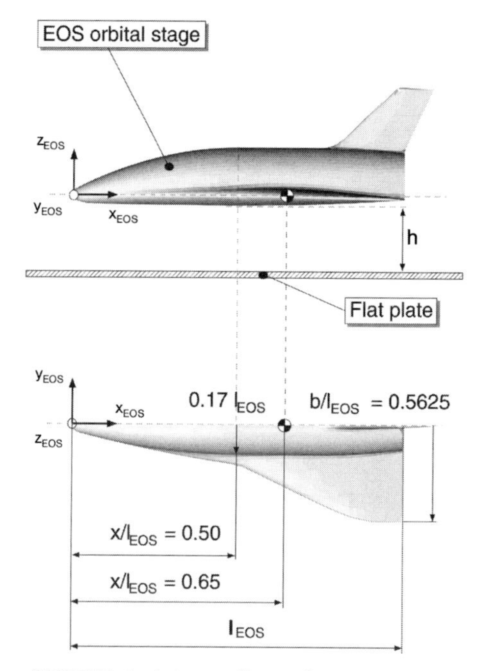


Figure 4.3.3: Geometry of EOS/flat plate configuration.

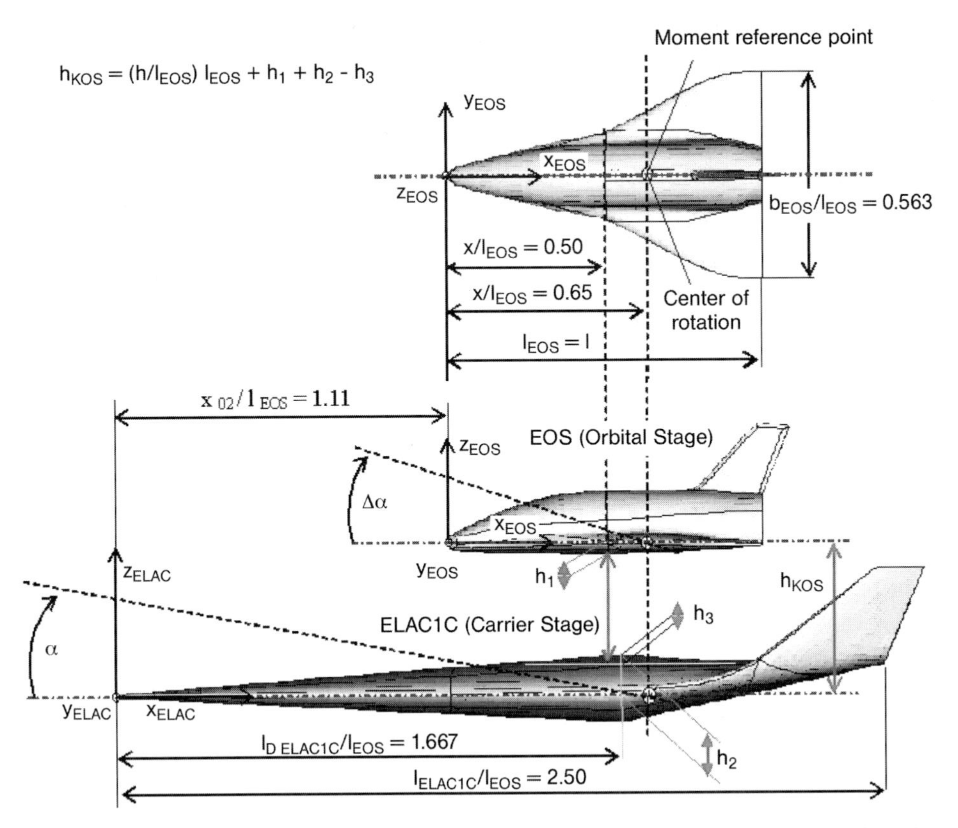


Figure 4.3.4: Geometry and geometric reference values of carrier stage ELAC1C and orbital stage EOS.

### 4.3.3 Numerical Simulation

The numerical method used is based on a finite-volume approximation to the integral form of the unsteady Euler and Navier-Stokes equations, respectively.

#### 4.3.3.1 Flow Solver

The time-dependent flow vector for multi-dimensional problems is calculated by using the Strang-type of fractional step. The convective fluxes are calculated by the modified AUSM method (Advection Upstream Splitting Method) [13]. This method was employed as it has proved to be very stable for hypersonic applications. The modified AUSM scheme represents a hybrid approach between the van Leer flux vector splitting scheme and the original AUSM scheme [14, 15]. The AUSM scheme is based on the idea of regarding the convection and acoustic waves as physically distinct processes and thus defining the fluxes as a sum of the convective and pressure terms [15]. In order to guarantee high order accuracy in the spatial domain, the left and right states at the cell interfaces are obtained with the MUSCL approach [16]. The van Albada limiter function is chosen to extrapolate the state values and thus provides high-order fluxes in smooth regions. At discontinuities the function switches to first order accuracy to ensure optimal shock-capturing features [17]. Regarding the Navier-Stokes code the viscous fluxes are calculated following the method of Chakravarthy [18]. The algebraic Baldwin-Lomax turbulence model [19] with the modification of Degani and Schiff [20] is introduced to study the influence of turbulent boundary layers. The time integration is performed using either an explicit forth order Runge-Kutta scheme or the implicit LU-SSOR (lower upper symmetric successive over-relaxation) scheme [21]. Especially for unsteady simulations, the implicit method in conjunction with the dual time stepping approach leads to significantly reduced calculation times.

Impermeable wall characteristic boundary conditions are applied to evaluate the primitive variables. At the farfield boundaries, the flow variables are set to their freestream values for hypersonic inflow conditions, whereas for outflow conditions, the flow variables are extrapolated by employing the solution of the computational domain. The accuracy of the Euler and Navier-Stokes code has been proven by evaluating several standard test cases for subsonic, supersonic and hypersonic flow [7, 22]. In addition, grid convergence studies are carried out for steady cases [10, 22]. The comparison of numerical and experimental results substantiates that the dominant flow features are represented correctly by the grid resolutions employed.

#### 4.3.3.2 Grid Generation

The computational meshes are based on a structured multi-block approach [7, 22]. The block topology construction allows to set up several boundary condition types at any block face. This segmentation provides larger vector lengths leading to higher calculation efficiency within both the grid generator and flow solver. The use of a Poisson algorithm results in smoothing the initial grid in order to achieve small cell deformations and continuous cell growth. The connections between adjacent blocks are organized by mother-child relations where the grid points located at block connections are allowed to move during the iteration process. The source terms are determined at the solid body by inverting the Poisson equation. Mirror points with a fixed distance from the solid body wall are employed to calculate the derivatives at block boundaries. The convergence criterion for sufficient smoothness is fulfilled if the change in the source strengths is below $10^{-5}$. The Laplace equation is used to spread the source terms into the calculation domain. The elliptic smoothing applied for each block is also used over block boundaries. The mechanism described presents a stable and flexible tool avoiding cell singularities or overlaps. The distribution of mesh points is performed to represent adequately the geometrical shape of the body and to concentrate them in regimes with high gradients of both the body contour and the flow variables.

Unsteady calculations are performed by a dynamic adaptation of the grid to the actual body position for each time step. Due to the motion induced mesh deformations near the body, it is necessary to smooth the grid again using the Poisson algorithm. The costs for the elliptic smoothing process with respect to flow solver costs are typically about 3%–5%. The unsteady transformation of the Euler and Navier-Stokes equations, respectively, takes into account the velocity of the mesh as well as the deformation of the cells.

## 4.3.4 Experimental Simulation

### 4.3.4.1 Models and Facility

Models of 1/150-scale are used based on the hypersonic ELAC1C/EOS vehicles (cf. Fig. 4.3.4). ELAC1C represents the carrier (lower) stage while EOS is the orbital (upper) stage [12]. ELAC1C has a delta-wing planform with a leading-edge sweep of 75°, an elliptical forebody and afterbody, the latter with integrated winglets. The forebody elliptical cross sections have a major-to-minor axis ratio of 4 (upper side) and 6 (lower side), respectively. The junction between forebody and afterbody is at two thirds of the total axial length. There is also a bay (cavity) for embedding the upper stage EOS in the carrier when testing the entire TSTO system. The EOS vehicle is a wing-body configuration with a centre-line fin. The configuration consists of an elliptical nose and forebody, a centre body with half cylindrical upper side, wing strakes of elliptical cross sections, and a highly swept main wing and fin composed of standard NACA profiles.

The tests were carried out in the T-313 wind tunnel at the Institute of Theoretical and Applied Mechanics (ITAM) of the Russian Academy of Sciences, Siberian Branch (SB RAS) in Novosibirsk, Russia [12]. The blowdown wind tunnel operates on dried air, which is compressed in large gas holders. The facility is composed of input pipelines with pressure adapter, plenum chamber, nozzle block, pressure chamber with test section, supersonic diffuser, two ejectors, and output pipeline, from which the air is fed to a noise damping chamber before it is released into the atmosphere. The size of the rectangular test section is 0.6 m×0.6 m×2 m. The test section Mach number can be varied within $M_\infty$ = 2–6 by a plane contoured nozzle with inserts of different throat sizes. A rather wide range of Reynolds numbers and dynamic pressures can be adjusted using ejectors and maximum plenum chamber pressure of about 1.18 MPa. The measurement time reaches typically 10 min. The degree of non-uniformity of the velocity field in the test section is quantified by the relative root-mean-square (rms) deviations of the Mach number. At standard model positions the deviations are less than 0.5% for $M_\infty$ = 2–4 and less than 1% for $M_\infty$ = 5–6.

a)

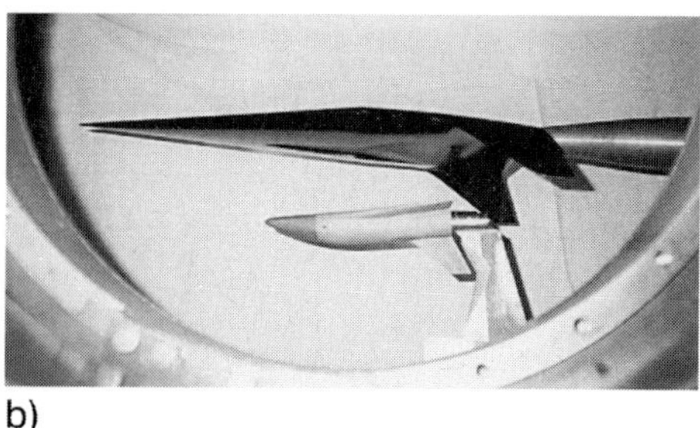
b)

Figure 4.3.5: Wind tunnel models mounted in the ITAM T-313 wind tunnel test section; (a) EOS/flat plate configuration (flow direction from right), (b) ELAC1C/EOS configuration (flow direction from left).

The two-stage separation is simulated in a quasi steady manner by mounting the upper stage at different vertical positions $h/l_{EOS}$ as well as at different angles of attack $\Delta\alpha$, sideslip $\Delta\beta$, and bank $\Delta\Phi$ relative to the lower stage. The separation test bench is described in detail in [12]. Fig. 4.3.5 depicts the models of the EOS/flat plate configuration and the full TSTO ELAC1C/EOS configuration mounted in the T-313 wind tunnel test section.

#### 4.3.4.2 Measurement Technique and Test Programme

An automatic four-component mechanical balance is employed to measure the loads acting on the carrier stage model ELAC1C. This balance is fitted with an arm system to decompose the total forces and moments of the balance coordinate system into the aerodynamic components lift and drag force and pitching and rolling moment. The instrumental error of the mechanical balance is less than 0.1% of the highest load [12]. An internal six-component strain-gauge balance is used to acquire the loads acting on the orbital stage model EOS. The construction of this balance is made in form of a tail sting measuring all components of aerodynamic forces and moments (Fig. 4.3.5 a). The instrumental error of the strain-gauge balance is less than 0.05% of the highest load [12]. The flow patterns around the separating models are visualized using a standard interfering shadowgraph to observe and register the density gradients of the air flow in the plane of the velocity vector. A specific device connects a CCD camera to the shadowgraph instrument through which digitized images of the Schlieren pictures are sampled by using frame grappers installed in a PC. Flow patterns are registered at each wind tunnel run for angles of attack of the entire configuration of $\alpha=0°$, $3°$, and $6°$.

The test conditions are as follows: The freestream Mach number is hold constant at a value of $M_\infty=4.04$. The corresponding Reynolds number based on the total axial length of the orbital stage is then $Re_{lEOS}=9.6\times10^6$. The angles of attack of the entire configuration are adjusted for the load measurements at $\alpha=0°$, $1°$, $2°$, $3°$, $4°$, $5°$, and $5.7°$. Simulating the stage separation, the orbital stage EOS is placed at three vertical positions relative to the flat plate or carrier stage ELAC1C, namely at $h/L_{EOS}=0.225$, 0.325, and 0.450 (Fig. 4.3.5). At each vertical station the angle of attack of EOS relative to ELAC1C is varied discretely with values of $\Delta\alpha=0°$, $2°$, and $5°$. In addition, effects of the lateral motion are included changing discretely the EOS sideslip and bank angle relative to the ELAC1C carrier stage using values of $\Delta\beta$ and $\Delta\Phi$ of $0°$, $2°$, and $4°$, respectively.

## 4.3.5 Steady State Flow

### 4.3.5.1 Dominant Flow Phenomena

The flow physics at stage separation flow conditions are discussed using results obtained for the two- (2D) and three-dimensional (3D) orbital stage generic configuration interfering with a flat plate simulating the carrier influence. The computational domain of the 2D case employs 6 blocks arranged in a hybrid C-H topology [11] (Fig. 4.3.6). The grid consists of 9000 cells for both Euler- and Navier-Stokes calculations. For the inviscid case, the distance of the first off-body grid line is fixed at $3\times10^{-3}$ L. For the viscous case, this distance is reduced to $3\times10^{-4}$ L to ensure an appropriate spatial resolution in boundary layer regions. The 3D-case is set up by a 9 block topology with 140 000 cells for the standard mesh and 900 000 cells for a refined mesh used for grid convergence studies [10]. The results presented herein correspond to a freestream Mach number of $M_\infty=6.8$, an incidence of the orbital stage of $\Delta\alpha=0°$, and a distance between the orbital stage and the carrier stage of $h/L=0.225$. A steady state solution is reached if the residual of the density does not exceed $10^{-6}$.

#### *4.3.5.1.1 Inviscid Case – 2D and 3D Simulations*

Contours of constant Mach number and constant pressure are shown for the 2D case in Fig. 4.3.7 and for the symmetry plane of the 3D case in Fig. 4.3.8. The main flow field features are as follows: A bow shock ahead of the blunt nose of the orbital stage is formed. The bow shock is closely located to the nose, where, in the vicinity of the stagnation point, a local subsonic region is present (Fig. 4.3.8b). The lower part of the bow shock hits the surface of the carrier stage, where it is reflected. For the 2D case, the reflected shock wave strikes again the lower side of the orbital stage. This shock impingement at the tail of the orbital stage is not observed in the 3D solution because the 3D flow results in a significantly smaller bow shock angle shifting the point of shock reflection on the carrier stage downstream. In the wake zone of the orbital stage, a recom-

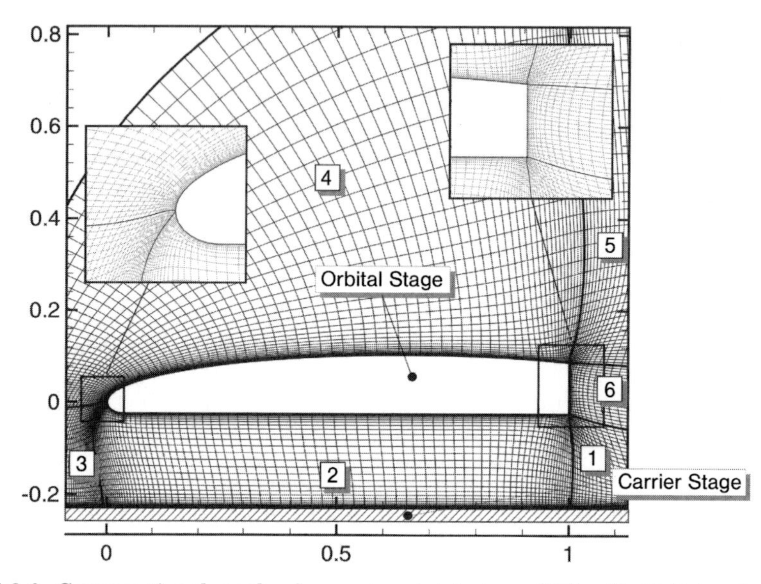


Figure 4.3.6: Computational mesh of generic orbital stage (2D) with C-H topology using 9000 cells arranged in 6 blocks (blocks are marked by numbers).

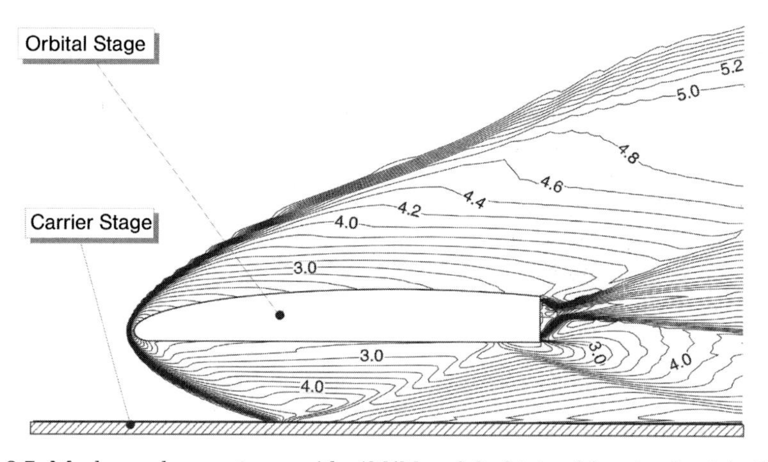


Figure 4.3.7: Mach number contours with $\Delta M/M_{\infty}=0.2$ obtained by steady state 2D Euler calculations for the case of the generic orbital stage with carrier stage interference at $M_{\infty}=6.8$, $\Delta\alpha=0°$, and $h/L=0.225$.

pression system is formed interfering with the shock wave reflected from the orbital vehicle. With respect to the orbital stage angle of attack of $\Delta\alpha=0°$ a negative lift force is obtained indicating an overall lower pressure for the lower side and a higher pressure for the upper side. Considering only the rear part of the vehicle, the pressure level increases at the lower side due to the impingement of the reflected bow shock wave. Hence, the wake region is deflected up-

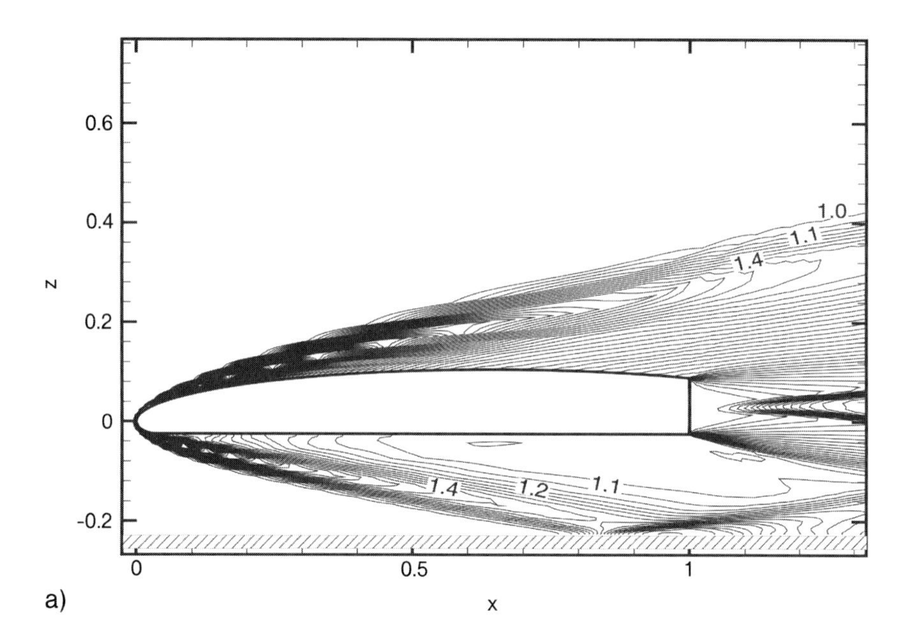


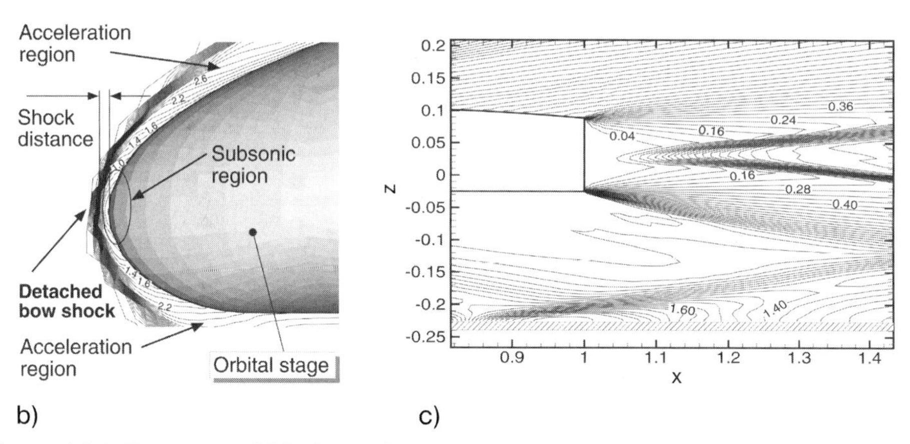


Figure 4.3.8: Pressure and Mach number contours obtained by steady state 3D Euler calculations using a refined mesh with 9 blocks and 900 000 cells for the case of the generic orbital stage with carrier stage interference at $M_\infty=6.8$, $\Delta\alpha=0°$, and $h/L=0.225$; (a) pressure contours in symmetry plane of the orbital stage with $\Delta p/p_\infty=0.05$, (b) Mach number contours in nose region of the orbital stage with $\Delta M/M=0.2$, (c) pressure contours in the wake region with $\Delta p/p=0.04$.

wards. As the 3D case exhibits a smaller shock reflection angle there is no shock impingement in the rear part of the orbital stage and no wake zone deflection is visible.

*4.3.5.1.2 Viscous Effects – Laminar and Turbulent Flow*

The Navier-Stokes calculations show basically the same flow field features as known already from the Euler results. More particularly, the 2D laminar case indicates some important viscous effects (Fig. 4.3.9 a). The reflected bow shock wave affecting the lower rear part of the orbital stage leads to a strong increase in wall pressure which causes a relatively large region of separated flow. Near

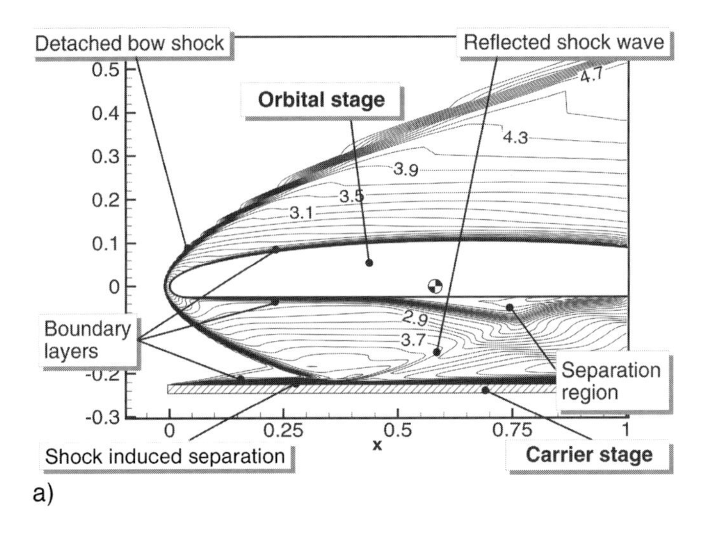


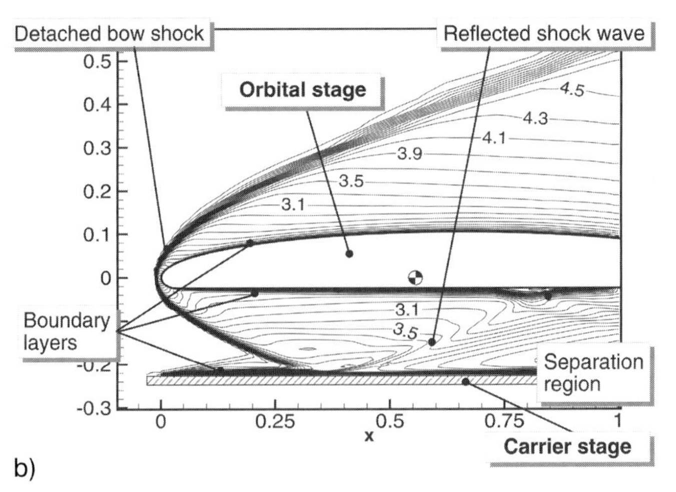


Figure 4.3.9: Mach number contours with $\Delta M/M_\infty = 0.2$ obtained by steady state 2D Navier–Stokes calculations for the laminar and turbulent flow case of the generic orbital stage with carrier stage interference at $M_\infty = 6.8$, $Re = 10^6$, $\Delta\alpha = 0°$, and $h/L = 0.225$; (a) laminar flow, (b) turbulent flow.

the base flow acceleration due to the expansion fan emanating from the edge forces the flow to re-attach. A shock induced separation is also found on the flat plate due to the bow shock impingement. Regarding turbulent boundary layers the region of separated flow on the orbital stage lower side is markedly reduced in size and it is slightly shifted downstream (Fig. 4.3.9 b). The separation bubble on the flat plate vanishes completely. The macroscopic fluctuations of the turbulent boundary layer result in a higher kinetic energy level to sustain the adverse pressure gradient over a longer distance before flow separation occurs.

### 4.3.5.2 Comparison of Experimental and Numerical Results

To quantify the accuracy of the flow solver when applied to complex geometries Navier-Stokes calculations are performed for the EOS/flat plate configuration (Fig. 4.3.3). The numerical results are compared with the data of the experiments explained in section 4.3.5. The numerical simulations are performed on a mesh consisting of 9 blocks with about 795 000 cells in total (Fig. 4.3.10). Figure 4.3.11 depicts the flow field patterns obtained for a freestream Mach number of $M_\infty=4.04$, a Reynolds number of $Re_{l\mathrm{EOS}}=9.6\times10^6$, an orbital stage angle of attack of $\Delta\alpha=0°$, and a distance between the orbital stage and the flat plate of $h/l_{\mathrm{EOS}}=0.1812$. The comparison between the Schlieren picture of the

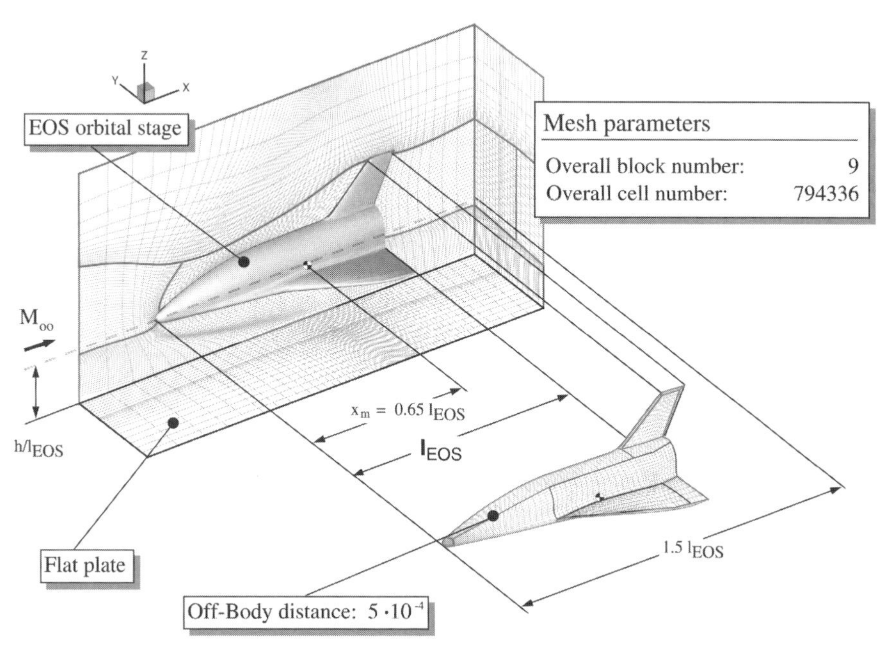


Figure 4.3.10: Computational mesh of EOS/flat plate configuration with about 795 000 cells arranged in 9 blocks.

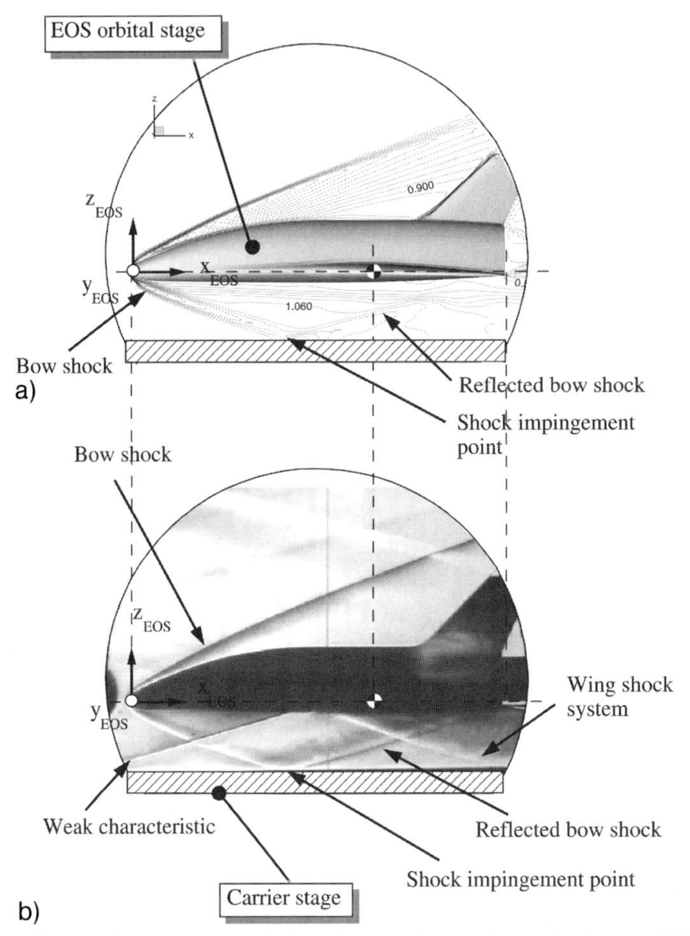


Figure 4.3.11: Comparison of numerical and experimental results for the EOS/flat plate configuration at $M_\infty = 4.04$, $Re_{lEOS} = 9.6 \times 10^6$, $\Delta\alpha = 0°$, and $h/l_{EOS} = 0.1812$; (a) Schlieren picture visualizing density gradients, (b) density contours with $\Delta\rho/\rho_\infty = 0.08$ in the symmetry plane achieved by Navier–Stokes calculations.

experiments and the calculated density distribution of the symmetry plane shows a very good agreement for the bow shock angle and shock wave propagation, the point of shock reflection on the flat plate and the shock impingement location on the orbital stage rear part. An excellent agreement between numerical and experimental results can also be found for the wall pressures taken on the flat plate in the orbital stage symmetry plane (Fig. 4.3.12). The results shown give strong evidence that the developed Euler/Navier-Stokes code predicts the dominant flow phenomena on an adequate level of accuracy to obtain precise aerodynamic data.

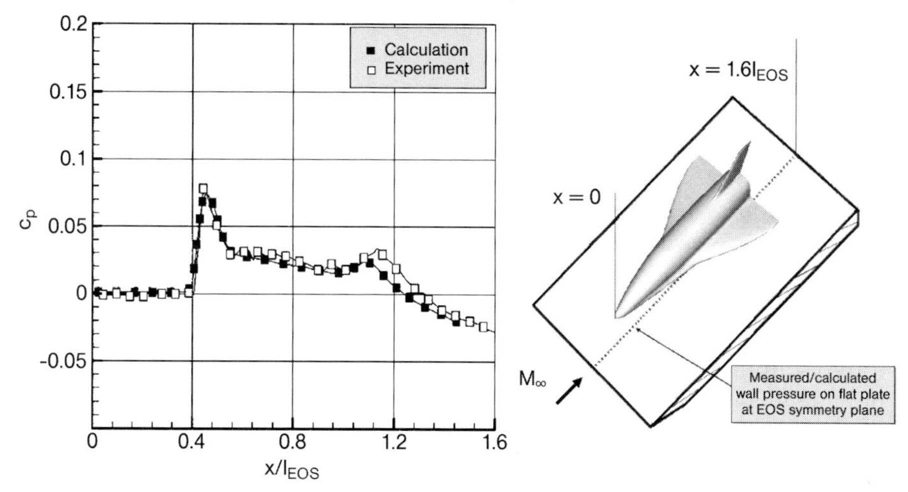


Figure 4.3.12: Comparison of numerically and experimentally determined pressure coefficient $c_p$ of the wall pressure taken on the flat plate in the symmetry plane for the EOS/ flat plate configuration at $M_\infty = 4.04$, $Re_{l\mathrm{EOS}} = 9.6 \times 10^6$, $\Delta\alpha = 0°$, and $h/l_{\mathrm{EOS}} = 0.1812$.

## 4.3.6 Unsteady Aerodynamics

### 4.3.6.1 Longitudinal Motion – Dynamic Separation

In general, the vertical distance between the stages, the orbital stage angle of attack as well as the pitch rate and separation speed changes continuously during the separation manoeuvre. Focusing on systematic investigations, basic parts of such a manoeuvre are studied separately in form of translatory and rotatory motions including also harmonic oscillations. A typical example is the translatory case explained below.

Starting from an initial position at $h/L = 0.225$ ($M_\infty = 6.8$, $\Delta\alpha = 0°$) the orbital vehicle performs a translatory motion in vertical direction. For three different vertical positions, i.e. $h/L = 0.155$, $h/L = 0.225$, and $h/L = 0.295$, Fig. 4.3.13 shows both Mach number contours and orbital stage surface pressure distributions achieved by Navier-Stokes calculations. Considering the closest distance between the orbital stage and the carrier stage ($h/L = 0.155$) the bow shock reflected from the carrier stage hits the orbital stage at $x/L \approx 0.5$ (Fig. 4.3.13 a). A small region of separated flow is visible resulting in a nearly constant pressure in this area on the lower side of the orbital stage. An increase of the separation distance shifts the shock impingement point in streamwise direction ($x/L \approx 0.65$ for $h/L = 0.225$ (Fig. 4.3.13 b)). At a distance of $h/L = 0.295$ no shock impingement occurs on the lower side of the orbital stage (Fig. 4.3.13 c). The effect of unsteadiness is clearly obvious for the separation distance at $h/L = 0.225$ comparing steady and unsteady results. In comparison to the steady case, with the flow separation at about $x/L \approx 0.75$ (Fig. 4.3.9 a), this region is located much

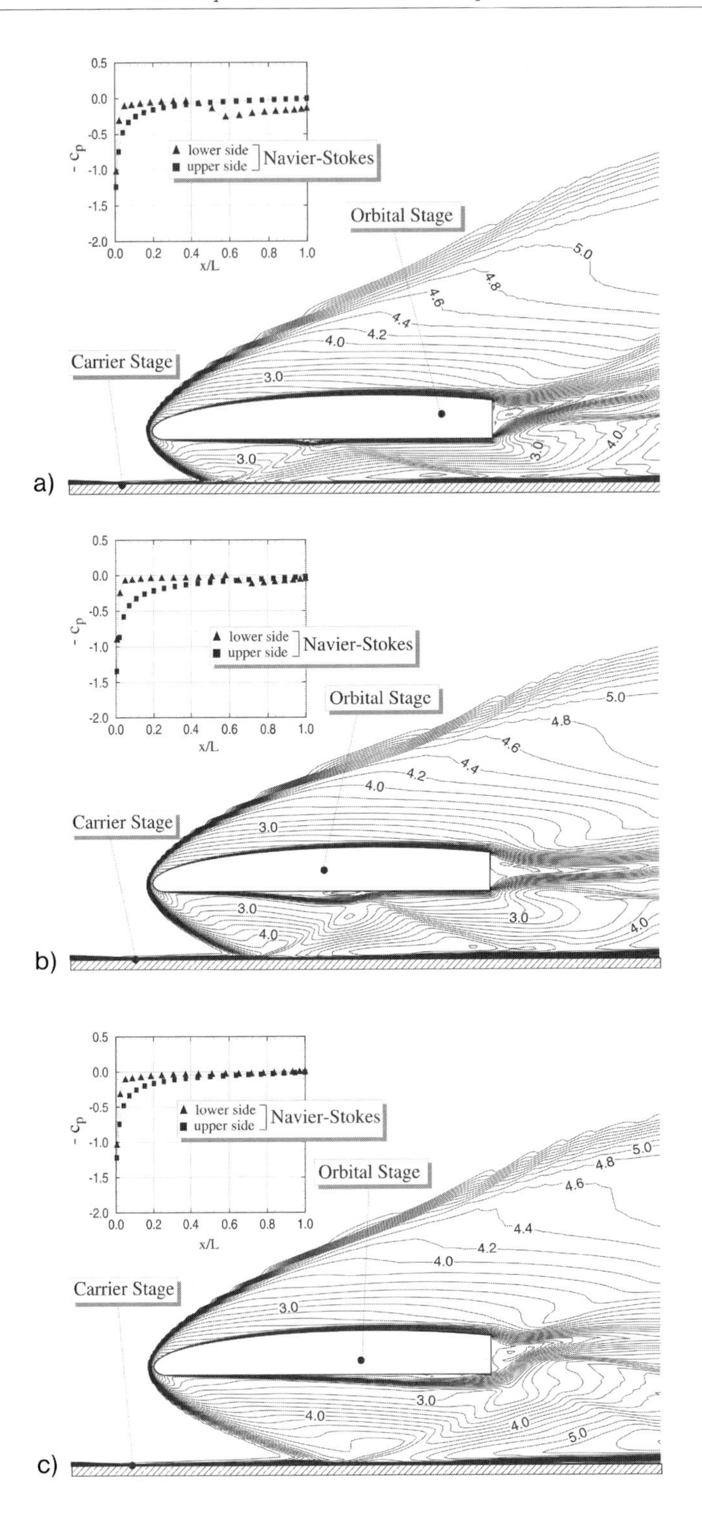
a)
b)
c)
Carrier Stage
Orbital Stage
lower side
upper side
Navier-Stokes
- $c_p$
x/L
0.5
0.0
-0.5
-1.0
-1.5
-2.0
0.2
0.4
0.6
0.8
1.0
3.0
4.0
4.2
4.4
4.6
4.8
5.0

more upstream considering the moving body (Fig. 4.3.13 b). Also, during lift-off the region of separated flow is smaller. The shock wave reflected from the carrier stage hits the lower side of the orbital stage in the middle part resulting in a discontinuous pressure increase. Consequently, in comparison to the steady state, the reflected shock wave system is shifted upstream in the unsteady case. This shift is caused by the motion induced vertical velocity which results in a decrease of the pitch angle. In addition, the displacement of the wake upwards is reduced in the unsteady case.

The orbital stage lift and pitch moment coefficient resulting from the Navier-Stokes calculations are shown in Fig. 4.3.14. Steady as well as unsteady results related to a low and high value of the reduced frequency $k$, describing the cycle time of the translatory harmonic motion, are presented. Starting from the initial position, the orbital stage lifts off and reaches the upper turning point and then moves downwards. Hereafter, the orbital vehicle passes the closest distance to the carrier stage, then it reaches the initial position again. During the upward movement a zone of separated flow on the lower side of the orbital stage is observed which is shifted from $x/L \approx 0.5$ to $x/L \approx 0.65$ (Fig. 4.3.13). Moving downwards, this zone decreases and finally disappears at $h/L \approx 0.225$. During this phase no shock wave reflections occur on the lower side of the orbital stage. At $h/L \approx 0.190$, a sudden increase of the lift and decrease of the moment coefficient can be detected. This significant change in lift and moment coefficients is mainly caused by the complex shock wave system reflected from the carrier stage which is formed during the motion. The results for the fast motion ($k = 1.0$) exhibit significant shifts in amplitude and phase while the slow motion values ($k = 0.05$) are close to the steady state values. As the lift decreases with increasing distance the orbital stage angle of attack has to be raised to perform the separation. Consequently, a practical separation manoeuvre forces a simultaneous change of distance and orbital stage incidence determined by separation speed and pitch rate. Within a (physical) time period of about 5 s the distance may raise from $h/L = 0.05$ to $h/L = 0.85$ at a nearly linearly increasing vertical speed up to $w/U_\infty \approx 0.005$ in conjunction with an increase of orbital stage angle of attack from $\Delta\alpha \approx 3^\circ$ to $\Delta\alpha \approx 12^\circ$ at a linearly increasing pitch rate up to 3.5 °/s.

#### 4.3.6.2 Lateral Motion – Disturbance Effects

Beside the pure longitudinal motion extensive numerical simulations were performed for lateral motions including harmonic oscillations in yaw and roll [8, 10]. Such motions can be evoked by inhomogeneous freestream conditions

Figure 4.3.13: Mach number contours with $\Delta M/M_\infty = 0.2$ obtained by unsteady 2D Navier–Stokes calculations (laminar flow) for the case of a translatory separation manoeuvre (harmonic motion with amplitude of $\Delta z/L = 0.07$ and reduced frequency of $k = 1.0$) of the generic orbital stage with carrier stage interference at $M_\infty = 6.8$, $Re = 10^6$, and $\Delta\alpha = 0^\circ$; (a) $h/L = 0.155$, (b) $h/L = 0.225$, (c) $h/L = 0.295$.

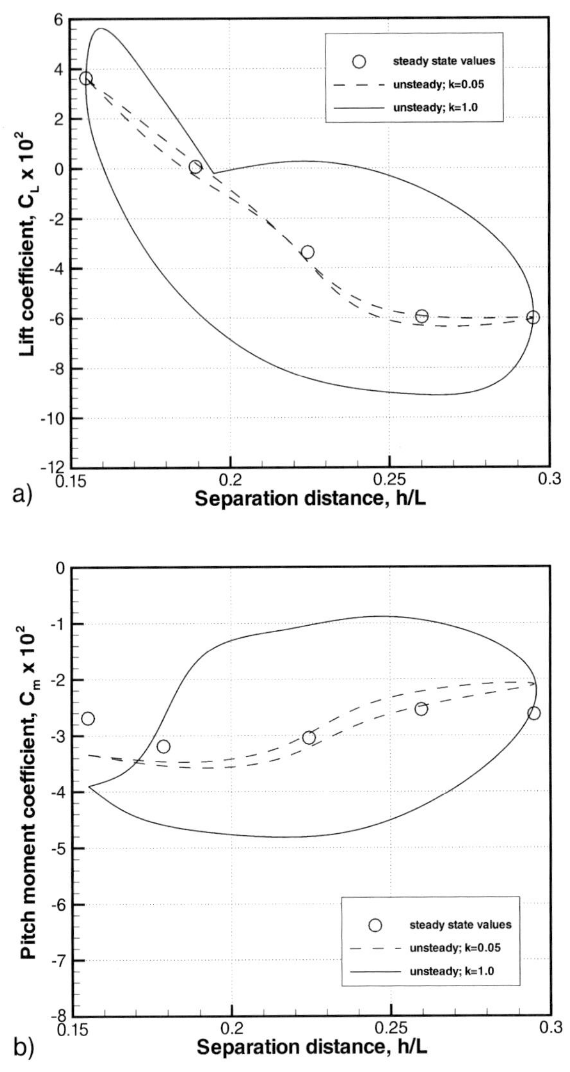


Figure 4.3.14: Aerodynamic coefficients obtained by unsteady 2D Navier–Stokes calculations (laminar flow) for the case of a translatory separation manoeuvre (harmonic motion with amplitude of $\Delta z/L=0.07$ and reduced frequency of $k=1.0$ and $k=0.05$) of the generic orbital stage with carrier stage interference at $M_\infty=6.8$, $Re=10^6$, and $\Delta\alpha=0°$; (a) Lift coefficient $C_L$ $(h/L)$, (b) Pitch moment coefficient $C_m$ $(h/L)$.

and/or aerodynamics and control disturbances occurring during stage separation which may severely endanger a successful separation manoeuvre. The calculated aerodynamic data are therefore used to evaluate dynamic stability derivatives needed for flight mechanics and flight control simulations.

## 4.3.7 Detailed Two-Stage-to-Orbit Configuration

The most complex case treated both experimentally and numerically is the ELAC1C/EOS configuration, cf. Fig. 4.3.4 and 4.3.5 b. The dominant flow field phenomena occurring typically at strong stage interference are visualized by the Schlieren picture of Fig. 4.3.15. Main effects are marked and numbered from 1 to 8 [12]. Primary factors of stage interference influencing the orbital stage aerodynamic characteristics are as follows: (i) the carrier stage bow shock and the disturbed flow behind it (1), (ii) the reflected orbital stage bow shock striking again on the orbital stage tail part at minimum vertical distance (6), (iii) the expansion fan at the inflection of the surface in the lower stage cross section of maximum thickness affecting the orbital stage tail part at small vertical distances (4), and (iv), the expansion fan due to flow turning at the front of the orbital transport cavity (2).

To further analyze the flow fields, calculations for this very complex geometry applying Euler and Navier-Stokes simulations are conducted comparing also numerical and experimental results. The computational mesh used is

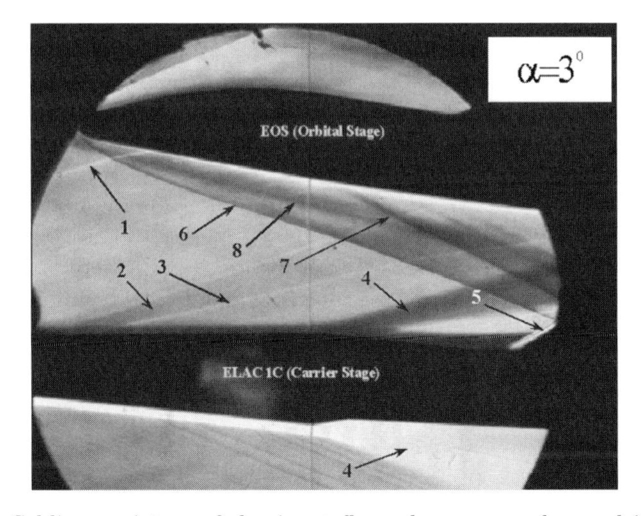


Figure 4.3.15: Schlieren picture of dominant flow phenomena observed in wind tunnel tests for the ELAC1C/EOS configuration at $M_\infty = 4.04$, $Re_{l\mathrm{EOS}} = 9.6 \times 10^6$, $\alpha = 3°$, $\Delta\alpha = 5°$, and $h/l_{\mathrm{EOS}} = 0.325$;

1. bow shock wave of ELAC1C,
2. expansion wave caused by flow turning to the cavity for EOS,
3. shock wave caused by flow turning along the cavity for EOS,
4. expansion fan caused by flow turning in the cross section of maximum thickness of ELAC1C,
5. shock wave from the winglet (fin) of ELAC1C,
6. bow shock wave of EOS,
7. shock wave of the inflection of the leading edges of EOS wings,
8. weak characteristic from the nozzle/test section junction (no effect on model interaction).

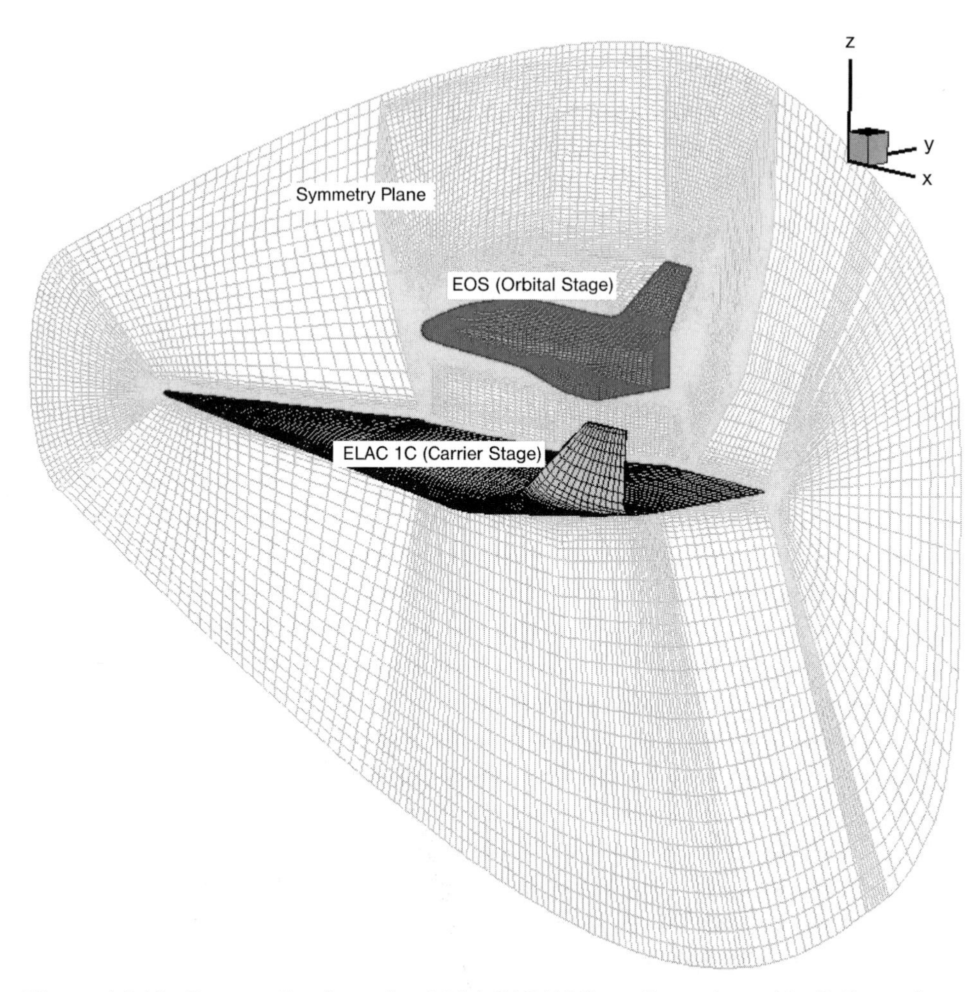


Figure 4.3.16: Computational mesh of ELAC1C/EOS configuration with O-O topology using about 1.1 million cells (half configuration) arranged in 45 blocks.

shown for a view of the symmetry plane in Fig. 4.3.16. The mesh is composed of 45 blocks arranged in O-O topology with about 1.1 million cells describing the half configuration. The comparison of results of numerical simulations and wind tunnel experiments indicates a very good agreement analyzing the locations, propagation and interaction of shock waves and expansion fans (Fig. 4.3.17 and Fig. 4.3.18). It can be summarized that the numerical simulations

Figure 4.3.17: Comparison of experimental and numerical results for the ELAC1C/EOS configuration at $M_\infty=4.04$, $\alpha=0.0°$, $\Delta\alpha=5.0°$, and $h/l_{EOS}=0.225$; (a) Schlieren picture visualizing density gradients, (b) density contours in the symmetry plane achieved by Euler calculations.

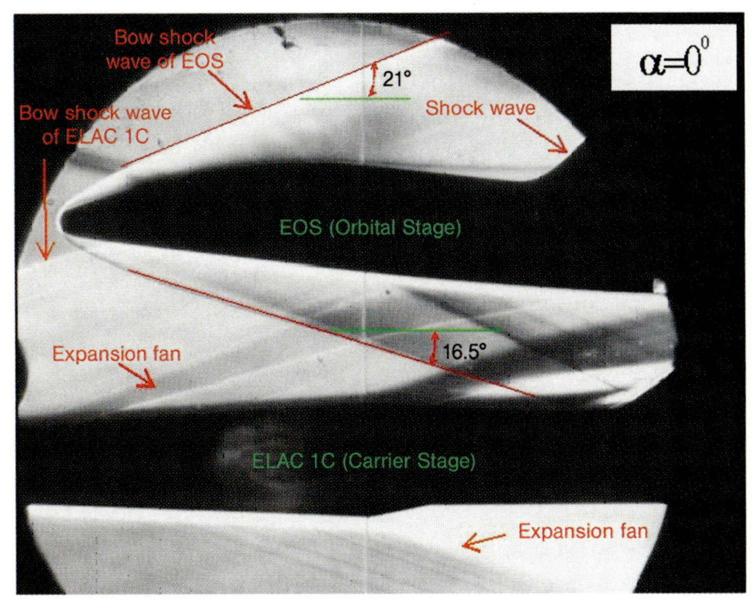

Bow shock
wave of EOS
α=0°
21°
Shock wave
Bow shock wave
of ELAC 1C
EOS (Orbital Stage)
Expansion fan
16.5°
ELAC 1C (Carrier Stage)
Expansion fan


a)

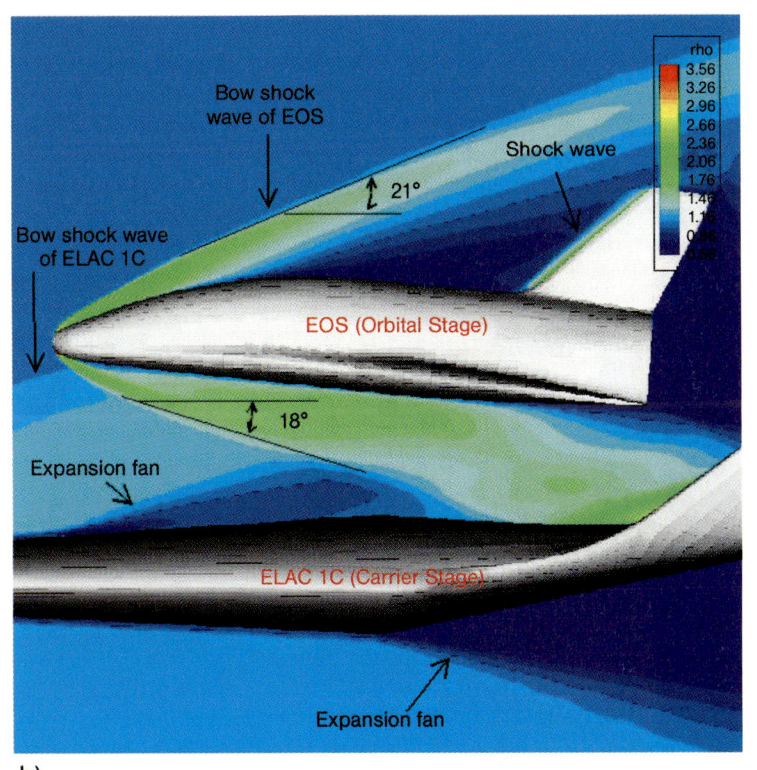

rho
3.56
3.26
2.96
2.66
2.36
2.06
1.76
Bow shock
wave of EOS
Shock wave
21°
Bow shock wave
of ELAC 1C
EOS (Orbital Stage)
18°
Expansion fan
ELAC 1C (Carrier Stage)
Expansion fan


b)

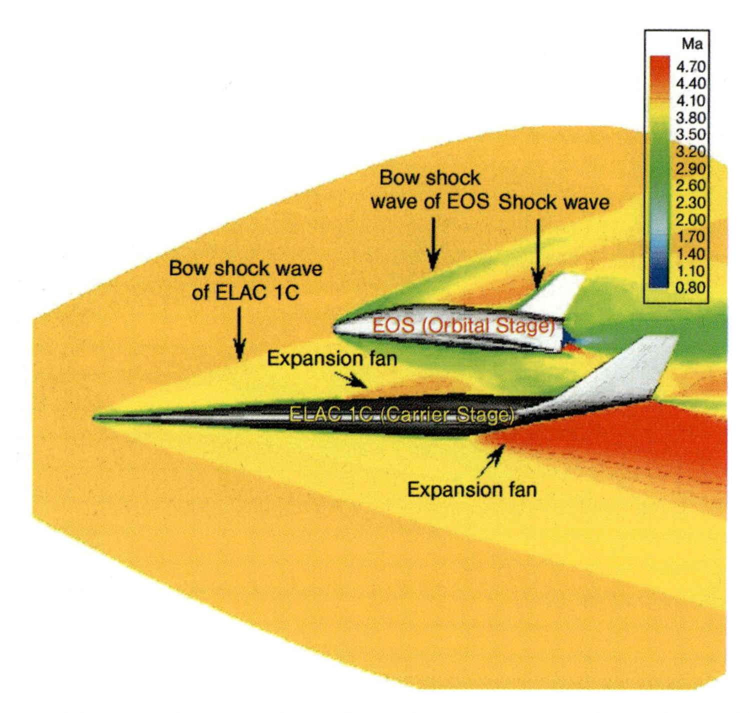


Figure 4.3.18: Mach number contours with $\Delta M/M_{\infty}=0.3$ obtained by Navier–Stokes calculations for the ELAC1C/EOS configuration at $M_{\infty}=4.04$, $Re_{l\mathrm{EOS}}=9.6\times10^{6}$, $\alpha=0.0°$, $\Delta\alpha=5.0°$, and $h/l_{\mathrm{EOS}}=0.225$.

carried out even on very complex configurations provide a high level of accuracy for the flow field data obtained.

### 4.3.8 Conclusions and Outlook

Comprehensive numerical and experimental simulations were performed on generic as well as on detailed two-stage-to-orbit configurations to study both the flow physics and the aerodynamic characteristics of the stages occurring at stage separation. During the separation manoeuvre complex systems of incident and reflected shock waves and expansion fans are present leading to strong shock/shock and shock/boundary layer interactions. The propagation of shock waves and expansion fans depends on the distance between the two stages and the incidence of the stages. Considering the dynamic case, distance and incidence of the orbital stage are a function of the motion due to separation speed and pitch rate. Thus, the points of shock impingement and reflection and expansion fan propagation, respectively, change markedly which causes strong unsteady airloads on the orbital stage affecting the safety of the separation manoeuvre. Instant changes in forces and moments of the carrier stage and to a much greater extent of the orbital stage might lead to collisions and damage or even loss of the vehicles if no proper flight control system is applied. The results of the developed Euler/Navier-Stokes code are in very good agreement with the data of qualified wind tunnel experiments. Consequently, the numerical method provides data to analyze the stage separation flow physics in detail and to obtain the vehicle unsteady aerodynamic quantities. The high fidelity flow solver will be improved further in its mathematics and physical modeling enhancing efficiency and performance, and in particular, introducing modules dealing with transport equation turbulence models and real gas effects.

## References


1. Freeman, D., Reubush, D., McClinton, C., Rausch, V., Crawford, L.: The NASA Hyper-X Program. IAF-97-V.4.07, 48th International Astronautical Congress, Turin, Italy, Oct. 1997.
2. Reubush, D.: Hyper-X Stage Separation – Background and Status. AIAA Paper 99-4818, AIAA 9th International Space Planes and Hypersonic Systems and Technologies Conference, Norfolk, Virginia, Nov. 1999.
3. Burkhardt, J., Gräßlin, M., Schöttle, U.M.: Impact of Mission Constraints on Optimal Flight Trajectories for the Lifting Body X-38. AIAA Paper 99-4167, AIAA Atmospheric Flight Mechanics Conference, Portland, Oregon, Aug. 1999.
4. Brücker, H.: Das deutsche Raumtransportprogramm ASTRA (Ausgewählte Systeme und Technologien für zukünftige Raumtransportsystem-Anwendungen). Deutscher Luft- und Raumfahrtkongress, DGLR-2002-044, Stuttgart, 23.–26. Sept. 2002.
5. Hirschel, E.H.: The Technology Development and Verification Concept of the German Hypersonics Programme. Aerothermodynamics and Propulsion Integration for Hypersonic Vehicles, AGARD-R-813, Oct. 1996, pp. 12-1-12-15.
6. Grallert, H.: Synthesis of a FESTIP Air-Breathing TSTO Space Transportation System. AIAA Paper 99-4884, AIAA 9th International Space Planes and Hypersonic Systems and Technologies Conference, Norfolk, Virginia, Nov. 1999.
7. Breitsamter, C., Laschka, B.: Instationäre Hyperschallströmungen – Zwei- und dreidimensionale Berechnungsmethoden. Sonderforschungsbereich 255, Transatmosphärische Flugsysteme, Grundlagen der Aerothermodynamik, Antriebe und Flugmechanik, Teilprojekt A1, Arbeits- und Ergebnisbericht, Juli 1998 – Juni 2001, 2001, pp. 23–94.
8. Breitsamter, C., Cvrlje, T., Laschka, B., Heller, M., Sachs, G.: Lateral-Directional Coupling and Unsteady Aerodynamic Effects of Hypersonic Vehicles. AIAA Journal of Spacecraft and Rockets, Vol. 38, No. 2, March-April 2001, pp. 159–167.
9. Weiland, C.: Stage Separation Aerodynamics. Aerothermodynamics and Propulsion Integration for Hypersonic Vehicles, AGARD-R-813, Oct. 1996, pp. 11-1–11-28.
10. Cvrlje, T., Breitsamter, C., Laschka, B.: Numerical simulation of the lateral aerodynamics of an orbital stage at stage separation flow conditions. Aerospace Science and Technology, Vol. 4, No. 3, April 2000, pp. 157–171.
11. Cvrlje, T., Breitsamter, C., Weishäupl, C., Laschka, B.: Euler and Navier-Stokes Solutions of Two-Stage Hypersonic Vehicle Longitudinal Motions. AIAA Journal of Spacecraft and Rockets, Vol. 37, No. 2, March-April 2000, pp. 242–251.
12. Kharitonov, A., Brodetsky, M., Vasenyov, L., Adamov, N., Breitsamter, C., Heller, M.: Investigation of Aerodynamic Characteristics of the Models of a Two-Stage Aerospace System During Separation. Final Report, Institute of Theoretical and Applied Mechanics, Russian Academy of Sciences, Siberian Division, and Institute of Fluid

Mechanics and Institute of Flight Mechanics and Flight Control, Technische Universität München, Nov. 2000.
13. Radespiel, R., Kroll, N.: Accurate Flux Vector Splitting for Shocks and Shear Layers. Journal of Computational Physics, Vol. 121, 1995, pp. 66–78.
14. Liou, M. S.: On a New Class of Flux Splittings. Lecture Notes in Physics, Vol. 414, 1992, pp. 115–119.
15. Liou, MS., Steffen, C.: A New Flux Splitting Scheme. Journal of Computational Physics, Vol. 107, 1993, pp. 23–39.
16. van Leer, B.: Towards the Ultimate Conservative Difference Scheme. V. a Second-Order Sequel to Godunov's Method. Journal of Computational Physics, Vol. 32, 1979, pp. 101–136.
17. van Albada, G., van Leer, B., Roberts, J.: A Comparative Study of Computational Methods in Cosmic Gas Dynamics. Astron. Astrophysics, Vol. 108, 1982, pp. 76–86.
18. Chakravarthy, S.-R.: High Resolution Upwind Formulations for the Navier-Stokes Equations. van Karman Institute Lecture Series on Computational Fluid Dynamics, VKI 1988-05, 1988, pp. 1–105.
19. Baldwin, B., Lomax, H.: Thin-Layer Approximation and Algebraic Model for Separated Turbulent Flows. AIAA Paper 78-0257, 1978.
20. Degani, D., Schiff, L. B.: Computation of Turbulent Supersonic Flows around Pointed Bodies Having Crossflow Separation. Journal of Computational Physics, Vol. 66, 1986, pp. 173–196.
21. Yoon, S., Jameson, A.: An LU-SSOR Scheme for the Euler and Navier-Stokes Equations. AIAA Journal, Vol. 26, No. 9, 1988, pp. 1025–1026.
22. Cvrlje, T.: Instationäre Aerodynamik des Separationsvorgangs zwischen Träger und Orbiter, Dissertation, Technische Universität München, 2001.

## 4.4 DNS of Laminar-Turbulent Transition in the Low Hypersonic Regime

Axel Fezer, Markus Kloker*, Alessandro Pagella, Ulrich Rist*, and Siegfried Wagner

### 4.4.1 Introduction

Laminar-turbulent transition in super- and hypersonic boundary layers does not only have strong influence on wall-shear stresses and heat flux, but also on other flow phenomena like shock-wave/boundary-layer interaction and flow separation, and can therefore influence the global flow field and the aerodynamic drag substantially. For airbreathing propulsion systems of the lower stage of space vehicles a thin laminar boundary layer on the forebody, designed to compress the air before the flow enters the engine intake, is clearly favourable.

* Universität Stuttgart, Institut für Aerodynamik und Gasdynamik, Pfaffenwaldring 21, 70550 Stuttgart; kloker@iag.uni-stuttgart.de, rist@iag.uni-stuttgart.de

Shock-wave/boundary-layer interaction itself is one major area of concern in technical applications at hypersonic speeds. It can result in high aerodynamic loads, engine inlet performance loss and increase of drag, to name only a few examples.

Direct Numerical Simulations (DNS) of high-order accuracy and resolution performed on super computers can enable undreamed-of insight in fundamental microscale or high-frequency flow phenomena like the transition process. A paradigm is the simulation of the physically unstable, dynamical processes during laminar-turbulent transition in boundary-layer flows forming along aerodynamical bodies in flight. For fundamental studies generic laminar base flows on simple bodies like the flat plate, wedges or cones are used, and transition is triggered by a well-defined disturbance input. This mimics so-called controlled experiments where the background disturbance level of the experimental facility is extremely low and special disturbance devices are used. Only results of such kind of experimental or numerical simulations enable lasting scientific progress.

Here, two kinds of investigations are presented. The first part is concerned with the simulation of transition in boundary layers over flat plates and sharp cones up to $M=6.8$, whereas the second part deals with transitional shock-wave/boundary-layer interaction at $M=4.8$.

## 4.4.2 Numerical Approach

The simulations presented here allow the investigation of the spatial instability and the laminar-turbulent transition of hypersonic boundary-layer flows along a flat plate or a sharp cone. Timewise periodic two- and three-dimensional disturbance waves with defined frequency and amplitude are triggered in a two-dimensional, laminar and steady-state base flow. The reactions of the base flow to these disturbances, i.e. the downstream evolution of the excited disturbance waves, is simulated within the integration domain by numerical solution of the complete, three-dimensional, unsteady, compressible Navier-Stokes equations.

The simulations are performed on the supercomputers NEC SX-4 and SX-5 of the High Performance Computing Centre Stuttgart, HLRS, which are perfectly capable of dealing with the complex and challenging computations.

In this chapter the common numerical aspects of the numerical methods are described; special features for the treatment of the different applications are specified in the respective subchapters. For a more thorough description of the numerical scheme refer to [1].

#### 4.4.2.1 Governing Equations

The complete, three-dimensional Navier-Stokes equations for unsteady, compressible flows, supplemented by the continuity and energy equation, are in conservative formulation:

$$\frac{\partial \rho}{\partial t} + \nabla \cdot (\rho \vec{u}) = 0 \tag{1}$$

$$\frac{\partial (\rho \vec{u})}{\partial t} + \nabla \cdot (\rho \vec{u} \vec{u}) + \nabla p = \frac{1}{Re} \nabla \vec{\sigma} \tag{2}$$

$$\frac{\partial (\rho e)}{\partial t} + \nabla \cdot (p + \rho e)\vec{u} = \frac{1}{(\kappa - 1) Re Pr Ma^2} \nabla \cdot (\vartheta \nabla T) + \frac{1}{Re} \nabla \cdot (\vec{\sigma} \vec{u}) \tag{3}$$

where

$$\vec{\sigma} = \mu \left[ \nabla \vec{u} + \nabla \vec{u}^T ) - \frac{2}{3} (\nabla \cdot \vec{u}) I \right] \tag{4}$$

is the viscous stress and

$$e = \int c_v(T) dT + \frac{1}{2}(u^2 + v^2 + w^2) \tag{5}$$

is the internal energy per mass unit. The fluid is considered as a non-reacting, perfect gas, for which the equation of state

$$p = \frac{1}{\kappa_\infty Ma_\infty^2} \cdot \rho T \tag{6}$$

is valid. The thermodynamic properties are approximated as calorically perfect gas as well as thermally perfect gas. In case of the calorically perfect gas the specific heats $c_P$ and $c_V$ and the Prandtl number $Pr = 0.71$ are assumed constant with the specific heat ratio $\kappa = c_p / c_v = 1.4$. The dynamic viscosity $\mu$ is determined from Sutherland's law and the thermal conductivity $\vartheta$ is set proportional to the viscosity. In case of the thermally perfect gas the specific heats and the Prandtl number are dependent on the temperature and are determined using tabulated values.

For the simulations, lengths are nondimensionalized by a reference length $L$, which appears in a global Reynolds number $Re = \rho_\infty \cdot u_\infty \cdot L / \mu_\infty = 10^5$. Another Reynolds number used here is $R_x = \sqrt{x \cdot Re}$. Time t is normalized by $L / u_\infty$. The specific heat $c_v$ is normalized with $u_\infty^2 / T_\infty$. Flow quantities with subscript "$\infty$" denote free-stream values. Density, temperature and viscosity are nondimensionalized by their respective free-stream values.

#### 4.4.2.2 Spatial and Time Discretization

We assume a numerical grid with $N \times M$ gridpoints in streamwise and wall-normal direction, as well as $k$ harmonics in spanwise direction. In streamwise direction the solution has a wave character in the presence of disturbance waves, which are either amplified or damped. Compact finite differences are able to resolve this kind of solution in an appropriate manner. They are applied here in a split-type form [2]. This form has some damping characteristics with respect to small-scale oscillations. In wall-normal direction either compact finite differences of $6^{th}$ order (cone calculations) or standard finite differences of fourth order accuracy (shock-wave/boundary-layer interaction) are used. Split-type differences are used to calculate convective terms, while viscous terms are computed by central differences. In spanwise direction periodic boundaries allow a spectral approximation with Fourier expansion. Time integration is performed with a standard 4-step Runge-Kutta scheme of fourth-order accuracy.

#### 4.4.2.3 Initial and Boundary Conditions

The first step is the calculation of a laminar base flow which satisfies the basic equations in steady, two-dimensional form, using pseudotime-stepping for integrating the time-dependent equations to a steady state. This base flow constitutes the initial condition for the disturbance calculation at $t=0$. Then, for $t>0$, disturbances are introduced, and the real-time downstream development of the disturbance waves is calculated solving the full equations. For this purpose the method of disturbance flow formulation is used, i.e. each flow quantity is decomposed in its base flow and disturbance part ($\phi = \phi_B + \phi'$). It is assumed that there is no alteration of the base flow in time. Note that a change of the time-mean of the flow is represented by $\langle \phi' \rangle \neq 0$.

The inflow variables result from calculations of the compressible boundary layer equations [3], and are held constant during simulation, i.e. disturbances are zero at the inflow boundary. At the outflow boundary, base-flow field variables are calculated neglecting second streamwise derivatives. Disturbance values are treated using a buffer domain [4] where all values are smoothly damped to zero in order to avoid reflections from the boundary.

At the wall, the no-slip condition and vanishing normal velocity component are assumed. Disturbances are introduced within a disturbance strip located at the wall by varying the wall-normal momentum disturbance $(\rho v)'$ simulating periodic blowing and suction. A thorough description of the disturbance function applied can be found in [5] and [1]. The wall-temperature can be modelled as constant ($T_B = \text{const}, T' = 0$), adiabatic $\left(\frac{\partial T}{\partial y} = 0\right)$ or radiation-adiabatic (see below).

The free-stream boundary conditions are different for the specific applications and are therefore discussed in the respective subchapters.

## 4.4.3 Transition on Flat Plate and Sharp Cone

In the rare wind-tunnel experiments transition is usually caused by uncontrolled background disturbances in the oncoming flow. From a historical point of view, transition has been observed at first to take place earlier on the flat plate than on the cone in 'cold' experiments – with a stagnation temperature equal to about the ambient temperature and under uncontrolled conditions –, contrary to predictions obtained by Linear Stability Theory (LST). Nowadays, investigations in 'quiet' tunnels and in-flight experiments indicate that this was caused by the wind-tunnel noise in connection with strong disturbance receptivity on the flat plate's leading edge, representing a line compared to the apex of the cone.

### 4.4.3.1 Application-Specific Details of the Numerical Method

*Integration Domain and Coordinate System*

The flow is considered in an integration domain on the flat plate and the sharp cone ($7^\circ$ semivertex angle), respectively, not containing any shock wave induced by the leading edge. Figure 4.4.1 shows a sketch of the integration domain as used on the cone.

The equations are solved in body-fitted coordinate systems, i.e. a rectangular domain $(x, y, z)$ on the flat plate and a conical system $(x, y, \varphi)$ on the cone, where $x$ is the axis in streamwise direction, $y$ is normal to the wall and $z/\varphi$ is the spanwise/azimuthal coordinate.

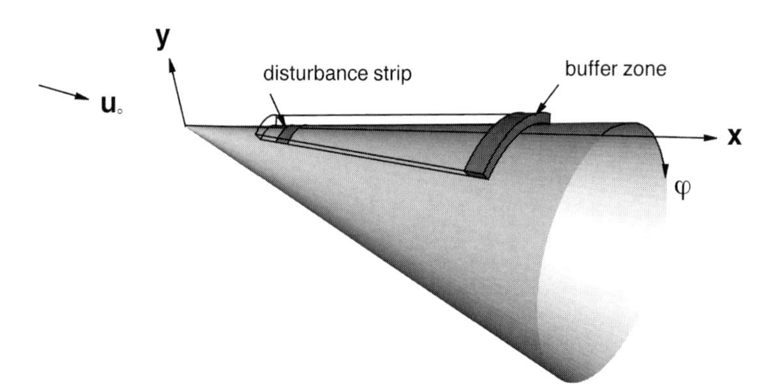


Figure 4.4.1: Integration domain and coordinate system on the sharp cone, angle of attack equal to zero.

*Boundary Conditions*

In contrast to the flat plate there is no self-similar boundary-layer solution for the cone which could be used as an inflow condition for the calculation of the base flow. However, neglecting body curvature the compressible boundary layer on the sharp cone at zero angle of attack can be put down to the self-similar boundary layer on the flat plate using the Mangler-Levy-Lees transformation [6] (with $x_{cone} = 3 \cdot x_{plate}$). Therefore, we use transformed flat-plate boundary-layer profiles as inflow boundary conditions. Additionally, these profiles are modified by multiplication with the inviscid solution resulting from the Taylor-Maccoll equation [7] to account for the non-constant values in wall-normal direction outside of the boundary layer.

For the base flow, a free-stream boundary condition is used which is based on the spatial, two-dimensional theory of characteristics. Along the characteristic directions the (planar or conical, respectively) compatibility equations are used to calculate the values at the boundary. For the disturbance calculations, a non-reflecting boundary condition according to [8] is used.

*Wall-Temperature*

A major goal of our simulations is to predict the wall temperature in the transitional regime in a realistic manner. Often the condition $T'=0$ is used, leading to no temperature alteration even in the near-turbulent regime compared to the laminar case. The other limit is the adiabatic condition where the wall temperature immediately follows the unsteady values near the wall. In reality we expect a behaviour between these two extremes. To this end we use a radiation-adiabatic wall-temperature boundary condition, where the heat flux into the wall by thermal conduction is in balance with the heat flux emitted by radiation:

$$\vartheta \frac{\partial T}{\partial y}\bigg|_{y=0} = \varepsilon \sigma T^4 \bigg|_{y=0} \tag{7}$$

with the Stefan-Boltzmann constant $\sigma$ and the emissivity $\varepsilon$ of the wall surface.

*Computing Details*

The typical grid sizes are $2500 \times 169 \times 32(64)$ points in x-, y- and z-($\varphi$-) direction, respectively. In the y-direction zones with successively halved $\Delta y$ can be used; typically two zones near the wall are applied. A simulation on the NEC SX-5 takes about 3.8 μs per grid point and full time step on a single processor. The memory requirement is about 300 bytes per point.

### 4.4.3.2 Results: Simulation of a Controlled Experiment

The first controlled experiment on a cone in hypersonic flow has been performed by Maslov et al. 2000 [9] at ITAM, Novosibirsk. The measurements were done on a 7° half-angle sharp cone model of 500 mm length in the T-326 hypersonic blow-down wind tunnel at $M_e$=5.3 (Mach number at boundary-layer edge, after shock). The disturbances were triggered by a high-frequency glow discharge system, containing frequencies of first ($\approx$ 80 kHz) or second mode disturbances ($\approx$ 280 kHz). The temperatures are $T_0 \approx 395$ K, $T_w \approx 320$ K, and $Re_{1e} \approx 15.5 \cdot 10^6$ $^1/_{\text{m}}$.

The glow discharge actuator has been modelled by a harmonic point source exciting the whole spectrum of spanwise wavenumbers. Additionally, various frequencies can be disturbed simultaneously.

An extensive presentation of the method and the results is available in [10]. Here we give a short description of the result for the first mode case at $f$=78 kHz.

Figure 4.4.2 shows the azimuthal wavenumber spectrum at two positions far downstream of the disturbance source, which were measured by rotating the cone in the experiment. For the DNS two lines are shown: The first represents the results obtained when one frequency ($f$=78 kHz) is disturbed. The main peak and the overall distribution as well as the amplitude growth from the first to the second station are in very good agreement; note that the slight anti-symmetry in the experiment is not present in the DNS because of the rigid compliance with the symmetric conditions. However, the experimental data show additional, smaller side peaks at $\beta \approx \pm 0.7$ that do not occur in the DNS.

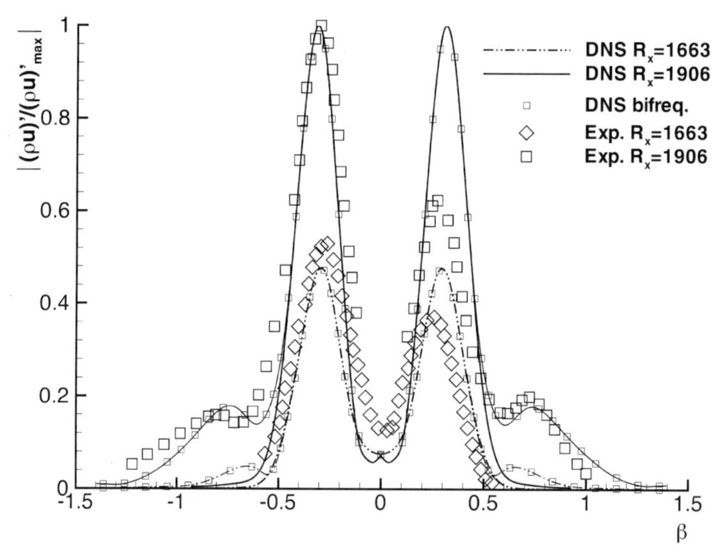


Figure 4.4.2: Azimuthal wavenumber spectrum at two positions downstream of the point-like disturbance source; symbols – experiment, lines – DNS.

Inspired by former investigations [11] the idea arose that these peaks could be the effect of some nonlinear subharmonic interactions. Keeping in mind that the glow discharge consists of periodically recurring short pulses containing higher-harmonics we disturbed one additional frequency, namely the double frequency $2f$ (in this context $f$ is the subharmonic to $2f$). Now, even the side peaks can be reproduced perfectly as can be seen from the second line (with small symbols) in Fig. 4.4.2, and an excellent overall agreement between experiment and DNS is achieved. The reason for this phenomenon are nonlinear interactions between fundamental ($2f$) and subharmonic ($f$) disturbances which create subharmonic disturbances with higher amplitudes at wave numbers where the additional peaks appear.

#### 4.4.3.3 Results: Flat Plate and Cone at $M=6.8$

The investigations presented in this chapter have several aims. First, different generic bodies are regarded that are in practice employed as basic forebody geometries of space vehicles. Second, we will distinguish flow parameters which typically are present in wind-tunnel experiments (WT) and in flight conditions (FC). Finally, we are interested in the identification of basic transition mechanisms and the underlying transitional structures.

The Mach number at the boundary-layer edge is $M_e=6.8$ in all cases. We define wind-tunnel conditions as 'cold' flow with $T_\infty=50$ K and an adiabatic wall temperature $T_w=T_{rec}\approx 437$ K. For flight conditions we choose $T_\infty=220$ K and a radiation-adiabatic cooled wall with $T_w\approx 0.5\cdot T_{rec}\approx 950$ K.

Investigations using Linear Stability Theory (LST) deliver an overview of the amplification of small-amplitude disturbance waves. It reveals for Mach numbers higher than 2.2 an amplified, so-called acoustic disturbance mode (or $2^{nd}$ mode) in addition to the vorticity mode ($1^{st}$ mode) known from incompressible flow. This acoustic mode is most amplified if it is 2-d, grows stronger than the vorticity mode for a Mach number higher than 4, and is destabilized by wall cooling. Different from the incompressible case, $1^{st}$ mode disturbances are most amplified as 3-d waves, and are stabilized by wall cooling.

Figure 4.4.3 shows integral amplification rates of discrete frequencies (dashed lines) according to LST which can be understood as the exponential growth a disturbance wave of a specific frequency experiences up to a certain x-position, and their envelope ($N$-factors, solid line). Chart (a) shows the situation for 3-d disturbances at wind-tunnel conditions. They are similarly amplified for flat plate and sharp cone, and the relevant frequencies are about the same. A specialty on the cone is that the (basic) azimuthal wavenumber is defined by $\beta=n/r$ with $n$ constant and $\beta_0=n/R_0$; $n$ represents the spanwise (integration-domain) wavelength as an integer fraction of the circumference: $n=2$ means that the wavelength covers half the circumference. $r$ denotes the radial distance of a point from the cone axis, thus $\beta=\beta_0 R_0/r$. Going downstream, all considered azimuthal wave numbers and the respective wave propagation angles reduce with the increasing radius, i.e. all modes tend to become 2-d. At flight conditions (b) exclusively 2-d modes are crucial because of the wall cooling. Disturbances on the cone appear to be stronger amplified at certain x-positions, and the relevant frequencies are higher.

Figure 4.4.4 shows the development of the maximum disturbance amplitudes (over y) of $u'$ for two cases. In the oblique-type transition scenario at wind-tunnel conditions (a) the disturbed symmetrical pair of waves $(1,\pm 1)$ has vorticity-mode type (1st mode) and the same frequency parameter $F=2\cdot 10^{-5}$ and obliqueness angle $60^\circ$ (referred to the location of the disturbance strip) for flat plate and cone, respectively. The characteristic, nonlinearly generated steady longitudinal vortex mode (0,2) has smaller amplification rates in the cone case, although the growth of the generating mode is almost the same. This is not caused by curvature or body divergence effects since simulations with different cone angles showed nearly the same result. It may be the effect of the slower-growing boundary layer which hinders the growth of vortices. Besides that, the mode development and the physical structures are quite similar. It turns out that oblique-type transition is a robust mechanism, active on both

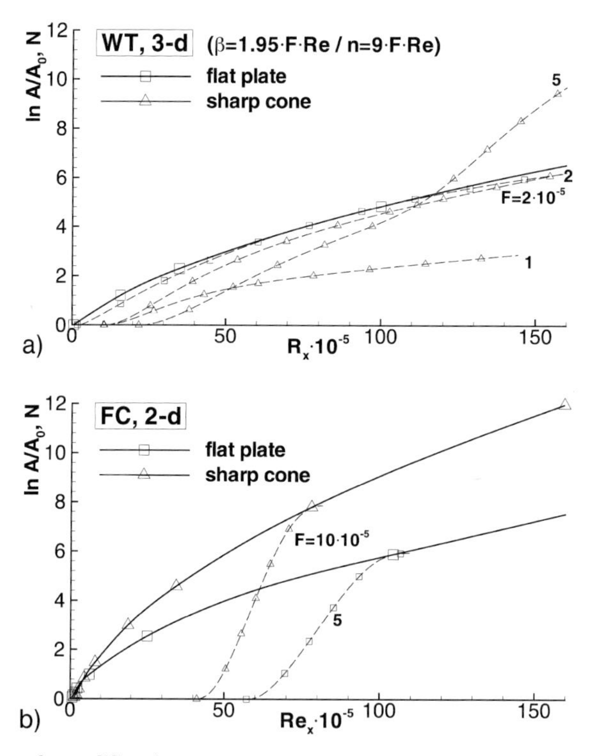


Figure 4.4.3: Integral amplification curves as obtained from Linear Stability Theory; (a) 3-d disturbances, wind-tunnel conditions, (b) 2-d disturbances, flight conditions. Dashed line: specific frequency, solid line: envelope. $F=2\pi f/(u_\infty \cdot \mathrm{Re})$.

the flat plate and the cone. This holds also for flight conditions. However, the higher temperature at hot flow and the wall-cooling stabilize the 1$^{st}$ mode. Historically, this led to the conclusion that transition takes place more upstream in cold wind-tunnels than at flight conditions which is only true for this type of scenario.

In the fundamental-type transition scenario at flight conditions an acoustic 2-d primary disturbance (1,0) and a secondary 3-d mode (1,±1) with the same frequency, but smaller amplitude, are excited. The frequency parameters are $F=5\cdot10^{-5}$ for the flat plate and $F=10\cdot10^{-5}$ for the cone, respectively, so that the resulting $N$-factors are roughly the same (cf. Fig. 4.4.3). Again, the maximum amplitudes are displayed in Fig. 4.4.4 (b). The 2-d primary wave (1,0) grows due to the primary instability. At a threshold amplitude, the phase speed of the secondary 3-d wave (1,±1) starts to synchronize with the primary wave; after adaptation resonant growth of (1,±1) and (0,1) takes place. Due to the stability properties of the 2-d 2$^{nd}$-mode disturbance the threshold amplitude is reached much earlier at the cone (the disturbance strip is in both cases located

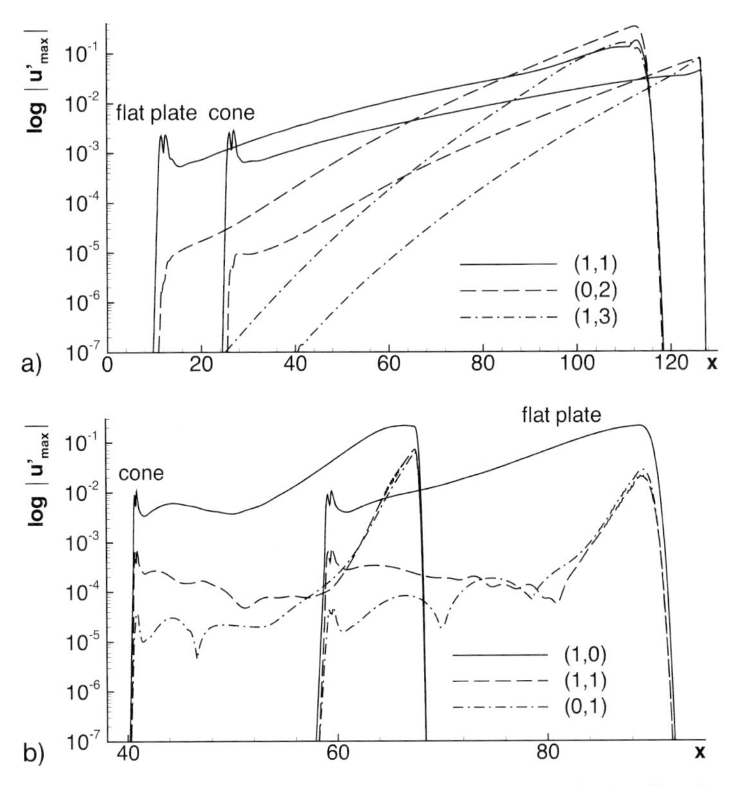


Figure 4.4.4: Downstream development of the maximum $u$-velocity disturbance amplitudes; (a) wind-tunnel conditions, oblique-type breakdown, (b) flight conditions, fundamental-type breakdown. $(h,k)$ refers to a disturbance mode with frequency $hf$ and spanwise wave number $k\beta$.

at the beginning of the unstable region and the disturbance amplitudes are identical). Since the 2-d and 3-d $2^{nd}$-mode waves tend to be the same going downstream on the cone due to the natural reduction of the azimuthal wave number, their phase speeds do similarly. This renders the necessary threshold amplitude on the cone to be only half that on the flat plate: it is $u'/u_\infty \approx 4\%$.

For the likeliness of the different transition scenarios at flight conditions with wall-cooling we find the following: Oblique-type transition is a robust transition mechanism for both the flat plate and the cone, also at cold wind-tunnel conditions (recall that wall-cooling stabilizes the vorticity modes). Important is the beginning of the unstable region. This is quite early for 3-d vorticity modes, and three times earlier on the flat plate than on the cone. Since relevant 2-d $N$-factors, decisive for fundamental breakdown, are reached only late on the flat plate, oblique breakdown is expected to be the most relevant mechanism there. On the cone, however, sufficiently large 2-d $N$-factors are reached earlier and the behaviour of the secondary 3-d waves accelerates the fundamental mechanism as described above. Here, fundamental transition caused by 2-d acoustic modes significantly destabilized by the wall-cooling, is likely to be the dominant mechanism.

The possibility of the appearance of fundamental breakdown on the cone has an important consequence in practice. It is a mechanism caused by acoustic disturbances which have their maximum located near the wall. Therefore, the heating of the wall in the transitional regime is much larger than in the oblique-type scenario. This is verified by the examination of the wall temperatures computed using a radiation-adiabatic wall-temperature condition for the instantaneous total flow. The time-averaged wall-temperature alteration for the fundamental case on the cone is shown in Fig. 4.4.5. The heating in the com-

Figure 4.4.5: Time-averaged wall-temperature alteration. Sharp cone at flight conditions, fundamental-type breakdown with radiation-adiabatic wall cooling.

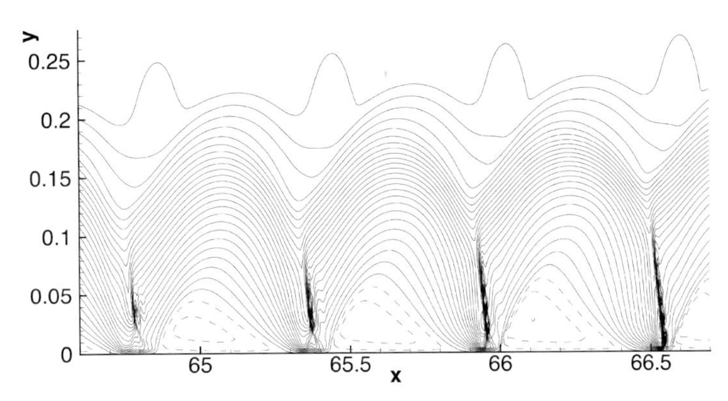


Figure 4.4.6: Isocontours of instantaneous velocity component $u$ in a longitudinal flow cut. Flight conditions, fundamental-type breakdown; the increment is 0.04 and negative values are dashed.

puted area reaches locally 250 K, which is more than 25% of the wall temperature of the laminar base flow ($\approx 950$ K). We note that the heating in the transitional regime is known to be higher than in the turbulent regime [12].

High-amplitude acoustic modes cause complex physical phenomena. These disturbances have the special characteristic that they travel by more than the speed of sound faster than the base flow in a relative supersonic region near the wall. However, this only holds if the amplitudes are small. As the amplitudes increase, we observe the formation of shocklets. They can be clearly seen in Fig. 4.4.6 where isolines of $u$ are plotted. Dashed lines correspond to negative values, so strong local flow separation areas are also visible. They are caused by the presence of the strong 2-d wave with its near-wall maximum. The simulation of these phenomena needs an enormous computational effort, and the simulations can not yet be easily extended into later stages. Therefore, some interesting questions are unresolved so far, for instance: Are there any $\Lambda$-vortices like in subsonic flow? We incessantly work on the resolution. More details of our DNS of super-/hypersonic boundary-layer transition can be found in [10, 11, 13, 14].

### 4.4.4 Transitional Shock-Wave/Boundary-Layer Interaction at $Ma$=4.8

We now move on to shock-wave/boundary-layer interactions, which form the second part of this paper. The base flow properties of shock-wave/boundary-layer interactions for simple geometries, such as an impinging shock wave on a flat-plate boundary layer, a forward or rearward facing step and the compression ramp problem are well-understood and have been thoroughly explained in the many publications. A good summary can be found in [15]. Early experimental results, such as experiments referred to in [15] indicate, that flows with shock-wave/boundary-layer interaction do behave in a similar manner in major parts, independent of the cause of separation (impinging shock, ramp or step). This led to the derivation of some correlation laws, which are known as the free interaction concept.

The pressure gradient of an impinging shock wave thickens the boundary layer. Due to the displacement of the flow, compression waves are formed up- and downstream shock-impingement. Provided the shock is strong enough, the boundary layer separates. Farther away from the boundary layer, the compression waves near separation and reattachment merge into the separation- and reattachment-shock, respectively. On its way into the boundary layer, the impinging shock gets steeper until it ends as an almost vertical shock at the sonic line. There it is reflected as a system of expansion waves. For a compression ramp the following physical phenomena occur: The change in direction of the wall due to the ramp forces the boundary layer to follow the contour, which causes a pressure gradient yielding the boundary layer thickness to increase. Depending on such parameters like the ramp angle, the Reynolds number, the wall temperature conditions and the boundary layer thickness, a complicated system of compression waves occurs. Provided the ramp angle is large enough, the boundary layer separates. Compression waves form upstream the corner, which is caused by an initial turn of the flow at separation. Well outside the boundary layer, those coalesce to the separation shock. At reattachment, additional compression waves are present, which merge with the separation shock. Another possible scenario is the formation of a reattachment shock before the compression waves reach the separation shock. The two shocks will then meet at the so-called triple or bifurcation point. Although the base flow properties have been intensively studied, much less is known about the transitional behaviour of such flows with shock-wave/boundary-layer interaction. Transition to turbulence is of high significance in practical hypersonic flows. Depending on the flight state the transition zone can have a comparable length like the fully turbulent flow on the body of a vessel [3]. Also, temperature disturbance peaks can reach even higher values than in turbulent flow, which has to be considered in structural design. In the present paper, we will compare the results obtained for a flat plate with impinging shock with a compression ramp flow. The flow parameters of both configurations correspond to each other. The ramp angle was chosen to induce a separation bubble which has the same length at the wall compared to the separation bubble in the case with impinging shock.

#### 4.4.4.1 Application-Specific Details of the Numerical Method

*Grid Mapping*

In the case of an orthogonal integration domain the three-dimensional, unsteady, compressible Navier-Stokes equations in conservative formulation can be written as

$$\frac{\partial Q}{\partial t} + \frac{\partial F}{\partial x} + \frac{\partial G}{\partial y} + \frac{\partial H}{\partial z} = 0 \tag{8}$$

with the conservative variable vector $Q$ and the fluxes $F$, $G$, $H$ in streamwise, normal and spanwise direction, respectively.

An appropriate transformation is necessary for simulating more complicated bodies, such as the compression ramp, of which results will be presented in a later section of this paper. The idea is to map a contour-fitted numerical grid in physical space onto an equally spaced in numerical space. For more details refer to [16]. Here, only transformations in streamwise and wall-normal direction are considered. According to [16], streamwise derivatives are replaced by

$$\frac{\partial}{\partial x} = \frac{1}{J}\left[\frac{\partial}{\partial \xi}\frac{\partial y}{\partial \eta} - \frac{\partial}{\partial \eta}\frac{\partial y}{\partial \xi}\right] \tag{9}$$

while wall-normal derivatives become

$$\frac{\partial}{\partial y} = \frac{1}{J}\left[\frac{\partial}{\partial \eta}\frac{\partial x}{\partial \xi} - \frac{\partial}{\partial \xi}\frac{\partial x}{\partial \eta}\right] \tag{10}$$

with the determinant of the Jacobian

$$J = \frac{\partial x}{\partial \xi}\frac{\partial y}{\partial \eta} - \frac{\partial y}{\partial \xi}\frac{\partial x}{\partial \eta} \tag{11}$$

The transformed flux-vectors then become

$$F_T = F\frac{\partial y}{\partial \eta} + G\frac{\partial x}{\partial \eta} \tag{12}$$

and

$$G_T = -F\frac{\partial y}{\partial \xi} + G\frac{\partial x}{\partial \xi} \tag{13}$$

The transformed Navier-Stokes equations are

$$J\frac{\partial Q}{\partial t}+\frac{\partial F_T}{\partial \xi}+\frac{\partial G_T}{\partial \eta}+J\frac{\partial H}{\partial z}=0 \tag{14}$$

*Free-Stream Boundary Conditions*

At the free-stream boundary, a characteristic boundary condition [1] is applied for the base flow calculation. For the disturbance calculations, a non-reflecting boundary condition according to [8] is used. In the case of the compression ramp, the boundary condition is transformed appropriately as described before. The shock is prescribed at the free-stream boundary using Rankine-Hugoniot relations to calculate the flow variables behind the shock with the chosen shock angle. For several grid-points up- and downstream of the shock at the upper boundary, flow variables are held constant. A steady shock then establishes itself within the flow field during calculation.

*Filtering*

As long as step-sizes are fine enough, the above discretization reliably works for oblique shocks in the computation of the base flow. However, as the shock approaches the sonic line in the boundary layer it turns towards a vertical shock and sawtooth oscillations in streamwise direction (similar to the Gibbs phenomenon) occur around the shock wave because of insufficient resolution in x-direction. To stabilize the computation, a fourth-order accurate implicit, spatial filter [17] is used (*in the base flow calculations only!*). Step size and integration domain variations proved grid-independency.

*Grid Generation*

For the case with impinging shock on a laminar boundary layer a transformation of the governing equations is not necessary. Therefore, the physical and the computational grids are identical.

Figure 4.4.7 gives the body-fitted grid used in simulations of the compression ramp. The grid is based on the analytical formulation by Adams [18]. A more thorough description of the grid-generation can be found in [19]. $X_0$ refers to the location of the beginning of the integration domain; the corner is located at $X_C$. $\Phi_w$ is the physical ramp angle, $y_1$ the height of the integration domain and $L_{gr}$ the length. The indices "$w$" and "$\infty$" refer to the wall and the free-stream boundary, respectively. The lower picture shows the computational space, where the equations presented in the previous section are solved.

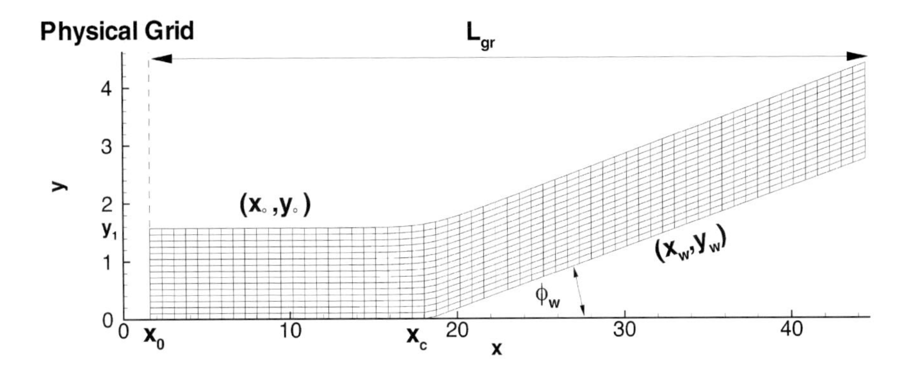


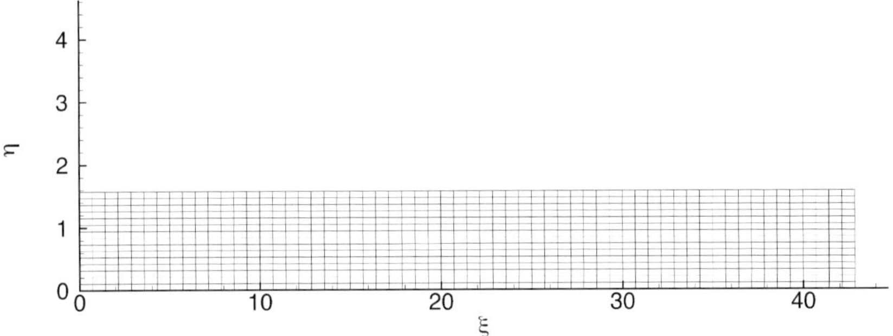


Figure 4.4.7: Physical (upper part) and numerical grid (lower part) used in simulations. Figures are stretched in wall-normal direction.

#### 4.4.4.2 Results: Impinging Shock on a Flat Plate vs. Compression Ramp at $Ma=4.8$

*Base Flow Properties*

In the following, results for the compression ramp will be shown and compared to the flat-plate boundary layer with impinging shock. In both cases, the free stream Mach number is $Ma=4.8$, free stream temperature $T_\infty=55.4$ K and the wall temperature is held constant at $T_w=270$ K. For the case with impinging shock on the flat plate the shock angle with respect to the wall is $\sigma=14^\circ$. For this case results were published in [5]. The ramp angle with respect to the horizontal direction is $\Phi=6^\circ$. As a first approximation for turbulent boundary layers a corner angle half the angle of the impinging shock gives a similar wall-pressure distribution in both cases, according to Knight et al. [20]. The ramp angle of $\Phi=6^\circ$ was obtained by iteratively changing the wedge-angle until maximum agreement in terms of the skin friction distribution could be achieved. In Fig. 4.4.8 the density field for both cases is given. Also shown are selected stream-

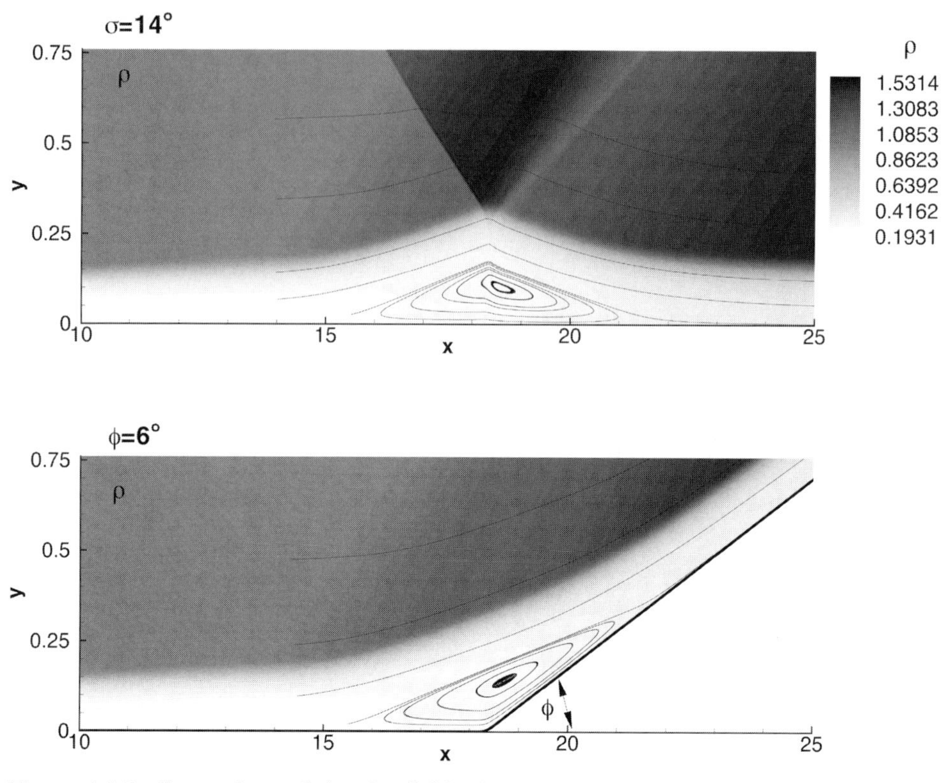


Figure 4.4.8: Comparison of density fields: impinging shock wave (upper figure) and compression ramp (lower figure). Figures are stretched in wall-normal direction.

lines, which visualize the flow in the free stream and the boundary layer, where a separation bubble can be observed. It will be shown in the following, that the two flows are very similar.

When the flow field in the ramp case downstream of the corner is mapped on the flat plate, which means a rotation with its negative ramp angle the resulting flow field matches with the impinging shock. In Fig. 4.4.9 such a construction is given for the density fields. As it can be seen, the two density fields look very similar except for the impinging shock and its reflection as an expansion wave at the sonic line, which is not present in the compression corner. The contour levels of both configurations agree quantitatively. For all other flow variables the results compare equally well.

We now take a closer look at the skin friction distributions, which are given in Fig. 4.4.10. It can be seen, that the skin friction of the compression ramp perfectly matches the skin friction of the case with impinging shock. Also shown in Fig. 4.4.10 are results of resolution studies (dotted lines with symbols, according to the legend). They prove the grid-independence of our simulations. According results for the case with impinging shock can be found in [5].

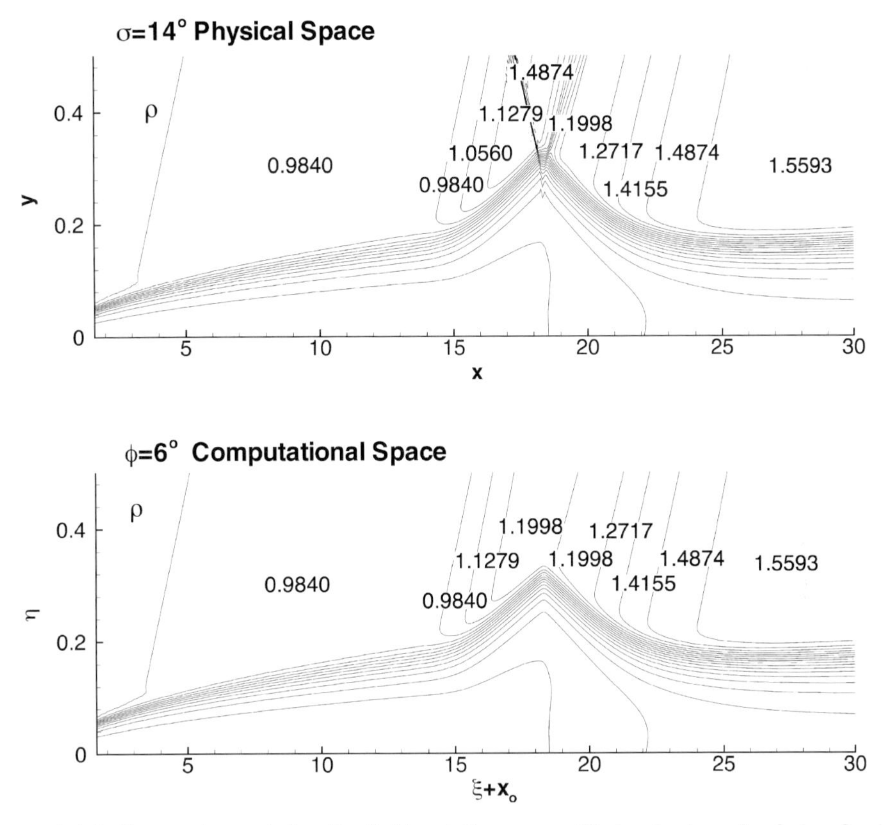


Figure 4.4.9: Comparison of density fields of the case with impinging shock in physical space with the compression corner in computational space. Figures are stretched in wall-normal direction.

Base flow calculations presented here are in full agreement with free-interaction theory, which states, that the boundary layer behaviour in shock-wave/boundary-layer interaction is largely independent of its origin. This holds even quantitatively, as it can be seen in the present comparisons.

*Small-Amplitude Disturbance Behaviour*

We now compare compressible linear stability theory results obtained from investigations of both the impinging shock and the compression ramp. Our linear stability theory results are based on the scheme developed by Mack [21]. Non-parallel effects are not taken into account here. The stream-wise velocity and temperature profiles, which are used by linear stability theory, were extracted locally from the direct numerical simulations of the base flow. Figure 4.4.11 compares the stability diagrams obtained by linear stability theory for the two

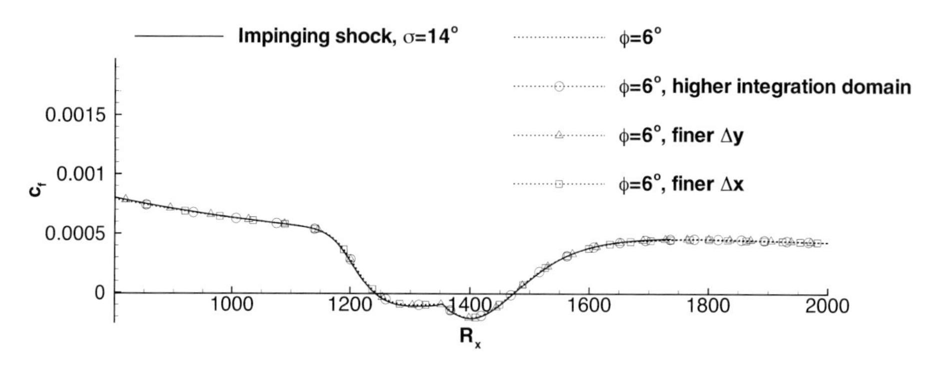


Figure 4.4.10: Skin friction coefficient distribution $c_f$ versus square root of local Reynolds number $R_x$. Comparison of case with impinging shock (solid line) and compression corner (dotted lines). Results of grid-refinement/enlargement studies represented by dotted lines with symbols.

cases. $F = (2 \cdot \pi \cdot f^* \cdot L)/(u^*_\infty \cdot Re)$ is the disturbance frequency, darker shadings indicate larger amplification rates $\alpha_i = (-\partial \ln(A(x)/A_0))/\partial x$. $A(x)/A_0$ refers to an amplitude ratio with respect to an initial amplitude of an arbitrary flow variable. $f^*$ is the disturbance frequency. Variables with an asterisk * and the reference length $L$ are dimensional. As it can be seen, the two figures are virtually identical. Due to the influence of the shock, the first instability mode is stabilized and the second mode is destabilized and locally shifted to lower frequencies. New instabilities at higher frequencies, which belong to a third mode according to the zeros in the pressure eigenfunction are formed near $R_x = 1350$ and $F = 0.00012$.

We now compare direct numerical simulations with controlled, small-amplitude disturbances. Because the two base flows are not identical in all aspects, i.e. there is no impinging shock and reflected expansion wave in the compression corner and curvatures of streamlines in physical space are different, differences between the two configurations might exist, which are not taken into account by linear stability theory. Therefore, a more detailed investigation with direct numerical simulation seems appropriate. Figures 4.4.12 and 4.4.13 show the maximum temperature disturbance amplitudes and the wall pressure amplitudes versus the downstream coordinate, respectively. For the direct numerical simulations, they were obtained by a Fourier analysis of the flow variables over one disturbance period. The wall pressure and temperature were chosen here because of results obtained in [5]. These two quantities have the largest and smallest non-parallel effects of all primitive flow variables in our simulations. Non-parallel effects highly influence the disturbance behaviour, which was quantified in [5], as well.

The differences between the amplitudes of the direct numerical simulation and linear stability, which can be seen in Figures 4.4.12 and 4.4.13, were identified and quantified as an effect of non-parallelism for the case with impinging shock in [5]. The strongest non-parallel effects appear in the maximum pres-

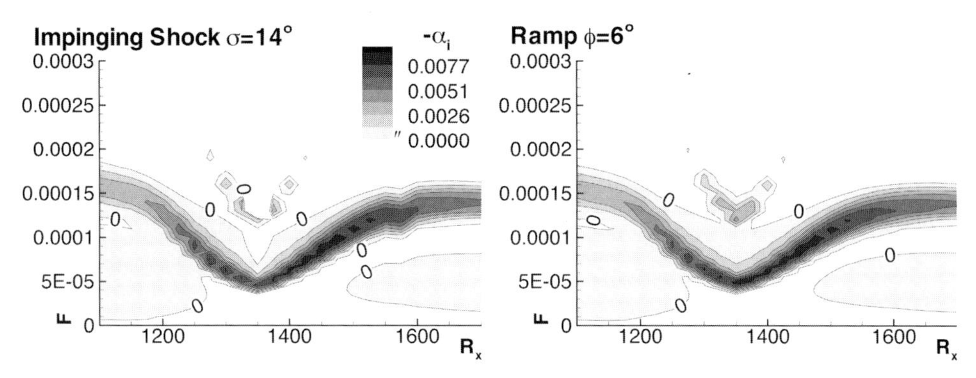


Figure 4.4.11: Linear stability theory results. Disturbance frequency $F$ versus local Reynolds number $R_x$. Darker shadings indicate larger amplification rates $a_i$. White represents negative/neutral amplification.

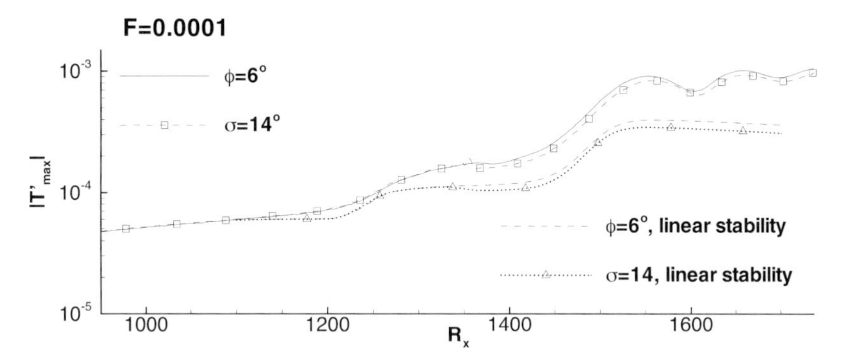


Figure 4.4.12: Maximum temperature disturbance amplitudes of both direct numerical simulations and integrated amplification rates from linear stability theory versus $R_x$.

sure disturbance, given in Fig. 4.4.13. Disturbance behaviour of the case with impinging shock and the ramp look very similar for both the direct numerical simulation and linear stability. It is identical until shock-impingement at $R_x \approx 1350$. Downstream of $R_x = 1350$ the amplitudes both for the maximum temperature amplitude and the wall pressure remain smaller in the case with impinging shock. As it could already be concluded from Fig. 4.4.11, amplification rates in the case with impinging shock are slightly damped near shock-impingement, which results in smaller disturbance amplitudes as it can be seen in Figures 4.4.12 and 4.4.13. However, compared to the non-parallel effects present in the DNS solutions such differences are marginal. For instance, the predicted amplification of linear stability theory between $R_x = 1100$ and $R_x = 1550$ would be $A/A_0 \approx 7$. This is far less than the actually observed amplification of the temperature maxima and the wall pressure fluctuations, which are $\approx 15$ and $\approx 68$, respectively. At $R_x = 1625$ the amplitude of the wall pressure is again very close to linear stability theory, because of considerable spatial variations.

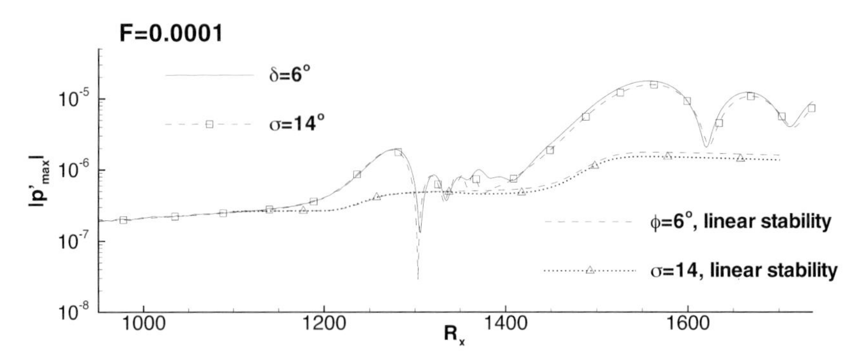


Figure 4.4.13: Maximum pressure disturbance amplitudes of both direct numerical simulations and integrated amplification rates from linear stability theory versus $R_x$.

*Larger-Amplitude Disturbance Behaviour*

We now turn to the development of larger disturbances yielding non-linear behaviour. In the following, $(h, k)$ represents a mode of the frequency $F_0$, where $F_0$ is the fundamental disturbance frequency and a spanwise wave number $k \cdot \beta$, with $\beta$ as the fundamental spanwise wave number. The disturbance frequency $F_0$ determines the streamwise wave number $a_r$ via the dispersion relation of the disturbances. Thus, the obliqueness angle $\psi$ of the disturbance is given by $\tan \psi = k \cdot \beta / a_r$. For the cases presented here, a fundamental frequency $F_0 = 1 \cdot 10^{-4}$ is chosen. In the oblique scenario, of which results are shown in Fig. 4.4.14, two waves (1,1) and (1,–1) interact with each other. Within this section, only results of the oblique scenario will be considered because simulations with other non-linear scenarios, such as subharmonic or fundamental breakdown showed very similar behaviour. Because of symmetry, only (1,1) without its symmetric counterpart (1,–1) is given. (1,1) is represented by solid, the directly generated (0,2) and (1,3) by dashed and dotted lines, respectively. The case without shock is represented by the lines with square symbols, while the case with impinging shock has triangles and the compression ramp no symbols at all. It can clearly be seen, that the disturbance amplitudes of the cases with shock-wave/boundary-layer interaction strongly exceed the amplitudes of the pure flat plate. For the (1,1), the amplitude-behaviour can qualitatively be explained with linear theory for this three-dimensional disturbance wave. Due to shock-wave/boundary-layer interaction amplification is increased for $F_0$. As explained in the section before, the compression ramp constitutes larger disturbance amplitudes compared to the case with impinging shock. This is shown in Fig. 4.4.14, as well. The impinging shock locally stabilizes the flow, which can be observed in all disturbance modes shown in Fig. 4.4.14.

In Tab. 4.4.1, amplitude-ratios are given. The highest absolute amplitude is reached by (0,2) in the ramp-case. For the impinging shock, this mode seems to become saturated downstream the damping zone, while still appearing to grow for the ramp. In earlier investigations with impinging shock at $Ma = 4.5$

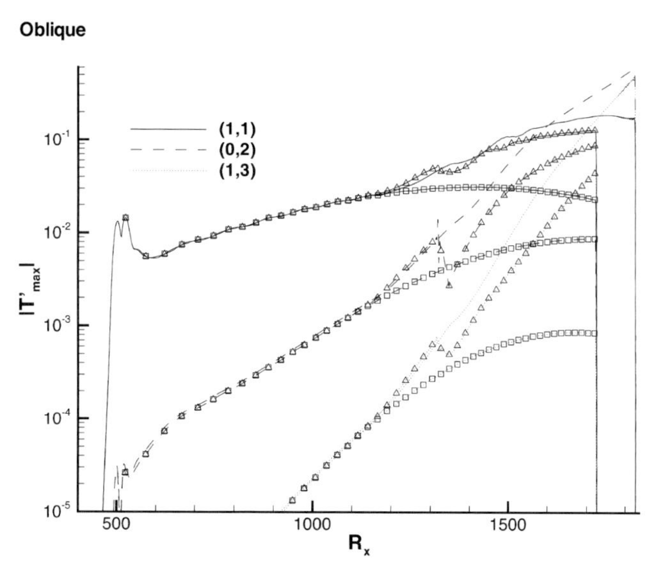


Figure 4.4.14: Maximum temperature disturbances versus $R_x$. Comparison of results from the compression ramp (no symbols), impinging shock (triangles) and the flat plate without shock-boundary layer interaction (square symbols). Oblique scenario.

Table 4.4.1: Comparison of disturbance amplitudes for oblique case at $R_x$ = 1700. Reference amplitude in the case without shock-boundary layer interaction ("Plate only") is set to 1. Values for impinging shock and ramp represent ratio of their respective amplitude to reference amplitude.

| Mode | Plate only | Impinging shock | Ramp |
|---|---|---|---|
| (1,1) | 1 | 5.3 | 7.4 |
| (0,2) | 1 | 9.6 | 32.4 |
| (1,3) | 1 | 43.4 | 153.1 |

and $Ma$=6, such a dominance of the (0,2) in the oblique scenario could be observed, too. Since these modes are linked to the formation of vortices, it may be possible, that a Görtler-type instability mechanism plays a key role in the growth of those so-called streak or vortex modes. In general, Görtler-instability is triggered by concave curvature of the body. In the case with shock-wave/ boundary-layer interaction, the curvature of the body might be replaced by the thickening of the boundary layer near impingement, causing a concave curvature of the streamlines. However, this is difficult to quantify and still remains to be proven.

#### 4.4.4.3 Conclusions

For the base flow and small disturbance amplitudes it was shown, that the boundary layer under the influence of an impinging shock and along a compression corner are practically identical. Therefore, conclusions drawn in [5] can be applied to the compression ramp, too. This result lies in accordance with earlier experimental observations and analytical concepts stating that the physics of such flows are not determined by the type of shock-wave/boundary-layer interaction, but rather determined by the flow-field properties at the onset of interaction. Although the qualitative behaviour of the case with impinging-shock and the compression corner is still comparable, more significant differences exist in the case with larger disturbance amplitudes. For the ramp, the highest amplitudes are reached, which are significantly larger compared to the impinging shock, and, of course, exceed the reference case without shock-wave/boundary-layer interaction even more so. The impinging shock locally stabilizes the boundary layer explaining the smaller disturbance amplitudes in this case. This stabilizing effect of the shock wave is also predicted by linear stability theory.

## Acknowledgements

We greatly acknowledge the former contributions of W. Eissler and H. Bestek who prepared the ground for this work. We thank the DFG for the long-time support within SFB259-C4/C11, and HLRS (High Performance Computing Centre Stuttgart) for competently providing the necessary computer resources within the project package LAMTUR, as well as INTAS who sponsored scientific communication with partners from Russia and Europe within INTAS-2000-0007.

## References

1. Eissler, W., Numerische Untersuchungen zum laminar-turbulenten Strömungsumschlag in Überschallgrenzschichten, Dissertation, Stuttgart University, 1995.
2. Kloker, M. J., A Robust High-Resolution Split-Type Compact FD Scheme for Spatial Direct Numerical Simulation of Boundary-Layer Transition, Applied Scientific Research, Vol. 59, pp. 353–377, 1998.
3. Anderson Jr., J. D., Hypersonic and High Temperature Gas Dynamics, McGraw-Hill, 1989.
4. Kloker, M., Konzelmann, U., Fasel, H., Outflow boundary conditions for spatial Navier-Stokes simulations of transition boundary layers. AIAA Journal 31, pp. 620–628, 1993.
5. Pagella, A., Rist, U., Wagner, S., Numerical investigations of small-amplitude disturbances in a boundary layer with impinging shock wave at $Ma=4.8$, Phys. Fluids, Vol. 14 (7), pp. 2088–2101, 2002.
6. White, F. M., Viscous Fluid Flows, $2^{nd}$ ed. Series in Mechanical Engineering. McGraw-Hill, New York, 1991.

7. Anderson Jr., J.D., Modern Compressible Flow, McGraw-Hill, New York, 1990.
8. Thompson, K.W., Time dependent boundary conditions for hyperbolic systems, J. Comput. Phys., Vol. 68, pp. 1–24, 1987.
9. Maslov, A.A., Shiplyuk, A.N., Sidorenko, A.N., Study of hypersonic boundary-layer instability on a cone using artificial disturbances. Proc. Internat. Conf. on the Methods of Aerophysical Research (ICMAR), Novosibirsk-Tomsk, July 2000, Publishing House SB-RAS, ISBN 5-7692-0303-6, pp. 132–137, 2000.
10. Fezer, A., Kloker, M., DNS of point-source induced transition in a hypersonic cone boundary layer. Submitted for publication in J. Fluid Mech. (2005).
11. Fezer, A., Kloker, M., Spatial direct numerical simulation of transition phenomena in supersonic flat-plate boundary layers. In Laminar-Turbulent Transition. Proc. IUTAM Symp., Sedona, Az, USA, 1999 (eds. Fasel, H., Saric, W.), Springer, 2000, pp. 415–420.
12. Lachowicz, J.T., Chokani, N., Wilkinson, S.P., Boundary-layer stability measurements in a hypersonic quiet tunnel. AIAA Journal, 34, 1996.
13. Fezer, A., Kloker, M., Transition processes in Mach 6.8 boundary layers at varying temperature conditions investigated by spatial DNS. In New Results in Numerical and Experimental Fluid Dynamics II (eds. Nitsche, W., Heinemann, H.-J., Hilbig, R.), Proc. $11^{th}$ AG STAB/DGLR Symposium, Berlin, 1998. Notes on Numerical Fluid Mechanics 72, 138–145, 1999, Vieweg.
14. Fezer, A., Kloker, M., DNS of transition mechanisms at Mach 6.8 – flat plate vs. sharp cone. Proc. West East High Speed Flow Fields 2002 (eds. Zeitoun, D.E., Periaux, J., Desideri, A., Marini, M.), Marseille, © Cimne, Barcelona, Spain, 2003, pp. 434–441.
15. Délery, J., Marvin, J.G., Shock-Wave Boundary Layer Interactions, AGARDograph, 280, 1986.
16. Anderson Jr., J.D., Computational Fluid Dynamics, McGraw-Hill, 1995.
17. Lele, S.K., Compact Finite Difference Schemes with Spectral-like Resolution, J. Comp. Phys., Vol. 103, pp. 16–42, 1992.
18. Adams, N., Direct simulation of the turbulent boundary layer along a compression ramp at $M=3$ and $Re_{\Theta}=1685$, J. Fluid Mech., Vol. 420, pp. 47–83, 2000.
19. Pagella, A., Babucke, A., Rist, U., Two-dimensional numerical investigations of small-amplitude disturbances in a boundary-layer at $Ma=4.8$: compression corner versus impinging shock wave. Phys. Fluids 16 (7), 2272–2281 (2004).
20. Knight, D., Yan, H., Panaras, A.G., Zheltovodov, A., Advances in CFD prediction of shock wave turbulent boundary layer interactions, Progress in Aerospace Sciences, Vol. 39, pp. 121–184, 2003.
21. Mack, L.M., Boundary layer stability theory, Jet Propulsion Laboratory, Report No. 900-277, 1969.

## 4.5 Numerical Simulation of High-Enthalpy Nonequilibrium Air Flows

Farid Infed, Markus Fertig*, Ferdinand Olawsky, Panagiotis Adamidis, Monika Auweter-Kurtz, Michael Resch, and Ernst W. Messerschmid

Flight through the Earth's atmosphere imposes an enormous stress on a space vehicle. This is true in both directions: during ascent after launch, and during re-entry at the end of the mission.

The speeds associated with spaceflight are very high: at least 7.9 km/s is necessary to reach low Earth orbit. Space vehicles suffer considerable buffeting and heating. The heat that the space vehicle creates is mostly due to the pressure wave that the vehicle creates in front of it as it moves at extreme speeds within atmosphere. It is commonly thought that this heat is due to friction but the heat caused from friction is really not what causes the extreme temperatures on the vehicle. The Ideal Gas Law for example shows where the extreme temperatures come from. As the pressure increases, the temperature must increase also to balance the equation. The volume does change, but not enough to compensate for the dramatic increase in pressure. The temperatures become so extreme that they facilitate in creating plasma in front of the vehicle as it is flying through the atmosphere. The resulting thermo-chemical processes are dominating the composition and the behaviour of the flow. Beside high-temperature effects, rarefaction effects take place due to the low density of the atmosphere at the height where re-entry effectively occurs.

### 4.5.1 Aerothermodynamic Aspects of Re-Entry Flows

In high-enthalpy gases, numerous phenomena arise and interact among each other: excitation of the internal degrees of freedom, dissociation, ionization and radiation of the gas species.

The thermal processes occur because the energy is conserved in different degrees of freedom of the gas as well as in different reactive processes, which significantly influence the chemical composition of the gas. The vibrational excitation of the air molecules starts at a flow speed of 1.0 km/s. When the flow reaches a speed of 2.5 km/s the vibrational excitation is at a maximum so that the dissociation of oxygen begins. The dissociation of oxygen is terminated at a speed of 5 km/s followed by the initial dissociation of nitrogen. Finally, at flow velocities higher than 10 km/s the nitrogen dissociation ends followed by ionization.

* Universität Stuttgart, Institut für Raumfahrtsysteme, Pfaffenwaldring 31, 70550 Stuttgart; fertig@irs.uni-stuttgart.de

All thermal and chemical processes are invoked by collisions and radiative exchange processes in the gas and seek to relax to an equilibrium state. The process takes place during a finite amount of time, the so-called relaxation time, which depends on the collision frequency. The collision frequency is considered to be low for low densities and high temperatures. Both processes, thermal and chemical, may have separate relaxation times. If thermal and chemical relaxation time are smaller by many magnitudes of order then the time, in which the convective processes occur, the flow is an equilibrium state. Otherwise the flow is in a state of nonequilibrium.

Using a unit less parameter, the Knudsen Number *Kn*, the re-entry of a space vehicle can be divided into regimes, where only the gas kinetics apply, and into another one, where continuum gas dynamics is valid as well. The Knudsen number denotes the relationship between mean free path length and the characteristic length of the re-entry vehicle. Along a re-entry trajectory of a space vehicle the following regimes can be observed:

- free molecular regime $10 < Kn$
- transition regime $10^{-1} < Kn < 10$
- slip stream regime $10^{-2} < Kn < 10^{-1}$
- continuum regime $Kn < 10^{-2}$

Since a re-entry trajectory traverses areas with low and high air densities, the simulation of the resulting nonequilibrium flows has to consider kinetical gas dynamics and continuum flow behaviour. The resulting governing equations, the Navier-Stokes equations, which are accurate for $Kn < 0.1$, include both.

#### 4.5.1.1 Inviscid Fluxes

Outside of the boundary layer, where flow speed is high while gradients in temperature, gas composition as well as velocity are rather low, air flow is dominated by inviscid terms. Simulation of supersonic flows requires sophisticated methods to describe the inviscid fluxes with 2$^{nd}$ order accuracy [16]. As usual, discretization of the inviscid fluxes of the governing equations is performed in the physical space by a Godunov-type upwind scheme, whereas the viscous fluxes are discretized in the transformed computational space by central differences using formulas of second order accuracy. For simulation of nonequilibrium re-entry flow Gas Kinetic Flux Vector Splitting (KVFS) [2], an approximate Riemann solver following Roe [38] and AUSMDV [32] have been implemented.

#### 4.5.1.2 Thermal Relaxation

Rotational excitation of the molecules usually starts at cryogenic temperatures. Yet below room temperature rotational degrees of freedom are fully excited. Perfect gas relations account for rotational excitation by different specific heat capacities for atoms and molecules. This implies that rotational excitation is always in equilibrium with translational excitation, i.e. only one temperature is sufficient for the determination of the partition function. The exchange between translation and rotation was theoretically investigated in 1959 by Parker [35]. He found that relaxation of rotational excitation with regard to translational energy requires about 5 intermolecular collisions. Lordi and Mates found by revision of Parker's model [26] that rotational relaxation of air species requires about 23 intermolecular collisions if nonequilibrium is strong. Within the slip flow regime where air density is low, equilibrium assumption for translational energy with rotational excitation is invalid in the post shock region in front of a re-entry vehicle [17, 21]. Other investigations imply that rotational relaxation could be decelerated close to equilibrium [31] and that rotational relaxation may require more collisions than vibrational relaxation behind strong shock waves [34].

Usually, about 1000 collisions are required in order to relax vibrational degrees of freedom. Accounting for vibrational nonequilibrium is therefore mandatory at high altitudes [24, 25]. Empirical [30] as well as theoretical [28] models exist for the determination of characteristic relaxation times for vibrational relaxation. Up to now, differences between these models are rather big [23]. Since the spacing between vibrational energy levels differs, vibrational excitation has to be considered for every molecular species separately. Hence, an additional exchange mechanism arises due to the vibrational exchange between different molecular species [1, 24]. At low translational temperatures, which do occur in the vicinity of cold surfaces and in wakes [21], a significant decrease of the V-V transition probability with decreasing translational temperature arises [36]. Up to now, only a little number of experiments is available for validation of the exchange models [7].

#### 4.5.1.3 Electronic Excitation

Modelling of electronic excitation requires much computational work, since electronic energy levels are not Boltzmann distributed in general [33]. For significantly ionized flows the influence of the electronic excitation on the ion-chemistry of atoms can be modelled via the Quasi-Steady-State (QSS) theory of Park [33]. The theory uses the fact that the change of the population density of an energy level occurs much slower than the involved collision processes. The Quasi-Steady-State theory can be implemented consistently into a multi-temperature Navier-Stokes code which assumes Boltzmann distributions for the other energies [17]. The influence of the electronic excitation on the post shock ionization of atomic nitrogen and oxygen which is modelled by nonequilibrium factors has been investigated for the FIRE II experiment [7].

#### 4.5.1.4 Thermochemical Relaxation

Typically, 100 000 collisions are required for chemical relaxation. Usually, only very little collisions exchange enough energy to dissociate a molecule in rotational and vibrational ground state. Excited internal degrees of freedom lower dissociation energy considerably such that reaction probability of excited molecules is significantly higher. Hence, a strong coupling exists between thermal relaxation and chemistry. This coupling can be described by so-called state selective models [37] where molecules with different internal energy states are considered as different species. Disadvantage of this approach is the large number of species continuity equations associated with much more computational effort. Up to now, application of state selective models is limited to one dimensional or simple inviscid two-dimensional flow problems. If internal energy is Boltzmann distributed multiple-temperature models can be derived which can be used for the simulation of complex three-dimensional flows as well. So-called multi-temperature models were proposed by several authors [21, 25, 27, 33]. The models of Knab [22, 25] and Kanne [21] were derived from gas kinetics most rigorously.

Starting from state-selective rates, multiple temperature rates depending on translational temperature as well as on vibrational and rotational temperatures are obtained by averaging over the distribution function of internal energies [19]. The model can be used for all types of reactions which are important in high-enthalpy flows like dissociation, exchange and associative ionization reactions. Consistent with the modelling of the production rates, the production rates of vibrational and rotational energies due to chemical reactions are computed. Using quasi-classical trajectory calculations, the parameters of the CVCV-model have been calibrated for dissociation and exchange reactions [20]. With these improvements nonequilibrium flows are dissociated higher because the dissociation energy is lowered by the rotational excitation. The kinetics of the Zeldovich reactions is now in accordance with quasiclassical trajectory theory. The improved NO-chemistry modelling in the URANUS code was validated indirectly by means of a spectrometer simulation performed with the PARADE radiation code [18] and the HERTA radiation transport code [9] for the Bow Shock Ultraviolet Experiment II, where spectra in the wavelength range of NO have been measured, see [17]. With the exception of large altitudes where weak radiation signals do occur, good agreement was obtained between measured and simulated spectra along the trajectory.

#### 4.5.1.5 Transport Coefficients

In dissociated and ionized re-entry flows strong gradients can be observed in densities, temperatures and velocities. Multiple models exist in order to describe the exchange of mass, momentum and energy under these conditions [4, 12, 13]. In addition to the widely used transport coefficient model of Yos [12] Chapman-Cowling's first approximations for the transport coefficients translational thermal conductivity of heavy particles, viscosity and mass diffusion were implemented as well as the second approximation of thermal conductivity of electrons [4]. Results of models have been compared with the simplified model of Gupta and Yos for the Stardust capsule flow [4]. It was found that the latter model gives accurate results for dissociated flows but not in partly ionized flows. Nevertheless, simplified models can not be recommended for flows in the slip flow regime, since viscous effects have a significant influence on post shock relaxation.

#### 4.5.1.6 Turbulence

Analysis of the free stream in plasma wind tunnel flows showed that turbulence is an important issue. Algebraic models have been implemented due to robustness and simplicity [11]. After testing several models, it was found that the model of Ferri, Libby and Zakkay [29] is the most appropriate for the simulation of turbulent free streams in plasma wind tunnels.

#### 4.5.1.7 Electrical Discharge

In order to take electrical discharge effects into account the Maxwell equation and Ohm's Law have to be considered for the physical modelling. With the solution of these equations the magnetic induction and the current density are given [11]. These terms are necessary for a consistent coupling with the flow source terms of the Navier-Stokes equations.

#### 4.5.1.8 Gas-Surface Interaction Modelling

The finite-rate catalytic behaviour of a TPS-surface influences the surface heat flux significantly. Gaskinetic gas-surface interaction models are implemented into the URANUS code in order to take into account near surface rarefaction effects in leeward side and base flows and generally in re-entry flow at large altitudes. Detailed catalysis model for $SiO_2$- and SiC-surfaces have been implemented by Daiß [3] and Fertig [5] into the URANUS code which distinguish adsorption, desorption and recombination reactions according to the Eley-Rideal and Langmuir-Hinshelwood mechanisms as well as dissociative adsorption reactions. Furthermore, an expression for the estimation of the chemical energy accommodation coefficient of chemical reactions has been derived. These detailed models which consist of elementary reactions are able to describe the non-Arrhenius behaviour of the recombination coefficients and have been successfully validated for the Shuttle re-entry [3] and the MIRKA re-entry (see section 7.2) from the slip flow range close to the perfect gas range.

Future reusable spacecrafts like X-38 will be smaller in size than the US Shuttle Orbiter. Therefore, the maximum thermal loads at the surface will be significantly higher. Hence, higher surface temperatures arise which require improved heat shield materials as compared to the US Shuttle Orbiter. Currently, SiC-based materials suit best the need for advanced materials. However, at temperatures of about 2000 K oxidation reactions between air and SiC may arise, which lead to high material loss rate. The oxidation behaviour was modelled by Fertig [5], cf. section 7.2.

The finite-rate catalytic behaviour can also be predicted by a purely phenomenological or simplified global model, which tries to match the available experimental data obtained in high-enthalpy facilities in a large surface temperature range by a surface temperature-dependent expression for the overall recombination coefficients of the atomic species. In this simple single step model the chemical energy accommodation coefficient is assumed to equal one. It has been shown in the case of the Shuttle windward side flow that detailed and global catalysis models nearly predict the same surface heat flux. Since $SiO_2$-coated surfaces generally behave weakly catalytic in a large surface temperature range and there is only a small leeward side surface temperature variation catalysis can even be modelled with sufficient accuracy by constant overall recombination coefficients.

The catalytic behaviour of C/SiC and SiC-surfaces changes from weakly to strongly catalytic in a large surface temperature range, see [15]. Furthermore, in high surface temperature areas such as nose and body-flap regions the surface temperature may change significantly and hence also the catalytic behaviour. Therefore, the non-Arrhenius behaviour of recombination coefficients and overall recombination coefficients has to be taken into account by detailed and global catalysis models, respectively.

#### 4.5.1.9 Radiative Exchange at the Surface

Every surface, which has a temperature that exceeds 0 K, emits a radiative flux in all directions. Thus, if a surface element is able to see another one, it receives a part of the flux emitted by the other one.

Every body emits an energy which depends on its temperature. Hence, a surface $S_1$ emits energy to surface $S_2$. The radiation emitted by $S_1$ in the direction of $S_2$ leads to an increase of surface temperature $T_2$. The same phenomenon applies to $S_2$ such that the temperature $T_1$ of $S_1$ increases as well. As a consequence, modification of $T_1$ changes the flux emitted from $S_1$. To evaluate the radiative flux $q_{i,rad}$ received by a surface $S_i$ from all other surfaces $j$, one may write

$$q_{i,rad} = \sum_j F_{i,j} q_{j,rad} \tag{1}$$

where $F_{i,j}$ is the view factor to see the surface $S_j$ from the surface $i$ and $q_{j,rad}$ is the radiative flux emitted by the surface $S_j$. According to Stefan-Boltzmann's law of radiation it follows

$$q_{j,rad} = \varepsilon_j \cdot \sigma \cdot T_j^4 \tag{2}$$

where $\sigma$ is the Stefan-Boltzmann constant and $\varepsilon_j$ is the emission coefficient of surface $j$. The emission coefficients for the widely used TPS materials (C/SiC and $SiO_2$) are temperature dependent and so they are considered in the model. For complex and non-convex geometries the radiative exchange could influence significantly catalytic reactions and thus the surface temperature distribution [15].

#### 4.5.1.10 Heat Conduction within TPS Materials

TPS materials have a heat conduction behaviour which influences the temperature distribution at the surface, on one hand, and thus the gas state, on the other hand, significantly. One approach to take this phenomenon into account is to take the computed heat load distribution assuming radiation equilibrium at the surface from the flow solver as initial input for an external heat conduction solver for the TPS, which calculates then a temperature distribution at the surface [6]. This distribution will be then put back into the flow to recalculate a new gas state and thus a new heat load distribution at the surface. The procedure will be repeated until convergence is achieved for the coupling of both solvers. This method is very time-consuming and not practical for large 3D simulations.

So, a finite element-based model, which gives a new temperature distribution at the surface taking thermal conduction within the TPS material into account has been developed for the URANUS code [15]. We modify the governing heat conduction equation by introducing the concept of shape functions to have a relation between the TPS volume data and the data at the nodes which define the TPS control volume element. The advantage of this approach is that any kind of grid and element constellation of the grid can be treated. Once one has derived theory of the heat conduction model the obtained equations are valid for any element constellation. Hence, knowledge about the elements shape function and the grid points which define the element is sufficient. URA-

NUS generates automatically a TPS grid. Starting from the flow mesh, the surface geometry of re-entry vehicle is given. Using few input parameters such as element type, cell number and TPS material's width, the mesh will be calculated in the pre-processing phase. Both flow mesh and TPS grid share the same nodes at the surface.

To ensure accurate modelling, experimentally gained and temperature dependent emissivity, conduction coefficients and heat capacity coefficients for real TPS materials (C/SiC and $SiO_2$) are used.

## 4.5.2 Numerics and Parallelization

Within the framework of the SFB 259 two nonequilibrium Navier-Stokes codes namely SINA (Sequential Iterative Nonequilibrium Algorithm) and URANUS (Upwind Relaxation Algorithm for Nonequilibrium Flows of the University of Stuttgart) have been developed. SINA has been developed in order to simulate the plasma generation in a plasma wind tunnel as well as the flow within the plasma wind tunnel including flow around a probe. Since details about the SINA code (Sequential Iterative Nonequilibrium Algorithm) can be found elsewhere [11], numerics and parallelization strategies will be described for the URANUS code. URANUS is a fully implicit Navier-Stokes code which has been developed in order to simulate the loads onto re-entry vehicles during descent accurately.

### 4.5.2.1 Conservation Equations

In the URANUS nonequilibrium Navier-Stokes code the governing equations are solved in finite volume formulation fully coupled [8]. The Navier-Stokes equations

$$\frac{\partial \vec{Q}}{\partial t} + \frac{\partial(\vec{E} - \vec{E}_v)}{\partial x} + \frac{\partial(\vec{F} - \vec{F}_v)}{\partial y} + \frac{\partial(\vec{G} - \vec{G}_v)}{\partial z} = \vec{S} \tag{3}$$

consist of ten species continuity equations, three momentum equations, the total energy equation, three vibrational energy equations and one rotational energy equation for the molecular species as well as one equation for the electron energy, where

$\vec{Q}$ is the conservation vector:

$$\vec{Q} = \left[\rho_i, \rho u, \rho v, \rho w, \rho e_{tot}, \rho_k e_{vib,k}, \sum_{l=1}^{6} \rho_l e_{rot,l}, \rho_e e_e\right]^T, \ [i = 1, 10, k = 1, 3] \tag{4}$$

$\vec{E}, \vec{F}, \vec{G}$ are the inviscid fluxes in x-, y- and z-direction,
$\vec{E}_V, \vec{F}_V, \vec{G}_V$ the viscous fluxes and
$\vec{S}$ the source terms vector for chemical reactions and energy exchange [16, 21].

4.5.2.2 Solver

To calculate the steady state solution of the Finite Volume Navier-Stokes equations

$$V\frac{\partial Q}{\partial t} = R(Q) \tag{5}$$

with the conservation variables Q, the volumes $V$ and $R(Q)$ the sum of the inviscid and viscous fluxes and the source terms, the implicit Euler time differencing with the usual Taylor series linearization for $R$ is applied. The resulting linear system

$$\left(\frac{V}{\Delta t} - \frac{\partial R(Q)}{\partial Q}\right)\Delta Q = R(Q^n) \tag{6}$$

with

$$\Delta Q = Q^{n+1} - Q^n \tag{7}$$

has to be solved for each time step. For

$$\Delta t \to \infty \tag{8}$$

the scheme is exactly Newton's method. It is not necessary for Newton's method to compute the Jacobian exactly. Approximations can be made to reduce memory requirement. For this reason, the Jacobian is only computed with first order inviscid fluxes and a thin shear layer approximation for the viscous fluxes [32]. So, the resulting matrix consists of seven block diagonals with $19\times19$-blocks. Further reduction of the memory requirement can be achieved by storing the Jacobian using single precision, while computing and storing the fluxes and source terms using double precision. Hence, the memory requirement of the Jacobian is halved.

The linear system is solved with the Jacobi-Line-Relaxation-Method (JLR). After preconditioning with the inverse of the main blockdiagonal, four blockdiagonals of the matrix corresponding to two grid directions are transferred to the right hand side of the linear system. The resulting tridiagonal system is solved by vectorized LU-decomposition. Thereafter, the step will be repeated with the other grid directions. This procedure is iterated until convergence is achieved. The linear systems are solved on each block separately. After this solving step the resulting data at the subdomain boundaries are exchanged and another solving step follows.

To accelerate the code, the Krylov subspace methods GMRES, BiCGstab, CGS, TFQMR and QMRCGstab with a vectorizable ILU-preconditioner have been implemented [32].

#### 4.5.2.3 Multiblock

The newly designed Parallel Multiblock version of 3D URANUS code is able to deal with nearly any kind of structured multiblock meshes, which enables to deal with complex re-entry vehicles' geometry. In a multiblock mesh, beside the surface boundary, physical boundaries (inflow, outflow, symmetry) may occur at each of the six block sides.

Furthermore, each block can have neighbouring blocks at each block side. When using GridPro meshes, there is exactly one block connected to one block side as a neighbour, but using other multiblock meshes there may be more than one neighbour at one block side, which is the case for the mesh used for the X-38 simulations presented here. The URANUS code is able to handle data exchange with several neighbours at one block side. Moreover, it is able to handle any combination of neighbourhood, e.g. a block can have at two of its sides a boundary to one and the same side of a neighbour block. Such layouts exist in current multiblock meshes.

The data exchange in P-MB URANUS is performed by MPI. Hence, the portability of the code on widely used supercomputers like NEC SX-5, CRAY T3E or Hitachi SR8000 is guaranteed [8].

In order to maintain second order in discretization and extrapolation, the data exchange occurs at the block boundary over an intersection zone (domain overlapping) of two cells.

Different blocks may have different local coordinate directions. This has to be considered when exchanging data between two block neighbours. The necessary data conversion is done during the communication and is hidden from the application.

To obtain the full performance of the system used, the load balancing tool JOSTLE was implemented into the P-MB URANUS code. Applying load balancing, several cases are possible when using multiblock meshes for flow calculations: Large blocks may exist. Calculating them using one processor will lead to an immense loss of performance if all the other processes are ready and have to wait for one or two large processes to finish. These blocks are cut into several pieces, depending on their size. The resulting "new" blocks are then handled as separate blocks. Furthermore, there may be also blocks which are very small in size, i.e. they include only a small number of control volumes and therefore they do not fully utilize a CPU. Therefore, the code has been extended to be able to compute several blocks together on one processor [14].

The attached load balancing tool is able to find an optimal distribution of the blocks to the available processors in most cases.

#### 4.5.2.4 Metacomputing

Metacomputing stands for using virtually networked high performance computers, which use ordinary networks with low data bandwidth to exchange data, to perform large simulations. The URANUS code has been successfully used for metacomputing test applications. The URANUS code has received the High Performance Computing Award 1999. The award honours the code for the simulation of the re-entry of HERMES on three Cray T3Es located in Manchester, Pittsburgh and Stuttgart involving 1536 processors.

#### 4.5.2.5 Adaptive Grids

The idea behind the parallel multilevel multiblock method is to coarsen the mesh of each block and to start the computations on the blocks with coarse meshes. After some iteration on the coarse grid the algorithm switches to a finer level and the computations are carried out there. This means that the system of equations is solved in the first phase on the coarse meshes of each block in parallel, and then after an interpolation step the solutions of each block are mapped onto a finer level until the last level, which is the original fine grid, is reached. The problem under consideration is three-dimensional. The coarsening procedure is flexible regarding the direction in which the grid is going to be coarsened. This means, that it is possible to coarsen the grid in all three directions independent of each other. It is also possible to choose how coarse each block should be for each level. So, the coarsening is done by specifying two parameters. The first determines, how many levels are desired, and the second determines how coarse the mesh of each block and in each level should be. An interesting matter, from the parallelization point of view, is when to do the domain decomposition. In this algorithm the initial mesh of each block is coarsened first, and the domain decomposition is done on the lowest level. After the solution step has finished on the coarsest level, the conservative variables have to be mapped onto the fine level. This is done by a three-dimensional volume interpolation, which consists of three steps. In the first step the conservation vectors of both grids are assigned to coordinates. This is done by calculating the cell centres of each cell of the coarse and of the fine grid. In this way, the dual grids are constructed. In the second step, the algorithm goes through each cell of the dual grid of the coarse mesh and searches for the nodes of the fine dual mesh, which reside in the interior of the coarse cell. Once these have been found, the local coordinates of the fine mesh into the coarse mesh are computed. For each of the eight nodes of the coarse cell a shape function is defined in local coordinates. In the third step, the values of the conservative variables, which reside on the vertices of the coarse cell, are being interpolated to the nodes of the finer mesh using the shape functions. In this way, the value of the conservative variable of each node of the fine mesh is a combination of the values on the vertices of the coarse cell.

### 4.5.3 Results

Two representative solutions for URANUS and SINA are presented.

#### 4.5.3.1 Simulation of the Re-Entry of the X-38

The re-entry of the X-38 demonstrator has been simulated for three different trajectory points [14]. First point is at an altitude of 74.6 km. The simulation is performed for the upstream conditions $Ma=25$. The second trajectory point, which is shown in Fig. 4.5.1 is at 64.6 km of altitude with $Ma=19.8$. Both trajectory points are part of NASA cycle 8. The third point is computed in accordance with the X-38 re-entry trajectory simulations for NASA cycle 9 worked out within TETRA-cooperation at the IRS [10]. The trajectory point is at an altitude of 55.8 km with $Ma=16.4$. At all three trajectory points the X-38 has an angle of attack $\alpha=40^\circ$ and a body flap deflection of $\delta=20^\circ$. The surface temperature distribution is computed for the real TPS, which is based on SiC- and $SiO_2$ technology. Nose cap, nose skirt, chin panel and the body flap are made of C-SiC and so they are considered for the simulations. The remaining part of the vehicle's surface is equipped with $SiO_2$ tiles, which is also taken into account for the simulations discussed here. The emissivity of all TPS tiles was assumed to be 0.87 and constant over all temperature ranges. Neither radiative exchange among the surfaces components nor heat transfer within the TPS has been taken into account for the computation of the surface temperature distributions presented in [14]. Non- and full catalytic design assumptions have been also simulated for all trajectory points.

For the here presented simulations of the X-38 a DLR multiblock mesh, which contains wake flow regions, is used. This mesh has 1.02 million cells and consists of 18 initial blocks. The near surface resolution is within the scope of mean free path length ($10^{-6}$ m) and thus sufficient for the prediction of the surface heat flux. Unfortunately, the shock wave and the wake flow region are poorly resolved. The sizes of the blocks differ strongly, the smallest blocks have 9200 cells and the largest block contains 260700 cells. Most of the small sized mesh blocks were necessary to describe the geometry of the body flap deflection of the X-38 vehicle. The simulations were performed on the NEC SX-5 at HLRS using 6 CPUs requiring 17.8 GB of memory. The large discrepancy in the number of cells of each block caused at the very beginning of the simulations a

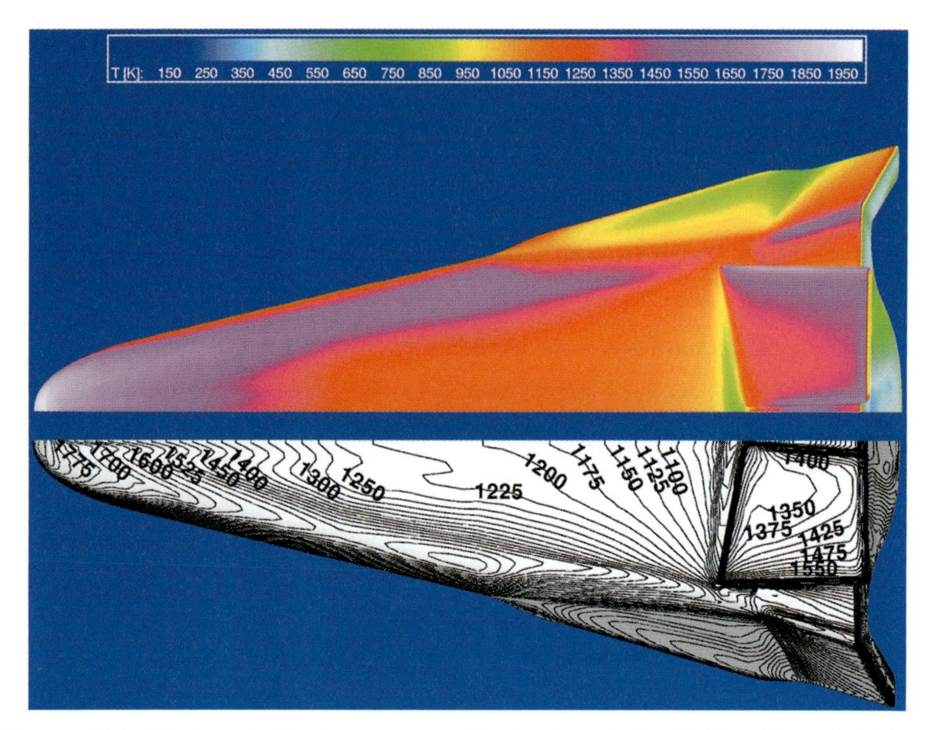


Figure 4.5.1: Windward-side temperature distribution of the X-38 with real TPS at 64.4 km of altitude.

large unbalance of the loads among the processors even though using the advanced balancing tool JOSTLE. This did not agree with our previous experiences with this tool, which input options have been embedded into the URANUS code. After several modifications in the way that the input parameter of the balancer became part of the general input parameters of the code, the previous unbalance of 38% could be reduced to a range of 8% to 10%. It was remarkable, that even after finding a good distribution of the load on the processors for one run of the simulations, the continuation runs could not maintain this distribution for the same case. This problem has been solved by increasing the coarsing/reduction threshold – the level at which graph coarsing ceases. Two main effects were observed: The first is that it speeds up the partitioning and the second is that since coarsing gives a more global perspective to the partitioning, it should reduce the amount of data that needs to be migrated at each repartition. It was necessary to move away from the default value 20 to higher values up to 200.

Finally, the simulations ran with a total performance of 8.4 GFLOPS on 6 NEC-SX5 processors by an average vector operation ratio of 95.4%. CFL numbers up to 100 have been reached for the computations with $2^{nd}$ order flux discretization of the complex X-38 geometry; higher CFL numbers were possible for $1^{st}$ order discretization. On average, each iteration took about 45 seconds,

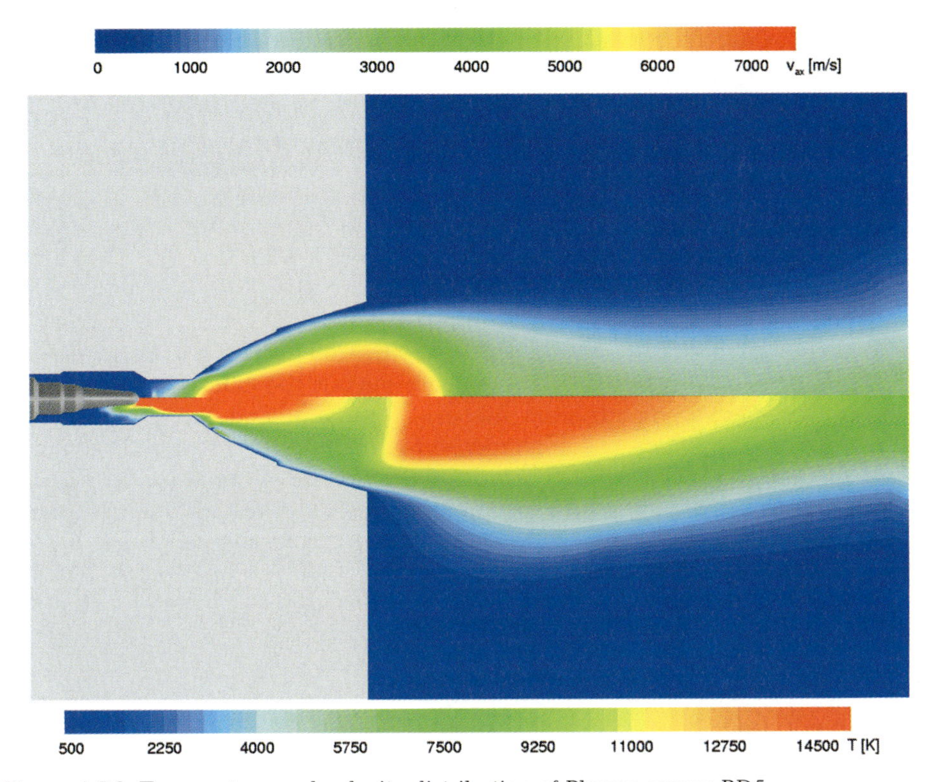


Figure 4.5.2: Temperature and velocity distribution of Plasma source RD5.

where up to 18% of the computational time was spent on communication among the processors. For all trajectory points the residual dropped six orders of magnitude which is very satisfying for such kind of complex geometry. A better convergence could have been obtained if the mesh, especially in the wake flow region, was of better quality. However, 600 iterations were necessary to reduce the residual by one order of magnitude in Newton phase of convergence. We have to mention that it took lots of iterations to place the shock wave such that the Newton phase of convergence could begin. After all, on average each simulation of the three trajectory points with different modeling of the surface (non-, full and real TPS surface) took about 21 to 25 hours of computation. Hence, computations in about one day are possible.

Figure 4.5.1 shows the temperature distribution computed for 64.4 km of altitude of the X-38 with real TPS. As expected, high thermal loads are observed at the body flap region, where a thin boundary layer, as a result of angular geometry, causes temperatures of up to 95.5% of the stagnation point temperature.

At the conjunction zone between body flap and the fore body on the windward side a recirculation bubble originates and strongly influences shape and size of the local boundary layer, which is mainly responsible for the thermal load distribution. While the bubble size is small for the lowest trajectory point, at the highest trajectory point a larger bubble arises. Hence, energy transfer of the gas flow is hindered. A reduction of the local temperature at the mentioned conjunction zone is the result.

#### 4.5.3.2 Simulation of the Plasma Source RD5

Within the plasma source RD5 only nitrogen is injected along the cathode in the combustion chamber to create a plasma stream. The oxygen is injected in the divergent part of the nozzle to avoid erosions at the cathode. This mechanism requires high demands from the grid generator and the solver. Figure 4.5.2 shows the simulation of the temperature and velocity distribution of an air plasma stream generated by RD5 using the SINA code. To perform this simulation, a cold stream simulation for air mixture consisting of a nitrogen stream of 1.6 g/s and an oxygen stream 0.4 g/s has been complemented by a plasma stream computation for an electrical stream of $I = 1200$ A. Good agreement between experimental measurements and simulation results has been found [11].

## Acknowledgements

Thomas Bönisch, Andreas Daiß, Hans-Heiner Frühauf, Alfred Geiger, Uwe Gerlinger, Thomas Grau, Thomas Gogel, Stefan Jonas, Sebastian Kanne, Oliver Knab, Eberhard Schöll as well as all staff members of the Institute of Space Systems, the High Performance Computing Centre of Stuttgart and many students are gratefully acknowledged by the authors for their contributions.

## References

1. Candler, G.V., MacCormack, R.W.: The Computation of Hypersonic Ionized Flows in Chemical and Thermal Nonequilibrium, AIAA-Paper 88-0511, 1988.
2. Chou, S.Y., Baganoff, D.: Kinetic Flux-Vector Splitting for the Navier-Stokes Equations, Journal of Computational Physics 130, pp. 217–230, 1997.
3. Daiss, A., Frühauf, H.-H., Messerschmid, E.W.: Modelling of Catalytic Reactions on Silica Surfaces with Consideration of Slip Effects, AIAA Journal of Thermophysics and Heat Transfer, Vol. 11, No. 3, July–September 1997.
4. Fertig, M., Dohr, A., Frühauf, H.-H.: Transport Coefficients for High Temperature Nonequilibrium Air Flows, AIAA Journal of Thermophysics and Heat Transfer, Vol. 15, No. 2, 2001.
5. Fertig, M., Frühauf, H.-H., Auweter-Kurtz, M.: Modelling of Reactive Processes at SiC-Surfaces in Rarefied Nonequilibrium Airflows, AIAA-Paper 2002-3102, 8th AIAA Joint Thermophysics and Heat Transfer Conference, St. Louis, Missouri, USA, June 24–27, 2002.

6. Fertig, M., Frühauf, H.-H.: Strömungs-Strukturwechselwirkung bei der X-38 Klappe, IRS, Universität Stuttgart, IRS-02 P 08, Oktober 2002.
7. Frühauf, H.-H., Fertig, M., Kanne, S.: Validation of the Enhanced URANUS Nonequilibrium Navier-Stokes Code, AIAA Journal of Spacecraft and Rockets, Vol. 37, No. 2, March–April 2000.
8. Frühauf, H.-H., Fertig, M., Olawsky, F., Infed, F., Bönisch, T.: Upwind Relaxation Algorithm for Re-entry Nonequilibrium Flows, High Performance Computing in Science and Engineering 2000, Springer, Berlin, pp. 440–445, 2001.
9. Gogel, T. G., Numerische Modellierung von Hochenthalpieströmungen mit Strahlungsverlusten, Dissertation, Fakultät für Luft- und Raumfahrttechnik, Universität Stuttgart, 1994.
10. Gräßlin, M., Walz, S.: X-38 V201 Re-entry Trajectory Simulations for NASA Cycle 9, TET-IRS-22-TN-5305.
11. Grau, T. G.: Numerische Untersuchungen von Plasmawindkanalströmungen zur Wiedereintrittssimulation, Dissertation, Universität Stuttgart, Germany, 2000.
12. Gupta, R. N., Lee, K. P., Thomson, R. A., Yos, J. M.: A Review of Reaction Rates and Transport Properties for an 11-Species Air Model for Chemical and Thermal Nonequilibrium Calculations to 30000 K, TM 85820, NASA, 1990.
13. Hansen, C. F.: Approximations for the Thermodynamic and Transport Properties of High-Temperature Air, Technical Report R-50, NASA, 1959.
14. Infed, F., Auweter-Kurtz, M.: Simulation of Hypersonic Flows in Thermochemical Nonequilibrium Around the Re-entry Vehicle X-38 with URANUS Code, Journal of Space Technology, Vol. 24, 2, pp. 81–93, 2004.
15. Infed, F., Olawsky, F., Auweter-Kurtz, M.: Stationary Coupling of Hypersonic Nonequilibrium Flows and Thermal-Protection-System, Journal of Spacecraft and Rockets, Vol. 42, 1, January–February, 2005.
16. Jonas, S.: Implizites Godunov-Typ-Verfahren zur voll gekoppelten Berechnung reibungsfreier Hyperschallströmungen im thermo-chemischen Nichtgleichgewicht, Dissertation, Fakultät für Luft- und Raumfahrttechnik, Universität Stuttgart, VDI Fortschrittsbericht, Reihe 7, Nr. 221, 1993.
17. Kanne, S., Frühauf, H.-H., Messerschmid, E. W.: Thermochemical Relaxation Through Collisions and Radiation, AIAA Journal of Thermophysics and Heat Transfer, Vol. 14, No. 4, 2000.
18. Kanne, S., Gogel, T. G., Dupuis, M., Messerschmid, E. W., Roberts, T. P.: Simulation of Radiation Experiments on Re-entry Vehicles Using the New Radiation Database PARADE, AIAA Paper 97-2562, 1997.
19. Kanne, S., Knab, O., Frühauf, H.-H., Messerschmid, E. W.: The Influence of Rotational Excitation on Vibration-Chemistry-Vibration-Coupling, AIAA-Paper 96-1802, 1996.
20. Kanne, S., Knab, O., Frühauf, H.-H., Pogosbekyan, M., Losev, S. A.: Calibration of the CVCV-Model against Quasiclassical Trajectory Calculations, AIAA-Paper 97-2557, 1997.
21. Kanne, S.: Zur thermo-chemischen Relaxation innerer Freiheitsgrade durch Stoß- und Strahlungsprozesse beim Wiedereintritt, Dissertation, Fakultät für Luft- und Raumfahrttechnik, Universität Stuttgart, VDI Fortschrittsbericht, Reihe 12, Nr. 429, 2000.
22. Knab, O., Frühauf, H.-H., Messerschmid, E.: Theory and Validation of the Physically Consistent Coupled Vibration-Chemistry-Vibration Model, AIAA Journal of Thermophysics and Heat Transfer, Vol. 9, No. 2, April – June 1995.
23. Knab, O., Frühauf, H.-H., Messerschmid, E.: URANUS/CVCV-Code Validation by Means of Thermochemical Nonequilibrium Nozzle Airflow Calculations, Proceedings of the Second European Symposium on Aerothermodynamics for Space Vehicles and Fourth European High-Velocity Database Workshop, ESTEC, Noordwijk, The Netherlands, November 22–25, 1994.
24. Knab, O., Jonas, S., Frühauf, H.-H.: Fundamental Investigation on Post-Shock Thermodynamic Nonequilibrium Effects of Air Including Ionization, Orbital Transport,

Technical, Meteorological and Chemical Aspects, Third Aerospace Symposium, Braunschweig, Germany, August 1991, Springer, 1992, pp. 245–255.
25. Knab, O.: Konsistente Mehrtemperatur-Modellierung von thermochemischen Relaxationsprozessen in Hyperschallströmungen, Dissertation, Fakultät für Luft- und Raumfahrttechnik, Universität Stuttgart, VDI Fortschrittsbericht, Reihe 7, Nr. 300, 1996.
26. Lordi, J. A., Mates, R. E.: Rotational Relaxation in Nonpolar Diatomic Gases, The Physics of Fluids, Vol. 13, No. 2, 1970, pp. 291–308.
27. Marrone, P. V., Treanor, C. E.: Chemical Relaxation with Preferential Dissociation from Excited Vibrational Levels, The Physics of Fluids, Vol. 6, No. 9, pp. 1215–1221, 1963.
28. Mattes, P.: Berechnung von Translation-Vibration Relaxationszeiten in Luft mit Hilfe der SSH-Theorie, Studienarbeit, Universität Stuttgart, Institut für Raumfahrtsysteme, IRS 94-S-12, 1994.
29. Messerschmid, E. W., Fasoulas, S., Gogel, T. G., Grau, T.: Numerical Modeling of Plasma Wind Tunnel Flows, Journal of Flight Sciences and Space Research 19, pp. 158–165, Springer 1995.
30. Millikan, R., White, D.: Systematics of Vibrational Relaxation, Journal of Chemical Physics, Vol. 39, No. 12, 1963, pp. 3209–3213.
31. Moreau, S., Bourquin, P. Y., Chapman, D. R., MacCormack, R. W.: Numerical Simulation of Sharma's Shock-Tube Experiment, AIAA-Paper 93-0273, 1993.
32. Olawsky, F., Infed, F., Auweter-Kurtz, M.: Preconditioned Newton-Method for Computing Supersonic and Hypersonic Nonequilibrium Flows, AIAA-Paper 2003-3702, 16th AIAA Computational Fluid Dynamics Conference, Orlando, Florida, USA, June 23–26, 2003.
33. Park, C.: Nonequilibrium Hypersonic Aerothermodynamics, John Wiley & Sons, New York, 1989.
34. Park, C.: Rotational Relaxation of N2 Behind a Strong Shock Wave, AIAA-Paper 2002-3218, 2002.
35. Parker, J. G.: Rotational and Vibrational Relaxation in Diatomic Gases, The Physics of Fluids, Vol. 2, No. 4, July 1959, pp. 449–462.
36. Rapp, D.: Interchange of Vibrational Energy between Molecules in Collisions, Journal of Chemical Physics, Vol. 43, 1965, pp. 316–317.
37. Riedel, U.: Numerische Simulation reaktiver Hyperschallströmungen mit detaillierten Reaktionsmechanismen, Dissertation, Universität Heidelberg, Germany, 1992.
38. Roe, P. L.: Approximate Riemann Solvers, Parameter Vectors and Difference Scheme, Journal of Computational Physics, Vol. 43, pp. 357–372, 1981.

## 4.6 Flow Simulation and Problems in Ground Test Facilities

Uwe Gaisbauer, Helmut Knauss, Siegfried Wagner, Georg Herdrich *, Markus Fertig, Michael Winter, and Monika Auweter-Kurtz

### 4.6.1 Introduction

Experimental flow simulations necessary in both the design process and for the qualification of aerospace technologies comprise a large variety of flow types. This becomes plausible from the velocity altitude map for all sorts of flight trajectories (in the ascent and descent phase of the carrier and/or the orbiter) shown in Fig. 4.6.1. The range of flow velocity is covering almost 3 orders of magnitude. Combined with atmospheric change of state from continuum to highly rarefied conditions caused by gasdynamic, thermodynamic and real gas effects (chemical reactions and thermal nonequilibrium), various flow phenomena arise around the vehicle with increasing speed and flight altitude. Between take-off/landing and space flight, sub-, trans-, super- and hypersonic flow regimes are crossed (through). A bulk of similarity parameters are to be duplicated for a transferable simulation capturing all specific phenomena, typical for the individual flight conditions. For the totality of these parameters we can form two groups: group I comprises all parameters necessary to duplicate flow conditions without real gas effects and group II contains all additional ones essential to take account for all aero-thermodynamic phenomena of a high-enthalpy flow. We can assume, that group I covers all flight conditions along the trajectory till stage separation, i.e. are sufficient for flow simulations of the carrier, while group II joins for corresponding flow simulation along the orbiter's ascent and descent trajectories.

Group I includes Reynolds-, Mach- and Prandtl-number, duplicating viscous-, compressible-effects and properties of the test gas respectively. For a non-reacting gas it is a function of temperature only. If real gas effects apply, similarity parameters also depend on the local physicochemical gas state. For a duplication of the temperature boundary conditions the wall to free stream temperature ratio $T_w/T_\infty$ is a further similarity parameter which has to be kept. When a radiation adiabatic wall condition shall be simulated the Thring number must be equivalent in the wind tunnel simulation and original. The ratio of specific heats $\kappa$ of the test gas should also be duplicated. In addition to temperature, real gas effects make $\kappa$ a function of composition and excitation of internal degrees of freedom.

* Universität Stuttgart, Institut für Raumfahrtsysteme, Pfaffenwaldring 31, 70550 Stuttgart; herdrich@irs.uni-stuttgart.de

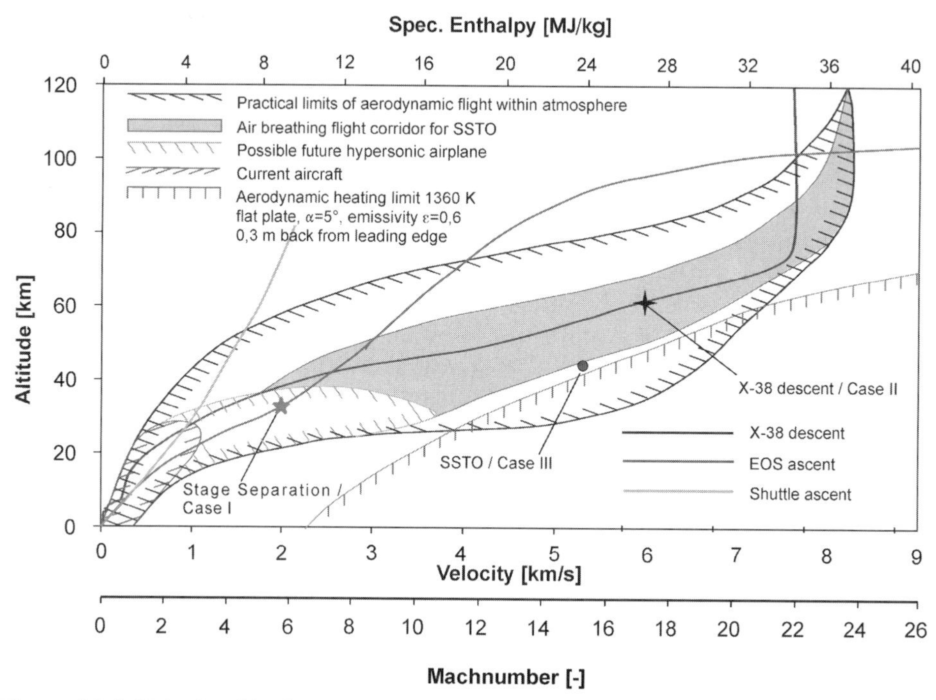


Figure 4.6.1: Velocity altitude map for typical trajectories.

Group II includes Damkoehler number a very crucial similarity parameter. It represents either the ratio of a chemical reaction time (relaxation time) to a characteristic flow time or the ratio of corresponding lengths belonging to these periods of time. Consequently, when the time factor will be added, there is no longer an obvious way to scale. Hence, only single aspects of the flow can be simulated with a scaled model. Plasma wind tunnels e.g. are aero-thermodynamic facilities where most often the stagnation point is simulated such that rather Group II of the similarity parameters is of concern. Some other parameters are the Knudson number describing the degree of rarefaction existing in the flow. The local Knudsen number close to the surface unambiguously defines the wall conditions (i.e. free molecular flow, slip and non-slip conditions). Schmidt number and Lewis number are further similarity parameters for a duplication of mass transfer by diffusion and combined heat and mass transfer in gas surface interaction, respectively. Due to the different scaling laws of the parameters no scaling is possible if all aspects of the flow should be simulated simultaneously.

The dimensionless similarity parameters, to be kept in the simulation, define the "operational conditions" of the test gas in the ground test facility. The generation of prevailing stagnation conditions in pressure and temperature extending over about 4 orders of magnitude results in large efforts and comes up against limiting factors, being partially beyond actual state of the art. Therefore, it is not surprising that practically any flow simulation, no matter whether

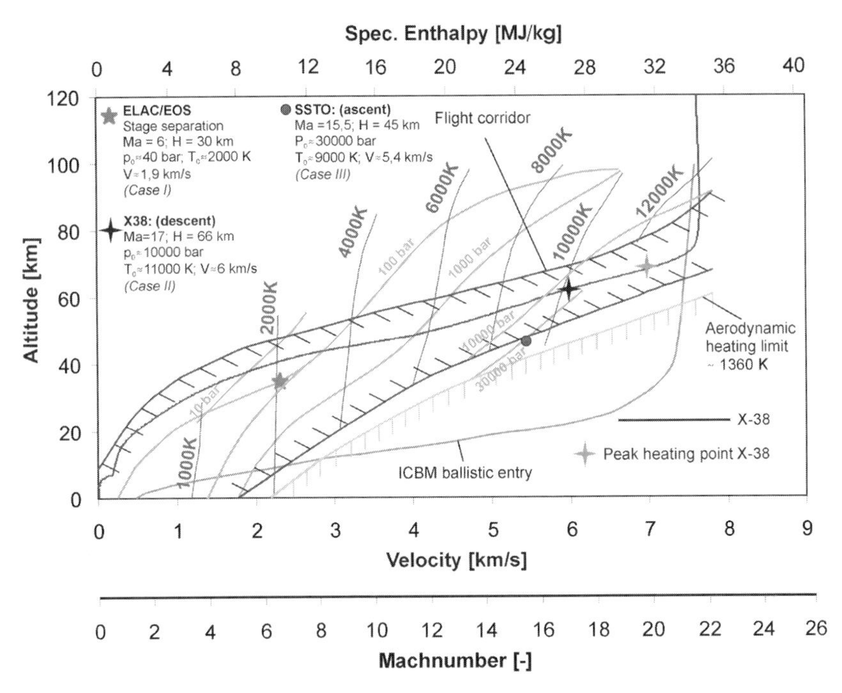


Figure 4.6.2: Velocity altitude map for typical trajectories including required reservoir conditions.

for low or high enthalpy flow, for different reasons, a complete keeping of similarity condition (similitude) is not possible. Compromises, neglects as well as leading on support measures have to be taken to circumvent such incompleteness and to interpret the information gained in an optimal sense.

In the following sections 4.6.2 and 4.6.3 two different problems of simulation are considered, situated almost at the lower and upper edge of the velocity altitude map. Hence, it is reasonable to mention at least some fundamental problems and limits of hypersonic flow simulation, assigned to flow conditions in the linking trajectory field of the two special situations (positions) of discussion. Three trajectory positions are illustrated in Fig. 4.6.2 as well in order to clarify the fictitious reservoir conditions. Case I concerns the stage separation as the upper limit of the carrier trajectory, case II pertains to typical flow conditions of the X-38 rescue vehicle (comparable to the shuttle re-entry path conditions) and case III belongs to a trajectory point of a hypothetical SSTO vehicle in ascent [1, 2]. A complete simulation only of the free flight conditions at 5400 m/s ($M = 15.5$) at 45 km altitude in case III would require a tunnel supply temperature of 9000 K and a supply pressure of about 30 000 bar. Operation at such high temperature is difficult and operation at such a high pressure is not feasible at present. But also a perfect duplication of case II free flow conditions is strictly speaking not feasible. Due to the kinematic reversal in flow generation the test gas is not in equilibrium owing to its rapid expansion from a highly ionized and dissociated state in the reservoir. This is an inherent handicap in all hyper velocity tunnels and slightly reduced in expansion tubes. On the other hand a stagnation pressure of 10000 bar is still too high to be realized, so free stream pressure, i.e. Reynolds number is not perfectly duplicated.

A perfect generation of case I stagnation conditions is possible in a hyper velocity wind tunnel. However, measuring time of such facilities is of the order of 1 s at the very best [3, 4]. Duplication of the similarity parameter $T_w/T_\infty$, e.g. assigned to radiation adiabatic wall conditions, in a steady case, as a rule is not feasible. Here, compromises must and have been partially found by preheating of the model using some specially developed similarity laws [6, 7].

In conventional hypersonic wind tunnels, operating in a blow down mode and of sufficient measuring time, Mach and Reynolds number are duplicated. But generally it is not possible to duplicate velocity or temperature in these types of facility because of too low stagnation temperature. Therefore, global temperature- and heat transfer distributions measured on a model configuration are on a too low level and have to be extrapolated to actual flight heating levels by specially developed computational tools [5]. A further consequence of the too low stagnation temperature especially in the hypersonic Mach range is that it is not possible to maintain similarity of boundary layer (BL) profiles between wind tunnel and flight. Consequently, BL transition simulation from laminar to turbulent is affected by this handicap and cannot be duplicated. Additional disturbances in the test section flow of conventional hyper- and supersonic wind tunnels generate mainly acoustic noise beside some other perturbations. From these, further distorting effects arise for the transitional process because of the receptivity of the BL to such disturbances.

A possible approach to obtain a solution to this dilemma is to take a combination of analytical and experimental studies: To the extent possible, experiments should be conducted to define the phenomenon, to compare with theory, to validate the computational method and to evaluate the difference that occur because of a non duplicated BL.

At lower Mach numbers i.e. in the supersonic regime the influence of minor stagnation temperatures and of disturbances correlating with Mach number are somewhat reduced but still existing. Subject of the following article 4.6.2 is to investigate a supersonic wind tunnel with regard to such disturbances distorting the natural transitional process.

Full similarity of the flow can only be reproduced in the full 1:1 scale experiment e.g. with a re-entry vehicle. However, the simulation of plasma-surface-interaction is possible using PWT. Here, the point of similarity considerations is limited to the stagnation point as a sectional part to be tested in the facility. Hence, rather group II of the above described similarity parameter is of concern as these parameter take the aero-thermodynamic parameter into consideration. For the qualification of TPS materials e.g. the reproduction of wall temperature $T_W$ and total pressure $p_{tot}$ is often sufficient. However, full similarity is not possible in these ground test facilities. With regard to heat exchange and diffusion processes in the stagnation point boundary layer in front of the TPS material, the situation is described by BL theories where both similarity of

Lewis and Schmidt number is taken into account [8, 9]. The corresponding relations enable the description of the material catalysis. Nevertheless, the scaling factors by themselves are quite difficult to be handled as the flow conditions usually are in thermochemical non-equilibrium.

Research of the plasma-surface-interactions requires a reproduction of the boundary layer with regard to concentration, temperature and velocity profiles in addition. Here, rather the wall temperature, the total pressure, the total enthalpy and the mass flow rate at the local section have to be reproduced. In fact, none of these parameters is a direct control parameter of the facilities and, in addition, each of the parameters depends on the other. But it becomes evident that the enthalpy-pressure envelopes of the facilities as depicted in Fig. 4.6.1 are a binding assumption for the possibility to obtain similarity. The Figure shows typical ascent and descent trajectories together with corridors derived from boundary conditions. The upper right area shows the enthalpy-pressure zone in which the different IRS plasma wind tunnels are operated. The corresponding zones are shown in Fig. 4.6.15 of the following section 4.6.3 where the IRS PWT are described. The difficulty to find feasible simulation points led to simplified similarity theories where the thermochemical similarity in the BL in front of the stagnation point is produced [10]. This is of major concern for research work as a direct extrapolation to flight trajectories became possible and as the situation on ground can be used of the investigation of the thermochemical behaviour of the TPS material in flight.

One of the best established theories in this field is the thermochemical similarity theory of Kolesnikov which has successfully been used e.g. for the investigation of the catalytic behaviour of ceramic TPS materials. Here, the specific enthalpies, shown in Fig. 4.6.1, the total pressures and the material probe geometry are the input parameter for the model. Apart from the enthalpy similarity a set of extrapolating transformations between ground facility and flight is made: The flight velocity is extrapolated from the PWK enthalpy, the density and hence the altitude is obtained via the stagnation pressure. Using the PWK flow velocity together with the model radius $R_m$ the nose radius can be derived. The effective model radius $R_m^*$ is known [10].

With the example of the heat flux simulation performed with the IRS plasma wind tunnel PWK1 the different levels of similarity are outlined (see Fig. 4.6.15). In [11] wall temperature measurements are presented. It could be shown that in for trajectory times smaller equal 400 s a pretty good correspondence between the expected stagnation point temperatures during flight and the measured wall temperatures during the trajectory simulation could be achieved. With larger $t > 600$ s active oxidation appeared (section 3.6) which is not included in the stagnation point models used for the trajectory simulation. Subject of the following section 4.6.3 is to outline the IRS PWT with regard to their design and function together with experimental examples.

## References to Section 4.6.1

1. Bussing, T., Scott, E.: Chemistry Associated with Hypersonic Vehicles, AIAA Paper No. 87-1292, 1987.
2. Tauber, M. E., Meneses, G. P.: Aerothermodynamics of Transatmospheric Vehicles, AIAA Paper No. 86-1257, 1986.
3. Schulz, D. L., Jones, T. V.: Heat-Transfer Measurements in Short-Duration Hypersonic Facilities, AGARD-AG-165, 1973.
4. Pélissier, C., Traineau, J. C., Kharitonov, A. M., Lapygin, V. I., Gorelov, V. A.: Review of Aerothermodynamic Facilities in Europe, The high enthalpy Facilities Example, ICMAR, Novosibirsk, 2002.
5. Micol, J. R.: Langley Aerothermodynamic Facilities Complex: Enhancement and Testing Capabilities, AIAA No. 98-0147, 1998.
6. Henckels, A., Maurer, F.: Hypersonic Wind Tunnel Testing with Simulation of Local Hot Wall Boundary Layer and Radiation Cooling, ZFW 18-160-166, 1994.
7. Heynatz, J. T.: Modellähnlichkeit der strahlungsadiabaten Wand in drei Aufsätze zur Strömungsmechanik, Verlag Dieter Thomas, Fürstenfeldbruck, 1997, ISBN 3-931776-12-3.
8. Goulard, R.: On Catalytic Recombination Rates in Hypersonic Stagnation Heat Transfer, Jet Propulsion, pp. 737–745, Nov. 1958.
9. Fay, J. A., Riddell, F. R.: Theory of Stagnation Point Heat Transfer in Dissociated Air, Journal of the Aeronautical Sciences, pp. 73–85, Vol. 25, No. 2, Feb. 1958.
10. Kolesnikov, A. F.: Extrapolation from High Enthalpy Tests to Flight Based on the Concept of Local Heat Transfer Simulation, Institute for Problems in Mechanics RAS, VKI Lecture, 1999.
11. Herdrich, G., Auweter-Kurtz, M., Hartling, M., Laux, T.: PYREX-KAT38, Temperature Measurement System for the X-38 Nose Structure TPS, 2nd Int. Symposium Atmospheric Reentry Vehicles and Systems, Arcachon, France, March 2001.

### 4.6.2 Validation of a Short Duration Supersonic Wind Tunnel for Natural Laminar Turbulent Transition Studies

#### 4.6.2.1 Introduction to the Problem

In the design of hypersonic vehicles boundary layer and transition play a dominant role because of their decisive influence to drag and especially to the local heat load distribution onto the structure. Considering a generic hypersonic configuration, to obtain an order of magnitude estimate, Burns et al. [1] found that skin friction contributes over 30% of the overall vehicle drag for fully turbulent and corresponding 10% for the fully laminar flow. It is quite likely that transition occurs at some point on the vehicle even at highest Mach number. Transition on the lower (compression) surface of the vehicle is especially likely due to the adverse pressure gradient there. Its location just here should be known, to handle shock boundary layer interaction more downstream in the ramp flow of the air breathing propulsion system. Clearly, transition location can be a significant source of uncertainty in vehicle heat load, drag and other local boundary layer characteristic predictions. However, the mechanisms leading to transition, are still not completely understood, because they are very complex! The transition process is initiated through the growth and development of disturbances originating on the body or in the free stream environment. For a realistic transition simulation the so-called transition Reynolds number ($R_{Trans}$) as a similarity parameter must be duplicated in the wind tunnel test. But free flight transition experiment on pointed circular cones at zero angle of attack show $R_{Trans}$-values up to one order of magnitude higher compared with corresponding cone experiments in conventional wind tunnels [2-57; second number is reference number in [2]]. Stetson [2-1] concludes from extensive experiments "one should not expect transition Reynolds number obtained in any wind tunnel to be directly relatable to flight". Hereby he expresses more or less evidently the current level of the simulation possibility. It is reasonable that an internal flow system, such as a wind tunnel, has a number of sources to generate velocity, temperature and pressure (acoustic) fluctuations, not present in the atmosphere at high altitude. To avoid disturbance fields in a super- or hypersonic test section flow of a wind tunnel great efforts have been undertaken since the seventies in the US with NASA Langley [2-6] and NASA Ames [2-7], but nowhere in Europe. So-called quiet wind tunnels have been developed to reduce these "disturbance modes" especially the acoustic noise in the free stream, originating either in the pre-history of the flow generation (in regulating flow valves) or radiated from the turbulent boundary layer (BL) of the test section walls.

The acoustic disturbance field does not only accelerate the transitional process (i.e. reduce $R_{Trans}$-values) but also distorts its spatial extent. But of a special meaning is the anomaly found in the transitional data of the flat plate and conical BL in noisy and quiet flow. They show under quiet condition a higher $R_{Trans}$ for the flat plate as for the cone BL, just the opposite to the noisy flow. Therefore, a designer relying on the conventional tunnel data would seek to create a conical fore body for a hypersonic vehicle in order to delay transition although a more 2D wedge shaped body is to be preferred (and as has been realized in the design of the NASA hyper X 43 vehicle). Reference is made to [2-53] where a detailed survey about the effects of tunnel noise on parametric trends is given. In summary, noise levels can have dramatic effects on transition and consequently experimental measurements should be carried out at conditions comparable to those in free flight.

#### 4.6.2.2 The Shock Wind Tunnel at Stuttgart University

The shock wind tunnel (German: *S*tosswind*k*anal SWK) is a large short duration supersonic facility with nominal, fixed Mach numbers 1.75, 2.54, 3.6 and 4.5. Because of its large rectangular test section (1,2×0,8 m) Reynolds numbers close to original flight conditions can be simulated on large models. Maximum loading pressure of 6 bar in the driver tube corresponds to a maximum unit Reynolds number ($R$) ranging from $20\times10^6/\text{m} < R < 80\times10^6/\text{m}$ depending on Mach number. The working principle is like in a Ludwieg tube. For the actual arrangement shown in Fig. 4.6.3 two steady flow states of about 120 ms each and of slightly different stagnation conditions are given.

Here, only some peculiarities of the tunnel shall be mentioned which are promoting for a reduction of the different kind of disturbances expected. One is the use of rapid expansion nozzles (like in the NASA quiet wind tunnel [2-57]) which have a much lower emission in acoustic radiation from turbulent BL on the nozzle when compared with a conventional design [2-9]. The kind of nozzle block suspension on the side walls, forming a bypass at the outer side (Fig. 4.6.4) and pealing off the driver tube BL, can bring further positive aspects for influencing the BL development on the nozzle contour. The pure size of the SWK yields some other advantages because as shown in [2-10, 2-11] there is a decrease of intensity and spectral bandwidth in acoustic disturbances with an increasing Reynolds number $R_D$ based on the effective diameter $D$ of the nozzle exit cross section. For more details about SWK reference is made to [2, 2-8]. All these distinctive factors and especially a projection study in [2-17] for an enlargement of the NASA quiet tunnel gave motivation and have initiated the performance of qualifying measurements in the test section flow, consisting of the detection and analysis of disturbances and the determination of $R_{\mathrm{Trans}}$-values on a cone for a comparison with corresponding free flight experiments.

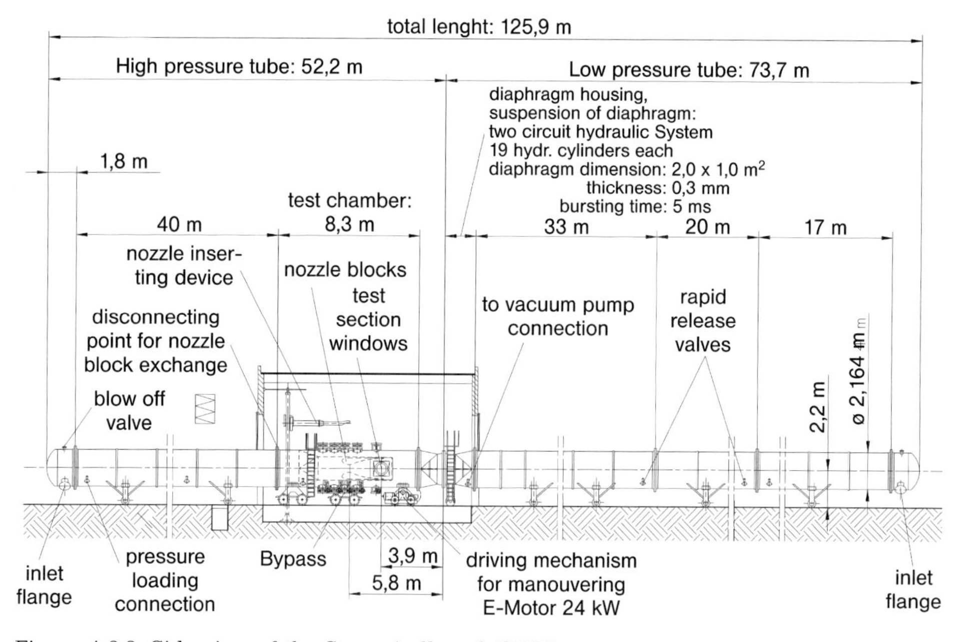


Figure 4.6.3: Side view of the Stosswindkanal (SWK).

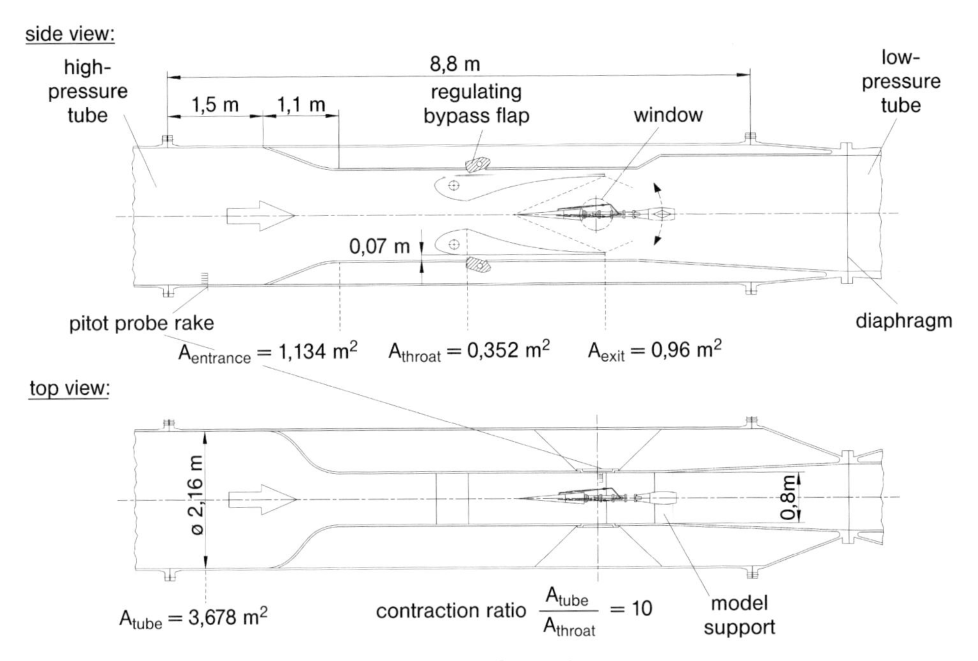


Figure 4.6.4: Details of test section and main dimensions.

*Definition of the Different Disturbance Fields*

Kovasznay [2-34, 2-35] and Morkovin [3] showed that a turbulent field in compressible flow is composed of three fluctuating "modes" which are independent of each other when their amplitude is small. This results from a linearization of the Navier-Stokes equations for a compressible, viscous and heat conductive gas and is represented in 3 differential equations:

$$D(u')_{rot}/Dt = 0 \tag{1}$$

$$D(T_0')_s/Dt = 0 \tag{2}$$

$$\frac{1}{a}\frac{D^2p'}{Dt^2} - \nabla^2 p' = 0 \tag{3}$$

Eq. (1) refers to the "vorticity mode", which is a variation of the rotational component of the velocity field whereas Eq. (2) describes the entropy mode (also defined as entropy or temperature spottiness) and refers to a variation of entropy, density and temperature at constant pressure. These 2 first modes can be considered as "frozen pattern" converted by the free stream. Eq. (3) is the wave equation and describes sound waves travelling within the flow (isentropic variation of pressure, density, temperature, and irrotational part of the velocity). It corresponds to the "acoustic mode". The intensities of the 3 modes are defined by their normalized RMS-value of the fluctuation value.

$$\text{Vorticity mode: } \langle |(u'_{rot})| \rangle = \frac{(|(u')_{rot}|)_{rms}}{|\overline{u}|} = \sqrt{\frac{\overline{(u')^2_{rot}}}{|\overline{u}|}} \, ,$$

$$\text{Entropy mode: } \langle (T_0)_s \rangle = \frac{((T'_0)_s)_{rms}}{\overline{T_0}} = \frac{\sqrt{\overline{(T'_0)^2_s}}}{\overline{T_0}} \, ,$$

$$\text{Acoustic mode: } \langle p \rangle = \frac{(p')_{rms}}{\overline{p}} = \sqrt{\frac{\overline{p'^2}}{\overline{p}}} \, .$$

*Origin and Character of Expected Disturbances in the Test Section Flow*

The free stream disturbance field in the SWK consists of a contribution from each of these 3 modes. Because propagation mechanisms and origin of the disturbances are different, the 3 turbulent modes are assumed to be uncorrelated.

The supersonic flow in the test section is generated by an unsteady expansion wave in the driver tube and a successive steady expansion in the Laval nozzle. This isentropic process does not produce disturbances in the core flow of the tube. It becomes obvious, that a short duration facility with such a gas dynamic principle must have low disturbances concerning these modes, due to the smooth acceleration from inherent constant loading conditions [2-15, 2-16]. Vorticity and entropy disturbances are reminiscent of the filling process of the driver tube and should therefore be small if the air allowed to equilibrate before each wind tunnel run. But a perfectly settled "state of quiescence" is however improbable, and this existing initial condition in the driver tube will be converted almost uninfluenced downstream into the test section flow. Because the residual disturbances from this filling process are unique and stochastic by nature, their amplitudes are supposed to change slightly from run to run. Therefore it is of great importance to determine the order of magnitude of such unavoidable and uncontrollable vorticity and entropy fluctuations within *one* run to have a proof for a reproducibility of the flow conditions including the superimposed disturbance fields.

The intensity of the acoustic noise in the free stream of a test section flow as a rule is by far the highest of all disturbance modes. Laufer [2-4] showed, that the turbulent BL along the nozzle and tunnel walls is the origin of this sound field. Sound waves are produced by transitional and turbulent BL and radiated in the free stream in form of fluctuating Mach waves. This includes the truly aerodynamic sound in the sense of Lighthill [4-20] and referred as "eddy Mach waves" (EMW) as well as "shivering" or "shimmering Mach waves" (SMW) produced by reflection or diffraction of normally steady pressure gradients by the turbulent BL and which can have their origin in the imperfection of the nozzle contour or in surface irregularities. It is noteworthy that these two types of fluctuating Mach waves have different characteristics, which allow an identification of both in the course of their experimental detection. Concerning the EMW only some features are given: In [2-4] was found that EMW in the BL are also moving with respect to the tunnel walls. For a free stream Mach number of $M=2.5$ between the so-called source velocity of the EMW and free stream velocity a ratio $u_s/u_\infty= 0.36$ was found. Furthermore, EMW production is a strong function of the Mach number and acoustic intensity $\langle p\rangle^2 \propto M^4$: Measurements in [2-9] showed that an increase in $R$ reduces the EMW radiation by reducing the nozzle BL thickness, but at the expense of a shift to higher frequencies in the radiation.

In contrast SMW have a fixed position independent on $R$ because their origin are steady pressure gradients, caused by imperfections of the boundary surfaces either in the nozzle curvature or in an insufficient surface quality. In [2-63, 2-64] a strong degree of correlation between mean and fluctuating Mach waves has been found and according to [2-2], SMW dominate in many wind tunnels and result in lower frequency fluctuations compared with EMW. Consequently a reduction of the acoustic noise level will only be achieved, as a rule, by a laminarization of the BL on the test section walls and an extremely high wall surface quality which avoid EMW and SMW respectively. In some wind tunnels [2-57], [2-7] such measures have been realized and resulted in a noise level of $\langle p\rangle \leq 0.1\%$. But this low level is only attained in the upstream portion of the Mach rhombus in the so-called "quiet core" which is bounded upstream by the limiting characteristics defining the Mach rhombus and downstream by the EMW originating in the transitional region of the BL on the side wall and nozzle contour.

### 4.6.2.3 Detection Techniques for Flow Disturbance Fields

*Pitot Pressure Fluctuation Measurements*

Because acoustic waves form the bulk of free stream disturbances mostly Pitot pressure probes for the determination of the acoustic noise level ([2-56, 2-15, 2-16, 2-66]) have been used. Also first extensive Pitot pressure fluctuation (PPF) measurements have been performed in the SWK up to $R=35 \cdot 10^6$/m, where a fragile hot wire would not survive [2].

Besides, concerning the detection of temperature fluctuations at higher $R$-values only, a new sensor was tested successfully in the SWK [6]. But dynamic PPF measurements are not free of drawbacks. First, a Pitot probe is much larger than a HW, so spatial resolution is inferior. Second, although pressure transducers benefit from a large bandwidth, their transfer function is probe dependent and not easy to determine. Third, the relation between Pitot pressure fluctuations $\langle p_t\rangle$ and static pressure fluctuation $\langle p\rangle$ cannot be derived unless certain assumptions about the character of the flow itself are made or known from complementary detection techniques ([2-66, 4]). Strictly speaking an analysis of acoustic properties is limited when using this technique only.

*Hot Wire Technique and Modal Analysis*

The more meaningful procedure for flow disturbance detection is given by means of the hot wire (HW) anemometry combined with the so-called "modal analysis" or "fluctuation diagram technique" ([2-34, 2-3, 3]). Presently this is still the only reliable procedure allowing a detection of all possible, existing disturbance modes like vorticity, entropy spottiness and sound waves simultaneously with a high degree of temporal and spatial resolution. The principle of HW signal decomposition into various "modes" is based on dissimilar ratios $r = -F(\tau)/G(\tau)$ (the so-called modal parameter) of the "sensitivities" $F(\tau)$ and $G(\tau)$ for the heat exchange of the probe element in reference to velocity, density and temperature of the flow, when overheating parameter defined by the "temperature loading" $\tau = (T_w - T_e)/T_0$ is varied during the flow measurement. Subscript "$w$" pertains to conditions of heated wire probe, $T_0$ and $T_e$ are assigned to stagnation and recovery conditions. In a short duration facility one gets in some difficulties and compromises had to be made to realize this fluctuation diagram technique. Sufficient information must be captured within one flow event (run).

*The Constant Temperature Anemometer Scanning Device and Procedure*

A constant temperature anemometer (CTA) device was developed and manufactured in a cooperation with IAG, ITAM Novosibirsk and COSYTEC Elektronik GmbH to operate the HW in a sequence of different overheat parameters within the available short time.

For detailed description about its main technical features and dynamic behaviour reference is made to [5, 8, 10]. But the automatic rapid scanning of wire overheat yielded new problems in the experimental procedure and the most challenging has been the optimization of the anemometer's frequency response. To eliminate such dilemma with all its consequences an efficient new fast procedure has been developed to determine a transfer function of the anemometer system for the individual overheat ratios enabling the performance of a post correction of the non-optimized frequency responses [5, 9, 10]. This method enables also a determination of the signal to noise ratio. Besides, the handling of the inherent non-linear behaviour of the CTA to temperature fluctuation at low overheat ratio is described in [5, 7]. Exemplary, a typical scanning sequence is shown in Fig. 4.6.5, with signal conditioning DC/AC.

*Data Reduction: The Fluctuation Diagram Technique*

The basic dynamic equation, relating instantaneous HW fluctuation signal $e'$ to corresponding fluctuation quantities $(\rho u)'$ and $T_0'$ is given in [2-36, 2-37, 5]. It is important to notice that a usual fluctuation diagram enables only the evaluation of the normalized mass fluctuation $\langle \rho u \rangle$, total temperature fluctuation $\langle T_0 \rangle$ and

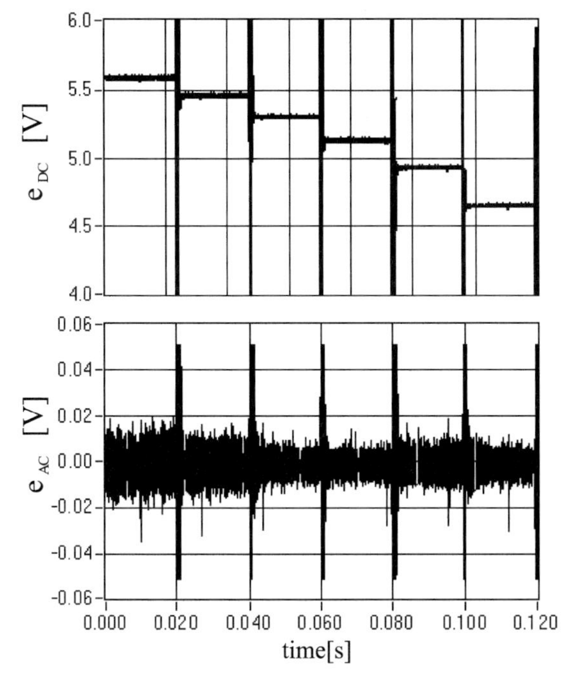


Figure 4.6.5: Typical DC/AC time traces of the CTA scanning system; sequence of 6 wire temperature loadings. $0.4 \leq \tau \leq 0.9$.

their correlation coefficient $R_{\rho u,T_0}$. A further splitting of the data to obtain the magnitude of the three turbulent uncorrelated modes (vorticity, entropy and acoustic mode) requires further assumptions which depend on flow conditions.

For the flow in the free stream of the SWK test section, two special cases are of interest. Case I happens, when the acoustic mode is dominant. In [2-4] it is shown that $R_{\rho u,T_0}$ in a pure acoustic field is equal to –1, i.e. in this case the mode diagram simplifies to a linear function.

$$\Theta = \langle \rho u \rangle r + \langle T_0 \rangle \tag{4}$$

Moreover, following relations between the normalized pressure and velocity fluctuation are valid:

$$\langle \rho u \rangle = \frac{1}{\gamma}\langle p \rangle - \langle u \rangle;\ \langle T_0 \rangle = a(\gamma - 1)\left( M^2 \langle u \rangle - \frac{1}{\gamma}\langle p \rangle \right);\ a = \frac{1}{1 + \frac{\gamma - 1}{2} M^2} \tag{5}$$

So when a linear fluctuation diagram is obtained by measurements, $\langle \rho u \rangle$ and $\langle T_0 \rangle$ are defined by the slope and the interception of the straight line and $\langle p \rangle$ is determined using (4). When the acoustic sources are stationary, i.e. SMW are existing, $\langle T_0 \rangle = 0$ and the straight line passes to the origin. On the other hand in the case of EMW, where the acoustic sources are converted downstream, $\langle T_0 \rangle \neq 0$. It is possible to calculate the mean propagation velocity $u_s$ of the sources and the most probable wave orientation $\theta$ using the equation

$$\frac{\overline{u_s}}{\overline{u}_\infty} = 1 + \frac{1}{M \cdot \cos\theta} = 1 - \frac{\langle p \rangle}{\gamma M^2 \langle u \rangle} \tag{6}$$

The special case II involves the uncorrelated modes $\langle p \rangle$, $\langle u_{rot} \rangle$ and $\langle (T_0)_s \rangle$, which form as a rule, according to [2-2], the fluctuation field in a supersonic wind tunnel flow. To extract their magnitude from the fluctuation diagram a relation derived in [2-35] for the case of SMW has been extended in [5] for EMW.

$$\theta^2 = (L_1 r + L_2)\langle p \rangle^2 + (\beta - r)^2 \langle u_{rot} \rangle^2 + \left(1 + \frac{r}{a}\right)^2 \langle (T_0)_s \rangle^2 \tag{7}$$

By means of $\in (\Theta^2, r)_{i, 1 \leq i \leq N}$ measured values, a fitted second order polynomial function can be defined

$$\Theta^2 = a_0 + a_1 r + a_2 r^2 \tag{8}$$

where coefficients $a_0$, $a_1$ and $a_2$ result from the general least square algorithm. The three unknowns $\langle p \rangle$, the rotational velocity $\langle u_{rot} \rangle$ and the non-isentropic total temperature fluctuation $\langle (T_0)_s \rangle$ are determined by a comparison of the coefficients in Eq. (7) and Eq. (8). L1 and L2 are function of the source velocity according to [5].

#### 4.6.2.4 Free Stream Disturbance Measurements in the Shock Wind Tunnel

*Introductory Remarks*

Extensive PPF measurements [2] have set the pattern and have been the decisive factor for further specific disturbance measurements by modal analysis.

HW measurements were performed along the test section centre line (y=z=0) between x=696 mm (coinciding with Mach rhombus tip) and x=2036 mm (at about nozzle exit), where the reference system (x=0) is positioned on centre line in the plane of the nozzle throat.

The nominal Mach number was $M=2.54$ and $10 \times 10^6/\text{m} \leq R \leq 25 \cdot 10^6/\text{m}$. Measurements have been restricted to $R=25 \cdot 10^6/\text{m}$ because of the fragile HW and to the first quasi-steady supersonic flow state.

*Parameter Setting for Modal Analysis*

To realize a mode analysis in *one* tunnel run, because of the short available time, a compromise between the number of steps of overheat and time span per overheat defining the integration time to estimate $\langle e \rangle$ had also to be made. Six overheat steps of 20 ms each were used, but only the last 15 ms for integration. For an accurate determination of the three turbulent modes a relative low value of $r(\tau) = -F(\tau)/G(\tau)$ is necessary, corresponding to a low overheat ratio too. In a first step modal analysis has been carried out using 6 overheat ratios of relative low values ranging from $\tau$=0.05–0.5. As it appeared that the acoustic mode is largely dominant, successive distribution measurements along centre line axis were performed with $\tau$=0.4–0.9 to benefit from a larger bandwidth and a better signal to noise ratio [5-31, 5-34].

*Measured Correlated and Uncorrelated Disturbance Modes*

In Fig. 4.6.6 fluctuation diagrams are depicted for the 2 centre line positions x=1771, 2036 mm close to the nozzle exit at 4 different $R$ values.

For each tunnel run a "settling time" of at least 15 min for the compressed air in the driver tube has been provided. Wind tunnel runs at different settling times have shown that such a time span was sufficient after the charging process for a settling down of the compressed air. All determined fluctuation values in Fig. 4.6.6 (derived from scanning sequences as in Fig. 4.6.5) seem to be well fitted by straight lines, thus demonstrating the dominance of acoustic disturbances in the free stream. Of course this can strictly speaking only be an approximation: vorticity and entropy spottiness generated during the loading process are certainly still present in the flow even though on a small scale. But the analysis of a data sample $\in (\Theta^2, r)$ at x=2036 mm, with the assumption of uncorrelated modes, when using Eq. (7) and Eq. (8) revealed for the rotational velocity fluctuation $\langle u_{rot} \rangle$=0.017% and the non-isentropic total temperature fluctuation $\langle (T_0)_s \rangle$=0.015%, whereas the acoustic pressure fluctuation resulted in $\langle p \rangle$=0.39%. So these uncorrelated modes originating in the driver tube are at least one order of magnitude lower than $\langle p \rangle$! This fact is of importance in a broader sense: on one hand it is legitimizing the linearization of the mode diagram. On the other hand it allows the conclusion, that the consequence of a small variation of these uncontrollable disturbances, originating in the loading process and by stochastic nature, will have no remarkable influence onto the reproducibility of the unavoidable disturbance fields, superimposed the test section flow.

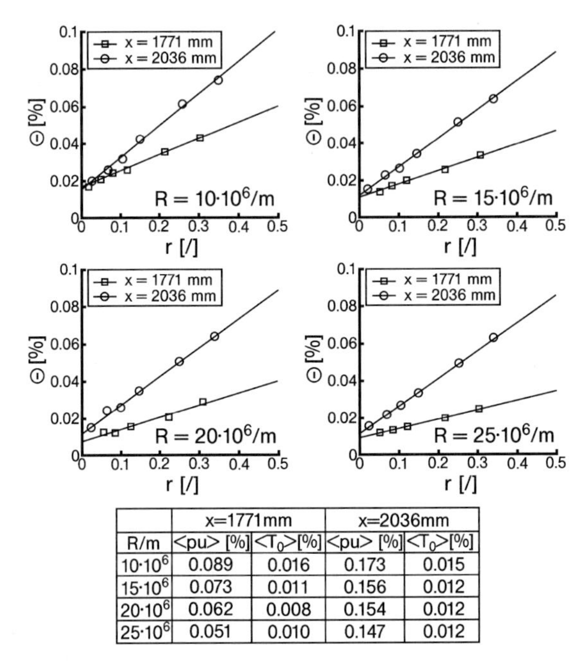


| | x=1771mm | | x=2036mm | |
|---|---|---|---|---|
| R/m | <pu> [%] | $\langle T_0 \rangle$[%] | <pu> [%] | $\langle T_0 \rangle$[%] |
| $10 \cdot 10^6$ | 0.089 | 0.016 | 0.173 | 0.015 |
| $15 \cdot 10^6$ | 0.073 | 0.011 | 0.156 | 0.012 |
| $20 \cdot 10^6$ | 0.062 | 0.008 | 0.154 | 0.012 |
| $25 \cdot 10^6$ | 0.051 | 0.010 | 0.147 | 0.012 |

Figure 4.6.6: Typical mode diagrams with results for $0.05 \leq \tau \leq 0.5$ in 2 points at centre line axis.

*Acoustic Field Properties*

Some properties in this dominating acoustic field could be deduced following the method in [2-4] when using Eq. (5) and Eq. (6). Pressure fluctuation $\langle p \rangle$ and source velocities of the EMW has been determined and depicted in Fig. 4.6.7.

It is seen that sound intensity is decreasing in a more upstream position and with increasing $R$, in accordance to [2-9]. But the large local difference in the sound level (more than a factor 2) between x=1771 und x=2036 mm should be caused neither by a difference of the local Mach number at the acoustic origin nor in the BL growth on the nozzle contour. It is assumed that SMW (caused by wall imperfection) are superimposed on the EMW. This assumption could be supported by different facts, deduced from the character of the mode diagram in Fig. 4.6.6 and the source velocity $u_s$ for both positions. An analysis described in [5] revealed that 44% of the total noise in x=2036 mm is contributed by SMW.

*Noise Level Distribution along the Centre Line Axis*

Because BL development at nozzle contour and side wall is responsible for the noise level progression and lowest disturbance levels are existing only in the upstream part of the rhombus, measurements along centre line axis has been per-

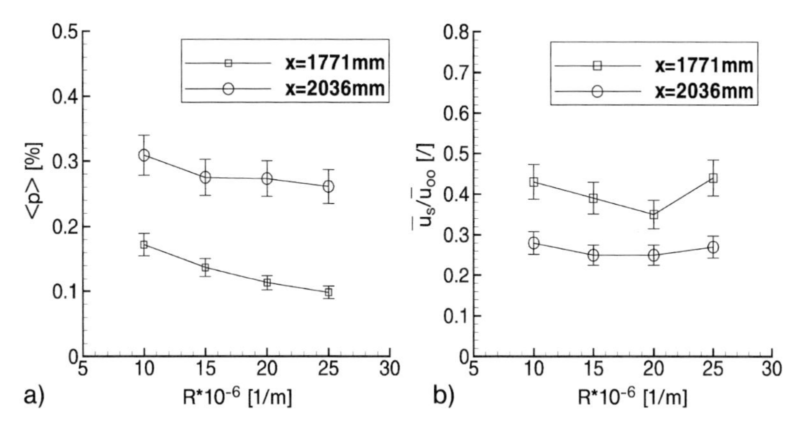


Figure 4.6.7: Acoustic fluctuations: (a) intensity level, (b) source velocity.

formed from the Mach rhombus tip in downstream direction. The resulting data were fitted to linear mode diagrams since dominance of acoustic disturbances had already been shown above in Fig. 4.6.6. In Fig. 4.6.8 the resulting pressure fluctuation course $\langle p \rangle$ is presented for 4 different $R$-values. Although most of the values are above the "quiet" limit, defined as $\langle p \rangle = 0.1\%$, some of them are close to this borderline, even far downstream at x=1771 mm for $R = 25 \times 10^6$/m. Most noticeable however is the irregularity of the development in streamwise direction: both, very strong peaks as well as calmer regions can be observed.

The 3 peaks (1) to (3) in Fig. 4.6.8 correspond to very low frequency disturbances and are correlating with temporal mean irregularities detected already in [11] in static pressure measurements on a pointed cone. We can deduce, that a large part of the detected fluctuating pressure disturbances along the axis, possessing irregular large intensity, are in fact steady compression waves "shivered", i.e. excited by the turbulent BL to oscillations of low frequency. Possible reasons for such irregularities and their avoidance, reference is made to [2-3, 2]. An estimation about the noise field along centre line axis without SMW contribution, only caused by EMW, is presented in Fig. 4.6.9. The establishing of such a hypothetic curve [5] results in a course which incorporates more or less the lower envelope of the lowest measured pressure fluctuations and allocated to intensity values, which are assigned to a 100% contribution from EMW radiation. This approximation must fail of course in the upstream part of the Mach rhombus x<1000 mm, where the noise level is close to or in the quiet region.

More significant is a decrease of the fluctuation level with increasing $R$ which is typical for a turbulent BL [2-4, 2-17]. No sign of laminar-turbulent transition of the nozzle BL (which would produce a remarkable qualitative change in the progression $\langle p(\mathrm{x}) \rangle$ with increasing $R$ like in [2-17]), can be observed. Although of a turbulent BL, the disturbance level is low, because of a moderate Mach number, the rapid expansion nozzle [2-9] and the large test section [2-10, 2-11, 2-17].

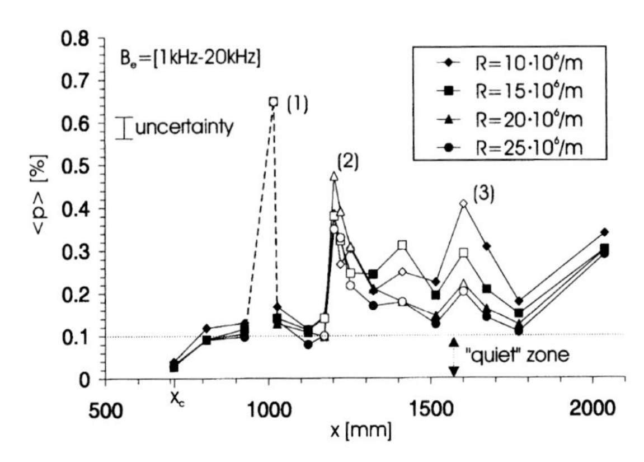


Figure 4.6.8: $\langle p \rangle$ fluctuation along centre line axis; solid symbols with scanning $0.4 \le \tau \le 0.9$, open symbols with $\tau = 0.9$.

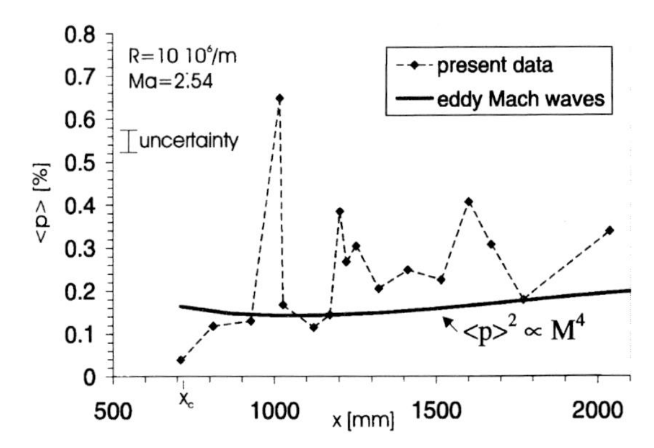


Figure 4.6.9: Estimation of EMW proportion in the measured total disturbance level.

*Spectral Analysis of the Acoustic Disturbance Field*

Not only the sound intensity but also its spectral distribution is of special importance. The degree of receptivity to the incidence of acoustic disturbances on the model BL to be simulated can be more or less distinctive because it is a selective process. Stetson [2-1] argues, that most of the acoustic noise in the free stream of the test section flow is not important for instability and transition studies; the only part of significance is said to be that, which is at frequencies similar to those of the dominant instability modes on the model leading to transition. Therefore, it makes sense to identify the prevailing frequencies in the acoustic disturbance field and, if possible, estimate the dominant most unstable frequency of the relevant instability mode, be expected.

Some representative post-corrected power spectra without any averaging are presented in Fig. 4.6.10, directly obtained by a standard FFT of the measured time trace. The dashed line "electronic noise" with a slope equivalent to

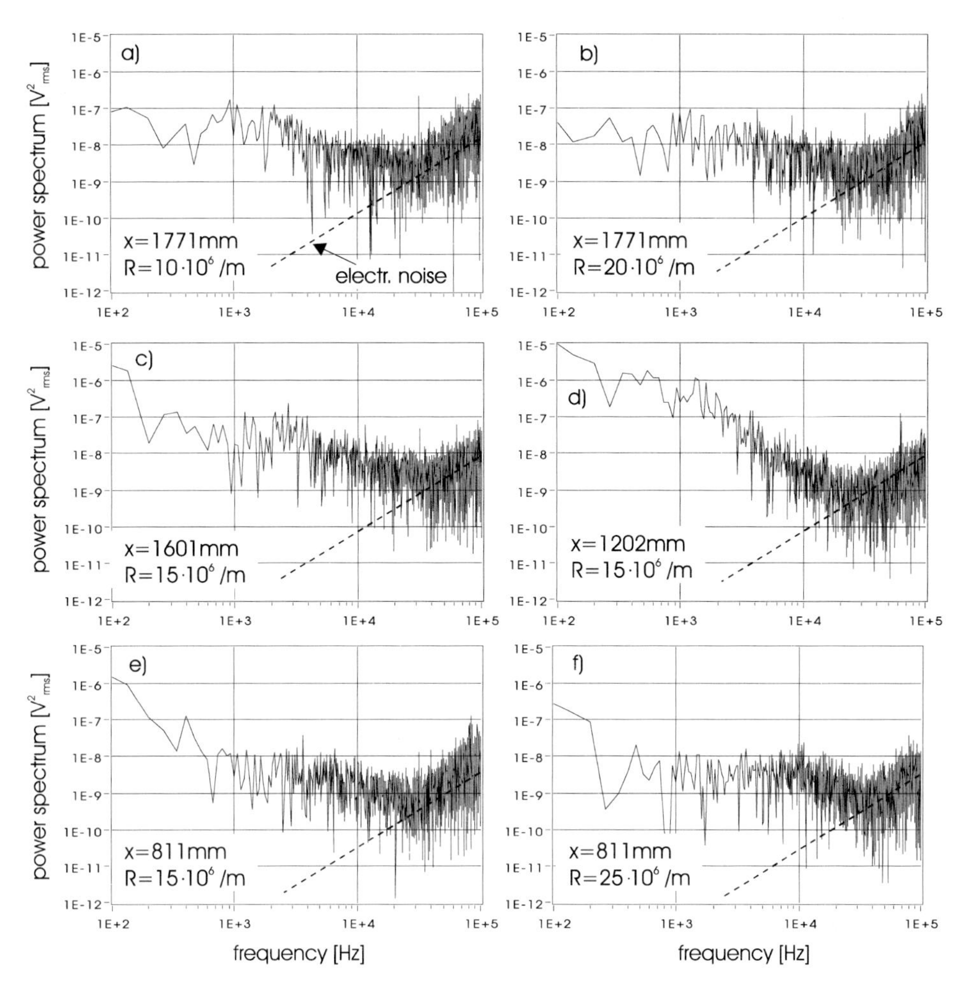


Figure 4.6.10: Representative power spectra, on different stations along centre line axis.

a square dependency on frequency indicates, where coinciding with the spectrum, the signal to noise ratio 1, i.e. the useful bandwidth of the CTA system, which is about 20–30 kHz. The spectra (a) and (b) have their maximum energy approximately at 2–3 kHz. This is in good agreement with the universal law, found in [5-57] which predicts in a rough estimation for the present conditions ($Ma$=2.5, $R$=10·10$^6$/m, BL thickness $\delta \approx 1$ cm) about 4 kHz. In both spectra (c) and (d) equivalent to the peak positions (3) and (2) of diagram 4.6.8 resp., low frequency oscillations are superimposed on the previous distributions in (a) and (b). They confirm that at these positions local SMW are superimposed on the bulk of EMW and total energy content is shifted to lower frequencies, as predicted in [2-2]. Comparing respectively spectra (a) with (b) and (e) with (f) the influence of the $R$ on EMW radiation can be observed. With increasing $R$, the thickness of nozzle BL decreases, thus leading to a slightly lowering of intensity and shifting to higher radiating frequencies. Furthermore, because BL thickness is much smaller close to the throat region, corresponding peak frequency at x=811 mm is higher than at x=1601 mm (at $R=15\times10^6$/m each). At frequencies where maximum energy was found, electronic noise is one order of magnitude lower, i.e. the bandwidth was sufficient to capture the dominating frequency range of the disturbances. Of special interest is a comparison of these maximum energy frequencies with the most relevant instability modes under present conditions. For a cone or a slender generic hypersonic configuration at $M=2.5$ the 1st mode (TS instability) will be relevant. A rough $e^N$-estimation by [13] reveals most unstable frequencies in a conical BL of about 22 and 44 kHz at $R=25\times10^6$/m, and almost 80 kHz at $R=50\times10^6$/m. These are remarkable aspects for the validation of the test section flow and with regard to transition studies of generic bodies.

*Review of the Measured Disturbance Fields*

The low level of the dominating acoustic field, originating from EMW, with low frequency disturbances, relatively independent from $R$, is particularly advantageous for laminar turbulent transition experiments. On the other hand, one handicap in the SWK test section flow is (up to now) the presence of these highly localized mean flow disturbances, which result in a large amount of SMW and in an irregular progression of the noise field in streamwise direction. These perturbations can trigger the transition process on the model BL, reducing the $R_{\mathrm{Trans}}$-value. They can also induce an "apparent" unit Reynolds number dependency (i.e. a variation of $R_{\mathrm{Trans}}$ with $R$), since SMW have a fixed position and do not scale with $R$ whereas the model BL do. The elimination of these mean flow disturbances caused mainly by nozzle contour and surface imperfection will improve the flow quality considerably.

#### 4.6.2.5 Transition Experiments in the Test Section Flow

*Common Remarks and Detection Techniques*

A common established test procedure to validate the test section flow of a supersonic wind tunnel are transition measurements on a slender pointed circular cone (5 degree half apex angle) at zero angle of attack. Under these conditions and an actual free stream Mach number of $M=2.5$, the 1st mode (TS instability) will be relevant for transition. For such a model specification a bulk of $R_{\mathrm{Trans}}$-values exists over a wide range of Mach and unit Reynolds numbers from free flight and low disturbance wind tunnel experiments for a comparison. Besides a verification of a comparable high $R_{\mathrm{Trans}}$-value at necessarily high $R$-value also the $R$ independency should be confirmed. Extensive measurements with 3 different transition detection techniques: 1. The thin- (hot)-film (HF) technique in cone I, 2. The "thin wall" technique combined with a use of thermocouples in cone II and 3. a Preston tube probe in combination with cone I have been carried out at 3 different model positions in the Mach rhombus in a range of $6\times10^6$/m≤R≤$25\times10^6$/m. About the complete measurements, instrumented cones and developed detection techniques etc. detailed information is given in [2, 8, 11]. Here only main results of the measurements are depicted.

*Presentation of Cone Transition Experiments*

All here presented $R_{\mathrm{Trans}}$-values have been determined in the test section at a cone position most upstream close to the Mach rhombus tip. For other cone locations more downstream in the region of the nozzle exit, $R_{\mathrm{Trans}}$-values found, seemed not to be relevant because only comparable with conventional tunnel results [2].

In cone I with a total length of about 1 m, five HF sensors (2 failed during experiments) have been applied on a surface line far behind the cone tip at x=475, 540 and 605 mm. The protruding caused by HF installation in the cone surface could be minimized to about 1–2 µm. But with a cone surface *rms* roughness of about 0.2–0.3 µm, these sensor regions had still to be regarded as single roughness elements almost one order of magnitude larger, which could cause a roughness induced transition process. This has been the very reason for using still two other techniques.

The determination of a transition Reynolds number by means of the 3 HF gauges is based on the movement of transition location when $R$ is changed. In the actual situation transition moves upstream from the rear end of the cone when $R$ is increased in a range between $5.65\times10^6/\mathrm{m}<R<13\times10^6/\mathrm{m}$. The state of transition is characterized by extremely high fluctuations in all flow quantities therefore also captured by the individual sensor when this event is passing over its position.

In Fig. 4.6.11 the *rms* fluctuation courses of the 3 sensors are plotted versus corresponding Reynolds numbers $Re_{pr,i} = [(\rho_\infty \cdot u_\infty)/\eta_\infty] \cdot x_{pr}$ based on free stream condition and the individual probe position $x_{pr,i}$. The measured effective fluctuation values $e'_{rms}$ are normalized by $e'_1$, which is a corresponding $e'_{rms}$ value measured for lowest $R$-value of the respective HF gauge. Therefore, all 3 curves have their onset in a normalized fluctuation value of unity. As seen in Fig. 4.6.11 the 3 peak values 3, 4 and 6 are different because of different sensitivity of the individual sensors, but their position collapses fairly well in a range of $R_{\mathrm{Trans}}=4\text{–}4.2\times10^{6}$, equivalent to a mean value of $R_{\mathrm{Trans}} \approx 4.07\times10^6$.

Simultaneously with HF experiments, wall PPF measurements have been performed by means of 3 Preston tubes, mounted at $x_{pr}$=630 mm in 3×120° angular position, twisted 20° off to the reference line where the HF have been installed. Fig. 4.6.12 shows the PPF distribution in an analogue presentation to Fig. 4.6.11. Also here the fluctuation curves are matching fairly well in their peak positions between $R_{\mathrm{Trans}}=4.15\text{–}4.38\times10^6$, equivalent to an averaged value

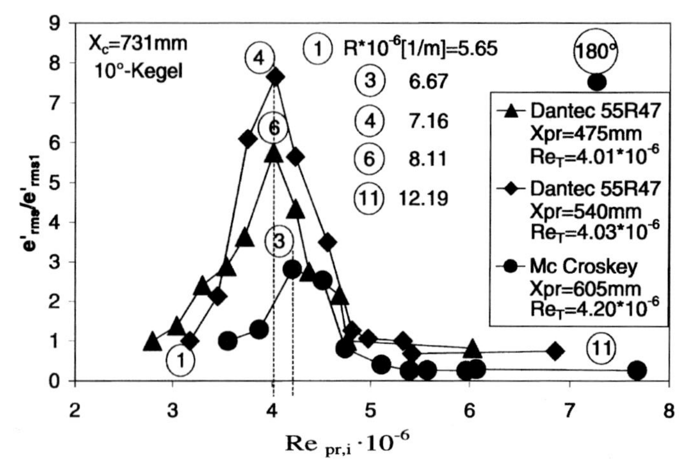


Figure 4.6.11: Normalized rms-values of HF fluctuations signals in the cone BL vs $Re_{\mathrm{pr,i}}$.

of $R_{\mathrm{Trans}} \approx 4.3 \times 10^6$. The averaged $R_{\mathrm{Trans}}$-value found by this method is unambiguously slightly above the corresponding $R_{\mathrm{Trans}} = 4.07 \times 10^6$ of the HF technique. But to compare averaged $R_{\mathrm{Trans}}$-values assigned to different $R$-values is only justified when there is no "unit Reynolds number effect", influencing the transition Reynolds number $R_{\mathrm{Trans}}$.

To answer this question, transition measurements have been carried out with cone II, a model manufactured in thin wall technique, where 19 thermocouples are completely integrated into the thin cone wall (0.15–0.2 mm thickness) along a cone generator line without producing artificial roughness in the surface. This DLR model was already used for corresponding measurements at the RWKG [12]. Because of limited model strength of the thin wall construction

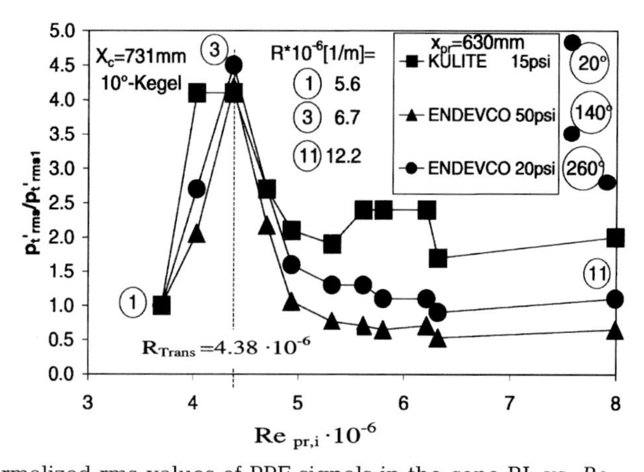


Figure 4.6.12: Normalized rms-values of PPF signals in the cone BL vs. $Re_{\mathrm{pr,i}}$.

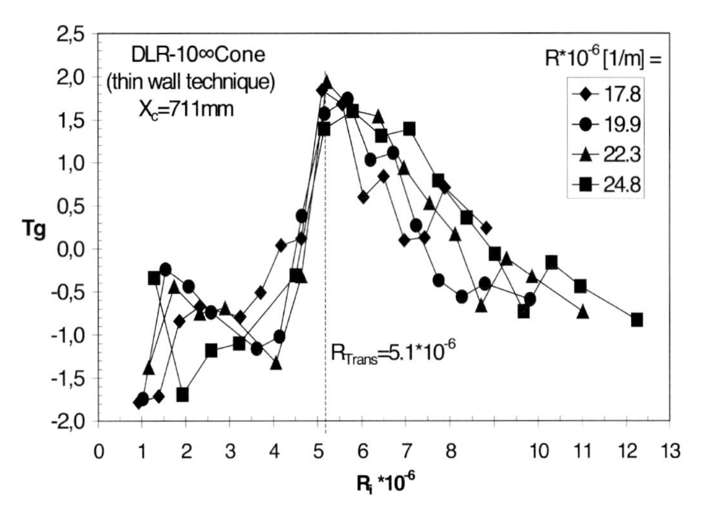


Figure 4.6.13: Transition detection by the thin wall technique for different R-values.

concerning differential pressure [2], experiments had to be limited to a maximum of $R=25\times10^6$/m.

Electric disturbances in the thermocouple signals (but only during flow), a low stagnation temperature $T_0$, i.e. according ambient condition and therefore a low heat flux in the thin wall too, resulted in a low signal to noise ratio. Only by a statistical evaluation process, a significant local change of a normalized temperature time gradient $Tg$ in the cone wall, which indicates transition, could be detected. The course of these dimensionless gradients defined by the relation $Tg_i = ((dT/dt)_i - (dT/dt)_{\text{mean}})/(dT/dt)_{rms}$ with the dimensional local gradient $(dT/dt)_\mathrm{i}$ of the thermocouple "$i$", the average $(dT/dt)_{\text{mean}}=(\Sigma(dT/dt)_i)/N$ and the $rms$ value $(dT/dt)_{rms}=((\Sigma((dT/dt)_i-(dT/dt)_{\text{mean}})^2)/N)^{0.5}$ is shown in Fig. 4.6.13 for $17.8\times10^6/\text{m}\leq R\leq 24.8\times10^6$/m.

The peaks of the four plots belonging to different $R$-values coincide at $R_{\text{Trans}} \approx 5.1\times10^6$ and demonstrate $R$ independence.

But this is in contradiction to the $R_{\text{Trans}}$-values 4.07 and $4.3\times10^6$ found in the range of $6.6\times10^6/\text{m}\leq R\leq 13\times10^6$/m by the other two techniques, where transition location is situated more downstream of the cone tip. A clarification of this discrepancy is given by the presumption expressed in Chapter 4.8. The superimposed SMW are triggering the transitional process. It was proved just recently [12] and not shown here, that there is a stepwise increase of $R_{\text{Trans}}$-values, when transition region is moving more upstream and has passed these 3 mean disturbances marked (1) to (3) in Fig. 4.6.8. For $R>18\times10^6$/m transition location is in an area upstream of the peak position (2) marked in Fig. 4.6.8 and maximum $R_{\text{Trans}}$-values of about $5\times10^6$ are obtained and remain constant.

An extension of transition measurements to $R=50\times10^6$/m attainable in the SWK are on a go by means of the HF technique. It is noteworthy that no other transition techniques like pressure sensitive paint or the liquid crystal method

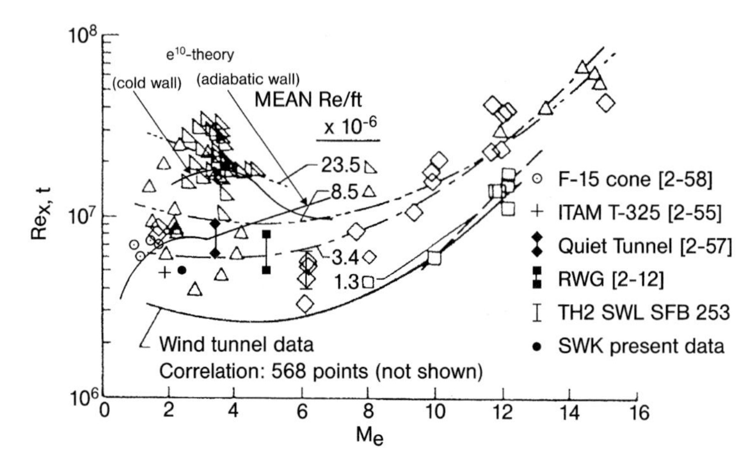


Figure 4.6.14: $R_{\text{Trans}}$-values on pointed 5° half cones [2-57], enlarged by other wind tunnel results and present measurement.

can be used, because no optical access is available in this most upstream cone position.

*Comparison with Other Transition Measurements*

For a validation of the SWK results, a comparison with relevant values from free flight and quiet tunnel experiments is necessary. Figure 4.6.14 shows the well-known collection of Beckwith [2-57] for such transition data plotted against *Me* (i.e. BL edge Mach number).

The diagram is extended by the actual data and some results found in other facilities. The diagram clarifies that $R_{\text{Trans}}$-values of conventional wind tunnels, summarized in a correlation, are in average up to one order of magnitude smaller than highest flight data. Only transition data of the NASA Langley ($M=3.5$) quiet tunnel [2-57] and of the DLR ($M=5$) Ludwieg tube [2-12] are lying unequivocally in the mid range of free flight data. Additionally in these 2 measurements an $R$ independence was found up to $R=30\times10^6$/m and $50\times10^6$/m, respectively.

Most relevant SWK result is the transition Reynolds number $R_{\text{Trans}}=5.1\times10^6$ found for higher $R$-values and with no dependence in the range of $17\cdot10^6/\text{m}\leq R\leq25\times10^6/\text{m}$. The upper limit of this range was defined by the confined rigidity of the thin wall cone. But there is no reason, not to believe that this $R$ independency shall not exist up to a maximum of $R=50\times10^6$/m, because once noise level will be further reduced and no SMW have been detected in this region of the Mach rhombus where transition location will be situated under these conditions. The size of the $R_{\text{Trans}}$-value is in reference to the F-15 free flight cone data, measured in the Mach range $M=1.5$–2.,0, about 70–85% and consequently in the flight data range and significant above the correlation of conventional wind tunnel data.

But wall temperature condition has also an influence to $R_{Trans}$. In [2-58] by means of the $e^N$ computational method ($N=10$) it was proved, that for adiabatic wall as well as cold wall conditions there is a good agreement with F-15-, quiet tunnel and other flight data as shown in Fig. 4.6.14. A cold wall (i.e. $T_w < T_r$ and a heat flux from the flow into the wall) has a stabilizing effect and explains the high $R_{Trans}$-values. In the actual SWK situation we have a slightly reverse situation i.e. $T_w > T_r$ in consequence of the working principle of the tunnel [2], around 30 K. So, a heat transfer from the wall to the flow can give a trend to more instability in the BL which could promote earlier transition.

#### 4.6.2.6 Conclusion and Aspects

To validate the SWK test section flow, disturbance field and cone transition measurements have been performed. The first time, by means of a new developed technique in a short duration facility, a short-term modal analysis within *one* run has been performed. It was found, that the acoustic noise is the dominating disturbance mode in the free stream of the test section flow. Vorticity and entropy spottiness, both of stochastic nature in each individual tunnel run, originating in the filling process of the driver tube are modes found one order of magnitude smaller compared with the acoustic noise. With a settling time of 15 min after end of the charging process, a reproducibility of the unavoidable existing disturbance field superimposed on the test section flow has been confirmed. The acoustic noise level found is low in comparison with conventional wind tunnels. The available results comprising the low noise level particularly with the accompanying spectral energy distribution, concentrated at low frequencies, as well as the $R_{Trans}$-value lying clearly in the range of flight data, have confirmed predictions made by Beckwith et al. [2-17], maintaining that there can exist a so-called "turbulent quiet test core" where $R_{Trans}$-values can be attained in the range of flight data, providing a large enough test section. Essential reason is the existing disturbance spectrum in a frequency range which does not include (measurable) relevant critical frequencies of the dominating instability (first mode at $M=2.5$) in the BL of a conical or common generic body. Also the assertion of Stetson [2-1] (that only disturbance frequencies matching the most critical unstable frequencies will be relevant for transition) is confirmed by these results too.

Though there is still one handicap in the test section flow up to now. Highly localized mean flow disturbances caused by imperfections in the nozzle and wall surfaces as well as in irregularities of the nozzle contour and corner flow effects result in a large amount of low frequency shivering Mach waves of considerable high intensity levels along the centre line axis. It is assumed, that these disturbances are triggering the transition process and have led to low $R_{Trans}$-values at lower $R$-values, where transition location is still downstream of all these perturbations. To eliminate these almost stationary pressure waves, the surface quality of the test section walls and the continuity of the nozzle curvature will be improved and has been already performed on the nozzle blocks for higher Mach numbers.

In summary we can say even with a turbulent BL on the test section walls, the SWK will be remarkably advantageous for the global simulation of natural laminar turbulent transition studies on generic hypersonic configurations in the supersonic Mach range.

References to Section 4.6.2

1. Burns, K.A., Deters, K.J., Haley, C.P., Khilken, T.A.: Viscous effects on complex configurations. Air Force Wright Laboratory Technical Report WL-TR-3059,1995.
2. Knauss, H., Riedel, R., Wagner, S.: The Shock Wind Tunnel of Stuttgart University. A Facility for Testing Hypersonic Vehicles. AIAA paper 99-4959, 1999.
3. Morkovin, M.V.: Fluctuations and Hot Wire Anemometry in Compressible Flows. AGARDograph 24.
4. Gaisbauer, U., Knauss, H., Wagner, S., Weiss, J.: The meaning of disturbance fields in transition experiments and their detection in the test section flow of a short duration wind tunnel. In S. Wagner, U. Rist, J. Heinemann and R. Hilbig, editors, New Results in Numerical and Experimental Fluid Mechanics III, volume 77 of Notes on Numerical Fluid Mechanics, pages 403–410. Springer, 2000.
5. Weiss, J.: Experimental Determination of the Free Stream Disturbance Field in the Short Duration Supersonic Wind Tunnel of Stuttgart University, Dissertation, Institut für Aerodynamik und Gasdynamik, Universität Stuttgart, 2002.
6. Knauss, H.*, Gaisbauer, U.*, Wagner, S. (*IAG); Buntin, D.[2]; Maslov, A.[2]; Smorodsky, B. ([2]ITAM); Betz, J. (Fortech): Calibration Experiments of a New Active Fast Response Heat Flux Sensor to Measure Total Temperature Fluctuations, Part I-III. Proceedings of XI International Conference on the Methods of Aerophysical Research, ICMAR, Part III, 2002 Part 1–3.
7. Weiss, J., Knauss, H., Wagner, S.: On the total temperature sensitivity of constant-temperature Anemometers. Proceedings Part III, International Conference on Methods of Aero-Physical Research, ICMAR, 2002, Novosibirsk.
8. Gaisbauer, U., Knauss, H., Wagner, S., Weiss, J.: Measurement techniques for the detection of flow disturbances and transition localization in a short duration wind tunnel. In Kharitonov, A.M., editor, Proceedings of the International Conference on the Methods of Aerophysical Research, pages 54–67, Russian Academy of Sciences, 2000, Novosibirsk, Russia.
9. Weiss, J., Knauss, H., Wagner, S.: Method for the determination of frequency response and signal to noise ratio for constant-temperature hot-wire anemometers. Review of Scientific Instruments, 72(3):1904–1909, 2001.
10. Weiss, J., Kosinov, A.D., Knauss, H., Wagner, S.: Constant temperature hot-wire measurements in a short duration supersonic wind tunnel. The Aeronautical Journal, 105(1050):435–441, 2001.
11. Riedel, R.: Erprobung eines Stoßwindkanals zur Untersuchung des laminar-turbulenten Umschlags in Überschallgrenzschichten. Dissertation, Institut für Aerodynamik und Gasdynamik, Universität Stuttgart, 2000.
12. Gaisbauer, U.: Untersuchungen zur Stoß-Grenzschicht-Wechselwirkung an Doppelrampen unter verschiedenen Randbedingungen. PhD-thesis, Stuttgart University, 2004.
13. Kloker, M.: Private communication. Institut für Aerodynamik und Gasdynamik.

### 4.6.3 Plasma Wind Tunnels

The term plasma wind tunnel is to be understood as a steady state test facility in which a high-enthalpy flow is produced with the help of a plasma generator. For the air system the simulation regimes of the PWKs at IRS are shown in Fig. 4.6.15 together with some typical re-entry trajectories [1]. Here, pressure ranges and local specific enthalpies in the plasma jets are related to corresponding velocities and altitudes of the space vehicle.

Three different vacuum chambers can be operated with different types of plasma generators. Fig. 4.6.16 shows a photograph of the vacuum chambers PWK1 and PWK2 which are 5 m long steel tanks with a diameter of 2 m. The housing of PWK1 is divided into three cylindrical segments, PWK2 contains only one segment, all of which have a water cooled double-wall.

The plasma source is flanged from the outside to the plate which covers the tank on a cone-shaped element which prevents deformation under vacuum conditions. Each tank is equipped with a 4-axis positioning system on which the different probes and the specimen support system can be mounted. This al-

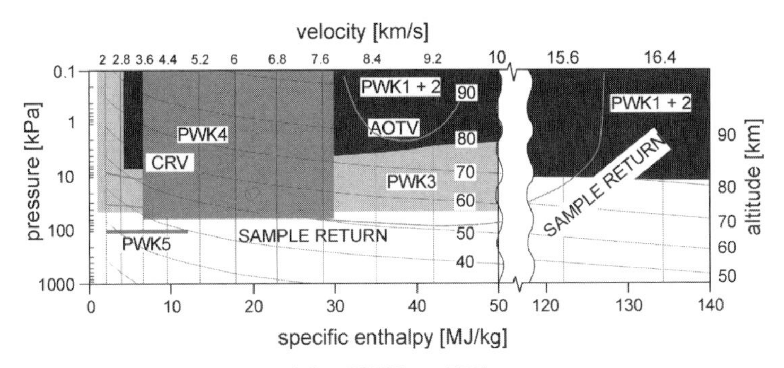


Figure 4.6.15: Simulation regimes of the PWKs at IRS.

Figure 4.6.16: The plasma wind tunnels PWK1 (left) and PWK2 (right).

lows the simulation of parts of the re-entry trajectories by moving the specimen in different plasma flow regimes. The whole plasma jet range is accessible by optical methods. Windows with optical glass allow for pyrometric temperature measurements on the front side of the specimen at distances from the plasma source between about 50 mm and 1 m. Moreover, optical measurements perpendicular to the plasma jet axis are possible through three movable flanges of three optical glasses each, which are located on both sides of the tank opposite each other and on the top (Fig. 4.6.16). As long as the described vacuum tanks are equipped with magnetoplasmadynamic generators (MPG), these plasma wind tunnels are called PWK1 or PWK2. If, instead, a thermal arcjet plasma generator (TPG) is used, they are called PWK4.

PWK3 consists of the inductively heated plasma generator (IPG) and the vacuum chamber with a size of about 2 m in length and 1.6 m in diameter. Material support systems and mechanical probes can be installed onto a moveable platform (2-axis) inside the tank. The plane lid of PWK 3 carries the plasma generator and the external resonant circuit, which contains the capacitors with the connection to the IPG. Again, windows allow optical access to the plasma jet and also to the inside of the generator.

The plasma wind tunnels PWK 1–4 are connected to a central gas supply system and to the central vacuum system with a total suction power of 6000 $m^3/h$ at atmospheric pressure and about 250 000 $m^3/h$ at 10 Pa. Apart from PWK3, they are all connected to the central power system with a maximal power up to 6 MW at a maximal current of 48 kA or a maximal voltage of 6000 V.

### 4.6.3.1 Plasma Generators

With inductively heated and arc-heated plasma wind tunnels, two generally different concepts of plasma generation are operated at IRS. Again, the latter use two basically different types of plasma generators, namely thermal (TPG) and magnetoplasmadynamic (MPG) generators.

#### *4.6.3.1.1 Arc-Driven Plasma Generators (TPG and MPG)*

In TPGs the test gas is heated by means of an electric arc and accelerated through a nozzle. For MPGs, additional electromagnetic forces are used to further accelerate the plasma. Both devices are water-cooled with the exception of the 2% thoriated tungsten cathode which reaches more than 3000 K during the steady-state operation. Since contact between the oxygenic part of the test gas and the cathode has to be avoided to minimize electrode erosion, nitrogen and oxygen are supplied separately for re-entry simulation. Only the nitrogen is guided along the cathode and the oxygen is fed downstream. Instead of oxygen other gases such as $CH_4$ and $CO_2$ can also be mixed into the main flow. In

PWK2, for example, the qualification tests were carried out for the heat shield of Huygens for the entry into the atmosphere of Saturn's moon Titan [12]. In operation, the current is controlled and the voltage adjusts itself. Further adjustable parameters are the mass flow rate, the ambient pressure and the distance of the measurement probes to the exit of the plasma source.

Both devices heat the gas by an arc but the MPGs use the nozzle itself as anode while the TPGs have a settling chamber and the nozzle is electrically separated from the anode. This enables higher settling chamber pressures and consequently higher velocities and pressures for the TPGs at the cost of specific enthalpy. Fig. 4.6.17 shows examples for the different plasma generators.

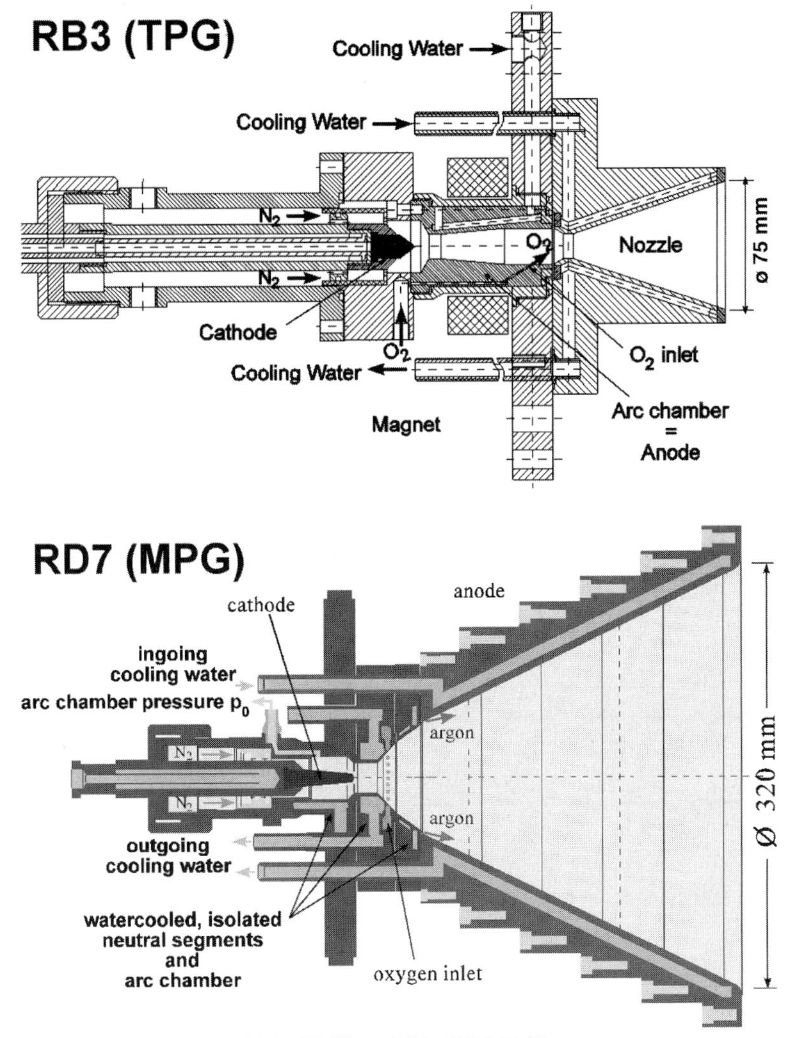


Figure 4.6.17: Plasma sources RB3 (TPG) and RD5/7 (MPG).

Self-field MPGs of different sizes have been developed over the past decades at IRS, they are employed successfully in the IRS plasma wind tunnels PWK1 and PWK2. The nozzle-type MPG plasma generators consist of two coaxial electrodes, separated by neutral, water-cooled copper segments. The arc is ignited by a Paschen breakdown. The current passes through the expansion nozzle from the cathode tip to the end of the nozzle. The test gas is dissociated and partly ionized. The magnitude of the magnetic acceleration force strongly depends on the current level of operation. The nozzle exit, which is also a water-cooled segment, forms the anode. The tungsten cathode is mounted in the center of the plenum chamber. The high-temperature and the low-work function of the cathode result in a diffuse arc attachment and, consequently, a very low cathode erosion rate at the order of sublimation. This results in operation periods of the MPG of hundreds of hours. The oxygen needed for the duplication of high-enthalpy air flows is fed in radially at a high velocity at the supersonic part of the nozzle but still within the arc region. To avoid a spotty arc attachment on the anode, which would cause a contamination of the plasma flow, a small amount of argon can be injected tangentially along the anode contour. This method has been shown experimentally to remarkably decrease anode erosion [5]. For testing large TPS structures the MPG generator RD7 with an enlarged nozzle has been built, which enables the investigation of TPS materials up to 400 mm in diameter [10].

In the TPG, the anode is a water-cooled copper cylinder which is electrically insulated against the nozzle. The nitrogen is passed along the cathode into the plenum chamber and the oxygen is injected at the downstream end of the anode towards the nozzle throat. To ensure a good mixing of the nitrogen and the oxygen, the injection point is positioned in the subsonic part of the TPG such that a backflow of oxygen into the cathode region cannot be ruled out completely. However, tests have shown that the cathode erosion rate due to the bi-throat design for RB3 is as low as observed in the MPGs operated in PWK1 and PWK2 and in the order of the sublimation rate [1, 5, 10]. To avoid anode erosion caused by spotty arc attachment, a coil is used to generate an axial magnetic field which moves the arc rapidly around. Table 4.6.1 sum-

Table 4.6.1: Characteristic operation parameters for MPGs and TPGs at IRS.

| No. | Generator | $P_{(\text{inf})}$ [Pa] | $P_{\text{tot}}$ [Pa] | $\dot{m}$ [$10^{-3}$ kg/s] | I [A] | x-Pos. [mm] | $T_{\text{surface}}$ [K] |
|---|---|---|---|---|---|---|---|
| I | RD5 | 1050 | 1400 | 8.0 $N_2$/2,0 $O_2$ | 1400 | 575 | C/C-SiC: 1893 |
| II | RD5 | 290 | | 450 1.6 $N_2$/ 0.4 $O_2$ | 1200 | 117/300/ 467 | C/C-SiC: 1673 |
| VI | RD2/5 | 290 | | 1.3 $N_2$ | 750 | 50/300 | |
| VII | RD5 | 490 | 800 | 6.4 $N_2$/1.6 $O_2$ | 1235 | 368 | SiC: 1973 |
| VIII | RD5 | 2770 | 3500 | 14.4 $N_2$/3.6 $O_2$ | 880 | 346 | SiC: 1853 |
| IX | RB3 | 330 | 3500 | 7.2 $N_2$/1.8 $O_2$ | 950 | 300 | SiC: 1853 |

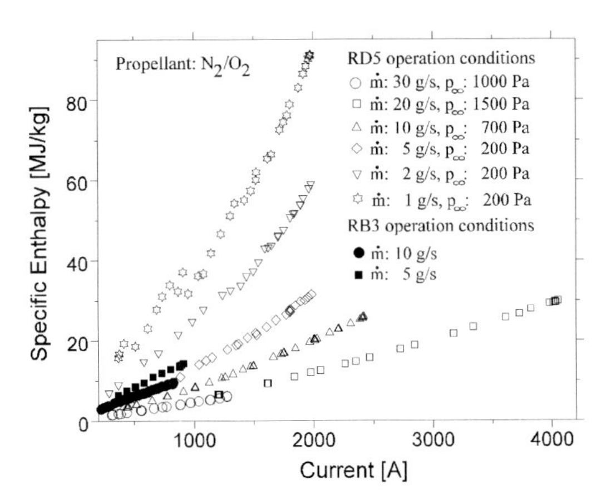


Figure 4.6.18: Specific enthalpy vs current for RD5 and RB3 in the moderate enthalpy range [10].

marizes some typical characteristic parameters (ambient, total and settling chamber pressure $p_\infty$, $p_t$, and $p_0$, nozzle exit diameter $D_{\text{nozzle}}$ mass flow rate $\dot{m}$, current $I$, average specific enthalpy $\overline{h}$, electrical power $P_{el}$ and heat flux on copper $\dot{q}$) of both TPGs and MPGs.

In Fig. 4.6.18 the mean specific enthalpy $\overline{h}$ at the nozzle exit of the plasma generator as a function of the current intensity is shown for various mass flows for the RD5 including data of RB3. Low mass flow and high current intensity make it possible to reach specific enthalpies of up to 150 MJ/kg which can, e.g., appear during return missions to Earth or during aerobreaking manoeuvres. The specific enthalpies attainable with an MPG are higher than those reached with a TPG due to additional acceleration of the gas through magnetic Lorentz forces, lower radiation losses and heating the gas up to the nozzle exit.

In the early phase of SFB 259, a number of test cases were assessed [4] based on HERMES trajectory data (No I and II) and in cooperation with NASDA (No VII–IX) with respect to HOPE-X [2]. Recently, these conditions had to be adapted to specific problems (e.g. for the investigation of passive and active oxidation or for catalytic effects a large number of different conditions are necessary). Nevertheless, the environmental conditions of the originally assessed conditions are given in Tab. 4.6.2 since various diagnostic methods will refer to these points. A more detailed list can be found in [11].

#### 4.6.3.1.2 Inductively Heated Plasma Generators (IPGs)

Due to the cathode sublimation of the TPG and MPG generators, plasma pollution may be present, yielding unwanted chemical reactions in front of the material probe. Therefore, the electrode-less, inductively heated plasma generators IPG3, IPG4 and IPG5, which enable the generation of metallic particle free

Table 4.6.2: Environmental data for selected plasma conditions.

| $p_{(\text{inf})}$ [hPa] | $p_t$ [hPa] | $p_0$ [hPa] | $D_{\text{nozzle}}$ [mm] | $\dot{m}$ [g/s] | I [kA] | $\overline{h}$ [MJ/kg] | [ $P_{el}$ [kW] | $\dot{q}$ [MW/m$^2$] |
|---|---|---|---|---|---|---|---|---|
| MPG | | | | | | | | |
| 0,5–70 | 1–100 | 20–400 | 125/320 | 0.3–50 | 0.2–4 | 2–150 | ≤1000 | up to 100 |
| TPG | | | | | | | | |
| 0,5–1000 | 1–1000 | 400–2000 | 75 | 2-50 | 0.05–1.2 | 1–30 | ≤500 | 0.1–5 |

plasma flows, have been developed at IRS. The inductively heated plasma generators (IPG) also belong to the group of thermal plasma generators. With this electrode-less design the plasma is produced by inductive heating using a triode-driven radio frequency power supply. The behaviour of the probe and the plasma in front of it can be considered to be unadulterated. Even rather reactive gases (oxygen, carbon dioxide) which would produce strong electrode erosion in the arc-heated devices, are applicable. The influence of the single gas components can be investigated e.g. with regard to the catalytic behaviour of TPS material and to a comparison with the arc-heated plasma. In addition, pure atmospheres of celestial bodies like Mars or Venus can be simulated.

An inductive plasma generator basically consists of an induction coil surrounding a quartz tube and capacitors. This resonant circuit is fed by an energy supply. The alternating current in the coil induces a mostly azimuthal electric field inside the quartz tube. This field initiates an electric discharge in the gas

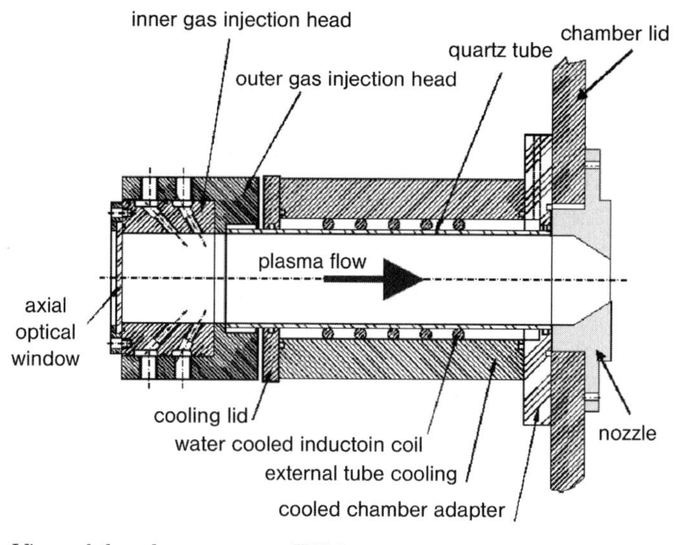


Figure 4.6.19: View of the plasma source IPG4.

which is injected at one side into the tube (Fig. 4.6.19). The produced plasma is expanded into the vacuum chamber. The electric discharge in the plasma is carried by mostly azimuthal currents. The current amplitude and thus the Ohm heating strongly depends on the electric conductivity of the plasma and the frequency of the electric circuit.

The plasma generator IPG4 is shown in Fig. 4.6.19. The gas injection head enables different gas injection angles. The quartz tube contains the plasma, which leaves the generator through the water-cooled vacuum chamber adapter. The induction coil is connected to the external resonant circuit. Furthermore, both the tube and the coil are surrounded by the external tube cooling, which protects the tube from overheating. The total length of IPG4 is about 0.4 m, its diameter about 0.08 m.

The Meissner type resonant circuit is supplied by the DC anode power $P_A$, which is calculated from the measured anode voltage $U_A$ and the anode current $I_A$ during the operation of the device. The anode voltage is controlled. Hence, the anode current results from the load of the resonant circuit (plasma) and the accompanying operating conditions.

#### 4.6.3.2 Heat Flux Simulation for X-38 Using PWK1 as Example for PWK Investigation

As an example for typical tasks in plasma wind tunnels, the functional qualification and acceptance tests for PYREX-KAT38 FM are described which were performed under the X-38 heat flux with the magnetoplasmadynamically driven PWK1 using RD5 [8]. The stagnation point heat flux according to the so-called cycle 8 was used. Fig. 4.6.20 demonstrates the good correspondence.

These results were achieved by programming the sample holder position in PWK1 using a numerically controlled four-axis table. The calibration was performed with a measured axial profile using a calorimetric copper based heat

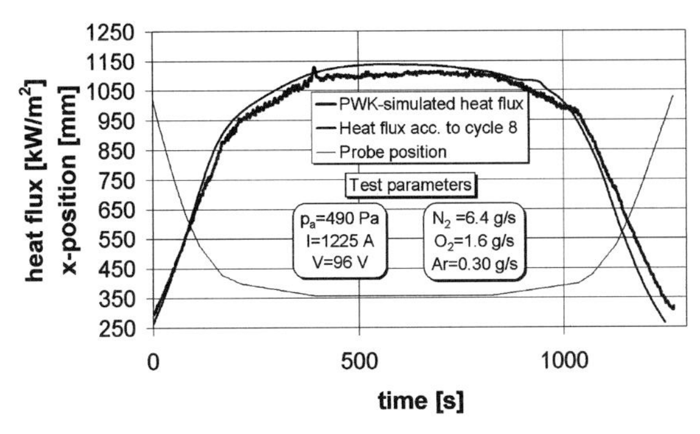


Figure 4.6.20: PWK-simulated X-38 heat flux profile (IRS-PWK1) and PWK-related test parameters.

flux sensor with the same geometries. All tests of PYREX-KAT38 were done with similar sections as in the X-38 thermal protection structure.

The set-up of the PYREX-KAT38 FM with all components was integrated in the plasma wind tunnel. The PWK tests additionally provided information to estimate front side temperatures of X-38 using a sample of the same thickness. Test geometries were according to the X-38 nose cap area to reproduce the heat conductivity features of the zone. Thermocouples were placed in the insulation, along the sensor head (SH) SiC tube and on the upper flange of the SH. With the FM test original C/C-SiC material samples (DLR-Stuttgart) were used. The measured data can be used as information for the front side temperature encountered by the X-38 vehicle. After the test, the calibration of the tested channel was checked. PYREX-KAT38 FM withstood this test successfully. For both tests all functional requirements were fulfilled very well. Both the temperature profile (TCs) through the structure itself (TPS-material, insulation, Al-structure) and the profile along the sensor could be verified. An important temperature was the maximum value on the upper flange of the sensor head which did not exceed 160 °C.

Acknowledgements

The authors are grateful to R. Riedel and J. Weiss for their contributions to section 4.6.2. Jennifer Baer-Engel, Patrizia Dabalá, Pia Endlich, Helmut Früholz, Manfred Hartling, Gerhard Hilfer, Harald Habiger, Stefan Laure, Torsten Laux, Wolfgang Röck, Karl-Heinz Schneider, Edgar Schreiber, Thomas Stöckle, Thomas Wegmann as well as all staff members of the Institute of Space Systems and many students are gratefully acknowledged by the authors for their contributions, either in form of scientific results or by assistance for the development, manufacturing and operation of the experimental set-ups and facilities.

References to Section 4.6.3

1. Auweter-Kurtz, M., Wegmann, T.: Overview of IRS Plasma Torch Facilities, RTO-EN-8. AC/323(AVT)TP/23, RTO Educational Notes 8, April 2000; von Karman Institute for Fluid Dynamics, RTO AVT/VKI Special Course on Measurement Techniques for High Enthalpy Plasma Flows, Belgium, Oktober 1999.
2. Auweter-Kurtz, M., Habiger, H., Hilfer, G., Laux, T., Wegmann, Th., Winter, M.: Comparison of a Magnetoplasmadynamic and a Thermal Plasma Wind Tunnel as Tools for the Investigation of Heatshield Materials for Space Transportation Systems, Final Report for the DARA Grant, FKZ: 50 TT 96163, 2000.
3. Auweter-Kurtz, M., Bauer, G., Behringer, K., Dabalá, P., Habiger, H., Hirsch, K., Jentschke, H., Kurtz, H., Laure, S., Stöckle, T., Volk, G.: Plasmadiagnostics within the Plasma Wind Tunnel PWK, Zeitschrift für Flugwissenschaften und Weltraumforschung, Vol. 19, No. 3, Springer, June 1995, 166–179.

4. Auweter-Kurtz, M., Dabalá, P., Feigl, M., Habiger, H., Kurtz, H., Stöckle, T., Wegmann, Th., Winter, M.: Diagnostik mit Sonden und optischen Verfahren, Arbeits- und Ergebnisbericht 1998 zum Sonderforschungsbereich 259 Hochtemperaturprobleme rückkehrfähiger Raumtransportsysteme, Stuttgart, Germany, July 1998.
5. Auweter-Kurtz, M., Kurtz, H.L., Laure, S.: Plasma Generators for Re-Entry Simulation, Journal of Propulsion and Power, Vol. 12, No. 6, 1996, 1053–1061.
6. Herdrich, G., Auweter-Kurtz, M., Endlich, P.: Mars Entry Simulation using the Inductively Heated Plasma Generator IPG4, Engineering Note for the Journal of Spacecrafts and Rockets Vol. 41, September–October 2003.
7. Herdrich, G., Auweter-Kurtz, M., Fertig, M., Laux, T., Pidan, S., Schöttle, U., Wegmann, Th., Winter, M.: Atmospheric Entry Experiments at IRS, 3rd Int. Symposium Atmospheric Re-entry Vehicles and Systems, Arcachon, France, March 2003.
8. Herdrich, G., Auweter-Kurtz, M., Hartling, M., Laux, T.: PYREX-KAT38: Temperature Measurement System for the X-38 Nose Structure TPS, 2nd Int. Symposium Atmospheric Re-entry Vehicles and Systems, Arcachon, France, March 2001.
9. Herdrich, G., Auweter-Kurtz, M., Kurtz, H., Laux, T., Winter, M.: Operational Behavior of Inductively Heated Plasma Source IPG3 for Entry Simulations, Journal of Thermophysics and Heat Transfer, Vol. 16, No. 3, July–Sept. 2002.
10. Laure, S.: Experimentelle Simulation der Staupunktströmung wiedereintretender Raumflugkörper und deren Charakterisierung mittels mechanischer Sonden, Dissertation, Universität Stuttgart, Mai 1998.
11. Laux, T.: Die Testbedingungen in den IRS-Plasmawindkanälen. 2. Ausgabe, interner Bericht, 2001.
12. Röck, W.: Simulation des Eintritts einer Sonde in die Atmosphäre des Saturnmondes Titan in einem Plasmawindkanal, Dissertation, Universität Stuttgart, Dezember 1998.

## 4.7 Characterization of High-Enthalpy Flows

Monika Auweter-Kurtz, Markus Fertig, Georg Herdrich, Kurt Hirsch, Stefan Löhle, Sergej Pidan, Uwe Schumacher, and Michael Winter*

The accuracy of the simulation of re-entry conditions strongly depends on the ability to determine the flow conditions. The minimum set of parameters which have to be duplicated during the tests for material qualification are the specific enthalpy of the gas, the stagnation pressure and the surface temperature in the case of a radiation-cooled material, or rather the heat flux for an ablative material (compare Section 4.6.). A whole series of probes and non-intrusive techniques were developed to determine these parameters.

* Universität Stuttgart, Institut für Raumfahrtsysteme, Pfaffenwaldring 31, 70550 Stuttgart; winter@irs.uni-stuttgart.de

Typical re-entry plasma flows are in thermal and chemical non-equilibrium. For building a relation between the plasma state in a ground test facility and a determined trajectory point in real flight and for validating theoretical models and numerical simulations, as many quantities as possible have to be determined. Obviously, a material sample to be investigated has to be exposed to the plasma flow and the sample mount has to fulfill various constraints especially concerning heat transfer to and from the material sample. The high-enthalpy plasma flows themselves are typically characterized by the total pressure, the species concentrations, the velocities and the different temperatures which define the internal energies of the plasma components. If these values can be determined experimentally, the degrees of ionization and dissociation and the specific enthalpy of the flow can be deduced. In addition, the measurement method should not influence the flow field if possible. If boundary layer effects and gas surface interactions are investigated, non-intrusive methods (e.g. optical measurements) are strictly required.

Within the evaluation of some basic physical models even separate information from only one measurement method may be sufficient but, generally, the simultaneous application of different methods is essential. Reported plasma diagnostic methods are optical measurements such as emission spectroscopy, laser induced fluorescence, laser absorption, Raman spectroscopy, coherent anti Stokes Raman spectroscopy (CARS), laser-based velocity measurements, Thomson scattering, Schlieren methods, Fabry-Perot interferometry, thermography and pyrometry. But also intrusive measurements like electrostatic probes, static and dynamic pressure probes, wedge type probes, heat flux probes and mass spectrometry are in use and necessary to complete the picture of the flow field. The applicability of each method depends on the plasma condition and therefore also on the type of ground test facility [3].

Having a sufficient understanding of the flow may even make it possible to gain valuable information from selected measurements in combination with a CFD analysis, e.g. when flight experiments are conducted [4].

### 4.7.1 Intrusive Measurement Methods

As the name already suggests, each of these diagnostic methods is based on a suitably designed probe being placed in the plasma stream which is to be investigated. This differs from the optical measurement techniques which will be described in the next section. All of these probes can be installed at the IRS on four-axis platforms inside the plasma wind tunnels. The mechanical probes are among the most important instruments for plasma-diagnostic measurements. They can essentially be divided into five groups according to the parameters to be investigated. We differentiate between material sample support, Pitot pressure probes, aerodynamic wedge probes, heat flux probes, enthalpy probes and solid-state electrolyte probes [3].

The wedge probe is used to ascertain the static pressure and the Mach number of the supersonic flow. A solid-state electrolyte probe can determine the oxygen partial pressure. With a mass spectrometer probe, particle densities and the energy distributions of the particles can be investigated. With the exception of the Mach number and the oxygen partial pressure measurements [26], all of the measuring techniques with mechanical probes are ultimately based on the measurements of pressure, flux and/or temperature whereby the attempt must be made to ascertain these parameters with the highest possible accuracy. In practice, these supposedly simple measurements are already very difficult [3]. A more or less complex theory is hidden between determining the basic parameters and ascertaining the parameters of particular interest, especially in the case of enthalpy and heat flux. Besides the enthalpy probes, all mechanical probes have one common denominator – they do not actively influence the plasma, but rather quasi passively register the effects of the plasma flow in an appropriate way. This information is then used to determine the important parameters.

Besides the so-called mechanical probes, the mass spectrometry, electrostatic and radiation probes also belong to the group of intrusive measurement techniques. A mass spectrometer probe can be used with low pressure plasmas to investigate the composition of the plasma [6, 24, 25]. Due to catalytic effects and the large pressure reduction which is necessary, the ability of this system to be applied at pressures above 1 hPa is limited and a two-stage system is needed. However, a mass spectrometer probe can be used advantageously for on-line investigations of erosion behaviour [6] and the catalytic behaviour [28, 29] of thermal protection materials. Important information about the erosive behaviour, catalytic efficiency and the failure of surface protection layers has been obtained using a mass spectrometer positioned directly behind the material sample within the plasma wind tunnel [6].

Electrostatic probes are used to ascertain the plasma potential, electron density and temperature, energy distribution of the electrons, ion temperature, flow velocity and flow direction. The measurement principle is based on an active influence on the plasma boundary layer which forms on the probe [11, 12].

Table 4.7.1: Different kinds of electrostatic probes and determinable plasma parameters [12].

| Probe | Parameters |
|---|---|
| Single probe | Electron temperature $T_e$, electron density $n_e$, plasma potential, electron energy distribution |
| Double probe | Electron temperature $T_e$, electron density $n_e$ |
| Triple probe | Electron temperature $T_e$, electron density $n_e$ |
| Time of flight probe | Plasma velocity $v_{pl}$ |
| Angle probe | Plasma flow line |
| Rotation probe | Plasma flow line |
| Crossed probe | $v_{pl}$ at known ion temperature, ion temperature at known $v_{pl}$ |

The use of radiometric probes is unavoidable when the radiation heat flux can not be neglected compared to the convective part. This is the case when during sample return missions the entry speed into the Earth's atmosphere is especially high or when the atmosphere of another celestial body which is to be entered contains strongly radiating species, as for example the atmosphere of Titan [22].

Due to their importance for investigation of the catalytic and oxidation behaviour (compare section 7.2) the material sample support system and the heat flux probe are discussed more intensely. All other measurement methods are explained in great detail in [3].

#### 4.7.1.1 Material Sample Support System

The material sample support system minimizes the heat exchange of the sample with the supporting structure in axial as well as in radial direction. The rear side temperature of the sample can be determined with a linear pyrometer via a fibre optic connection [3]. Figure 4.7.1 shows the standard sample support for ceramic material in a stagnation point configuration and a sketch of the pyrometric measurement system. The sample is captured by a glowing SiC cap which yields a smooth temperature distribution in radial direction. Behind the sample, a strongly isolating material prevents from heat conduction to the probe.

This kind of probe is designed to block all heat flux from the sample yielding a material temperature which adapts to the particular plasma state. Unfortunately, many processes like catalysis and oxidation behaviour show a strong dependence on temperature. Even if the material temperature can be changed by choosing another probe position with respect to the plasma source or by readjusting the plasma source parameters, a change in the plasma state will follow immediately. Usually, it is rather complicated to separate the influence of material and plasma properties.

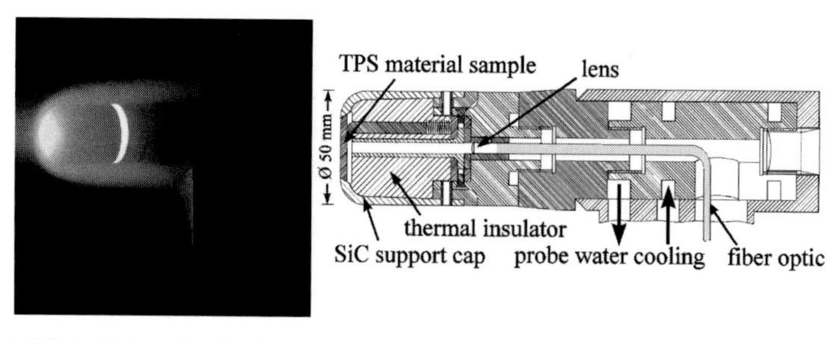


Figure 4.7.1: Material probe in the plasma flow and sketch of probe with miniaturized pyrometer.

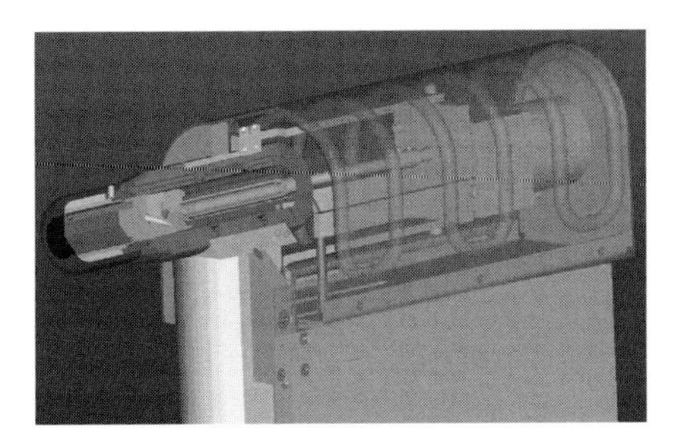

Figure 4.7.2: Probe with regeneratively cooled material sample.

Therefore, a recent development at IRS in the field of intrusive plasma diagnostics is a material probe equipped with the variable regenerative sample cooling. With this probe, it is possible to change the temperature of the hot material sample in the range of about 300 K without moving the probe into the plasma beam or changing the test conditions. To achieve the cooling effect a movable molybdenum alloy cooling body is used combined with water-cooled copper bars (see Fig. 4.7.2). The copper bars and the cooling body can be moved together towards the rear side of the material sample during the test until the cooling body touches the sample. Due to controlled movement of the bars, they can cover a variable part of the cooling body, which leads to adjustable cooling of the material sample. This probe can be engaged for the accurate temperature determination of the passive-active transition in the oxidation process on materials containing the silicon carbide and, e.g. in combination with LIF or emission spectroscopy, for the investigation of the temperature dependent catalytic efficiency of heat shield materials. The probe can also be used for conventional material testing up to material temperatures of 2300 K, if an insulation is applied instead of the water-cooling.

### 4.7.1.2 Heat Flux Measurements

The measurement of the total heat flux to a probe is one of the most important tasks for material investigations. The thermal loads on a material sample include the convective and radiative heat load from the plasma as well as the energy released by atom recombination at the surface. Typically, the heating by gas radiation can be neglected at the densities of interest for re-entry problems. The loads must be balanced by heat removal processes e.g. by heat conduction to the rear side (which is not desired for a heat shield material) or in surface direction, by endothermic chemical reactions which destroy the heat shield and by surface radiation. Obviously, the last process is desirable for re-useable heat shields. For the total heat flux measurement the material of the sample is of great importance since thermal loads vary depending on the catalytic efficiency of the material which – in a non-equilibrium situation (compare section 7.2) – influences the recombination rates at the surface. Furthermore, the surface temperature of the probe affects recombination rates and emission coefficient of the sample as well as the convective heat transfer from the plasma.

Recently, four types of heat flux probes have been applied to the re-entry plasmas. The first type uses a stationary calorimetric method where the corresponding surface is water-cooled. The heat flux is determined from the temperature difference of the in- and outflowing cooling water. As a standard reference material for heat flux probes mostly copper is used as it has a relatively high catalytic efficiency [18]. However, copper is not an ideal material as a standard reference because, in an oxidizing atmosphere, it forms two different oxides with different catalytic efficiency. The second type of stationary measurements is done with so-called Gardon Gage sensors. Here, the heat flux is determined from the temperature gradient which forms between the centre and the edge of a circular foil. This foil consists of constantan and is mounted on the front side of a hollow cylinder made of copper which is kept at constant temperature and works as a heat sink. The connection between foil and cylinder forms one thermocouple, a copper wire which is mounted to the centre of the foil builds another one. The difference between the thermocouples voltages is proportional to the heat flux applied to the sensor. The compact design of only a few mm in diameter makes them ideally suited to be integrated in a probe body. At IRS, these heat flux sensors are equipped with surface layers of different materials to study catalytic effects via the heat flux reduction versus constantan [29] (compare Section 7.2.2).

The third method of heat flux measurement uses the heating phase of a probe. If a radial heat transport is avoided the axial heat flux can be calculated from the one dimensional Fourier equation. According to the transient thin wall method the heat flux is proportional to the measured temperature rise. Here, emissivity, density, specific heat capacity and thickness of the material are needed for the evaluation. Usually, the temperature is measured by thermocouples or by a pyrometer from the outside of the PWK [20].

Recently, a pyrometric double probe – also called catalytic probe – was designed. The pyrometric heat flux probe is a combination of both transient and steady state heat flux measurement techniques. The probe is equipped with two PYREX mini-pyrometers for the transient temperature measurements of the

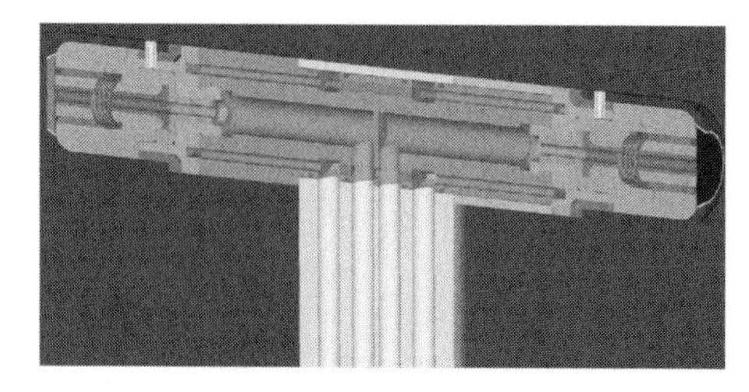

Figure 4.7.3: Pyrometric Double Probe (Catalycity Probe).

sample rear side. Figure 4.7.3 shows the probe that can be turned by 180° during a test to enable the measurement of different materials with different catalytic efficiencies under the same plasma condition [20].

In Fig. 4.7.4, typical results with this probe in a PWK environment with a Sintered Silicon Carbide (SSiC) sample are depicted. Just like with the instationary heat flux probe a heat flux can be determined from the temperature gradient. With the pyrometric double probe, only the rear side temperature is measured so the front side temperature must be measured with an external pyrometer since it is needed to calculate the radiative heat flux emitted by the sample front side. The total heat flux to the probe is the sum of the instationary and the radiative heat flux plus conductive losses where the insert is mounted to the probe. Up to now, these structural losses are not included into the determination of the total heat flux as depicted in Fig. 4.7.4 [10]. Therefore, the steady state heat flux was assumed to equal the radiative one emitted by the sample front surface.

Shortly after the probe is moved into the plasma beam a maximum in total heat flux can be noticed. Then the total heat flux slowly decreases to a steady state value. Up to that time, the sample temperature still rises. When the sample temperature is low the convective heat flux from the plasma should be at its maximum. Instead, the measured data shows smaller heat flux values prior to the maximum around $t = 12$ s which are caused by the temperature distribution inside the sample which is not yet readapted to the thermal load. The decrease from the maximum to the steady state value can be explained by a decrease of convective heat flux and an increase of structural losses both due to

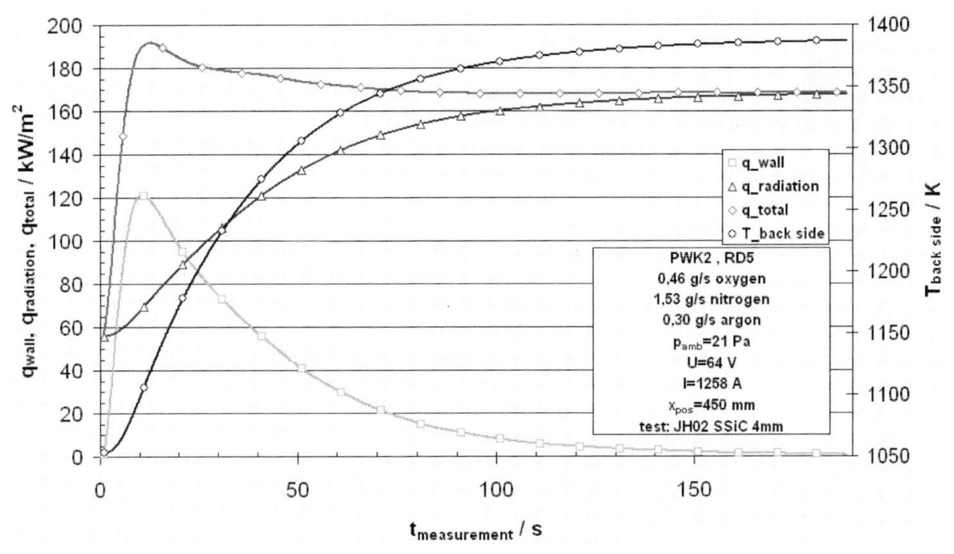


Figure 4.7.4: Heat flux measurement and rear side temperature history for SSiC using catalytic probe.

rising sample temperature. Furthermore, the surface catalytic efficiency usually is a function of temperature such that the amount of heating by recombining particles may change with surface temperature. A separation of these effects is going on.

### 4.7.2 Non-Intrusive Techniques

Most non-intrusive measurement techniques use optical methods and focus on the investigation of the plasma except for pyrometric measurements which investigate the material sample surface with respect to surface temperature, local damage and overheating effects. To characterize the high-enthalpy flows under investigation the methods of emission spectroscopy, laser induced fluorescence, Thomson scattering, Fabry Perot interferometry and laser absorption spectroscopy have been applied [1–3, 27]. Some of these methods are well suited to investigate even the boundary layer very close to the material probe surface although others are not. Table 4.7.2 summarizes the individual methods, their ability to be used for investigation of the boundary layer and the thermodynamic quantities that can be determined from the measurements. Options that are generally possible but were not applied are mentioned in brackets. Details and exemplary results are found in the corresponding subchapters.

Table 4.7.2: Measurement methods and measured physical quantities.

| Measurement method | Measured quantity | Boundary layer/ free stream |
|---|---|---|
| Emission spectroscopy (EMS) | $T_{ex}$, $T_{rot}$, $T_{vib}$, $n_{ex}$, $[n_{gs}]$, $n_e$, $T_e$, $n_{Si}$, $T_{Si}$ | Both |
| Laser induced fluorescence (LIF) | $T_{rot}$, $[T_{vib}]$, $n_{gs}$, $n_e$, $T_e$, $[v]$ | Both |
| Thompson scattering | $n_e$, $T_e$, | Free stream |
| Fabry Perot interferometry (FPI) | $T_{trans}$, $v$ | Free stream |
| Laser absorption spectroscopy (LAS) | $n_{ex}$, $T_{trans}$, $[v]$ | Free stream (boundary layer) |
| Pyrometry spatially resolved: | material sample surface temperature local defects and overheating zones | n.a. (sample surface) |

Physical quantities: $T$=temperature, $n$=particle density, $V$=velocity
Suffixes: $_{trans}$=translational, $_{vib}$=vibrational, $_{rot}$=rotational, $_{el}$=electronic, $_{Si}$=silicon, $_{ex}$=excitation, $_{ex}$=excited state, $_{gs}$=ground state, $_{e}$=electron

#### 4.7.2.1 Emission Spectroscopy

Emission spectroscopy accesses the spontaneous emission emitted by excited states of atoms and molecules. Qualitative information about the presence of different species is directly available from the measured spectra e.g. providing information about the boundary layer shape [3, 21]. In order to obtain information about temperatures and particle densities, knowledge of the distribution function of the excited states is required, except when methods based on evaluation of the line broadening are used [16, 33]. If components of the plasma are in equilibrium, a Boltzmann distribution can be used. Although the plasma states of concern are usually in non-equilibrium, at least partial validity of such a distribution function can often be found, e.g. for the rotational energy of the molecules [3] or between some of the electronically excited states of atoms. Typically, the assumption of a Boltzmann distribution does not hold for the ground state densities which consequently require rather sophisticated methods such as non-equilibrium collision radiative modelling [16] or utilization of processes like inverse pre-dissociation [32] to be determined from emission spectroscopic measurements. For plasma wind tunnel application of high enthalpy air plasmas the wavelength range under investigation reaches from the UV up to the near IR. If this whole range is to be covered, a typical set-up uses mirrors instead of lenses to inhibit chromatic aberration. Figure 4.7.5 shows a typical overview spectrum in front of a water-cooled copper probe and in front of a glowing SiC material probe. The spectrum is dominated by the emission of NO, $N_2$, $N_2^+$, N and O as well as the erosion products Si and CN.

Atomic excitation temperatures were determined from the radiation of atomic oxygen and nitrogen using a Boltzmann plot [32]. The same principle was used to determine rotational temperature of $N_2^+$ from spectrally highly re-

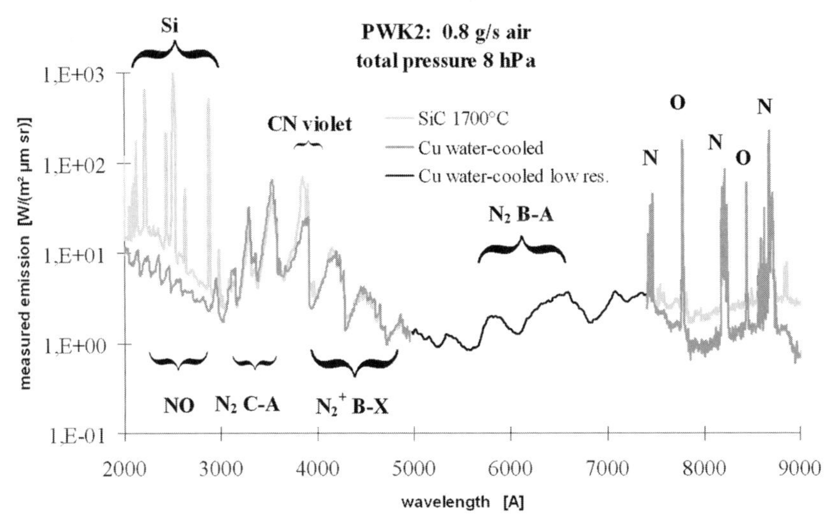


Figure 4.7.5: Overview spectra of the PWK plasma in front of a water-cooled copper surface and a glowing SiC material probe.

solved measurements but the method was found to be too time-consuming for application to the boundary layer investigation. Instead, a theoretical simulation of the molecular radiation was established and compared to measurements in low spectral resolution to determine vibrational and rotational temperatures with a two-parameter fit [3]. Recently this method was extended to contain the main radiating systems of air molecules such as $N_2$ $2^{nd}$ pos., $N_2^+$ $1^{st}$ neg., five bands of NO and $O_2$ Schumann Runge [32]. All data is available across the boundary layer in front of a water-cooled copper probe and in front of a glowing SiC material sample as well as for the free stream at the nominal probe position. As an example for the obtained results, Fig. 4.7.6 shows temperature dis-

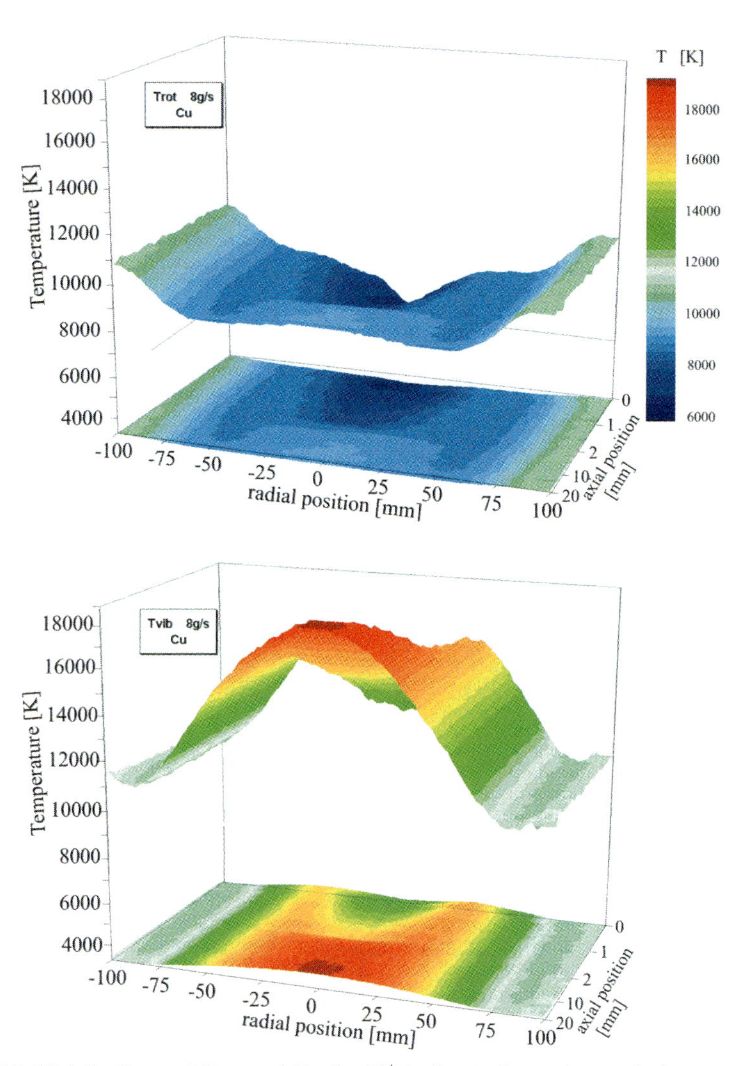


Figure 4.7.6: Distributions of $T_{\text{vib}}$ and $T_{\text{rot}}$ for $N_2^+$ in front of a water-cooled copper probe.

tributions of rotational and vibrational temperatures of $N_2^+$ in front of a water-cooled copper probe. For these measurements, a spatial resolution of 0.1 mm in axial direction could be achieved. The total number of measurement points is about 3000, each of them providing one UV, one VIS and one IR spectrum. The probe surface is located at the rear side of the diagrams, the scaling of the distance to the probe is non-linear. All results are converted from line of sight measurements to spatially resolved values by Abel inversion.

#### 4.7.2.2 Laser-Induced Fluorescence

This method uses a laser beam to excite discrete states of atoms or molecules to a higher electronic level. By measuring the subsequent fluorescence emission, information about the particle density and the rotational and vibrational temperature of the absorbing state (e.g. the ground state) as well as the translational temperature can be gained. Depending on the energy that must be provided for the desired excitation and the possible wavelength region of the applied laser, a one (e.g. molecules like NO or SiC) [8, 9, 19] or two photon excitation (e.g. for atoms like O and N) can be performed [7]. The laser pulse is generated by a dye laser which is pumped by an excimer laser. The laser light can be frequency doubled, detection can be done either by a gated photo multiplier tube or by an intensified CCD camera. No basic equilibrium assumptions are necessary to access the ground state population. If the resonance fluores-

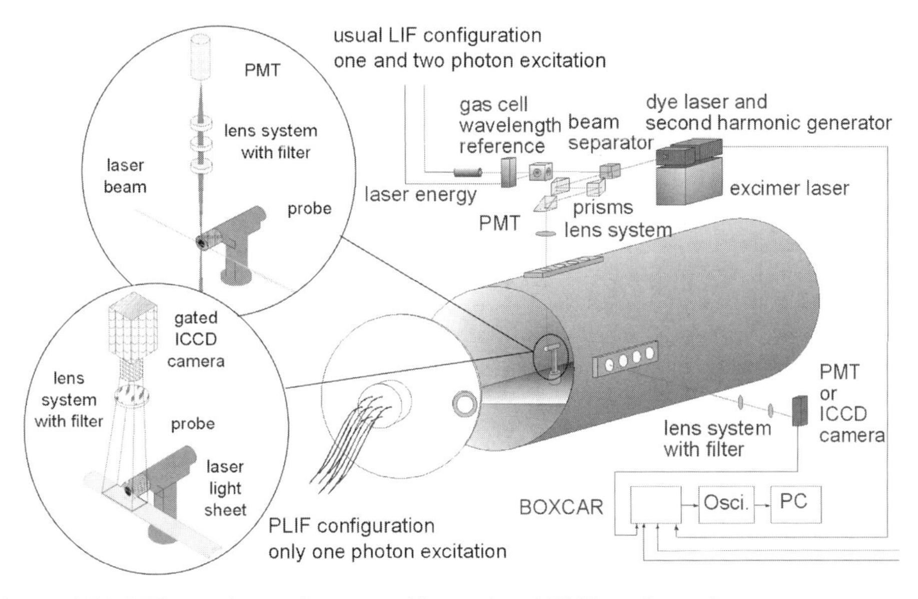


Figure 4.7.7: LIF experimental set-up with usual and PLIF configuration.

cence is measured perpendicular to the laser beam, the method itself shows an excellent spatial resolution without the necessity of an Abel inversion. Previous measurements used the detection on top of the tank and the laser in a horizontal plane, the recent set-up shows the laser on top of the vacuum chamber. Planar measurements (PLIF) can be conducted by producing a laser sheet using

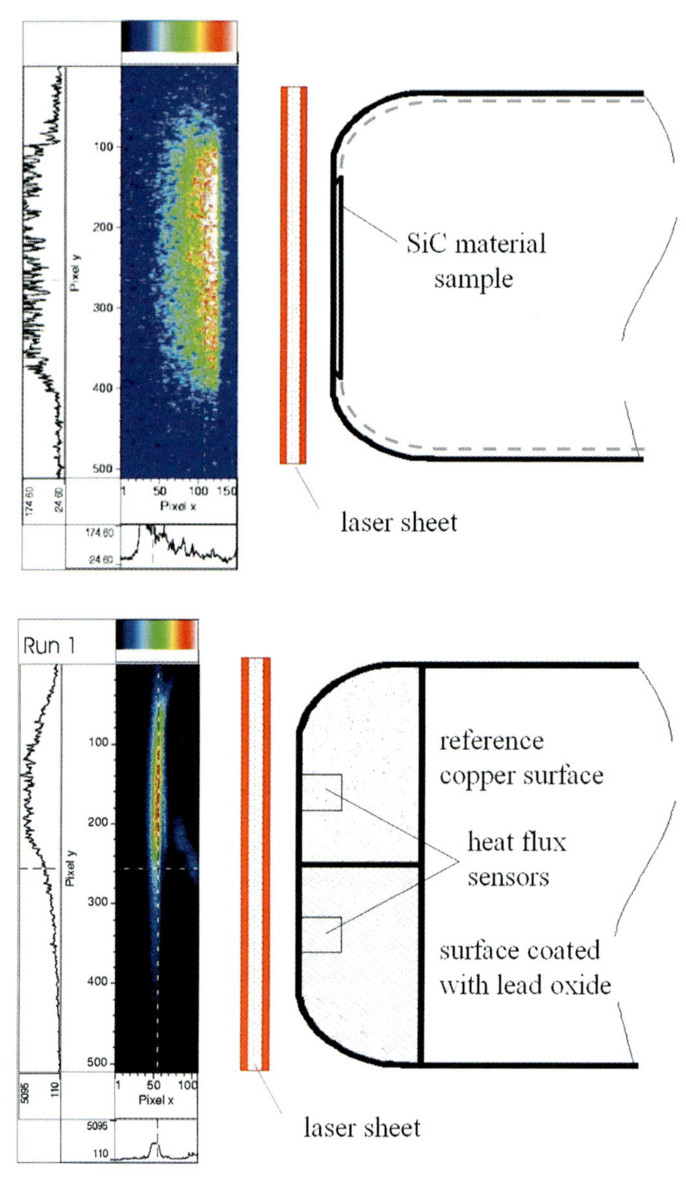


Figure 4.7.8: PLIF signals – SiO in front of a SiC material probe and NO in front of copper and lead oxide surfaces.

two cylindrical lenses. Since the laser energy in every point is much less than with the conventional LIF, this method is usually applied to one photon excitation processes. A camera is used for detection of the fluorescence signal distributions so that a whole area can be investigated and spatial distributions of the species under investigation can be obtained. Figure 4.7.7 shows the experimental set up and the two basic configurations.

Examples for PLIF results are shown in Fig. 4.7.8 for measurements of SiO as an erosion product of the glowing SiC material sample which indicates active oxidation [9] and of NO in front of a water-cooled probe [19]. One half of the probe surface was coated with lead oxide the other half was blank copper. In front of the copper surface the NO emission is remarkably stronger than in front of the coated part which indicates a higher recombination due to the different catalytic efficiency of copper and lead oxide. Both phenomena will be explained in more detail in chapter 7.2.

### 4.7.2.3 Thomson Scattering for Electron Temperature and Density Determination

Thomson scattering of intense laser light on the plasma electrons represents the direct method for electron temperature and density determination, which does not interfere with the plasma. Since the Thomson scattering cross section $\sigma_{\mathrm{Th}} = (8\pi/3)\ [e^2/(m_e c^2)]^2 = 6.65 \times 10^{-29}\ \mathrm{m}^2$ is extremely small, the successful application of this diagnostic asks for high laser intensities for the light source or for the development of sensitive detection methods. Using a frequency doubled Nd:YAG laser (delivering 0.3 J at the wavelength of 532 nm with a pulse repetition rate of 10 Hz) and an intensified CCD camera the first time-averaged results obtained for the position near the plasma jet entrance [15] were substan-

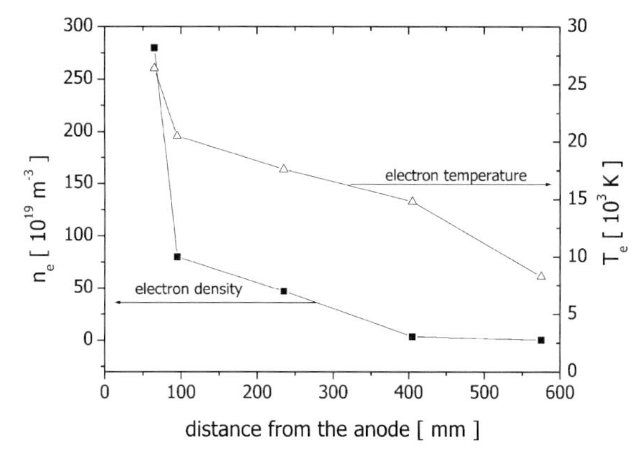


Figure 4.7.9: Axial electron density and temperature distributions of the plasma wind tunnel jet at the test case SFB I.

tially extended to single laser shot results along the plasma jet axis from the jet entrance down to the position of the probe, which is given in Fig. 4.7.9 [23]. The electron density determination covers a range from $1.3\times10^{21}\,m^{-3}$ to $3\times10^{18}\,m^{-3}$, while the electron temperature decreases from a value of 26 000 K near the plasma source down to 8000 K at the position of the probe. The low density limit is close to the goal reached by complicated diagnostic methods like cavity-ring-down spectroscopy.

Thomson scattering results are compared to those of in-situ visualization techniques and of different spectroscopic measurements from Stark broadening of emission lines and of their intensities using a collisional radiative model [14, 16, 30].

#### 4.7.2.4 High-Resolution Spectroscopy and Fabry Perot Interferometry

These methods take advantage of the Doppler effect which shifts the emission of plasma particles according to their velocity component in the direction of the optical axis. Since the translational temperature can be expressed as an undirected movement of the plasma particles with a mean thermal velocity, each particle can have a different velocity component in the direction of the optical axis. If the emission of all particles is superposed, a broadened emission line is obtained. Under the assumption of a Maxwellian distribution of the particle velocities, the translational temperature can be determined from the measured line broadening [12, 21, 33]. The significant advantage of this method is that

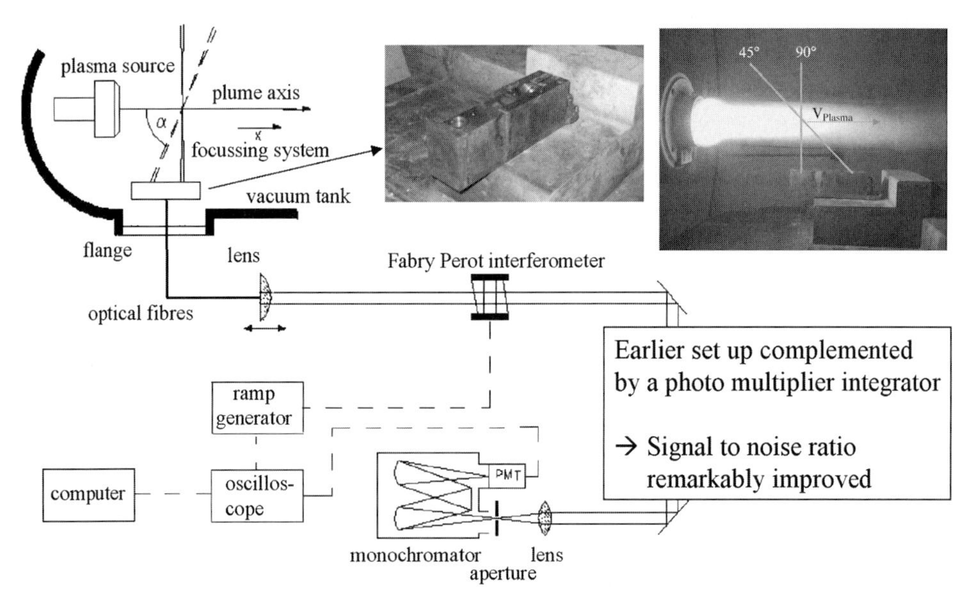


Figure 4.7.10: Experimental set-up for Fabry Perot interferometry.

apart from this very basic assumption, no other equilibrium distributions are necessary. A disadvantage is the often poor spatial resolution and the line of sight measurement which require sophisticated methods to obtain spatially resolved results. With the use of an Echelle spectrometer and a Fabry Perot interferometer two different methods have been applied.

The Fabry Perot interferometer allows for a resolution of up to $3 \times 10^5$. Translational temperatures determined from the emission of atomic nitrogen show good agreement with ion temperatures determined with electrostatic probes [12]. The set-up also allows plasma velocities to be determined. If an emission line of the plasma is detected at different angles (e.g. perpendicular and at 45° to the plasma jet) the broadened line profile at 45° will be shifted due to the plasma velocity while the emission line detected perpendicular to the jet remains unshifted. This difference yields the plasma velocity [12]. To access the measurement under different angles an optical probe to be placed inside the plasma wind tunnel was designed which collects simultaneously two different signals at 90° and 45° to the plasma axis. The signals are transferred to the FPI by fibre optics. Figure 4.7.10 shows the recent experimental set-up at IRS [33].

The Echelle spectrometer with an extremely high resolution of $10^5$ allows for the investigation of all radiating species in the plasma jet and in the boundary layer. Measurements of a translational temperature of 8000 K +/– 500 K at the simulation point SFB I based on the emission of N, O and $N_2^+$ lines show an excellent agreement with electron temperatures determined by Thomson scattering.

A comparison between non-intrusive and intrusive techniques is presented in [2] where results of spectroscopic and Fabry Perot interferometric measurements are compared to electrostatic probe measurements for the simulation point SFB II (compare section 4.6.3 – The IRS Plasma Wind Tunnels).

### 4.7.3 Flight Instrumentation (PYREX, RESPECT, PHLUX, COMPARE)

Ground testing facilities and numerical codes are applied to generate and simulate the atmospheric entry of spacecrafts. These entry simulating tools provide an excellent opportunity to develop and qualify radiation-cooled materials for re-useable spacecraft and are much less expensive than space flights. However, such ground tests and computer simulations cannot replace space flights completely. The demand for reusable spacecraft, equipped with radiation-cooled fibre-reinforced ceramic thermal protection shields (TPS), requires re-entry experiments for the validation of the ground-based experiments and computer codes [4, section 4.6]. Material test diagnostic tools and TPS material tests have been developed and qualified. Parallel to these methods, additional measurement techniques were applied for the investigation of plasma flows and material behaviour during the tests [3]. As a result of the gathered experience, flight experiments like the PYRometric Entry EXperiment PYREX (capsules EXPRESS

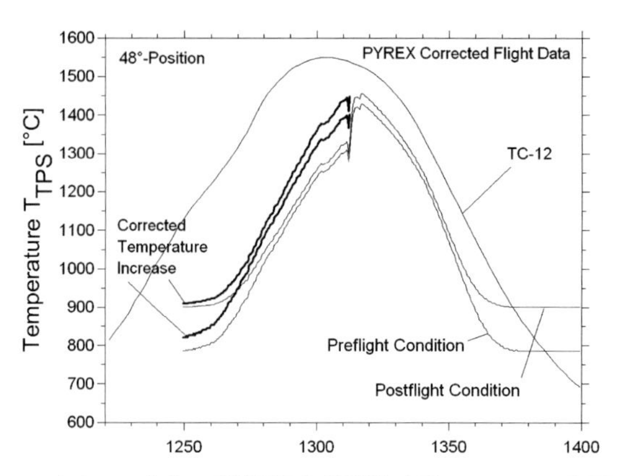


Figure 4.7.11: Measured data of the PYREX-MIRKA 48° sensor and HEATIN TC 12.

and MIRKA) and HEATIN (HEATshield INstrumentation on MIRKA) were developed, qualified and successfully flown [5, 26]. Figure 4.7.11 shows results for PYREX-MIRKA and HEATIN [4]. After dismounting the PYREX 48° sensor, minor pollution was found in the optical path. This led to a decrease of measured radiance intensity, which was corrected numerically by the simulation of the calibration curves by means of variation of aperture [13].

Flight experiments such as RESPECT (RE-entry SPECTrometer) and PYREX on HOPE-X are in the conceptual phase. An experiment to estimate the dissociation degree during an entry using the measured heat fluxes on different catalytic surfaces (PHLUX), a boundary layer probe to determine the boundary layer thickness and pressure using a retractable pressure sensor as well as a radiometer experiment for high-enthalpy re-entries are being developed. Detailed information can be found in [4].

#### 4.7.3.1 Description of PYREX-KAT38 (Pyrometric Entry Experiment)

PYREX-KAT38, measuring the temperature histories at five positions of the TPS rear side in the X38 nose structure, was contributed by the Space Transportation Division of IRS [4]. The information provided by the system is pertinent to several fields of interest, i.e. the temperature distribution in the nose structure during entry, statements on the behaviour of the TPS material and the heat flux distribution [4]. A sixth channel is used as reference channel to survey the system performance. Additionally, a comparative temperature and heat flux measurement using two of the sensors each measuring zones of different catalytic efficiencies was intended. Hereby, information on atomic species concentrations in front of the heat shield was expected. Unfortunately, no suitable material of sufficiently low catalytic efficiency could be found within the scheduled time. The sensors do not touch the TPS and consist of an optical path, a flange and a lens system. The mass is about 0.25 kg for each sensor, the diameter and height are roughly 50 mm. The fibre optics transmit the radiation to the sensor unit containing the electronics. From there data and power are transmitted to the X38 Vehicle Analysis Data Recording System (VADR).

The most important supplementations to the former PYREX-systems [4, 13] are the independent memory bank of the sensor unit and the overall system integration (VADR). The fibre optics can be dismounted from the sensor unit and the sensors. Hence, the whole nose structure can be dismounted from the remaining X38 vehicle. The sensor unit contains the optical system in front of the five photodiodes. Its mass is 2 kg, the size is $0.1 \times 0.13 \times 0.2\ m^3$. A 630 nm filter system is placed between the fibre optics and the photodiode. The wavelength is a useful compromise between signal and sensitivity [4].

The qualification procedures already performed for PYREX have partially been accepted for the PYREX-KAT38 qualification. The system was calibrated with the IRS black body source between 1150 K and 2700 K. After that, a qualification under the X38 heat flux with PWK 1 [3, 4] was performed. Here, good correspondence was achieved [3]. The tests of the components in the PWK were done with similar geometries and sections as in the X38 TPS. They additionally provided information to estimate front side temperatures using a sample of the same thickness for the test of the engineering model and an original nose cap sample for the approval test of the flight model. System function, thermal behaviour, data transfer and independent data acquisition were checked. All functional requirements were fulfilled very well. Additionally, the assembly of a qualification model of the nose structure including PYREX was required [4]. The corresponding fit check was successful. After the finalization of instrumentation, the qualification model thermomechanical tests were performed. PYREX-KAT38 withstood the tests and the system worked very well [4]. The mechanical loads for X38 were lower than those for EXPRESS and MIRKA missions [4, 5, 26]. Here, the sensor unit was tested successfully. The system was integrated in the X38 nose structure (see Fig. 4.7.12). The calibration was

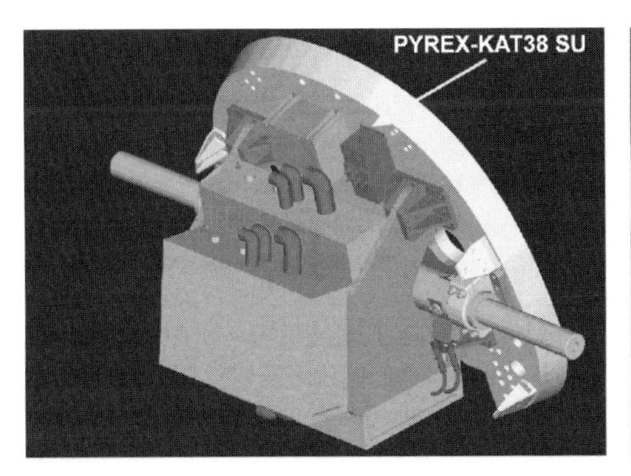

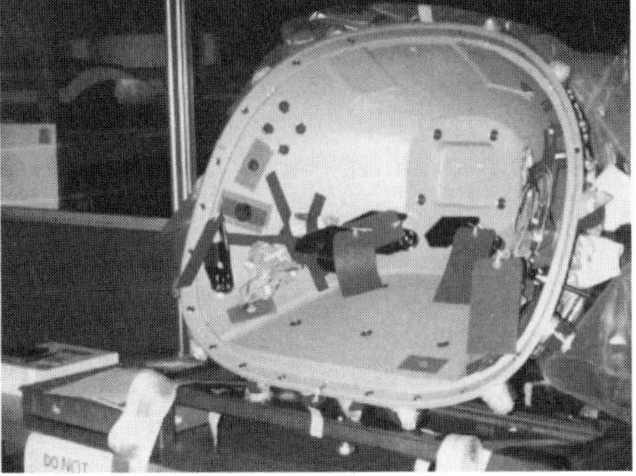

Figure 4.7.12: PYREX-KAT SU mounted to X38 front bulkhead (left), X38 rear side of nose structure together with integrated PYREX-KAT38 sensors (right).

checked using a calibrated lamp. Both the test and the ongoing integration were successful.

#### 4.7.3.2 RESPECT (Re-Entry SPECTrometer)

The goal for RESPECT is to obtain detailed information about the plasma state in the post-shock regime of an entry vehicle by measuring the spectrally resolved radiation. The obtained database will be compared with the results of numerical simulations. One of the codes is the URANUS Navier-Stokes Code in combination with a radiation transport code HERTA and the radiation data base PARADE [17]. The flight data will be compared with the simulated spectra based on the flow field data provided by the URANUS code and with the measured data from plasma wind tunnel experiments. The investigation requires pre-flight calculations, functional qualification of the spectrometer in plasma wind tunnels [3], data collection during the entry and a post-flight analysis.

Experimental data which can be used to verify the numerical simulation are poor. Existing experimental data concentrate on single species e.g. NO radiation [4]. One way to gain information about multiple species is given by emission spectroscopic measurements during entry. Fig. 4.7.13 shows the principle of the experiment. Due to the integrating character of the measurement, a direct extraction of temperatures or densities does not seem possible, however, the desired information can be obtained by comparison with numerically simulated data. This requires a successful validation of the chemical models implemented in the codes. Here, discrepancies will provide information concerning

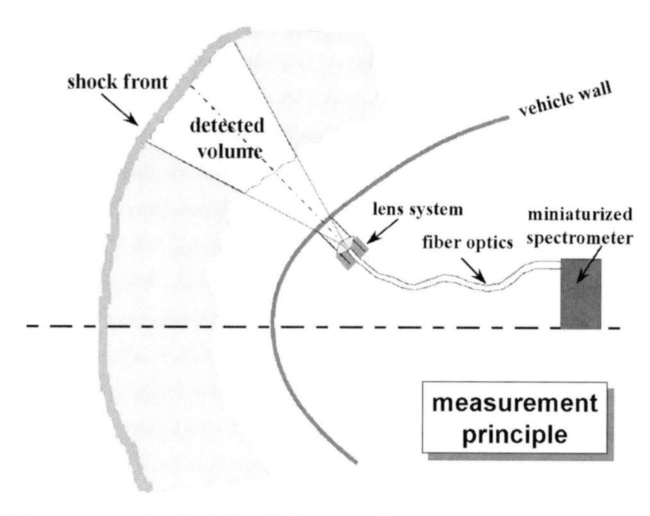


Figure 4.7.13: Measurement principle of the experiment.

necessary code improvements. The obtained data will cover a minimal wavelength range from 200 to 800 nm with a resolution of about 0.5 nm. It contains the emission of relevant radiating air species. Emission spectroscopic experiments have already been carried out to investigate the boundary layer in front of a blunt body in subsonic high-enthalpy plasma flows. The ratios of the emission of different species were found to be a valuable indicator for chemical processes, especially in the boundary layer [31].

A comparison with numerically simulated data will provide information about the thermochemical processes in the plasma. Since several flight experiments have been developed at IRS [4], a lot of experience in space qualified electronic and data acquisition systems is available. The operational requirements have to be defined depending on the flight mission. The verification matrix of the measurement system for the spectrometer is also defined by the wavelength range, allowing the main radiating species in the flow to be detected as presented in [4]. If weight and size restrictions allow for a multi-channel version, various wavelength regions can be covered by different channels. The dynamic range of the detector has to be chosen as high as possible to detect both varying intensities within one spectrum and the variation along the flight trajectory even if some control of the overall signal strength is given by the detection time. At the same time, a sufficient readout frequency has to be guaranteed to minimize the integration along the flight path. In addition, the measurement rate has to face the time scale of the re-entry trajectory in combination with the required amount of data sets. A feed-through is required for the coupling of the measurement system via the TPS structure. Most specifications for the coupling to the space vehicle will have to be specified by the requirements of the flight mission. The transmission of the radiation to the spectrometer inlet has to be done using optical fibres.

#### 4.7.3.3 COMPARE

For missions to Mars or Venus or hyperbolic re-entries radiation heat fluxes have to be considered for the TPS design. With increasing nose radius the radiative heat fluxes increase such that for future crewed missions, e.g. to Mars, radiation has to be taken into account. Within the Archimedes mission a concept for a TPS experiment has been developed. This experiment combines pyrometric and radiometric measurement in the stagnation point region [4]. Additionally, total pressure can be measured via the optical path of the radiometer. The experiment will enable a separation of the radiative heat flux from the total heat flux and it will enable the stagnation point specific enthalpy to be determined. In fact, one aspect of the design was the optimization with regard to the measurement system volume and power consumption. The short entry period allows the Peltier cooling system to be disposed of such that power consumption goes below 5 W. In addition, the simulation of the calibration curves enabled a reduction of the required measurement ranges of the pyrometric sensor.

28. Stöckle, T., Winter, M., Auweter-Kurtz, M.: Simultaneous Spectroscopic and Mass Spectroscopic Investigation of Surface Catalytic Effects in High Enthalpy Gas Flows, AIAA-98-2845, 7th AIAA/ASME Joint Thermophysics and Heat Transfer Conference, Albuquerque, NM, 1998.
29. Stöckle, Th.: Experimentelle und numerische Untersuchung der Oberflächenkatalyzität metallischer und keramischer Werkstoffe in Hochenthalpieströmungen, PhD Thesis, Institute of Space Systems, Universität Stuttgart, 2000.
30. Summers, H. P., Wood, L.: Atomic Data and Analysis Structure (ADAS), User Manual JET-R(94)06.
31. Winter, M., Auweter-Kurtz, M., Kurtz, H. L.: Spectroscopic Investigation of the Boundary Layer in Front of a Blunt Body in a Subsonic Air Plasma Flow, AIAA-97-2526, 32nd AIAA Thermophysics Conference, Atlanta, GA, USA, June 1997.
32. Winter, M., Auweter-Kurtz, M., Park, C.: Determination of Temperatures and Particle Densities in a Subsonic High Enthalpy Air Plasma Flow From Emission Spectroscopic Measurements, 32nd AIAA Plasmadynamics and Lasers Conference, Anaheim, CA, USA, Juni 2001.
33. Winter, M., Auweter-Kurtz, M.: Translational, Electronic and Molecular Temperatures in a Subsonic Air Plasma Flow Determined by Optical Methods and Langmuir Probes, AIAA-2003-3489, 36th AIAA Thermophysics Conference, Orlando, FL, USA, 23.–26. Juni 2003.

## 4.8 Numerical Simulation of Flow Fields Past Space Transportation Systems

Andreas Henze*, Wolfgang Schröder, and Matthias Meinke

### Abstract

Several numerical investigations in sub- and supersonic flow fields around the space configurations ELAC and EOS are presented. Turbulent flow was simulated for subsonic flow over ELAC and the piggyback system ELAC/EOS. Laminar flow was simulated for supersonic flow over a simplified separation system consisting of EOS and a flat plate. The numerical results are obtained by solving the Navier-Stokes equations using a finite volume integration scheme based on node-centered block-structured and unstructured grids in curvilinear coordinates. Algebraic and one-equation turbulence models were used in fully turbulent assumed flows. The results presented are computed at different Mach numbers and angles of attack. The findings include distributions for the field variables, wall streamlines and velocity profiles. The numerical data of the present investigation is juxtaposed with experimental data from different facilities. The comparison shows satisfactory agreement.

* RWTH Aachen, Aerodynamisches Institut, Wüllnerstr. 5–7, 52062 Aachen; andreas@aia.rwth-aachen.de

## 4.8.1 Introduction

The present paper deals with the numerical simulation of three-dimensional flows around reusable space transportation systems at high and low Mach numbers. The development of such vehicles requires accurate numerical investigations for the determination of the overall flow structure and the computation of aerodynamic forces and moments. The first configuration of the investigation is the delta-wing shaped lifting body ELAC. The analysis was accomplished for subsonic flows at high Reynolds numbers. Generally, the models used in wind tunnels are much smaller than real configurations. Full similarity cannot be reached between the model and the full scale configuration. The influence of the Reynolds number cannot be investigated correctly. Therefore, to achieve a more realistic behaviour of the flow field, experiments with a large scale model of ELAC have been performed in the German Dutch Wind tunnel (DNW). The numerical results will be compared with these measurements.

The second geometry of the investigation is the piggyback system of ELAC and the upper stage EOS. Again, the analysis was performed for subsonic flows at moderate high Reynolds numbers. The numerical results will be compared with experimental data obtained in the Aerodynamisches Institut.

A simplified stage separation configuration consisting of EOS over a flat plate is the third geometry of the investigation. Previous experimental and numerical analyses on stage separations showed a significant influence of the vehicles. The flow field between the upper and the lower stage consists of interacting shocks, vortices, and boundary layers. For a stable stage separation it is necessary to have accurate data about the influence of these phenomena on flight mechanics and thermal loads on the wall surfaces. The analysis was performed for supersonic Mach numbers at laminar flow conditions.

## 4.8.2 Numerical Scheme

### 4.8.2.1 Basic Equations

In the mathematical description the index notation of tensor calculus is applied. In an arbitrary coordinate system the components of the velocity vector read $v^k (k = 1, 2, 3)$. The indices are both latin $(i, j, \ldots = 1, 2, 3)$ and greek $(\alpha, \beta, \ldots = 1, 2, 3)$. The Navier-Stokes equations can be expressed by the vector of conservative variables $q = (\rho, \rho\vec{v}, \rho E)^T$ and the flux $F^k (k = 1, 2, 3)$ in the following form

$$Sr \frac{\partial q}{\partial t} + F^{\alpha}_{|\alpha} = 0$$

All values are dimensionless and the reference values are the stagnation variables. Decomposing $F^k$ into the advective part $F^k_A$, consisting of the convective transport and the pressure term, and the diffusive part, describing viscous fluxes and the heat conduction term, the total fluxes are

$$F_A^k - F_D^k = \begin{pmatrix} \rho v^k \\ \rho v^k v^j + p\delta_{jk} \\ v^k(\rho E + p) \end{pmatrix} + \frac{1}{Re_0} \begin{pmatrix} 0 \\ \lambda g^{jk} v^\beta_{|\beta} + \mu[g^{ja} v^k_{|a} + g^{k\beta} v^j_{|\beta}] \\ \frac{\kappa g^{ka} T_a}{(\gamma - 1)Pr_0} + \lambda g^{k\delta} v^\delta v^a_{|a} + \mu v^\delta [g^{\delta a} v^k_{|a} + g^{k\beta} v^\delta_{|\beta}] \end{pmatrix}$$

with the first and second viscosity coefficients $\mu$ and $\lambda$. Neglecting the bulk viscosity they are related by $\lambda = -2/3\mu$. The coefficients $g$ describe the coordinate transformation from an cartesian coordinate system. The equations are closed using the thermal and calorical equations of state for an ideal gas

$$p = \frac{T}{\gamma} \rho R \, .$$

Using perfect gases the viscosity is determined by a potential law $\mu/\mu_0 = (T/T_0)^{\overline{\omega}}$ with $\overline{\omega} = 0.72$ for air. Assuming a constant Prandtl number $Pr = \mu c_p/\lambda$ the dimensionless heat conductivity equals the dimensionless viscosity $\lambda/\lambda_0 = \mu/\mu_0$. Besides $Pr$ the Reynolds number $Re = \rho u_\infty L/\mu$ and the Strouhal number $Sr = L/u_\infty \tau$ appear in the fundamental equations.

For flows at moderate angles of attack the Baldwin/Lomax model [9] is used. Its benefit is the efficiency of the method that is a local approach. It can be applied without knowledge about the edge of the boundary layer, but lacks like transport or diffusion properties. Hence, the Fares/Schröder [10] model was used for higher angles of attack, where strong vertical flows are expected. It is a one-equation model that has been derived from the $k$-$\omega$ model and validated against experimental and analytical solutions. The findings showed good results for numerous wall bounded and free shear flows.

#### 4.8.2.2 Initial and Boundary Conditions

The system of equations described in the previous section represents an initial boundary value problem. A parallel flow field serves as initial condition.

$$\rho = \rho_\infty, \; \vec{v} = \vec{v}_\infty, \; p = p_\infty$$

The boundary conditions are divided into far field and wall conditions. In the far field the conditions are to be computed from the theory of characteristics. Since the geometry of the configurations computed is symmetrical and no yaw angle is taken into account, the flow field is simulated only for the half model. At the symmetry plane the following conditions are valid:

$$\frac{\partial u, v}{\partial z} = 0, \; w = 0, \; \frac{\partial \rho}{\partial z} = 0, \; \frac{\partial T}{\partial z} = 0,$$

where z is the coordinate normal to the symmetry plane. A Dirichlet boundary condition is used in supersonic flow at the inlet boundary by defining the free stream values.

$$\rho = \rho_\infty, \; \vec{v} = \vec{v}_\infty, \; p = p_\infty \, .$$

In subsonic flow the boundary of the integration regime has to be far upstream of the body. The one-dimensional theory of characteristics gives the following boundary inflow boundary conditions

$$\rho = \rho_\infty, \; \vec{v} = \vec{v}_\infty, \; \frac{\partial p}{\partial n} = 0 \, .$$

At supersonic outflow the Neumann condition for the conservative variables is applied

$$\frac{\partial \rho}{\partial n} = \frac{\partial \vec{v}}{\partial n} = \frac{\partial p}{\partial n} = 0 \, .$$

At subsonic outflow the boundary conditions are formulated according to the eigenvalues of the flux vector

$$\frac{\partial \rho}{\partial n} = 0, \; \frac{\partial \vec{v}}{\partial n} = 0, \; p = p_\infty \, .$$

At solid walls the no-slip condition is used

$$\vec{v} = 0 \, .$$

Due to the boundary layer character of the flow in the vicinity of solid walls the normal pressure gradient in the wall normal direction is $\partial p / \partial n = 0$. The thermal wall boundary condition reads

$$\vec{q} = 0 \rightarrow \frac{\partial T}{\partial n} = 0 \, .$$

#### 4.8.2.3 Spatial Discretization in Structured Grids

The governing equations are discretized on a node-centered block-structured grid in general curvilinear coordinates with a finite volume method. The convective fluxes are computed using a modified AUSM-scheme [1]

$$F_\beta^C = (F_1^C, F_2^C, F_3^C)$$
$$F_1^C = \frac{M_\beta}{2}((\rho a)^L + (\rho a)^R) + \frac{|M_\beta|}{2}((\rho a)^L - (\rho a)^R)$$
$$F_2^C = \frac{F_1^C}{2}(u_\alpha^L + u_\alpha^R) + \frac{F_1^C}{2}(u_\alpha^L - u_\alpha^R) + \left(\frac{p^L + p^R}{2} + \chi(p^L u_\alpha^L - p^R u_\alpha^R)\right)\delta_{\alpha\beta}$$
$$F_3^C = \frac{F_1^C}{2}\left(\frac{(\rho e)^L + p^L}{\rho^L} + \frac{(\rho e)^R + p^R}{\rho^R}\right) + \frac{|F_1^C|}{2}\left(\frac{(\rho e)^L + p^L}{\rho^L} - \frac{(\rho e)^R + p^R}{\rho^R}\right).$$

Superscripts $^L$ and $^R$ denote left and right interpolated variables which are obtained via a quadratic MUSCL interpolation [2] of the primitive flow variables. The quantity $M_\beta = \frac{u_\beta}{a}$ is the average of the left and right interpolated Mach number. The speed of sound $a$ is computed as an average of the left and right interpolated values. The parameter $\chi$ is used to control the amount of numerical dissipation. To avoid oscillations near extrema the flux-limiter of van Albada has been implemented in the approach [3].

The viscous stresses are discretized using central differences of second-order accuracy with a modified "Cell-Vertex" scheme [4].

#### 4.8.2.4 Spatial Discretization in Unstructured Grids

The convective flux across a triangular surface of a control volume is

$$\vec{F_C} = |\vec{S_f}| \cdot \left(\frac{1}{2}[Ma(\Phi_L + \Phi_R) - |Ma|(\Phi_R - \Phi_L)] + p\right),$$

with $\vec{S_f} = (S_x, S_y, S_z)$ representing the dimensionless normal vector of the surface. The Mach number and pressure splitting is the same as in the discretization in structured grids with $\Phi = (\rho a, \rho a\vec{v}, \rho a H)$. For a linear reconstruction of the left and right values the average vector of gradients is computed using Greens formula.

$$\int_V \nabla u dV = \int_\Omega u\vec{n} d\Omega \rightarrow \nabla u = \frac{1}{V}\int_\Omega u\vec{n} d\Omega \rightarrow \nabla u_i = \frac{1}{V_i}\sum_{f=1}^{n} \overline{u}_f \vec{S_f}$$

$V_i$ is the control volume, $n$ is the number of surface elements enclosing the control volume, and $\overline{u}_f$ represents the average values of the corresponding quantity (density, pressure, velocity component). In order to avoid oscillations in the vicinity of discontinuities the formulation of Barth and Jespersen [5] is used.

#### 4.8.2.5 Structured/Unstructured Coupling

The structured and unstructured grids are used in different parts of the flow field. From the numerical point of view it is necessary for flows around complex geometries like the combination ELAC-EOS configuration to use smooth grids as well as robust and accurate numerical methods. The generation of structured grids for such geometries is still a difficult and time consuming task. An easier method is the discretization of the integration domain using unstructured grids. Nevertheless, in the boundary layer adjacent to the wall it is still necessary to use small cells in the wall normal direction, while the tangential step size can be much larger. Pure unstructured grids using triangles or tetrahedrons are strongly stretched and a sufficient accuracy for the computation of the viscous terms cannot be attained. In order to achieve higher accuracy close to the wall the use of structured grids using hexahedrons is recommended. For flows with shock waves adaptive grid generators for unstructured grids are useful. This adaptivity is much easier to achieve in unstructured grids than in structured grids, since single points can easily be added or eliminated.

To take the advantage of both approaches different combinations are possible. Several hybrid methods can be found in literature with various control volumes. In the different parts of the integration regime either structured or unstructured grids are used, but the two parts are connected with each other and data have to be exchanged at the boundaries between the two regimes. The exchange was realized using a Message-Passing-Library (PVM) which enables a parallelization of the procedure. The data that have to be exchanged are located in an overlapping regime that is defined in the structured area and in the unstructured areas.

The shared area consists of a certain number of layers of hexahedrons depending in the accuracy of the discretization. In the second-order upwind discretization used in this study three layers are used. The overlapping area is triangulated in the unstructured approach and inserted into the unstructured part of the grid. In two-dimensional problems a quadrangle is divided into 2 or 4 triangles. In the latter case an additional point arises which does not occur in the structured grid. Its flow variables are interpolated from its neighbours. In the three-dimensional case a hexahedron is divided into 5 or even 12 tetrahedrons, while the latter case turned out to be easier to implement. The conservative variables $q = (\rho, \rho\vec{v}, \rho E)^T$ have to be exchanged and for three layers two lines of data have to be known in the structured and in the unstructured part, respectively.

#### 4.8.2.6 Temporal Integration

The integration in time of the differential equation is carried out using a 5-step low storage Runge-Kutta method. The scheme is adapted to a maximum Courant number of 4 for central and 3.5 for upwind schemes. The coefficients in the Runge-Kutta steps are chosen $a_1 = (0.059, 0.145, 0.273, 0.5, 1.0)$. They are optimized for maximum stability for an upwind scheme. The temporal accuracy is $O(\Delta t^2)$.

## 4.8.3 Results

### 4.8.3.1 Geometry of the Two-Stage System

The lower stage looks like a slender delta-wing with an aspect ratio of $1 \cdot 1$, lateral stabilizers with rudders at the wing tips, trailing edge flaps for pitch and roll control (Fig. 4.8.1), and an airbreathing engine. The cross section consists of two semi ellipses with equal major and different minor axes. For the upper side the ratio of the semi axes is 1:4, and for the lower side 1:6. The sweep angle is $75^\circ$. Downstream of the rounded nose the minor semi axes increase linearly to $x/l = 0.63$, where the maximum thickness of ELAC occurs, and then they decrease linearly to the sharp trailing edge. The engine is located on the lower side of the first stage whose forebody serves as compression ramp. The rear part of ELAC's lower surface is formed as an expansion ramp. The fins have a NACA 0010 profile, and their dihedral angle is $65^\circ$.

At first sight the upper stage possesses a Shuttle-like shape (Fig. 4.8.2). Conic sections such as ellipse, circle, parabola etc. are used to define the geometry of EOS. For example, the nose is part of an ellipsoid and the adjoined fuse-

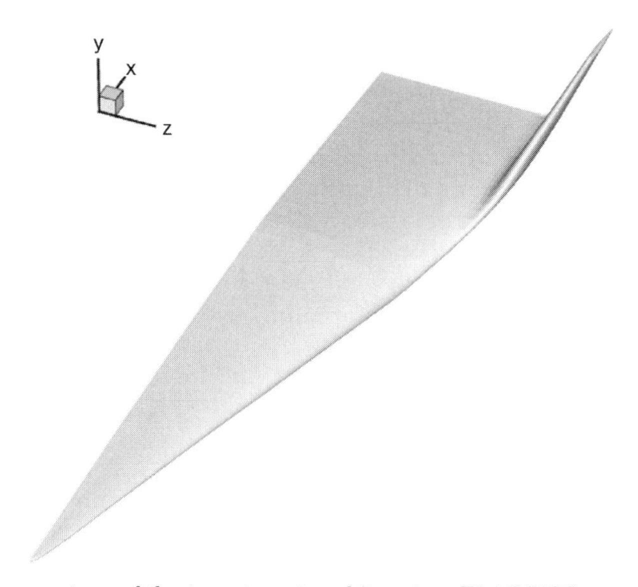


Figure 4.8.1: Lower stage of the two-stage-to-orbit system ELAC/EOS.

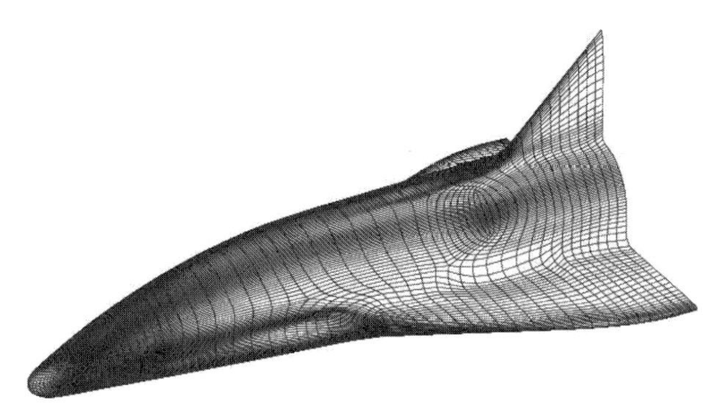

Figure 4.8.2: Upper stage of the two stage to orbit system ELAC/EOS.

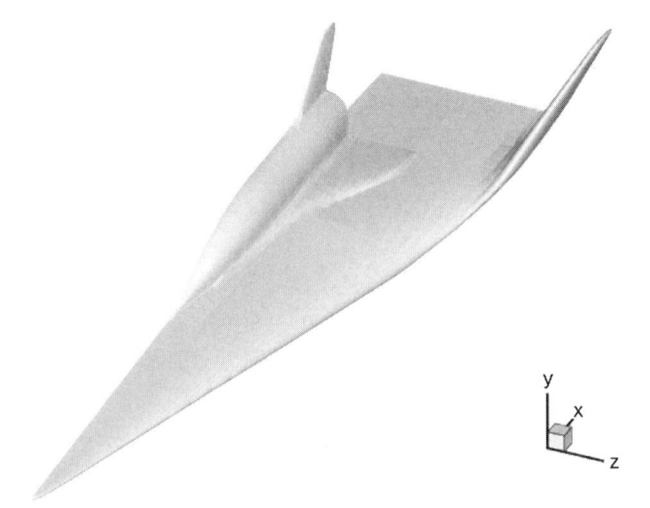


Figure 4.8.3: Piggyback system ELAC/EOS.

lage consists on the upper side of a half cylinder that is connected with the ellipse of the lower side by vertical surfaces. The front part of the delta-wing, that is defined by a parabola in the horizontal plane, extends from $x/l=0.125$ to $x/l=0.5$. Downstream of $x/l=0.52$ the delta-wing possesses a straight leading edge with a sweep angle of $65^\circ$. Using a third order polynomial the maximum span is reached at $x/l=0.94$. The profile of the wing is given by the NACA 0009 profile. The angle of the fin's leading edge is $45^\circ$ and that of the trailing edge $65^\circ$. The shape of the fin is defined by the NACA 0010 profile. ELAC-1c is a piggyback system of both configurations (Fig. 4.8.3). More details on the geometry can be found in [6].

### 4.8.3.2 Flow Past ELAC

In the following results for the computation of the subsonic flow field around ELAC are presented for different flight conditions. The velocity in the turbulent boundary, the pressure distribution as well as the three-dimensional structure of the flow field will be analyzed. The results will be compared with experimental data coming from the German-Dutch windtunnel [7, 8].

For the computation of the turbulent flow around ELAC (Fig. 4.8.1) a block structured multiblock grid with 10 blocks and about 1 000 000 grid points was used. For a good resolution of the boundary layer at a Reynolds number of $4\times10^7$ the minimum step size at the wall is $\Delta y = 6\cdot10^{-6}(y^+ \approx 2)$. The different flow conditions for ELAC are shown in Tab. 4.8.1.

The computations were performed under the assumption of a fully turbulent flow and adiabatic wall conditions. In most of the simulations the turbulence model of Baldwin and Lomax [9] was used. At an angle of attack of

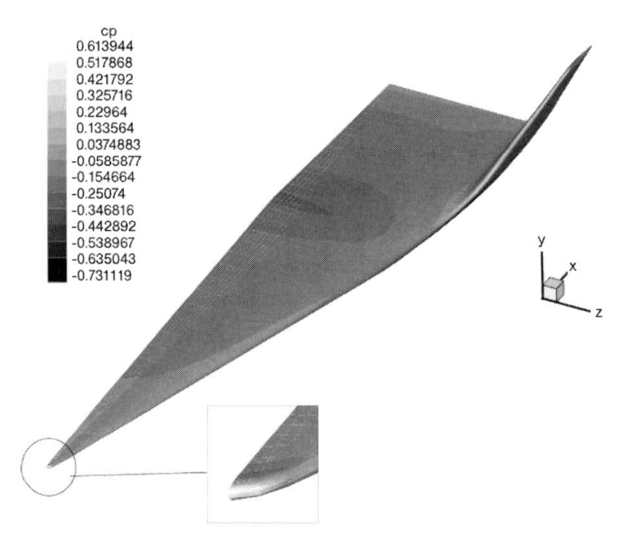


Figure 4.8.4: Pressure contours on the surface at $Ma_\infty = 0.29$, $Re_\infty = 40\times10^6$, $a = 0.07°$.

Table 4.8.1: Flow parameters of the computations for ELAC.

| $MA_\infty$ | $Re_\infty$ | $a$ |
|---|---|---|
| 0.29 | $40\times10^6$ | 0.07 |
| 0.29 | $40\times10^6$ | 4.26 |
| 0.29 | $40\times10^6$ | 7.91 |
| 0.29 | $40\times10^6$ | 15.73 |
| 0.14 | $20\times10^6$ | 4.06 |
| 0.14 | $20\times10^6$ | 8.23 |

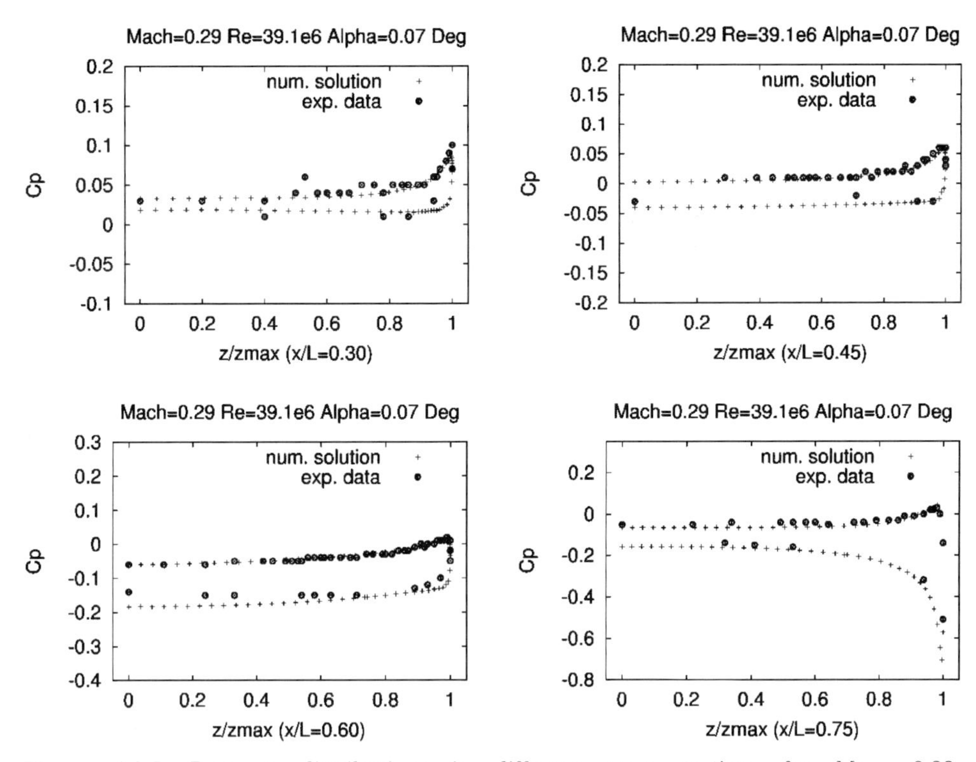


Figure 4.8.5: Pressure distributions in different cross sections for $Ma_\infty = 0.29$, $Re_\infty = 40 \times 10^6$, $a = 0.07^\circ$.

$15.73^\circ$ the Fares-Schröder [10] model was applied since the leeward vortices are much more pronounced for larger angles of attack.

In Fig. 4.8.4 the pressure distribution on the surface of ELAC at $Re_\infty = 4 \cdot 10^7$, $Ma_\infty = 0.29$ and $a = 0.07^\circ$ is shown. Due to the slenderness of the contour no strong gradients are visible. In the nose region, along the leading edge and at the transition between the leading delta wing and the rear part of the contour slight pressure gradients are visible. Fig. 4.8.5 shows the pressure distribution in different cross sections at 30, 45, 60, and 75% of the overall length. The pressure distribution is compared with experimental data from the DNW-experiments described in [7]. Due to the acceleration at the leading edge the pressure gradient has its maximum in this region. In all cross sections a good agreement between the numerical and the experimental results can be found. For the same flow conditions the measured velocity profiles in the symmetry plane are available. They are compared with the numerical data in Fig. 4.8.6 (left). Since the Reynolds numbers of the experiment and the numerical simulation are slightly different, the dimensionless profiles are depicted. Again, a good agreement between experimental and numerical data can be detected. Supplementary, in Fig. 4.8.6 (right) the turbulent viscosity $\mu_t$ is depicted at dif-

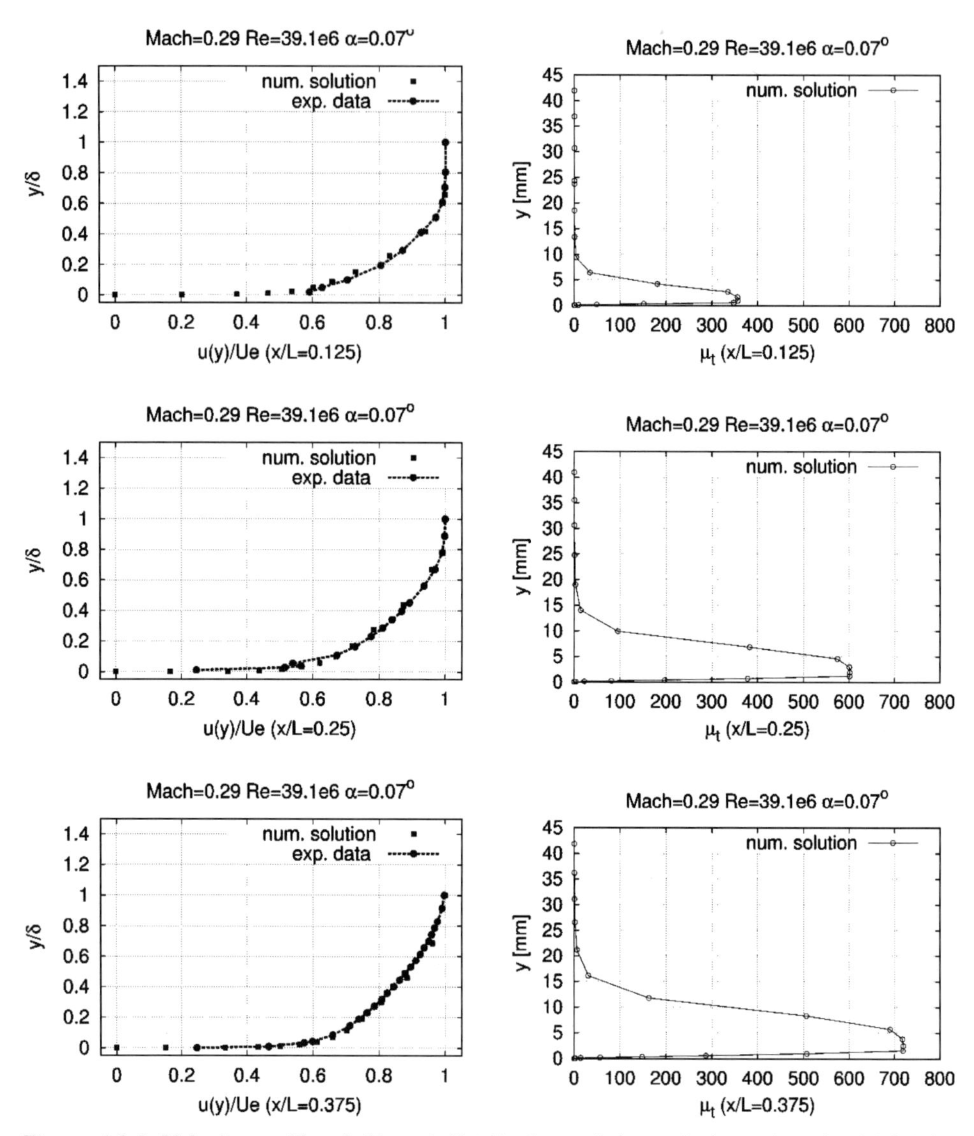


Figure 4.8.6: Velocity profiles (left) and distributions of the turbulent viscosity (right) for $Ma_\infty = 0.29$, $Re_\infty = 40 \times 10^6$, $a = 0.07^\circ$.

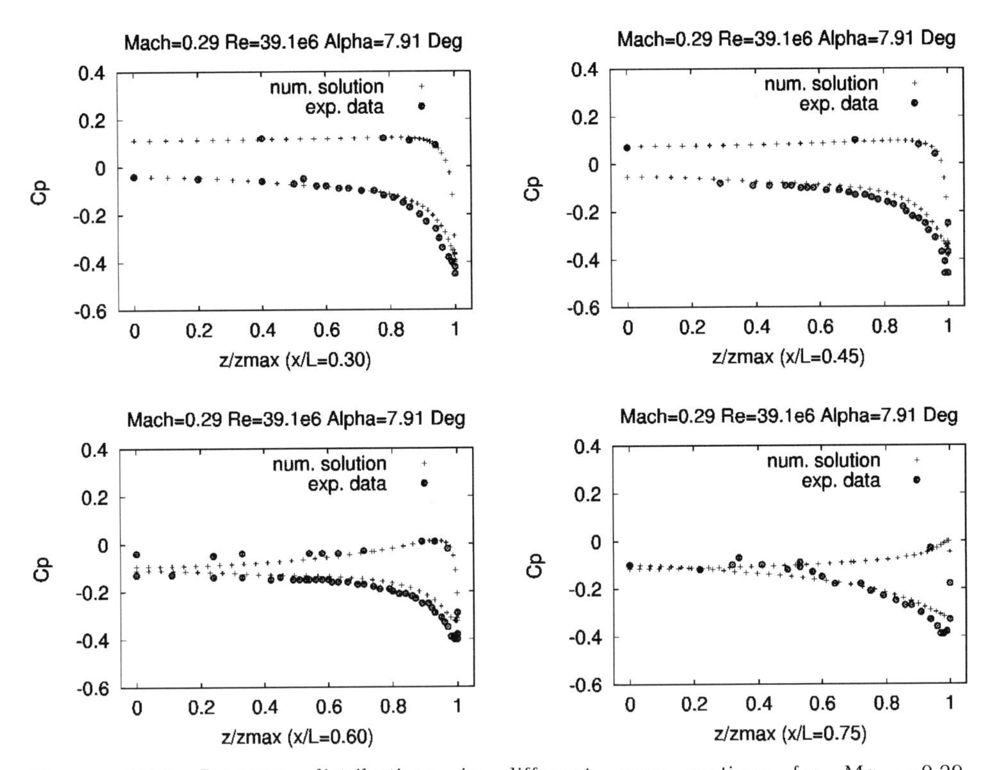


Figure 4.8.7: Pressure distributions in different cross sections for $Ma_\infty = 0.29$, $Re_\infty = 40 \times 10^6$, $\alpha = 7.91^\circ$.

ferent points on the symmetry plane. The pronounced peak near the wall and the increase of its maximum value at growing *x/l* is clearly visible.

The next results presented belong to the same Reynolds and Mach number, but to a higher angle of attack $\alpha = 7.91^\circ$. In Fig. 4.8.7 again the numerical pressure distributions in different cross sections are depicted together with the experimental data. Again, the agreement is excellent. Comparing the results of the lower and the higher angle attack, it becomes clear that in the former case the stagnation line is on the upper side of the contour while it moves to the lower side when the angle of attack is increased. Nevertheless, at the higher angle of attack no primary vortex can be observed on the lee side. In Fig. 4.8.8 the turbulent viscosities as well as the numerical and experimental velocity profiles at different points at the symmetry plane are plotted. The angle of attack in the experiments is $\alpha = 12.0^\circ$. Due to this difference the numerical data slightly deviate from the reference data.

At higher angles of attack flow separation and a vortex system are expected. In order to scrutinize the influence of the turbulence model on the solution the zero-equation Baldwin/Lomax model [9] as well as the one-equation Fares/Schröder model [10] were used. Figure 4.8.9 shows the simulated wall

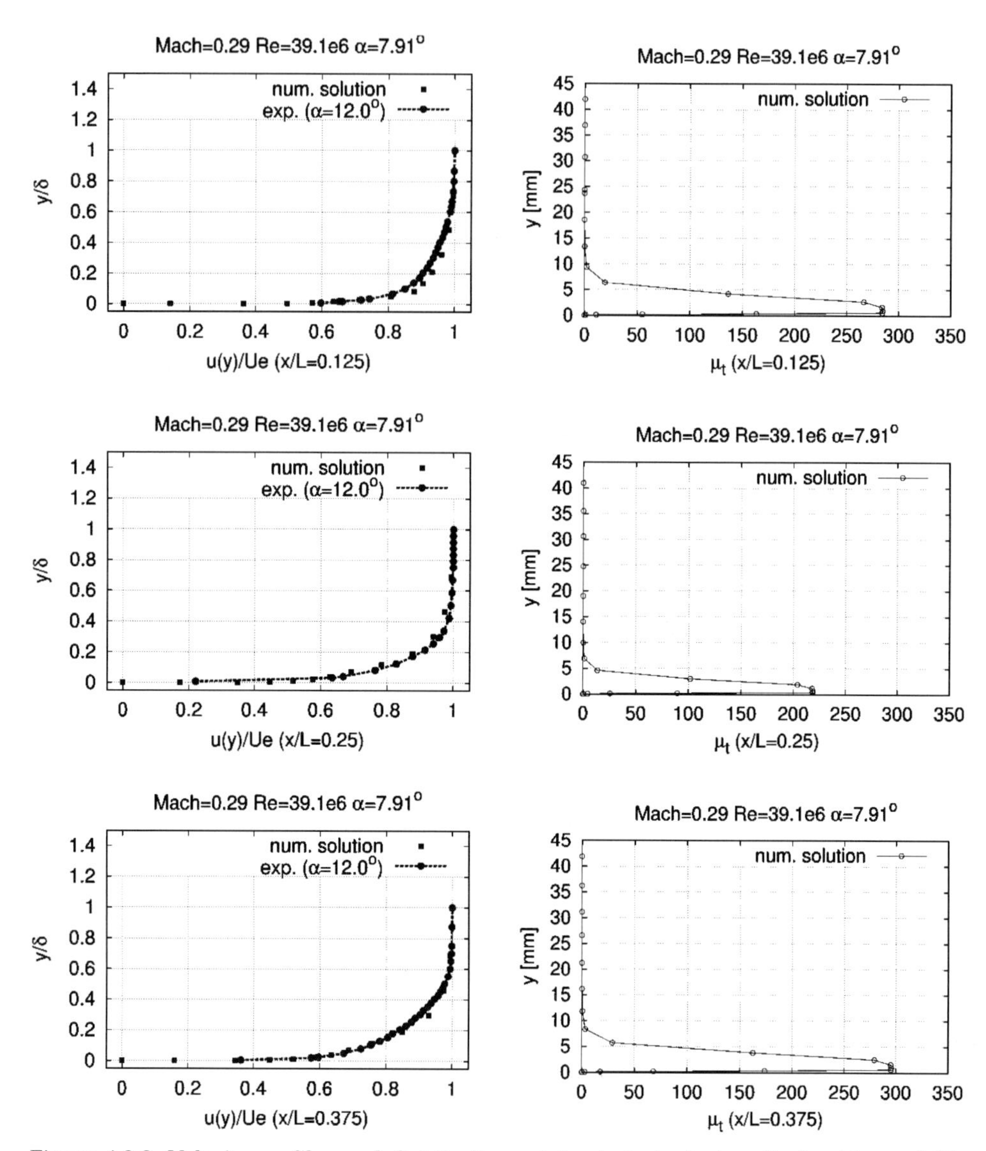


Figure 4.8.8: Velocity profiles and distributions of the turbulent viscosity for $Ma_\infty = 0.29$, $Re_\infty = 40 \times 10^6$, $\alpha = 7.91°$.

streamlines for both computations at $\alpha = 15.73°$. At this angle of attack the stagnation line is completely on the lower side and flow separation occurs on the lee side. The primary and secondary separations as well as the secondary attachment line become clearly visible. It becomes apparent that the flow topology is more or less alike in both cases, but the position of the limiting streamlines changes a bit. In Fig. 4.8.10 the pressure distributions in different cross sections for both models and in comparison to experimental data is presented.

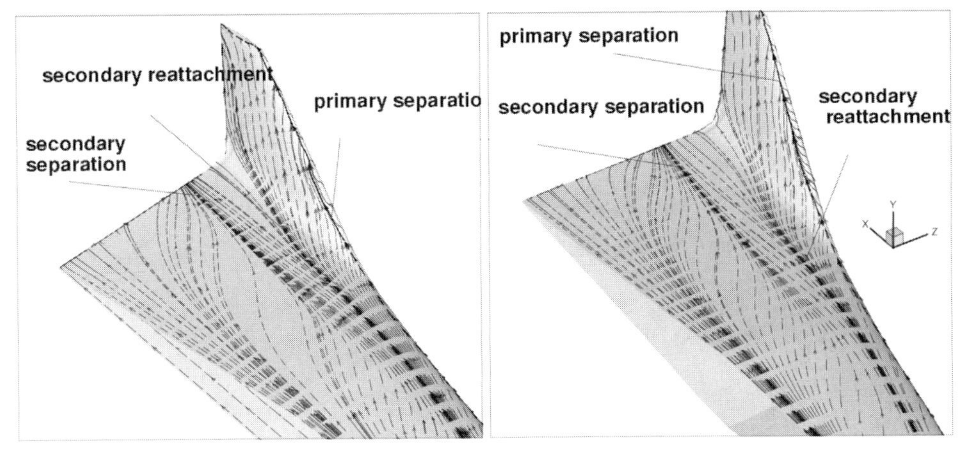


Figure 4.8.9: Wall streamlines for. Left: Baldwin/Lomax, Right: Fares/Schröder $Ma_\infty = 0.29$, $Re_\infty = 40 \times 10^6$, $\alpha = 15.73^\circ$.

In the region of the main vortex the experimental data show a strong decrease in the wall pressure. This decrease cannot be found in the Baldwin/Lomax solution whereas the Fares/Schröder solution at least indicates this although no quantitative agreement is achieved. All things considered the agreement between the experimental data is better but not excellent when the Fares/Schröder model is used. Further investigations concerning the transition, the grid refinement, and the turbulence model itself are necessary. Additional results can be found in [11].

### 4.8.3.3 Flow Past ELAC-1c

In the following the computations for the piggyback configuration ELAC-1c are presented for different flow conditions. The flow parameters are given in Tab. 4.8.2. Experimental results from the Aerodynamisches Institut are compared with numerical results.

Table 4.8.2: Flow parameters of the computations for ELAC-1c.

| $Ma_\infty$ | $Re_\infty$ | $\alpha$ |
|---|---|---|
| 0.4 | $2.6\times10^6$ | 0.0 |
| 0.4 | $2.6\times10^6$ | 5.0 |
| 0.4 | $2.6\times10^6$ | 10.0 |
| 0.6 | $3.6\times10^6$ | 0.0 |
| 0.6 | $3.6\times10^6$ | 5.0 |
| 0.6 | $3.6\times10^6$ | 10.0 |

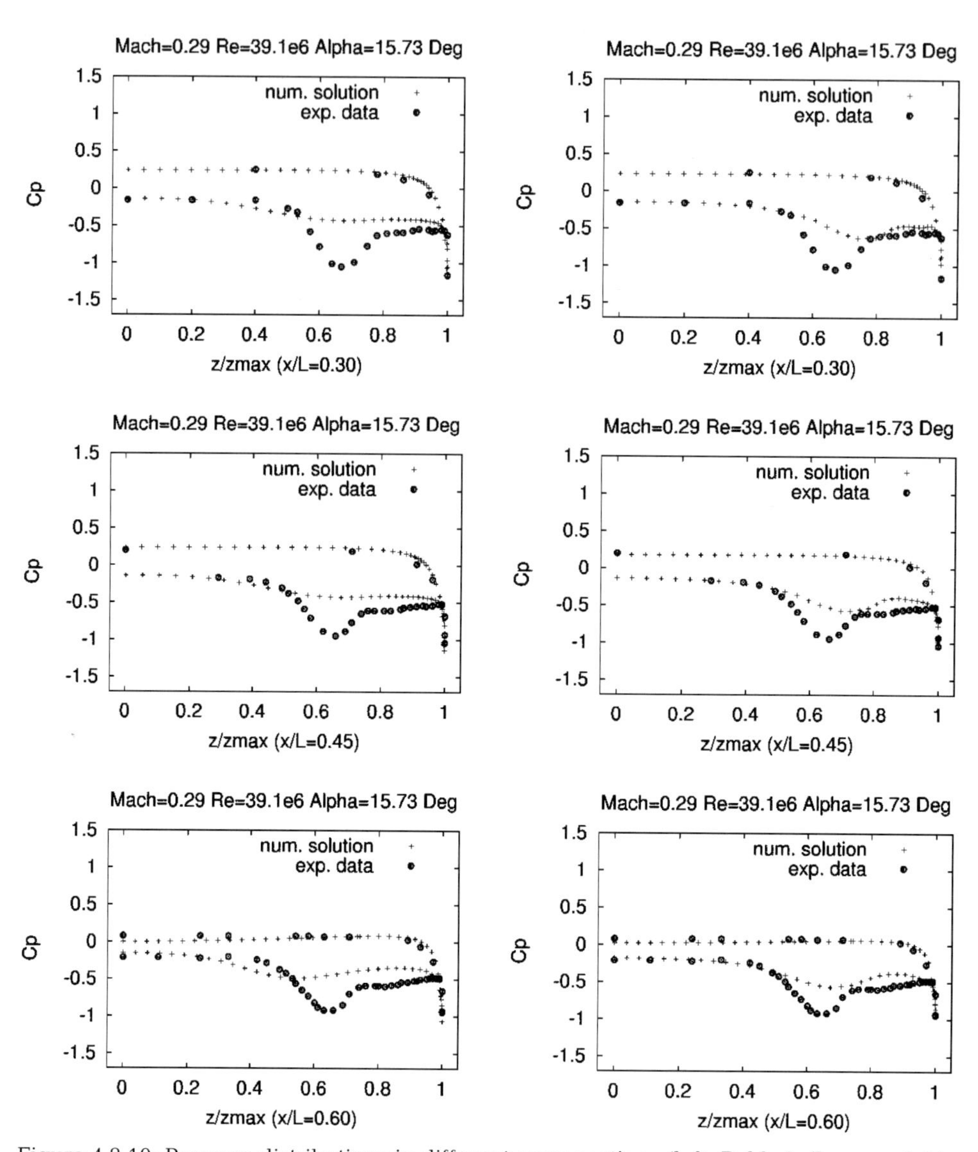


Figure 4.8.10: Pressure distributions in different cross sections (left: Baldwin/Lomax, right: Fares/ Schröder) for $Ma_\infty = 0.29$, $Re_\infty = 40 \times 10^6$, $\alpha = 15.73^\circ$.

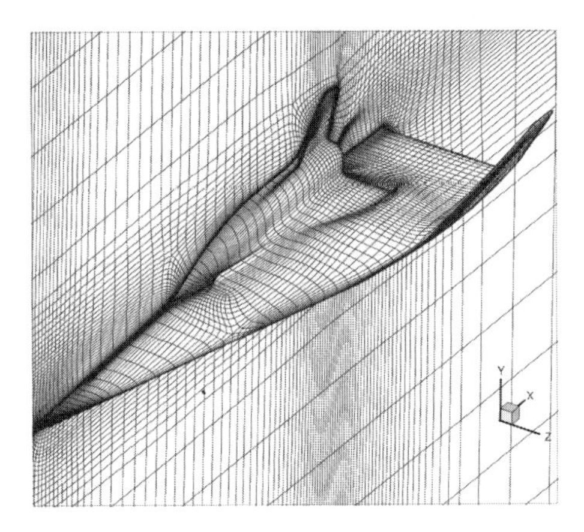


Figure 4.8.11: Computational grid for the piggyback configuration.

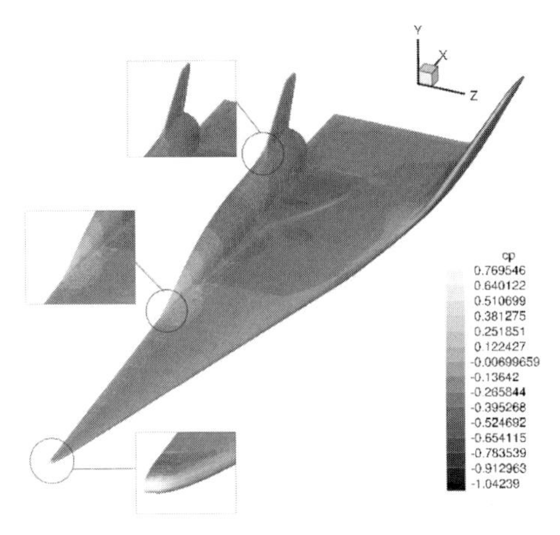


Figure 4.8.12: Pressure contours on the surface at $Ma_\infty = 0.4$, $Re_\infty = 2.6 \times 10^6$, $\alpha = 0^\circ$.

Figure 4.8.11 shows the numerical grid which was used for the computations. It contains 32 blocks with approx. 1 000 000 points. The computations assume a fully turbulent flow and adiabatic walls. Only at $\alpha = 10^\circ$ the Fares/Schröder model was applied and otherwise the Baldwin/Lomax model was used.

In Fig. 4.8.12 the pressure distribution on the wall of ELAC-1c at an angle of attack $\alpha = 0^\circ$ is shown. Compared with the solution of the flow around ELAC no remarkable changes can be observed on the lower side of the lower stage and on the upper side in the nose and the leading edge region. However, due to the pressure increase in the nose region of the upper stage the pressure distribution on the lower stage is massively influenced. This becomes clear when the wall streamlines are plotted. In Fig. 4.8.13 they are juxtaposed with an oil-flow picture taken at the same flow conditions [12]. The wall streamlines in the vicinity of the upper stage are displaced outboard. A separation occurs down-

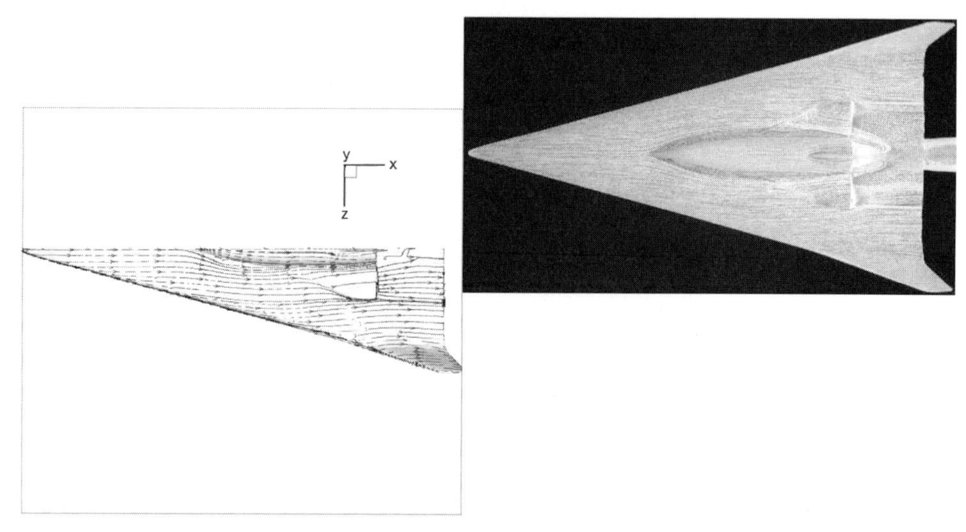


Figure 4.8.13: Numerical and experimental oil flow pattern at $Ma_\infty = 0.4$, $Re_\infty = 2.6 \times 10^6$, $a = 0°$.

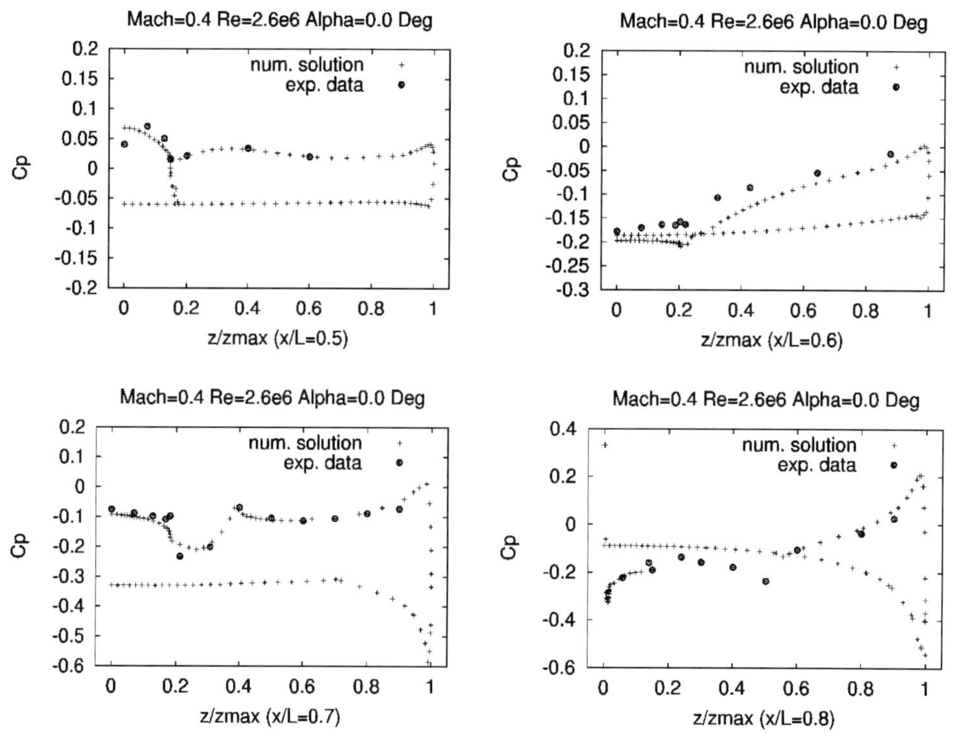


Figure 4.8.14: Pressure distributions in different cross sections for $Ma_\infty = 0.4$, $Re_\infty = 2.6 \times 10^6$, $a = 0°$.

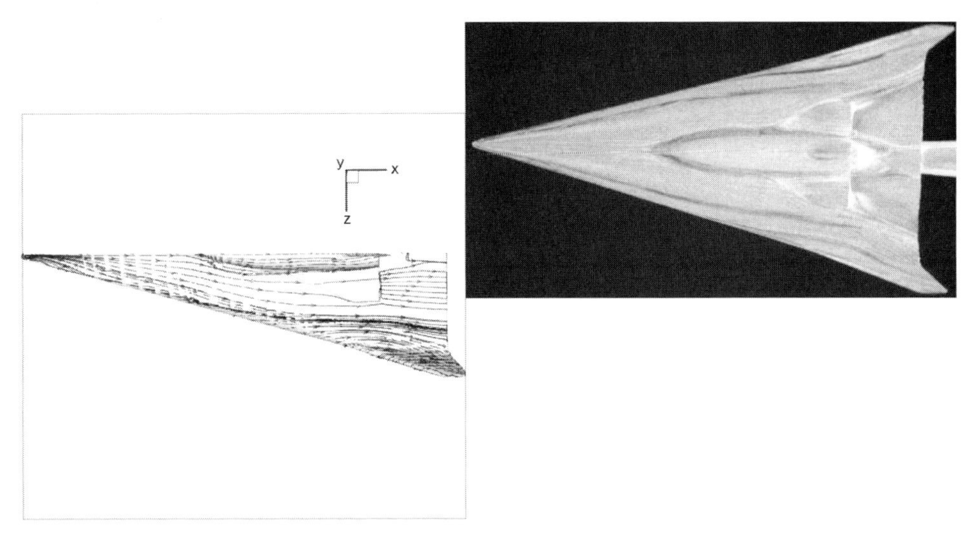


Figure 4.8.15: Numerical and experimental oil flow pattern at $Ma_\infty = 0.4$, $Re_\infty = 2.6 \times 10^6$, $a = 0^\circ$.

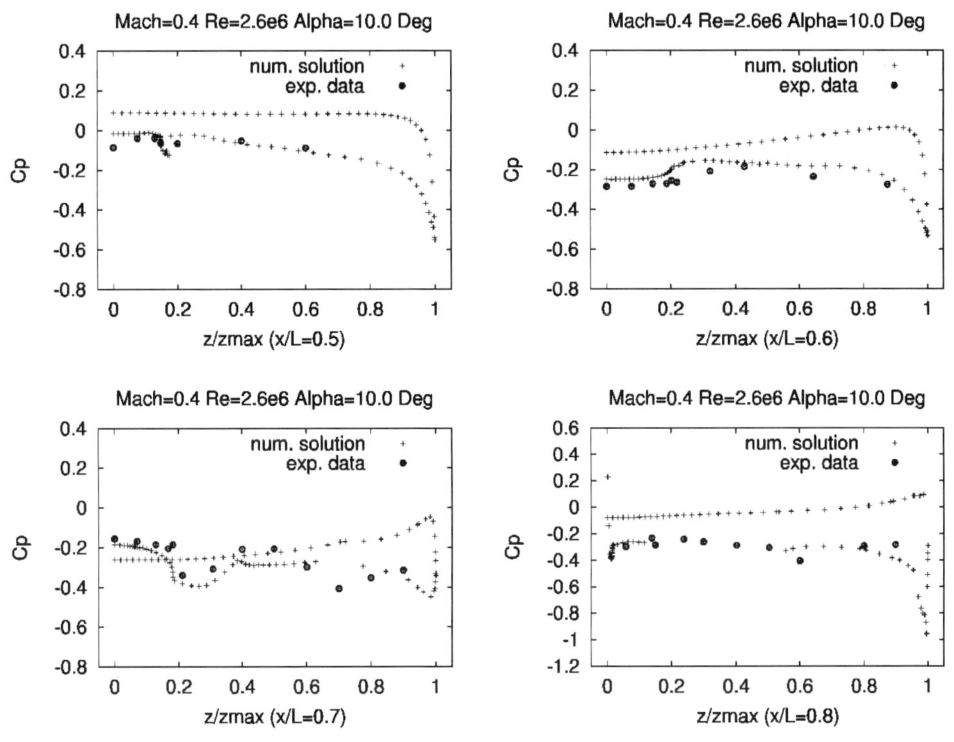


Figure 4.8.16: Pressure distributions in different cross sections for $Ma_\infty = 0.4$, $Re_\infty = 2.6 \times 10^6$, $a = 0^\circ$.

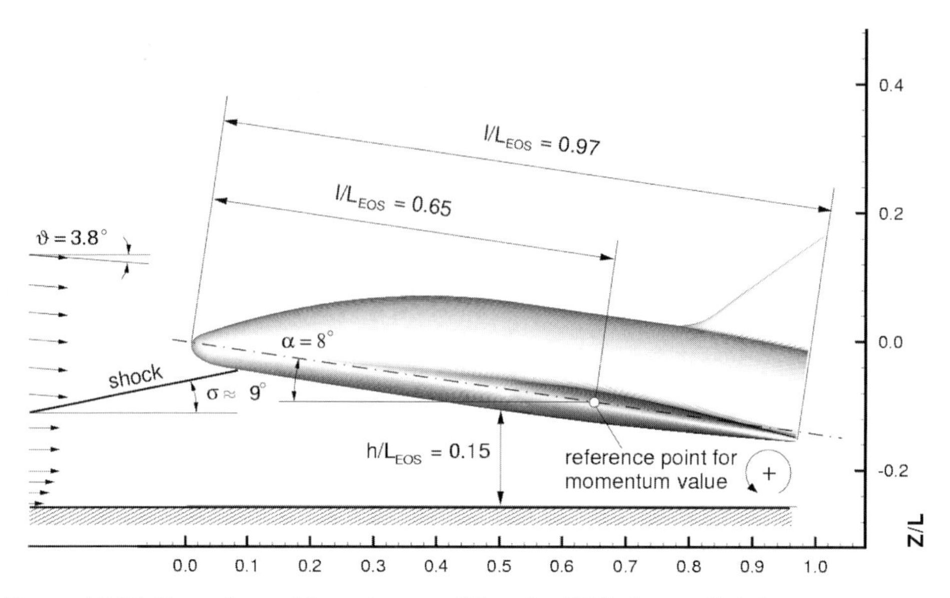


Figure 4.8.17: Geometry and boundary conditions for EOS above a flat plate.

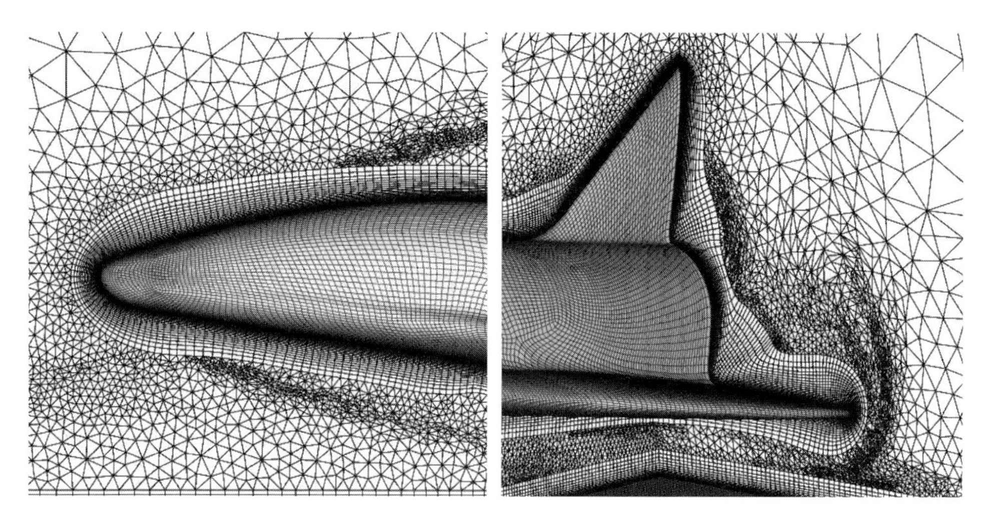

Figure 4.8.18: Hybrid adapted grid.

stream of the upper stage. The results are also compared with pressure measurements (Fig. 4.8.14). A good agreement was found in all cross sections. The increase of the angle of attack to $\alpha = 10°$ leads to a weak separation on the upper side of the lower stage. This separation is also influenced by the upper stage as evidenced in Fig. 4.8.15. The agreement between the experimental and the numerical data in Fig. 4.8.16 is again convincing.

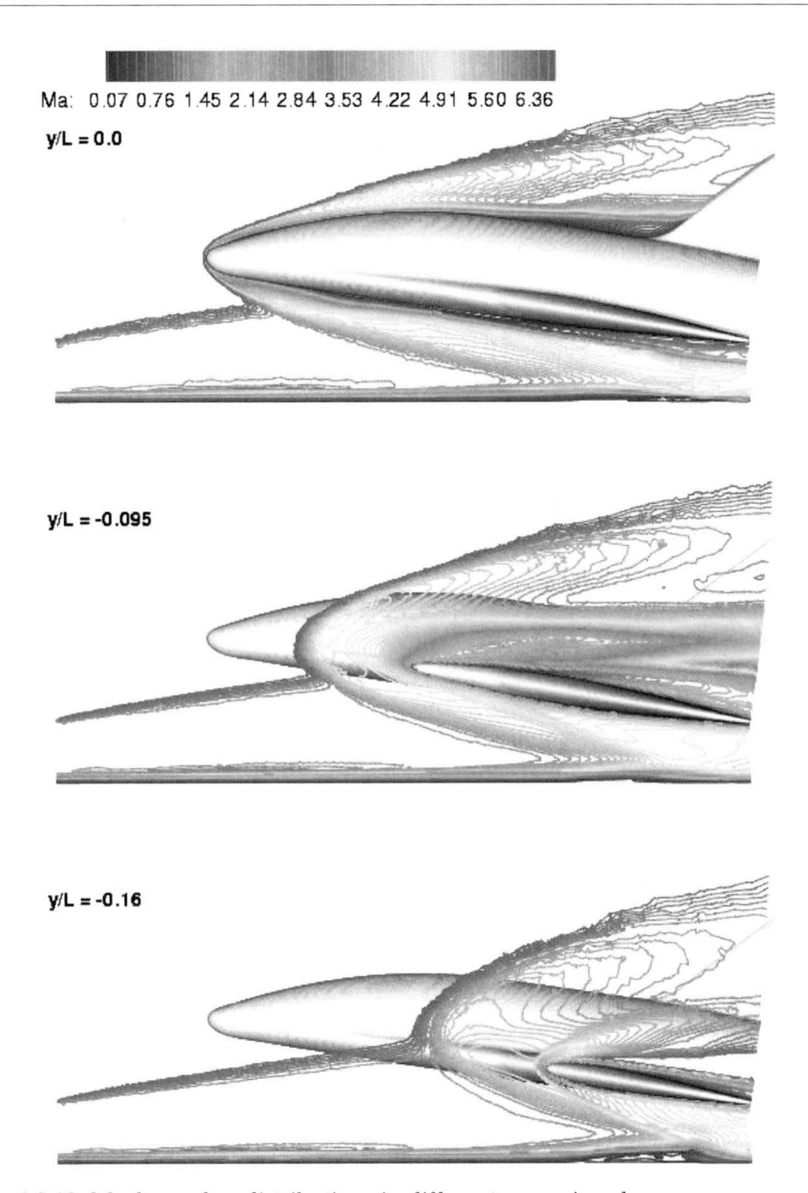


Figure 4.8.19: Mach number distributions in different spanwise planes.

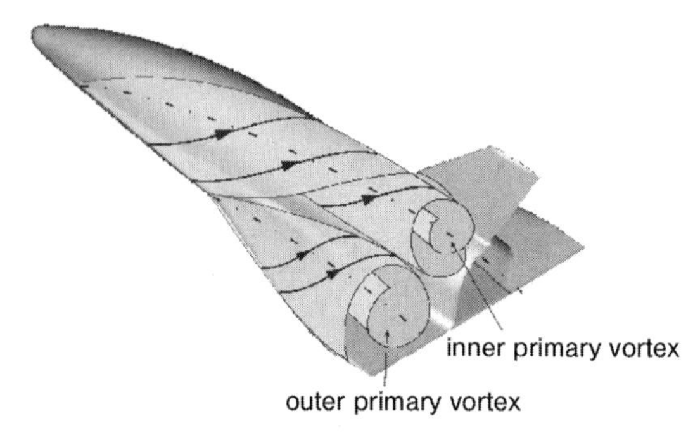


Figure 4.8.20: Sketch of the vortex system.

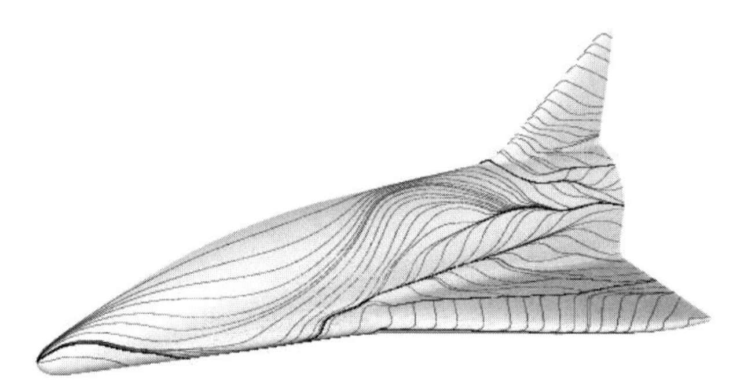

Figure 4.8.21: Numerically simulated wall stream lines.

#### 4.8.3.4 Simplified Stage Separation

We now turn to the results of the flow around the upper stage EOS in close proximity to a flat plate. The geometry of the system is shown in Fig. 4.8.17. The flow conditions are $Ma_\infty = 6.8$ and $Re_L = 1 \cdot 10^6$. The reference length is the total length of the orbital stage. At the inlet boundary a self similar solution of the boundary layer equations for the flow along a flat plate at $Re_L = 0.8 \cdot 10^6$ together with an impinging shock was applied. The grid consists of approx. 1 000 000 points in 1 unstructured and 8 structured blocks, the former of which was adapted to the solution (Fig. 4.8.18).

Figure 4.8.19 shows the Mach number distribution in different spanwise cross sections parallel to the symmetry plane. In the symmetry plane $y/L = 0$ itself the bow shock and the impinging shock are clearly visible. They interact and the deflected impinging shock causes a flow separation on the lower side of EOS, and is reflected. In the other planes the bow shock from the leading edge of the wing is evident. This shock interacts with the boundary layer and causes a large separated region. The wing geometry of EOS consists of a double delta-wing. The first part generates a vortex that moves along the fuselage and finally reaches the trailing edge in the junction between the fuselage and the fin. The second delta-wing causes another vortex that covers mainly the outboard part of the wing. In Fig. 4.8.20 this situation is sketched. Figure 4.8.21 shows the numerically simulated wall streamlines. The flow separation on the lower side due to the interaction of the EOS bow shock with the shock of the inclined flat plate and the impact of the two vortices becomes apparent together with the secondary structures in the neighbourhood of the trailing edge. A detailed analysis of the flow field can be found in [13].

## 4.8.4 Conclusions

Surface pressure distributions, velocity profiles, and flow visualizations were obtained using numerical solutions of the Navier-Stokes equations for three-dimensional sub- and supersonic flows. Both laminar and turbulent flows were computed. The results are obtained for subsonic flows around ELAC and the piggyback system ELAC+EOS and for supersonic flows around a simplified stage separation system consisting of EOS and a flat plate. The comparison with measurements from the German Dutch Windtunnel and the Aerodynamisches Institut of the University of Aachen shows that the overall agreement is good.

## References

1. Y. Wada, M.S. Liou: A flux splitting scheme with high-resolution and robustness for discontinuities, 32nd Aerospace Sciences Meeting and Exhibit, 1994.
2. B. van Leer: Towards the ultimate conservative difference schemes, V. A second-order sequel to Godunov's method, Journal of Computational Physics, 32:101–136, 1979.
3. G. van Albada, B. van Leer, J. Roberts: A comparative study of computational methods in cosmic gas dynamics, Astron. Astrophysics, Vol. 108, 76–86, 1982.
4. M. Meinke: Numerische Lösung der Navier-Stokes Gleichungen für instationäre Strömungen mit Hilfe der Mehrgittermethode, Dissertation, Aerodynamisches Institut, RWTH Aachen, 1993.
5. T.J. Barth, D.C. Jespersen: The design and application of upwind schemes on unstructured meshes, AIAA-89-0366, 1989.
6. E. Krause: German university research in hypersonics, Fourth International Aerospace Planes Conference, 1–4 December 1992, Orlando, Florida.
7. G. Neuwerth, U. Peiter, F. Decker, D. Jacob: Reynolds number effects on the low-speed aerodynamics of a hypersonic configuration, Journal of Spacecraft and Rockets, Vol. 36, No. 2, March–April 1999, pp. 265–272.
8. E. Krause, R. Abstiens, S. Fühling, V.N. Vetlutsky: Boundary layer investigations on a model of the ELAC I configuration at high Reynolds numbers in the DNW, EJMB, vol. 19:745–764.
9. B.S. Baldwin, H. Lomax: Thin layer approximation and algebraic model for separated turbulent flows, AIAA-paper 78–257, 1978.
10. E. Fares, W. Schröder: A general one-equation model for near-wall and free shear turbulence, AIAA 2003-3742, 2003.

11. F.W. Zhou: Numerische Simulation des Strömungsfeldes des zweistufigen Raumflugzeugs ELAC, Dissertation, Aerodynamisches Institut, RWTH Aachen, 2003.
12. C.-C. Ting: Strömungs- und Wärmeübergangsmessung für das zweistufige Raumtransportsystem ELAC, Dissertation, Aerodynamisches Institut, RWTH Aachen, 2003.
13. C.M. Brodbeck: Entwicklung eines strukturiert/unstrukturierten Verfahrens zur Lösung der Navier-Stokes-Gleichungen, Dissertation, Aerodynamisches Institut, RWTH Aachen, 2003.

## 4.9 High-Speed Aerodynamics of the Two-Stage ELAC/EOS-Configuration for Ascend and Re-entry

Martin Bleilebens, Christoph Glößner, and Herbert Olivier*

### Abstract

An overview is given over the experimental work accomplished at the Shock Wave Laboratory within the projects A2 and B3 of the Collaborative Research Centre 253. Experiments in hypersonic flow were performed for the overall configurations ELAC and EOS as well as for basic ramp configurations representing control surfaces or engine intakes. The working principle of the hypersonic wind tunnel TH2/TH2-D and the special measurement techniques needed in a short duration facility like this are outlined shortly, with emphasis on the special needs of measurements with controlled surface temperature of the wind tunnel models. Pressure and heat flux distributions as well as aerodynamic coefficients are given for both lower and upper stage of the reference configuration. The comparison with numerical data from other projects shows good agreement between experimental results and numerical calculations. The effects of surface temperature on ramp flows are measured and show also good agreement with numerical and theoretical results.

### 4.9.1 Introduction and Experimental Conditions

If ground testing of hypersonic or re-entry vehicles is to be representative of real in-flight conditions, the flow around the wind tunnel model has to duplicate flight velocity for elevated densities. One means to establish a flow meeting these requirements is the use of a shock tunnel such as the TH2/TH2-D at the Shock Wave Laboratory of the RWTH Aachen University [1]. In this facility,

* RWTH Aachen, Stoßwellenlabor, Templergraben 55, 52062 Aachen; olivier@swl.rwth-aachen.de

the investigation of pressure and heat flux distributions as well as the measurement of aerodynamic forces and moments is possible. Experiments are performed with complete configurations such as ELAC or EOS and with more basic setups like ramp or intake models, to investigate in detail the flow over e.g. deflected flaps or inlets typical for hypersonic propulsion systems.

Close cooperation with other projects within the Collaborative Research Centre 253 is mandatory. Special requirements on the reference configuration and its trajectory – represented by the test conditions used in the shock tunnel – were coordinated with project A1 and A4, while the experimental results obtained in TH2 are cross-checked with the numerical results of projects A3 and C3, respectively. Contact to the other Collaborative Research Centres and external research facilities, as e.g. the Institute of Theoretical and Applied Mechanics (ITAM) of the Russian Academy of Sciences in Novosibirsk, is also very important.

The fundamental setup of the shock tunnel is shown in Fig. 4.9.1. It can be driven by a standard helium driver or by an upstream detonation driver, generating the shock wave travelling into the air-filled low pressure section either by releasing the pressurized helium in the high-pressure section by bursting diaphragms or by igniting a detonation in a stoichiometric oxyhydrogen mixture, diluted with helium or argon [2]. The incident shock wave in the driven section reflects at the end wall generating a reservoir of high-pressure and high-temperature test gas. This test gas expands through a convergent-divergent nozzle (divergent part of conical shape with 5.8° half apex angle) into an evacuated dump tank, thus being accelerated to hypersonic speed. The model is mounted inside the dump tank in front of the nozzle exit, exposed to the high-enthalpy flow. Windows lateral to the test section allow optical access for flow visualization methods like Schlieren technique etc. Different test conditions are realized by adjusting the filling pressures in the different tube sections and if necessary the fraction of the diluting gas in the detonation driver. For the experiments presented in the following the conditions shown in Tab. 4.9.1 were employed. A major part of the methods and the corresponding measurement equipment used in the tunnel operation was developed within the HERMES- and MSTP-programmes initiated by ESA. For the evaluation of the

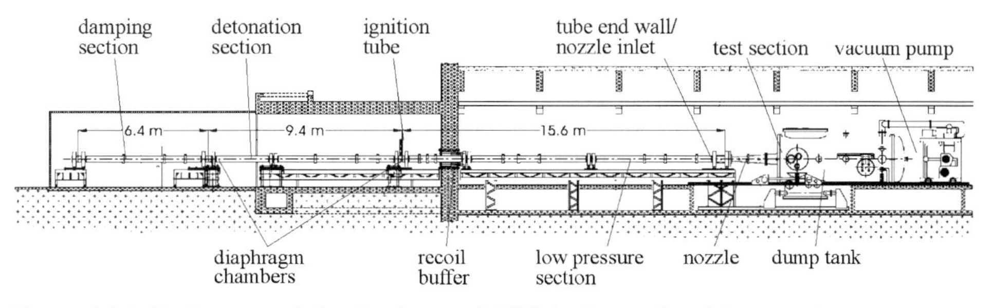


Figure 4.9.1: Basic setup of the shock tunnel TH2 in detonation driven mode.

free stream conditions for example, the stagnation point heat flux of a sphere [3] and the static pressure in the flow [4] are measured.

## 4.9.2 Measurement Equipment

### 4.9.2.1 Pressure Measurement

The short testing times in the shock tunnel and the relatively low pressures require the use of commercial piezo-resistive pressure sensors (Kulite) of short response time, high natural frequency and of high sensitivity. To protect the sensors from the harsh flow conditions, they are installed behind small cavities connected with the model surface by thin pressure taps. The sensor temperature should not exceed 150 °C, otherwise active cooling of the sensors is necessary. Pressures in the shock tube are measured by piezo-electric pressure gages (Kistler).

### 4.9.2.2 Temperature and Heatflux Measurement

Thin film sensors, thermocouples and infrared thermography are used to measure time dependent temperature distributions of the model surfaces which allows to deduce the heatflux distribution [5]. Thin films are suitable to measure small heat fluxes below about 5 W/cm$^2$. For higher values coaxial thermocouples are in use. The thermocouples are also capable to withstand elevated surface temperatures whereas thin films can not. Lately infrared thermography is in operation to determine two-dimensional temperature (270 Hz) and heat-flux distributions near the model centreline (9400 Hz). For this a high speed camera (Phoenix, Indigo Systems) is used with an array of 320×256 pixels. Non-uniformity correction of the sensor array for different exposure times, filters and temperature ranges is realized at the SWL with a commercial black body calibration source. Calibration curves of the measured intensity of the infrared camera as function of the model surface temperature are acquired within the evacuated test section with thermocouple measurements as reference. These calibration curves can be directly used to deduce temperature and heat flux distributions.

### 4.9.2.3 Force and Moment Measurement

The measurement of aerodynamic forces and moments is performed with a sting-mounted strain gage balance especially designed for the use in the TH2 facility. High natural frequencies of the balance, necessary for short duration measurements, are obtained by a very compact design of high rigidity, and the use of semiconductor strain gages is required to obtain an acceptable signal to noise ratio. On the other hand, this results in increased interferences and non-linearities of the balance, which have to be considered for its calibration and the data reduction of the measurements [6, 7]. The measuring of the loads in all 6 derees of freedom allows for a correction of the interferences and non-linearities. By thorough calibration a precision of ≤1% for lift, drag and pitching moment, and ≤2% for the other components is reached. The mass of the model oscillating on its support leads to additional inertial loads acting on the balance. The influence of these oscillating loads can be eliminated by an acceleration compensation [8]. The acceleration induced inertial loads are deduced from the measured model accelerations and the acceleration sensitivity of the model-balance system, determined by excitation tests [9]. Adding these to the total signals yields the quasistatic aerodynamic loads. In the present case, only the normal force and pitching moment are compensated, for all other components a gliding averaging is employed or the total loads are of negligible magnitude. Up to now, compensation of inertial loads is possible only for a narrow band of low frequencies, averaging or filtering eliminates higher frequencies in the force and acceleration signals.

#### 4.9.2.4 Flow Visualization

For flow visualization, a colour Schlieren setup is used. By comparing a visualized density gradient with a reference object it is possible to recognize it as a compression or an expansion, respectively. Thus, important information even

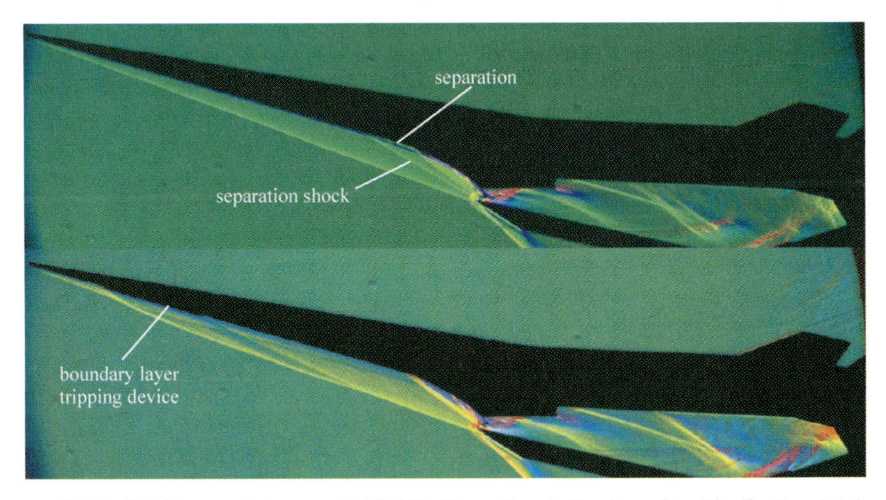


Figure 4.9.2: Schlieren pictures of ELAC 1b without (top) and with (bottom) tripped boundary layer (condition I, $M_\infty = 7.9$, $\alpha = 15°$).

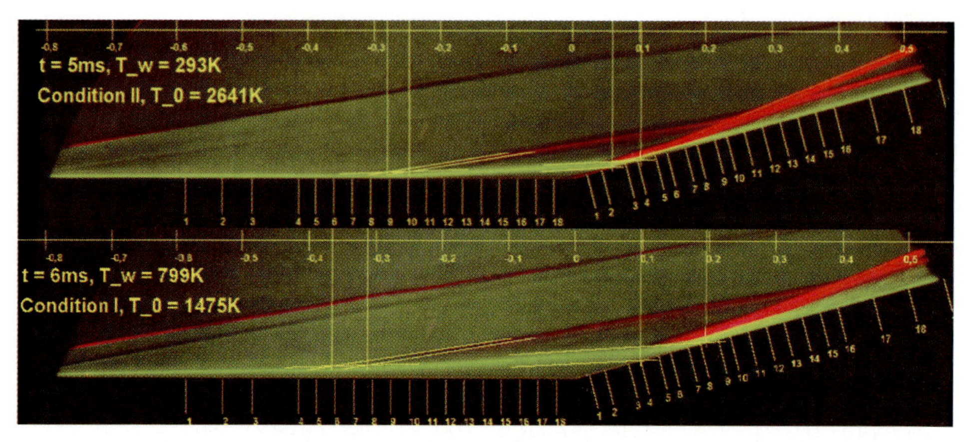


Figure 4.9.3: Schlieren pictures showing the influence of the temperature ratio $T_w/T_0$ on the extent of the separation bubble at a ramp configuration.

about complex flow fields can be obtained. Figures 4.9.2 and 4.9.3 give examples of colour Schlieren pictures taken in experiments with different configurations and models in the TH2 facility.

### 4.9.3 Measurements on the ELAC- and EOS-Configurations

#### 4.9.3.1 Pressure and Heat Flux Measurements on the ELAC-Configuration

Multiple tests were performed on the lower stage of the reference configuration ELAC, investigating the flow field around the vehicle with and without engine intake simulation, at moderate angles of attack. Due to the small scale of the

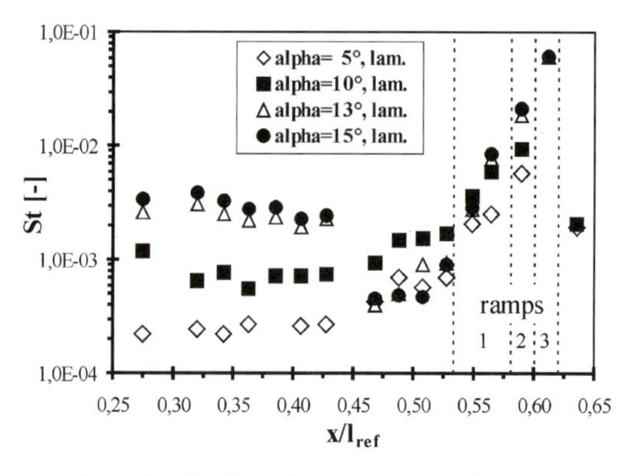


Figure 4.9.4: Stanton number distribution along windward centerline of ELAC 1b for laminar boundary layer.

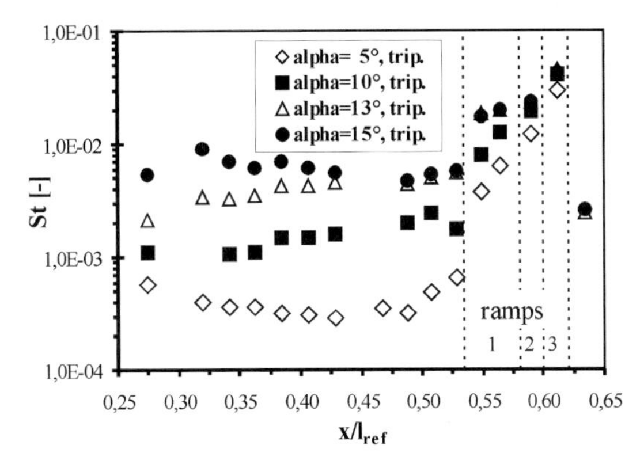


Figure 4.9.5: Stanton number distribution along windward centerline of ELAC 1b for tripped boundary layer.

model (1:240, model length 300 mm) the available space for instrumentation and therefore the number of pressure and heat flux gages is limited. Exemplarily, the Stanton number distribution along the windward centreline of the ELAC-model is shown, achieved at condition I for laminar and tripped boundary layer (see Figs. 4.9.2, 4.9.4 and 4.9.5) [10]. The boundary layer is tripped by an artificial roughness at the nose of the vehicle. As expected, for the tripped case the Stanton numbers are higher than for the laminar one, and increase with increasing angles of attack in both cases. For the experiments with laminar boundary layer, a steep decrease in the Stanton number between $x/l_{\text{ref}}=0.43$ and 0.47 for high angles of attack indicates a separation region in front of the first compression ramp, at low angles of attack the beginning of the separation bubble is located upstream of the first sensor position and therefore cannot be detected. No separation at all occurs in case of a tripped boundary layer (see also Fig. 4.9.2). The heat fluxes on the third ramp seem to be independent on the angle of attack and are of the same order of magnitude for both laminar and tripped boundary layer.

### 4.9.3.2 Force and Moment Measurements on the ELAC-Configuration

For the ELAC-configuration without engine inlet simulation, the aerodynamic forces and moments were investigated at condition I for angles of attack between $-2^{\circ}$ and $8^{\circ}$ [7]. In general, it is recommended to keep the mass and the moments of inertia of the model small with respect to the balance reference point, in order to reduce the additional inertial loads during the experiments and increase the eigenfrequencies of the model-balance system. This facilitates the elimination of the acceleration induced parts of the balance signals and can be achieved by using a lightweight model with its centre of gravity close to the balance reference point. For the ELAC-model, at least the latter requirement

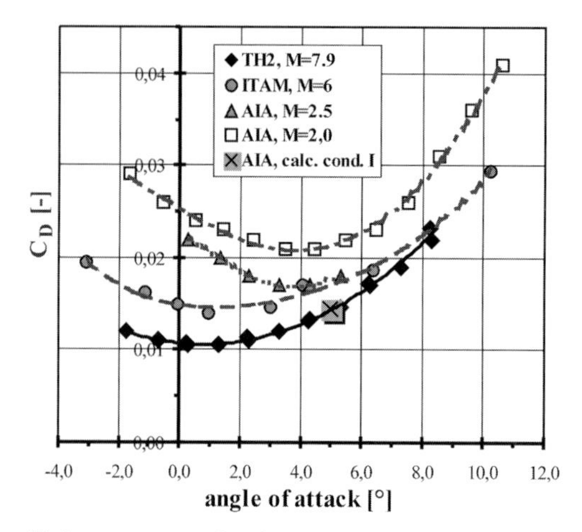


Figure 4.9.6: Drag coefficient over angle of attack for ELAC 1a, condition I.

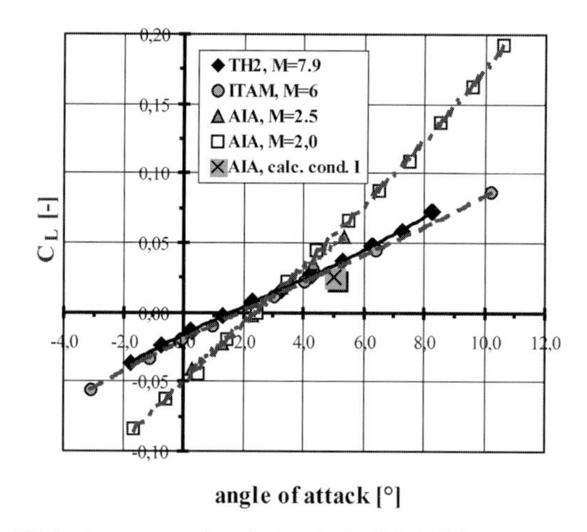


Figure 4.9.7: Lift coefficient over angle of attack for ELAC 1a, condition I.

cannot be accomplished, since the model fuselage is not large enough to accommodate the balance. Despite using a carbon fibre model with a total mass of only approx. 220 g, the necessity of using a sting to mount the model on the balance leads to large Steiner parts of the moments of inertia and thus to low eigenfrequencies of the model-balance system. Nevertheless, the methods used in the data reduction process are effective enough to provide reliable results, as proved by the consistency of especially the force coefficients, shown in Figs. 4.9.6 and 4.9.7. The experimentally deduced coefficients are compared with results obtained in the hypersonic blow-down tunnel T-313 at the Institute of Theoretical and Applied Mechanics (ITAM) of the Russian Academy of Sciences in Novosibirsk [11], at a Mach number of 6 and a Reynolds number of

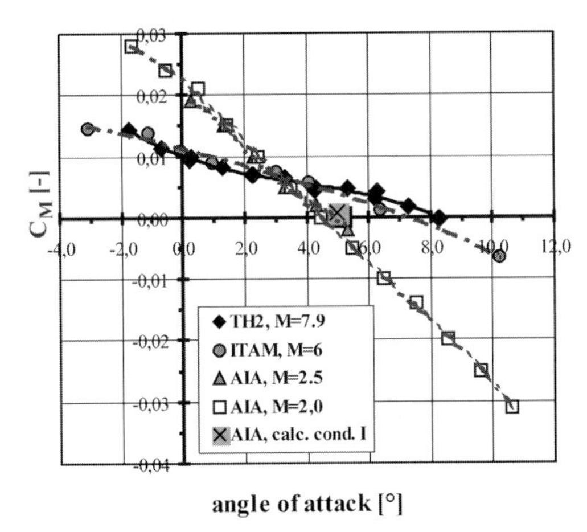


Figure 4.9.8: Pitching moment coefficient over angle of attack for ELAC 1a, condition I.

$3\times10^6$ referring to the model length of 300 mm. Additionally, experiments were performed at the Aerodynamisches Institut AIA of RWTH Aachen University [12]. Two sets of experiments at $M_\infty=2.0$ and $M_\infty=2.5$, and Reynolds numbers of $3.8\times10^6$ and $3.0\times10^6$, respectively, are shown. A numerical calculation for cond. I is available for $\alpha=5^\circ$ [13]. Figure 4.9.6 shows a strong dependency of the drag coefficient on the Mach number: the minimum $C_D$-value decreases with increasing Mach number and also the correspondent angle of attack is decreasing. The numerical value fits well with the TH2-measurements. The lift coefficient in Fig. 4.9.7 shows in a first order a linear dependency on the angle of attack for all Mach numbers considered. The difference between the experiments at $M_\infty$= 6 and at $M_\infty$= 7.9 is small, and the numerical result for $\alpha=5^\circ$ is in good accordance with the experiments.

In Fig. 4.9.8, the pitching moment coefficient $C_m$ is shown as a function of $\alpha$. The calculated trimming angle (approx. 5°) is considerably different to the measured one in the TH2 (approx. 8°). This discrepancy between measurements and numerical results also exists for lower Mach numbers [12, 13], so obviously the proper determination of the pitching moment in super- or hypersonic flow is still a task which is not easily accomplished, neither numerically nor experimentally, where the consistency of the experimental data for $M_\infty$= 6 and $M_\infty$= 7.9 may indicate a problem of the numerical simulation.

#### 4.9.3.3 Pressure and Heat Flux Measurements on the EOS-Configuration

Pressure and heat flux data is available for the lower surface of the upper stage of the reference configuration, EOS, for three typical phases of its mission: stage separation, ascent, and atmospheric re-entry at high angles of attack and in high-enthalpy flow [9]. In the following, results are shown exemplarily for

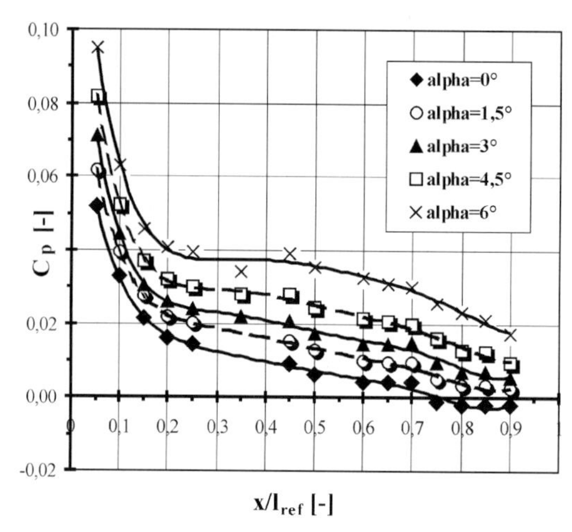


Figure 4.9.9: Pressure coefficient distribution, EOS windward centreline, ascent phase, cond. D-I.

ascent and stage separation. In the experiments for the ascent phase, performed at cond. D-I at angles of attack between 0° and 6°, the distribution of the pressure coefficients along the windward side is as expected (see Fig. 4.9.9): the steepest gradients are observed in the nose region up to $x/l_{\mathrm{ref}}=0.3$, where the biggest change in the local inclination of the body contour occurs, whereas downstream of this region, where the curvature of the surface is small, there is also only a small change in the pressure coefficients.

Stage separation is simulated by placing a flat plate, aligned to the free stream direction (angle of attack of the flat plate $\alpha_P=0°$), underneath the upper stage at different vertical distances scaled with the EOS body length ($h/l_{\mathrm{ref}}=0.2$, 0.15, 0.05 and 0.035). Angle of attack of EOS is $\alpha=0°$ for all distances. Further experiments were performed with various angles of attack of EOS ($0° \leq \alpha \leq 4.5°$) at a constant distance to the flat plate of $h/l_{\mathrm{ref}}=0.05$. Reference point for the measured distance is the lowest point of the body contour at 50% $l_{\mathrm{ref}}$. The results of the experiments for the stage separation phase are represented exemplarily by the $C_p$-distributions in Fig. 4.9.10. The data for three different angles of attack of EOS at a constant distance to the flat plate of 0.05 $l_{\mathrm{ref}}$ are shown. The data for the case without flat plate are also shown to document the changes. Two effects are displayed in this diagram: first, a step in the pressure distribution, growing in height with increasing angle of attack, and second, a backward shift of this step for increasing $\alpha$. The first effect can be explained by the bow shock of EOS, becoming stronger at higher angles of attack, so that the reflexion on the flat plate, impinging on the lower surface of EOS, results in higher pressures. The Schlieren pictures in Fig. 4.9.11 show the second effect, i.e. that the reflected shock hits the lower surface of EOS further backwards for higher angles of attack. Further investigations on stage separa-

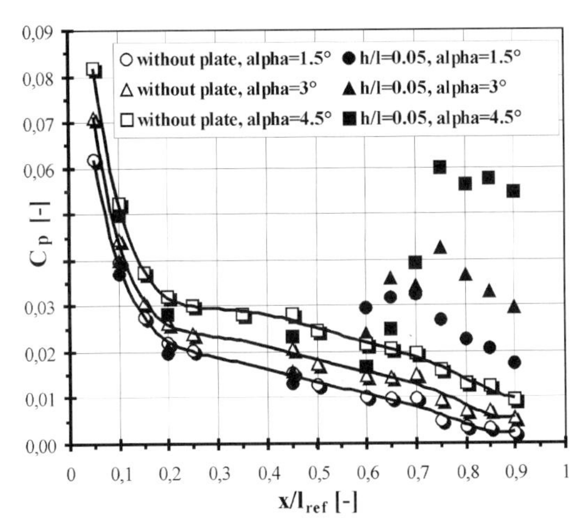


Figure 4.9.10: Pressure coefficient distribution, EOS windward centreline, stage separation phase, different angles of attack, condition D-I.

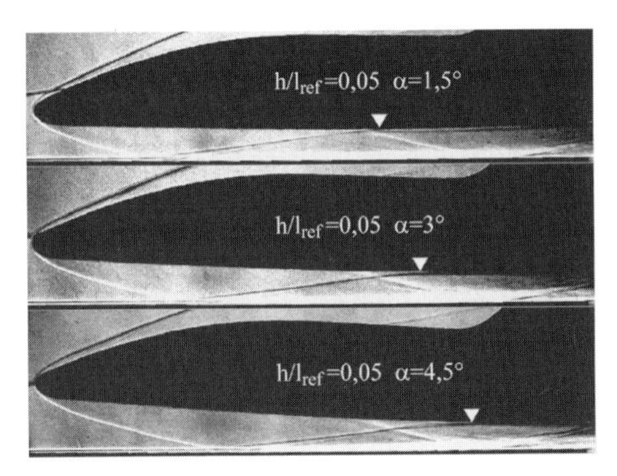


Figure 4.9.11: EOS Schlieren pictures, stage separation phase, different angles of attack, cond. D-I.

tion were performed within the Collaborative Research Centre 255 together with ITAM in Novosibirsk (see Chapter 4.3).

### 4.9.3.4 Force and Moment Measurements on the EOS-Configuration

The experiments were performed with a carbon fibre model of 1:100 scale with a length of 288 mm and a weight of approx. 250 g. The fuselage of this model is wide enough for the balance to be mounted internally, close to the centre of gravity. Thus, the eigenfrequencies of the model's oscillations on the balance

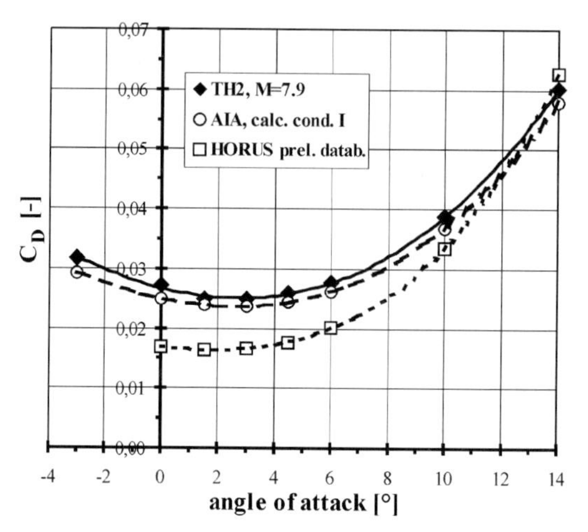


Figure 4.9.12: Drag coefficient over angle of attack for EOS, ascent phase, condition I.

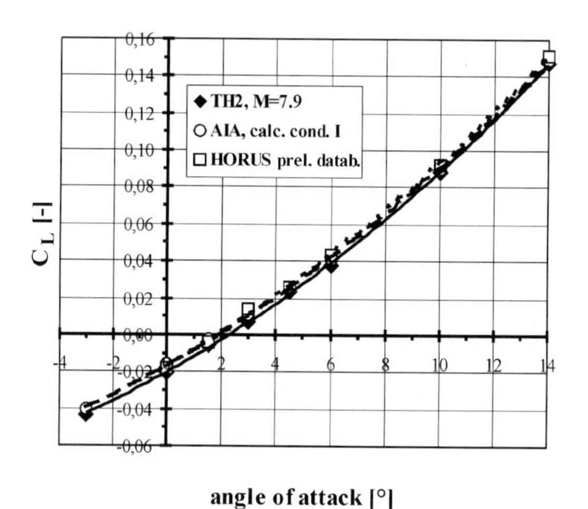


Figure 4.9.13: Lift coefficient over angle of attack for EOS, ascent phase, condition I.

were kept as high as possible. Two different phases of the mission were investigated, the ascent and the re-entry phase, the latter for high-enthalpy flow to try to evaluate the influence of real gas effects. The ascent phase was investigated with cond. I, the angle of attack ranging from $\alpha$=–3° to $\alpha$=14°. Drag and lift coefficient, shown in Figs. 4.9.12 and 4.9.13, are compared with numerical results [14]. A Navier-Stokes code without turbulence modelling was used, the assumed base pressure was set to the static pressure of the free stream. The wall/stagnation temperature ratio of the experiments was maintained for the calculations at a wall temperature of 293 K. Additionally, for further information the values of the HORUS configuration, the SÄNGER upper stage, are plotted in the diagrams. They were obtained with a panel method based on Newtonian theory and give preliminary estimations for the HORUS aerodynamics [15]. Though slightly different in geometry, the EOS and HORUS configurations are comparable referring to the qualitative aerodynamic features. For EOS, both drag and lift coefficient show a very good agreement between the experimental and the numerical data. The small offset observed in the drag coefficient may be due to the base pressure influence, but the resulting error is less than 4%. For the lift coefficient, the agreement is even better. The angle of zero lift is numerically determined to $\alpha_0 = 1.9°$ and experimentally to $\alpha_0 = 2.3°$.

During atmospheric re-entry of real space vehicles dissociation and ionization processes take place, which have influence on the aerodynamic coefficients. The detonation driven condition D-IV provides a flow with a stagnation temperature of about 7400 K, where the oxygen is completely dissociated and the nitrogen to a fraction of about 12 %. One aim of the experiments was to try to evaluate the influence of these real gas effects on the EOS aerodynamics. To rule out viscosity effects in the comparison of data with and without real gas effects, cond. I-f was chosen, providing a flow of similar Mach and Reynolds number (see Tab. 4.9.1), but with considerably lower stagnation temperature and thus without real gas effects. A clean configuration, i.e. without deflected body flap, was studied at angles of attack of $\alpha = 20°$, $25°$ and $30°$. Unfortunately, the small order of magnitude of the flow parameters as well as of the measured forces for condition If induces a comparatively large data scatter, blurring the anyway small expected effects, especially the change in pitching moment. Despite the fact that for cond. I the Reynolds number is about five times larger than for conds. If and D-IV, available numerical results are given for cond. I for comparison. For all conditions considered laminar flow establishes around the windward side of the model, and since the total enthalpy $h_0$ for conds. I and If is comparable, the agreement between the numerical and experimental results for the low-enthalpy cases is not surprising. No difference could be detected neither in drag nor in pitching moment coefficient between low- and high-enthalpy flow for the three angles of attack. Only the lift coefficient, shown in Fig. 4.9.14, is slightly increased for the high-enthalpy experiments, which is

Table 4.9.1: Test conditions of experiments performed within this study.

| Test cond. | Reservoir cond. | | Free stream conditions | | | | | | |
|---|---|---|---|---|---|---|---|---|---|
| | $p_0$ [MPa] | $h_0$ [MJ/kg] | $p_{t2}$ [kPa] | $p_\infty$ [kPa] | $T_\infty$ [K] | $u_\infty$ [m/s] | $M_\infty$ [-] | $Re_{\infty,1}$ [$10^6$/m] | $\Delta t_{test}$ [ms] |
| I | 7.1 | 1.68 | 59 | 0.8 | 125 | 1750 | 7.7 | 4.2 | 2.0 |
| II | 37.0 | 3.4 | 142 | 2.0 | 250 | 2350 | 7.4 | 4.0 | 2.0 |
| If | 1.7 | 2.0 | 24 | 0.4 | 210 | 2070 | 6.7 | 0.8 | 3.0 |
| D-I | 27.3 | 2.4 | 184 | 2.3 | 180 | 2120 | 7.9 | 7.6 | 8.0 |
| D-III | 26.3 | 9.0 | 185 | 3.0 | 830 | 3950 | 6.9 | 1.3 | 3.4 |
| D-IV | 23.6 | 14.2 | 168 | 2.9 | 1250 | 4840 | 6.7 | 0.8 | 2.0 |

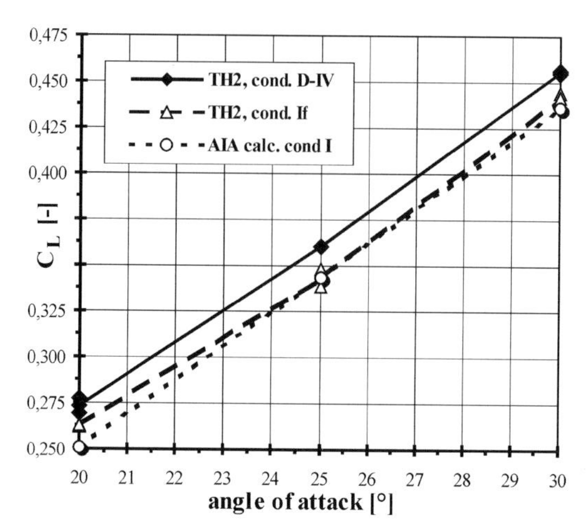


Figure 4.9.14: Lift coefficient over angle of attack for EOS, re-entry phase.

consistent with the fact that for a high-enthalpy flow the surface pressure is slightly higher than for a low-enthalpy case. Considering the fact that nitrogen dissociation is far from being complete and ionization rates are negligible, a distinct difference can not be expected for the EOS vehicle in clean configuration. The measurement techniques and methods developed and improved in this project for the ELAC/EOS configuration were applied in a recent test campaign on the HOPPER configuration within the framework of the ASTRA programme. Very interesting results could be obtained in the experiments performed at re-entry conditions in high-enthalpy flow [16].

### 4.9.4 Detailed Measurements on Ramp-Configurations

#### 4.9.4.1 Laminar and Turbulent Shock-Wave/Boundary-Layer Interactions

Shock-wave/boundary-layer interactions take place at different parts (e.g. flaps and engine inlets) of vehicles flying at high Mach number. Their behaviour is mainly influenced by the deflection angle $\alpha$, Mach number $Ma_\infty$ and Reynolds number $Re_\infty$ of the incoming flow. At low Reynolds numbers the interactions are purely laminar, at high Reynolds numbers transitional and turbulent interactions occur. In this context the total temperature $T_0$ of the fluid and surface temperature $T_w$ play also an important role. At high total temperatures real gas effects start to influence the interaction behaviour.

The main part of the work related to shock-wave/boundary-layer interactions performed at the Shock Wave Laboratory was to study the influence of the following parameters on the overall flow behaviour:

- Reynolds number $Re_{\infty}$,
- Total temperature $T_0$,
- Surface temperature $T_w$.

In this report the main interest is related to the last two parameters. Ramp models to generate a quasi two-dimensional flow were used to simulate the flow around deflected flaps or engine inlets. The experiments were performed at different flow conditions (see Tab. 4.9.1) to study total temperature effects. With heatable model surfaces it was also possible to investigate the influence of the surface temperature.

### 4.9.4.2 Theoretical Considerations

Figure 4.9.15 shows schematically a shock-wave/boundary-layer interaction for a ramp configuration. Due to a high deflection angle inducing a positive pressure gradient the flat plate boundary layer separates and a separation shock develops. This shear layer reattaches at the ramp surface. The main ramp shock is displaced downstream the ramp hinge line. The characteristics of the areas of attached flow can be described by oblique shock theory and boundary layer theory. Separated areas are characterized by free interaction theory and scaling laws for the size of the separation bubble.

The static pressure upstream of the separation of the boundary layer mainly corresponds to the free stream pressure $p_{\infty}$. The static pressure $p_{III}$ downstream the interaction is given by oblique shock theory. The Stanton number distribution for attached boundary layer flow can be described by the flat plate solution extended by the reference temperature method [17]

$$St(x) = 0.5 \cdot K \cdot Pr^{-2/3} \cdot Re_x^{-n} \left(\frac{T^*}{T_{\infty}}\right)^{-n(1-\omega)} \left(\frac{p(x)}{p_{\infty}}\right)^n$$

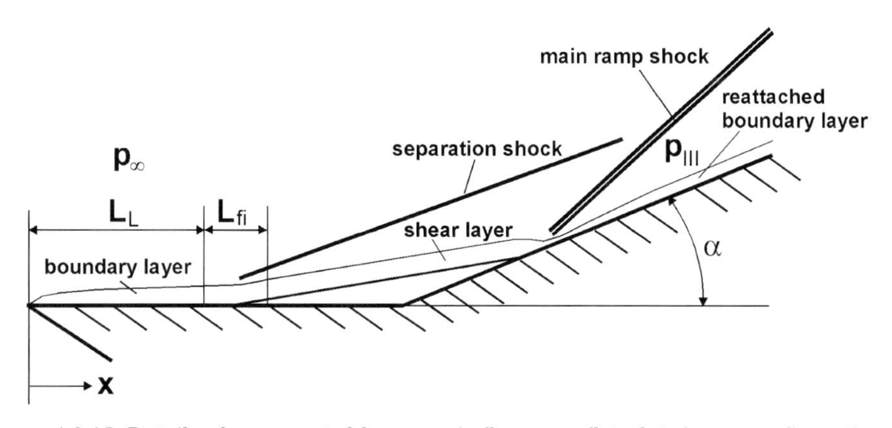


Figure 4.9.15: Details of a separated hypersonic flow on a flat plate/ramp configuration.

where $K$ and $n$ are dependent on the boundary layer flow (laminar, turbulent). The factor $\omega$ is defined by the approximation of the viscosity law (air: $\omega=0.75$). $Pr$ is the Prandtl number and $Re$ the Reynolds number. With the pressure ratio $p_{III}/p_\infty$ the Stanton number of the attached boundary layer downstream of the interaction can also be calculated.

In general, the subsonic part of the flat plate boundary layer in a supersonic flow can be influenced by disturbances generated downstream. If the ramp angle exceeds a certain value $\alpha_{sep}$ the boundary layer cannot follow any longer the pressure increase induced by the ramp shock and separates. This disturbance then propagates upstream in the subsonic part of the boundary layer. The behaviour within the separation region is independent of the ramp angle and generally independent of the cause of the disturbances. The concept of this "free interaction" was first developed by Chapman [18]. The pressure within the separation region is mainly dependent on the flow parameters upstream the separation and a function $F$, which was determined experimentally [18].

$$p(x) = f\left(F\left(\frac{x - L_L}{L_{fi}}\right), x, c_f, Ma_I\right) \cdot q_I + p_I$$

This function approaches a certain value depending on whether the boundary layer is laminar or turbulent. This value determines the pressure within the separation bubble.

The geometry of the separation bubble e.g. beginning of free interaction region, reattachment position etc. as a function of the ramp angle $\alpha$ was subject of many investigations [18]. The flow parameters varied cover Mach number $Ma_\infty$, Reynolds number $Re_\infty$ of the incoming flow and total temperature $T_0$ of the flow. The length of the separation bubble is mainly scaled with the boundary layer thickness and parameters depending on the state of the boundary layer at the separation point. It was found that with increasing ratio wall temperature to total temperature $T_w/T_0$ the length of the separation bubble increases. If the separation length $L_{sep}$ is normalized by the boundary layer thickness or displacement thickness for laminar flow this influence disappears. A correlation is given to calculate the length of the separated region [19] where surface temperature effects are partially accounted for.

#### 4.9.4.3 Ramp Flows with Variation of Surface Temperature

The experimental simulation of hypersonic flows involves a decrease in testing time with increasing total temperatures or flow velocities. In shock tunnels the typical testing time is in the order of a few milliseconds. Within this short time period usually the model surface temperature does not reach the temperatures of free flight e.g. the adiabatic wall temperature $T_{aw}$ or the radiation adiabatic wall temperature $T_{raw}$ [20]. As key figures the ratios $T_w/T_0$ or $T_w/T_{aw}$ can be uti-

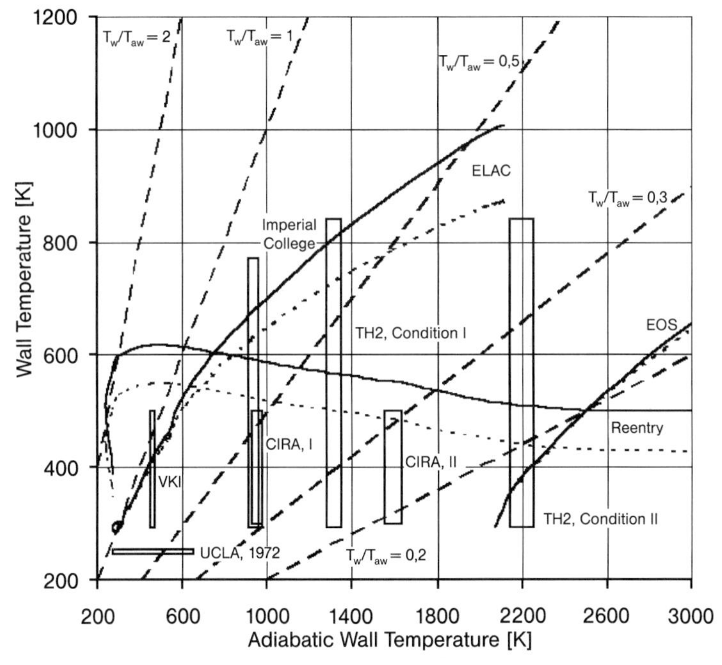


Figure 4.9.16: Surface temperatures $T_w$ versus adiabatic wall temperatures $T_{aw}$ achieved in different facilities.

lized to characterize the temperature influence. Although ramp flows are investigated for decades, the effects of surface temperature on the flow field are not completely understood. Basic work was done in 1970 starting again in 1996 with improved surface materials and sensors. Fig. 4.9.16 shows surface temperatures $T_w$ and adiabatic wall temperatures $T_{aw}$ which were achieved in earlier reports [21–24] compared to the present study (TH2, condition I and II). For comparison estimations of the wall temperature near a flap (EOS) or engine inlet (ELAC) at ascent and values for a typical Shuttle descent are added.

Because internal heating of the models is necessary for application in the shock tunnel, the first aim was to test different heating techniques and their ability to generate a uniform wall temperature distribution of the heated surfaces. For this, within the ESA-FESTIP programme a ramp model with a ramp angle of 15° was provided by former DaimlerChrysler Aerospace GmbH. The aerodynamic surfaces consisted of quartz-glass plates (transmissivity $\varepsilon \approx 0.7$) which were coated with a high-temperature resistant paint. This thin layer was heated up to $T_{w,\max} \approx 1000$ K within two minutes by infrared radiators. Temperature distributions of the surfaces were measured by infrared thermography and separation lengths by colour Schlieren pictures [25]. Fig. 4.9.17 shows a typical temperature distribution of the aerodynamic surfaces immediately before the experiment. The geometry is scaled with the length of the flat plate $L_1$ and the temperature is scaled with the maximum temperature $T_{w,\max}$ reached at

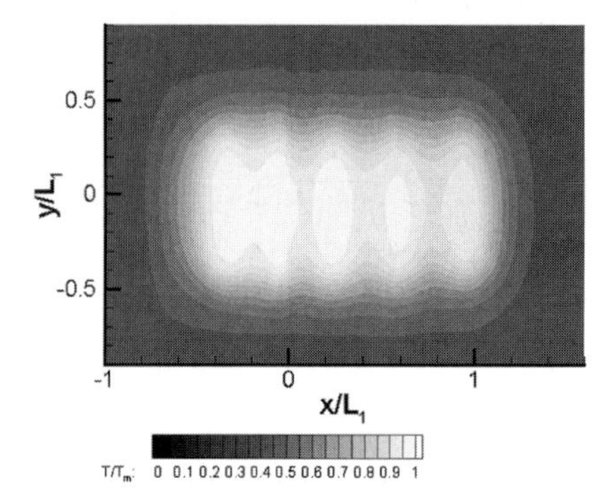


Figure 4.9.17: Temperature distribution of aerodynamic surfaces heated by infrared radiation.

the surface. The ramp hinge line is located at $x/L_1=0$. Regions near the centreline were heated up to maximum temperatures whereas lateral regions remained at lower temperatures. Therefore, an improved heating technique yielding a more uniform temperature distribution was desirable. Furthermore, the experiments showed that due to the high surface pressures tests at high total temperatures with quartz-glass surfaces were not possible.

Based on these experiments for detailed measurements it was decided to build an improved model which includes the following features:

- Electrical resistance heating elements to achieve an improved temperature distribution,
- Suitable model geometry to achieve two-dimensional ramp flow at the centreline,
- Possibility to change ramp angle and to use the model as "flat plate configuration",
- Installation of suitable thermocouples for high surface temperatures,
- Static wall pressure measurements at elevated surface temperatures,
- Possibility to use Schlieren photography and infrared thermography with the same model.

#### 4.9.4.4 Description of Ramp Model

As surface temperatures increase, the use of intrusive measurement techniques (pressure probes, thermocouples) gets more difficult. Although non-intrusive optical methods (Schlieren pictures, infrared thermography) can be used, both approaches were followed during the experiments. Due to the results of preliminary investigations the ramp model was equipped with 8 electrical heating circuits (totally 2.3 kW heating power) with PID controllers to achieve a temperature distribution as uniform as possible. The heating elements are mounted into

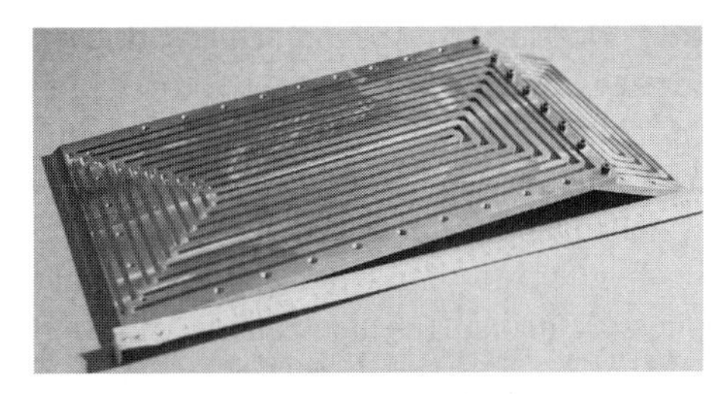

Figure 4.9.18: Backside of heated surfaces.

the backside of the aerodynamic surfaces (Fig. 4.9.18). To assure a two-dimensional ramp flow the ratios of model width $W$ and ramp length $R$ to the forebody length $L$ are close to one ($W/L=0.92$; $R/L=1.18$) which is in agreement with the criteria found by Lewis and Kubota [26].

Thermocouples (K-type) with short response time for use at higher temperatures have been manufactured at the Shock Wave Laboratory. The pressure gages have been installed into two seperate cooling devices to allow for a change of the ramp angle without any change of the sensor installation and pressure taps. Water cooling of the sensor installation unit allows to keep the gages at room temperature while heating the surfaces up to $T_{w,\max} \approx 840$ K. The pressure taps are connected to the sensors by stainless steel capillary tubes avoiding big cavities. The model is equipped with 20 thermocouples and 42 pressure gages to achieve a high spatial resolution. During heating the length of the aerodynamic surfaces increases about 1 to 2 mm due to thermal expansion. Since thermal movement of the surfaces takes place at the hinge line, it is checked and carefully sealed before every experiment. It was found from Schlieren pictures that due to the heating the ramp angle does not change more than $\pm 0.1^\circ$. The model can be rotated to get either a Schlieren picture or a two-dimensional infrared picture or an infrared line-scan of high frequency.

Figure 4.9.19 shows a typical temperature distribution of the new ramp model for a ramp angle of $15^\circ$. Comparison with Fig. 4.9.17 shows a significant

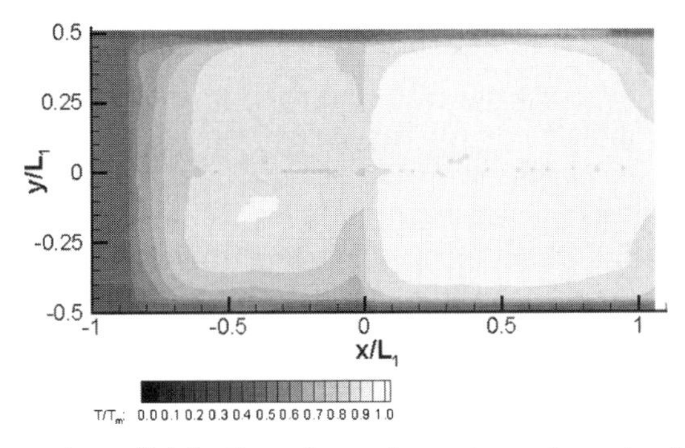


Figure 4.9.19: Temperature distribution of aerodynamic surfaces heated by electrical resistance heating.

improvement of the temperature uniformity. Nearly the whole surface is heated up to about 90% of the maximum temperature. Especially the area close to the leading edge is more uniformely heated up.

### 4.9.4.5 Schlieren Pictures and Position of Separation

Figure 4.9.3 exemplarily depicts two Schlieren pictures which clearly show the changes in the geometry of the separated region caused by different ratios of wall to total temperature. The separating and reattaching boundary layer, separation shock and main ramp shock are clearly visible. The positions of the pressure gages are indicated. For $T_w/T_{0,min}=0.11$ and $T_w/T_{0,max}=0.54$ respectively, the pictures show the influence of the main scaling parameter of interest keeping Reynolds number and Mach number almost constant (Tab. 4.9.1). The Schlieren pictures were taken at the end of the measurement time to ensure a complete establishment of the separation bubble. This characteristic time scale can be estimated from the time dependent pressure coefficients $c_p$ measured within the separation region. An example of this is given in Fig. 4.9.20 for the experiment shown in the upper Schlieren picture (condition II). Two different $c_p$-levels are visible: Upstream of the separation all pressure gages measure the free stream pressure ($c_p \approx 0$) whereas downstream of the positions 13 to 15 the pressure plateau level within the separated region ($c_p \approx 0.019$) is registered. As the separation bubble starts growing, more and more pressure taps upstream the hinge line are successively covered by the separation bubble (see Fig.

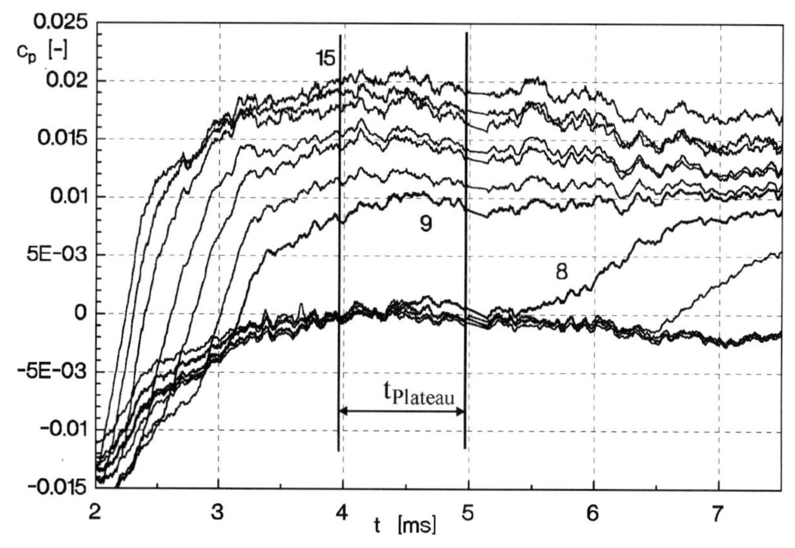


Figure 4.9.20: Time dependent pressure coefficients and determination of measurement time.

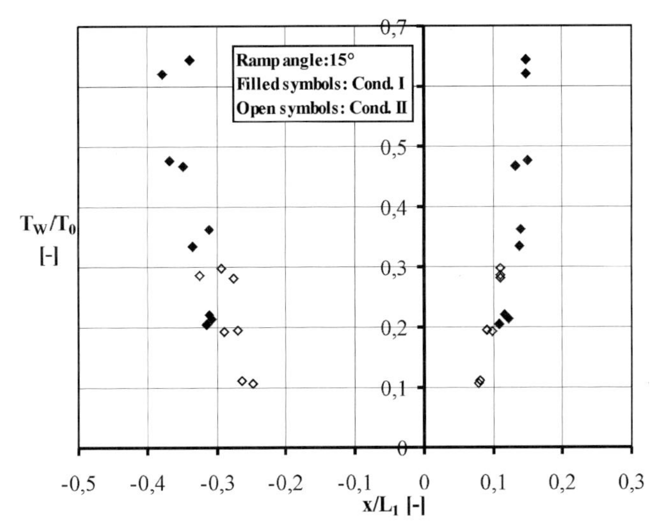


Figure 4.9.21: Separation and reattachment positions obtained from Schlieren pictures for 15° ramp angle.

4.9.20). At $t \approx 4$ ms all positions have reached a plateau value holding it for about 1 ms so that a fully developed separation bubble can be assumed up to 5 ms. At $t \approx 5.5$ ms the free stream pressure starts to decrease limiting the usable measurement time. The separation position derived by the Schlieren picture and by the pressure measurements match and in this case it is located between the positions 8 and 9.

Figure 4.9.21 shows for different wall to stagnation temperature ratios the measured separation and reattachment positions taken from Schlieren pictures. The dependence is clearly visible. Although the ratio $T_w/T_0$ is kept almost constant at two levels, the separation bubble seems to be a little smaller for higher stagnation temperature $T_0$. This was also observed in different studies (e.g. [27]).

The measured separation length $L_{sep}$ normalized by the boundary layer thickness at the separation position $\delta_{sep}$ is shown in Fig. 4.9.22. In [19] Katzer gives the original correlation for the separation length:

$$\frac{L_{\text{sep}}}{\delta^*_{x0}} = 4.4 \cdot \frac{p_3 - p_{\text{inc}}}{p_1} \cdot \sqrt{\frac{Re_{x0}/C}{Ma^3_\infty}}$$

Katzer used the boundary layer displacement thickness and the Reynolds number at the hinge line which in our case gave no good results with the flow conditions and ramp geometries used [28]. Therefore, the boundary layer thickness and the Reynolds number at the separation position were employed in the formula above (modified Katzer). The results for the new 15°-ramp model and the quartz-glas model used within the FESTIP-programme show a good agreement (see Fig. 4.9.22). The values calculated with the modified formula but with the

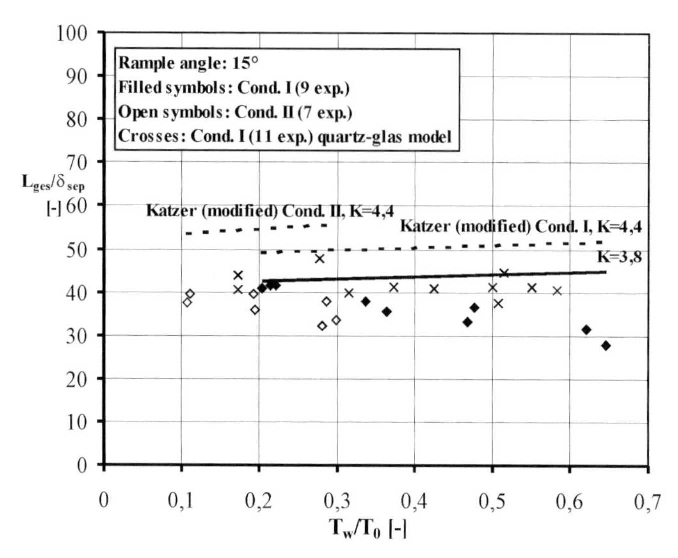


Figure 4.9.22: Comparison of measured dimensionless separation length with the modified Katzer correlation.

constant of the original Katzer formula are higher than the experimental values. With a constant of about $K=3.8$ there is quite a good agreement of theory and measurements.

#### 4.9.4.6 Determination of Pressure Coefficients

The mean values of the pressure coefficients have been taken at the same time as the Schlieren pictures. Figure 4.9.23 shows a comparison of the $c_p$-distributions for the same ratios $T_w/T_0$ as for the two Schlieren pictures (two experiments per ratio). Both clearly show the change in separation and reattachment position which correspond to the positions in the Schlieren pictures. To carefully compare the pressure measurements with theory, Figs. 4.9.24 and 4.9.25 show the results for conditions I and II respectively. For information, the pressure levels expected by free interaction and oblique shock theory are also given. Deviations between these levels at different conditions occur because of slightly different flow conditions which are considered in the calculation of the theoretical distributions. For the lowest pressures which occur at condition I, measurements in flat plate configuration have been performed to test the response characteristic of the pressure taps (circles in Fig. 4.9.24). An increase in response time was observed for all positions with increasing surface temperature due to the decreasing density of the inflowing gas. For both conditions the pressure level within the separation bubble is nearly the same as expected by free interaction theory. A pressure-overshoot is visible for both cases, which for

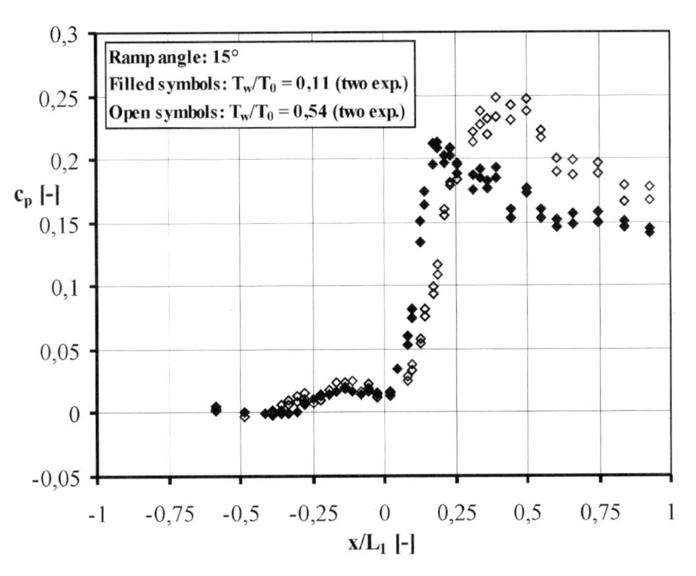


Figure 4.9.23: Pressure coefficient distributions for minimum and maximum value of temperature ratio $T_w/T_0$.

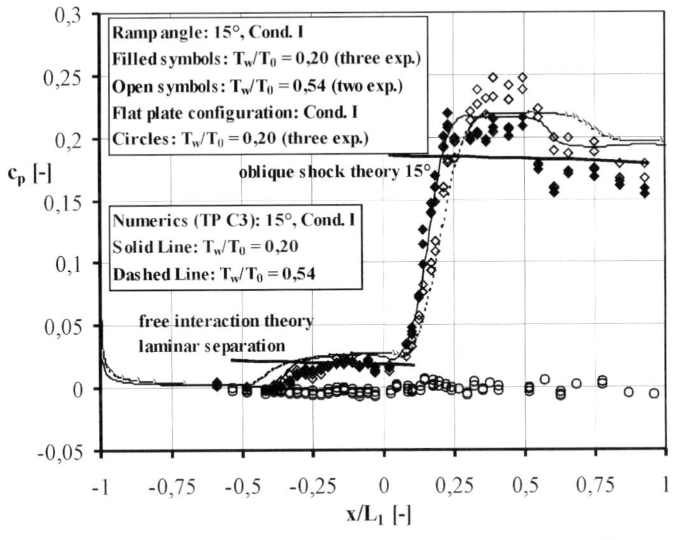


Figure 4.9.24: Comparison of calculated and measured pressure coefficient distributions for experiments performed at test condition I.

the higher wall temperature is more pronounced than for the lower wall temperature case. This was also observed at higher ramp angles [22]. The pressures downstream of the interaction are also in good agreement with the theoretical values.

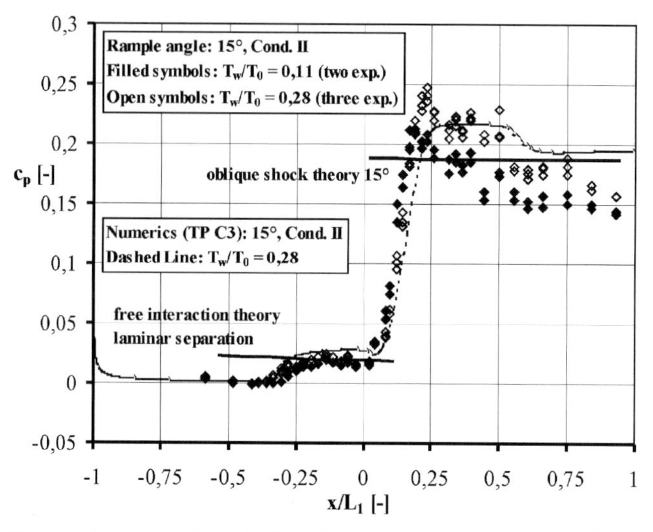


Figure 4.9.25: Comparison of calculated and measured pressure coefficient distributions for experiments performed at test condition II.

Numerical results of Navier-Stokes simulations provided by project C3 [29] have been added to Figs. 4.9.24 and 4.9.25. They show a good agreement with both the theoretical and experimental results. The positions of separation and reattachment are predicted very good for condition II. Even the large pressure gradients within the reattachment regions are in a good agreement. Although for condition I the calculated reattachment behaviour is also in good agreement with measured pressure coefficients the separation positions are not. The free stream gradients in static pressure and Mach number induced by the nozzle flow have not been considered in the numerical simulations, which may be of influence on the remaining deviations between the measured and calculated $c_p$-values.

#### 4.9.4.7 Determination of Stanton Numbers

In Fig. 4.9.26 the Stanton numbers deduced from thermocouple measurements at condition II are compared with theoretical and computational (laminar) results. Due to the small signals, heat flux measurements with thermocouples were only possible on the ramp. Because of smaller wall temperature gradients the measured Stanton numbers at all positions decrease as $T_w/T_0$ increases. The measured Stanton number levels on the ramp surface show a good agreement with theoretical values for a turbulent boundary layer. This means a laminar/ turbulent transition is possibly triggered at the reattachment position. A laminar/turbulent transition upstream the separation position can be excluded because according to free interaction theory a turbulent separation is not pos-

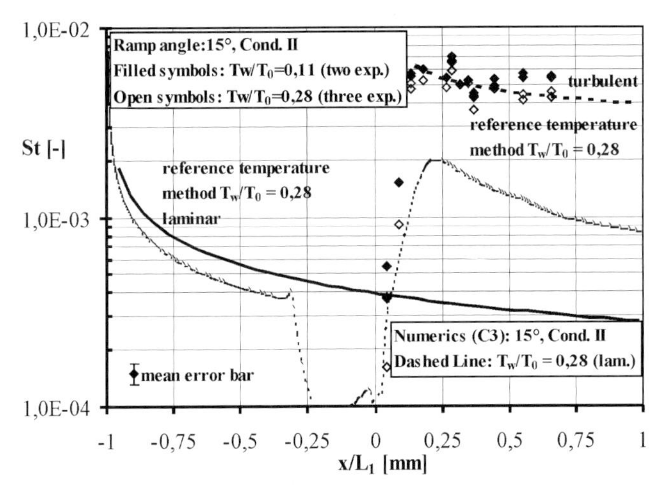


Figure 4.9.26: Comparison of calculated and measured Stanton number distributions for experiments performed at test condition II.

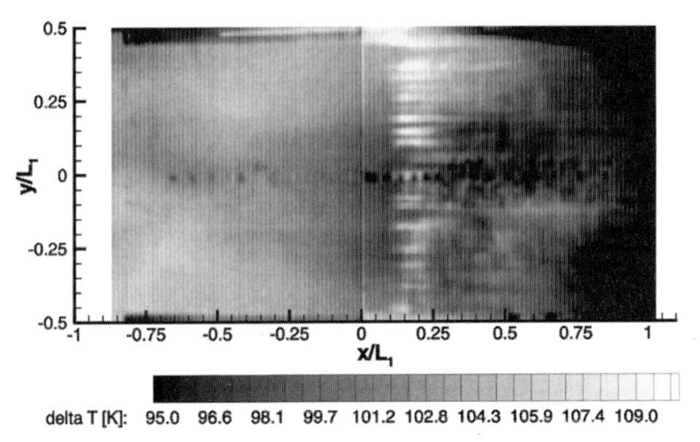


Figure 4.9.27: Temperature distribution at the end of the measuring time for unheated model wall.

sible for a 15°-ramp but the pressure measurements clearly show a separation bubble which indicates a laminar separation. Some details within the reattachment region were observed by two-dimensional infrared thermography (see Fig. 4.9.27 for $T_w$=293 K and Fig. 4.9.28 for $T_w$=733 K). They show the local temperature increase of the surface at the end of the measuring time caused by the flow. Within the reattachment region for both cases streamwise striations of the surface temperature became visible. These are caused by spanwise deviations of the heat transfer coefficient from a mean level which have been observed many times within reattachment regions of weak shock-wave/boundary-layer interactions [17]. For a boundary layer flow with concave curvature, like e.g. the reattachment region on the ramp instabilities occur as Görtler vortices within the boundary layer which are visible in the infrared pictures. Infra-

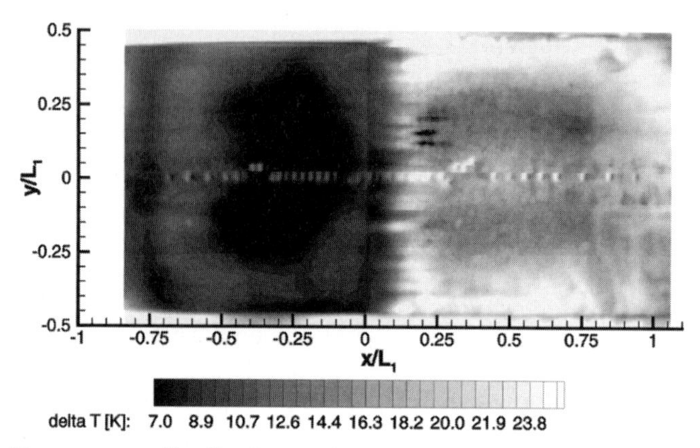


Figure 4.9.28: Temperature distribution at the end of the measuring time for heated model wall, $T_w$=730 K.

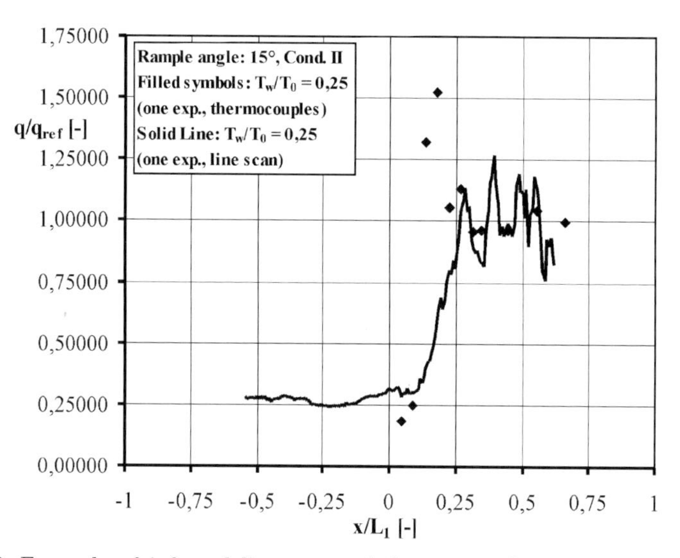


Figure 4.9.29: Example of infrared linescan and thermocouple measurements and influence of Görtler vortices.

red line scans ($f$=9.4 kHz) which were compared with thermocouple measurements show a different behaviour downstream of the hinge line (see Fig. 4.9.29; $T_w$=733 K). The measured heat fluxes are nondimensionalized by a reference value chosen on the ramp. The deviation between the two longitudinal distributions is due to a shift between the lateral positions of the line scan and the thermocouples. In this case especially the thermocouples close to the hinge line seem to be located in a region of high heat flux caused by the Görtler vortices.

### 4.9.5 Conclusions

The results presented above give an overview of hypersonic flow phenomena investigated in the shock tunnel TH2/TH2-D within the Collaborative Research Centre 253. Both stages of the reference configuration were examined in detail regarding pressure distribution, heat flux distribution and aerodynamic loads. Numerical results obtained by other projects of the Collaborative Research Centre show in general a good agreement with the experimental data. The experiences made in the test campaigns performed with the reference configuration were used in industrial research projects such as investigations on the Atmospheric Re-entry Demonstrator (ARD, within the MSTP programme) and – recently performed – re-entry experiments for the HOPPER configuration (within the ASTRA programme). Special measurement techniques for use in wind tunnel models with controlled surface temperature were developed and applied in ramp configurations representing control surfaces or engine intakes. The measured pressure and heat flux distributions were compared with numerical and theoretical results to show the surface temperature effects on such flows in short duration facilities.

### References

1. H. Grönig, H. Olivier: Experimental Hypersonic Flow Research in Europe. JSME International Journal, Series B, Vol. 41 (1998), No. 2, pp. 397–407.
2. M. Habermann, H. Olivier, H. Grönig: Operation of a High Performance Detonation Driver in Upstream Propagation Mode for a Hypersonic Shock Tunnel. Proceedings of the 22nd Int. Symposium on Shock Waves, Imperial College, London, 18–23 July 1999, pp. 447–452.
3. H. Olivier, H. Grönig: An Improved Method to Determine Free Stream Conditions in Hypersonic Facilities. Shock Waves Vol. 3 (1993), No. 2, pp. 129–139.
4. G. Kindl, H. Olivier: Development of a Static Pressure Probe. MSTP Final Report, Shock Wave Laboratory, RWTH Aachen 1996.
5. D.L. Schultz, T.V. Jones: Heat Transfer Measurements in Short Duration Hypersonic Facilities. AGARDograph 165, NATO 1973.
6. C. Jessen, H. Grönig: Six-component Force Measurements in the Aachen Shock Tunnel. In: R. Brun, L. Dumitrescu (Eds.), Shock Waves @ Marseille, Springer Berlin Heidelberg 1995, pp. 288–292.
7. V. Störkmann, H. Olivier, H. Grönig: Force Measurements in Hypersonic Impulse Facilities. AIAA-Journal Vol. 36 (1998), No. 3, pp. 342–348.
8. L. Bernstein: Force Measurements in Short-duration Hypersonic Facilities. AGARDograph 214, AGARD, London 1975.
9. C. Glößner, H. Olivier: Aerothermodynamics of a Winged Re-entry Vehicle in Hypersonic Flow. Proceedings of the 4th European Symposium on Aerothermodynamics for Space Vehicles, Capua/Italy, 15–18 October 2001, ESA SP487, pp. 233–240.
10. J.-H. Schulte-Rödding, H. Olivier: Experimental Investigations on Hypersonic Inlet Flows. AIAA 98-1528, Proceedings of the AIAA 8th International Space Planes and Hypersonic Systems and Technologies Conference, Norfolk/Va., April 27–30 1998.
11. A. Kharitonov, M. Brodetsky, E. Krause, M. Jacobs: An Experimental Investigation of ELAC 1 Model at Supersonic Speeds. Technical Report A-96-10 SFB 253, RWTH Aachen 1996.

12. A. Stromberg: Experimentelle Untersuchungen an der Hyperschallkonfiguration ELAC 1 bei Unter- und Überschallanströmung. Doctoral thesis, RWTH Aachen (1994).
13. D. Hänel, A. Henze, E. Krause: Supersonic and Hypersonic Flow Computations for the Research Configuration ELAC 1 and Comparison to Experimental Data. Journal of Flight Sciences and Space Research, Vol. 17 (1993), No. 2, pp. 90–98.
14. A. Henze, W. Schröder, M. Meinke: Numerical Analysis of the Supersonic Flow around Reusable Space Transportation Vehicles. Proceedings of the 4th European Symposium on Aerothermodynamics for Space Vehicles, Capua/Italy, 15–18 October 2001, ESA SP487, pp. 191–197.
15. G. Cucinelli: Sänger-Systemstudie – Aerodynamischer Datensatz HORUS 8-88. Technical Note TN/HYP/71, Messerschmitt-Bölkow-Blohm GmbH, München 1988.
16. C. Glößner, H. Olivier: Force and Moment Coefficients of the HOPPER/PHOENIX-Configuration in Hypersonic Flow. Proceedings of the 3rd Int. Symposium on Atmospheric Re-entry Vehicles and Systems, Arcachon/France, 24–27 March 2003.
17. G. Simeonides: Experimental, Analytical, and Computational Methods Applied to Hypersonic Compression Ramp Flows. AIAA Journal Vol. 32 (1994), No. 2, pp. 301–310.
18. J. Delery, J.G. Marvin: Shock-Wave Boundary Layer Interactions. AGARDograph 280, NATO 1986.
19. E. Katzer: On the Lengthscales of Laminar Shock/Boundary-Layer Interaction. Journal of Fluid Mechanics Vol. 206 (1989), pp. 477–496.
20. E.G. Hirschel: Thermal Surface Effects in Aerothermodynamics. Proceedings of the 3rd European Symposium on Aerothermodynamics for Space Vehicles, Noordwijk/ The Netherlands, 24–28 November 1998, ESA SP-426, pp. 17–31.
21. F.W. Spaid, J.C. Frishett: Incipient Separation of a Supersonic, Turbulent Boundary Layer, Including Effects of Heat-Transfer. AIAA Journal Vol. 10 (1972), No. 7, pp. 915–922.
22. G.M. Elfstrom: Turbulent Hypersonic Flow at a Wedge Compression Corner. Imperial College, Aeroreport 71-16, London 1971.
23. F. Scortecci, F. Paganucci, L. d'Agostino: Experimental Investigation of Shock-Wave/ Boundary-Layer Interactions over an Artificially Heated Modell in Hypersonic Flow. AIAA 98-1571, Proceedings of the AIAA 8th International Space Planes and Hypersonic Systems and Technologies Conference, Norfolk/Va., April 27–30 1998.
24. J.M. Charbonnier: Hot Wall Testing in High Speed Flows. Project Report VKI Reprint 1998-24, Von Karman Institute for Fluid Dynamics, Brussels 1998.
25. H. Olivier, M. Bleilebens: Hot Experimental Testing. Technical Report ESA WP5510, Shock Wave Laboratory, RWTH Aachen 1999.
26. J.E. Lewis, T. Kubota, T. Lees: Experimental Investigation of Supersonic Laminar, Two-Dimensional Boundary-Layer Separation in a Compression Corner with and without Cooling. AIAA Journal Vol. 6 (1968), No. 1, pp. 7–14.
27. S.G. Mallinson, S.L. Gai, N. R. Mudford: The Interaction of a Shock Wave with a Laminar Boundary Layer at a Compression Corner in High-Enthalpy Flows Including Real Gas Effects. Journal of Fluid Mechanics Vol. 342 (1997), pp. 1–35.
28. M. Bleilebens, H. Olivier: Surface Temperature Effects on Shock-Wave Boundary-Layer Interaction of Ramp Flows. In: S. Wagner, U. Rist, J. Heinemann, R. Hilbig (Eds.): New Results in Numerical and Experimental Fluid Mechanics III (Notes on Numerical Fluid Mechanics, Vol. 77), Springer, Berlin Heidelberg 2002, pp. 161–168.
29. B. Reinartz, J. Ballmann: CFD-Results for the Heated Ramp Model. Private communication 2003.

# 5 Propulsion

## 5.1 PDF/FDF-Methods for the Prediction of Supersonic Turbulent Combustion

Stefan Heinz and Rainer Friedrich *

### Abstract

The paper presents an overview of recent developments of PDF/FDF-methods for the prediction of turbulent combustion with special emphasis on work performed within the SFB 255/TP A2. On the basis of a discussion of available methods for the calculation of turbulent reacting flows, significant conceptual shortcomings of currently applied PDF/FDF-methods are highlighted and new concepts to overcome these problems are described. A main purpose of developing PDF/FDF-methods is to use them for the solution of real technological turbulent combustion problems. However, this goal could not be achieved until now due to the unavailability of appropriate codes. It is reported in which way this fundamental problem has to and will be solved in the near future.

### 5.1.1 Introduction

The reliable prediction of turbulent combustion processes represents a need for many technological developments (for example, to enhance the performance of airbreathing propulsion systems). Measurements can be used for that only in a limited way because they are often laborious and prone to errors. In addition to that, they usually provide only an incomplete picture of the underlying physics.

* Technische Universität München, Fachgebiet Strömungsmechanik, Boltzmannstr. 15, 85748 Garching; r.friedrich@lrz.tum.de

*Basic Research and Technologies for Two-Stage-to-Orbit Vehicles*
DFG, Deutsche Forschungsgemeinschaft

ISBN: 3-527-27735-8

Hence, predictions of turbulent combustion processes have to be based, essentially, on results of numerical methods. In particular, one needs numerical methods which are available for prospective customers, applicable to a broad range of combustion problems, efficient and accurate.

From a conceptual point of view, so-called hybrid PDF-methods offer at this stage of development the most promising way to find an appropriate solution to that problem because they combine two important advantages: they are able to take nonlinear chemical conversion processes exactly into account, whereas they are (in contrast to other methods) realizable from a computational point of view, this means they can be used for technological applications. However, these methods apply different approximations for turbulent transport processes, which determine, therefore, the accuracy of predictions. With regard to this, currently applied hybrid PDF-methods are faced with significant problems: they are inconsistent by the use of different models for the same physical process, simulate turbulent mixing partly in contradiction to the observed physics and describe the turbulent transport of energy on the basis of not well-founded, empirical assumptions. Apart from that, one has to see that there are (in contrast to the main purpose of developing such methods) no applications of PDF-methods to the solution of real technological turbulent combustion problems at present: such codes only exist within a few groups of specialists at universities, and there is hardly a way for other people (potential industrial users) to make use of them. Accordingly, the work performed within the SFB 255/TP A2 was devoted to two basic tasks: to develop new solutions to overcome conceptual problems of hybrid PDF/FDF-methods, and to push the use of corresponding numerical methods for predictions of technological turbulent combustion processes. The realization of these tasks will be described here.

The paper contains an overview of basic methods for turbulent reacting flow calculations in Section 5.1.2. This results in the conclusion that hybrid PDF/FDF-methods are particularly appropriate for turbulent reacting flow calculations at this stage of development. Section 5.1.3 deals with a discussion of shortcomings of currently applied PDF/FDF-methods. Solutions that were developed to overcome these problems are described in Section 5.1.4. The development of corresponding numerical methods for turbulent combustion prognoses is the concern of Section 5.1.5. Finally, Section 5.1.6 deals with a discussion of future developments that can be expected in this field.

### 5.1.2 Methods for Turbulent Reacting Flow Calculations

The problem related to the calculation of turbulent reacting flows is that (due to the memory capabilities and speed of high-performance computers) there are clear limitations for applying direct numerical simulation (DNS), i.e., the direct numerical integration of the basic equations of fluid and thermodynamics. At present and in the foreseeable future one can only use DNS for the calculation of flows with relatively small Reynolds numbers (flows with a weak or moderate turbulence intensity) [1–6]. However, flows of technological and environmental relevance are usually characterized by high Reynolds numbers, and the consideration of Reynolds number effects is clearly a requirement with regard to reacting flow simulations [7].

### 5.1.2.1 Basic Methods

Table 5.1.1 describes in an idealized way basic methods that are available for the calculation of turbulent flows with high Reynolds numbers. PDF- and FDF-methods [1–20] generalize RANS- and LES-methods, respectively: they do not only describe the dynamics of filtered variables, but they also provide the dynamics of fluctuations around these variables. The significant advantage of the extension of the set of variables considered is that nonlinear chemical reaction rates are given in a closed form, so that the consideration of chemical reactions does not require modelling [1]. Therefore, the use of PDF- and FDF-methods offers the best chance to develop numerical methods which are applicable to a wide range of flows and capable of providing reliable predictions of flows with complex chemistry.

The difference between PDF- and FDF-methods is given by the fact that PDF-methods represent a special case of FDF-methods. FDF-methods provide deviations from spatially-filtered values, where a suitable value for the filter width has to be chosen. In contrast to that, one works with a large filter width in PDF-methods, which is of the order of the correlation lengths of variables considered [2]. In general, FDF-methods are applied so that modelling assumptions do only concern the treatment of small-scale turbulent motions, this means one applies a small filter width [15–18]. Thus, one may expect that such predictions are more accurate than those of PDF-methods. However, the problem of the FDF approach consists in its computational costs, which do not allow the use of these methods for technical applications at present.

Table 5.1.1: Basic methods for the calculation of turbulent reacting flows with significant turbulence: large-eddy simulation (LES), Reynolds-averaged Navier-Stokes (RANS), filter density function (FDF) and probability density function (PDF) methods.

| Equations for filtered variables | Hybrid methods | Equations for stochastic variables |
|---|---|---|
| LES-methods | Velocities: LES-methods<br>Scalars: FDF-methods | FDF-methods |
| RANS-methods | Velocities: RANS-methods<br>Scalars: PDF-methods | PDF-methods |

#### 5.1.2.2 Hybrid PDF/FDF-Methods

The stochastic treatment of both velocities and scalars (mass fractions of species and temperature) in full PDF/FDF-methods is expensive and often the reason for significant problems, in particular with regard to the calculation of compressible flows. Some details of stochastic velocity equations for such flows are still unclear at present, and the numerical solution of such equations is very complicated [2]. An alternative is given by the use of hybrid PDF/FDF-methods. Within that approach, one calculates velocity fields by RANS/LES-methods so that stochastic simulations can be restricted to the treatment of scalar transport, see Fig. 5.1.1 for an illustration. The advantage of this approach is that chemical reactions still can be handled formally exactly, and one does not have a significant problem with regard to the calculation of velocity fields: well-developed methodologies can be used for that. However, the full utilization of the capabilities of such hybrid methods requires their consistent and physically correct formulation. Unfortunately, this constraint is not satisfied for the currently applied methods, as shown in Section 5.1.3. This may be the cause for significant errors and hampers remarkably the assessment of differences between full and hybrid PDF/FDF-methods.

### 5.1.3 Some Deficiencies of Existing Hybrid PDF-Methods

Some deficiencies of existing hybrid methods will be pointed out in this Section. For simplicity, this discussion will be limited to the consideration of hybrid PDF-methods because FDF-methods are faced with the same problems.

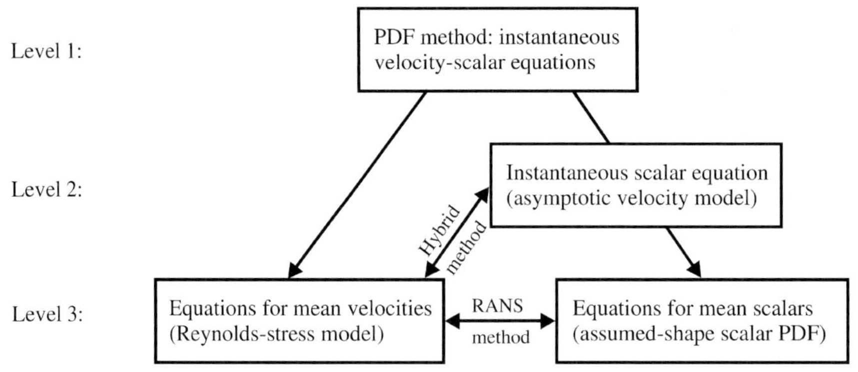


Figure 5.1.1: An illustration of the reduction of PDF-methods (stochastic equations for velocities and scalars) to RANS- and hybrid methods. Equations for mean velocities must be closed by algebraic or transport equation models for Reynolds stresses. The direct use of equations for mean scalars requires (assumed-shape) models for scalar PDFs to close mean reaction rates. A hybrid method is given by combining RANS equations for velocities with stochastic equations for scalars. The latter involve an asymptotic velocity model to calculate the transport of scalars in physical space. In consistency with that, one has to use an algebraic Reynolds stress model in RANS equations for velocities.

#### 5.1.3.1 The Transport Problem

From the view point of a full PDF-method for velocities and scalars, the use of a hybrid method represents a simplification where (instead of equations for instantaneous velocities) deterministic velocity equations are solved, see Fig. 5.1.1. This reduction corresponds to the assumption that velocities are considered in a somewhat coarser scale where details of the acceleration statistics are not resolved [2].

The treatment of velocities and scalars on the basis of differently structured models then leads to constraints with respect to the consistency of these models, for example, due to the fact that both models require knowledge about velocity correlations (to close the turbulent transport term in equations for the averaged velocity field and the diffusion coefficient in stochastic equations for the transport of scalars in physical space). The application of a consistent hybrid method (see level 2 in Fig. 5.1.1) then requires the use of the same model for the Reynolds stress tensor in deterministic equations for velocities and stochastic equations for scalars. However, the latter is not guaranteed in existing methods. Anisotropy effects are considered in velocity equations whereas one applies isotropic diffusion coefficients in scalar equations. It is known that such imbalances may cause significant errors.

#### 5.1.3.2 The Mixing Problem

As pointed out above, the significant advantage of using PDF-methods is the exact treatment of chemical reactions. This allows the reduction of the complex turbulence-chemistry interaction to the problem of describing the turbulent mixing of scalars appropriately. An example for such a mixing model represents the so-called "interaction by exchange with the mean" (IEM) model, which is (due to its simplicity) used in most of the PDF simulations of turbulent reacting flows. Within the frame of this model, a realization of the probability density function (PDF) of mass fractions $m_a$ ($a=1, N$; $N$ is the number of species considered) changes according to

$$\frac{d}{dt} m_a = -\frac{C_\varphi}{2\tau} (m_a - \overline{m}_a) + S_a \,. \tag{1}$$

The overbar refers to a mass density-weighted ensemble mean (a Favre mean), and $S_a$ represents a chemical reaction rate, which follows from the reaction scheme considered. The mixing frequency $C_\varphi/(2\tau)$ is proportional to the mean frequency $\tau^{-1}$, which characterizes the inertial range of turbulence. $C_\varphi$ is a constant with a standard value $C_\varphi=2.0$ [1].

However, the use of the model (1) is related to different problems. One problem is that this model is unable to simulate the initial stage of complex mixing processes in detail, which may be important with respect to accurate predictions of non-premixed combustion. Another problem is that this model is

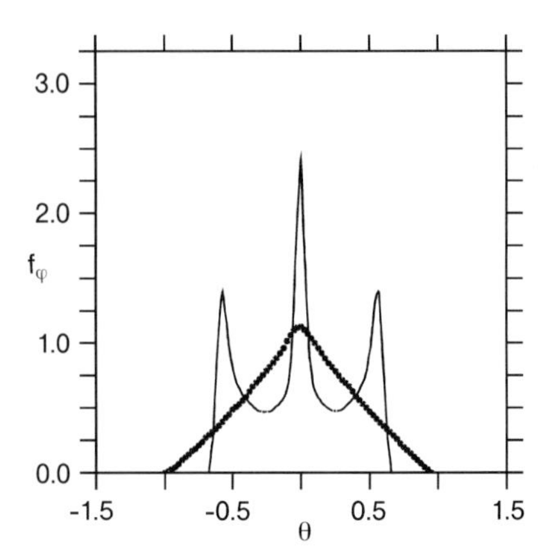


Figure 5.1.2: The dots present DNS data of the PDF $f_\varphi(\theta)$ of a passive scalar (for $\Phi_T$=0.6, see Fig. 5.1.3). The thin solid line shows the corresponding predictions of the IEM-model for the evolution of the scalar PDF at this stage.

unable in general to calculate the correct asymptotic shape of scalar PDFs. It was shown by DNS data that scalar PDFs evolve (independent of the initial PDF shape) towards a Gaussian PDF in statistically homogeneous isotropic and stationary turbulence [21]. Instead, the PDF calculated by the IEM-model depends on the initial distribution also for very long times, see the illustration in Fig. 5.1.2. Such predictions are in contrast to the physics of turbulent mixing, such that significant errors in turbulent combustion simulations may occur.

#### 5.1.3.3 The Energy Problem

The stochastic treatment of reacting scalar transport requires the incorporation of temperature in order to calculate instantaneous reaction rates in equations for species mass fractions. Furthermore, the consideration of energy variables (internal energy, enthalpy or temperature) is a requirement to calculate the pressure in compressible flows.

There exists a variety of ways to formulate equations for the transport of energy in compressible reacting flows: one can consider equations for the internal energy, enthalpy or temperature [22–24]. The way to be preferred currently consists in a combination of the stochastic model (1) with an equation for the enthalpy $h$. This equation can be written

$$\frac{d}{dt}h = -\frac{C_\varphi}{2\tau}\left(h - \overline{h}\right) + S_h\,. \tag{2}$$

$S_h$ is a source term which follows from the basic equations of fluid and thermodynamics as

$$S_h = 2\nu S^d_{jk} S^d_{kj} + \frac{1}{\rho}\frac{dp}{dt} \,. \tag{3}$$

$\nu$ represents the kinematic viscosity, $S^d_{jk}$ is the deviatoric part of the rate-of-strain tensor, $p$ the pressure and $\rho$ the mass density. $S^d_{jk}$, $p$ and $\rho$ are instantaneous (fluctuating) variables, and the problem is given by the fact that they appear as unknowns within the frame of combined mass fraction-enthalpy equations. Therefore, the model (2) is unclosed. To overcome this problem, one currently applies the following approximation [24]

$$S^{cl}_h = \overline{\frac{1}{\rho}\frac{dp}{dt}} \approx \frac{1}{\langle\rho\rangle}\left\langle \frac{\partial p}{\partial t} + U_k \frac{\partial p}{\partial x_k} \right\rangle = \frac{1}{\langle\rho\rangle}\left[ \frac{\partial \langle p\rangle}{\partial t} + \langle \bar{U}_k\rangle \frac{\partial \langle p\rangle}{\partial x_k} + \left\langle u_k \frac{\partial p}{\partial x_k} \right\rangle \right] ; \tag{4}$$

this means one neglects the first term on the right-hand side of relation (3), which simulates an increase of internal energy due to viscous friction, and replaces the pressure change by the corresponding mean value (the symbol $\langle\ \rangle$ refers to an ensemble mean). The resulting problem to explain the last term of relation (4) by means of variables that are known is then solved on the basis of usual parametrizations, which are applied within the framework of turbulence modelling [24].

This previously applied way to involve energy variables in PDF-methods for reacting flows is characterized by different disadvantages. First, one neglects fluctuations of dp/dt, which may be important in supersonic flow simulations [22]. This approximation also disagrees with the goal of PDF-methods to describe the dynamics of scalar fluctuations in a consistent way (the neglect of fluctuations of dp/dt corresponds to the neglect of temperature fluctuations, which appear, for instance, in instantaneous reaction rates). Furthermore, one produces additional closure problems in this way (see the last term in (4)), which cannot be solved appropriately for inhomogeneous reacting flows. The closure (4) also has disadvantages from a numerical point of view, because the calculation of the substantial derivative of the mean pressure causes several problems [25]. Due to these reasons, the currently applied way to involve energy variables in hybrid PDF-methods cannot be seen as a satisfactory solution to this problem and requires methodological improvements.

### 5.1.4 New Theoretical Concepts

The work performed within the SFB 255/TP A2 to overcome the conceptual problems of hybrid PDF-methods pointed out in Section 5.1.3 will be described next.

### 5.1.4.1 The Transport Problem

A solution to the transport problem pointed out above may be obtained by reducing the underlying transport equation for the joint velocity-scalar PDF, see Fig. 5.1.1. In this way, one obtains a generalization of methods applied usually, where contributions due to anisotropy and shear (which appear in general in anisotropic and inhomogeneous flows) are involved in calculations of scalar transport in physical space [2]. The result is a consistent hybrid method where the same sub-models are used in equations for velocities and scalars.

### 5.1.4.2 The Mixing Problem

An important goal of the work performed within the SFB 255/TP A2 was the improvement of models for the turbulent mixing of scalars. Such an improvement of the IEM-model can be obtained on the basis of the so-called projection operator technique [2, 19, 26]. In this way, the IEM-model can be generalized (in a non-expensive way) so that complex mixing processes within the initial stage of turbulent mixing and the evolution of scalar PDFs can be simulated correctly in agreement with the boundedness constraint [1].

Some essential properties of this new model are described in the following by considering a binary mixing process. The test case considered is given by DNS data of Juneja and Pope [21] who investigated the mixing of passive scalars in statistically stationary isotropic and homogeneous turbulence. Results obtained by means of the new mixing model and a simplified version are given in Fig. 5.1.3 in comparison to the corresponding DNS data. This comparison demonstrates the good performance of the new mixing model. The evolution of the PDF in the initial and asymptotic stages of mixing is described in agreement with DNS data. Figures 5.1.3/c1 and 5.1.3/c2 show that the simplified version of the new mixing model also enables good predictions, if a very detailed resolution of complicated mixing processes in the initial stage of turbulent mixing is not essential. The simplified mixing model represents, basically, a stochastic variant of the IEM-model. Thus, it is easy to use and does hardly require more computational effort than the IEM-model.

### 5.1.4.3 The Energy Problem

As pointed out above, the presently applied formulation of the stochastic energy transport on the basis of an enthalpy equation is related to the appearance of a variety of problems. A way to overcome these problems has been developed as part of the work performed within the SFB 255/TP A2. The basic idea is to use a temperature formulation of the energy transport (a temperature equation) in conjunction with the consistent use of approximations that are a

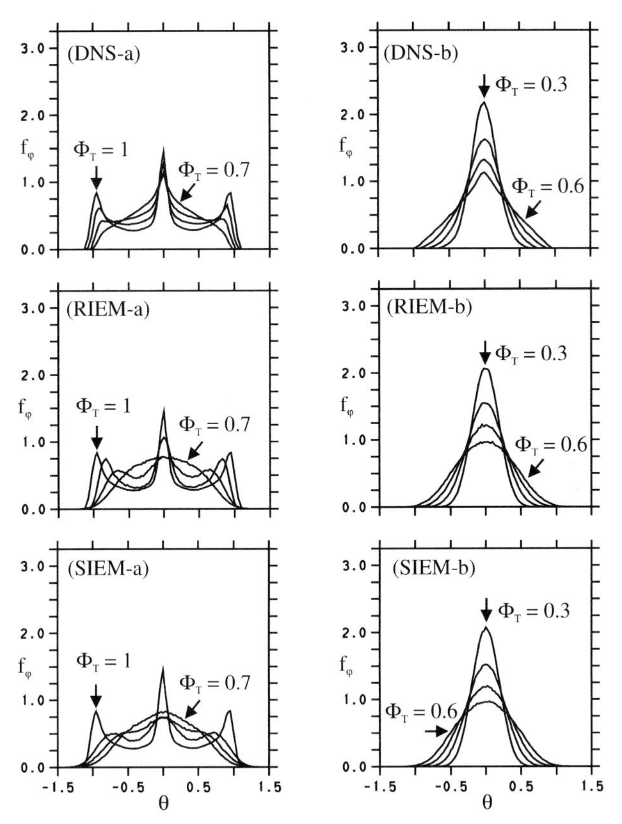


Figure 5.1.3: The evolution of a scalar PDF $f_\varphi(\theta)$ as function of the progress variable $\Phi_T$, which runs from unity to zero in time. The DNS data of Juneja and Pope [21] are given in (DNS-a) and (DNS-b), and the corresponding prognoses of the new mixing model in (RIEM-a) and (RIEM-b). The Figs. (SIEM-a) and (SIEM-b) show the performance of a simplified version of the new model.

consequence of reducing a transport equation for the joint PDF of velocities and scalars to a hybrid method, see Fig. 5.1.1 for an illustration. The simplicity and consistency of this concept offers significant advantages compared to the use of other methods [2].

## 5.1.5 The Use of PDF Combustion Codes

### 5.1.5.1 The Current Use of PDF/FDF-Methods

One goal related to the development of PDF/FDF-methods is to investigate the physics of turbulent flows with high Reynolds, Schmidt and Damköhler numbers, which cannot be assessed by means of DNS. FDF-methods are particularly appropriate for that due to the fact that large-scale motions are treated without modelling. However, their application to such studies is still in an initial stage. In contrast to that, PDF-methods, which represent a special case of FDF-methods (see Section 5.1.2.1), are already applicable to such investigations. Due to their close relationship to FDF-methods, such PDF-studies are well appropriate to prepare the use of corresponding FDF-methods.

Another very important goal of the development of PDF/FDF-methods is their use for predictions of technological combustion processes. Due to their lower computational costs, PDF-methods are more appropriate for that than FDF-methods. However, despite of important conceptual advantages of PDF-methods compared to RANS-methods, one has to see that these methods are not used at present for the solution of technological problems. The main reason for that is given by their requirements with regard to the memory and speed of computers, which are significantly higher than those related to the use of conventional turbulence models, which are based on deterministic equations. The constraints for the application of PDF-methods can be illustrated in the following way. The calculation of industrial flows often requires a number of grid points which is of the order of $100^3$. Without adopting techniques for the reduction of statistical errors, one needs $10^4$ or $10^6$ realizations (particles) in each box of the grid to keep the statistical error of calculations below 1% or 0.1%, respectively. The treatment of the evolution of at least $10^{10}$ particles then represents a huge challenge. To handle this problem, the use of PDF-methods for the solution of technological problems requires the careful analysis of strategies for the numerical solution of stochastic equations and error sources in conjunction with the development of efficient techniques for the reduction of errors in order to limit the significant computational costs and memory requirements. These questions were solved recently only within a few research groups of specialists at universities who apply PDF codes for scientific investigations. Such codes, which are the result of long-standing investments, are not made accessible to potential industrial users in general, which would be related to an enormous effort to explain the use of codes and treatment of problems. Apart from that, one finds that such developments often do not correspond to technological requirements with regard to computing facilities, generation of grids and turbulence modelling.

This lack of convincing technological applications of PDF-methods has several negative consequences. Due to the fact that the development of FDF-methods is still in its initial stage, it supports doubts regarding the advantage of developing such methods, which results in an unsatisfactory status of research funding in this area so that these developments are carried out by only a few groups in the world. This leads to the problem to keep pace with modern developments, for instance with regard to the generation of unstructured grids, the improvement of turbulence models and the tabulation of chemistry. Another problem is the implied limited number of applications (the availability of PDF codes for many users at universities and the industry offers the chance to obtain a much broader knowledge regarding the advantages and disadvantages of different models).

### 5.1.5.2 New Developments

Fortunately, there are promising new activities of Fluent Inc. (a software company that offers a computational fluid dynamics (CFD) code called FLUENT) to overcome the problem related to the lack of using PDF-methods for technological applications. Fluent Inc. was the first firm that involved PDFs in combustion simulations. That concerned an advanced RANS-method where the shape of the scalar PDF has to be provided to close mean reaction rates in equations for the scalar transport (so-called assumed-shape PDF-models, see the illustration in Fig. 5.1.1). However, one has to see that such assumed-shape PDF-methods do not contain the required physics in order to simulate the processes that take place in non-premixed flames [2]. Therefore, one finds significant shortcomings of assumed-shape PDF-models [27], for example, concerning the calculation of non-premixed hydrogen and methane turbulent jet flames [28, 29].

The natural next step was the integration of a hybrid PDF-method (see Fig. 5.1.1) in FLUENT [30], which was performed in the beginning of 2003 in collaboration with Professor S. B. Pope (Cornell University). It is worth noting that the incorporation of ISAT-routines [31] in FLUENT offers the possibility to calculate then complex chemical reactions in a very efficient manner. Nevertheless, it has to be pointed out that this code still does not offer solutions for the transport, mixing and energy problems described above. Furthermore, the code is not tested until now with regard to its numerical accuracy and performance as combustion model.

### 5.1.5.3 Common Activities to Develop a New Combustion Code

Obviously, the new development of Fluent Inc. offers the great chance to overcome the problems pointed out at the end of Section 5.1.5.1. In particular, it turned out that the activities at the Technical University of Munich (Fachgebiet Strömungsmechanik) and Fluent Inc. can be combined in a way that this development is made much more efficient. The basic configuration of this collaboration is that Fluent Inc. implements the methodological improvements reported in Section 5.1.4 in the PDF code of FLUENT. This allows to test the relevance and performance of the new models on the basis of various applications. From the view point of Fluent Inc., this work is also very helpful because it supports the code testing and provides evidence for the performance of the new PDF code as a combustion model. Accordingly, the Technical University of Munich (Fachgebiet Strömungsmechanik) and Fluent Inc. Germany agreed in corresponding common activities to test and improve the existing FLUENT code.

In a first application, the PDF code of Fluent Inc. is used to simulate a three-dimensional supersonic turbulent channel flow where a passive scalar is injected from one wall and removed from the other wall. For that flow, DNS data are available (at different Reynolds and Mach numbers) for comparisons, which were produced within a project funded by the DFG [32, 33]. One goal of these comparisons between PDF simulations with DNS data is to validate the numerical accuracy of the PDF code. Due to its simplicity, the flow considered is very appropriate for that. A relevant question concerns, for instance, the number of particles that are required within the Monte Carlo simulation, which is essential to the efficiency of calculations. From a scientific point of view these comparisons are also relevant since they can be used for the assessment of the relevance of the methodological improvements described in Section 5.1.4. It is, for example, possible to get a better insight into the relevance of anisotropy on the transport of scalars in physical space and the influence of compressibility on the PDF of scalars. Further, the DNS data are very appropriate to investigate the advantages of the temperature formulation of the energy transport in comparison to the enthalpy formulation applied previously. The results of these investigations are currently prepared for publication. It is also worth noting that such studies of the dynamics of scalars are also relevant to the construction of FDF equations, where very similar questions have to be answered [2].

Further applications of this PDF code are planned in the years 2005–2006. This concerns the simulation of different subsonic and supersonic flames. These investigations will be done to reach two goals. The first one is to assess the performance of the improved PDF code (the relevance of methodological improvements). The second one is to provide evidence for the applicability and numerical accuracy of the PDF combustion code developed by Fluent Inc.

### 5.1.6 Prospects for Further Developments

Questions related to the further development of computational methods will be addressed now. This concerns, for example, the question whether (in the light of the development of FDF-methods) the further development of PDF-methods still will be in agreement with the general trend of development that can be expected. This is the case, as will be shown in Section 5.1.6.1. This results then in the question about the future relation of PDF- and FDF-methods, which will be discussed in Section 5.1.6.2. More detailed explanations of these questions may be found elsewhere [2].

#### 5.1.6.1 The Current and Future Use of Computational Methods

To understand needs for further methodological developments, let us have a closer look at current and future applications of turbulence models for solving industrial (technological and environmental) problems.

The current use of computational methods is illustrated in Fig. 5.1.4, which closely follows an analysis of Pope [3]. Obviously, most applications are performed on the basis of the simplest (RANS) models. In particular, the k-$\varepsilon$ model

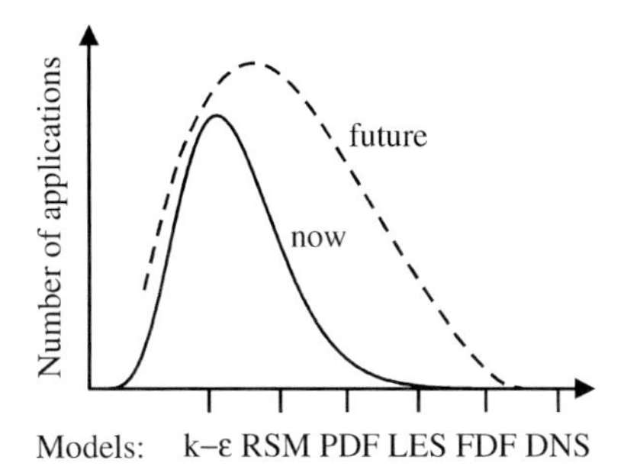


Figure 5.1.4: A slightly extended prognosis of Pope [3], who considered the current and future (next 10–20 years) use of turbulence models for industrial applications. RANS-models are split into algebraic ($k$-$\varepsilon$) and Reynolds stress models (RSM) which apply transport equations to calculate Reynolds stresses.

and (less often) Reynolds-stress models (RSM), which make use of transport equations to obtain the Reynolds stresses, are the models which are usually employed. As pointed out above, the use of RANS equations for reacting flow calculations is faced with serious problems regarding the closure of mean reaction rate terms. This problem can be solved by extending RANS to PDF-methods.

Regarding the future use of computational methods for industrial applications, an important conclusion from previous developments is that a simple displacement of the distribution in Fig. 5.1.4 towards more advanced methods (an abandonment of RANS in favour of LES) will probably not take place. The general trend is to use the simplest models as much as possible. Thus, it is likely that the contributions of PDF-, LES- and possibly FDF-methods grow, but most of the applications will be still performed on the basis of RANS(PDF)-methods.

### 5.1.6.2 Some Challenges

A significant problem related to the use of RANS-models for industrial applications is the assessment and optimization of the model performance. In general, the use of DNS is too expensive for that purpose. Measurements are also very expensive and provide only partial (and often relatively inaccurate) information. With regard to this, it is essential to note that possible deficiencies of RANS-models are closely related to the underlying relative coarse filtering [2]. The most natural way to assess the suitability of RANS-models, therefore, is to investigate the consequences of applying (by the use of a RANS-method) a large filter width $\Delta$. This can be done by adopting LES predictions for comparisons, provided the LES equations recover the RANS-model considered in the large-$\Delta$ limit. The latter is not guaranteed, however, in the majority of existing LES-models. This leads to the need for the construction of unified models that can be used, depending on the resolution, as either LES- or RANS-methods, see

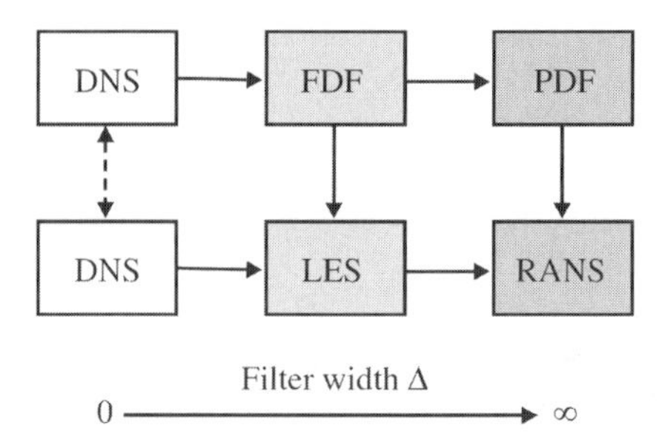


Figure 5.1.5: An illustration of unified turbulence models. Equations for filtered variables are given below. They can be applied as DNS-, LES- or RANS-models, depending on the choice of $\Delta$. The upper line shows corresponding stochastic models (SGS fluctuations vanish for $\Delta \to 0$ such that FDF-methods reduce to DNS).

Fig. 5.1.5 for an illustration and [34]. It is relevant to note that the requirement of developing unified turbulence models is not restricted to the need to represent RANS and PDF equations as large-$\Delta$ limits of LES and FDF equations, respectively, but a corresponding deepening of the relations between DNS and LES/FDF equations will also be very helpful [2].

On the other hand, it is obvious that the computational realization of unified turbulence models requires solutions for several new questions. Instead, the advantage of such efforts may be enormous: it offers the chance for a significant reduction of expensive experiments, which are still needed to assess the performance of numerical prognosis methods.

## References

1. S. B. Pope: Turbulent Flows, Cambridge University Press, Cambridge (2000).
2. S. Heinz: Statistical Mechanics of Turbulent Flows, Springer-Verlag, Berlin, Heidelberg, New York, Tokyo (2003).
3. S. B. Pope: A Perspective on Turbulence Modeling, In: Modeling Complex Turbulent Flows, edited by M. D. Salas, J. N. Hefner, L. Sakell, Kluwer, 53–67 (1999).
4. R. O. Fox: Computational Methods for Turbulent Reacting Flows in the Chemical Process Industry, Revue de l'Institut Français du Pétrole 51, 215–243 (1996).
5. J. Baldyga, J. R. Bourne: Turbulent Mixing and Chemical Reactions, John Wiley & Sons, Chichester, New York, Weinheim, Brisbane, Singapore, Toronto (1999).
6. N. Peters: Turbulent Combustion, Cambridge University Press, Cambridge (2001).
7. S. Heinz, D. Roekaerts: Reynolds Number Effects on Mixing and Reaction in a Turbulent Pipe Flow, Chem. Eng. Sci. 56, 3197–3210 (2001).
8. P. A. Nooren, H. A. Wouters, T. W. J. Peeters, D. Roekaerts: Monte Carlo PDF Modeling of a Turbulent Natural-Gas Diffusion Flame, Combustion Theory Modeling 1, 79–96 (1997).
9. P. A. Nooren: Stochastic Modeling of Turbulent Natural-Gas Flames, PhD thesis, TU Delft (1998).
10. H. A. Wouters, P. A. Nooren, T. W. J. Peeters, D. Roekaerts: Simulation of a Bluff-Body Stabilized Diffusion Flame Using Second-Moment Closure and Monte Carlo Meth-

ods, In: Twenty-Sixth Symp. (International) on Combust., The Combustion Institute, Pittsburgh 177–185 (1996).
11. H.A. Wouters: Lagrangian Models for Turbulent Reacting Flow, PhD thesis, TU Delft (1998).
12. M. Muradoglu, P. Jenny, S.B. Pope, D.A. Caughey: A Consistent Hybrid Finite-Volume/Particle Method for the PDF Equations of Turbulent Reacting Flows, J. Comput. Phys. 154, 342–371 (1999).
13. B.J. Delarue, S.B. Pope: Application of PDF-methods to Compressible Turbulent Flows, Phys. Fluids 9, 2704–2715 (1997).
14. B.J. Delarue, S.B. Pope: Calculation of Subsonic and Supersonic Turbulent Reacting Mixing Layers Using Probability Density Function Methods, Phys. Fluids 10, 487–498 (1998).
15. P.J. Colucci, F.A. Jaberi, P. Givi, S.B. Pope: Filtered Density Function for Large Eddy Simulations of Turbulent Reactive Flows, Phys. Fluids 10, 499–515 (1998).
16. F.A. Jaberi, P.J. Colucci, S. James, P. Givi, S.B. Pope: Filtered Mass Density Function for Large-Eddy Simulation of Turbulent Reacting Flows, J. Fluid Mech. 401, 85–121 (1999).
17. X.Y. Zhou, J.C.F. Pereira: Large Eddy Simulation (2D) of a Reacting Plane Mixing Layer using Filter Density Function Closure, Flow, Turb. Combust. 64, 279–300 (2000).
18. L.Y.M. Gicquel, P. Givi, F.A. Jaberi, S.B. Pope: Velocity Filtered Density Function for Large Eddy Simulation of Turbulent Flows, Phys. Fluids 14, 1196–1213 (2002).
19. S. Heinz: On Fokker-Planck Equations for Turbulent Reacting Flows. Part 1. Probability Density Function for Reynolds-averaged Navier–Stokes Equations, Flow, Turb. Combust. 70, 115–152 (2003).
20. S. Heinz: On Fokker-Planck Equations for Turbulent Reacting Flows. Part 2. Filter Density Function for Large Eddy Simulation, Flow, Turb. Combust. 70, 153–181 (2003).
21. A. Juneja, S.B. Pope: A DNS Study of Turbulent Mixing of two Passive Scalars, Phys. Fluids 8, 2161–2184 (1996).
22. W. Kollmann: The PDF Approach to Turbulent Flow, Theoretical and Computational Fluid Dynamics 1, 249–285 (1990).
23. P. Eifler, W. Kollmann: PDF Prediction of Supersonic Hydrogen Flames, AIAA Paper 93-0448, Reno (1993).
24. A.T. Hsu, Y.-L.P. Tsai, M.S. Raju: Probability Density Function Approach for Compressible Turbulent Reacting Flows, AIAA Journal 32, 1407–1415 (1994).
25. H. Möbius, P. Gerlinger, D. Brüggemann: Efficient Methods for Particle Temperature Calculations in Monte Carlo PDF-methods, ECCOMAS 98 John Wiley & Sons Ltd., 162–168 (1998).
26. S. Heinz: Nonlinear Lagrangian Equations for Turbulent Motion and Buoyancy in Inhomogeneous Flows, Phys. Fluids 9, 703–716 (1997).
27. U. Allgayer: Turbulent Combustion in Compressible Shear Layers (in German), Ph.D. Thesis, Dissertationsverlag NG Kopierladen GmbH, Munich.
28. G.M. Goldin, S. Menon: A Scalar PDF Construction Model for Turbulent Non-Premixed Combustion, Combust. Sci. Technol. 125, 47–72 (1997).
29. G.M. Goldin, S. Menon: A Comparison of Scalar PDF Turbulent Combustion Models, Combust. Flame 113, 442–453 (1998).
30. FLUENT 6.1 User's Guide, Fluent Inc., Lebanon, New Hampshire, USA.
31. S.B. Pope: Computationally Efficient Implementation of Combustion Chemistry Using In Situ Adaptive Tabulation, Combustion Theory Modeling 1, 41–63 (1997).
32. H. Foysi, R. Friedrich: DNS of Passive Scalar Transport in Turbulent Supersonic Channel Flow, Proc. 3rd Int. Symp. Turbulence and Shear Flow Phenomena, Sendai, Japan, June 25–27, 1121–1126 (2003).
33. H. Foysi, S. Sarkar, R. Friedrich: On Reynolds Stress Anisotropy in Compressible Channel Flow, Proc. 3rd Int. Symp. Turbulence and Shear Flow Phenomena, Sendai, Japan, June 25–27, 1103–1108 (2003).
34. C.D. Pruett, T.B. Gatski, C.E. Grosh, W.D. Thacker: The Temporally Filtered Navier–Stokes Equations: Properties of the Residual Stress, Phys. Fluids 15, 2127–2140 (2003).

## 5.2 Design and Testing of Gasdynamically Optimized Fuel Injectors for the Piloting of Supersonic Flames with Low Losses

Anatoliy Lyubar, Tobias Sander *, and Thomas Sattelmayer

### 5.2.1 Introduction

Airbreathing engines as a propulsion system for space transportation vehicles offer the advantage of a higher payload compared to rocket-based propulsion systems, as the oxygen from the atmosphere can be used for the oxidation of the fuel. However, the flight regime for these propulsion systems is limited by the low partial pressure of the oxygen at high altitude.

At supersonic flight conditions the efficiency of gas turbines decreases, because the shocks in the inlet ducting upstream of the compressor lead to tremendous losses of total pressure. Furthermore, the contribution of the compressor to the total pressure rise decreases with increasing flight Mach numbers. At a flight Mach number of 3 for example the compression in the turbomachine is less than 4%. Thus, the compression is achieved predominantly by the shock wave system under these conditions. Ramjet and SCRamjet propulsion systems don't require any rotating parts and basically consist of a diffuser, a combustor and a nozzle.

In the diffuser the Mach number of the supersonic flow is decelerated by shock waves. While in ramjet propulsion systems a normal shock decreases the Mach number below 1, the flow stays supersonic in the diffuser of SCRamjet propulsion systems. Merely oblique shock waves are induced by the geometry of the channel. In well designed SCRamjet combustors the flow predominantly stays supersonic except for regions of a high hydrogen concentration or of large heat release.

Because of the short residence time of the gas mixture in both ramjet and SCRamjet propulsion systems, a fundamental problem is the realization of mixing and ignition of the fuel. Furthermore, the ignition delay of the hydrogen-air mixture at low temperatures (900 K–1300 K) is long compared with the available residence time in the combustion chamber.

In order to obtain a complete burnout within the chamber, an adequate mixing of the fuel and oxidizer must be obtained and, in addition, self-ignition of the fuel must be achieved right after the injection. Both tasks must be accomplished by a proper injector design. On the one hand the injector must generate sufficient turbulence and a shock wave system that increases static temperature and pressure. On the other hand turbulence and the shock wave

* Siemens VDO Automotive AG, Postfach 100943, 93009 Regensburg; tobias.sander@siemens.com

system both lead to a loss of total pressure that reduces the efficiency of the propulsion system.

In the nozzle the flow in ramjet and SCRamjet engines is accelerated to a supersonic Mach number and thrust is generated.

The maximum flight Mach number of ramjet-propelled aircrafts is approximately 6. Above this limit dissociation of air decreases the efficiency and the high temperature causes thermal problems. In SCRamjet engines the flow stays supersonic, the static temperature is below the critical limit for dissociation of the air and material problems. Its flight Mach number range extends from 6 to approximately 12.

A main goal of the study was to evaluate the conventional injectors, which were used in previous investigations with respect to their applicability in a SCRamjet combustor. After the results indicated that none of the investigated injectors provides a low-loss flame stabilization or a complete burnout of the fuel within the combustor, respectively, a new injector was developed, which overcomes these deficiencies. With numerical and experimental results it is demonstrated that this injector offers promising properties for a low-loss flame stabilization in a SCRamjet and the possibility of a broadband application with respect to the flight Mach number.

## 5.2.2 Experimental Setup

### 5.2.2.1 Model SCRamjet Combustor

The model SCRamjet combustor is shown in Fig. 5.2.1. The filtered test air, which is supplied with pressures up to 13 bar, is mixed with oxygen in order to compensate for the oxygen loss in the hydrogen preheater. The total temperature in the preheater can be increased up to 1400 K and is measured with an accuracy of ±3%. The maximum mass flow through the channel was 300 g/s ±3% during experiments. The preheated flow is expanded subsequently in a Laval nozzle to a Mach number of 2.1 and enters the combustion chamber.

### 5.2.2.2 Preheater

The hydrogen required for the preheating of the test air ignites in the preheater immediately after its addition to the main flow and a diffusive flame is stabilized by flame holders (Fig. 5.2.2). The residence time of the gas in the preheater is rather short to obtain a perfect mixture as its length is only 380 mm. This limits the burnout that can be achieved before the gases leave the preheater. In order to obtain a higher burnout with low radical content in the exhaust, a vortex generator was installed 700 mm upstream of the preheater (Fig. 5.2.1). One

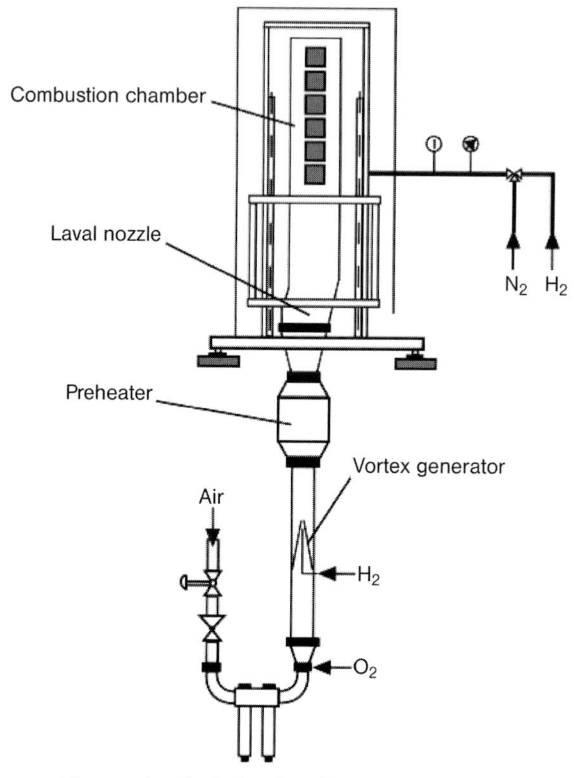


Figure 5.2.1: SCRamjet combustor and test rig.

half of the fuel is injected at the vortex generator and the other half is added at the original downstream injector to obtain partially premixed combustion. In this mode the flame holder acts as stabilization device for the piloted premixed flame. Since the mixing process is performed upstream in the supply tube, the preheater flame is shortened considerably and complete burnout is obtained upstream of the entrance of the SCRamjet combustion chamber.

#### 5.2.2.3 Combustion Chamber

The cross-sectional area at the entrance of the combustion chamber is 27.5 mm×25 mm. At a distance of 75 mm from the inlet plane the hydrogen fuel is injected using one of the two investigated injectors, which are described further below. The hydrogen mass flow can be varied up to 2.3 g/s by increasing the injection pressure, which is measured with an accuracy of ±2%.

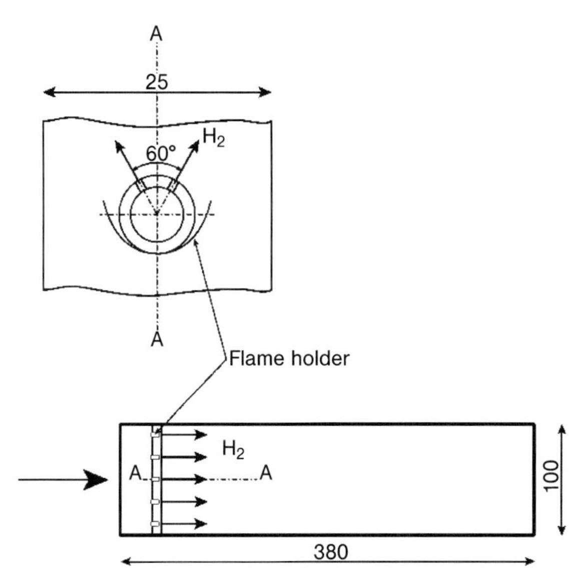


Figure 5.2.2: Geometry of the preheater.

145 mm downstream of the entrance, the flow in the combustion chamber is expanded continuously from 27.5 mm to 41.5 mm with an angle of 4° in order to prevent thermal choking [1]. In contrast to the modular construction of the chamber walls used in previous investigations, the experiments were made with single element walls in order to avoid disturbances of the flow caused by discontinuities between the modules (Fig. 5.2.3).

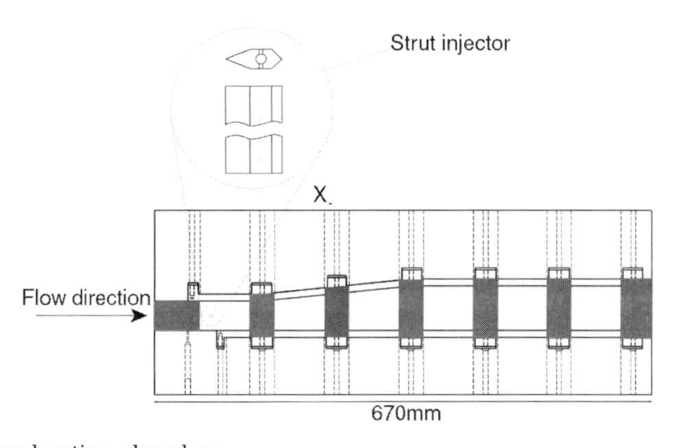


Figure 5.2.3: Combustion chamber.

### 5.2.2.4 Injectors

Two different injectors have been used in the tests. The first injector ("pylon injector") was developed by Grünig (Fig. 5.2.4). It represents a design optimized for quick ignition and intense reaction within the combustion chamber. It consists of a massive body, which produces a strong shock wave system in combination with secondary vortices generated by a ramp structure at its flanks. The hydrogen is injected at an angle of 30° to the main flow direction. Further details on this injector and the experimental results can be found in [1]. Because of the high pressure losses caused by the pylon, it was merely used to provide reference data for an injector with very robust flame stabilization.

The other injector, which was designed to achieve low total pressure loss with a smaller range of flame stability ("strut injector"), is depicted in Fig. 5.2.5. The slender cross section has a half angle of only 22.5° at its leading edge and a half angle of 45° at its trailing edge. This injector does not provide any features for the generation of secondary vortices. Furthermore, it causes a weaker shock wave system with lower pressure loss than the pylon injector. The fuel is injected at 90° to the main air stream through three orifices on each side of the strut. The orientation of the hydrogen jets relative to the air stream is shown in Figs. 5.2.4 and 5.2.5, respectively. The geometric and aerodynamic simplicity of this injector makes it particularly suitable for numerical simulations using CFD. Comparing the results of the two designs, the effects of the

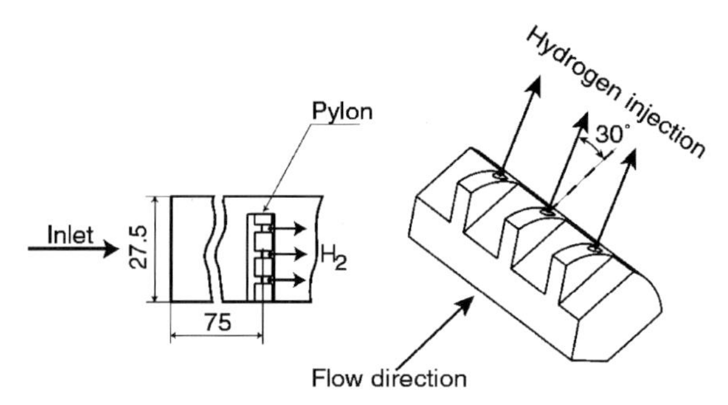


Figure 5.2.4: Pylon injector.

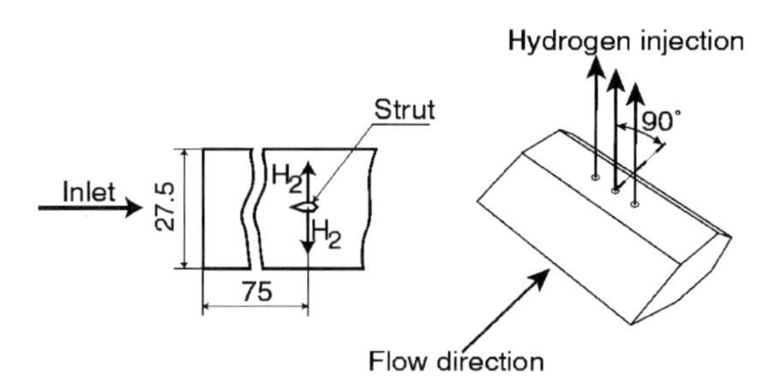


Figure 5.2.5: Strut injector.

inflowing radicals and the water vapor content on the ignition process will be illustrated.

### 5.2.3 Investigation Tools

#### 5.2.3.1 Shadowgraph Method

With this method, density gradients and, in particular, shocks in a supersonic flow can be visualized employing a change of the refractive index of a compressible transparent medium. The data delivers information of the flow integrated along lines of sight. In the presented investigations, a pulsed flash lamp (Nanolite) was synchronized with a CCD Camera (PCO, Flash Cam 335 CG 0045) with a resolution of 756×290 pixel.

#### 5.2.3.2 Rayleigh Scattering

In Rayleigh scattering the detected signals have the same frequency as the incident laser light and are a superposition from all existing molecules in the gas mixture stimulated. Thus, this technique cannot be used for individual species concentration measurements, but it offers the possibility to visualize a flow field with strong gradients in pressure and concentration of species with different Rayleigh cross section, e.g. air and $H_2$ [2]. The Rayleigh scattered power depends on both the total number density and the gas composition. Using a light sheet, two-dimensional planes in the flow can be investigated and a high spatial resolution can be obtained. A XeCl excimer laser (Lambda Physik, EMG201-204MSC) with a wavelength of 308 nm and a pulse energy of up to 400 mJ was used as light source. The elastic scattered light was detected with an intensified CCD camera (La Vision, Streak Star) and a Nikon lens (UV-Nikkor, 105 mm, 1:4.5).

#### 5.2.3.3 Raman Scattering

With the application of Raman scattering the static temperature and the concentration of the majority species in the flow can be measured. Raman scattering is an inelastic scattering process, which means that the light emitted from the molecules in the gas mixture is frequency shifted from the laser light. On one hand this offers the possibility of multispecies detection if the emitted light is resolved spectrally, on the other hand the Raman scattering is very weak (about three orders of magnitude lower than the elastic Rayleigh scattering), which complicates quantitative experiments enormously. As the population of rotational and vibrational energy levels of the molecules is temperature dependent, the Raman scattering can also be applied for the measurement of the static temperature. For the Raman measurements a Nd:YAG laser (Quantel YG 782 C 10) with a wavelength of 532 nm and a pulse energy of approximately 1000 mJ was used. The focal length of the focusing lens was 1000 mm. The scattered light was filtered with a Notch filter, in order to suppress the elastic scattered light, which would inhibit the detection of the comparatively weak Raman scattered light. The spectra was analyzed with a spectrograph (Acton Research, SpectraPro-275) and an intensified CCD camera (Princeton Instruments, ICCD-576 S/B).

#### 5.2.3.4 OH-LIF Measurements

With the laser induced fluorescence single species of interest may be detected in a gas mixture if an extremely narrow-banded light source is used for the excitation of the molecules. The detection limit of this technique is in the ppm-range. In hydrogen-air combustion, OH radicals are the only species that exist in sufficient concentrations for LIF measurements. They can be used as marker for the start of the reaction because of their fast formation during the ignition processes. In the measurements, a light sheet was used, which allowed the two-dimensional detection of the fluorescence signals. The molecules were excited at a wavelength of 283 nm by means of a dye laser (Lambda Physik, Scanmate 2 9703 F 2055, Coumarin 153), which was pumped by a XeCl excimer laser (Lambda Physik, EMG201-204MSC) at a wavelength of 308 nm. The corresponding energy transition was $A^2\Sigma^+(v'=1) \leftarrow X^2\Pi(v''=0)$. The fluorescence was detected with an intensified CCD camera (La Vision, Streak Star) and the already mentioned Nikon lens. In the experiments with the combustion chamber, the quartz windows caused reflection, which had to be suppressed by means of a band pass mirror filter, which was adjusted to transmit at 310 nm.

#### 5.2.3.5 Self-Fluorescence Measurements (Chemiluminescence)

The self-fluorescence technique offers the simple possibility to detect chemical reactions without using an excitation light source. In the experiments the signal was detected with an intensified CCD camera (La Vision, Streak Star) and, again, the Nikon lens. No band pass filter was used here, since the goal of the measurements was rather to distinguish between the different reaction types than to achieve a spectral selection of specific species in the reacting flow.

### 5.2.4 Numerical Modelling

#### 5.2.4.1 Numerical Simulation with the CFD-Code Fluent 5.5

The simulation of the supersonic combustor was carried out with the CFD-Code Fluent 5.5 on a Linux-Cluster consisting in 32 nodes. Fluent [3] uses a control-volume-based technique to convert the governing equations to algebraic ones. Both structured and unstructured meshes can be used for the simulation. Fluent also allows the user to refine or coarsen the grid based on the flow solution and thus to improve the simulation accuracy of the shocks without enormous increase of the number of cells. The Reynolds Stress Model [3] was applied for the modelling of turbulence. The non-equilibrium wall function was used for modelling the near-wall region [3]. The using of the near-wall modelling approach, which would be preferable for the simulation of flows with boundary layer separation, was not possible, because it requires a very fine grid resolution near the wall ($y^+ \approx 1$). Such a resolution near the wall is possible only for a two-dimensional simulation, because of the enormous increase of the number of cells and because of the limited computational resources. The governing equations were solved using a coupled solver with a second-order discretization scheme, which means that all values are second-order accurate.

A 2D structured grid with quadrilateral cells and a 3D structured grid with hexahedral cells were used for the simulation of the SCRamjet combustor. The average size of the cells was 0.1 mm and 0.25 mm and the number of cells was about 12,500 and 500,000 respectively. Additional transport equations were solved for all relevant species and the source terms of these equations were calculated using detailed hydrogen kinetics, whereas turbulence-chemistry interaction was not considered. Additional simulations of the flow in the Laval nozzle were carried out to obtain the profiles of the inlet parameters for the modelling of the flow in the chamber.

#### 5.2.4.2 Special Features of the Modelling of the Supersonic Combustion

The numerical description of the reaction in a supersonic flow has some special features compared to the subsonic combustion e.g. in a gas turbine. The main characteristic of the supersonic combustion is that the timescales of the flow and chemistry are of the same order. This is why the complete chemical kinetics have to be taken into account, which requires additional memory for the storage of the reaction intermediates (radicals) and cpu-time to solve the stiff ODEs of chemical kinetics. The simplest way for the modelling of the supersonic combustion with complete chemical kinetic is using of the laminar Finite-Rate model, which is available in the CFD-Code Fluent [3] and which uses the Arrhenius law to calculate the chemical reaction rates. The model is exact for laminar flames, but it is generally inaccurate for turbulent flames due to highly non-linear Arrhenius chemical kinetics. The laminar model is, however, acceptable for combustion with relatively slow chemistry and high local mixedness, such as in supersonic flames. Including the chemistry calculation of the hydrogen-air reaction with five reaction intermediates ($H$, $O$, $OH$, $HO_2$, $H_2O_2$) and 37 elementary reactions according to the reaction scheme of Warnatz [4] into the CFD-simulation slows down the computation about the factor of two and increases the memory requirements also about a factor of two. The using of the PDF-approach for coupling the chemistry with the mixing and the temperature fluctuations due to the turbulence is very expensive concerning the computational time and together with the chemical kinetics reduces strongly the maximum computational volume and the spatial resolution. For example, including the temperature-PDF transport equation from Gerlinger [5] and taking into account the temperature fluctuation for the computation of the chemical reaction rates slows down the simulation about a factor of 40. This means that only small two-dimensional geometries can be simulated using typical computational resources, if the PDF-coupling is included. For the simulation of a real 3-D chamber taking into account the turbulent fluctuations a supercomputer is needed. The available Linux-cluster, which was realized for the numerical investigations of the supersonic chamber, allows the 3-D simulation of the chamber using only the finite-rate combustion model, which was considered as a standard model. To speed up the simulations with the finite-rate model and to enable the use of the PDF-approach, a new technique of the reaction mapping using the multilinear polynomial was developed and will be presented below.

#### 5.2.4.3 Reducing the Number of Species

A sensitivity analysis according to Pilling [6] was carried out for a wide spectrum of conditions with the goal to decrease the number of the reaction intermediates, which are needed to describe the reaction process. According to this method, all species are divided into three groups: important, necessary and redundant. A species may be considered redundant, if its concentration change has no significant effect on the rate of production of important species.

The influence of the change of the concentration of species $i$ on the rate of production of a $N_{imp}$-membered group of important species can be evaluated as follows:

$$B_i = \sum_{n=1}^{N_{imp}} \left( \frac{\partial \ln \dot{\omega}_n}{\partial \ln [X_i]} \right)^2 ,$$

where $N_{imp}$ is the number of important species, $\dot{\omega}_n$ is the molar production rate of species $n$ and $[X_i]$ is the molar concentration of species $i$. The higher the $B_i$ value of species, the greater is its direct effect on important species. The groups of species have to be identified by an iterative procedure. From the analysis the following grouping was found:

- important species: $H_2$, $O_2$, $H_2O$,
- necessary species: $H$, $O$, $OH$,
- redundant species: $HO_2$, $H_2O_2$.

The variation of the $B_i$-values during the reaction in a constant pressure reactor relative to important species is shown in Fig. 5.2.6. The $B_i$-values of $HO_2$ and $H_2O_2$ are definitely much smaller than the $B_i$-values of necessary species and consequently their influence on the production and consumption of the important species is negligible. After elimination of the redundant species the reaction is described by 7 species ($H_2$, $O_2$, $H$, $O$, $OH$ and $H_2O$). However, it has to be mentioned that the species, which were identified as redundant, play an important role in the chain reactions at low temperatures (under 1100 K–1300 K) and their elimination can influence the ignition delay time in this temperature region. As the reduced chemistry was used mainly for the tabulation procedure, the redundant species were not eliminated from the reaction scheme and only their concentrations were set to zero after each iteration. Thus the effect of elimination of redundant species was reduced. The decrease of the number of species needed to describe the reaction was very important for the development of the reaction mapping method, which is presented below.

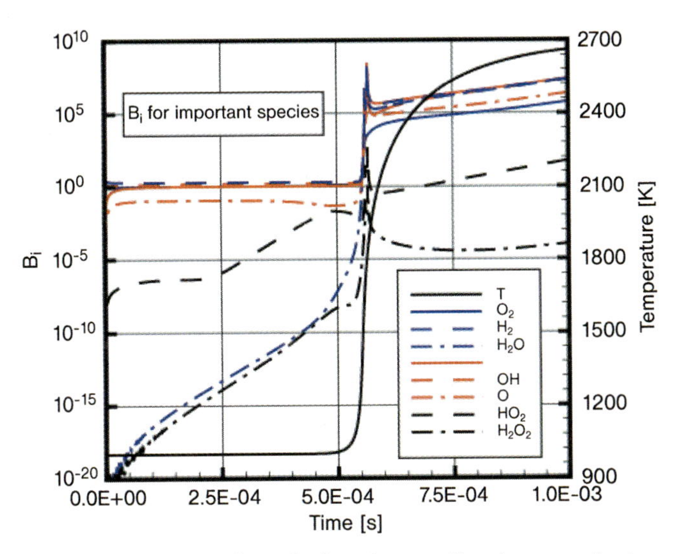


Figure 5.2.6: Variation of the $B_i$-values during the reaction in a constant pressure reactor relative to important species.

#### 5.2.4.4 Reaction Mapping by Using of the Polynomials

This reduced scheme for the description of the reaction was used for the construction of the lookup-table, where the thermochemical states of the mixture was tabulated. The thermochemical state of the mixture is described by the temperature and mole fractions of the seven species. The mole fractions of eight species (nitrogen is considered only as a third body) are calculated under the assumption that the mole fractions sum to unity. The assumption of a constant pressure of 1 bar was made to keep the dimensionality of the table as small as possible during the development phase. The thermochemical states are tabulated with a time step $\Delta t$, which means that an adiabatic reaction under constant pressure is calculated from the initial state $t=0$ during the time interval $\Delta t$ and the obtained thermochemical state is written into the table. The nodes in the table are interconnected by pointers forming rectangular cells in the multidimensional space. Such connectors allow producing the tables with an arbitrary form of the boundary and they provide a very efficient searching procedure. The structure of the node and of the table is described detailed in the report of the previous research phase [7]. Using this table decreases the simulation time because the time consuming calculation of the chemistry in the CFD-code can be avoided. The change of the thermochemical state is interpolated from the lookup-table.

A further increase of the efficiency of the simulation can be obtained through the fitting of the tabulated data to a set of polynomials. Different kinds of polynomials such as high-order orthogonal polynomials and multilinear polynomials were analyzed concerning their suitability to match the tabulated data. The analysis showed that the best performance can be achieved by using multilinear polynomials. In the case of multilinear polynomials one polynomial describes the data inside of one multidimensional rectangular cell. The accuracy of the polynomial corresponds to the accuracy of the interpolation from the lookup-table, but the search for the nodes needed for the interpolation and the evaluation of the weighting coefficients (the number of nodes and coefficients is 128 and the dimensionality is 7) are avoided. At the vertexes of the cell the multilinear polynomial delivers exactly the values of nodes placed there and no discontinuity is present from one polynomial to another. Because such discontinuities have a very negative effect on the numerical stability, this is one of the important features of the approach. The multilinear polynomial has the form:

$$f(x_1, \ldots x_7) = a_0 + a_1 x_1 + \ldots + a_7 x_7 + a_{12} x_1 x_2 + a_{67} x_6 x_7 + \ldots \\ + a_{1234567} x_1 x_2 x_3 x_4 x_5 x_6 x_7$$

The coefficients $a$ represent the corresponding partial derivatives and are evaluated numerically using the function values of the nodes. The evaluation of derivatives in discrete coordinates for a two-dimensional case is as follows (Fig. 5.2.7):

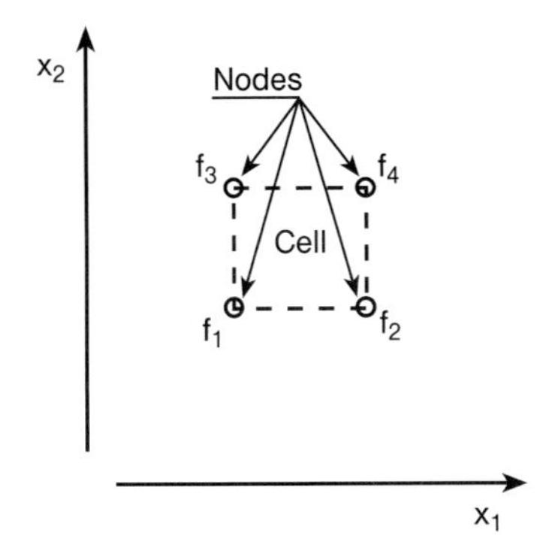


Figure 5.2.7: Simplified 2D illustration of the multidimensional cell.

$$a_0 = f_1 ,$$
$$a_1 = \frac{\partial f}{\partial x_1} = \frac{f_2 - f_1}{\Delta x_1} = f_2 - f_1 , \quad a_2 = f_2 - f_1, \ldots ,$$
$$a_{12} = \frac{\partial f^2}{\partial x_1 \partial x_2} = \frac{(f_4 - f_2) - (f_3 - f_1)}{\Delta x_1 \Delta x_2} = f_1 + f_4 - f_2 - f_3 , \quad \ldots . \quad \Delta x_i = 1$$

During the CFD-simulation only the used polynomials are stored in the memory. A cashing procedure allows keeping the usage memory required for the polynomials insignificantly low.

#### 5.2.4.5 Validation of the Modelling Approach with Polynomials

The polynomials were constructed using a lookup-table consisting in 226,663 nodes, which were transformed into 28,969 cells and consequently 28,969 polynomials. The time interval was 1e-8 s. The lookup-table covers an entire region of physically possible thermochemical states in the temperature range from 300 K to 3000 K. The limits for each quantity and the discretization steps are presented in Tab. 5.2.1. The boundary of the physically possible region is defined on the basis of the direction of the reaction evolution. This means that the mixture with a thermochemical state lying on the boundary due to the chemical reaction progresses towards the center of the physically possible region.

After the polynomials had been validated by the simulation of the ignition in a spatially homogeneous reactor, they were applied to simulate an experiment of Evans [8], where hydrogen is injected into a preheated vitiated air stream. The geometry of the experiment and the inlet conditions are shown in Fig. 5.2.8. The inlet profiles had been obtained by an additional simulation of the flow in an axis-symmetric tube of the same size. The simulation of the combustion process was realized for both models, the finite-rate model with full chemistry and the polynomial-based method with reduced chemistry. The re-

Table 5.2.1: Tabulated area and discretization steps

| Axis | Minimum | Maximum | Discretization step |
|---|---|---|---|
| T | 300 | 3000 | 150 |
| $H_2$ | 0 | 1 | 0.1 |
| $O_2$ | 0 | 0.24 | 0.08 |
| O | 0 | 0.02 | 0.02 |
| H | 0 | 0.1 | 0.02 |
| OH | 0 | 0.04 | 0.04 |
| $H_2O$ | 0 | 0.55 | 0.05 |

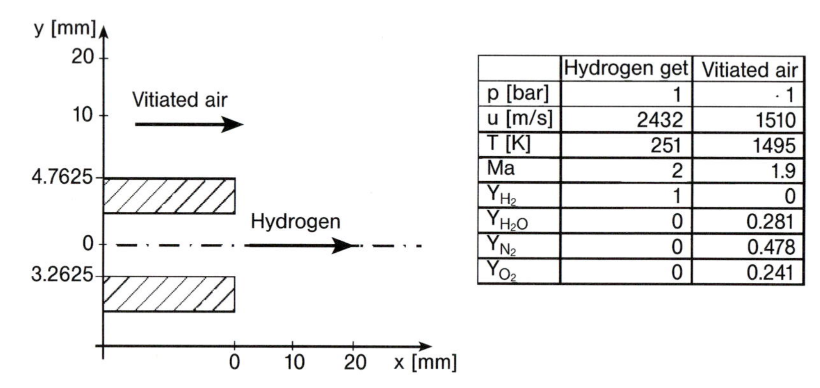


| | Hydrogen get | Vitiated air |
|---|---|---|
| p [bar] | 1 | 1 |
| u [m/s] | 2432 | 1510 |
| T [K] | 251 | 1495 |
| Ma | 2 | 1.9 |
| $Y_{H_2}$ | 1 | 0 |
| $Y_{H_2O}$ | 0 | 0.281 |
| $Y_{N_2}$ | 0 | 0.478 |
| $Y_{O_2}$ | 0 | 0.241 |

Figure 5.2.8: Geometry and inlet conditions for the axis-symmetric experiment of Evans.

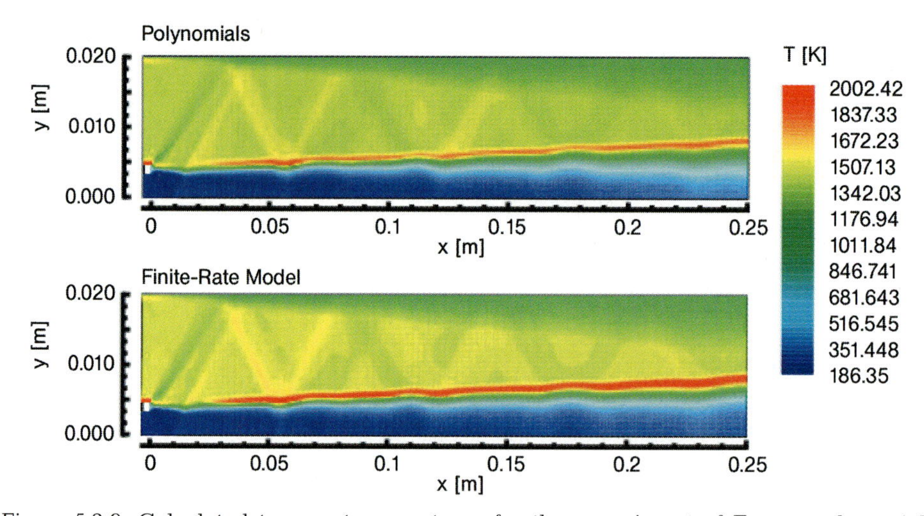


Figure 5.2.9: Calculated temperature contours for the experiment of Evans: polynomials (top) and finite-rate model (bottom).

sults of simulations, shown in Fig. 5.2.9, are almost identical, although the simulation time using the polynomials approximately equals the simulation time of the non-reacting flow. This is a remarkable asset of the approach. Although the reaction has a diffusive character with a wide variety of mixing ratios and temperatures, the number of used polynomials was very low. Interestingly, only approximately 180 polynomials were used during the simulation. 165 polynomials were stored in the memory and approximately 1–2 polynomials were read and deleted each time step. This low number of used polynomials allows an increase of the dimensionality of the interpolated data without the usual enormous growth of used memory and hard drive – memory traffic.

The high computational efficiency of the approach using polynomials is the prerequisite for using PDF-methods for technically relevant combustors with complicated geometries on computer platforms with limited power. Including additional variables into the description of the thermochemical state to represent the variances of temperature and species allows eliminating the time consuming integration procedure for evaluating of the averaged reaction rates. Consequently, the use of polynomials for the calculation of the chemistry reduces the computational demands of PDF-methods tremendously.

### 5.2.5 Two Stage Injector

The following part of the present work is focused on the design of an injector, which mixes fuel and oxidizer without large losses of total pressure and which provides stable ignition under various operation conditions. The concept was deduced from theoretical considerations and was investigated numerically and experimentally. In an iterative approach, several test injectors were designed, built and tested in order to optimize the flame stabilization potential.

In the cold mode of operation of the test rig the total temperature was 298 K and in the hot mode 900 K. The latter temperature was selected in the tests presented below, since the extension of the operation regime to low flight Mach numbers was the main target of the injector development. The total pressure in both cases was 7.5 bar.

#### 5.2.5.1 Theoretical Considerations

Both, the study of the literature and our former investigations of flame stabilization by means of struts or pylons in a supersonic chamber [9–11] point out that a subsonic area with high static temperature and turbulence is required to achieve a stable ignition. Oblique shocks do not influence the stabilization significantly, since the increase of the static temperature is too weak to induce ignition. This finding implies that using well known injector types the ignition is initiated as the result of irreversible total pressure losses, which are produced in recirculation areas. Such zones are usually formed in cavities in the chamber walls or behind pylons or struts [12, 13]. Although stable ignition can be achieved in these areas, burnout is insufficient at low flight Mach numbers due to low transverse flame propagation. Pylon as well as cavity stabilization concepts have very thin reaction zones and a very slow burnout [9–11]. Furthermore, the reaction in cavities or wall-near regions leads to an enormous thermal stress in the combustor.

The novel approach to the solution of this problem presented subsequently is based on the reduction of the self-ignition temperature and the ignition delay time and the acceleration of the flame propagation under the presence of radicals in the flow. As it was shown in several works, the same effect is present in experimental setups that use pre-combustion heating to simulate flight conditions and represents a major drawback of test setups with vitiators, if the vitiator does not achieve complete burnout and emits radicals far above the equilibrium concentration of the mixture burnt [12, 14]. Own numerical studies on the influence of radicals on the ignition of hydrogen-air mixtures showed that the ignition delay can be decreased up to microseconds at low temperatures of about 900 K. From our own experience it can be concluded that using zones with high radical concentrations in a supersonic chamber is much more efficient for achieving stable ignition at low temperatures and for increasing burnout than increasing the static temperature by means of a system of oblique shocks for the same purpose.

In the present work a novel injector type was designed to demonstrate the flame stabilization using the above mentioned radical-based mechanism. The scheme of the injector employed is shown in Fig. 5.2.10.

The test setup was designed to simulate one sector of a SCRamjet combustor and the side walls represent planes of symmetry in the real case of an implementation of the principle in a chamber with multiple injectors. The flow is divided into three parts: two outer or main flows and one inner or injector air flow. The latter is decelerated in the convergent duct of the diffuser by a system of oblique shocks and a final normal shock with a drop to a subsonic Mach number. The static pressure is increased accordingly and the static temperature

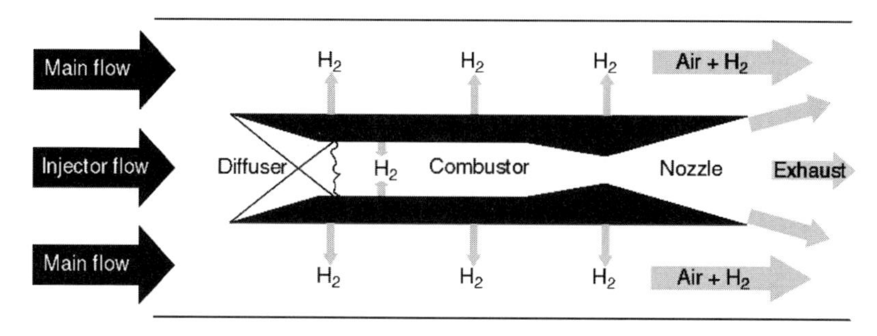


Figure 5.2.10: Scheme of the injector concept.

reaches the self-ignition temperature of hydrogen. The reaction of the hydrogen, which is injected into the subsonic region of the injector flow, is mixing-controlled. As a consequence, the exhaust gas, which is accelerated to a supersonic Mach number in the nozzle again, emits large concentrations of reaction intermediates from the injector end in downstream direction. Three additional hydrogen streams are injected into the two main flows. The reaction intermediates from the injector flow cause the ignition of the hydrogen-air mixture in the main flow and the reaction is stabilized over the entire cross section of the flow downstream of the injector.

The basic element of the injector is the supersonic diffuser, whose task is the deceleration of the flow with the smallest possible pressure loss. This means that a system consisting in several oblique shock waves and a concluding normal shock must be established inside the duct. If the diffuser works stably, the normal shock lies at the throat and the diffuser is said to run full, i.e. the streamlines at the intake cross section are parallel. If the shock propagates downstream from the desired position during operation, the thermodynamic conditions suitable for the self-ignition of the hydrogen are not achieved and the reaction will break down. If the back pressure of the flow is too large, the normal shock propagates upstream, stabilizes outside the diffuser and causes a decrease of the mass flow through the injector by bending the streamlines. This causes choking of the injector flow and the main flow with high total pressure losses reducing the efficiency of the engine.

Compatible with these basic considerations was the observation made in the numerical investigations as well as during the tests that the shock stabilization in the injector flow is a very challenging task: A normal shock is induced by the kink of the injector wall at the transition from the precombustor inside

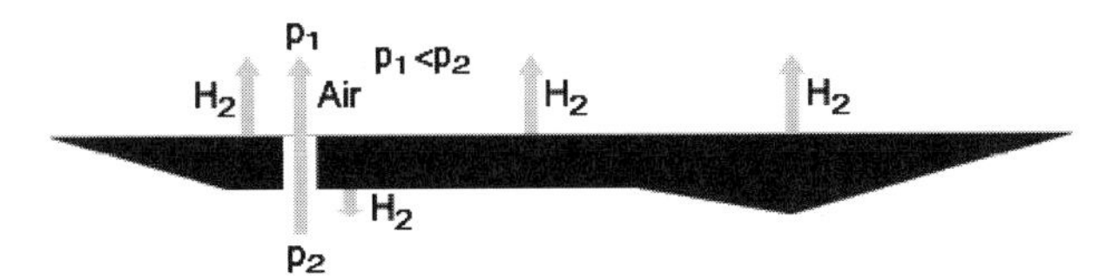


Figure 5.2.11: Injector with orifices.

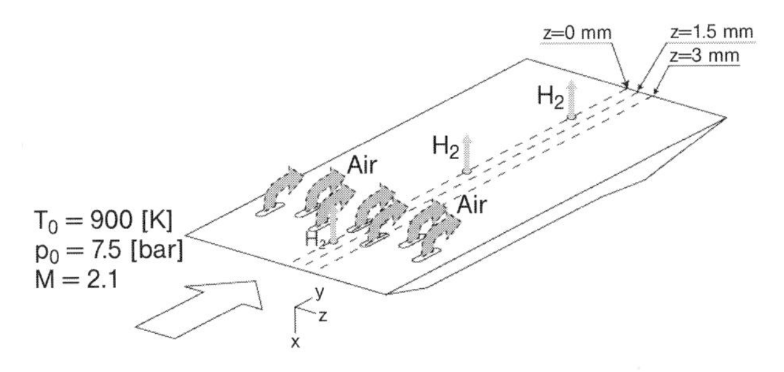


Figure 5.2.12: Injector with orifices.

the injector to the throat of the nozzle and propagates upstream up to the leading edge of the injector. In this case the injector channel is choked and the injector flow as well as the main flow become subsonic.

Taking into account the properties of a supersonic diffuser, the injector scheme, described in Fig. 5.2.10, was equipped with orifices, in order to overcome this difficulty. In Fig. 5.2.11 one half of the modified injector is shown.

The orifices are placed between the first two hydrogen jets and interconnect the injector flow of high static pressure $p_2$ with the main flow of low pressure $p_1$. This modification has two beneficial effects:

- The mass flow through the orifices leads to a decrease of $p_2$ and reduces the mass flow through the injector. Thus the normal shock stabilizes inside the injector flow.
- The flow of the air through the orifices intensifies the mixing of the hydrogen in the main flow and improves the ignition of the mixture at the trailing edge of the injector.

In Fig. 5.2.12 the shape and the position of the stabilization orifices and of the injection orifices are shown in a three-dimensional view of one injector half.

The orifices were machined as slotted holes with a cross section of 1 mm×4 mm. They are placed in two rows, which are shifted to each other by 1 mm, in order to achieve a better mixing of fuel and air in the main flow and to increase the region of a potential shock stabilization. The injection of hydrogen into the injector flow takes place through one orifice with a diameter of 0.66 mm and into the main flow through three orifices with a diameter of 0.5 mm each. All injection orifices are placed in the middle plane. Three planes were extracted from the results of the three-dimensional calculation of the flow for the purpose of illustration of the described effects. The location of these planes is shown in Fig. 5.2.16. The z-coordinates are $z=0$ mm, 1.5 mm and 3 mm, respectively, with $z=0$ mm at the centreline of the injector.

#### 5.2.5.2 Shock Stabilization

The numerical as well as the experimental investigations demonstrate that the position of the normal shock can be controlled with orifices. Subsequently, the injector version without orifices will be referred to as configuration 1 and the version with orifices as configuration 2.

*Version 1*

The computations of the flow field of version 1 were first carried out in cold mode. It was observed during the computations that the normal shock in the injector flow propagates upstream in the diffuser.

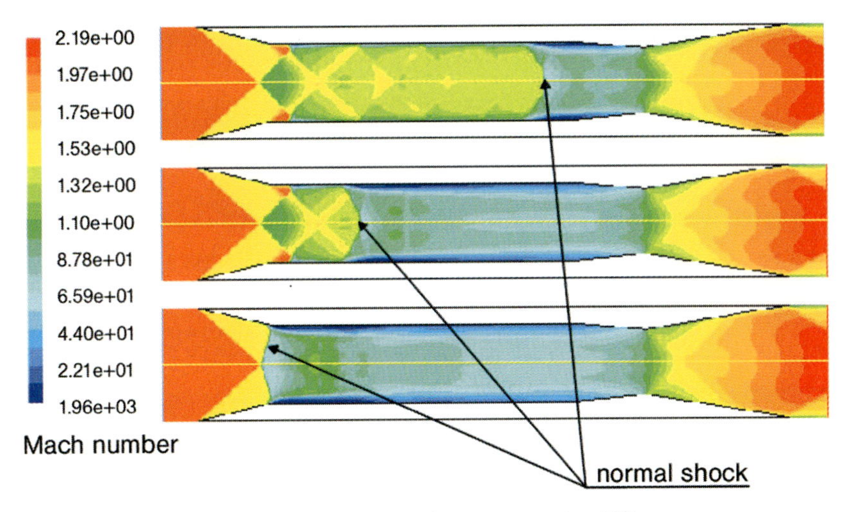


Figure 5.2.13: Propagation of the normal shock upstream the diffuser.

The three snapshots of the Mach number distribution shown in Fig. 5.2.13 illustrate the propagation of the normal shock from the transition of the pre-combustor to the throat upstream. During this propagation the Mach number upstream of the normal shock and the shock strength increases. The loss of total pressure reaches its maximum when the normal shock reaches the leading edge of the injector and the injector flow as well as the main flows become choked.

The experimental investigations of version 1 in cold mode showed the same behaviour. For reasons of a better optical access, a special injector without main flow on both sides was built. In Fig. 5.2.14 the result of the shadowgraph measurement is shown.

A dark curved line at the inlet of the diffuser can be seen, which indicates a strong pressure gradient with a three-dimensional structure. This pressure gradient is caused by a boundary layer separation that leads to a shock. Because of the integrative character of the shadowgraph method along the line of sight, it is not visible that this shock is located near the walls of the channel and does not extend over its whole depth. The flow in the diffuser still is supersonic. Shadowgraph sequences reveal that the normal shock (two vertical lines in Fig. 5.2.14) propagates upstream. It was observed that the final flow state is established when the normal shock reaches the leading edge of the injector and leads to choking of the flow (Fig. 5.2.15). The oblique shocks at the trailing edge of the nozzle indicate that an overexpanded supersonic flow is generated in the nozzle independently from the location of the normal shock during the shock propagation.

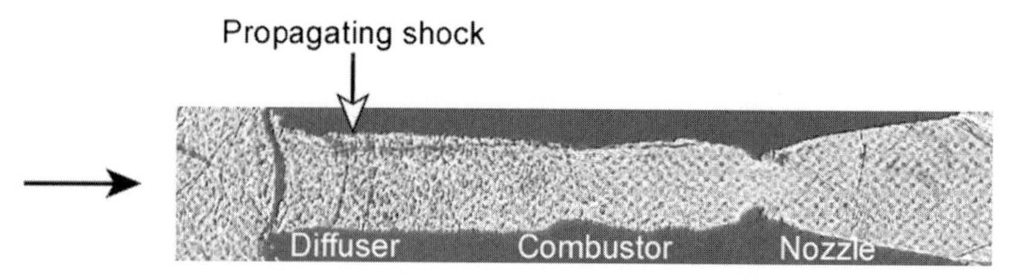


Figure 5.2.14: Injector without orifices, $Ma=2.1$, $p_0=7.5$ bar, $T_0=298$ K.

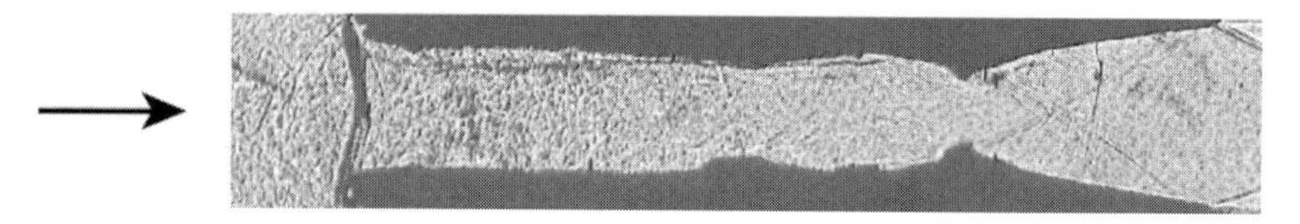

Figure 5.2.15: Injector without orifices, $Ma=2.1$, $p_0=7.5$ bar, $T_0=298$ K.

*Version 2*

In this section the results of the investigations of version 2 in cold mode are presented. The contour plots of the Mach number are shown in Fig. 5.2.16.

The locations of the presented planes are sketched in Fig. 5.2.12. Two planes penetrate the orifices (planes $z=0$ mm and $z=3$ mm) and the third one is in between two orifices (plane $z=1.5$ mm). The simulation of the flow predicts a reliable stabilization of the shock at the orifices. It is illustrated that the normal shock is fixed at the first row (plane $z=3$ mm) of orifices whereas the second row (plane $z=0$ mm) is in the subsonic region after the normal shock. The air flow through the orifices disturbs the main flow and leads to an increase of the boundary layer along the injector downstream of the orifices. The flow through the orifices of the second row influences the main flow much stronger than the orifices of the first row because of the larger pressure difference between injector flow and main flow near the second row.

The same stabilization pattern of the normal shock was detected experimentally (Fig. 5.2.17). The shock at the inlet of the diffuser is similar to the above mentioned cases. However, two oblique shocks are visible. These shocks lead to the normal shock, which decelerates the injector flow to a subsonic Mach number. In the configuration with orifices no propagation of the shock wave system as in version 1 occurred. Like in version 1 the flow is accelerated in the nozzle of the injector and the shocks at the trailing edge indicate an overexpanded flow.

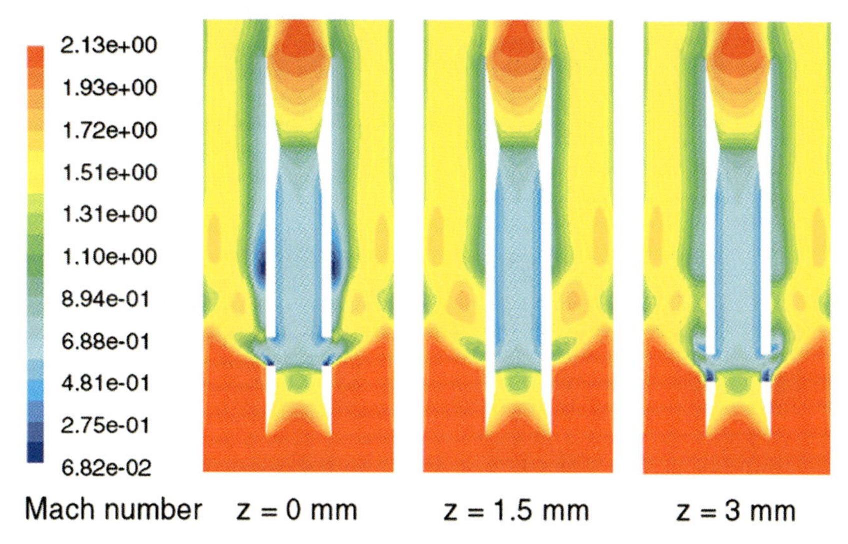


Figure 5.2.16: Injector with orifices, $Ma=2.1$, $p_0=7.5$ bar, $T_0=298$ K.

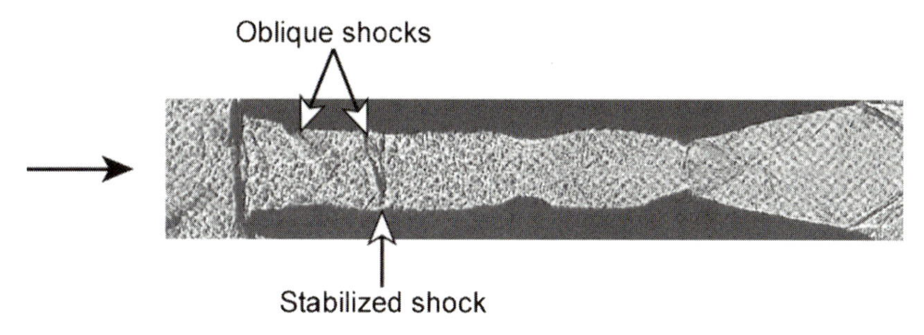


Figure 5.2.17: Injector with orifices, $Ma=2.1$, $\mathrm{p}_0=7.5$ bar, $\mathrm{T}_0=298$ K.

### 5.2.5.3 Combustion

*Ignition and Burnout*

In order to visualize the chemical reaction in the injector flow, version 1 of the injector was investigated with the self-fluorescence technique. As the gain of the ICCD camera was held constant during the experiments, the acquired data allows to directly compare the intensity of the reaction in the investigated cases. In the tests the total temperature was 900 K, the mass flow of injected hydrogen 1.27 g/s ($\Phi=0.35$ based on the injector air flow) and 3.8 g/s ($\Phi=1.06$) respectively. As in the previous experiment, the Mach number was 2.1 and the total pressure 7.5 bar. The result is shown in Fig. 5.2.18.

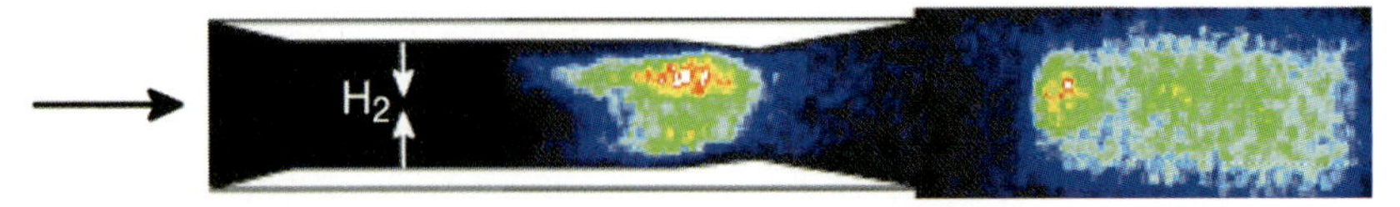


Figure 5.2.18: Injector flow with reaction, $Ma=2.1$, $p_0=7.5$ bar, $T_0=900$ K, $\Phi=0.35$.

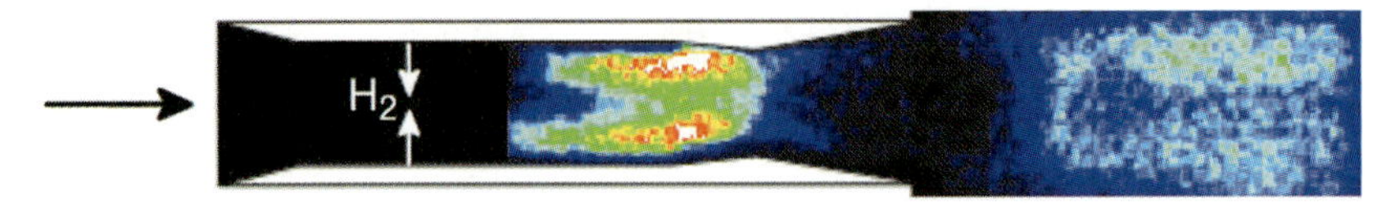


Figure 5.2.19: Injector flow with reaction, $Ma=2.1$, $p_0=7.5$ bar, $T_0=900$ K, $\Phi=1.06$.

The ignition of the hydrogen occurs with a spatial delay of about 11 mm downstream of the injection orifices and two diffusive flames are visible in the precombustor. The reaction rate is limited by the moderate mixing of the fuel with the hot air. The heat release accelerates the subsonic flow and critical flow conditions are achieved in the throat of the injector. Unburned fuel is still present in the flow, which is accelerated in the nozzle to a supersonic Mach number. The static temperature decreases and consequently the reaction is quenched. No fluorescence signal is detectable in this region. Due to the overexpanded character of the injector flow two oblique shocks at the trailing edge of the nozzle are generated (Figs. 5.2.13–5.2.17), which have a major influence on the heat release downstream of the injector. At the intersection of both oblique shocks 8 mm downstream of the trailing edge of the injector the temperature rises and re-ignites the mixture. Since the heat release zone is of a homogeneous character, it can be concluded that transverse mixing in the injector and the nozzle is intense enough to produce a uniform bulk flow of reactants, products and combustion intermediates, which are emitted from the injector exit and afterburn without major influence of the main flows.

In the second test series the mass flow of injected hydrogen was increased to stoichiometric conditions in the injector precombustor (3.8 g/s fuel); the other experimental conditions were identical to the previous investigation. In Fig. 5.2.19 the self-fluorescence of the flow is shown for the staged combustion mode under stoichiometric conditions in the injector flow. The ignition occurs 8 mm downstream of the injection orifices and the reaction is more intense than in the case with the smaller equivalence ratio. Again, the heat release causes an acceleration of the ignited mixture up to critical condition in the throat. The reaction is quenched in the nozzle and re-ignition occurs after the shock wave system, which is formed by the pressure equalization. The re-ignition occurs 8 mm downstream of the trailing edge of the injector at the same position as in the case with a smaller equivalence ratio. Whereas the reaction after re-ignition had a homogeneous appearance in the former case, now two streaks of high reaction rate are clearly visible. Numerical calculations reveal that the structure of the fuel jets has not yet vanished at the exit of the injector. This means that the reaction in the injector flow is limited to the shear layer between hydrogen and preheated air. In the stoichiometric case the larger amount of fuel causes an elongated time for mixing and additionally a decrease of the temperature. Consequently cold fuel rich areas exist at the exit of the injector and the shear layer between the injector flow and the ambient air influences the reaction by improving the mixture and decreasing the flow velocity, which leads to an increase of the static temperature. In the fuel lean case the mixture of fuel and air is more advanced and the effect of the shear layer is not observable.

*Air Split*

To date, version 1 of the injector was tested at two different global equivalence ratios of 0.35 and 1.06 respectively. Both, the high temperature of the injector flow and the temperature gradient between the relatively cool main and the hot injector flow, generates a hostile environment for the injector. For a safe and reliable operation of the injector a lean mixture with a lower reaction temperature in the injector flow is advantageous. Furthermore, an equivalence ratio profile in the main flow with a maximum near the injector walls cannot be avoided, as the hydrogen jets will always have a limited penetration. As a consequence, a rich inner region near the injector walls will develop. In the case of a lean global mixture in the injector flow, oxygen is still present downstream of the nozzle, which promotes the reaction in the fuel rich layer of the main flow.

*Expected Performance of Injectors with Shock Stabilizing Orifices*

At the present stage of the investigations data from combustion experiments is not yet available for version 2. The numerical and experimental study of injector 2 is subject to future activities. It is expected that the ignition behaviour of the injector flow doesn't differ significantly from that of version 1, as the self-ignition temperature of hydrogen is achieved by the deceleration of the flow to subsonic Mach numbers. However, the shock stabilization at the position of the orifices avoids the choking of the entire flow and the high losses of total pressure. Furthermore, the flow, which enters the main flow through the stabilization orifices, intensifies the mixing of the hydrogen and the air in the main flow. For these reasons, an improvement of the combustion performance is expected in comparison to version 1.

### 5.2.6 Conclusions

The chemical kinetics of the hydrogen-air reaction were analyzed and two redundant species were found. A lookup-table with a special structure was developed to tabulate the reaction in a seven-dimensional composition space. The efficiency of the look-up table was enhanced by using multilinear polynomials instead of the interpolation.

- The validation of the modelling approach using polynomials showed a good agreement of the results simulated with full chemical kinetics and polynomials.

- The efficient method of reducing of the simulation time with the help of polynomials can be applied to the simulation using PDF-approaches and thus enormous requirements of computational resources for such simulations can be avoided.

The investigations of the pylon and the strut injector in previous investigations revealed that reliable flame stabilization leads to intolerable losses of total pressure and that low-loss flame stabilization downstream of slender injectors causes an incomplete burnout. Both inherent properties reduce the efficiency of the propulsion system. For these reasons a new injector was developed and tested in the present work, which overcomes these deficiencies:

- Decelerating a small fraction of the overall airflow in a SCRamjet combustor to subsonic conditions provides the basis for a stable ignition zone in SCRamjet combustors. This is achieved in a novel injector type with a shaped inner surface, which produces a normal shock with a strong temperature rise and subsequent self-ignition of the fuel-air mixture.

- The stabilization of a normal shock in a supersonic diffuser is a complex problem. In the present work the normal shock is stabilized at orifices between injector flow and main flow. These orifices improve the fuel-air mixing in the main flow and favour the ignition of the mixture downstream of the injector.

- The reaction of the hydrogen-air mixture in the injector flow is mixing-controlled and not yet terminated at the throat. The reaction is quenched in the nozzle, where the flow is overexpanded to a supersonic Mach number.

- The shock wave system induced by the overexpanded flow at the trailing edge of the injector causes the unburned fuel to ignite and to stabilize the reaction downstream of the injector.

- In the main flows the static temperature is too low for the self-ignition of the fuel. The flame stabilization downstream of the injector is achieved by reaction intermediates stemming from the injector flow, which accelerate the ignition of the hydrogen-air mixture in the main flows dramatically and produce strong lateral flame propagation.

- The reaction in the injector flow can be stabilized already at low temperatures and leads to a reliable ignition of the hydrogen-air mixture of the main flow. Thus, the injector can cover a broad operation range concerning the flight Mach number.

## References

1. Grünig, C.: Gemischbildung und Flammenstabilisierung bei Pyloneinblasung in Überschallbrennkammern, Dissertation, Technische Universität München, 1999.
2. Eckbreth, A.C.: Laser Diagnostics for Combustion Temperature and Species, Combustion Science and Technology Book Series, Volume 3, chapter 5, 1996.
3. Fluent Inc.: Fluent 5.5, Documentation, 2000.
4. Maas, U., Warnatz, J.: Detailed Numerical Simulation of H2-O2-Ignition in Two-Dimensional Geometries, Internal Report, Preprint 90-01, Universität Heidelberg.
5. Gerlinger, P.: Investigation of an Assumed PDF Approach for Finite-Rate Chemistry, Proceedings of the 40th AIAA Aerospace Sciences Meeting and Exhibit, 2002, Paper AIAA 2002-0166.
6. Pilling, M.J.: Comprehensive Chemical Kinetics. Low-Temperature Combustion and Autoignition, Vol. 35, Elsevier, 1997.
7. Lyubar, A., Sander, T., Sattelmayer, T.: Selbstzündung, Flammenstabilisierung und Stoß-Flammen-Wechselwirkung in stoßinduzierten Überschallflammen, Arbeits- und Ergebnisbericht, Sonderforschungsbereich 255, Stuttgart, 2001.
8. Evans, J.S., Schexnayder, C.J., Beach, Jr. H.L.: Application of a Two-Dimensional Parabolic Computer Program to Prediction of Turbulent Reacting Flows, NASA TP 1169, 1978.
9. Lyubar, A., Sattelmayer, T.: Numerical Investigation of Fuel Mixing, Ignition and Flame Stabilization by a Strut Injector in a SCRamjet Combustor, Proceedings of the 11th International Conference on Methods of Aerophysical Research (ICMAR), Volume 2, pp. 122–127, 3.–7. July 2002, Akademgorodok, Novosibirsk – Denisova Cave, Altai region, Russia.
10. Sander, T., Sattelmayer, T.: Application of Spontaneous Raman Scattering to the Flowfield of a SCRamjet Combustor, Proceedings of the 11th International Conference on Methods of Aerophysical Research (ICMAR), Volume 2, pp. 143–147, 3.–7. July 2002, Akademgorodok, Novosibirsk – Denisova Cave, Altai region, Russia.
11. Lyubar, A., Sander, T., Sattelmayer, T.: Selbstzündung, Flammenstabilisierung und Stoß-Flammen-Wechselwirkung in stoß-induzierten Überschallflammen, Arbeits- und Ergebnisbericht, Sonderforschungsbereich 255, 2001.
12. Li, J., Yu, G., Zhang, Y., Li, Y., Qian, D.: Experimental Studies on Self-Ignition of Hydrogen/Air Supersonic Combustion, Journal of Propulsion and Power, Vol. 13, No. 4, 1997, pp. 538–542, American Institute of Aeronautics and Astronautics.
13. Ben-Yakar, A., Hanson, R.K.: Cavity Flameholders for Ignition and Flame Stabilization in Scramjets: Review and Experimental Study, AIAA/ASME/SAE/ASEE Joint Propulsion Conference & Exhibit, 34th, Cleveland, OH, July 13–15, AIAA Paper 98-3122, 1998.
14. Mitani, T., Hiraiwa, T., Sato, S., Tomioka, S., Kanda, T., Tani, K.: Comparison of SCRamjet Engine Performance in Mach 6 Vitiated and Storage-Heated Air, Journal of Propulsion and Power, Vol. 13, No. 5, 1997, pp. 635–642, American Institute of Aeronautics and Astronautics.

# 5.3 Hypersonic Propulsion Systems: Design, Dual-Mode Combustion and Systems Off-Design Simulation

## 5.3.1 Combustion Stability of a Dual-Mode Scramjet – Configuration with Strut Injector

Sara Rocci-Denis *, Armin Brandstetter, Dieter Rist, and Hans-Peter Kau

### Nomenclature

| | |
|---|---|
| $M_{CC}$ | Combustor entrance Mach number |
| $M_F$ | Flight Mach number |
| $p_0$ | Total pressure upstream the Laval nozzle |
| $p_{CC}$ | Combustor entrance static pressure |
| $p_{tH2\boldsymbol{A}}$ | Total pressure of hydrogen injected through system ***A*** |
| $p_W$ | Combustor wall static pressure |
| $T_{SS}$ | Flame-holder surface temperature |
| $T_{tCC}$ | Combustor entrance total temperature |
| $T_{tH2}$ | Hydrogen injection total temperature |
| $\phi$ | Equivalence ratio |

#### 5.3.1.1 Introduction

One of the challenges in the development of airbreathing hypersonic transport systems is the efficient combustion in supersonic flows. Although the high flight Mach number of the vehicle drastically reduces before the flow enters the propulsion system, the air stream still flows at high speeds into the combustor. If the flight is taking place at hypersonic or high supersonic conditions, the combustor will operate in a supersonic field (SCRamjet=Supersonic Combustion Ramjet). The flow deceleration is accompanied by a significant increase in temperature which depends on the speed: the stronger the deceleration, the higher will be the temperature in the propulsion unit. The supersonic flow conditions lead to a very short residence time of fuel and air inside the combustor. Within that short time fuel has to be injected, to mix with the supersonic flow and to burn completely. Its ignition and reaction times have, therefore, to be short. Hydrogen is one of the best fuels to fulfil these requirements.

* Technische Universität München, Lehrstuhl für Flugantriebe, Boltzmannstr. 15, 85748 Garching; rocci@lfa.mw.tum.de

Hypersonic transport systems have also to be suitable for operating at relatively low Mach numbers: the combustor could eventually operate in a subsonic field (Ramjet). At these conditions the temperature increase due to the flow deceleration will be lower, the relative speed of sound will reduce and residence time and mixing rate will improve. Unfortunately, ignition and reaction times increase exponentially with temperature decreases, counteracting completely the effect of lower speed [1]. A temperature reduction from 1300 K to 1000 K goes with a ten-time rise of the ignition time. Below 1000 K it is required to introduce a slower flow region to increase the recovery temperature and the residence time together with it. Suitable solutions are recirculation zones behind bluff bodies, wakes of injectors, wall caverns and downward-facing steps; all these alternatives have already been investigated and the pertaining results have been published [2–4]. For very low flight velocities also the corresponding stagnation temperature is below the hydrogen self-ignition temperature and an external ignition source has to be installed [5].

A fundamental role in the development of a combustor for the aforementioned vehicle is also played by the overall shock structure that varies with flight parameters, i.e. Mach number, temperature and pressure. These shocks heavily influence the flow conditions in the combustion chamber as well as the fuel injection, primarily because of their variable interaction with the shear and boundary layers [6–8].

After the general functionality of combustion in supersonic flows has been demonstrated in earlier projects [9, 10], the research at the Institute of Flight Propulsion of the Technische Universität München has been focused mainly on the feasibility of stable dual-mode combustion over a broad operation range, at stagnation temperatures lower than the self ignition temperature of hydrogen in air. The considered operation field is indicated in Fig. 5.3.1.1 by the thick, dashed line along the trajectory of the Hypersonic Transport System Munich (HTSM).

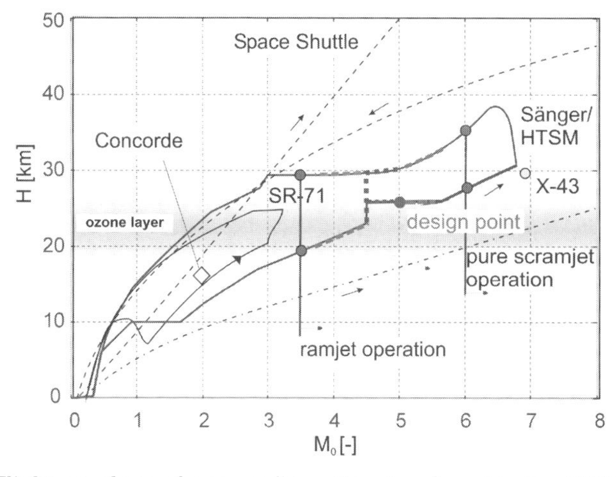


Figure 5.3.1.1: Flight envelope of one configuration developed in the HTSM project.

Adequate experiments have been planned and completed, starting in 1992 with the examination of a diverging combustion chamber and ramp injection with particular attention on the ignition at relatively low temperatures (about 1200 K) and on the combustion process at design conditions [9–11]. 1999 has seen an important turning point in the research with the introduction of a newly devised injection configuration that stabilized a pilot flame and thus the supersonic combustion. The experiments have been carried out up to a Mach number of 2.1 and at stagnation temperatures up to 700 K under free stream conditions and up to 1200 K inside the combustion chamber. Once the configuration has been optimized to operate at the design point, its functionality has been extended at off-design conditions, first remaining in the supersonic combustion field and later covering also subsonic conditions. Dual-mode operation has been performed in the ground experiments and the transition process has been controlled and described in terms of fuel injection pressure. Methane and ethylene have been positively tested as alternatives to hydrogen as fuel: the different combustion and heat release processes have been investigated and compared. Ceramic materials have been implemented to substitute steel elements of the system. Computational Fluid Dynamic (CFD) simulations have helped studying the shock structure inside the combustor and new geometries have been designed to reduce the shocks' interaction with boundary and mixing layers.

#### 5.3.1.2 Experimental Setup

*Test Facility*

The experiments have been performed in a direct connected test facility, schematically depicted in Fig. 5.3.1.2.

The supply pressure was 1 MPa. The air has been first electrically preheated up to 700 K. Higher stagnation temperatures, up to 1200 K, could be attained by means of a catalytic air heater implemented along the piping. If the temperature was kept under 700 K, long duration experiments could be performed: the only restriction was the exhaustion of the hydrogen supply, which lasted 30 to 50 minutes depending on the operative fuel injection pressure. At higher temperatures the test duration shortened because the combustion chamber walls have not been cooled: at values around 1200 K, the typical testing time was 15 s. A maximum flow rate of 0.5 kg/s stoked the combustor. The latter has been connected to the exit of an interchangeable Laval nozzle accelerating the flow up to supersonic conditions.

To properly simulate the transition between ramjet and scramjet modes, the Mach number at the combustor entrance ($M_{CC}$) has been varied during the experiments between 1.7 and 2.1, which was also the highest value attainable.

In free stream measurements, linear guide vanes allowed axial displacement of the injection group further downstream the nozzle and the Mach num-

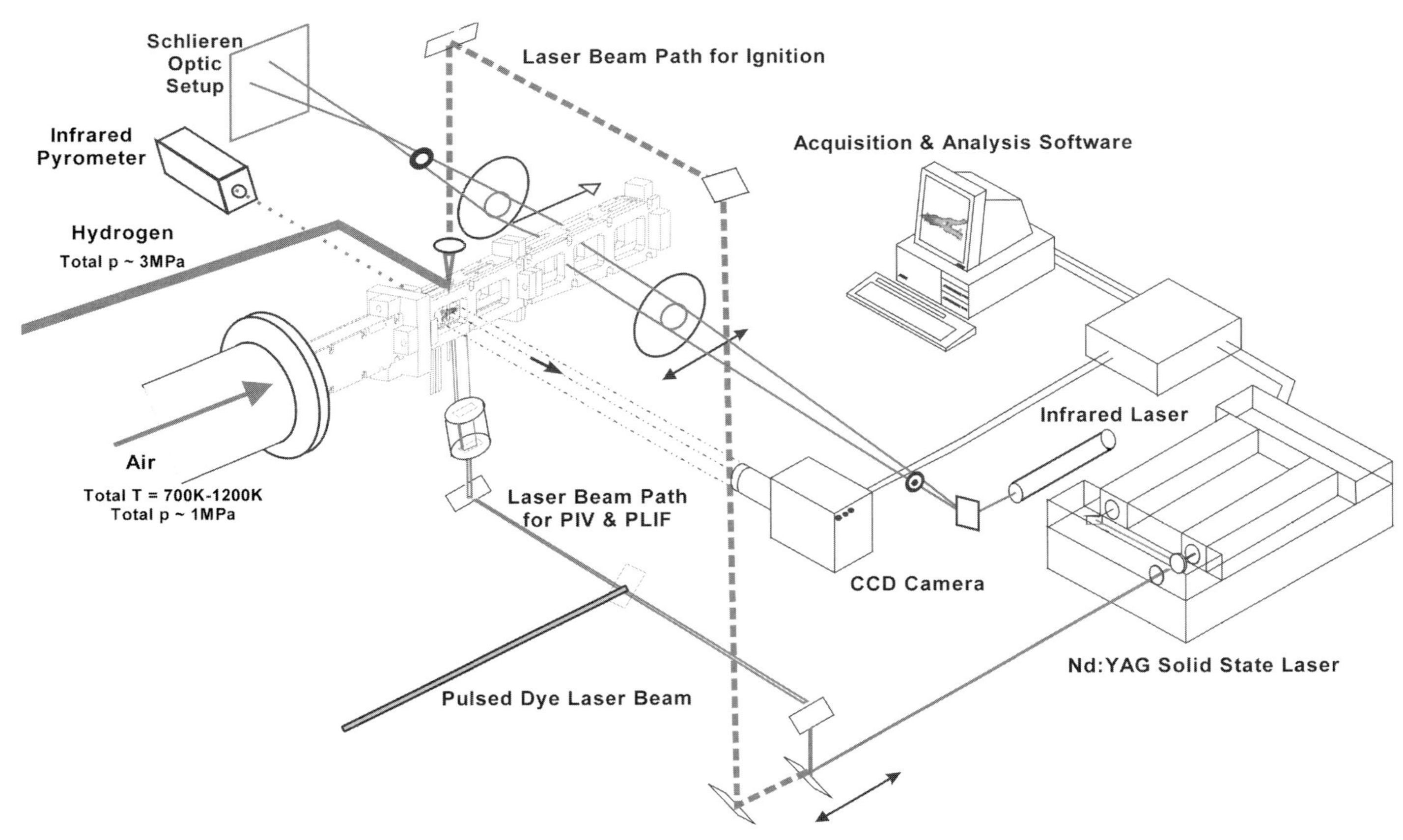


Figure 5.3.1.2: Schematic drawing of the test facility at the Institute of Flight Propulsion.

ber in the test section could be reduced to 1.5. Optical measurements equipment has been as well integrated to the linear guide vanes and could follow the group in its movement.

Gaseous hydrogen has been injected at 290 K with a maximum pressure of 3 MPa.

*Combustion Chambers*

Two different combustion chambers have been used: early experiments have been done with injection through a ramp, as shown in Fig. 5.3.1.3, while in the following experiments since 1999 a different chamber with strut injection has been adopted, as shown in Fig. 5.3.1.4.

First experiments have been carried out at the assumed design point (indicated in Fig. 5.3.1.1). The injection system consisted of a swept sidewall ramp, generating vortices and causing significant stagnation pressure losses. These total pressure losses and complicacies in flame stabilization have been a major downfall of the ramp injection design.

Main changes in the combustor geometry targeted the injection system: the ramp has been substituted with a strut coupled with a flame-holder. Addi-

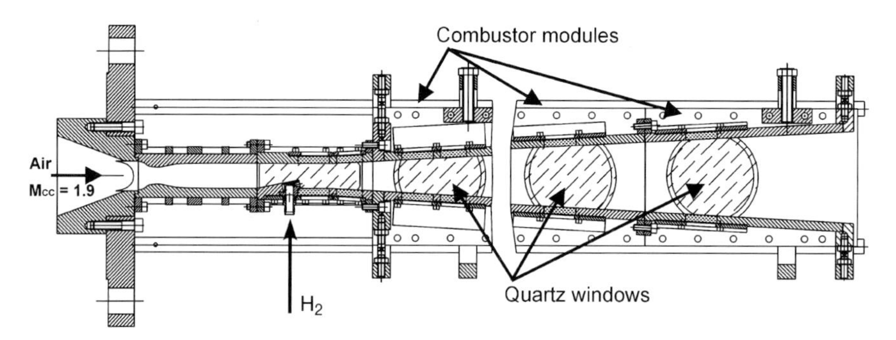


Figure 5.3.1.3: Side view of the modular combustion chamber with ramp injection.

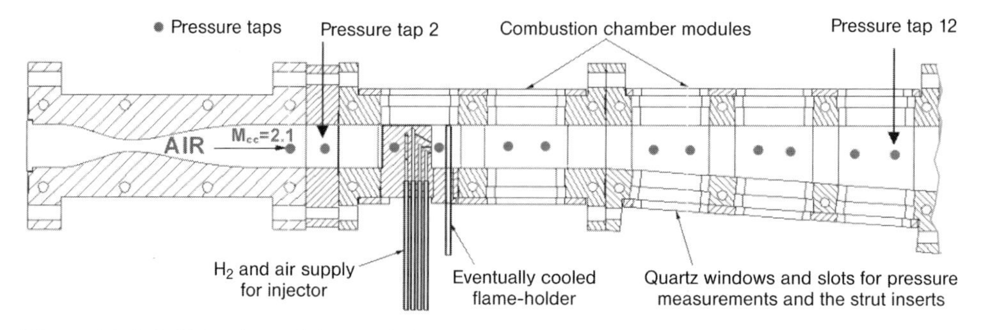


Figure 5.3.1.4: Side view of the modular combustion chamber with strut injection.

tionally, while the previous combustion chamber diverged almost along its whole length, the new combustor featured a longer constant-area section.

The subscale combustion chamber consisted of two modular sections: the first, hosting the injection group, had a constant cross section of 25×25 mm and length of 160 mm. The second module, 200 mm long, diverged with a 4 degrees angle to counteract the static pressure peak induced by combustion. All along the combustor length static pressure taps have been installed to get the wall static pressure distribution.

System flexibility has been achieved not only through the modular arrangement, but also due to uniformly distributed slots, notably five in each wall. The injection group could be consequently positioned either directly at the Laval nozzle exit or further downstream, and allowed for considering two different configurations. Additionally quartz windows could be positioned into the slots and guaranteed optical accessibility for non-intrusive laser measurements.

*Strut Injection System*

Injection has been performed through a strut coupled with a cylindrical pipe acting as a flame-holder. Views of the group are shown in Fig. 5.3.1.5.

Different versions of the basic configuration have been used to get more detailed information on the involved mechanisms. All models featured a 7.5 degrees apex half-angle, a 30 mm length and a 22 mm width. The versions under investigation differed in their thicknesses, respectively 3 mm and 5 mm. The flame-holder was initially a cylindrical steel pipe of 3 mm diameter positioned in the strut wake. In 2003 composite materials (Si and SiC) substituted steel. The intervening distance between strut and pipe has been optimized in early free stream measurements [12]. Nevertheless one injector still allowed for varying the flame-holder axial position so that different recirculation regions built up in the wake, to diversify the conditions for combustion.

Hydrogen has been fed at sonic speed to the combustor in two ways, exploiting the three holes **A**, **B** and **C**. Fig. 5.3.1.5 illustrates the concept: through system **A** hydrogen has been injected perpendicular to the supersonic flow and pre-mixed with air.

Systems **B** and **C** constitute an unlike-triplet impingement injection system [13] with 30 degrees inclined holes, chosen to fasten the fuel-oxidizer mixing. Through hole **B** hydrogen has been supplied into the wake of the strut. Additional air injection through hole **C** allowed for adjusting the fuel to air ratio in the wake, where the mixture has been ignited by means of a pulsed Nd:YAG laser focused to produce an ionization spark. The pulse rate has been varied from 1 Hz to 8 Hz.

As clarified in Fig. 5.3.1.5, cooling of the steel flame-holder needed to be implemented to preserve the material and retard its failure. The relatively cold, impinging jets from **B** and **C** partially fulfilled this function, but their cooling effect could not be controlled and did not allow for regulating the pipe surface temperature. For this purpose internal cooling has been performed by sending

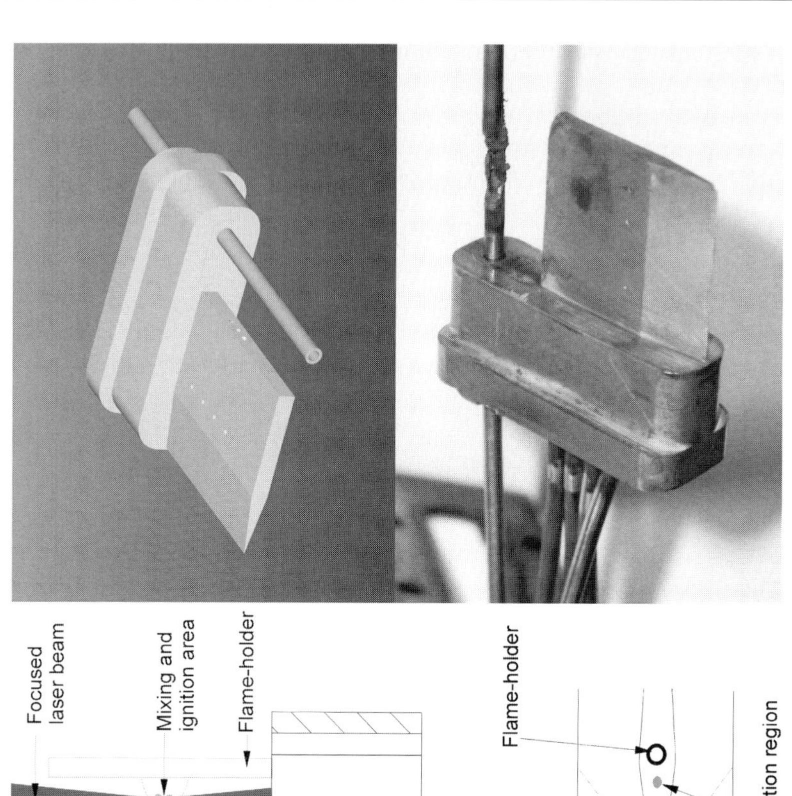


Figure 5.3.1.5: Injection system.

an air jet through the pipe. The air could either flow as a stand-alone jet or be pre-heated and then sent to the strut wake through **C**. The latter solution could be actuated as well with a hydrogen jet.

*Measurement Techniques*

Besides various digital image acquisition devices, thermocouples and pressure sensors, the test lab has been equipped with several non-intrusive measurement systems.

A Schlieren optic apparatus has been used to assess supersonic flow conditions both in free stream investigations and at the combustor exit in combustion experiments. A wedge has been inserted at significant flow positions and photographs have been acquired to get information on the shock structure, as well as a primary indication on the corresponding Mach number. The setup has been built of lenses and of an infrared laser as a light source.

The Schlieren images have been exploited also to calibrate the Particle Image Velocimetry (PIV) arrangement and to validate the pertaining results. PIV measurements have been carried out under free stream conditions to study the strut wake and the interactions with the surrounding flow [12, 14].

An infrared pyrometer calibrated for a range of 630 K to 2000 K provided information on the surface temperature of the flame-holder as well as on the combustor wall temperature.

The information obtained during the experiments required to be complemented with other data relative to the mixing and combustion processes. Therefore, a pulsed Dye laser and an excimer laser have been set up and adjusted to perform Planar Laser Induced Fluorescence (PLIF) measurements. The laser beam wave length could be tuned between 266 nm and 1064 nm. The intent was to detect the OH radicals' concentration and to investigate the planar distribution of reaction zones.

#### 5.3.1.3 Results and Discussion

The major aim of the research has been to succeed stable dual-mode combustion for a range of combustor inlet conditions below self-ignition conditions and representing different portions of the flight envelope of a hypersonic vehicle, mainly in its low speed range.

Since 1992 investigations have been carried out on the combustion chamber with swept ramp injection at a single operation point, represented from combustor entry conditions of $M_{CC}=1.9$, static temperature of 850 K and stagnation pressure of 0.96 MPa. At these conditions, mixture self-ignition would be possible in principle, but does not necessary occur [1, 5].

A positive influence on the mixture self-ignition can be exerted by flow zones having a reduced speed and thus a higher recovery temperature. In such a region, the combustion undergoes a higher mixing rate and contributes rising the local temperature; additionally the shocks impinging on the region itself can be exploited as well for the same purpose.

The idea of the ramp injection has been conceived on the basis of the aforementioned considerations. Behind the ramp, a significant recirculation zone is created, so that the recovery temperature should be high enough to enable self-ignition. Furthermore, the hot combustion products and chain-carriers generated inside this area should mix with the injected hydrogen and air and sustain the combustion. The ramp itself should initiate a shock system in which each shock causes an increase in static temperature and consequently supports self-ignition.

However, LIF measurements on the spontaneous emission of OH molecules showed that self-ignition only occurred in a small region immediately behind the ramp and close to the wall (see Fig. 5.3.1.6, left picture). The heat release has been minimal, likely due to the low temperature (290 K) of the injected hydrogen that cooled down the recirculation zone below the self-ignition capabilities. In the diverging section, immediately after the recirculation region, an expansion took place causing a temperature decrease which could not be counteracted by the temperature rise accompanying the oblique shocks that occurred further downstream.

Improvements have been made through an additional "shock" ramp, which has been installed on the top wall, just downstream the injection ramp (see Fig. 5.3.1.6, right picture). At the leading edge of the new ramp, an oblique shock propagated through the test section entraining several effects. A significant interaction between the wake of the ramp, the wall boundary layer

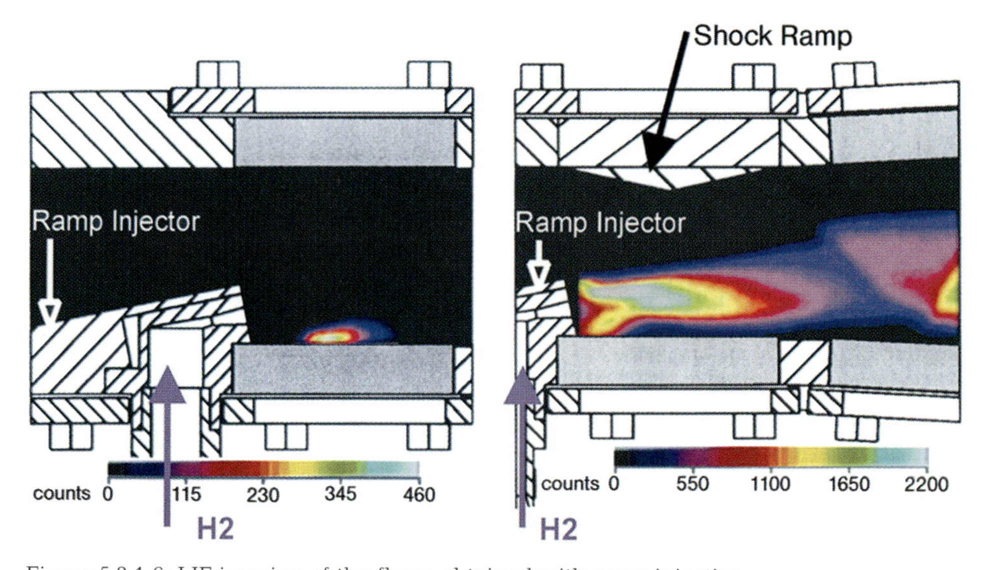


Figure 5.3.1.6: LIF imaging of the flame obtained with ramp injection.

and the shock has been recognized. The increased pressure difference between the area behind the shock and the wake influenced the recirculation zone and favoured the combustion of the mixture upstream the shock; additionally a small recirculation zone established at the base of the ramp. The high OH radicals' concentration indicated a high reaction rate and the flame covered completely the injected hydrogen flow. The overall effect has been significant, as observable in the Laser Induced Fluorescence (LIF) images in Fig. 5.3.1.6. Further downstream the flow has been accelerated by the expansion wave on the wedge and the combustion quenched. Burning of the mixture has been promoted again by the oblique shock originating at the wedge trailing edge.

The overall combustion stability in the ramp setup showed to be largely dependent on the location of the shock system, especially on its interaction with the wake and, as the configuration was based on self-ignition, on the temperature and pressure levels. Therefore, it has already been expected that the combustor would have worked only sufficiently close to the design parameters.

Based on the previous experience and on results published worldwide [2, 4, 15–21], a new configuration has been developed with injection through a small strut, which has been installed inside the flow. The strut has been conceived to generate a small wake in which a pilot flame could be stabilized. Inside this wake a small pipe has been mounted to work as flame-holding device. Geometrical details of this strut/flame-holder injector are described above.

This injection system has been suggested because of its peculiarities. Unlike wall or ramp injectors, it creates a low speed region in the centre of the supersonic flow. This entails a better mixing because of the extended interface between the flows and facilitates the entrainment of the low speed wake into the supersonic air stream which accelerates it. Other advantages are reduced thermal load on the combustor walls and minor pressure losses [3, 11, 12, 14, 22, 23].

Early free stream tests (Mach number of the flow hitting the strut varied between 1.5 and 1.7) showed the performances of the basic setup, i.e. a strut positioned directly behind the Laval nozzle and without flame-holder. As operation has been typically under the mixture self-ignition temperature, an external ignition source has been required. The pulsed Nd:YAG laser has been exploited to create an ignition spark. With the first setup combustion of the fuel did not continue after the laser was switched off. As soon as the laser induced spark has been interrupted, the reaction zone has been seen moving downstream with the main flow without leaving back any source for further ignition. It has been concluded that the wake behind the strut was unsuitable to recirculate the combustion products and that more detailed information about the flow was required.

By means of cold flow (i.e. without combustion) Particle Image Velocimetry (PIV) measurements the flow behind the strut could be visualized, see Fig. 5.3.1.7. The vectorial analysis indicates that the residence time of the particles inside the wake has been much higher than the reaction time. The results show as well that at those Mach numbers, the trailing edge of the strut did not generate a recirculation zone which could transport hot combustion products

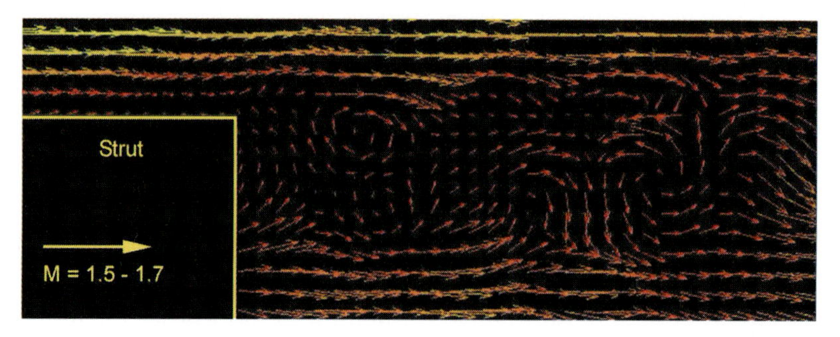


Figure 5.3.1.7: PIV visualization of the strut wake under free stream conditions.

upstream to initiate further chemical reaction. Instead, a well established street of vortices flowing down with the main stream has been identified. Further improvements have been done by positioning an additional pipe, acting as flame-holder, shortly downstream the strut. Its axial position could be varied for optimization purposes.

During free stream investigations not only the flame-holder placement, but also its surface temperature ($T_{SS}$) turned out to be a fundamental parameter. Flame stabilization has been augmented by the steel pipe, which absorbed energy released by ionization sparks and by following short reactions in the mixing region at the beginning of the ignition sequence. Once the pipe has been heated up to a temperature level suitable for mixture self-ignition in the wake, thermal energy released and promoted the creation of combustion radicals, so that the pilot flame could burn without external ignition source. The flame kept the pipe surface at the required temperature and therefore the system has been self-stabilizing.

As mentioned before, the flame-holder had to be cooled. Control of the cooling allowed regulating $T_{SS}$ and consequently the time required for the combustion to start. In order to reduce the mixture ignition time down to 6 s for example, the air jet impinging the wake through **C** (see Fig. 5.3.1.5) has not been activated from the beginning on.

Figure 5.3.1.8 presents the steel flame-holder surface temperature plot for a typical free stream experiment ($M_{CC}$=2.1, $p_{CC}$=1 MPa, $T_{tCC}$=500 K–700 K) with use of the steel pipe. After the ignition sequence has been performed, the pilot flame burned in a stable manner while hydrogen has been injected only through **B**, see Fig. 5.3.1.5.

The flame-holder started being cooled: its surface temperature lowered, but the flame did not blow off. Cooling has been then intensified by switching on the air injection into the wake, also in order to improve the mixing behind the strut; as expected, $T_{SS}$ lowered, but the flame persisted. The next step has been injection of additional hydrogen through the strut sides (**A** in Fig. 5.3.1.5): the sudden increase of the flame-holder temperature $T_{SS}$ proved that the main mixture has been ignited and burned. Finally, the right part of the diagram is indicative of the flame behaviour consequent to an axial displacement of the injection group

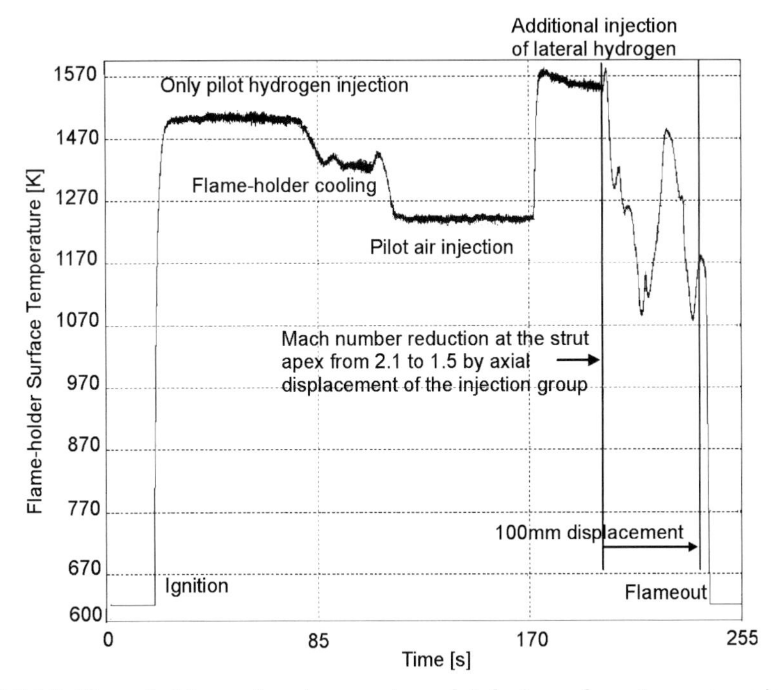


Figure 5.3.1.8: Flame-holder surface temperature plot during a free stream experiment.

further downstream the Laval nozzle. By this means, the shock structure of the flow hitting the strut has been varied and simulated a different Mach number condition. Although $T_{SS}$ featured wide oscillations, the flame continued burning and has been blown off only if the axial movement exceeded 100 mm, corresponding to a Mach number of 1.5. The data acquired with the infrared pyrometer demonstrate pilot flame stability under variable flow conditions.

As next step, the injection configuration has been mounted inside the combustion chamber and wall static pressure data have been obtained for different test modalities, as reported in Fig. 5.3.1.9, where $p_W$ is non-dimensioned with the total pressure upstream the Laval nozzle. The entry flow conditions have been kept constant at $M_{CC}$=2.1, $p_{CC}$=1 MPa and $T_{tCC}$=700 K.

The first measurements have been performed with cold flow, to clarify the way different fuel injection modes influence the flow. The dark-blue line describes the general pressure tendency: $p_W$ peaks due to the bow shock on the strut leading edge and then decreases at the trailing edge consequently to the Prandtl-Mayer expansion fan. The next pressure rise in the constant cross area section of the combustor is induced by reflected shocks [10]. Finally, the flow expands in the diverging section and the wall static pressure reduces.

A quite drastic effect resulted from the beginning of hydrogen injection into the wake: the jet at $T_{tH2}$=290 K cooled down the supersonic flow and a completely new shock configuration appeared and affected the pressure level

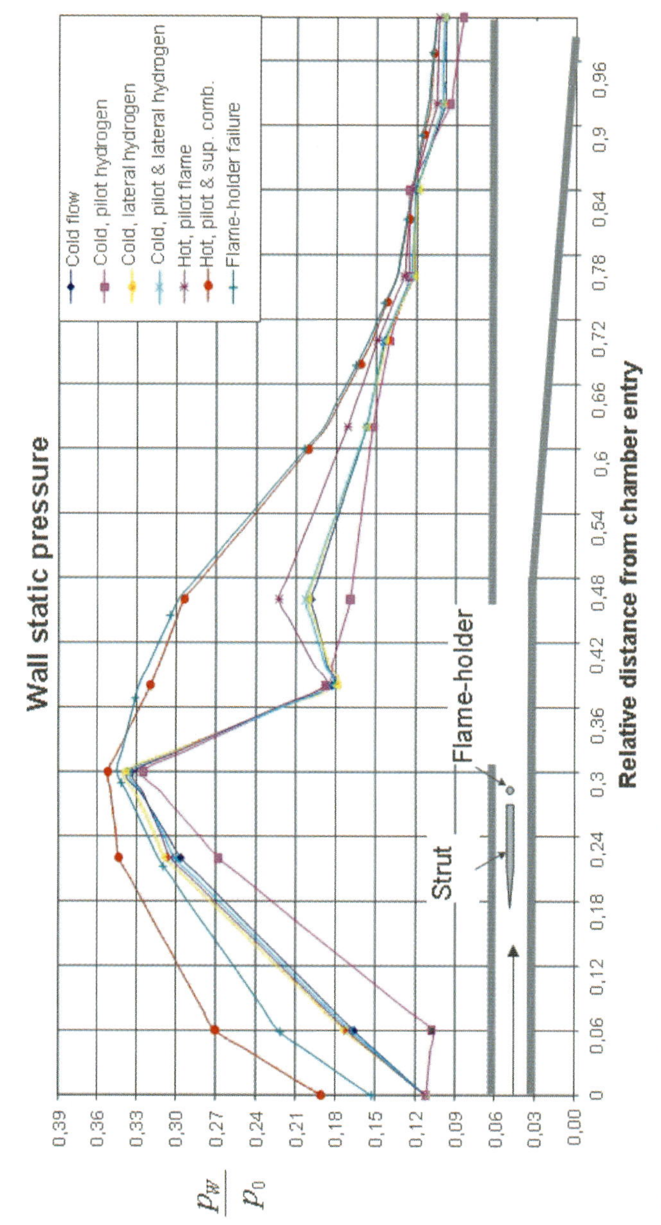


Figure 5.3.1.9: Wall static pressure distribution.

upstream the injector. The pink curve details the lower $p_W$ level, as well as the absence of pressure recovery downstream the injector in the constant cross section, due to the different shock reflection. Once these flow conditions had settled, the previous distribution could not be restored unless the total pressure of the supersonic flow has been lowered and then reset to the initial conditions.

With hydrogen injection through the strut sides (**A** in Fig. 5.3.1.5) the illustrated effect did not show up and the relative curve (yellow) matches the baseline. Moreover, if hydrogen was simultaneously injected through **A** and **B**, the dominant effect of **A** counteracted the pressure drop resulting from the pilot hydrogen injection, see pale-blue curve.

The next experiments have been carried out with combustion. The dark-red curve indicates the overall wall static pressure increment consequent to effective burning of the mixture; the rise upstream the strut led to the conclusion that a homogeneous fuel distribution inside the supersonic flow has been achieved through penetration and pre-mixing of hydrogen injected through **A**.

The last curve has been obtained during a test in which the flame-holder glowed until failure. The data of sensors downstream the injector demonstrate persisting of combustion, although the upstream $p_W$ values decreased because of lower heat release from the steel pipe. After a short while the configuration spontaneously recovered to stable operation, but again the previous pressure level could not be restored.

This occurrence opened the way for a twofold series of investigations. On one side the flame-holder failure has been intentionally provoked to study the system reaction and to figure out possible improvements. On the other side, the aptitude of this configuration to work in dual-mode has been extended.

Early experiments have been performed at conditions demanding continuous flame-holder cooling and device replacement after some tests, but still at heat release levels too low to achieve an efficient combustion. The total injection pressure of the hydrogen supplied by **A** ($p_{tH2A}$) has been increased to 4 MPa to bring the equivalence ratio (fuel-to-air ratio to fuel-to-air stoichiometric ratio) into the suitable range (=0.2–0.4). On account of published work treating mixing in supersonic flows [23–25], keeping $\phi$ inside the mentioned range will optimize the mixing process. Due to the higher heat release level, combustion has been leading to flame-holder failure within short times, typically 15 to 40 minutes for the usual $p_{tH2A}$ values.

Tests ($M_{CC}$=2.1, $T_{tCC}$=700 K and $p_{CC}$=1 MPa) have been filmed in which the well-established combustion led to pipe failure. In this occurrence the main combustion temporarily ceased, while the pilot flame kept burning. Within a short time suitable flow conditions have been spontaneously reset, pre-mixing of hydrogen and supersonic air stream became effective again and the pilot flame re-ignited the surrounding mixture. Figure 5.3.1.10, extracted from the video, documents the three mentioned phases. The first picture on the left shows the flow burning while the flame-holder glows and sustains combustion. The top right picture captures the moment when the flame-holder breaks, recognizable from the burning metallic particles blown out of the combustor. In the lower left picture, the flame-holder's broken sides still glow, but only the pilot flame

Figure 5.3.1.12: Free stream combustion of injected methane.

lease profile improved the system capability to prevent eventual thermal choking of the combustor.

As cooling of the flame-holder has not been enough to guarantee its integrity during the tests, new materials have been selected to withstand the temperatures attained in the combustor during its operation (typically 1400 K). Cylindric bars of silicon and silicon carbide have been tested with the usual experiment sequence and proved to be suitable for stabilizing the pilot flame as well as the steel pipe. The new flame-holder additionally simplified the system configuration, because internal cooling has not been required and the corresponding setup has been eliminated.

#### 5.3.1.4 Conclusions

For the purpose of stabilizing combustion in supersonic flows different kinds of injection systems and combustor configurations have been designed and tested. First experiments assessed the possibility of igniting a hydrogen sonic jet in a supersonic air stream. Tests performed with a ramp injector allowed fuel ignition and combustion at the design point, but the mixture has not been homogeneously burned and the configuration did not operate properly at off-design. Investigations on a combustor configuration with a novel strut injection system confirmed stable combustion and detected supersonic flow at the chamber exit within a broad operation range. Two injector versions, whose thickness differs by 2 mm, have been developed. While the injection pressure of normally injected hydrogen was varied, transition from ramjet to scramjet mode has been observed. The higher aerodynamic blockage of the thicker injector led to ramjet operation at low $\phi$ ($\sim 0.2$). None of the injectors caused reaction quenching in the combustor. A steel pipe coupled with the strut acted as a flame-holder and proved to be the key element for flame stabilization, which then enabled the main combustion. During the operation the pipe had to withstand extremely high temperatures and encountered failure even though it is cooled. Further experiments have been carried out both to better explain the way the flame-holder fulfilled its function and to improve its performance. For this purpose ceramic materials have been tested and proved to be suitable to substitute steel.

## References to Section 5.3.1


1. Huber, P.W., Schexnayder, C.J., Jr., McClinton, C.R.: Criteria for Self-Ignition of Supersonic Hydrogen-Air Mixtures, NASA Technical Paper 1457, August 1979.
2. Curran, E.T., Murthy, S.N.B. (Editors): SCRAMJET PROPULSION, Progress in astronautics and aeronautics, Volume 189, AIAA 2000.
3. Takashi, N., Kenichi, T., Hideaki, K., Susumu, H.: Flame stabilization characteristics of strut divided into two parts in supersonic airflow, Journal of Propulsion and Power, Vol. 11, No, 1, 1995.
4. Champion, M., Deshaies, B.: IUTAM Symposium on Combustion in Supersonic Flows, Proceedings, Poitiers, France, 1995.
5. Colket, M.B., Spadaccini, L.J.: Scramjet fuels auto ignition study, Journal of propulsion and power, Vol. 17, No. 2, March-April 2001.
6. Kim, J.-H., Huh, H., Yoon, Y., Jeung, I.S.: Effects of Shock Waves on a Supersonic Hydrogen-Air Jet Flame in a Model Scramjet Combustor, The 2nd Asia-Pacific Conference on Combustion, Taiwan, 1999.
7. Adamson, T.C., Messiter, Jr. A.F.: Analysis of Two-Dimensional Interactions between Shock Waves and Boundary Layers, Ann. Rev. Fluid Mech., 12, 103–138, 1980.
8. Jacobs, J.W., Collins, B.D.: Experimental study of the Richtmyer-Meshkov instability of a diffuse interface, 22nd International Symposium in Shock Waves, Imperial College, London 1999.
9. Hönig, R., Theisen, D., Fink, R., Kappler, G.: Experimental Investigation of a SCRAMJET Model Combustor with Injection Through a Swept Ramp Using Laser-Induced Fluorescence with Tenable Excimer Lasers, 26$^{th}$ Symposium on Combustion, The Combustion Institute, Naples, Italy, 1996.
10. Hönig, R., Theisen, D., Fink, R., Kappler, G., Rist, D., Andresen, P.: Diagnostics of Non-Reacting supersonic Flows in a SCRAMJET Model Combustor Using Non-Intrusive Spectroscopic Methods, 4$^{th}$ International Symposium on Special Topics in Chemical Propulsion, Stockholm, 1996.
11. Kau, H.-P., Rist, D., Brandstetter, A.: Teilprojekt B12 Arbeits- und Ergebnisbericht, Sonderforschungsbereich 255, Institute of Flight Propulsion, Technische Universität München, Munich, Germany.
12. Rocci Denis, S.: Basics of hydrogen combustion in supersonic flows and experimental investigation for a specific scramjet configuration, Diploma Thesis, Institute of Flight Propulsion, Technische Universität München, Department of Aeronautic&Aerospace Engineering of Politecnico di Torino, 2001.
13. Soller, S., Kau, H.-P., Kretschmer, J., Martin, P., Mäding, C.: Untersuchungen zur Verbrennung von Kerosin in Raketentriebwerken, DGLR Jahrestagung, Stuttgart, 2002.
14. Brandstetter, A., Rocci Denis, S., Rist, D., Kau, H.-P.: Flame Stabilization in Supersonic Combustion, 11th AIAA/AAAF International Conference, Space Planes and Hypersonic Systems and Technologies, Orleans, AIAA-2002-5224.
15. Anderson, G., Kumar, A., Erdos, J.: Progress in Hypersonic Combustion Technology with Computation and Experiment, AIAA 1990.
16. Philip Drummond, J.: Enhancement of Mixing and Reaction in High-Speed Combustor Flowfields, International Colloquium on Advanced Computation and Analysis of Combustion, Moscow, Russia.

17. Dessornes, O., Jourdren, C.: Mixing enhancement techniques in a scramjet, AIAA 98-1517 Norfolk, USA, 1998.
18. Billig, F.S.: Research on Supersonic Combustion, Journal of Propulsion and Power, Vol. 9, No. 4, p. 499–514, July–August 1993.
19. Archer, D.R., Saalas, M.: Introduction to Aerospace Propulsion, Prentice Hall, Upper Saddle River, New Jersey, 1996.
20. Brandstetter, A.: Auslegung und Konstruktion einer Modellbrennkammer zur Untersuchung der Wasserstoffverbrennung in Überschallströmungen, DGLR Annual Conference, Berlin, 1999.
21. Luo, K.H.: Combustion Effects on Turbulence in a Partially Premixed Supersonic Diffusion Flame, Combustion and Flame, 119:417–435, The Combustion Institute, 1999.
22. Brandstetter, A., Rocci Denis, S., Rist, D., Kau, H.-P.: Experimental Investigation of Supersonic Combustion with Strut Injection, 11th AIAA/AAAF International Conference, Space Planes and Hypersonic Systems and Technologies, Orleans, AIAA-2002-5242.
23. Rocci Denis, S., Brandstetter, A., Kau, H.-P.: Transition between Ramjet and Scramjet Modes in a Combustor for an Air-Breathing Launcher, AIAC Paper 2003-054.
24. Guirguis, R.H.: Mixing enhancement in supersonic shear layers: III. Effect of convective Mach number, AIAA Paper 88-0701, 1988.
25. Tishkoff, J.M., Philip Drummond, J., Edwards, T., Nejad, A.S.: Future directions of supersonic combustion research, Air Force/NASA workshop on supersonic combustion, May 1996.
26. Dimotakis, P.E., Catrakis, H.J., Fourguette, D.C.: Flow structure and optical beam propagation in high-Reynolds-number gas-phase shear layers and jets, Journal of Fluid Mechanics, 433, pp. 105–134, 2001.

## 5.3.2 Hypersonic Highly Integrated Propulsion Systems – Design and Off-Design Simulation

Hans Rick *, Andreas Bauer, Thomas Esch, Sebastian Hollmeier, Hans-Peter Kau, Sven Kopp, and Andreas Kreiner

### Nomenclature

| | |
|---|---|
| $A$ | Area |
| $B$ | Nozzle width |
| $c$ | Velocity |
| $c_{fg}$ | Gross thrust coefficient |
| $c_{fgx,z}$ | Gross thrust coefficient in x,z-direction |
| $c_m$ | Pitching moment coefficient |
| $F$ | Force vector |
| $F_I$ | Inlet force |
| $F_{Add,Lip}$ | Drag additive, lip |
| $F_{Cowl}$ | Pressure force on cowling |
| $F_{Base}$ | Base drag |
| $F_{G,Net}$ | Gross, net thrust |
| $H$ | Altitude |
| $M$ | Mass flow |
| $Ma$ | Mach number |
| $p$ | Pressure |
| $q$ | Dynamic pressure |
| $T$ | Temperature |
| $\alpha$ | Angle of attack |
| $\delta$ | Inlet ramp |
| $\delta_{1,2}$ | Inlet first, second movable ramp 1,2 |
| $\delta_{Base}$ | Nozzle flap angle |
| $\Pi$ | Pressure ratio |
| $\sigma$ | Thrust vector angle |
| HFD | Horizontal fuselage datum |
| HTSM | Hypersonic Transport System Munich |
| LPC | Low pressure compressor |
| HPC | High pressure compressor |
| HPT | High pressure turbine |
| LPT | Low pressure turbine |
| NASP | National Aerospace Plane |
| SERN | Single expansion ramp nozzle |
| TSTO | Two Stage to Orbit |

#### 5.3.2.1 Introduction

Investigating future space transportation systems, horizontal take-off and landing spacecraft have been found to meet the main requirements by all components and simplify ground operations and flight preparations considerably [1–3]. Due to lower thrust requirement compared to conventional rockets or single-stage-to-orbit (SSTO) systems, a two-stage-to-orbit (TSTO) concept allows the use of combined variable cycle airbreathing turbo-ramjet engines with subsonic combustion or in the higher Mach number range with supersonic com-

* Technische Universität München, Lehrstuhl für Flugantriebe, Boltzmannstr. 15, 85748 Garching, rick@lfa.mw.tum.de

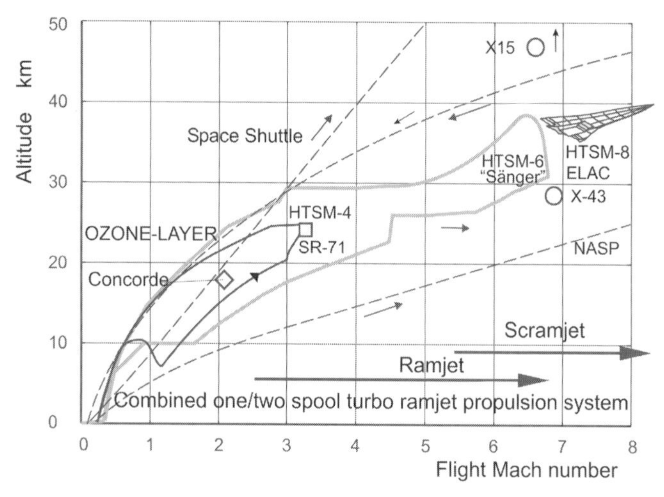


Figure 5.3.2.1: Reference ascent trajectory for HTSM-6 up to a separation Mach number of 6.8.

bustion like scramjet propulsion. Using parts of the airframe as parts of the propulsion system (e.g. precompression along the forebody of the aircraft, "Single Expansion Ramp Nozzle", SERN) can considerably reduce the propulsion system's size and weight. This leads to a strong coupling between the airframe and the engine making it impossible to consider both separately.

Within the Collaborative Research Centre (SFB) 255 this subject is studied by the Technische Universität München in cooperation with Daimler Benz Aerospace (DASA resp. EADS) and MTU München on the reference transport system **H**ypersonic **T**ransport **S**ystem **M**unich (**HTSM**). Figures 5.3.2.1 and 5.3.2.2 show the reference TSTO concepts [3]. The lower stage carries the

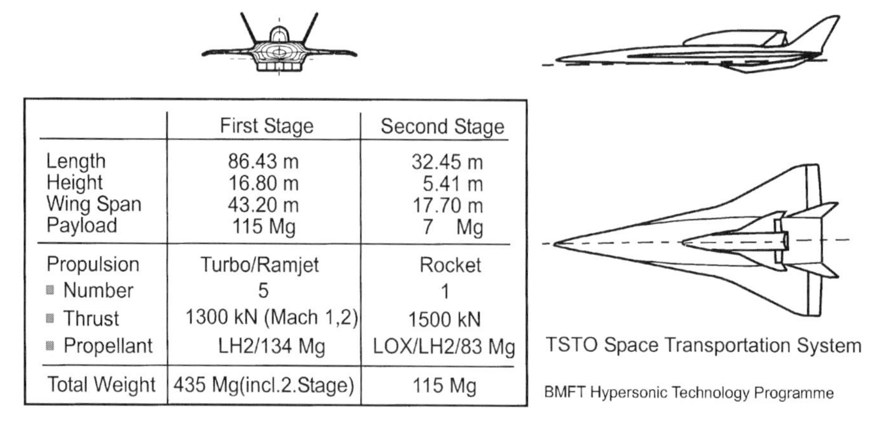

| | First Stage | Second Stage |
|---|---|---|
| Length | 86.43 m | 32.45 m |
| Height | 16.80 m | 5.41 m |
| Wing Span | 43.20 m | 17.70 m |
| Payload | 115 Mg | 7 Mg |
| Propulsion | Turbo/Ramjet | Rocket |
| ▪ Number | 5 | 1 |
| ▪ Thrust | 1300 kN (Mach 1,2) | 1500 kN |
| ▪ Propellant | LH2/134 Mg | LOX/LH2/83 Mg |
| Total Weight | 435 Mg(incl.2.Stage) | 115 Mg |



Figure 5.3.2.2 a: Reference TSTO transport system **H**ypersonic **T**ransport **S**ystem **M**unich (**HTSM**) with main characteristic data.

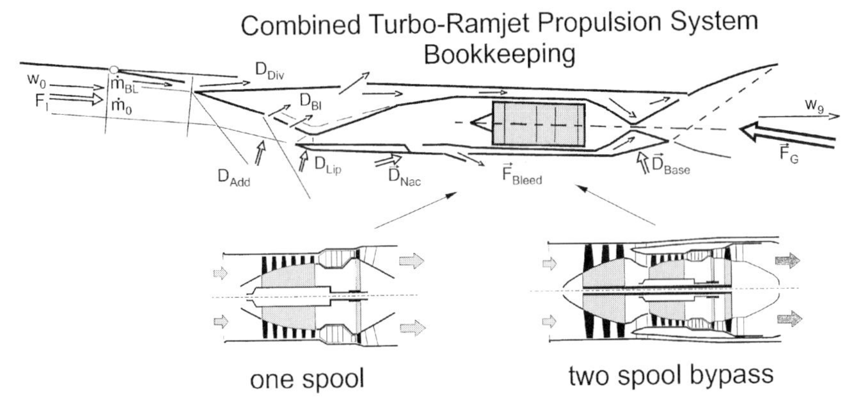


(a) One or two spool combined turbo/ramjet engine, coaxial position, flight Mach number range $Ma = 0\text{–}6+$, vehicle HTSM4, HTSM6

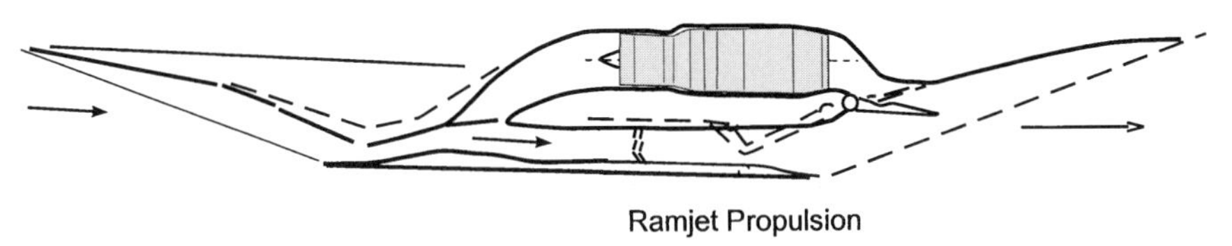


(b) Combined turbo/ramjet engine, parallel position, Mach number range $Ma = 0\text{–}6+$, vehicle HTSM4, HTSM6

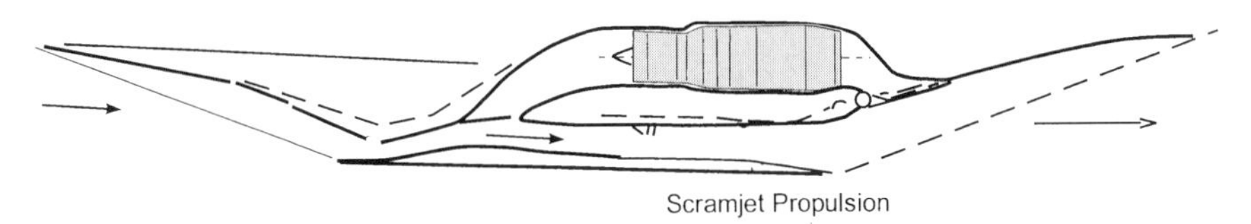


(c) Combined parallel turbo-ramjet resp. scramjet propulsion system for flight Mach number $Ma = 8$ for HTSM 8

Figure 5.3.2.2 b: Reference Propulsion systems for HTSM transport vehicle for different Mach number ranges.

upper stage up to a separation Mach number of about 6.8 in an altitude of 30–40 km (Fig. 5.3.2.2).

The objective of the present work is the prediction of the fully installed engine performance, with special emphasis on the propulsion system integration [4–6]. Figure 5.3.2.3 outlines the major aspects considered such as precompression, air inlet, nozzle/aftbody integration, control, and monitoring. Designing and controlling the single expansion ramp nozzle, the gross- and net-thrust vector angle especially in the transonic flight range had to be considered. The

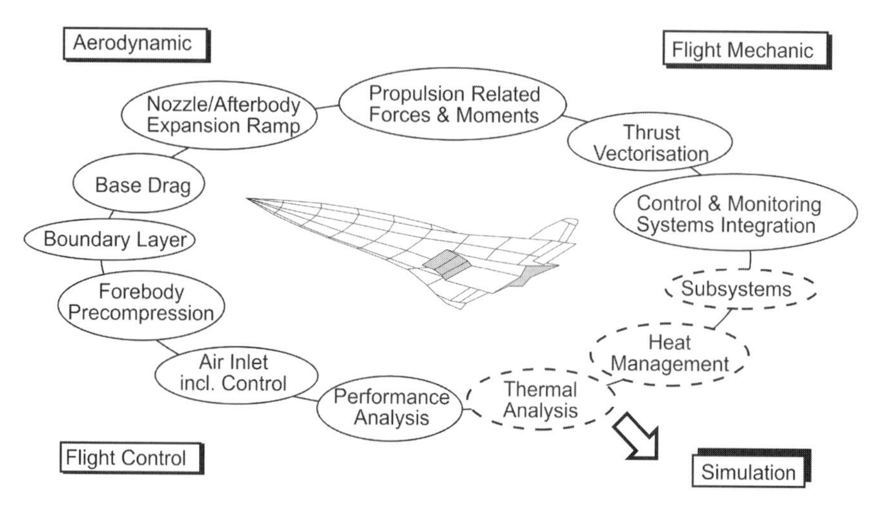


Figure 5.3.2.3: Integration and simulation model of a combined turbo-/Ramjet propulsion system.

present work does not include heat management, thermal analysis and secondary systems, however, these will have to be considered in the future.

### 5.3.2.2 Reference Propulsion System for the TSTO Concept

The reference propulsion system of the considered generic TSTO-concept is composed of a two-dimensional inlet with mixed external and internal compression, a diffuser with a cross-section change from rectangular to circular, a two-spool-low-bypass turbofan engine, a closure mechanism to seal off the turbo engine section during ramjet operation mode and shield it from extensive heat loads in the upper flight Mach number range, a reheat/ram combustor and a two-dimensional nozzle with an afterbody expansion ramp [3].

The quality of engine simulation depends on the accuracy of modelling the physical processes within the engine. Special emphasis was laid on the description of thrust nozzle and inlet characteristics at design and off-design conditions as well as failures. Figure 5.3.2.4 shows the engine modules as the basis for performance computation.

### 5.3.2.3 Engine Integration

To create a compatible formulation of all forces acting on airframe and propulsion system an overall bookkeeping system was defined. All effects caused by

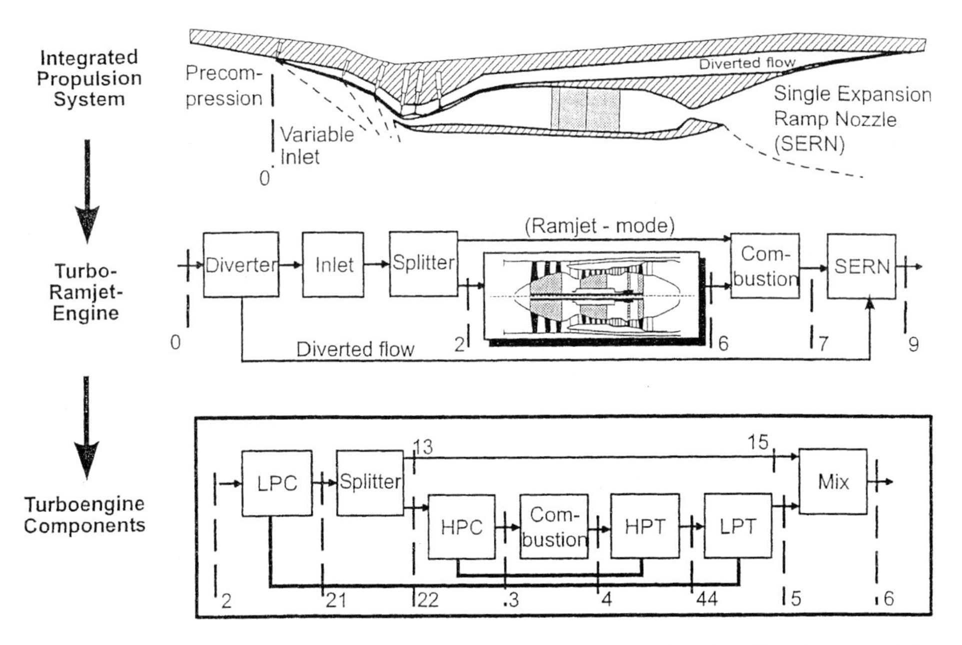


Figure 5.3.2.4: Integration and simulation model of a combined turbo-/Ramjet propulsion system.

the engines are summarized in the propulsion data set (with a resulting net thrust vector and a correlative momentum), while the forces acting on the defined airframe are contained in the aerodynamic data set [7].

*Inlet*

The two-dimensional ramp air inlet normally works with an external-internal five-shock-compression system. Figure 5.3.2.5 shows a calculated inlet flow field with contours of constant Mach number for an inlet entry Mach number of 5.0. The first inlet ramp is a shutter for a bypass duct diverting the aircraft forebody's boundary layer. The ramp cannot be fixed in an intermediate position, only the two modes "open" and "closed" are allowed. The air mass flow of the diverted boundary layer is added to the engine exhaust gas in the primary nozzle to influence nozzle performance.

The second and the third inlet ramps are movable. The angle $\delta$ denotes the geometric angle between the inlet ramp and HFD, a rather arbitrary horizontal, aircraft-fixed layer (cf. Figs. 5.3.2.1 and 5.3.2.5). As a result the angle of the oblique shock wave, the pressure loss, the inlet area ratio and finally the inlet mass flow ratio are changed as well.

During turbojet operation the matching of inlet air flow and turbo engine's demand is of main importance. The best possible inlet pressure ratio, an opti-

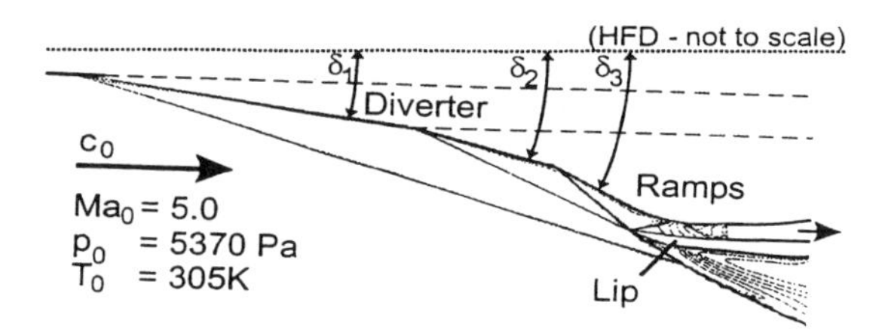


Figure 5.3.2.5: Regular inlet flow with lines of constant Mach number.

mized mass flow adaptation and a low level of compressor entry flow distortion is required. Therefore, the boundary layer diversion is active. Mass flow control and inlet-engine matching is obtained by controlling the movable inlet ramps and by an additional bypass diverting the surplus mass flow behind the inlet throat.

At ram-mode with increasing flight Mach numbers the mass flow determines inlet operation. The boundary layer diverter is closed, maximum mass flow is design issue.

In the bookkeeping system the inlet excludes the precompression along the forebody, which is part of the aerodynamic data set. A change of angle of attack during supersonic operation results in different precompression behaviour, thus in different inlet entry conditions. The inlet feels no change in flow direction. At the maximum inlet entry Mach number of 5.5 corresponding to a flight Mach number of 6.8 at an angle of attack of $5^\circ$ all external oblique shocks are focused on the inlet lip.

For the analysis of the inlet flow field two different wind tunnel models have been developed by DASA (Fig. 5.3.2.6). Boundary layer effects have been studied as well as airflow swallowing and choking characteristics. Both at the Technische Universität München and at DASA in München-Ottobrunn two- and three-dimensional Navier-Stokes calculations have been performed to analyse the flow fields of mixed compression inlets. These calculations have been verified with experimental data with good agreement [8, 9].

For an internal compression inlet side spillage is not possible without considerable pressure losses due to the breakdown of the external and internal shock system [10]. Hence an unstarted inlet should be avoided at hypersonic and high supersonic flight Mach numbers. However, inlet choking might occur and it is important to estimate its effects on aircraft operation. CFD-calculations have also been carried out to analyse the flow field of an unstarted air inlet. A one-dimensional model has been developed to determine the detached shock wave position and static pressure distribution on inlet ramps and across the capture area to calculate inlet forces reliably even in failure situations (Fig. 5.3.2.7). The captured airflow is assumed to pass a normal shock wave, whereas the spillage flow describes a flow past a blunt body formimg a detached shock wave which may be approximated by a hyperbola.

Figure 5.3.2.6: Inlet wind tunnel model ETM1 for hypersonic application tested by Daimler-Benz-Aerospace (DASA).

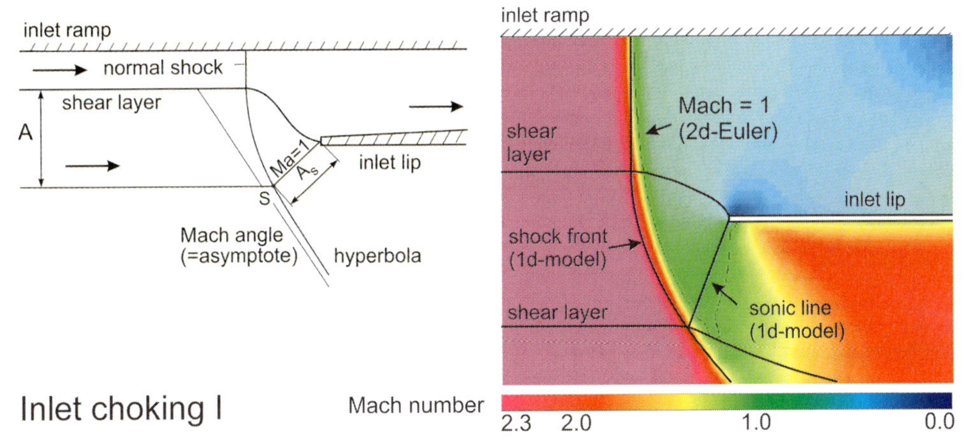



Figure 5.3.2.7: Flow field of an inlet unstart (63% of design massflow captured) in comparison with quasi one-dimensional model.

For inlet calculations two-dimensional flow has been assumed. In future, yaw-effects will have to be taken into account as additional oblique shock waves will be induced by the external inlet endwalls and flow dividers.

*Nozzle*

The exhaust nozzle for hypersonic aircraft has to work over a wide range of different operating conditions. From take-off to hypersonic speeds ($Ma>6$) the nozzle pressure ratio varies from 3 to about 800. For a complete expansion to ambient pressure large nozzle area ratios have to be realized. These can only be achieved by using parts of the airframe as part of the expansion nozzle. A two-dimensional-single expansion ramp nozzle is considered most suitable for hypersonic application [12].

High requirements are set on nozzle performance at low and high Mach numbers. At high flight Mach numbers even small changes in nozzle performance result in large deviations of net installed thrust. In the transonic flight range asymmetric thrust nozzles, designed for high flight Mach number operation, are known to produce large changes in gross thrust vector angle (of up to 40°). This is mainly caused by expansion losses of the nozzle flow and the increasing external drag of the lower nozzle flap. This behaviour considerably influences the flyability of the aircraft. However, the injection of the diverted forebody boundary layer in the divergent section of the nozzle at turbojet operation improves the nozzle performance considerably [13, 14].

For the study of the interaction between the main nozzle flow and the ambient flow a generic wind tunnel model was tested by DASA-MTU München in

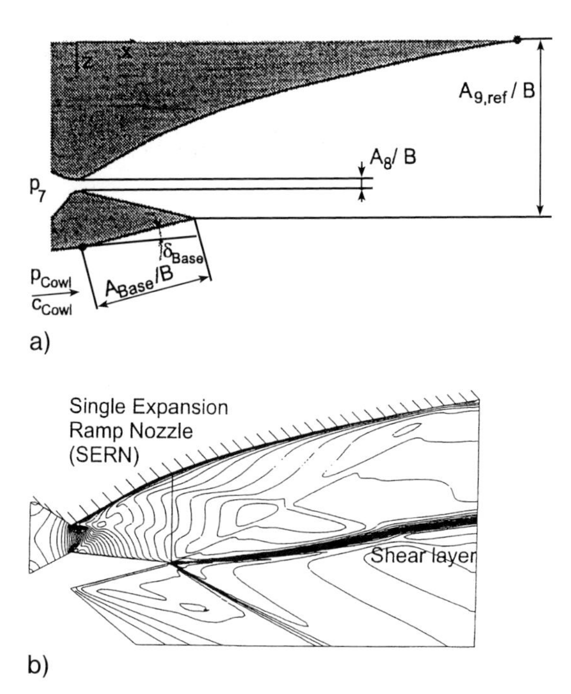


Figure 5.3.2.8: (a) Single expansion ramp nozzle (SERN) with variable geometrie; (b) Nozzle/Afterbody flow field interaction of a SERN. Flight Mach number $Ma=1.5$, $\Pi_{SERN}=19$.

the Trisonic Windtunnel and in the Hypersonic Windtunnel at DLR in Cologne for a joint research programme with the Institute of Aeronautics and Astronautics at the Technische Universität München. The experimental data have been used to validate Navier-Stokes calculations that were carried out over a wide range of different operating conditions (nozzle pressure $\Pi_{SERN}=3\ldots800$, Mach number $1.6\ldots6.0$, angle of attack $\alpha=0^\circ\ldots8^\circ$ (Fig. 5.3.2.8), [9, 15, 16].

As $\Pi_{SERN}>3$ throughout the mission, only critical nozzle flow occurs. However, in subsonic and transonic flight, subsonic speeds will be found in the nozzle region in the interactive flow around the lower engine cowl. Due to relatively low mass flow and static pressure in the main nozzle flow (compared to design point at ramjet-operation) the afterbody volume cannot be filled sufficiently. This effect results in considerable negative lift and a nose-down pitching moment.

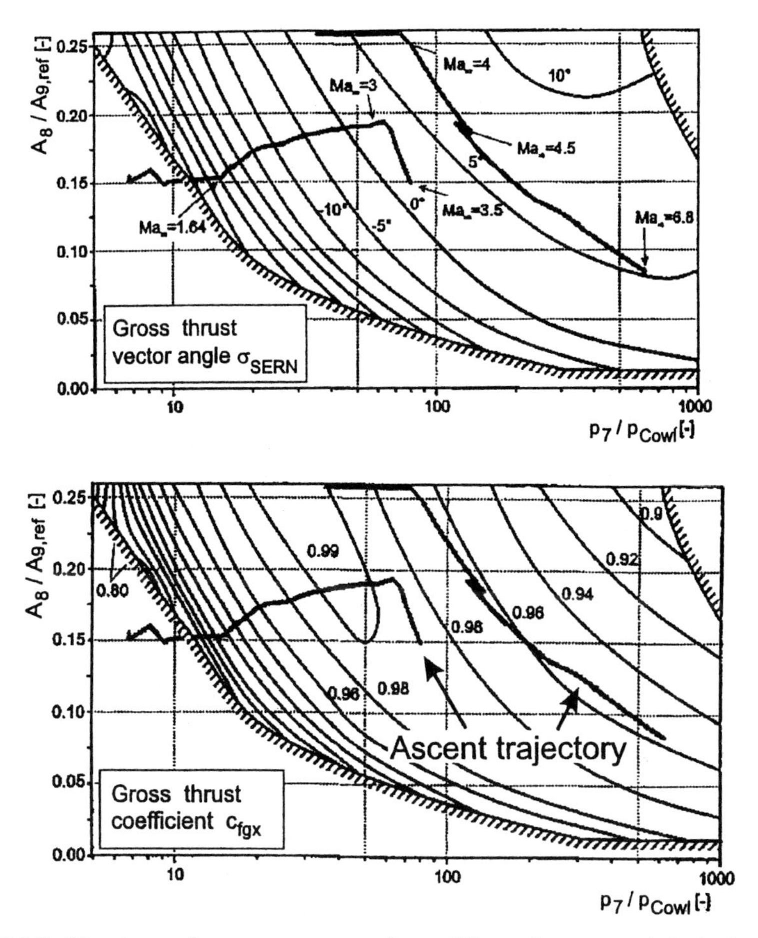


Figure 5.3.2.9: Nozzle performance map and condition along ascent trajectory: Gross thrust vector angle $\sigma_{SERN}$ and coefficient $c_{fgx}$ over nozzle pressure ratio and area ratio.

The single expansion ramp nozzle is designed to provide approximately axial thrust at design point. Figure 5.3.2.9 shows the nozzle performance map with gross thrust vector angle $\sigma_G$ and thrust coefficient $c_{fgx}$ over nozzle pressure ratio and nozzle area ratio, which basically is the throat area. It can be seen, that not only one design point with satisfying performance characteristics exists, but rather a combination of nozzle pressure and area ratios with axial gross thrust as well as good thrust coefficients. Additionally in Fig. 5.3.2.9 the operating conditions along the ascent trajectory are marked. The break at flight Mach nurnber Ma=3.5 is due to the change from turbo-mode to ramjet-operation as well as the closure of the inlet boundary layer diverter.

The nozzle map has been calculated with the method of characteristics assuming supersonic flow throughout the afterbody region. In subsonic and transonic flight (i.e. below a flight Mach number of about 1.4) prediction of the afterbody flow field is difficult. It depends mainly on aerodynamic flow characteristics around the airframe and its influence has to be included in the aerodynamic dataset. For calculation of the propulsion system performance a thrust vector angle is assumed as well as a constant thrust coefficient.

Especially in supersonic flight the thrust vector is affected by the variable lower nozzle flap angle (cf. Fig. 5.3.2.8 a). The influence of an active and dynamic geometry adjustment on off-design performance will be studied in the future.

Only two-dimensional flow aspects have been considered, sideslip flight control can not be performed with the current nozzle design.

### 5.3.2.4 Core Engine

Unlike the industry propulsion system concept, a one-spool turbojet, the reference propulsion system for the lower stage of the TSTO-concept is composed of turbofan engines for flight Mach numbers up to 3.5 and ramjet engines for Mach numbers up to 6.8 [1, 4, 11], Fig. 5.3.2.10.

The bypass ratio of the considered turbofan engine is 0.2, overall pressure ratio of the two compressors is 12 and the maximum turbine entry temperature is assumed to be 1850 K.

The turbofan engine has been favoured against a one-spool concept due to its lower specific fuel consumption at dry operation and subsonic cruise and due to its higher sensitivity on thrust nozzle control.

Due to turbine blade material limits, the temperature in the combustion chamber and therefore the fuel to air ratio is restricted. The unburned oxygen is used for reheat up to stoichiometric fuel-air ratio, beginning at a flight Mach number of 0.9.

The turbofan engine is shut off at flight Mach number 3.5, the turbo engine is bypassed and the afterburner is used as ramjet combustion chamber. In ram mode, the air is burned stoichiometrically up to the maximum nozzle entry temperature of $T$=2800 K. With increasing flight Mach numbers – after reach-

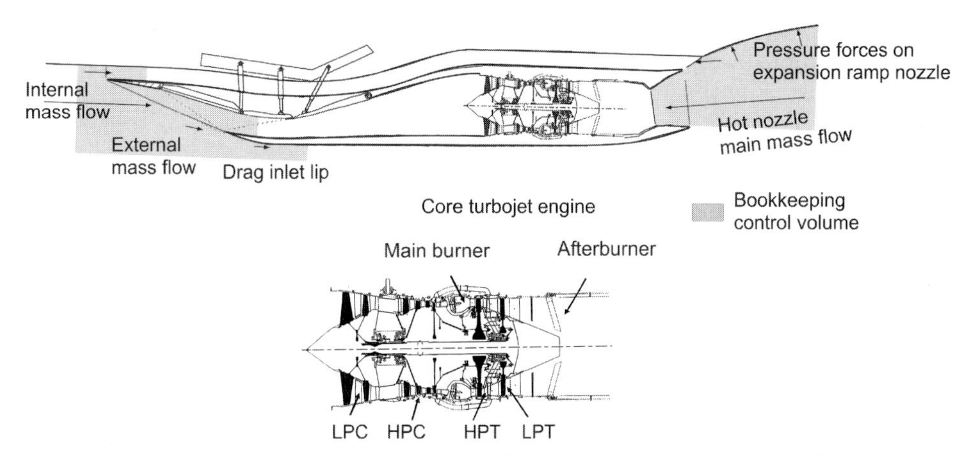


Figure 5.3.2.10: Highly integrated bypass turbo engine with some remarks upon the bookkeeping.

ing maximum temperature – the ram combustor works at over-stoichiometric fuel-air ratio.

### 5.3.2.5 Numerical Engine Simulation

#### *Performance Analysis*

Setting up a computational description of an engine, the propulsion system is subdivided into different components such as: inlet, compressor, combustion chamber, turbine, thrust nozzle, etc. (Fig. 5.3.2.4). Each component is characterized by a distinct physical behaviour. The interaction of the various engine parts and modules determines the steady state performance and transient operation of the complete propulsion system. Within each module the basic physical behaviour is described by an appropriate set of equations and/or characteristic maps. The engine components are coupled not only via the laws of mass-, momentum- and energy conservation, but also via the engine control system. This leads to a non-linear, coupled set of equations, which have to be solved through an iterative numerical procedure. The quality of the performance analysis depends strongly on an accurate modelling of the basic physical processes within a single engine component for design- and off-design conditions. For instance, the inlet and the nozzle of airbreathing hypersonic aircraft propulsion systems are characterized by the high degree of engine/airframe integration, by the high requirements on geometric flexibility, by the complex flow field phenomena and the wide range of operating conditions. Modelling these components extensively is too complex to be done within the performance analysis for real time simulation itself. Therefore, the performance- and operating-behaviour is commonly described by characteristic maps, which are based on data provided by experiments and/or time consuming theoretical methods, like CFD-analysis. Creating accurate component maps is a key issue in simulating hypersonic airbreathing engines.

In performance analysis characteristic component maps, for instance of compressors or turbines, frequently can be derived and scaled from known maps describing the component in similar design [4, 6, 10]. Off-design calculation iterates steady state and transient operation points within the component maps according to the laws of mass-, momentum- and energy conservation as well as the control system.

Details of the engine component models can be found in [18, 20, 21] and in the appendix.

*Real Time Simulation Environment*

The detailed computation of thermodynamic cycles is the basis not only for a comprehensive performance analysis, but also for real time dynamic engine simulations. Only some years ago special real time methods like state space models and transfer functions had to be applied and additional simulation codes had to be developed [18]. Due to increasing processor performance it is now possible to use the same methods for one-dimensional performance analysis and real time simulation even with a detailed description of thermodynamic properties and comprehensive component models even on a regular PC [11].

The achievable real time simulation sample time depends on model complexity: The quasi one-dimensional computation of the air inlet with its external and internal shock system, the iteration of oblique shock waves including shock reflection in the inlet duct and the matching of turbo engine, nozzle and inlet takes more than 50% of the total computing time. Heat transfer in turbo components, consideration of varying tip clearances in engine transients, active clearance control and inlet distortion (parallel compressor model) effect simulation time steps considerably. On a typical PC-processor with real time installed turbo engine simulation with full complexity can be achieved using time steps of less than 5 ms. This sample time is sufficient not only for flight mechanical investigations but also for the development of engine control systems and engine/aircraft integrated control systems.

#### 5.3.2.6 Thrust Vectoring

At the flight Mach numbers for hypersonic aircraft and airbreathing lower stages of TSTO-space transport systems it is essential to control the net thrust vector in value and direction. Additionally the application of aircraft rudders and flaps has to be minimized to reduce drag. However, as single expansion ramp nozzles suffer from poor off-design performance, thrust vectoring has to be taken into account in the transonic flight regime, too, e.g. by injection of forebody boundary layer mass flow.

For flight mechanical investigations concerning flyability and flying qualities of hypersonic vehicles the high Mach numbers are critical so that this report concentrates on this flight regime. The net thrust is basically the small difference between the large values of inlet drag and gross thrust, so it is very sensitive to any changes in propulsion system operating conditions.

Engine operating conditions are adjusted to the pilot's or auto-pilot's input by a control system, which commands the turbo engine respectively the ram-combustion as well as the inlet and the nozzle geometry. Controlling engine parameters are the diverter mode, two inlet ramp deflection angles, inlet air bleed, fuel flows of the turbo engine and afterburner, cooling air, nozzle throat area and nozzle flap geometry.

The digital control system observes the engine component limits and directs the dynamic transition from one operating condition to another. A variable maximum fuel flow increase, for example, prevents compressor surge and turbine overheating. The bypass nozzle adjusts the mass flow of the inlet and the turbo engine demand [19]. The component limitations include on the one hand control parameter limits (e.g., maximum inlet ramp deflection angle, fuel air ratio for the afterburner) and on the other hand maximum gas temperatures for compressor exit section, turbine and nozzle entry section, a maximum pressure for the ram combustion chamber etc.

Figure 5.3.2.11 (a, b) shows the maximum net thrust x-axis component $F_{Net,x}$ versus flight Mach number and altitude for a given angle of attack (AoA) as a result of steady state performance analysis. In addition lines of constant dynamic pressure $q_\infty$ and resulting net thrust vector angle $\sigma$ (the angle between the two-dimensional thrust vector and HFD) are included.

For a high speed flight case of Mach 6 in an altitude of 30,000 m some results shall be presented in the following (Figs. 5.3.2.12 and 5.3.2.13). Angle of attack (AoA) and ramp angles of the external inlet ramps have been varied while a functioning control system has been assumed. The inlet throat area has been adjusted in order to swallow the whole captured massflow and decelerate the flow to a minimum Mach number of 1.4. The ideal Mach number of 1.0 cannot be reached. Windtunnel experiments have shown, that there is a high risk of inlet choking below a throat Mach number of 1.4 [9, 20]. The ramjet combustion is assumed to be stoichiometric, i.e. the hydrogen air ratio is constant at 0.02916, maximum combustion chamber outlet temperature is granted. The nozzle throat area is adjusted to position the final normal inlet shock wave permanently at a Mach number of 1.6, i.e. Mach 0.2 over the inlet throat Mach number. A lower difference would cause an immediate risk of inlet unstart.

Under these assumptions, in Fig. 5.3.2.12 the absolute value of the net thrust vector is shown as a function of AoA. Additionally the first inlet ramp angle has been varied. Due to a higher AoA the total pressure loss increases over forebody precompression, but still static pressures increase. A higher airflow will be captured by the inlet and thus a higher absolute value of net thrust is

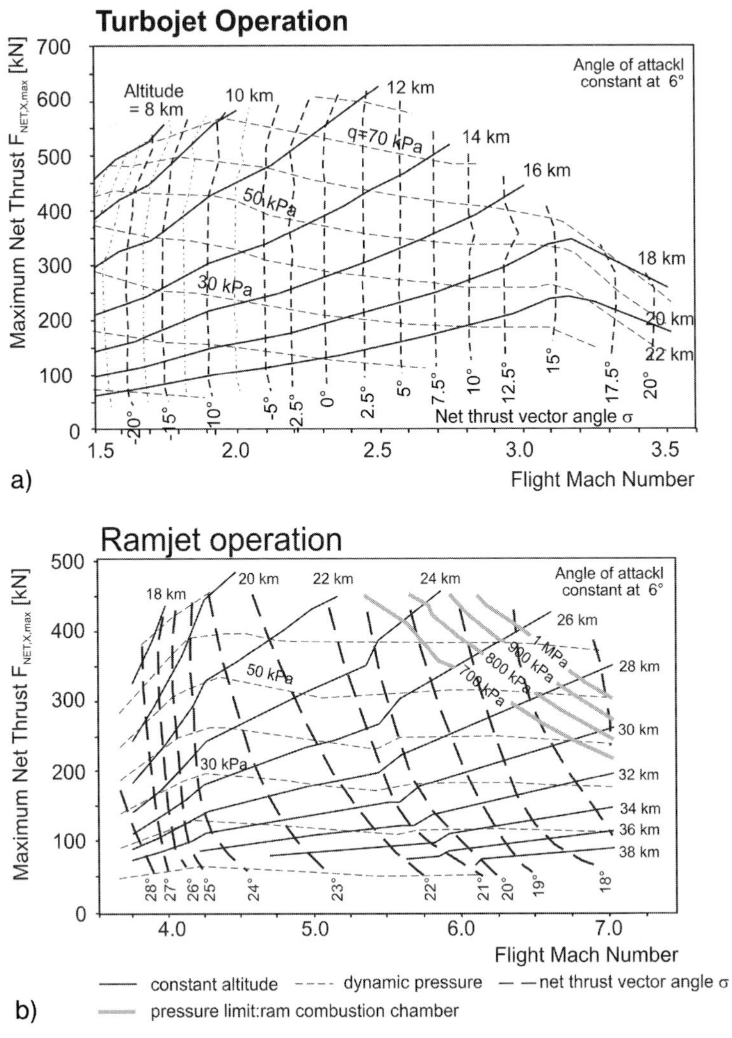


Figure 5.3.2.11: (a) Turbojet maximum net thrust and net thrust vector angle versus flight Mach number at constant angle of attack; (b) Ramjet maximum net thrust and net thrust vector angle versus flight Mach number at constant angle of attack.

obtained. As the net thrust vector turns, its x-component increases first but decreases at higher AoA.

Already in Fig. 5.3.2.12 can be seen that the net thrust vector greatly depends on inlet ramp positions. For Fig. 5.3.2.13 both movable external ramps have been varied at a constant AoA of 3°. Higher ramp angles cause increased induced inlet drag (which is lowest, ideally zero, when the external shocks are focused close to the inlet lip). Additionally the captured mass flow is decreased considerably. Increased losses over the external inlet shock waves and thus a

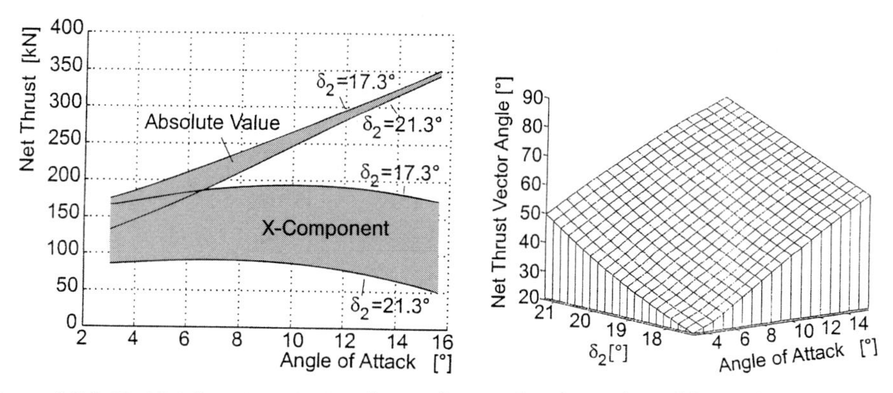


Figure 5.3.2.12: Net thrust vector angle varying angle of attack and first inlet ramp angle.

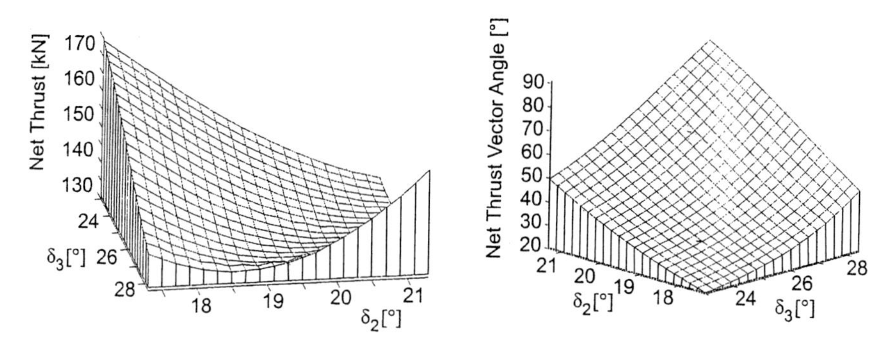


Figure 5.3.2.13: Net thrust and thrust vector angle varying both external inlet ramp angles.

lower nozzle pressure ratio has a minor effect on net thrust. The gross thrust vector angle varies between $-4^\circ$ and $+2^\circ$, while the sum of all inlet forces including inlet lip force varies between $8^\circ$ and $44^\circ$ due to higher static pressures. The resulting net thrust vector x-component decreases despite stoichiometric combustion to zero still with considerable lift (z-component) so that net thrust vector angle turns to more than $90^\circ$.

For flight-mechanical investigations of a hypersonic vehicle not only thrust and drag are important, but also the lift component of the net thrust and the resulting pitching moment. At transonic flight for example the gross thrust vector varies up to $40^\circ$, which produces a considerable nose-up pitching moment [21]. Figure 5.3.2.14 shows the pitching moment coefficients of the airframe, the engine and the total system along the ascent trajectory. The large pitching moments at transonic flight predominantly result from the large gross thrust vector angle.

With increasing angle of attack the nose-up moment of the inlet increases as well. However, the nozzle produces a larger nose-down moment. These two effects approximately level each other out at turbojet operation, at higher flight

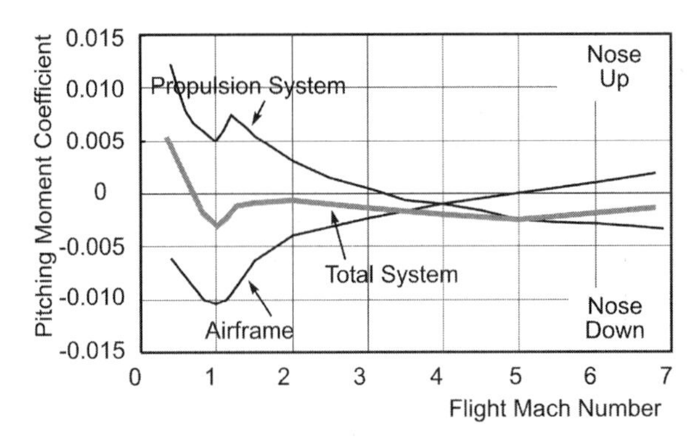


Figure 5.3.2.14: Pitching moment coefficient of airframe and propulsion system.

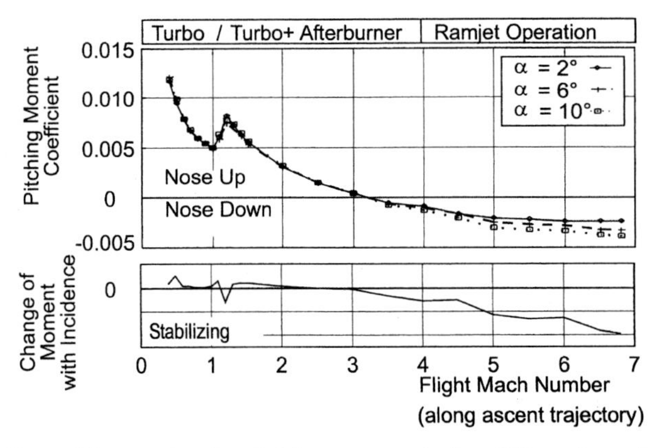


Figure 5.3.2.15: Propulsion system's pitching moments for different angles of attack and their derivative.

Mach numbers the nose-down moment of the asymmetric nozzle prevails. Figure 5.3.2.15 shows the engine's pitching moment coefficient for three different angles of attack and their derivative. An increasing nose-down moment with increasing angle of attack means a stabilizing effect of the propulsion system on the total aircraft.

It is not sufficient to evaluate the feasibility of a transport system by considering regular flight operation alone. The controllability of the aircraft at certain system failures has to be investigated as well. For the propulsion system two failure scenarios have been considered.

Figure 5.3.2.16 shows the engine force vectors and the resulting net thrust vector for a single two-dimensional engine at a flight Mach number of 6 and an angle of attack of 6° for nominal operation and various engine failures.

At flame out of the ram combustor the inlet would work normally, only the nozzle would produce less thrust due to the lower gas temperature. The resulting force would be small for drag and lift.

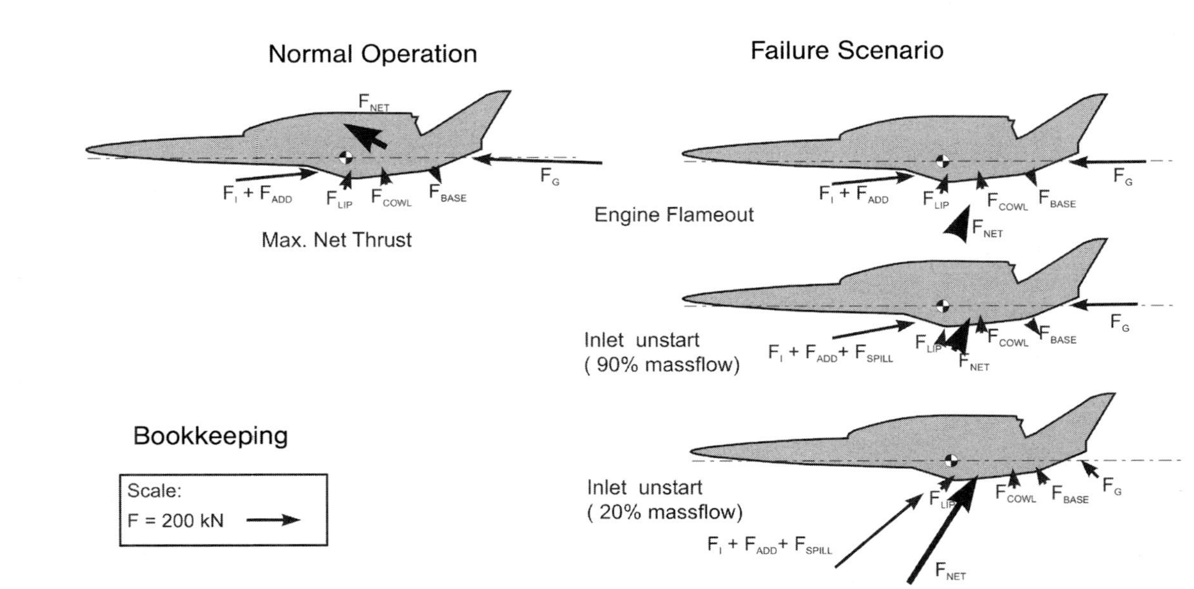


Figure 5.3.2.16: Propulsion system forces and net thrust vector at Mach 6 for nominal operation, engine flame out and two different degrees of inlet unstart.

Possibly due to a failure in the control system the inlet could unstart, i.e., the internal shock system is expelled and subsonic fore-spillage exists. 10% flow reduction and 80% flow reduction through the inlet were studied in a two-dimensional analysis. Dependent on the degree of choking this results in large inlet forces. Due to the normal shock in front of the inlet lip at a high local Mach number, total pressure losses are high. With a controlled immediate stop of combustion and a lower nozzle pressure ratio and mass flow the gross thrust decreases further. At a high degree of inlet choking ($m = 20\%$), the shock system leads to a high lift component in inlet, cowl and nozzle.

The disturbed flow field in front of an unstarted inlet might lead to a sympathetic unstart in the adjacent inlets. Although the three-dimensional forces acting on an engine are likely to be smaller than shown in this two-dimensional analysis an unstarted inlet has to be avoided.

#### 5.3.2.7 Real Time Flight Simulation

The above mentioned propulsion system simulation can be run in real time with time steps smaller than 5 ms on a micro processor. The engine simulation has been implemented in the research flight simulator of the Institute of Flight Mechanics and Flight Control – Technische Universität München (Fig. 5.3.2.17) together with the aircraft's aerodynamic, flight mechanics and flight control system. The flight-mechanical studies base on a six degree of freedom model. The non-linear control system of the total hypersonic flight system has been derived from a typical high speed aircraft [10].

The improved dynamic engine simulation serves for developing and testing advanced flight control systems and for the analysis of hypersonic flying qualities. Results of these flightmechanical studies are demonstated in chapter 6.

Figure 5.3.2.17: Research flight simulator at the Technische Universität München, Institute of Flight Mechanics and Flight Control.

#### 5.3.2.8 Conclusion

For the airbreathing lower resp. first stage of a two-stage-to-orbit transport system the main propulsion components air inlet and single expansion ramp nozzle have been analysed by Navier-Stokes calculations in comparison with experiments. Isolator and combustion chamber and the turbo jet engine, which is first used for low Mach number flight, have been described by means of typical component maps. Based on performance analysis a real time simulation of hypersonic propulsion systems has been developed for research flight simulators.

In a realistic flight mechanical analysis of hypersonic aircraft the steady state and dynamic propulsion system performance has to be considered not only with axial thrust and drag components, but also its significant lift and resulting pitching moments. At hypersonic speeds, the propulsion system has a stabilizing effect on the aircraft. Only small changes in ramp angle position and in angle of attack have immense effects on the net thrust vector. Therefore, a very detailed analysis of off-design engine characteristics has to be used to evaluate flight mechanical aircraft behaviour.

#### Acknowledgements

This study was carried out within the framework of the Collaborative Research Centre (SFB) 255 funded by the Deutsche Forschungsgemeinschaft (DFG). The authors are grateful for the support given by Daimler Benz Aerospace DASA München and MTU München, by the DLR Köln and especially the authors would like to thank their former colleagues in Technische Universität München for their contributions.

#### References to Section 5.3.2

1. Sonderforschungsbereich 255: Transatmosphärische Flugsysteme; Grundlagen der Aerodynamik, Antriebe und Flugmechanik. Arbeits- und Ergebnisbericht, 1995–1998, 1998–2001, Technische Universität München, Germany, 1998, 2001.
2. Dujarric, C., Caporicci, M., Kuczera, H., Sacher, P. W.: Conceptual Studies and Technology Requirements for a New Generation of European Launchers. IAF-97-V.3.03, 48th International Astronautical Congress, Turin, Italy, 1997.
3. Kuczera, H.: FESTIP System Study – An Overview, AIAA 8th International Space and Hypersonic Systems and Technologies Conference, Norfolk, Virginia, USA, April 27–30 1998.
4. Bauer, A.: Betriebsverhalten luftatmender Kombinationsantriebe für den Hyperschallflug unter besonderer Berücksichtigung der Triebwerksintegration. PhD thesis, TU München, Germany, 1994.
5. Sullins, G. A., Billig, F. S.: Force Accounting for Airframe Integrated Engines, AIAA-87-1965, San Diego, USA, 1987.
6. Herrmann, O., Rick, H.: Propulsion Aspects of Hypersonic Turbo-Ramjet Engines with Special Emphasis on Nozzle/Aftbody Integration. ASME 91-GT-395, Orlando, USA, 1991.

7. Weidner, J.P.: Propulsion/Airframe Integration Considerations for High Altitude Hypersonic Cruise Vehicles. AIAA-80-0111, Pasadena, USA, 1980.
8. Bissinger, N.C., Schmitz, D.M.: Design and Wind Tunnel Testing of Intakes for Hypersonic Vehicles. AIAA-93-5042, München, Germany, 1993.
9. Esch, T.: Zur Integration des Antriebs in ein Hyperschallflugzeug unter besonderer Berücksichtigung der Kräftebilanzierung. PhD thesis, TU München, Germany, 1997.
10. Hollmeier, S.: Simulation des Betriebsverhaltens von Antrieben für Raumtransporter/Hyperschallflugzeuge. PhD thesis, TU München, Germany, 1997.
11. Van Wie, D.M. et al.: Starting Characteristics of Supersonic Inlets. AIAA-96-2914, 32$^{nd}$ Joint Propulsion Conference, Lake Buena Vista, USA, 1996.
12. Dusa, D.J.: Exhaust Nozzle System Design Considerations for Turboramjet Propulsion Systems. ISABE 89-7077, Athens, Greece, 1989.
13. Berens, T.: Experimental and Numerical Analysis of a Two-Duct Nozzle/Afterbody Model at Supersonic Mach Numbers. AIAA-95-6085, AIAA 6$^{th}$ International Aerospace Planes and Hypersonics Technologies Conference, Chattanooga, USA, 1995.
14. Lindblad, I.A.A. et al.: A Study of Hypersonic Afterbody Flowfields. AIAA-97-2289, AIAA 15$^{th}$ Applied Aerodynamics Conference, Atlanta, USA, 1997.
15. Esch, T., Giehrl, M.: Numerical Analysis of Nozzle and Afterbody Flow of Hypersonic Transport Systems. ASME 94-GT-391, The Hague, Netherlands, 1994.
16. Niezgodka, J.: Druckverteilungsmessung zur Simulation der Interferenz des Triebwerksstrahls auf das Heck der Sänger-Unterstufe (MTU-Heckmodell). IB-39113-93C15, DLR Köln, WT-WK-KP, 1993.
17. Kreiner, A., Kopp, S., Preiss, A., Erhard, W., Rick, H.: Advances in Turbo Engine Real-Time Simulation for Modern Control System Development. International Council of Aerospace Sciences Congress 2000, Harrogate, GB, 2000.
18. Hollmeier, S., Kopp, S., Fiola, R., Rick, H.: Simulationsmodelle zur Regelung von Antriebssystemen für Hochgeschwindigkeitsflugzeuge. DGLR-Jahrestagung, München, Germany, 1997.
19. Morris, R.E., Brewer, G.D.: Hypersonic Cruise Aircraft Propulsion Integration Study, Vol. I, II, NASA CR-158926-1, Burbank, USA, 1979.
20. Esch, T., Hollmeier, S., Rick, H.: Design and Off-Design Simulation of Highly Integrated Hypersonic Propulsion Systems. 86$^{th}$ Symposium of Propulsion and Energetics Panel, AGARD CP-572, Seattle, USA, 1995.
21. Hollmeier, S., Esch, T., Herrmann, O., Rick, H.: Propulsion/Aircraft Integration and Dynamic Performance of Hypersonic TSTO Vehicles. 12$^{th}$ International Symposium on Airbreathing Engines, ISABE 95-7025, Melbourne, Australia, 1995.

# 5.4 Experimental Investigation about External Compression of Highly Integrated Airbreathing Propulsion Systems

Uwe Gaisbauer *, Helmut Knauss, and Siegfried Wagner

## 5.4.1 Introduction

For future reusable two-stage-to-orbit space transportation systems, the flight in the hypersonic velocity regime by using an airbreathing propulsion system is the main problem to be solved concerning the design and the overall vehicle conception. As mentioned above only the use of a scramjet-propulsion system meets all the aerodynamic and gasdynamic requirements. The scramjet, here in combination with all the necessary highly integrated components, is one of the key technologies for hypersonic flight, dominating the whole aircraft design concept.

For the flight under super- and hypersonic conditions the main part of the aerodynamic lift is provided by the front part of the lower surface of the necessarily slender vehicle, the so-called forebody.

At the same time the forebody must generate the boundary conditions in speed and pressure for the intake in the lack of compressor parts inside the scramjet engine itself. Therefore, due to the simultaneous increase of pressure and density it is obvious to place the whole engine in the rear part of the forebody to make use of the already precompressed airflow. So, the whole required compression for the scramjet can be split in two parts: The external compression just in front of the intake and an internal compression part, the so-called shock train, placed between intake and combustion chamber, see Section 5.7.

For the optimum in design, the flow should be decelerated with low loss while the achievable compression reaches a maximum. Thereby the drag should stay small and the stability of the whole vehicle should not be influenced [1, 2]. These different requirements are in a certain way controversial and demand to find a mission-depending compromise.

Keeping these requirements in mind, two different concepts are in discussion as far as the overall concept is concerned.

On the one hand the whole forebody can be designed as a more 2D-configuration, realized in the NASA HyperX-programme [3]. The geometrical consequence is to carry out the compression side of the forebody with multiple large ramps in combination with very small ramp-angles extending up to the intake. Configuring the whole lower side as a kind of "compression-field" one

* Universität Stuttgart, Institut für Aerodynamik und Gasdynamik, Pfaffenwaldring 21, 70550 Stuttgart; gaisbauer@iag.uni-stuttgart.de

must accept all the upcoming problems e.g. the question of stability in the case of an unpredictable burn-off.

On the other hand a 3D-design is possible, like the so-called Sänger-concept, where the whole necessary compression is focused on two or three combined short ramps with moderate ramp angles just at the position of the intake. Consequently, the point of load incidence of this slender configuration is near the vehicle's centre of gravity resulting in a quite good stability behaviour even in the case of burn-off. But it must also be considered that in the case of a more elliptical configuration some problems, albeit others, must be handled like cross-flow instabilities in the forebody boundary-layer and their consequence on the transition behaviour, only to mention one aspect. The general difference in wind tunnel testing between the two mentioned configurations is explained in Section 4.6.2.

#### 5.4.1.1 Focus on the Problem

A focus in the Collaborative Research Centres was on the latter mentioned design concept. Consequently, starting from a Sänger-configuration, the purpose was to design an highly integrated intake with two or three combined 2D-ramps under the condition of a not peeled-off incoming turbulent boundary layer to achieve the needed compression with less pressure loss compared with the single ramp case. The projected surface in flow direction of these compression surfaces should also be minimized to reduce total pressure drag of the vehicle.

In contrast to the 2D-vehicle design with long ramps, here the formed shocks are coupled and can only be treated as a common shock system with a strong mutual interaction. The combination of several shocks is a very complicate and sensitive structure with respect to small changes in the design point. This concur of a multiple shock structure with the incoming turbulent boundary layer leads to the problem of the so-called shock boundary layer interaction, focusing a double ramp configuration (shock/shock) with short as possible length and moderate ramp angles.

In the last 50 years big efforts were done to get more understandings of the phenomenon of the shock boundary layer interaction in supersonic flow [4]. A lot of experiments with different purpose were carried out: 2D-cases with big and small ramp angles, swept and un-swept, 3D-cases with different geometries (ramps, flaps etc.). Theories were developed describing the phenomenon of the interaction between shocks and boundary layer [5].

Also in the field of numerical computations there was a big success in simulating the above-mentioned experiments ([6, 7] and others).

In the past the main interest in the available literature, concerning the 2D, un-swept case, was focused on the single ramp with big ramp angles to achieve the desired compression in one step and to demonstrate clearly the visible upstream-interacting effect. Only a few work was done in the shock/shock case. An overview is given in [8, 9].

The theory describing the shock boundary layer interaction is called the Free Interaction Concept [5]. Here, due to experimental results it could be demonstrated that the major part of a supersonic interacting flow evolving towards separation does not appreciably depend on the agency at the origin of the separation itself. This agency being either a geometrical obstacle like a ramp(s), also steps, or an incident shock wave, as described in Chapter 4.4. That part of the flow independent of the downstream situation contains the concentrated compression as well as the development of a pressure plateau for largely separated flows. Everything happens as if the flow was entirely determined only by its properties at the onset of the interaction zone [5]. Deducing the corresponding conditional equations, a universal correlation function $F$ independent of Mach and Reynolds number must be found experimentally. Thereby, the most important normalizing factor is given by the upstream interaction length $L_0$, describing the upstream influence of the shock on the boundary layer. $L_0$ is used as a function of $L_0(M, Re_\delta, a)$, usually normalized by the boundary layer thickness $\delta \Rightarrow L_0/\delta = l_0$. Here, in the experimental studies $L_0$ is determined according to the method described in [10].

Unfortunately, up to now the theory is limited on the single ramp situation. So, the main idea is to analyse the influence on the parameter $L_0$ due to the geometrical double ramp configuration, specified by two-ramp angles $a_1$, $a_2$ and the distance $D$ between the two ramp kinks. This extension of parameters leads to the new functional dependency of the upstream interaction length $L_0$ at the ramp kink: $L_0(M, Re_\delta, a_1, a_2, D)$.

#### 5.4.1.2 Preliminary Measurements

Firstly, measurements have been done in a small suck down wind tunnel (HMMS) of IAG for different ramp angles but at fixed distance between the kinks. Tunnel runs were carried out for three different Mach numbers of $M=2$, 2.5 and 3 with a respectively change in unit Reynolds number. A more detailed description of the set-up and the conducted measurements is given in [11]. The main intention of this work was to understand much clearer the dependency of the upstream interaction length as a function of the angles $a_1$ and $a_2$. At the same time maximum ramp angles were searched leading to the wished two-shock system resulting in the intended compression.

Focusing on one researched case, measurements at $R/m=10\times10^6$ and $M=2.5$ showed that for a fixed length of $D=40$ mm ($D/\delta\approx5$) and an angle combination of $a_1\leq11^\circ$ and $a_2\leq9^\circ$ no separation of the second shock took place and the desired two-shock system was existent and could be measured – Fig. 5.4.1, left side. On the right side a fully separated case is shown. Accordingly, for the other two Mach and unit Reynolds numbers also a critical combination of ramp angles could be found [11] to clarify the angle-dependency of $L_0$ for still fixed $D$.

After finishing this first parameter analysis the next question was to find a second border line, i.e. the critical first ramp length $D$ (normalized also by the

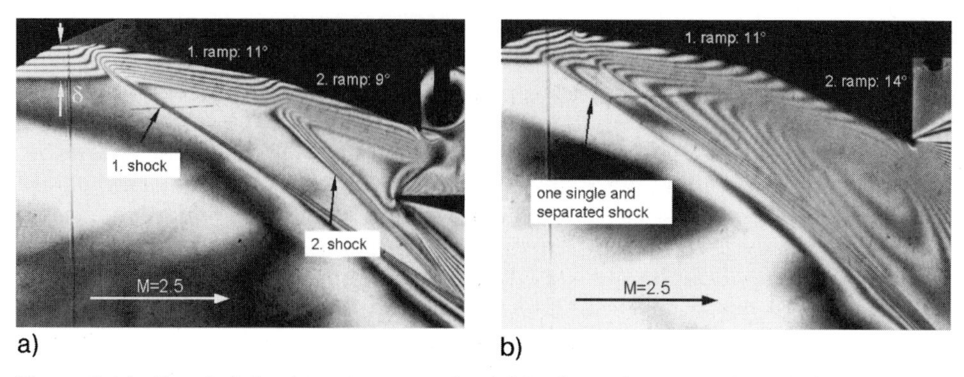


Figure 5.4.1: Coupled shock-system on a shock/shock-configuration; left: the desired two-shock system; right: separated case.

boundary layer thickness $\delta$ as $D/\delta = d$) so that the second shock is just not separated. While varying the parameter $D$ the unit Reynolds and Mach number as well as the ramp angles (with a focus on ramp angles of $\alpha_1 = 11^\circ$ and $\alpha_2 = 9^\circ$) are constant.

But first of all and of main interest for the experimental and numerical research in the field of shock/shock-phenomenon is to specify the boundary conditions of the whole experimental set-up. Generally, in this field not only the basic characteristics of the incoming boundary layer itself like thickness and the state of the boundary layer must be identified but also the flow conditions in the used wind tunnel (see section 4.6 and also [9, 12]).

Parallel to the experimental work, numerical simulations were made at the Institute of Theoretical and Applied Mechanics, Novosibirsk.

### 5.4.2 Experimental Facility

The experiments were carried out in the so-called shock wind tunnel, SWK, of IAG, Stuttgart University (see Fig. 4.6.1). This short duration facility is, according to the working principle, a kind of shock tube with a Laval nozzle. During one run there are two steady flow states of $2 \times 120$ ms. Because of the large test section ($1.2 \times 0.8$ m$^2$) investigations of models with large dimensions can be carried out from a Mach number range of $M = 1.75$–$4.5$ and a corresponding unit Reynolds number range of $R/m = 70 \times 10^6$–$30 \times 10^6$ [13, 14].

### 5.4.3 Wind Tunnel Models and Instrumentation

At the beginning of all measurements and numerical simulations within the framework of the shock/shock-subject, two basic problems appear: the determination of the basic flow on the flat plate and the investigation of the pressure distribution over the two ramps. To receive experimental data about these two main items, different models for different tasks were necessary. Consequently, three flat plate models with two different double ramp configurations were manufactured.

In all cases the basic model was a flat plate with a sharp leading edge. The size was $0.53 \times 0.2\ m^2$. Due to the manufacturing the surface was polished to reduce roughness. To simulate a real 2D-flow over the flat plate, the model was completed with trapezoidal side wings, designed with supersonic edges for $M=2.5$. The model was mounted on a special support to place the whole configuration outside the tunnel sidewall boundary layer at a position $x=1700$ mm downstream of the nozzle throat. To reduce the uncontrollable aerodynamic forces on the model during the flow establishing phase in the 2D-nozzle flow, the whole configuration was installed perpendicularly to the nozzle surfaces (see Fig. 5.4.2).

#### 5.4.3.1 Model 1

For the first investigations a flat plate model with static pressure tabs along the centre line equally spaced ($\Delta x=30$ mm) was manufactured. Additionally, in spanwise direction 4 rows in different distances from the leading edge gave the possibility to check the two-dimensionality in the pressure distribution over the whole flat plate in combination with the trapezoidal side wings. In the transverse direction the spacing was about $\Delta y=10$ mm.

The static pressure was determined with a fast responding pressure-scanning system (PSI 8400-system with two scanner modules, 34 ports each, $\Delta p=\pm 45$ psid) which allows 250 averages of the 64 pressure ports within the 120 ms SWK run.

For the determination of the transition Reynolds number ($Re_t$) wall Pitot-pressure measurements were done along the centre line with an adjustable Preston-tube, bearing flat on the surface [15]. Here a Kulite XCQ-062.3.5D differential pressure transducer was used integrated close to the head of the Preston-tube to minimize the overall tube length between probe tip and transducer itself to enable the determination of the mean pressure as well as its dynamic part. Accordingly, the rms-values could be computed and normalized with the mean (pressure-) data, a general procedure in the data processing in the field of transition [13].

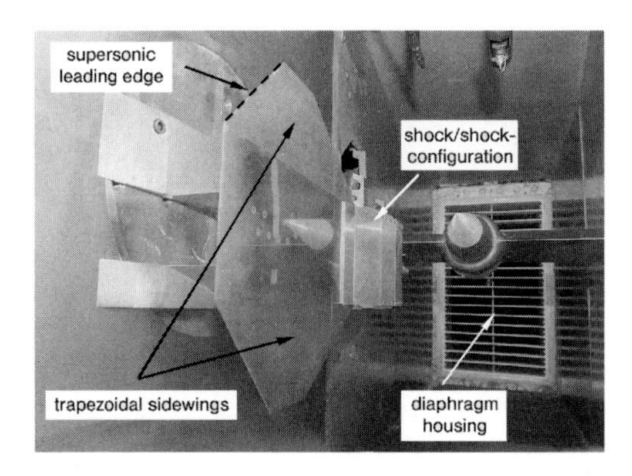


Figure 5.4.2: Model 2 with installed double ramp mounted at the tunnel sidewall.

5.4.3.2 Model 2

The second model, Fig. 5.4.2, was instrumented with three parallel to the centre line shifted rows of McCroskey hot film sensors, driven in the CCA-mode with an overheat ratio of $a=0.6$.

Their spacing was $\Delta x=15$ mm. The hot films were applicated on adjustable inserts, adapted to the plate surface by turning over in manufacturing to avoid artificial roughness [13]. This configuration was used to measure the development of the boundary layer to find out the location of transition. Again, the rms voltage output signal of each hot film was normalized with the mean value preconditioned by the used custom-made bridge amplifier.

Consequently, the transition Reynolds number could be detected and also the state of the boundary layer at the position of the later analysed double ramp configuration could be determined by two different and independent measuring methods.

It must be mentioned here as a kind of general remark that both methods are measuring different physical quantities. While the information of the Preston tube measurement is not based on direct observation of the turbulent spots (characterizing the "onset of transition" in the very early stage which can be detected by the first mentioned technique) but rather on the evolution of some macroscopic parameters like skin friction (by means of the measured wall Pitot-pressure, see section 5.4.5) [16], values of the transition Reynolds number based on the so-called "onset of transition point", are not, in general, comparable to the above mentioned techniques. It is pointed out in [16] and [17] that the hot film technique indicates the lowest "onset" $Re_t$-values because it is the most sensitive one in detecting turbulent spots i.e. instantaneous changes in the local heating rates, in the very early beginning of transition development. This is not obtained by the other used method. Therefore, to have a better comparability of $Re_t$-values, detected by the different methods, in correspondence to [18], the $Re_t$-values obtained by thin film technique and by the Preston tube are defined by the position of the distinctive peak in the rms-fluctuation values. This peak also coincides with an intermittency value of $\Gamma=0.5$ as could be verified and shown in [14].

Consequently, the identical procedure in data processing leads to smaller transition Reynolds numbers by using flush-mounted sensors compared to the wall Pitot-probe measurements due to the different frequency response of the two sensor types.

Additionally to the hot films, the first double ramp model was installed at a position of x=450 mm from the leading edge to get a first impression of the influence of the new parameter $D$ on the shock boundary layer interaction. The upper ramp with an angle of $\alpha_2=9^\circ$ was movable on the first ($\alpha_1=11^\circ$) to realize the variation of the distance between the two kinks. On the double ramp model the distance between the pressure ports was about $\Delta x=2.5$ mm. Again the above-mentioned PSI 8400 system with two pressure scanner modules came into operation. The set-up of the static pressure scanner system was identically in all cases.

#### 5.4.3.3 Model 3

The last used model was constructed modular to give the possibility of a variable sensor installation.

Various sensors like different kinds of hot films, the new developed Atomic Layer Thermo Pile (ALTP) [19] or an insert with static pressure ports were put in. In the measured configuration at a distance of x=162 mm from the leading edge a 3×3 matrix with flush-mounted sensors was installed with an equidistant spacing of $\Delta x=32$ mm (see Fig. 5.4.3). Again the determination of the transition location was intended by these nine sensors. For the first time in the SWK one ALTP-sensor was used for the transition measurement and a second for the determination of the wall temperature after extensive calibration and validation in another research programme [20].

Moreover, a new $11^\circ$-ramp/plate insert was built with a spacing of $\Delta x=1.25$ mm between the pressure tabs. So it was possible to measure the static pressure distribution in front of the first kink and on the ramp simultaneously with a very high resolution, ±1 mm in the vicinity of the kink, and without mechanical perturbation at the changeover between plate and first ramp. The upper $9^\circ$-ramp was identical to that one used at the plate model 2.

The position of the double ramp configuration was at x=420 mm due to constructive and gasdynamic requirements. Again, the PSI 8400 system was used to measure the static pressure distribution.

Additional a height adjustable Pitot-probe with a resolution of $\Delta y=0.04$ mm normal to the plate surface was installed to measure the boundary layer profiles. The probe was placed at x=380 mm from the leading edge (40 mm upstream of the first ramp), to ensure no upstream influence of the double ramp model on the boundary layer measurements (Fig. 5.4.3).

A very comprehensive and detailed description of the used models, the different installed sensors and the specific measurement techniques is given in [21].

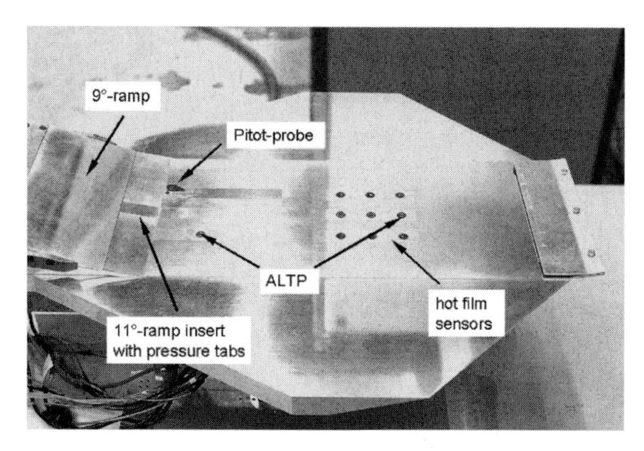


Figure 5.4.3: Plate model 3 with different installed sensors: ALTP, hot films and Pitot-probe.

## 5.4.4 Numerical Model

The computations were performed on the basis of Favre-averaged Navier-Stokes equations closed by a two-equation $k$-$\omega$ turbulence model by Wilcox [22]. The computation domain is restricted by inlet and outlet sections chosen far enough from the interaction zone, solid surface on the bottom and free surface on the top boundary. No-slip condition for velocity and adiabatic condition for temperature are set on solid surfaces. So-called simple wave conditions are used on the upper boundary. Profiles of all gasdynamic and turbulence parameters obtained using boundary layer approach are specified at the inlet section and "soft" conditions are used at the outlet section. The four-step implicit finite-difference scheme of splitting according to the space directions realized by scalar sweeps was used for time approximation. TVD-type scheme based on Van Leer Vector Flux Splitting method [23] was used to construct the high resolution approximation of inviscid fluxes space derivatives. Viscous terms were differenced to the second order accuracy in a centred manner. The computational method is described in details in [24]. This original numerical algorithm has been used to compute various 2D turbulent supersonic separated flows and has shown a good potential to predict the properties of the investigated flows [25–28].

## 5.4.5 Measurements and Results

In the present work the first main problem was to qualify the incoming flow field and the resulting boundary layer on the flat plate as a kind of basic flow, with great importance for the experiment and for the numerical simulation.

All the experimental work in the SWK has been done at a constant Mach number of $M=2.53$ but with a variation in unit Reynolds number, overall between $3\times10^6 \leq R/m \leq 34\times10^6$. For the studies in the shock/shock-field the measurements were done for two relevant cases: $R/m\times10^{-6}=9.82$ and 12.41 with an accuracy of 1.4%. The experiments were carried out in two steps: First the flow respectively the boundary layer was qualified for the required cases and second the measurements in the field of the shock boundary layer interaction with the double ramp models were done.

#### 5.4.5.1 Determination of the Boundary-Conditions

To qualify the incoming flow on the flat plate, in a first step studies have been done without installed double ramp in a unit Reynolds number range of $R/m\times10^{-6}=4.9$, 9.5, 14.4 and 18.8. The intention was to check the conformity of flat plate boundary layer. The corresponding measurements of the static pressure distribution were made with the above-described model 1. For two unit Reynolds numbers of about $R/m\times10^{-6}=4.9$ and 18.8 the static pressure distributions along the centre line of the plate are plotted in Fig. 5.4.4.

In both plots just a weak perturbation emanating from the kink in the leading edge, spreading in "Mach direction", is detectable, leading to a disturbance in the distribution at a fixed position of $x\approx233$ mm. Downstream, beginning at a position of $x=295$ mm the influence of the transition is perceivable. In the case of $R/m\times10^{-6}=4,9$ at the first measuring point also the effect of the very weak compression wave caused by this discontinuity in the leading edge can be seen by a small increasing of the static pressure.

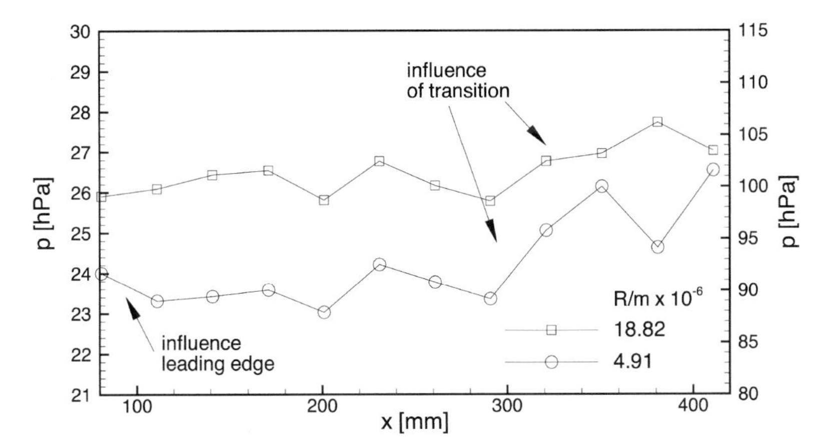


Figure 5.4.4: Static pressure distribution along the center line for two different unit Reynolds numbers.

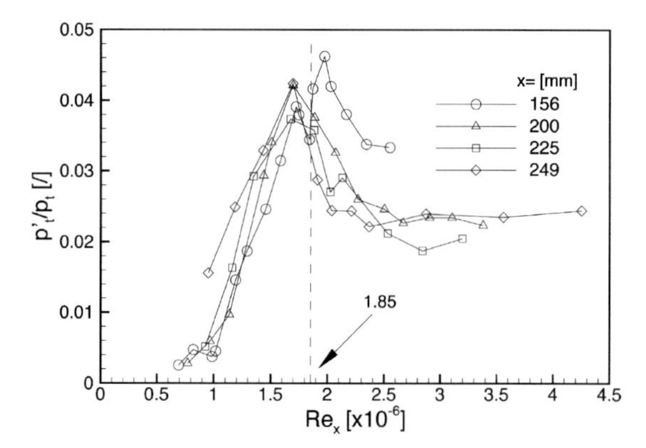


Figure 5.4.5: Normalized effective wall Pitot-pressure distribution for different Reynolds numbers $Re_x$.

Summarizing, the detectable variations in the static pressure distribution (on a length of 330 mm) have an overall order of magnitude of about 0.8% resp. 1.6%. In spanwise direction except the influence of the transition only very small changes in the range of about 1–1.5% could be found. Consequently the static pressure field over the flat plate model can be described as quasi two-dimensional.

In the next step studies have been carried out with all three models to determine the transition Reynolds number $Re_t$.

Therefore, first measurements with a Preston-tube at different positions along the centre line of model 1 have been done. For a unit Reynolds number range of $R/m \times 10^{-6} = 4.5–14.1$ in Fig. 5.4.5 the normalized effective pressure fluctuation values are plotted versus the Reynolds number calculated for each single probe position. Here a mean $Re_t$-value of about $Re_t \times 10^{-6} = 1.85$ can be detected [15].

The hot film measurements with model 2 for a $R/m$-range of about $R/m \times 10^{-6} = 4.5–37.6$ with 31 steps in between, verified the above made studies approximately. According to the general remarks made in section 5.4.3 the investigation of the region with the maximum effective fluctuation level $U_{rms}/U_{mean}$ (hot film output signal) leads to $Re_t \times 10^{-6} = 1.76$ – a slightly smaller value than measured before. After a careful analysis of the measured data an increasing of the $Re_t$-values up to $Re_t \times 10^{-6} = 3.2$ for $R/m \times 10^{-6} > 12$ can be detected. Remarkably, this enhancement of the transition Reynolds number proceeds in two discrete steps and not continuously like caused by a unit Reynolds number effect. A detailed description, a comparison with cone measurements and a first explanation are given in [21].

After all, a conclusive determination of the transition Reynolds number was made for the two relevant $R/m$-values of $R/m \times 10^{-6} = 9.82$ and 12.4 by using an ALTP-sensor installed in model 3. In Fig. 5.4.6 the maximum peak in the plot of the normalized effective rms-fluctuation output signal versus $Re_x$ results

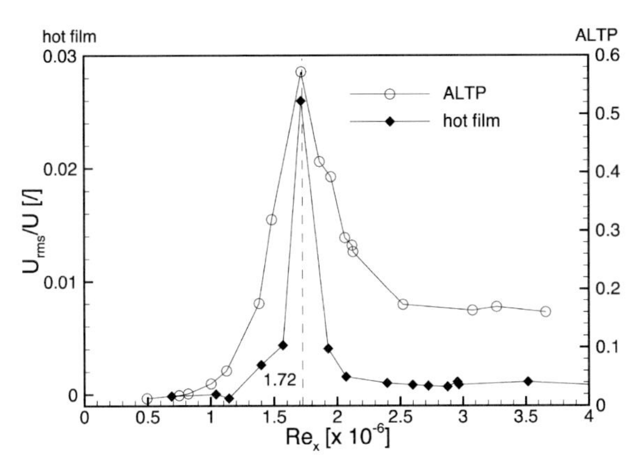


Figure 5.4.6: Normalized effective rms-output signal versus $Re_x$ of the ALTP in comparison with a custom-made hot film sensor.

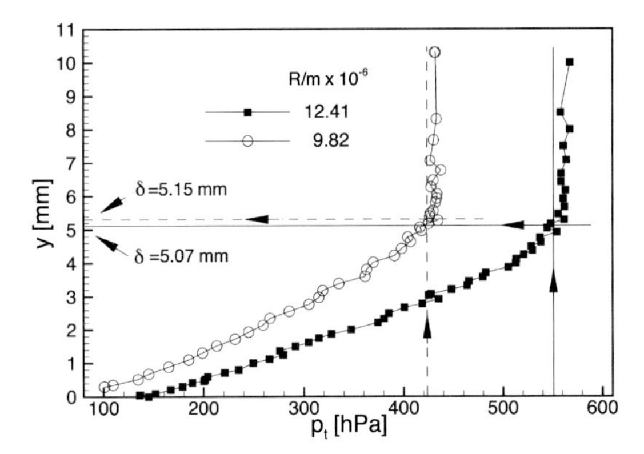


Figure 5.4.7: Measured Pitot-pressure profile with derived boundary layer thickness.

in $Re_t \times 10^{-6} = 1.72$, again less than the hot film result due to the higher frequency resolution of the used sensor.

Here, for the established $R/m$-values it can be concluded that at the position of the double ramp model at x=450 mm resp. x=420 mm the boundary layer is turbulent.

Finally the mean boundary layer profiles like Pitot-pressure, Mach number, velocity and dimensionless velocity $u^+(y^+)$ were determined to finalize the boundary conditions. From the measured Pitot-pressure profile (Fig. 5.4.7) the boundary layer thickness was determined as $\delta_{12.41} = 5.07$ mm resp. $\delta_{9.82} = 5.15$ mm by using the Pitot-Rayleigh relation and by calculating the Mach number profile. The pure shape of the profile leads also to the conclusion that the boundary layer is turbulent. It must be mentioned, that for each of the shown data points one single run of the SWK was necessary.

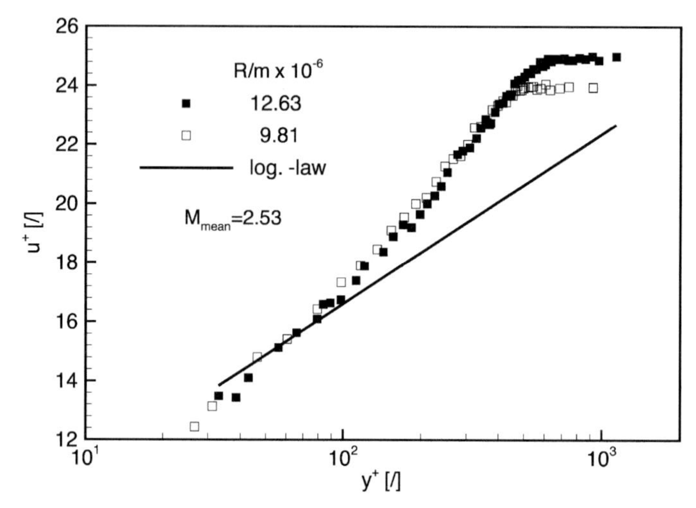


Figure 5.4.8: Nondimensional velocity profile $u^{+}(y^{+})$.

Further on, under the assumption of a quasi-adiabatic wall condition and $T_0$=const (perpendicular through the boundary layer) the velocity distribution was determined by using the Crocco-Busemann relation. Finally the dimensionless velocity distributions $u^{+}(y^{+})$ could be calculated. In Fig. 5.4.8, $u^{+}(y^{+})$ and the log-law derived from the computed $c_f$-values [29] of $c_{f;\,9.82}=1.977\times10^{-3}$ and $c_{f;\,12.41}=1.808\times10^{-3}$ are plotted with a quite good agreement between the measured and calculated data. The turbulent character of the boundary layer is again evident and the above made assumption can be verified. Consequently, as the first main results the incoming boundary layer on the flat plate in front of the first ramp kink can be qualified as turbulent and the boundary conditions for the advanced experiments and the numerical simulations can be specified.

#### 5.4.5.2 Measurements in the Field of Shock Boundary Layer Interaction

To get a first impression how $D$ effects the coupled shock system, parallel to the hot film measurements on model 2, the first ramp/ramp-configuration was installed at x=450 mm. Measurements of the static pressure distribution on the flat plate and on the two ramps were made simultaneously along the centre line for $R/m\times10^{-6}$=3.3–23. The distance $D$ between the two ramp kinks was varied for each unit Reynolds number. For $R/m\times10^{-6}=9.2$ in Fig. 5.4.9 the normalized pressure distribution for different distances of $D$ is plotted in the coordinate system of the flat plate. It seems that for $D\geq17$ mm the second shock starts to reattach. For $D\geq21$ mm the pressure plateau on the 11°-ramp starts to develop up to a constant value.

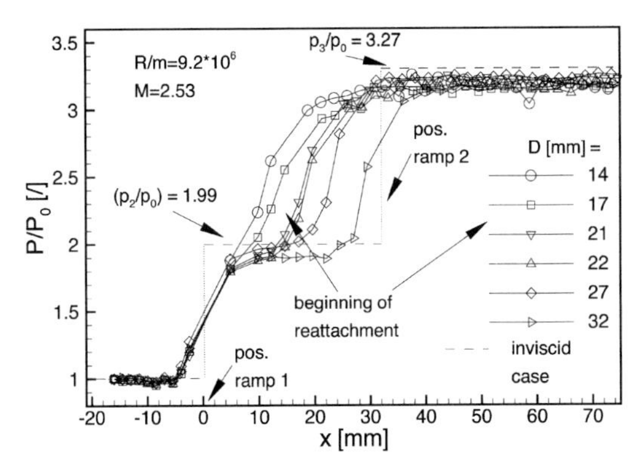


Figure 5.4.9: Normalized static pressure distribution of different distances of $D$ and constant $R/m$.

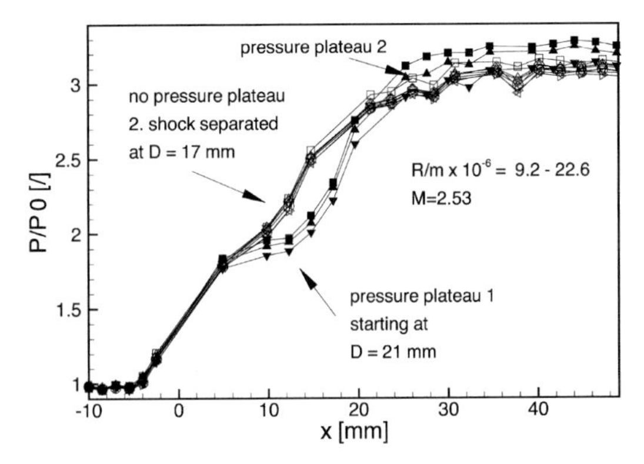


Figure 5.4.10: Influence of the $R/m$ on the pressure distribution for $D$=17 mm and 21 mm.

To clarify the influence of $R/m$ on the shock/shock-system, in Fig. 5.4.10 the pressure distribution for the two significant cases of $D$ are shown for $R/m \times 10^{-6}=9.2$–$22.6$. For both plots only a very small influence of the parameter $R/m$ is visible compared with the more decisive effect of $D$.

However, to get a final clarification of this problem the experimental spatial resolution in the pressure distribution, the spacing between the ports, was not high enough. Accordingly, the whole experiment had to be improved for the two relevant $R/m$-cases by using a new wind tunnel model and including the determined boundary conditions.

Consequently, with a much higher resolution, some further measurements have been carried out in the SWK for $R/m \times 10^{-6}=9.82$ and 12.41 with plate model 3, the second double ramp configuration installed.

Due to constructive reasons, the position of the first ramp was shifted to $x$=420 mm, but still situated in the well-developed turbulent boundary layer and not at the end of transition.

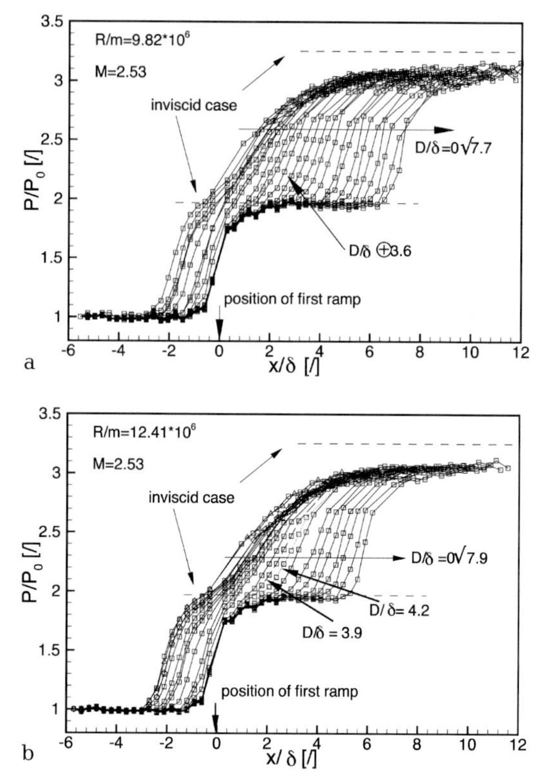


Figure 5.4.11: Normalized static pressure distribution for the shock/shock-configuration for $R/m \times 10^{-6} = 9.82$ and 12.41.

For both relevant $R/m$-ranges the variation of $D$ started at $D = 0$ mm, equal to a single ramp situation with $a_1 = 20°$, and went up to $D = 40$ mm, normalized with the respective boundary layer thickness.

For all values of $d = D/\delta$ in Fig. 5.4.11 the normalized pressure distributions are plotted versus $x/\delta$, in the plate's cartesian coordinate system, where $x/a = 0$ indicates the position of the first ramp.

Beginning from values of $d = 3.6$ resp. 3.9 the gradient in the normalized pressure distribution becomes smaller and a constant pressure level on the first ramp is detectable. The desired two-shock system is reattached and the second shock is now separated from the first. Concerning the overall compression, the ratio of the second shock is now independent from the first.

To get a much clearer understanding of the transition process between separated one-shock and attached two-shock system in Fig. 5.4.12 the normalized upstream interaction length $l_0 = L_0/\delta$ is plotted versus $d$.

In both relevant $R/m$-cases for $d \geq 3.2$ a constant value of $l_{01}$ can be established for the first ramp, independent of the double ramp parameter $d$. At this critical value of $d$ the second shock arises from fan-like isentropic compression

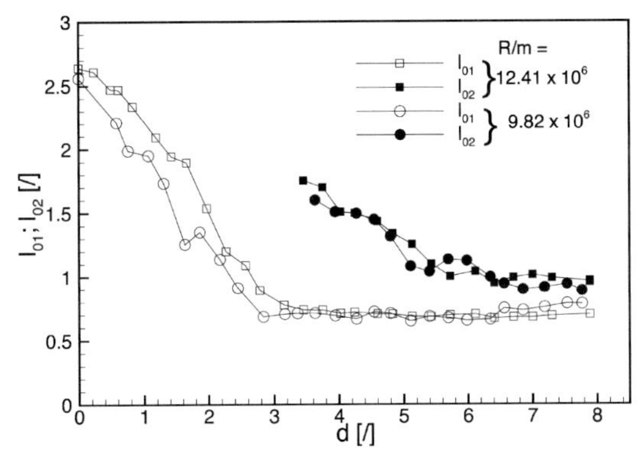


Figure 5.4.12: Dependency of the normalized upstream interaction length $l_0$ on the parameter $d$.

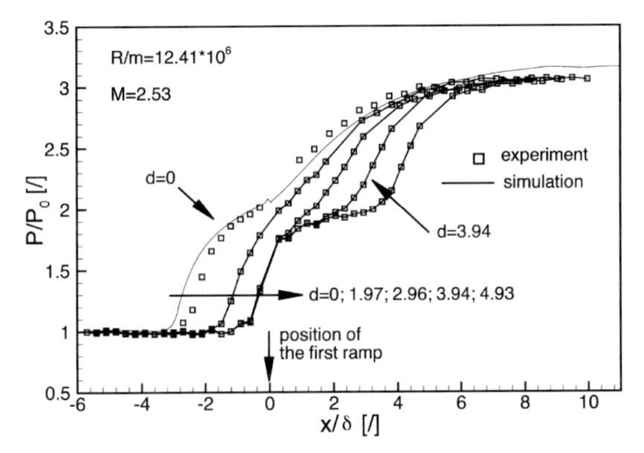


Figure 5.4.13: Comparison between measured and computed normalized static pressure distribution for $R/m \times 10^{-6} = 12.41$.

waves as a single gasdynamic phenomenon. From this point a status of flow is reached where the desired two-shock system develops.

Continuing from $d \approx 4$ a second interaction length $l_{02}$ can be well-defined and for all values $d \geq 5$, $l_{02}$ is only hardly dependent on $d$. Compared with the results of Fig. 5.4.11, this second critical value of $d = 5$ marks the beginning of an almost constant pressure level on the first ramp, i.e. the boundary layer starts to redevelop and the whole intake flow attains the working point, so that a compression takes place in two steps.

In Fig. 5.4.13, for 5 different values of $d$, the numerical results, computed with the measured boundary conditions, are compared with the experiment for $R/m = 12.41 \times 10^6$. The agreement in the prediction of the pressure distribution is quite good, even in the inflection points due to the ramp/ramp-configuration. Beginning from a ratio of $d = 3.9$ the gradient in the normalized pressure distribution becomes smaller and at $d = 4.9$ the constant pressure plateau is detectable.

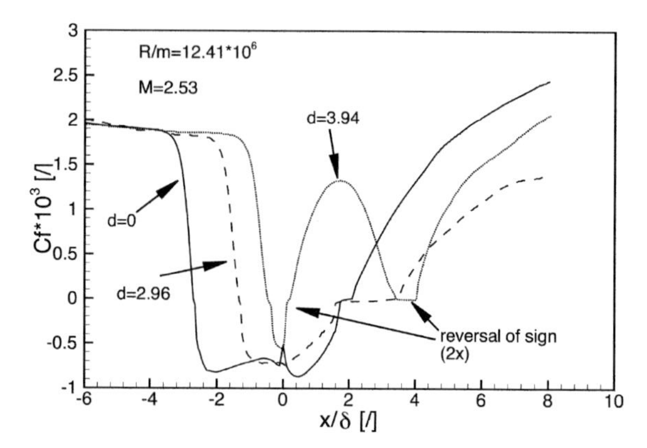


Figure 5.4.14: Computed skin friction coefficient for different values of $d$.

A confirmation is given in Fig. 5.4.14 by the skin friction coefficient distribution, determined by numerical simulations. Two small separation zones can be seen, beginning from $d \geq 4$, typical for a two-shock system. For smaller values of $d$ a wide range with negative sign in the $c_f$-distribution indicates one big separation zone with a separated second shock.

### 5.4.6 Conclusion and Outlook

In the presented work the influence of the parameter $D$ on the shock system of a double ramp configuration was investigated. On the basis of the free interaction theory especially the functional dependency of the most important parameter $L_0$ could clearly be shown. Therefore, an enlargement of the free interaction theory is suggested with respect to the "double ramp parameter" $d = D/\delta$. Moreover, a critical minimum of the first ramp length was found as $D/\delta \approx 5$ for the relevant $R/m$-ranges. For smaller values the pressure plateau as well as the boundary layer on the first ramp are not developed resp. the second shock is separated and merged with the first.

For future work three different investigations in the numerical (ITAM) and the experimental part (IAG) are planned. First on the base of the above obtained results, the unit Reynolds number will be varied about a much larger range. In a further step experiments concerning the shock oscillation itself and its correlation with the incoming boundary layer, described in [30, 31], will be realized. Additionally the effect of a real 3D-forebody boundary layer on the shock boundary layer interaction will be subject of future investigations.

## Acknowledgement

The authors are grateful to Dr. F. Schlaich for his contribution in this chapter.


## References


1. Ericson, L.E.: Effect of Boundary layer Transition on Vehicle Dynamics. Journal of Spacecraft, Vol. 6, No. 12, 1969.
2. Ericson, LE.: Transition Effects on Slender Vehicle Stability and Trim Characteristics. AIAA-Paper 73-126, Jan., 1973.
3. Rausch, V., McClinton, C., Sitz, J.: Hyper-X program overview. ISABE-99-7213.
4. Dolling, S.D.: Fifty Years of Shock-Wave/Boundary layer Interaction Research: What Next? AIAA-Journal Vol. 39, No. 8, August 2001.
5. Delery, J., Marvin, J.G.: Shock-Wave Boundary layer Interactions. AGARDograph 280, 1986.
6. Urbin, G., Knight, D.D., Zheltovodov, A.A.: Large Eddy Simulation of a Supersonic Compression Corner. AIAA-Paper 2000-0398, 2000.
7. Knight, D.D.: Numerical Simulation of Compressible Turbulent Flows Using the Reynolds-Averaged Navier-Stokes Equations. AGARD-Report 819, 1997.
8. Goldfeld, M.A., Nestoulia, R.V., Starov, A.V.: The Boundary layer Interaction with Shock Wave and Expansion Fan. Int. Journal of Thermal and Fluid Science, Vol. 9, No. 2, 2000.
9. Settles, G.S., Dodson, L.J.: Supersonic and Hypersonic Shock/Boundary layer Interaction Database. AIAA-Journal, Vol. 32, No. 7, 1994.
10. Settles, G.S., Bogdonoff, S.M., Vas, I.E.: Incipient Separation of a Supersonic Turbulent Boundary layer at High Reynolds Numbers. AIAA-Journal, Vol. 14, No. 1, 1976.
11. Schlaich, F.: Experimental investigations on the interaction between shock and turbulent boundary layer on a double ramp in supersonic flow. PhD-thesis, Stuttgart University, 1996 (in German).
12. Ardonceau, P.L.: The Structure of Turbulence in a Supersonic Shock Wave Boundary layer Interaction. AIAA-Journal, Vol. 22, No. 9, 1984.
13. Knauss, H., Riedel, R., Wagner, S.: The Shock Wind Tunnel of Stuttgart University – A Facility for Testing Hypersonic Vehicles, AIAA-Paper 99-4959, Norfolk, Virginia, 1999.
14. Riedel, R.: Erprobung eines Stoßwindkanals zur Untersuchung des laminar-turbulenten Umschlags in Überschallgrenzschichten. PhD-thesis, Stuttgart University, 2000.
15. Eberle, A.: Untersuchung einer transitionellen Plattengrenzschicht mittels einer Preston-Sonde im großen Stoßwindkanal des IAG. Studienarbeit, IAG, Stuttgart University, 2000.
16. Owen, F.K., Horstmann, C.C.: Hypersonic Transitional Boundary layers. AIAA-Journal, Vol. 10, No. 6, 1972.
17. Stainback, P.C., Wagner, R.D., Owen, F.K., Horstmann, C.C.: Experimental Studies of Hypersonic Boundary layer Transition and Effects of Wind Tunnel Disturbances. NASA TN D-7453, 1974.
18. Owen, F.K.: Transition Experiments of a Flat Plate at Subsonic and Supersonic Speeds. AIAA-Journal, Vol. 8, No. 3, 1970.
19. Gaisbauer, U., Knauss, H., Weiss, J., Wagner, S.: Measurement techniques for detection of flow disturbances and transition localization in a short duration wind tunnel. Proceedings of X International Conference on the Methods of Aerophysical Research, ICMAR, Part III, 2000.
20. Knauss, H., Gaisbauer, U., Wagner, S., Bountin, D., Maslov, A., Smorodsky, B., Betz, J.: Calibration Experiments of a New Active Fast Response Heat Flux Sensor to Mea-

sure Total Temperature Fluctuations – Part I–III. Proceedings of XI International Conference on the Methods of Aerophysical Research, ICMAR, Part III, 2002.
21. Gaisbauer, U.: Untersuchungen zur Stoß-Grenzschicht-Wechselwirkung an einer Doppelrampe unter verschiedenen Randbedingungen. PhD-thesis, Stuttgart University, 2004.
22. Wilkox, D.C.: Turbulence Modeling for CFD. DCW Industries, Inc., La Canada, California, 1993.
23. Van Leer, B.: Flux-vector splitting for the Euler equations, ICASE T. Rep. No. 82-30, N.Y., 1982.
24. Borisov, A.V., Fedorova, N.N.: Numerical simulation of turbulent flows near the forward-facing steps. Thermophysics and Aeromechanics, Vol. 4, No. 1. C 69–83, 1996.
25. Bedarev, I.A., Borisov, A.V., Fedorova, N.N.: Numerical Simulation of the Supersonic Turbulent Separated Flows in Vicinity of the Backward- and Forward-Faced Steps. Computational Fluid Dynamics Journal, Special Number, pp. 194–202, 2001.
26. Fedorova, N.N., Fedorchenko, I.A., Shuelein, A.: Experimental and numerical study of oblique shock wave/turbulent boundary layer interaction at M=5. Computational Fluid Dynamics Journal, Vol. 10, No. 3. pp. 376–381, 2001.
27. Fedorova, N.N., Kharlamova, Y.V., Gaisbauer, U., Knauss, H., Wagner, S.: Numerical and Experimental Investigation of Shock Wave/Turbulent Boundary layer Interaction on a double Ramp Configuration. PAMM, Proceedings in Applied Mathematics and Mechanics, GAMM-Tagung Augsburg, 2002.
28. Gaisbauer, U., Knauss, H., Wagner, S., Fedorova, N.N., Kharlamova, Y.V.: Shock/Turbulent Boundary layer Interaction on a Double Ramp Configuration – Experiments and Computations. Proceedings of XI International Conference on the Methods of Aerophysical Research, ICMAR, Part III, 2002.
29. Fenter, F.W., Stalmach, C.J.: The Measurement of Local Turbulent Skin Friction at Supersonic Speeds by Means of Surface Impact Pressure Probes. CM-712, DRL-30, Univ. of Texas, 1952.
30. Erengil, M.E., Dolling, D.S.: Correlation of Separation Shock Motion with Pressure Fluctuation in the Incoming Boundary layer. AIAA-Journal, Vol. 29, No. 11, Nov. 1991.
31. Erengil, M.E., Dolling, D.S.: Physical Cause of Separation Shock Unsteadiness in Shock Wave/Turbulent Boundary–Layer Interactions. AIAA-93-3134, Orlando, FL, 1993.

## 5.5 Experimental and Numerical Investigation of Lobed Strut Injectors for Supersonic Combustion

Peter Gerlinger *, Peter Kasal, Fernando Schneider,
Jens von Wolfersdorf, Bernhard Weigand, and Manfred Aigner

### Abstract

A lobed hydrogen strut injector is investigated for use in scramjet (supersonic combustion ramjet) engines at moderate combustor Mach numbers. The paper deals with the design of a strut injector that generates large-scale streamwise vortices. In case of a favourable chosen geometry counter-rotating vortices are induced that interact with the injected hydrogen and thus improve mixing. This concept achieves a rapid fuel-air mixing as well as an efficient combustion and therefore enables short combustor lengths. For cooling of the strut the injected hydrogen is used. A basic lobed strut injector is investigated experimentally for cold mixing as well as for combustion. Air flow and hydrogen injection Mach numbers are about 2. A comparison of measured and simulated hydrogen distributions will be given for a mixing test case. Based on these results modified injectors are investigated numerically. It will be shown that the lobed strut injector concept achieves good mixing and combustion performances at low losses in total pressure.

### 5.5.1 Introduction

At hypersonic flight the residence times for atmospheric air flowing from inlet to engine nozzle is in the order of a millisecond. Only a fraction of this time is available to achieve mixing of fuel and air and to complete combustion. Moreover, stable operating conditions are required over a wide range of flow Mach numbers. In some cases (usually at low flow Mach numbers) the combustion process is controlled by mixing but more often it is kinetically controlled because chemical and fluid mechanical time scales are in the same order of magnitude [1]. Besides short residence times in supersonic flow a good fuel/air mixing additionally suffers from inherently low mixing rates at high flow Mach numbers [2]. Increased compressibility causes reduced mixing layer growth rates and reduced levels of mixing. Thus efficient fuel injector concepts are a crucial point for the design of future scramjet propulsion devices.

* DLR, Institut für Verbrennungstechnik, Pfaffenwaldring 38–40, 70569 Stuttgart; peter.gerlinger@dlr.de

There are two widely used concepts of fuel injection into supersonic flow: Wall injectors (where the fuel is injected through the wall [3] or by some kind of ramp mounted to the wall [4, 5]) and strut [6–11] (or pylon [12, 13]) injectors which are located in the main flow and inject fuel directly into the core of the air flow. In some cases both kinds of injectors approach each other, e.g. if a ramp injector extends over significant parts of the channel height (e.g. [5]). The advantages and disadvantages of the different concepts are discussed in numerous papers. However, practical considerations like manufacturing and cooling problems are often neglected. Moreover, an injector concept should not only work well for one chosen set of inflow conditions but over a wide range of flow Mach numbers. In this case problems with flame stabilization and ignition delay increase strongly.

The present paper investigates a lobed strut injector with axial injection of hydrogen through the blunt end of the strut. In contrast to transverse fuel injections an injection in main flow direction may be achieved without the induction of strong shock waves if the strut is shaped in such a way that only weak oblique shock waves occur. Disadvantages of conventional streamwise injectors are its usually limited mixing capabilities. Thus techniques for mixing enhancement are required. An enhanced mixing is possible by using shock waves [14] or due to a special design of the strut [6–9, 11]. If a suited strut geometry is chosen changes in flow direction are caused which induce vortices in streamwise direction that considerably improve the fuel air mixing. A comparison between a planar and a lobed strut injector concept demonstrated (for cold hydrogen-air mixing) a significant improvement in mixing efficiency [11]. On the other hand most techniques for mixing enhancement cause additional losses in total pressure. Both factors have to be taken into account. In this paper a lobed strut is investigated for cold and hot inflow conditions, hydrogen/air mixing and combustion. All experimental results are obtained at the ITLR (Institute of Aerospace Thermodynamics, University of Stuttgart) supersonic combustion test facility [15–17]. The numerical simulations are performed at DLR Stuttgart (Institute of Combustion Technology) using the in-house TASCOM (Turbulent All Speed Combustion Multigrid) solver [18–20].

## 5.5.2 Experimental Setup and Measurement Techniques

Figure 5.5.1 shows the experimentally used mixing channel with a rectangular convergent divergent nozzle section. After expansion the duct has a constant cross section of 35.4 mm×40 mm (channel height×depth) and a length of 350 mm. The chosen inflow conditions produce under cold conditions a freestream Reynolds number of $5\times10^7$ per meter. Hydrogen is injected through the blunt end of a strut mounted at the channel centre line. The nozzle inside the strut is designed for an injection Mach number of about 2. In case of cold mixing experiments the side walls are made of optical glass or Perspex, respec-

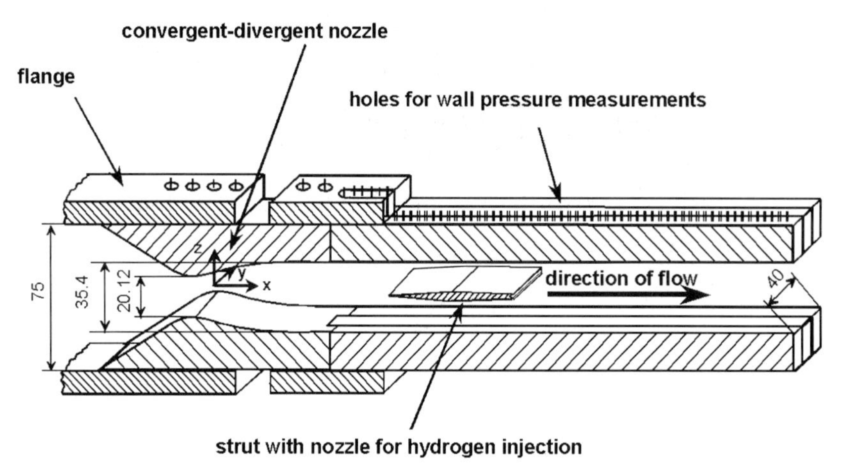


Figure 5.5.1: Supersonic mixing channel with planar strut injector (sizes in mm).

tively. For combustion experiments quartz side walls are inserted. Upper and lower walls are used as laser beam entrance and exit. The main air stream is fed by a continuously running screw compressor with a maximum mass flow of about 1.5 kg/s. Hot inflow cases are obtained by using an electric air heater which allows total temperatures of about 1400 K.

The experimental investigation of the flow field and hydrogen/air mixing is performed by:

- Schlieren technique,
- Static wall pressure measurements,
- Spontaneous 1D Raman scattering technique,
- LITA (Laser-Induced Thermal Acoustics).

For flow visualization the Schlieren technique is applied, giving a good overview about shock positions in case of planar flow fields. The upper wall is equipped with 40 holes for static wall pressure measurements. The spontaneous 1D Raman scattering technique is used to determine the mole fraction of hydrogen, oxygen and nitrogen along the laser beam simultaneously. The observed 1D length is 8 mm long. In areas where hydrogen is present, the temperature distribution could be determined from the rotational hydrogen Raman spectrum. Figure 5.5.2 shows the set-up of the 1D Raman measurement technique. A 20 W argon-ion-laser (Spectra Physic) at 514.5 nm is used for exiting of the Raman scattering. The laser light is focused over a lens FL ($f = 200$ mm) into the mixing channel. A $a/2$ retarding plate rotates the polarization for a maximum of scattered light intensity. The scattered light was collected through two achromatic lenses A1 and A2 and directed via an imaging correction unit in a polychromator (Acton AM-505F-S). A narrow band notch-filter NF, placed between both achromatic lenses, suppressed undesirable Mie and Rayleigh

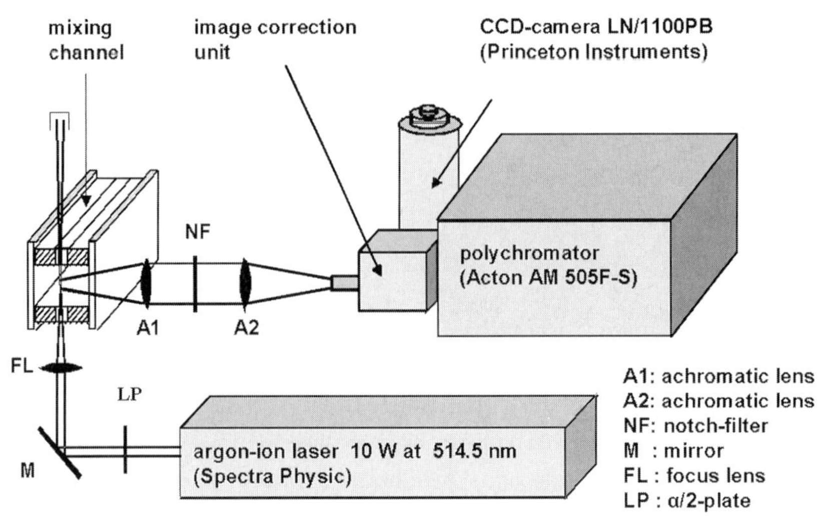


Figure 5.5.2: Set-up for 1D Raman measurement.

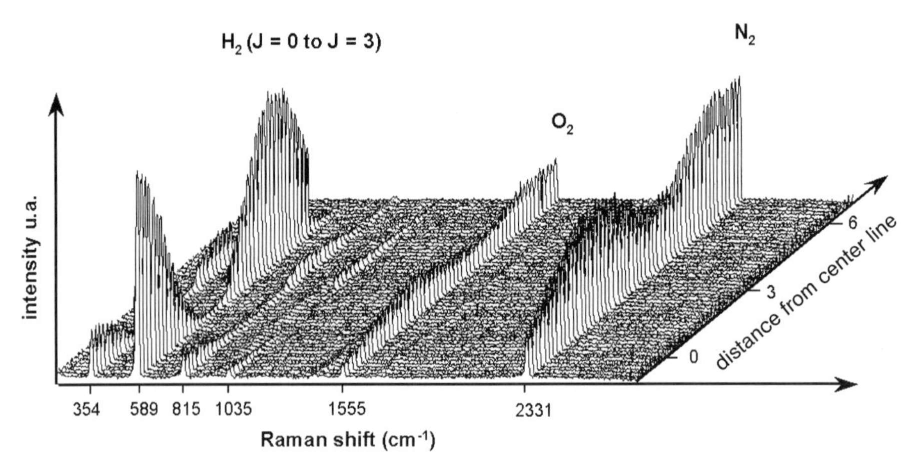


Figure 5.5.3: 1D Raman spectra 25 mm downstream of the injector.

scattered light. The Raman spectra of hydrogen, oxygen and nitrogen were recorded simultaneously with a CCD-camera (Princeton Instruments LN/CCD 1100PB). As an example Fig. 5.5.3 shows the complete 1D Raman spectra 25 mm downstream of the investigated lobed strut injector. The spectra contains the distribution of Raman intensities for four rotational lines ($J=0$ to $J=3$) of hydrogen and Q-branch intensities from oxygen and nitrogen.

### 5.5.3 Governing Equations and Numerical Simulation

The investigation of high speed turbulent combustion requires finite-rate chemistry because fluid mechanical and chemical time scales may be in the same order of magnitude. Two and three dimensional fully vectorized and parallelized codes have been developed for such simulations. While this paper only deals with applications of the developed numerical schemes a short overview about its capabilities is given. The TASCOM code solves the full Navier-Stokes, species, turbulence and variance transport equations which are given in three-dimensional conservative form by

$$\frac{\partial \boldsymbol{Q}}{\partial t}+\frac{\partial(\boldsymbol{F}-\boldsymbol{F}_v)}{\partial x}+\frac{\partial(\boldsymbol{G}-\boldsymbol{G}_v)}{\partial y}+\frac{\partial(\boldsymbol{H}-\boldsymbol{H}_v]}{\partial z}=\boldsymbol{S}. \quad (1)$$

Here

$$\boldsymbol{Q}=[\rho,\rho u,\rho v,\rho w,\rho E,\rho q,\rho\omega,\rho\sigma_T,\rho\sigma_Y,\rho Y_i]^T \quad (2)$$

is the conservative variable vector, $\boldsymbol{F}$, $\boldsymbol{G}$, and $\boldsymbol{H}$, are inviscid, $\boldsymbol{F}_v$, $\boldsymbol{G}_v$ and $\boldsymbol{H}_v$ are viscous fluxes in the x-, y-, and z-direction, respectively. $\boldsymbol{S}$ is a source vector resulting from chemistry. For turbulence closure a low-Reynolds-number $q$–$\omega$ turbulence model is employed [18, 21] using the turbulence variables $q=\sqrt{k}$ and $\omega=\varepsilon/k$ ($k$ is the turbulent kinetic energy and $\varepsilon$ its dissipation rate). The variables in Eq. (2) are the density $\rho$, velocity components $u$, $v$, and $w$, variance of temperature $\sigma_T$, mass fractions $Y_i$, and the sum of their variances $\sigma_Y$. The use of finite-rate chemistry causes the set of governing equations to be numerically stiff. This requires implicit numerical solvers. In the present case an implicit LU-SGS (Lower-Upper Symmetric Gauss-Seidel) algorithm is used for numerical time integration, which treats chemistry fully coupled with the fluid motion [19, 22–24]. As to account for turbulence effects on chemical production rates an assumed PDF (probability density function) approach with analytically formed Jacobians is available. In this case additional transport equations for the variance of temperature and the sum of species mass fraction variances have to be solved. These second order moments are used to define a clipped Gaussian PDF for temperature [20] and a multi-variate $\beta$-PDF [25, 26] for species mass fractions. Details concerning the assumed PDF closure as well as an analysis concerning its capabilities may be found in [27, 28]. Especially for complex combustor geometries, large reaction mechanisms and a large number of grid points we found this approach superior to transported PDF approaches due to much smaller computational times, the possibility to use methods for convergence acceleration and a good shock capturing. On the other hand in case of transported PDF approaches [29, 30] the shape of the PDF may freely evolve and no statistical independence between species and temperature fluctuations has to be assumed.

#### 5.5.3.1 Multigrid Convergence Acceleration

At high flow Mach numbers the combustion process is usually dominated by kinetic effects and an accurate prediction of ignition delay becomes a presumption for reliable scramjet combustor simulations. Thus finite-rate chemistry is required and a more or less large number of species transport equations has to be solved. As a consequence the numerical effort strongly increases in comparison to cases with fast chemistry or chemical equilibrium. Therefore, simulations of 3D scramjet combustors are extremely time consuming and in many cases grid convergence may not be achieved due to limitations of available computer resources. For this reason highly efficient numerical solvers are required that are able to deal with large numerical problems as well as with the numerous demands of high speed flows e.g. highly stretched grids with aspect ratios up to 10,000 or shock waves. One of the most efficient methods to solve large numerical problems is the multigrid technique [31–33]. In this case only the good abilities of iterative solvers (to damp out high frequency error components) are used while a sequence of coarser grid levels allows the reduction of low frequency error components at much lower cost. With this technique solutions of elliptic problems are possible with an effort that linearily scales with the number of grid points. In case of supersonic combustion a number of problems arises which endanger convergence of multigrid solvers or even make it impossible. The most important ones are caused by

- a set of governing equations which is of mixed type (e.g. elliptic hyperbolic),
- strong shock waves,
- grid alignment (main flow direction aligns with one direction of the computational grid),
- low-Reynolds-number turbulence closures,
- highly stretched grids,
- finite-rate chemistry.

In many practical scramjet calculations all mentioned difficulties appear at the same time. Nevertheless multigrid simulations and considerable reductions in CPU time are possible in this field too. However, modifications of the standard FAS (full approximation storage) multigrid procedure for non-linear problems [31] are required. Most of them concern restriction or prolongation. In the described TASCOM solver a full coarsening multigrid method is used for 2D simulations and a semi-coarsening for 3D cases. Due to the semi-coarsening problems of grid alignment and highly stretched grids (being more prominent in 3D cases) are reduced. As to deal with problems caused by low-Reynolds-number turbulence closures parts of the strongly non-linear source terms are kept frozen at all coarse grid levels [32, 33]. The simulation of strong (even detached) shock waves is enabled by a local damping of the restricted residual error in the vicinity of shock waves [32, 34]. The use of simple upwind prolongation techniques was found to be of minor importance. Multigrid simulations even of complex combustors with shock trains and separation have proven to be possible with these simple modifications. They are stable and reliable and achieve a considerable reduction in required CPU time. More complicated (and less reliable) are multigrid simulations using finite-rate chemistry. A local damping of restricted species residual errors in regions of high chemical activity works well for attached flames [19], but may fail for detached ones. Nevertheless strong CPU time reductions are demonstrated for attached and detached methane and hydrogen combustion and assumed PDF closure [20].

## 5.5.4 Strut Design and Performance Parameters

Planar and lobed strut injectors have been investigated experimentally and numerically. Due to the limited mixing capability of the planar version only the lobed injector is used for practical investigations. Fig. 5.5.4 shows a sketch of both injectors. The strongly improved mixing of the lobed injector is caused by streamwise vortices which are induced by the chosen geometry of the strut. In addition to the struts used in experiments a number of modified planar and lobed injectors have been investigated numerically [10, 11]. The aim of these studies is the design of a strut that allows to inject a defined mass flow rate of hydrogen, achieve a good mixing over a length which should be as short as possible, and at the same time keep the losses as small as possible. Obviously a compromise is required between the different demands. Losses in total pressure are caused by the mixing, by boundary layers and shock waves. Because the strut is located at the channel axis, shock waves are unavoidable but have to be kept as weak as possible. Additional compromises are required because of manufacturing and cooling demands.

Figure 5.5.5 shows a schlieren photograph of an experiment for which the planar strut has been used. Below the Schlieren photograph measured static wall pressures are given for both strut types. Because the geometry is more

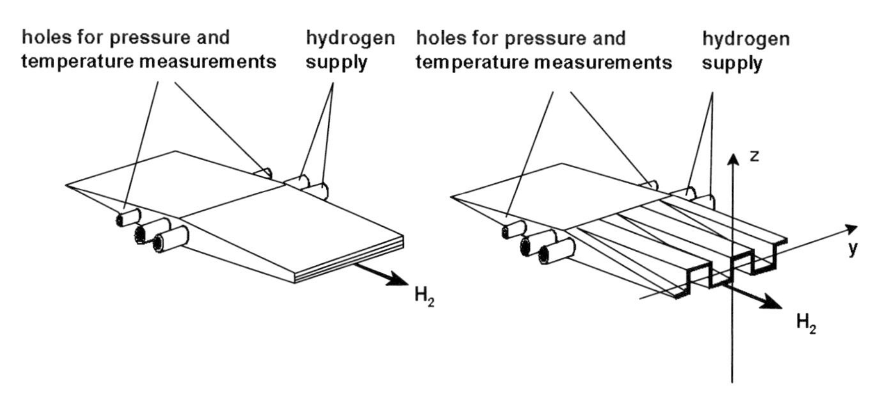


Figure 5.5.4: Geometries of the planar and lobed strut injector.

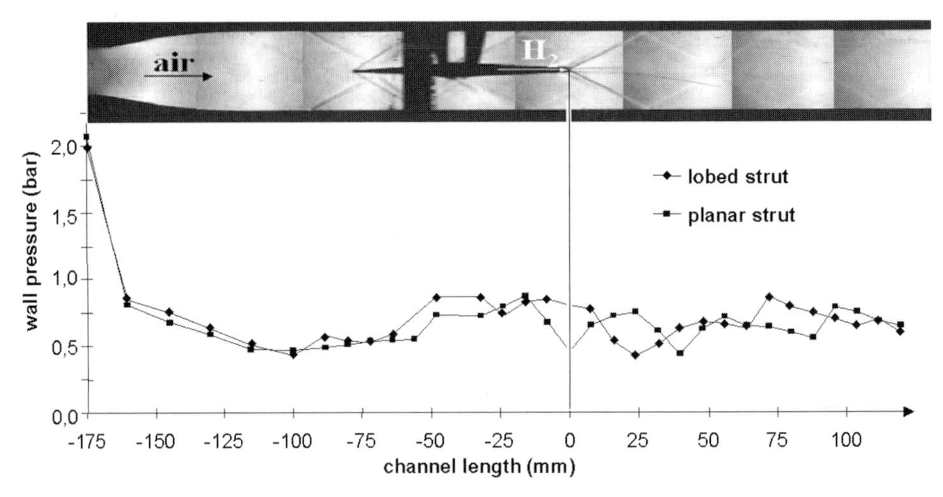


Figure 5.5.5: Schlieren photograph using the planar strut and wall static pressures along the upper wall.

complex and no longer planar in case of the 3D strut a Schlieren visualization gives no insight concerning the shock structure.

As to assess the mixing process, performance parameters are used for the numerical results. The mixing efficiency

$$\eta_{mix}(x) = \frac{\int_A \rho u Y_{H_2,r} dA}{\dot{m}_{H_2}} \tag{3}$$

is defined as the fraction of hydrogen mass flux in every cross section that could react due to available oxygen (if complete reaction took place) in relation to the total hydrogen mass flow. $Y_{H_2,r}$ is the hydrogen fraction that could react, and $A$ is the channel cross section. If the mixing efficiency is 1, all hydrogen could react because oxygen is available at least at a stoichiometric rate. If mixing does not take place in case of combustion it is due to finite-rate chemistry effects. Another important factor to evaluate a mixing or combustion chamber are losses in total pressure. The mass flux-averaged total pressure is calculated for every channel cross section by

$$\bar{p}_t(x) = \frac{1}{\dot{m}} \int_A p_t \rho u dA \tag{4}$$

where $p_t$ is the total pressure determined from local values of Mach number, gas composition, and static pressure.

## 5.5.5 Supersonic Mixing

In a first step both types of strut injectors (planar and lobed) are investigated for cold mixing of hydrogen and air. The experimental set up is depicted in Fig. 5.5.1. The corresponding 2D simulation of the planar test case uses the air nozzle throat (see Fig. 5.5.1) with sonic conditions as inlet while the 3D lobed injector simulations start at the tip of the strut to keep the number of grid points acceptable. The inflow conditions for the latter simulations are obtained from 2D calculations of the upstream channel part. In case of the planar strut the hydrogen nozzle inside the strut is included into the simulation, while for the lobed injector the hydrogen inflow is located at the nozzle exit. All simulations use as many symmetry conditions as possible. While the upper and lower wall are taken into account the side walls of the combustion chamber are neglected. Nevertheless 2.2 million grid points are required for the 3D simulations. The total pressures of air and hydrogen are 4 bar and 4.3 bar, respectively. The total temperatures are 295 K for air and 306 K for hydrogen.

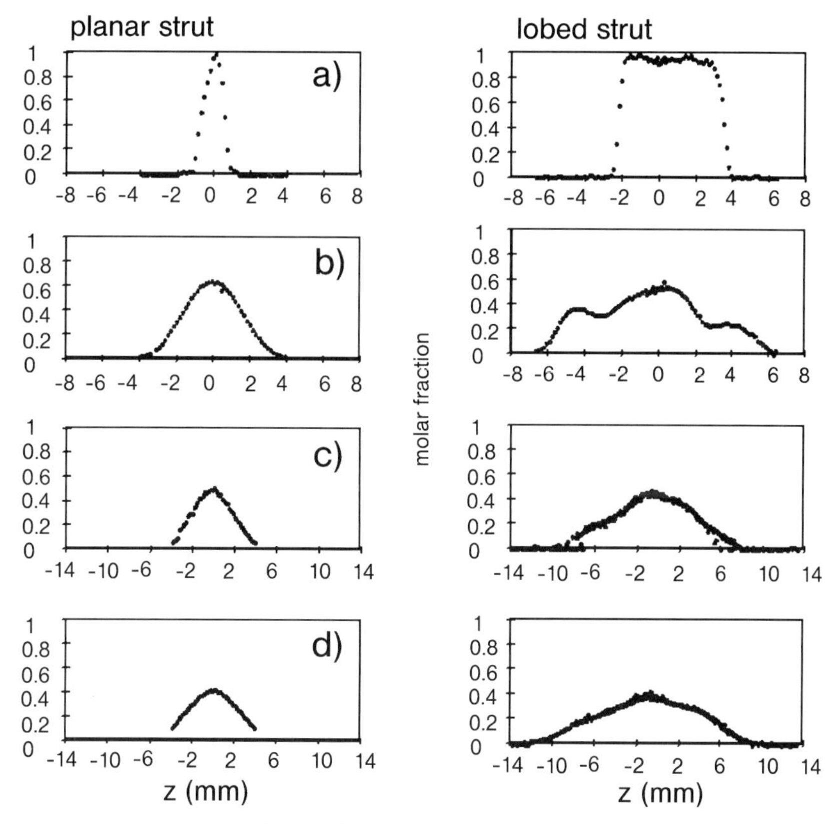


Figure 5.5.6: Experimentally obtained hydrogen molar fraction profiles for the planar (left side) and lobed (right side) injector (y=0 mm) at x=3 (a), 25 (b), 100 (c), and 150 mm (d).

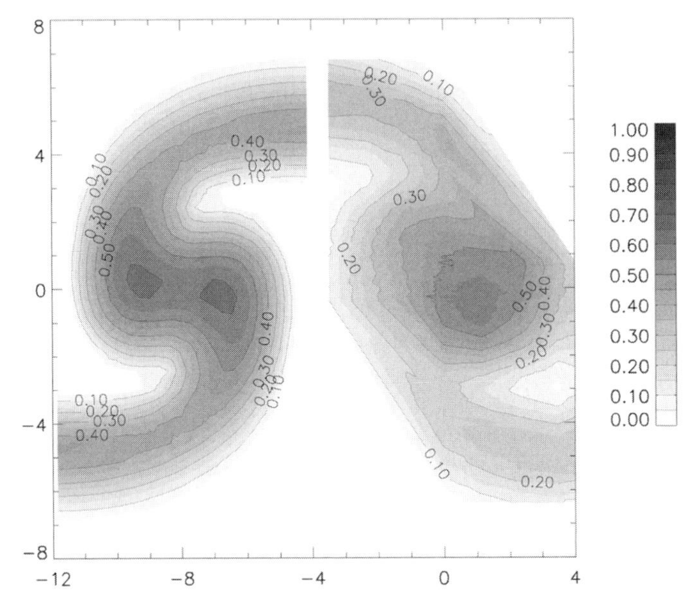


Figure 5.5.7: Calculated (left side) and measured (right side) hydrogen molar fractions in cross sections located at x=50 mm downstream of the injector.

Figure 5.5.6 shows experimentally obtained molar fraction profiles for both struts. A two-dimensional plot of measured and calculated hydrogen molar fractions is given in Fig. 5.5.7. This figure shows on the left side calculated and on the right side experimentally obtained distributions of hydrogen for a cross section located 50 mm downstream of the injector. The formation of streamwise vortices can be observed from these figures. These vortices are responsible for a more rapid decay of the maximum hydrogen concentration along the channel length and better mixing efficiencies in comparison to planar configurations. Even if the increased losses in total pressure are taken into account the lobed injector achieves better overall performances [11]. More details concerning these mixing investigations may be found in [10, 11].

## 5.5.6 Supersonic Combustion

In case of chemical reactions the chosen combustor geometry has to be changed in comparison to non-reactive flows as to compensate for density reductions due to heat release. In the present case the channel cross section is increasing in the main combustor section. This is achieved by a constant angle for the upper and lower channel walls to diverge. In case of kinetically stabilized flames the chosen angle is an important parameter which has to ensure a stable combustion without causing blockage of the main flow. The same strut

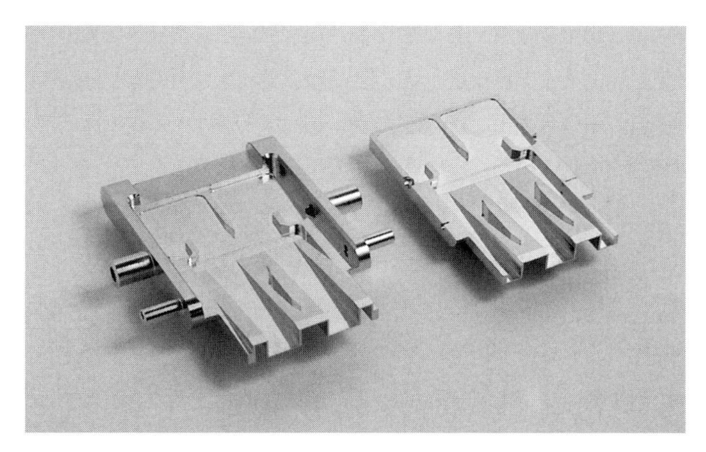

Figure 5.5.8: Inside contour of the lobed strut injector.

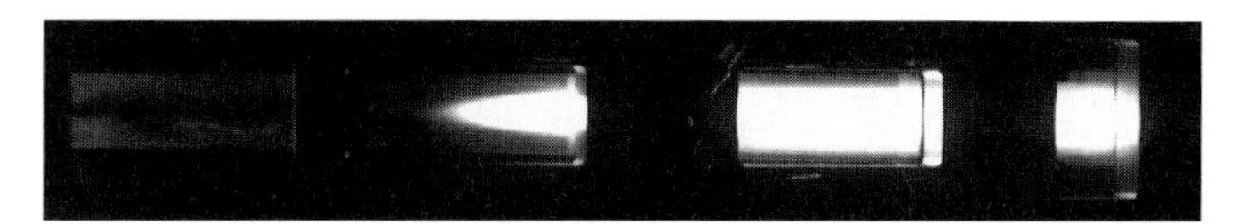

Figure 5.5.9: Combustion experiment with the planar strut injector: supersonic hydrogen/ air flame, $T_0 = 1364$ K, chamber opening angle 2°.

injectors are employed as for non-reactive mixing. Because long time experiments are performed, an active cooling of the strut is required. Fig. 5.5.8 shows the upper and lower parts out of which the lobed copper strut is manufactured. The injected hydrogen is used as a coolant and the struts inside is formed in such a way that the tip is efficiently cooled. Practical investigations have proven this concept to work reliably even in case of continuous operation. Fig. 5.5.9 shows a photograph of a kinetically stabilized flame using the planar strut for hydrogen injection. There is a considerable ignition delay between the end of the strut and the point where ignition takes place. For the planar strut injector this length is about 19.5 cm. It is strongly reduced if the hydrogen is injected by the lobed strut injector, mainly due to better mixing capabilities.

#### 5.5.6.1 Investigation of Different Lobed Strut Injectors

In order to optimize the experimentally chosen geometry of the combustion chamber and strut configuration a numerical study has been performed. These simulations are based on conditions which may be realized at the ITLR test facility. However, in some cases conditions which depart from these values would lead to better combustion efficiencies. For the simulation of hydrogen-air com-

Table 5.5.1: Inflow conditions for air and hydrogen, Mach number, static pressure, static temperature and species mass fractions.

| | *Ma* | *p* (bar) | *T* K) | $H_2$ | $O_2$ | $N_2$ |
|---|---|---|---|---|---|---|
| air | 1 | 2.11 | 1300 | 0 | 0.23 | 0.77 |
| hydrogen | 2 | 0.8 | 300 | 1 | 0 | 0 |

bustion a 9-species 20-step reaction mechanism is employed. Despite the use of finite-rate chemistry we have to keep in mind that for the lifted flames obtained the uncertainty of the reaction mechanism may be a source of error. A computational grid with 0.6 million volumes has been used for most of these simulations. This grid was found to be not fine enough to avoid grid dependencies of the solution. However, the comparison with a 3.2 million volumes simulation has shown that the basic features of the flow field as well as performance parameters like mixing efficiency and loss in total pressure are met pretty well by the coarser grid. Due to the extremely long computational times for a 3D, 9 species simulation with 3.2 million volumes the smaller grid is preferred. All simulations are parallelized by domain decomposition using MPI for message passing [35]. The inflow conditions for air at the nozzle throat ($Ma=1$) and hydrogen at the strut nozzle exit ($Ma=2$) are summarized in Tab. 5.5.1. The mass flow rate of air is 0.332 kg/s and for hydrogen between 3.41 g/s and 6.46 g/s depending on the chosen type of strut. Fig. 5.5.10 shows 3 numerically investigated types of lobed injectors from a back view. While the principal shape of the strut is the same, the hydrogen nozzle sections have different outlet areas. The upper strut (strut I) corresponds to the one being used in the experiment. The strut in the middle (strut II) is a modified version where only the horizontal parts of the nozzle are open for hydrogen injection. Finally strut III consists out of several injection parts that allow an air flow between the hydrogen jets and thus a further enhanced mixing. Because the inflow conditions are kept constant the injected hydrogen mass flow rates differ for the 3 struts. For strut I it is 6.46 g/s, for strut II 4.07 g/s, and for strut III 3.41 g/s, respectively.

As mentioned before the combustor geometry has to be diverging with beginning combustion. Because of the ignition delay it is advantageous to keep the channel cross section directly downstream of the injector constant up to the point where ignition takes place. From different simulations a length of about 5 cm (from the strut to the beginning of the diverging channel section) was found to work well and stabilize combustion. In a next step an angle of about 2.5° was identified to achieve best results concerning stable ignition and avoiding thermal choking. The use of these geometry parameters ensures stability of the combustion process even for slightly changed inflow conditions or changed types of injectors. Around the point of ignition a small subsonic zone is formed which is responsible for the stabilizing effect on combustion.

Figure 5.5.11 shows the temperature distribution in a cross section downstream of the injector. This simulation is performed for strut II. The length of

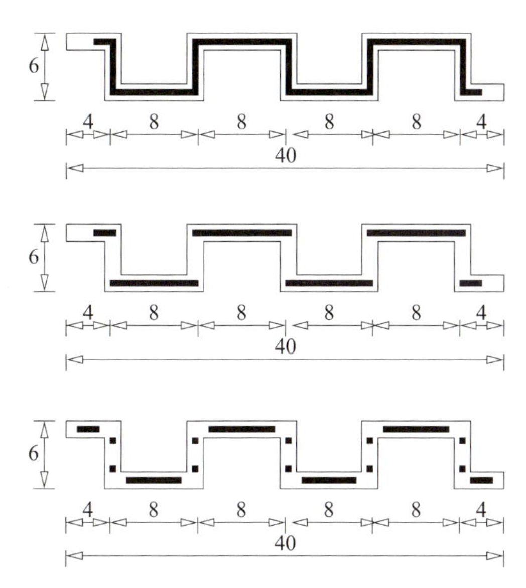


Figure 5.5.10: Back views of 3 different strut types (strut I, II and III from top to bottom). The active hydrogen nozzle sections parts are indicated.

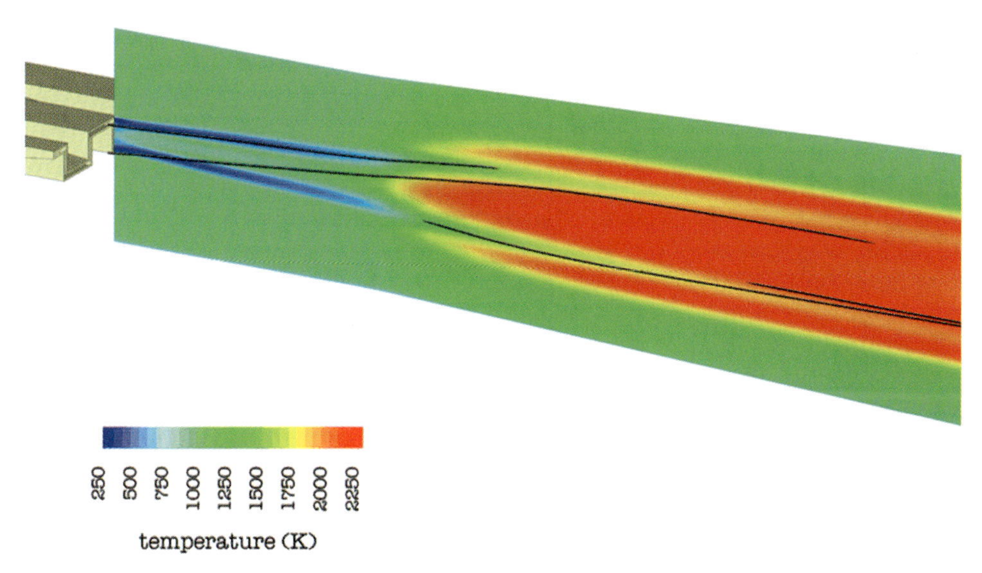


Figure 5.5.11: Streamlines and temperature distribution near the injector.

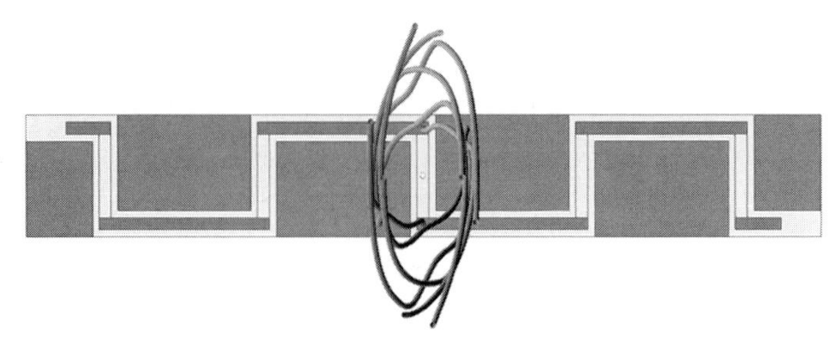

Figure 5.5.12: Streamlines starting at the injector exit (view from the back).

ignition delay is about 5 cm. The figure also shows the temperature rise for the cold hydrogen injected at 300 K into a surrounding air flow of about 950 K. It takes approximately 4.5 cm for the hydrogen to reach a temperature which is above ignition criterion. The rapid increase in temperature with ignition causes a reduction in Mach number and the occurrence of oblique shock waves. There is also a small subsonic bubble and a small normal shock in the centre of the combustion zone. Due to the diverging channel geometry the subsonic bubble disappears further downstream. The black lines in Fig. 5.5.11 indicate streamlines and demonstrate the formation of streamwise vorticity. This becomes even clearer from Fig. 5.5.12.

This figure shows streamlines starting at the end of the strut in a view from the back in upstream direction. The vortices induced by the strut are maintained up to the channel exit. These vortices are responsible for an enhanced mixing [11]. To provide a more complete view of the reacting flow field the distributions of $H_2$, $H_2O$ and OH mass fractions are plotted in Fig. 5.5.13 (from top to bottom) for the simulation using strut II. Again the ignition delay is clearly visible from these plots. The colour distribution for hydrogen is chosen in such a way that mass fractions below 0.2 are easy to indicate. This is important for a control of the hydrogen burnout. These figures show a problem which will be discussed later in more detail: In the core of the lobed structure there is still hydrogen available at the channel end while no oxygen is left.

For the evaluation of the combustion process performance parameters are calculated for struts I to III. Figure 5.5.14 shows the total pressures $\bar{p}_t$ and the hydrogen mass flow rates $\dot{m}_{H2}$ normalized with the corresponding inflow values for these struts. In addition the local mixing efficiency $\eta_{mix}$ is given which refers to values at the corresponding cross section. In all cases there is still unburnt hydrogen at the channel exit. This is mainly due to the lack of oxygen in the channel middle part. For the chosen air inflow conditions the injected hydrogen mass flow rates of strut I and II are too high as to achieve a complete burnout. Strut III injects less hydrogen and additionally offers better mixing capabilities. In this case about 12% of the hydrogen mass flux is still unburnt at the channel exit. However, in contrast to strut I and II the mixing efficiency is very high for strut III. Nearly half of the remaining hydrogen is mixed in such a way that it could react. If a complete burnout is not achieved for strut III, it may be due to

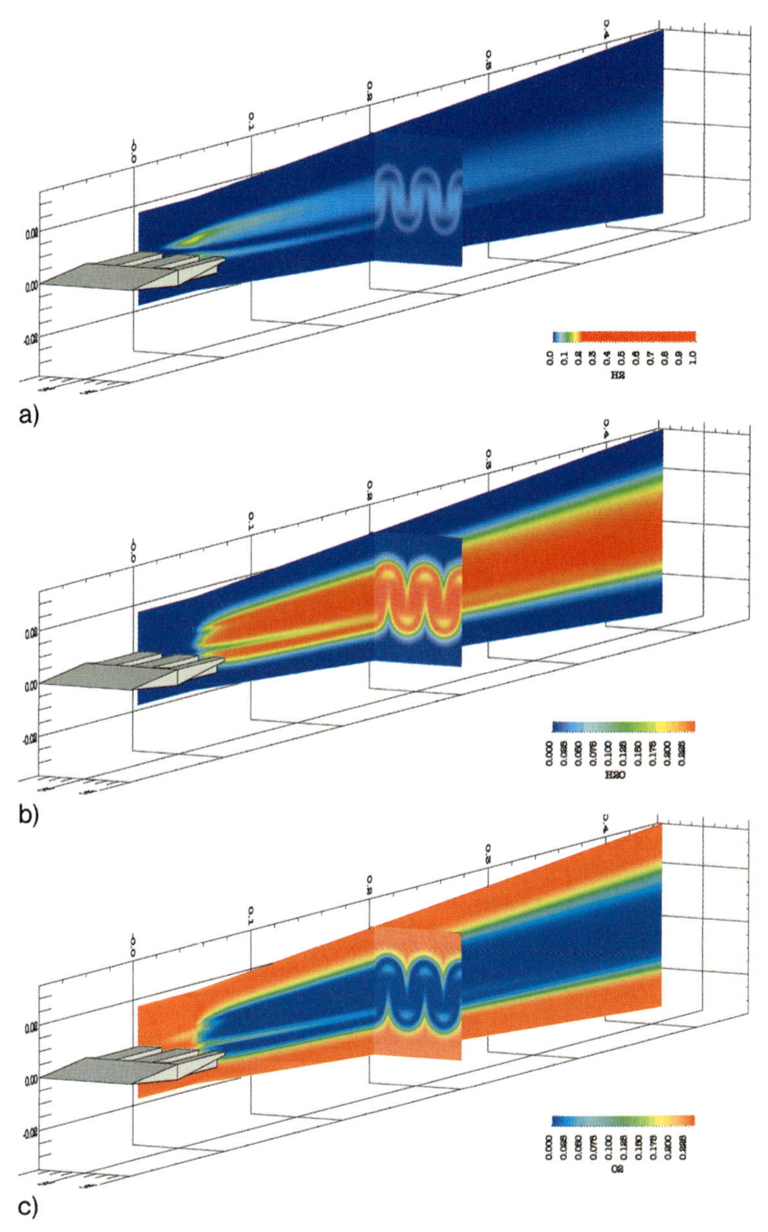


Figure 5.5.13: Distributions of hydrogen (top), water (middle), and oxygen (bottom) mass fractions obtained from a simulation of strut II.

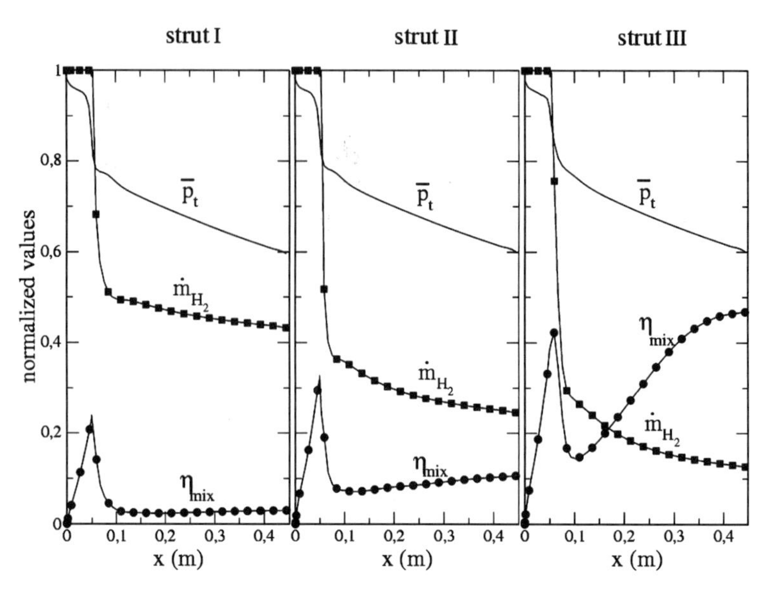


Figure 5.5.14: Performance parameters for strut I, II, and III (— normalized total pressure, ■ normalized hydrogen mass flux rate, ● mixing efficiency).

finite-rate chemistry effects or due to the diverging combustor geometry. Nevertheless strut III seems to be nearly optimized for the given air inflow conditions.

Better combustion efficiencies could be achieved for the struts I and II by an increase in air pressure (this value was defined by the experimental capabilities). A higher pressure would result in a better oxygen supply at the core of the channel and therefore enable a complete combustion. For practical realizations of a scramjet engine the injected hydrogen mass flow has to be controlled in accordance with the air inflow conditions. For injector III and the given inflow conditions a complete burnout should be possible in less than 50 cm combustor length.

## 5.5.7 Conclusions

Planar and lobed strut injectors have been investigated both experimentally and numerically. It is shown that the mixing process can be enhanced by choosing appropriate lobed injectors. Due to the induction of streamwise vortices mixing is accelerated and the ignition delay is reduced. Numerical simulations demonstrated a kinetic stabilization of the flame with beginning divergence of the channel. A small subsonic bubble is responsible for this effect. For the lobed strut injectors investigated an accurate choice of the hydrogen mass flow rate (in accordance with the air mass flow rate) is important to achieve a complete burnout. This is possible for all investigated types of lobed strut injectors. From the investigated types of struts, strut III offers the best mixing capabilities at comparable losses in total pressure.

## References


1. Drummand, J.P., Diskin, G.S., Cutler, A.D.: Fuel-Air Mixing and Combustion in Scramjets, AIAA Paper 2002-2378, 2002.
2. Glawe, D.D., Samimiy, M., Nejad, A.S., Cheng, T.H.: Effects of Nozzle Geometry on Parallel Injection from Base of an Extended Strut into a Supersonic Flow, AIAA paper 95-0522, 1995.
3. Belanger, J., Hornung, H.G.J.: Transverse Jet Mixing and Combustion Experiments in Hypervelocity Flows, J. Propul. Pow., 1996, 12, 186–192.
4. Riggins, D.W., McClinton, C.R., Rogers, R.C., Bittner, R.D.: Investigation of Scramjet Injection Strategies for High Mach Number Flows, J. Propul. Power, 1995, 11, 409–418.
5. Baurle, R.A., Fuller, R.P., White, J.A., Chen, T.H., Gruber, M.R., Nejad, A.S.: An Investigation of Advanced Fuel Injection Schemes for Scramjet Combustion, AIAA paper 98-0937, 1998.
6. Strickland, J.H., Selerland, T., Karagozian, A.R.: Numerical Simulation of a Lobed Fuel Injector, Phys. Fluids, 1998, 10, 2950–2964.
7. Charyulu, B.V.N., Kurian, J., Venugopalan, P., Sriamulu, V.: Experimental Study on Mixing Enhancement in Two Dimensional Supersonic Flow, Exp. Fluids, 1998, 24, 340–346.
8. Sunami, T., Wendt, M., Nishioka, M.: Supersonic Mixing and Combustion Control Using Streamwise Vorticity, AIAA paper 98-3271, 1998.
9. Sunami, T., Scheel, F.: Numerical Study on the Supersonic Mixing Enhancement Using Streamwise Vortices, AIAA paper 2002-5117, 2002.
10. Gerlinger, P., Brüggemann, D.: Numerical Investigation of Hydrogen Strut Injections into Supersonic Airflows, J. Prop. Power, 2000, 16, 22–28.
11. Gerlinger, P., Kasal, P., Stoll, P., Brüggemann, D.: Experimental and Theoretical Investigation on 2D and 3D Parallel Hydrogen/Air Mixing in a Supersonic Flow, ISABE paper 2001-1019, 2001.
12. Lyubar, A., Sattelmayer, T.: Numerical Investigation of Fuel Mixing, Ignition and Flame Stabilization by a Strut Injector in a Scramjet Combustor, Proc. Int. Conf. on Methods of Aerophysical Research (ICMAR), 2002, 122–127, Novosibirsk, Russia.
13. Sander, T., Sattelmayer, T.: Application of Spontaneous RAMAN Scattering to the Flowfield in a Scramjet Combustor, Proc. Int. Conf. on Methods of Aerophysical Research (ICMAR), 2002, 122–127, Novosibirsk, Russia.
14. Fernando, E.M., Menon, S.: Mixing Enhancement in Compressible Mixing Layers: An Experimental Study, AIAA J., 1993, 31, 278–285.
15. Brellochs, F., Fertig, M., Algermissen, J., Brüggemann, D.: Optical Measurements of Hydrogen Mixing in Supersonic Airflows, Fluid Mechanics and its Applications, Champion, M., Deshaies, B. (Eds.), 1997, 71–79.
16. Kasal, P., Boltz, J., Gerlinger, P., Brüggemann, D.: Raman Measurements of Supersonic Hydrogen/Air Mixing, Laser Application to Chemical and Environmental Analysis, OSA Technical Digest, 1998, 3, 91–97.
17. Kasal, P., Gerlinger, P., Walther, R., von Wolfersdorf, J., Weigand, B.: Supersonic Combustion: Fundamental Investigations of Aerothermodynamic Key Problems, AIAA paper 2002-5119, 2002.
18. Gerlinger, P., Algermissen, J., Brüggemann, D.: Numerical Simulation of Mixing for Turbulent Slot Injection, AIAA J., 34, 73–78, 1996.
19. Gerlinger, P., Stoll, P., Brüggemann, D.: An Implicit Multigrid Method for the Simulation of Chemically Reacting Flows, J. Comp. Physics, 146, 322–345, 1998.

20. Gerlinger, P., Möbus, H., Brüggemann, D.: An Implicit Multigrid Method for Turbulent Combustion, J. Comp. Physics, 167, 247–276, 2001.
21. Coakley, T.J., Huang, P.G.: Turbulence Modeling for High Speed Flows, AIAA paper 92-0436, 1992.
22. Jameson, A., Yoon, S.: Lower-Upper Implicit Scheme with Multiple Grids for the Euler Equations, AIAA J., 25, 929–937, 1987.
23. Shuen, J.S.: Upwind Differencing and LU Factorization for Chemical Non-Equilibrium Navier-Stokes Equations, J. Comp. Physics, 99, 233–250, 1992.
24. Stoll, P., Gerlinger, P., Brüggemann D.: Domain Decomposition for an Implicit LU-SGS Scheme Using Overlapping Grids, AIAA-paper 97-1896, 1997.
25. Girimaji, S.S.: A Simple Recipe for Modeling Reaction-Rates in Flows with Turbulent Combustion, AIAA-paper 91-1792, 1991.
26. Baurle, R.A., Alexopolous, G.A., Hassan, H.A.: Assumed Joint Probability Density Function Approach for Supersonic Turbulent Combustion, J. Propul. Power, 10, 473–484, 1994.
27. Gerlinger, P.: Investigation of an Assumed PDF Approach for Finite-Rate Chemistry, Combust. Sci. and Tech., 175, 841–872, 2003.
28. Gerlinger, P., Noll, B., Aigner, M.: Assumed PDF Modeling and PDF Structure Investigation Using Finite-Rate Chemistry, Progress Comput. Fluid Dynam., in press, 2005.
29. Pope, S.: PDF methods for turbulent reactive flows, Prog. Energy Combust. Sci., 11, 119–192, 1985.
30. Möbus, H., Gerlinger, P., Brüggemann, D.: Scalar and Joint Scalar-Velocity-Frequency Monte-Carlo PDF Simulation of Supersonic Combustion, Combust. Flame, 132, 3–24, 2003.
31. Brandt, A.: Multi-Level Adaptive Solutions to Boundary Value Problems, Mathemat. Comput., 1977, 31, 333–390.
32. Gerlinger, P., Brüggemann, D.: Multigrid Convergence Acceleration for Turbulent Supersonic Flows, Int. J. Numerical Methods in Fluids, 24, 1019–1035, 1997.
33. Gerlinger, P., Brüggemann, D.: An Implicit Multigrid Scheme for the Compressible Navier-Stokes Equations with Low-Reynolds-Number Turbulence Closure, J. Fluids Engineer., 120, 257–262, 1998.
34. Gerlinger, P., Stoll, P., Brüggemann, D.: Robust Implicit Multigrid Method for the Simulation of Turbulent Supersonic Mixing, AIAA J., 37, 766–768, 1999.
35. Gerlinger, P., Stoll, P., Schneider, F., Aigner, M.: Implicit LU Time Integration Using Domain Decomposition and Overlapping Grids, in High Performance Computing in Science and Engineering '02, Edts. E. Krause, W. Jäger, 2002, 311–322.

## 5.6 Experimental Studies of Viscous Interaction Effects in Hypersonic Inlets and Nozzle Flow Fields

Andreas Henckels and Patrick Gruhn *

### Abstract

Two experimental project studies concerning the aerodynamics of hypersonic airbreathing propulsion components were performed at the hypersonic wind tunnel H2K of DLR. Fundamental experiments on the inlet flow field revealed the influence of three dimensional flow structures, like Görtler vortices and a complex corner flow, on the inlet heat loads. Boundary layer bleed to reduce these heat loads was successfully implemented. Experiments on the SERN nozzle flow field gave detailed information about the boundary layer separation at the nozzle flap and contributed to the design of an optimized flap with increased nozzle performance. Further on, the influence of flow temperature effects on the thrust was shown by experiments with heated nozzle flow. Thus, the experiments gave substantial information to support the design of future hypersonic space vehicles.

### 5.6.1 Introduction

Since 1989 the DLR wind tunnel section of the Institute of Aerodynamics and Flow Technology in Cologne has been a partner of the University of Technology Aachen (RWTH Aachen) in frame of the "Sonderforschungsbereich 253" (Collaborative Research Centre 253) founded by the Deutsche Forschungsgemeinschaft (DFG). The programme objective is the development of "Fundamentals of Space Plane Design", with focus on the TSTO (**T**wo **S**tage **T**o **O**rbit) reference configuration ELAC (**EL**liptical **A**erodynamic **C**onfiguration). One main challenge was the design of the propulsion system for the lower stage, propelled by airbreathing engines. In case of ELAC, six combined co-axial dual mode turbojet/ramjet engines accelerate the vehicle along the ascent trajectory to the separation Mach number of 7.5. To guarantee the required thrust, all components of the propulsion unit, i.e. the adjustable air inlet, the combustion chamber and the adaptable nozzle have to act synchronized. Beyond this, a high degree of airframe integration as well as significant thermal and weight constraints have to be taken into account. In order to support feasible design solutions, the DLR wind tunnel section performed two project studies.

* DLR, Institut für Aerodynamik und Strömungstechnik, Linder Höhe, 51147 Köln; patrick.gruhn@dlr.de

The first project study was orientated towards the inlet and entitled "Experimental Optimization of a Hypersonic Inlet". One of the dominating flow phenomena inside an hypersonic inlet is the interaction of oblique shocks induced by the inlet ramps with the boundary layers developing on the walls. Here, viscous effects have an important impact on the flow field and by this, on the efficiency of the whole inlet system. The first objective of the study was to gain comprehensive information about the viscous influence on those interaction phenomena. Generic experiments on flat plate models with an impinging oblique shock addressed in particular shock induced boundary layer separation, generated heat loads, and boundary layer transition phenomena [1, 2]. In a more technically orientated step, methods to reduce interaction effects, for instance by bleed or surface curvature, were studied [3]. Finally, the entire inlet flow field, i.e. the external compression on the ramps as well as the internal compression process was evaluated by scaled wind tunnel models [4]. An important issue was the validation of numerical codes within another project of the same research programme (C3), which was supported by providing experimental data on the viscous corner flow structure, generated by the interacting shocks of the compression ramps and the boundary layer on the side walls.

In addition to the inlet, the thrust nozzle is a major aerodynamic propulsion component. A second project performed by DLR wind tunnel section, entitled "Investigation on a nozzle flow for hypersonic propulsion", was orientated towards a SERN (**S**ingle **E**xpansion **R**amp **N**ozzle) thrust nozzle. During the preliminary design phase special attention had to be paid to its geometry and the resulting drag in order to guarantee the required thrust for acceleration. Due to a high sensitivity of the net thrust against losses inside the nozzle at high Mach numbers [5] and due to strong losses at transonic Mach numbers [6], an optimization of the geometry based on an exact knowledge of the nozzle flow field is an essential task. To achieve feasible technical solutions for the nozzle design, the study focused on the integration of the nozzle into the flight vehicle and on the aerodynamic interaction effects between the ambient flow and the nozzle flow field. Here, particularly the flow around the flap, that guarantees the adaptation of the nozzle, was of interest, because of its substantial contribution to the nozzle drag [7, 8]. In a further step, the exhausted nozzle flow was heated up to demonstrate the influence of temperature effects, for example the changes in viscosity, on the nozzle flow field and the shear layers.

The design of propulsion components is mainly performed by the application of high performance CFD-simulations. Therefore, reliable experiments are essential for the validation of the numerical codes as well as to understand physical effects and their impact on the overall design. The long term experience of DLR in technology orientated hypersonic research in its facility H2K allowed it to provide a qualified contribution for the understanding of those aspects. The main goal of the experiments was to gain an insight into the complex supersonic and hypersonic flow phenomena governing the aerodynamics of the major aerodynamic propulsion components. Thus helpful information about technical capabilities and limitations were given in order to support the identification of feasible technical solutions for the design of propulsion components for advanced hypersonic space planes.

## 5.6.2 Experimental Techniques

### 5.6.2.1 Facility and Flow Diagnostics

The H2K facility at DLR Cologne is an intermittently working blow down tunnel with a free stream test section (Fig. 5.6.1). Since the stagnation pressure is limited to a maximum of 55 bar, the counter-pressure has to be decreased to a few mbar by a vacuum sphere to accelerate the flow to hypersonic speeds. Depending on the flow condition, a test duration up to 40 seconds can be achieved. The Mach number, which was 5.3 and 6.0 respectively for the presented experiments, is adjusted by the ratio of the geometrical cross sections of exchangeable axial symmetrically contoured nozzles with an exit diameter of 600 mm. To avoid air condensation and to operate the facility at high stagnation temperatures, electrical heaters with a power of 5 MW are integrated upstream the nozzle. Reynolds numbers in the range from $2.5\times10^6\ m^{-1}$ to $20\times10^6\ m^{-1}$ can be set by varying the stagnation pressure, while the stagnation temperature is kept constant. Additionally, an auxiliary electrical resistance heater of 400 kW power, equipped with twelve Kanthal-spirals of 24 mm diameter, is installed at the H2K facility to heat up secondary nozzle flows (Fig. 5.6.2). This heater provides dried air with mass flow rates of up to $0.3\ kg\times s^{-1}$ at pressures up to 60 bar and temperatures up to 1100 K. Before its injection into the test chamber over a hot gas valve the heated air flows into the ambient atmosphere until the desired steady state temperature for the test run is established.

The establishment of the flow behind the wind tunnel nozzle and around the model was controlled by optical flow field visualization and pressure measurements. Information gained by the Schlieren images were supported by sur-

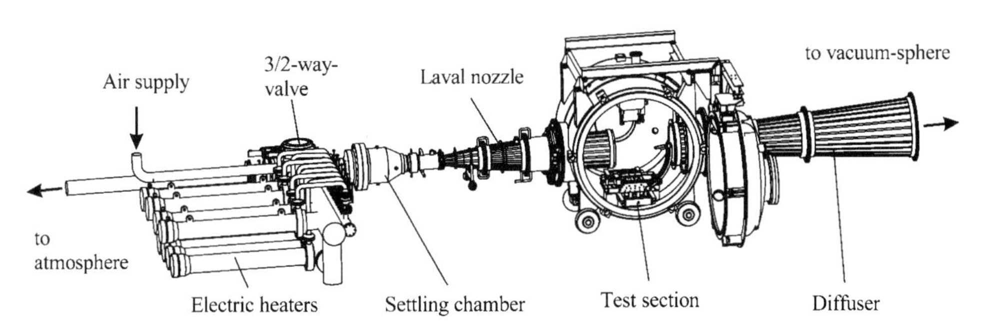


Figure 5.6.1: Components of H2K hypersonic wind tunnel facility.

Figure 5.6.2: Auxiliary heater to heat up secondary air flows with heating tube (only partially visible), hot gas valve and isolated pipe connected to the H2K test section.

face oil flow visualizations, performed with Silicon oil mixed with quantities of fluorescent pigments. The pressures at flow sections of particular interest were measured by a Pitot pressure rake with several probes mounted on a movable support inside the test section. Static pressure measurements were carried out on wall surfaces of the second inlet model and the nozzle/afterbody models. The measured pressures were recorded by the PSI-System 8400, which uses different piezoresistive pressure sensors of adequate sensitivity to convert the measured pressures into electrical signals. Thermal mapping of the PTFE model surfaces was performed by an infrared camera system. Heat flux rates at the wall were calculated from the gained temperature data, based on a thick wall analysis. The numerical evaluation procedure took radiation effects into account, as well as an instationary heat flux development due to rising surface temperatures on the model, which is typical for blow down facilities with their long test durations. Due to the technical progress of commercially available infrared systems, three different infrared systems were used. While early infrared studies were performed with an INFRAMETRICS 600, and later with an AVIO TVS 8000 system, recently a FLIR ThermaCAM SC 3000 is used at the H2K facility.

#### 5.6.2.2 Wind Tunnel Models

Preliminary studies on the shock boundary layer interaction were performed on generic flat plate models. Thereby, a wedge acting as shock generator produced an oblique shock front impinging on the boundary layer of the plate. Further generic models were equipped with various means to manipulate this interaction, for example with a bleed slot. In order to examine the efficiency of bleed slot configurations in three-dimensional flows, a side wall could be fixed to the plate. In order to allow the application of IR-techniques, all model walls consist of PTFE plates embedded in metal frames, which can easily be ex-

Figure 5.6.3: Wind tunnel model for external compression studies.

Figure 5.6.4: Scaled model of the ELAC inlet for combined pressure and infrared measurement.

changed and adjusted to provide various geometries. Subsequently, two scaled down models of the ELAC inlet (scale 1:21) were tested. The first one (Fig. 5.6.3) has three compression ramps in order to study the external compression process. Here, the inlet cowl was replaced by a short lip, in order to provide better optical access for the infrared thermography. The second model (Fig. 5.6.4) was equipped with 36 pressure tabs to gain information about the internal compression process. This model also featured boundary layer bleed at different possible positions. Both inlet models had side walls with integrated glass plates for optical access, to prevent lateral mass flow over the ramp sides. To study effects induced by the combustion chamber, the second model was mounted upstream a throttle that increased the pressure downstream of the inlet. When the pressure downstream of the inlet increases, the total pressure recovery (i.e. the ratio of the total pressures at the exit and the entrance of the inlet) was raised. At a certain pressure the inlet flow became choked and collapsed. Thus, typical throttle curves, representing the pressure recovery over the inlet mass flow rate, were gained.

For a first detailed investigation of the location of the boundary layer separation at the nozzle flap, a simplified model of the flap with different PTFE inserts representing different flap geometry was designed [9]. The pressure on the upper side of the model was increased by an oblique shock, generated by

Figure 5.6.5: Nozzle/afterbody model for experiments with heated nozzle flow mounted in the H2K test chamber.

the wedge shape of the model, and thus simulated the nozzle jet pressure at the tip of the flap. By this, the simplified configuration did not provide flow similarity of the total pressure ratio between the ejected and the ambient flow of a SERN nozzle, but provided similarity in the flow separation governing Euler numbers [10]. Tests were also performed on two scaled nozzle/afterbody models. The first one was originally developed by MTU-Munich and DLR in frame of the earlier "German Hypersonic Technology Programme" [11]. Here the adaptation of the nozzle flap was realized by installing different flap insets. Two side plates reduced disturbances resulting from the model strut and minimized three-dimensional effects. The nozzle jet plume was simulated by compressed air, supplied through a pipe inside the model strut. The internal model duct, the flap-insets and the ramp were instrumented by three rows of 96 static-pressure test ports in total. Due to design limitations of this model, a more temperature resistant model became necessary for experiments with heated nozzle flow (Fig. 5.6.5). Here, the total pressure and the temperature of the nozzle flow were measured inside the nozzle chamber. Additionally, the static pressure was measured at 13 positions, one located inside the nozzle chamber, five along the expansion ramp and seven at the nozzle flap, i.e. two at the upper and five at the lower flap side.

### 5.6.3 Inlet Studies

A governing flow phenomenon inside a hypersonic inlet is the oblique shock interaction, arising from the shocks induced by the ramps and the viscous boundary layer developing there. In literature, the oblique shock interaction is mostly treated as a purely planar flow problem, thus justifying the use of two-dimensional numerical simulations. On the other hand, preliminary project activities on experiments with generic flat plate models revealed the influence of three-dimensional shock interaction effects and their accompanying separation

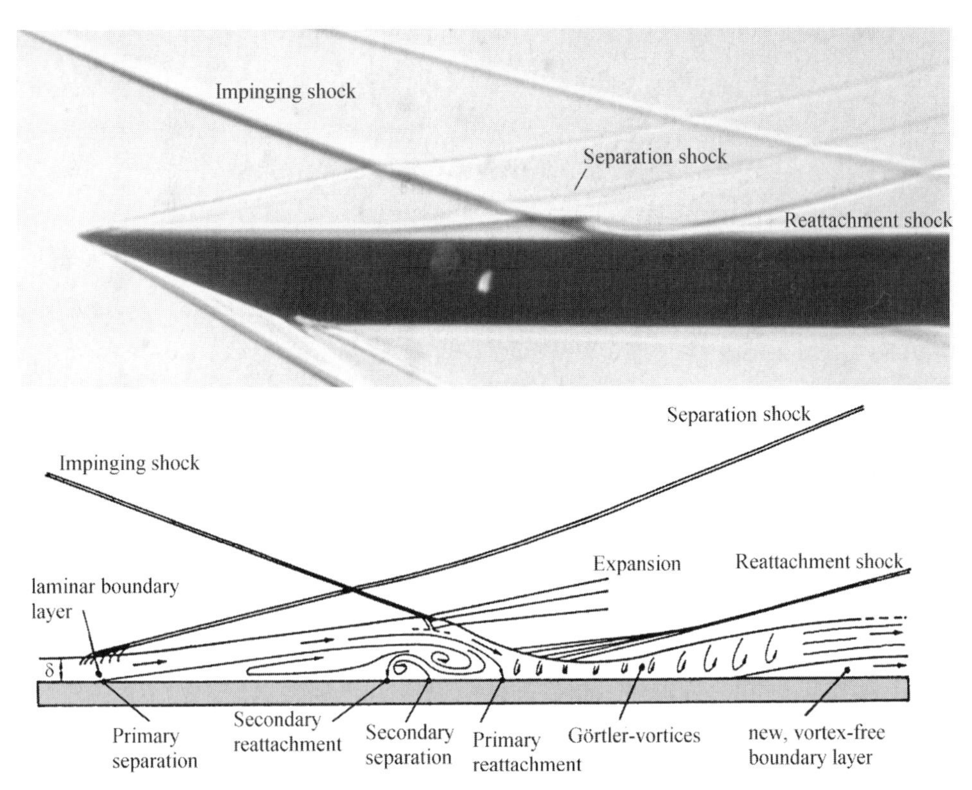


Figure 5.6.6: Schlieren image of plane shock boundary layer interaction on a flat plate and topological interpretation.

phenomena [12]. The impinging shock, generated by a wedge, induced boundary layer separation and reattachment phenomena (Fig. 5.6.6). The resulting flow field showed indications of longitudinal counter-rotating Görtler vortices downstream of the interaction region. The existence of these vortices was noticed by the interpretation of surface oil flow patterns and infrared images, which showed periodically changing skin friction and heat loads [1, 13].

In a subsequent parametric study, the effects of the impinging shocks strength and the boundary layer thickness on the development of these vortices were evaluated [2, 14]. It was found, that the longitudinal vortex diameter was approximately equal to the boundary layer thickness at the impingement location. The downstream development of the Görtler vortices and their effect on the stability of the laminar boundary layer was discussed. It was discovered that the formation of the vortices was not only directed downstream of the reattachment line, but also upstream towards the secondary separation, where they degenerate into a series of recirculating separation bubbles. Therefore, in contrast to previous interpretations from other authors, who treated this shock reflection problem as a purely two-dimensional case, it was shown, that depending on the Reynolds number a three-dimensional behaviour of the flow in the

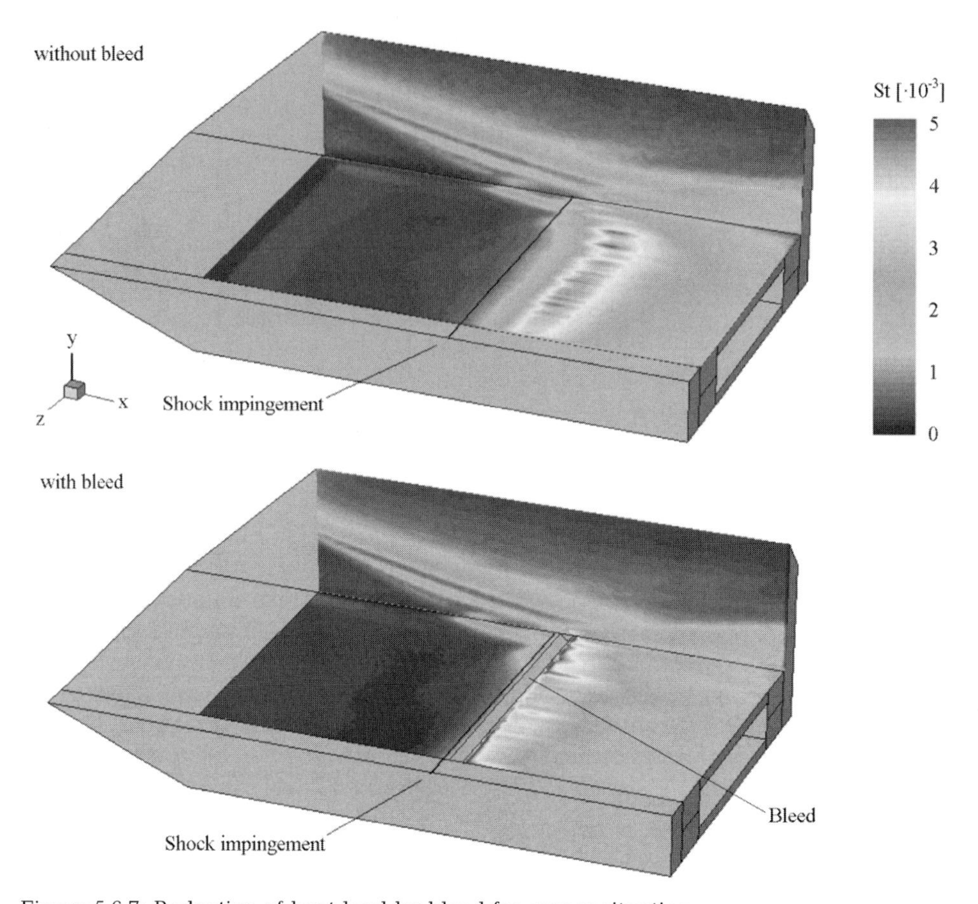


Figure 5.6.7: Reduction of heat load by bleed for corner situation.

interaction region is dominant. The gained experimental results stimulated numerical activities within the same research programme [15] and inspired further numerical studies in frame of the Sonderforschungsbereich 259 at Stuttgart [16]. The former simulations, based on Navier-Stokes calculations for two-dimensional as well as for three-dimensional flows, confirmed the existence of Görtler vortices due to the boundary layer interaction on a flat plate.

Shock induced boundary layer separation may disturb the desired inlet flow field, thus leading to a significant reduction of efficiency. The implementation of boundary layer bleed is a common method to reduce the boundary layer separation. For an effective bleed design, information about the governing parameters, i.e. the bleed position with regard to the interacting shock or the bleed mass flow rate, are needed. For the identification of favourable bleed designs, experiments were performed at the H2K on a generic flat plate model equipped with a planar bleed slot [17]. The wind tunnel model and the wind tunnel test procedure was optimized by using computational fluid dynamics.

An analytical method to predict the key parameters of the boundary layer bleed was also developed [18]. In case of the shock induced boundary layer separation the bleed implementation leads to a reduction of the separation bubble thickness by almost 50%. Further experiments dealt with the achievable reduction of heat loads on the wall surface depending on the amount of bled air mass and the position of the boundary layer bleed [3, 19]. These examinations were also extended to three-dimensional corner flows (Fig. 5.6.7). By this, favourable design parameters for the boundary layer bleed set-up were found. Finally, the results of the investigations using the generic models have been transferred to optimize a hypersonic inlet model. A significant increase of the inlet performance regarding the total pressure recovery was proven [20]. However, the use of boundary layer bleed in an hypersonic inlet to control shock/ boundary layer interactions depends strongly on the basic demands on the inlets purpose, as a gain in total pressure recovery is often combined with a slight loss in inlet mass flow.

The next step of inlet activities focused on the viscous flow phenomena occurring on the external compression ramps of the hypersonic RAM inlet for the TSTO configuration ELAC [4]. Here, the main design parameters for the ramps and the side walls are the pressure and the thermal loads. Therefore, these heat loads were determined from temperature measurements on an inlet model by infrared thermography, mainly at Mach 6.0. Additional information about the viscous ramp flow were gained by oil flow diagnostics, Schlieren optics and Pitot pressure measurements. The influence of model parts like side walls or cowl were demonstrated and separation regions were visualized with respect to the boundary layer state. The experimental results pointed out that the boundary layer has a strong influence on the flow field at the compression ramps. In agreement with boundary layer predictions from literature, laminar boundary layer separation was confirmed upstream of the second ramp for the ELAC inlet geometry, while turbulent boundary layer separation was found upstream of the third ramp. According to flight conditions, the most relevant test case was at the highest Reynolds number and with applied boundary layer tripping. Here, the formation of the separation bubble was found to be very sensitive on changes of the flow conditions or the model geometry.

In subsequent test runs a scaled model of the complete ELAC inlet (Fig. 5.6.4) with internal compression was tested at free stream conditions of Mach 5.3 and 6.0, which due to the shock at the front of the vehicle correspond to flight velocities of Mach 6.4 and 7.5, respectively. In order to reduce shock interaction effects of the inlet flow as well as to enhance the starting process, boundary layer bleed was implemented into the ramps at two different positions. But the gained data revealed that the maximum pressure recovery, which was measured at about 17%, was not significantly improved by bleed. For a test without bleed the static pressure distributions at the inlet wall were recorded (Fig. 5.6.8). On the ramps upstream of the internal compression, a good agreement between measured and analytically calculated pressures was found. Since the achieved Reynolds numbers in the wind tunnel were several times lower than those in real flight, the effects of the Reynolds number influence on

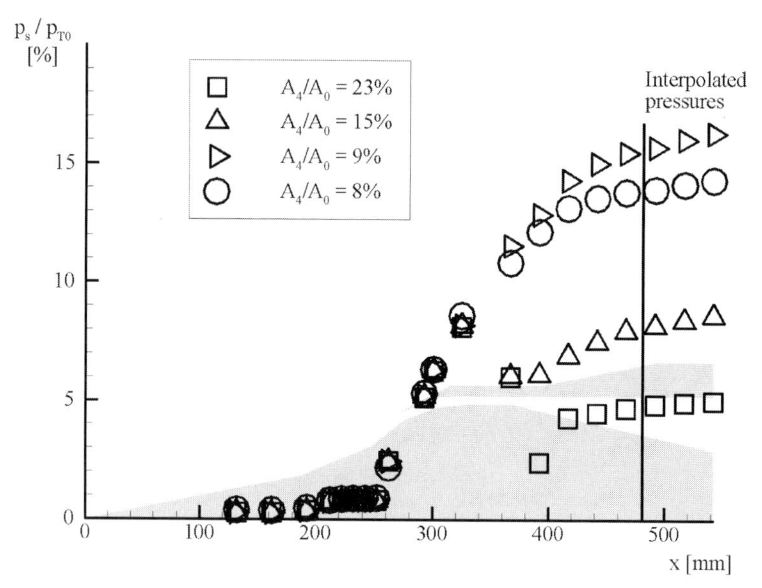


Figure 5.6.8: Pressure recovery at different throttle degrees.

the inlet flow had to be clarified. It was found that a change of the Reynolds number by a factor of about three led to a slight increase in the static pressures on the external compression ramps. On the other hand, the internal compression reacted sensitively on the counter-pressure produced by the throttle simulating the combustion chamber. With decreasing Mach number the possibility of a break down of the inlet flow at a constant throttle degree increased. The critical ratio between plug nozzle throat area and inlet capture area of the wind tunnel was found at approximately eight percent, a value which is assumed to lie closely to that of the inlet at flight condition.

Recent studies focused on the external compression of an hypersonic inlet featuring smaller ramp angles than the ELAC inlet, such as typical for SCRamjets. Therefore, the wind tunnel model (Fig. 5.6.3) was equipped with three compression ramps (5°, 5° and 10°) and tested at a free stream Mach number of 6.0. The tests were performed at different Reynolds numbers in order to gain information about the influence on the boundary layer transition as well as on the complex flow field in the vicinity of the corner. This corner flow field was caused by the ramp shocks glancing at the side walls of the inlet model, interacting with the boundary layer developing there. The interaction of multiple shocks and expansion waves is given by the oil flow visualizations shown in the upper parts of Figs. 5.6.9 and 5.6.10. The lateral views into the wind tunnel model visualizes oil traces of the ramp shocks at the side wall. The development of Görtler vortices, linked to transition phenomena, is indicated by the formation of a parallel oil flow pattern on the ramps. While these vortices are mainly initiated by the wedge of the third ramp at the lower Reynolds number, they are already initiated by the wedge of the second ramp at the larger Rey-

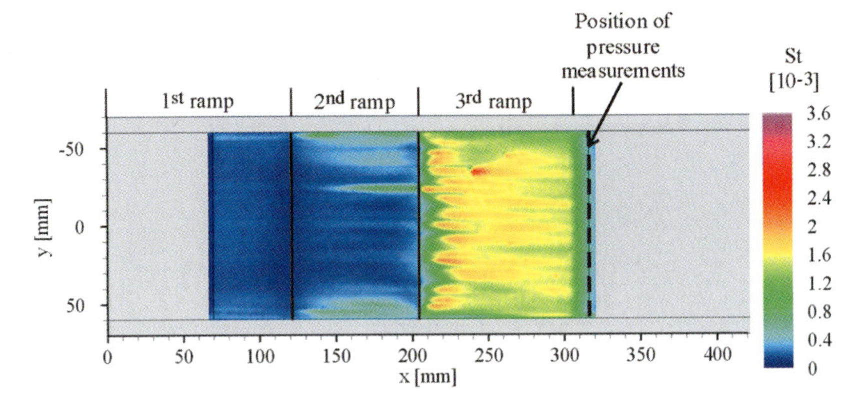


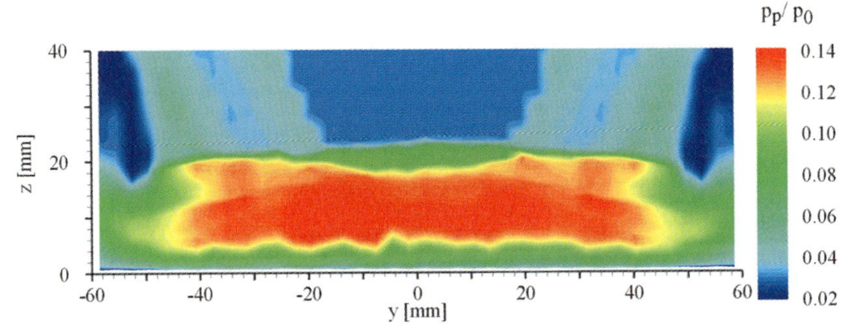


Figure 5.6.9: Oil flow visualization, heat flux distribution and pressure distribution at the external compression model ($Ma_\infty = 6.0$; $Re = 2.1 \times 10^6$; $l_{ref} = 0.42$ m).

nolds number. Furthermore, the parallel distance of the vortices decreases with increasing Reynolds number. Downstream of the third ramp, traces of the corner vortices propagate from the side walls towards the centre of the inlet due to the expanding flow.

On the ramps, the heat flux distribution in terms of Stanton numbers was calculated from the temperature data measured by the IR camera system with a

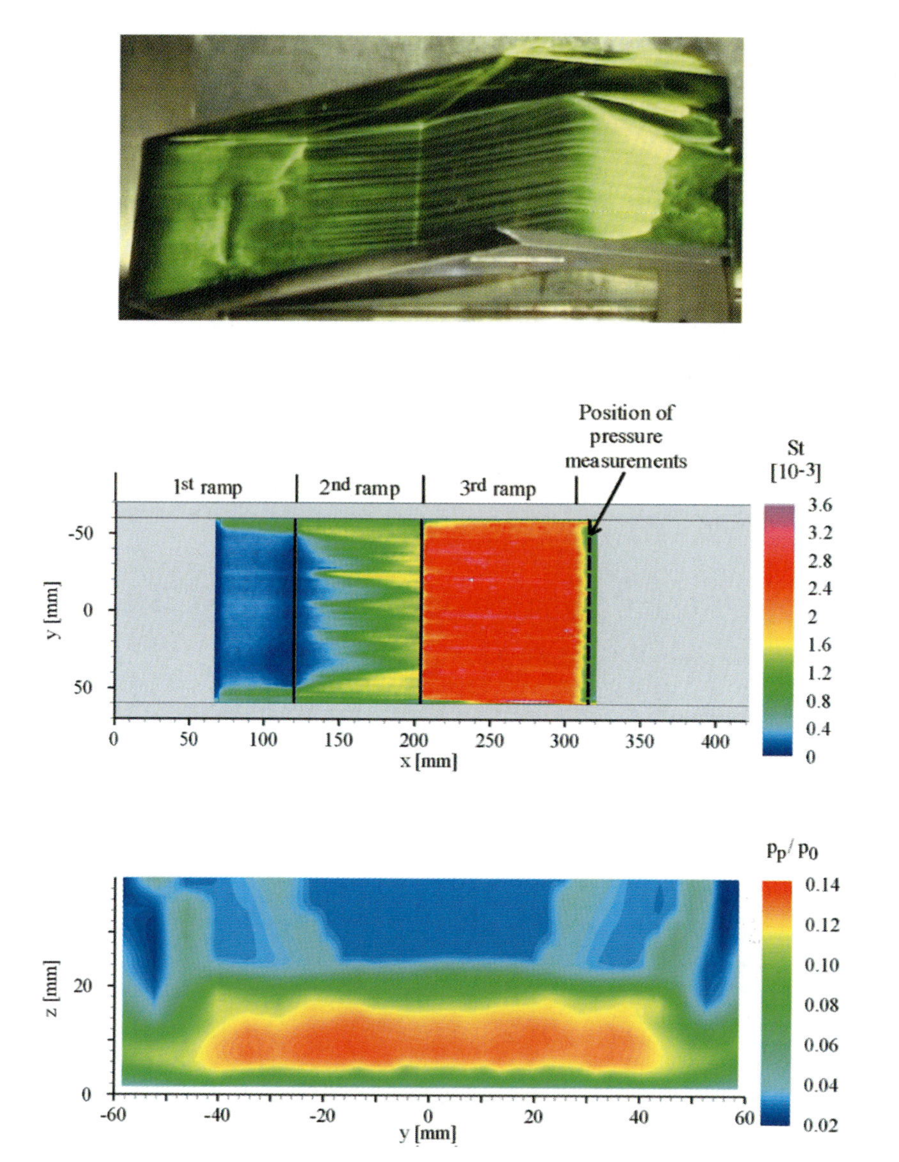


Figure 5.6.10: Oil flow visualization, heat flux distribution and pressure distribution at the external compression model ($Ma_\infty = 6.0$; $Re = 6.72 \times 10^6$; $l_{ref} = 0.42$ m).

spatial resolution of 1.8 mm per pixel (middle parts of Figs. 5.6.9 and 5.6.10). In accordance with the Reynolds analogy, the heat flux distribution resembles the established oil flow pattern and visualizes the lateral heat streaks of the Görtler vortices. As expected, the increase of the Reynolds number at identical stagnation temperatures leads to an increase of heat fluxes due to the thinner boundary layer. The lower part of the figures shows the pressure distribution orthogonal to the main flow direction in a plane downstream of the third ramp. The pressures were measured by a rake of 14 Pitot-probes, in order to provide a spatial resolution of 1 mm in y direction and 3 mm in z direction. The pressure contour plot reveals the complex shock structure in the vicinity of the corners, which is generated by the interaction of the ramp shocks and the shock caused by the displacement thickness at the side wall. It also provides first information about the magnitude of the external compression at the ramps. Due to an increased dissipation inside the turbulent boundary layer, the earlier transition at a larger Reynolds number leads to a reduced pressure recovery. In connection with the presented oil flow images the heat flux and pressure contour plots provide experimental data for comparison, validation and interpretation of numerical simulations of viscous interaction phenomena inside an hypersonic inlet.

### 5.6.4 Nozzle Studies

The flow field around the SERN nozzle for the ELAC vehicle was examined emphasizing the influence of the flap geometry on the nozzle performance. Combined wind tunnel experiments and numerical simulations were carried out in order to study the separation of the boundary layer at the flap cowl and to determine the thrust and the thrust vector angle. Early numerical studies on fourteen different contours for the flap cowl led to the conclusion, that a parabolic shaped cowl (i.e. a cubic polynomial), featuring only a slight expansion of the flow in the beginning, considerably reduces the vertical force compared to the original flap cowl [9, 21]. This leads to a smaller rotation of the thrust vector in the transonic and low supersonic Mach number regime. Experiments on a simplified flap model proved, that the location of the boundary layer separation is clearly shifted downstream for the parabolic shaped contour and that the separation bubble is therefore essentially smaller [10, 21]. The minimized size of the bubble reduces the energy losses of the flow and leads to an improved aerodynamic nozzle performance. Further wind tunnel experiments provided detailed information about the influence of the Reynolds number on the location of the boundary layer separation (Fig. 5.6.11). At laminar flow conditions the point of separation moved downstream by increasing the Reynolds number, mainly due to a rise of kinetic energy inside the boundary layer. At a certain Reynolds numbers the boundary layer became transitional and the point of separation moved slightly upstream again. Finally, at high Reynolds numbers the position of separation remained fixed, which indicated a fully turbulent boundary layer. Further tests with tripped turbulent boundary layers confirmed the same location of the separation also at lower Reynolds numbers.

The flow field at the nozzle/afterbody model was numerically simulated in order to determine the impact of an improved flap geometry on the nozzle efficiency [9, 21]. Based on the experience gained above, the cowl geometry was modified. Further on, the flap tip was shifted down and the nozzle throat was repositioned. The corresponding simulations demonstrated the importance of

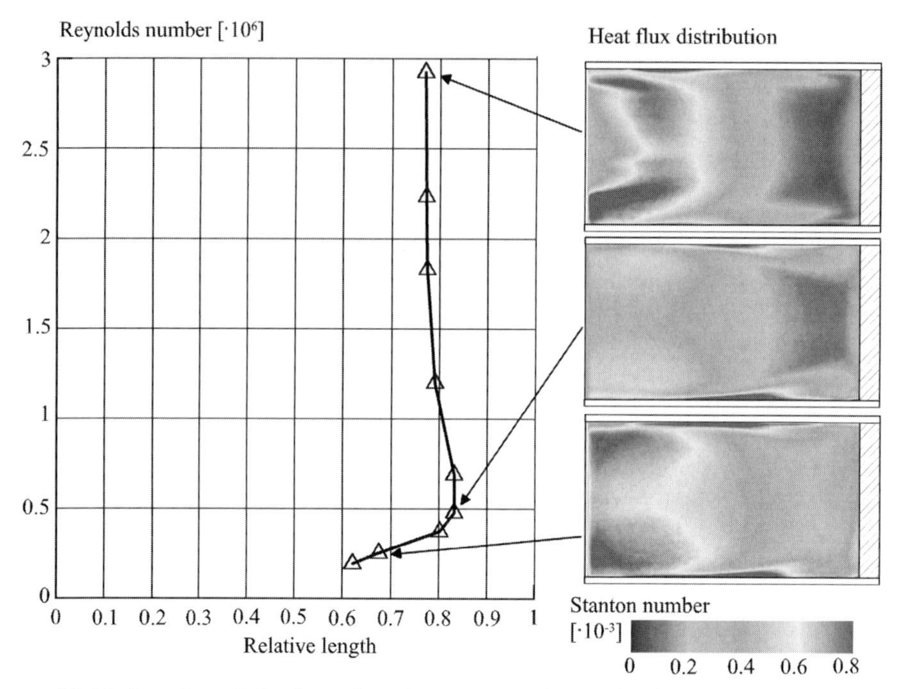


Figure 5.6.11: Location of the boundary layer separation at the flap cowl in dependence of the Reynolds number ($l_{ref}$=0.176 m, Boundary layer state indicated by IR-images).

shape variations of the nozzle flap to optimize a SERN nozzle. Particularly in the critical Mach regions, i.e. in the transonic and the high Mach region, the nozzle performance was improved. For the optimized flap with a parabolic cowl contour and a lowered tip, thrust improvement between 1 and 3.5 percent was achieved over the whole examined Mach number range between Mach 1.2 and Mach 7. At lower Mach numbers near the transonic Mach regime, the optimized flap (parabolic shape of flap cowl and lowered tip) achieved an improvement of the thrust vector angle up to 5°. Therefore, this optimized flap improved the nozzle performance in both critical areas, the transonic and the high Mach number regime. For another flap with a readjusted throat the thrust was increased even more (up to 5 percent) in the Mach range between Mach 1.64 and 4.5. However, this flap obtained no increase of thrust at higher Mach numbers, where an improvement of gross thrust is particularly desirable, because of the strong impact on net thrust.

Wind tunnel simulations at $M_\infty$=6 with an nozzle/afterbody model, equipped with the optimized flap, provided detailed information about the separation of the boundary layer at the flap cowl with respect to the Reynolds number and the nozzle pressure ratio $\Pi$, i.e. the ratio between the total pressure of the nozzle jet and the ambient pressure [22, 23]. Here, the Reynolds number showed only a small influence on the location of the separation. In contrast, an increase of the nozzle pressure ratio caused an upstream shift of the

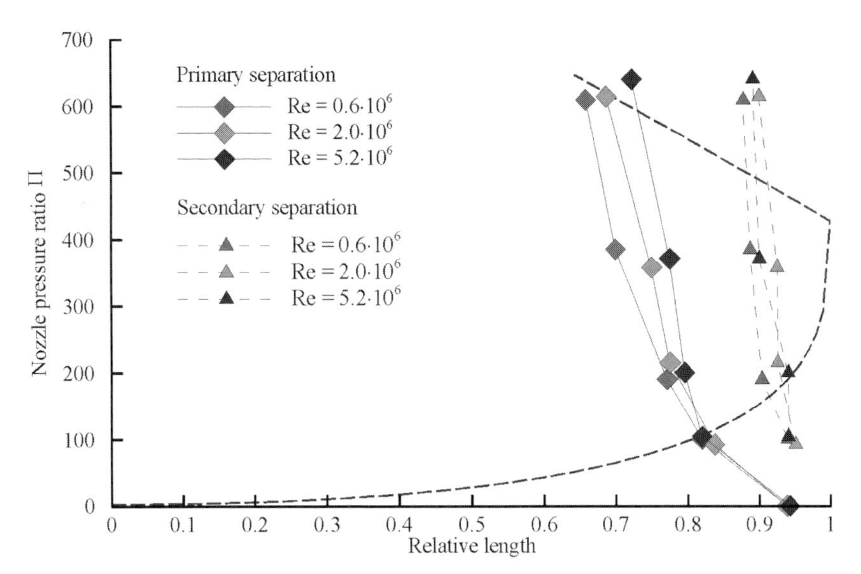


Figure 5.6.12: Location of the boundary layer separation at the flap cowl in dependence of the nozzle pressure ratio ($Ma_{\infty}=6.0$; $l_{\mathrm{ref}}=0.176$ m).

separation at the flap cowl (Fig. 5.6.12). Based on the experimental data, an empirical separation criterion was formulated in order to predict the separation point at the nozzle/afterbody model. For high nozzle pressure ratios, the thrust vector angle was found in an acceptable range between –5° and 5°. At a lower pressure ratio the thrust vector angle varied in a broader range from –20° up to 5°. It was also noticed that the Reynolds number influenced the thrust vector angle significantly at this lower pressure ratio. The axial thrust coefficient exceeded 0.95 in most of the experiments, except at lower nozzle pressure ratios where the deviation of the thrust vector angle lead to high angularity losses. A comparison with earlier experiments confirmed an increase of the axial thrust coefficient due to the optimized flap geometry. Again, the experimental results were compared to those from numerical simulations to gather further information on the flow and the structure of the separation bubble (Fig. 5.6.13). For a laminar boundary layer, the numerical simulations revealed an interpretation for a second separation point at the cowl as observed in the experiments. However, the location of the separation and the pressure distribution at the flap cowl matched better when assuming a turbulent boundary layer.

Finally, all gained numerical and experimental results were used to design a modified flap for the reference vehicle ELAC [22, 24]. For this new flap geometry, the contour of the cowl was adjusted, the nozzle flap tip was repositioned and a kinematic to adjust the nozzle throat was introduced. Comparative numerical simulations of the vehicle with the nozzle equipped with the original and the modified flap confirmed significant improvements (Fig. 5.6.14). The axial thrust coefficient was increased about 12 percent at $M_{\infty}=1.23$ and about 0.5 percent at $M_{\infty}=5.67$. Additionally the absolute value of the thrust vector

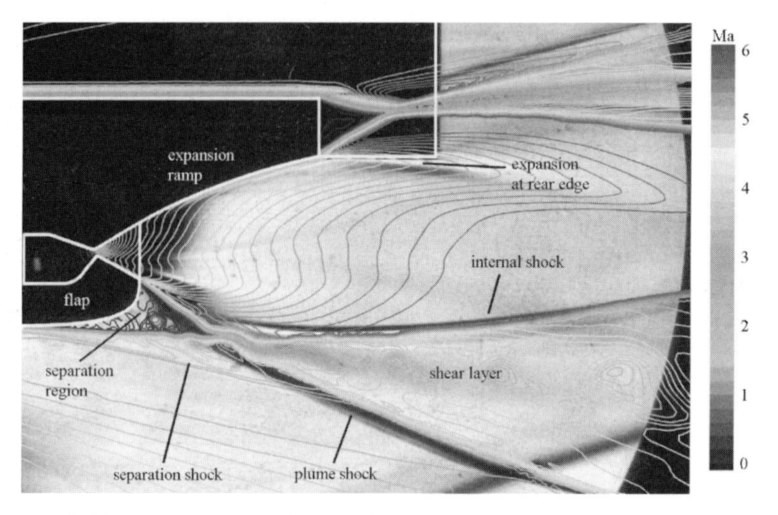


Figure 5.6.13: Schlieren picture of the nozzle with superimposed Mach number distribution from numerical simulations ($Ma_\infty=6.0$; $Re=5.1\times10^6$; $\Pi=600$; $l_{ref}=0.176$ m).

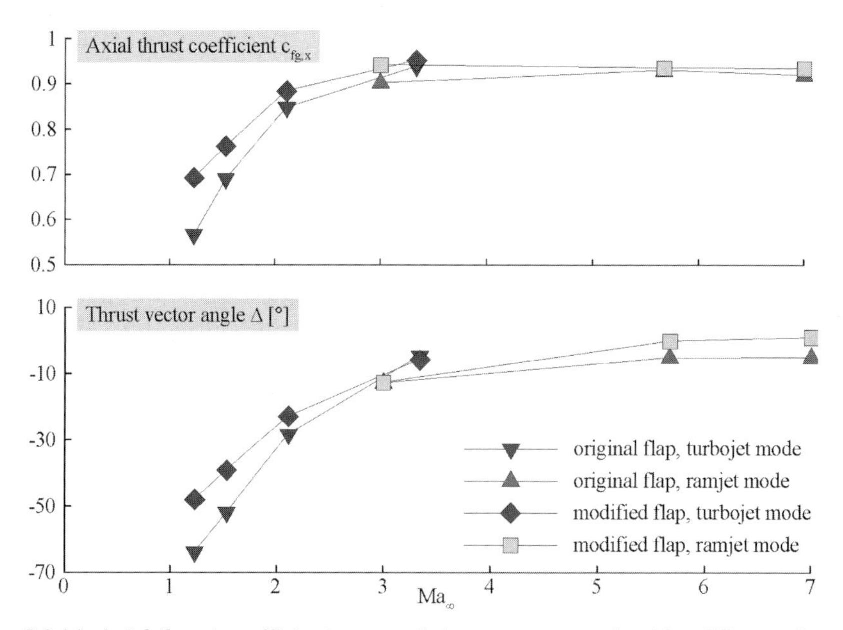


Figure 5.6.14: Axial thrust coefficient $c_{fg,x}$ and thrust vector angle $\Delta$ for different flap configurations for the ELAC nozzle.

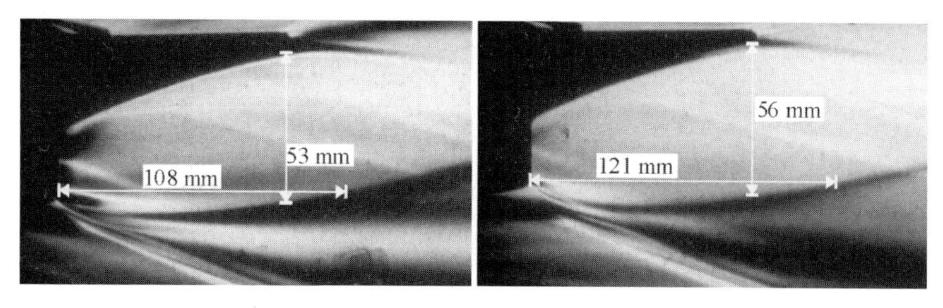


Figure 5.6.15: Extension of the nozzle flow field at $T_D$=298 K (left) and $T_D$=986 K (right). Re=$7.3\times10^6$ with $l_{ref}$=0.43 m and $\Pi$=615.

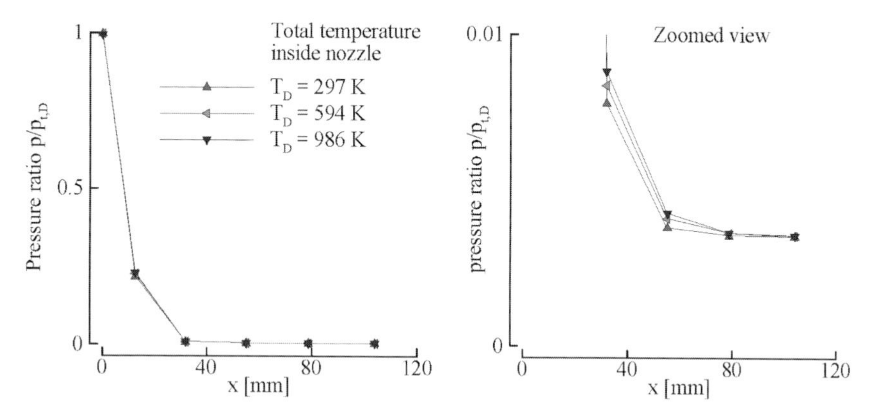


Figure 5.6.16: Ratio of static ramp pressure and total pressure inside the nozzle chamber at different total temperatures of the nozzle flow ($Re=7.3\times10^6$; $\Pi=615$; $l_{ref}=0.43$).

angle was decreased by up to 15° at low Mach numbers. Thus, the results showed a strong influence of the flow at the nozzle flap on the overall nozzle performance, especially in the transonic and low supersonic Mach range.

Recent experiments with a nozzle/afterbody model of the ELAC SERN-nozzle dealt with the influence of different temperatures on the flow field at the nozzle [25]. Hereby the nozzle exhaust flow was heated up by the auxiliary electric heater. A comparison of Schlieren images at the same nozzle pressure ratios showed, that the size of the nozzle jet grew with increasing temperature (Fig. 5.6.15). Hereby, the vertical reference length is defined as the span between the end of the expansion ramp and the interior shock, and the horizontal reference length is defined as the span between the flap tip and the interior shock. The quantitative information about the influence of different nozzle flow temperatures was provided by a comparison of the pressure distributions along the expansion ramp. The ratio between the static pressure along the expansion ramp and the total pressure inside the nozzle chamber at three different levels of the nozzle total temperature at a fixed Reynolds number and a fixed nozzle pressure ratio is given in Fig. 5.6.16. As illustrated, the pressure ratio grows

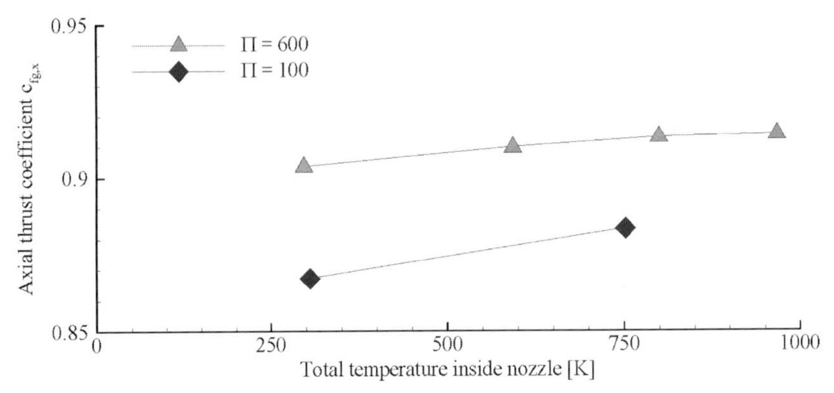


Figure 5.6.17: Axial thrust coefficient $c_{fg,x}$ at different nozzle pressure ratios ($Re = 7.3 \times 10^6$; $l_{ref} = 0.43$).

with the total temperature, which results in a higher nozzle thrust at increased temperatures for a given nozzle total pressure.

Since the static pressure is measured at only seven positions in the region that accounts for the nozzle thrust, the experimental data on its own were not sufficient to evaluate the nozzle thrust. Thus, the pressure and the momentum at the nozzle throat were calculated from the measured total conditions inside the nozzle chamber by one-dimensional gasdynamic equations. The pressures downstream of the expansion ramp and at the flap tip were extrapolated from measured pressures. Since the pressure gradients changed significantly in the region between the test points shortly downstream of the nozzle throat, so that a linear interpolation would overestimate the static pressures in this region, these pressures were approximated by a fitting Bezier curve. Thus, a pressure distribution at the nozzle was obtained, which was used to calculate the nozzle thrust and the axial thrust coefficient, which is defined as the ratio between the axial thrust and the ideal thrust. The comparison of the calculated thrust coefficients showed, that the nozzle thrust increased with an increasing nozzle temperature (Fig. 5.6.17). At a lower nozzle pressure ratio, the difference between the axial thrust coefficient at cold and heated nozzle flow was about 1.5 percent, at a higher nozzle pressure ratio the difference was about one percent. Considering the severe impact of nozzle thrust on the net thrust of airbreathing engines at hypersonic speeds, this study demonstrated the necessity of taking temperature effects into account for wind tunnel simulation of nozzle flow fields.

## 5.6.5 Conclusion

Fundamental experiments on the flow of an hypersonic airbreathing inlet have been performed to contribute to the design of the inlet and to increase its efficiency. A preliminary generic experiment of the oblique shock reflection on the surface of a flat plate, as being typical for inlet flow fields, visualized Görtler vortices that revealed the tree-dimensional character of this commonly two-dimensional treated flow problem. Further numerical and experimental studies showed, that these vortices are linked to the boundary layer transition by enhancing the heat flux downstream of the shock interaction. These vortices lead to high heat loads particularly at the third ramp of the ELAC inlet wind tunnel model. Boundary layer bleed was investigated successfully in order to reduce shock boundary layer effects, like the severe heating of the wall, and an optimized bleed configuration was found. Nevertheless, the bleed implementation at three distinct locations did not offer an increase in pressure recovery for the ELAC inlet model. It was found that the interaction of the glancing ramp shocks with the boundary layer developing on the inlet's side walls forces the development of a rather complex corner flow structure and thus affects the pressure distribution downstream of the external compression. The influence of the Reynolds number on this corner flow structure was demonstrated by wind tunnel experiments. These measurements were particularly performed to provide experimental data for the validation of numerical codes capable of simulating hypersonic flows with its severe viscous interactions.

In order to improve the performance of the ELAC SERN nozzle by an optimization of the nozzle flap geometry, the size of the separation bubble at the flap was minimized by combined numerical and experimental studies on boundary layer separation. In particular, the influence of the Reynolds number and the nozzle pressure ratio on the separation location was determined. Achieved improvements of the nozzle performance were verified by numerical simulations as well as experiments on a nozzle/afterbody wind tunnel model, featuring an exhausted jet flow. An upgraded version of the ELAC nozzle flap derived on basis of these results showed a significant increase in the thrust coefficient and a reduced shift of the thrust vector angle. In order to extrapolate these results to real, i.e. hot nozzle flow fields, the exhaust jet flow of a wind tunnel model of the ELAC nozzle was heated up to stagnation temperatures of about 1000 K. Thereby, an increase of thrust with increasing nozzle flow temperature was observed, thus pointing out the significance of such tests.

Finally, the experimental results on the hypersonic inlet flow and the efforts to improve the nozzle geometry provided a substantial insight into the aerodynamics of hypersonic propulsion components to support the design of future hypersonic space vehicles.

## References

1. A. Henckels, A.F. Kreins, F. Maurer: Experimental Investigation of Hypersonic Shock-Boundary Layer Interaction. Zeitschrift für Flugwissenschaften und Weltraumforschung, Nr. 17, Heft 2, Seite 116–124, Springer-, Berlin, 1993.
2. A.F. Kreins, A. Henckels, F. Maurer: Experimental Studies of Hypersonic Boundary Layer Separation. Zeitschrift für Flugwissenschaften und Weltraumforschung, Nr. 20, Heft 2, Seite 80–88, Springer, Berlin, 1996.

3. D. Schulte, A. Henckels, U. Wepler: Reduction of Shock Induced Boundary Layer Separation in Hypersonic Inlets Using Bleed. Aerospace Science and Technology, Vol. 2, No. 4, pp. 231–239, Elsevier, Paris, 1998.
4. R. Neubacher, A. Henckels, T.M. Gawehn: Experimental Investigation of a Hypersonic Inlet for the TSTO-Configuration ELAC. Beitrag zum 12. DGLR-Fach-Symposium der AG STAB, veröffentlicht in "New Results in Numerical and Experimental Fluid Mechanics III", ed. by S. Wagner et al., in der Reihe "Notes on Numerical Fluid Mechanics, Vol. 77, pp. 129–136, Springer, 2002.
5. R. Lederer, J. Hertel: Exhaust System Technology, 2. Space Course, SFB 255, München, 1993.
6. T. Berens: Thrust Vector Optimization for Hypersonic vehicles in the Transonic Mach Number Regime, AIAA-93-5060, AIAA/DGLR Fifth International Aerospace Planes and Hypersonics Technologies Conference, Munich, 1993.
7. T. Esch, A. Bauer, H. Rick: Simulation and nozzle/afterbody integration of hypersonic propulsion systems, Z. Flugwiss. Weltraumforsch. 19, pp. 19–28, 1995.
8. U.J. Fox: Finite-Elemente-Simulation von Hochtemperatur-Düsenströmungen, Dissertation RWTH Aachen, Shaker, 1999.
9. P. Gruhn, A. Henckels: Numerische Simulation der Strömung um die Heckklappe einer SERN-Düse für die Raumflugzeug-Konfiguration ELAC. DLR Interner Bericht, IB-39113-99A11, 1999.
10. S. Kirschstein, A. Henckels: Untersuchungen zum Ablöseverhalten der Hyperschall-Grenzschicht um eine SERN-Heckklappe mittels Prinzip-Windkanalmodellen. DLR Interner Bericht, IB-39113-99A24, 1999.
11. J. Niezgodka: Druckverteilungsmessungen zur Simulation der Interferenz des Triebwerksstrahls auf das Heck der Sänger-Unterstufe (MTU-Modell), DLR Interner Bericht, IB-39113-93C15, 1993.
12. A. Henckels, P. Herzog, F. Maurer: Experimental Study of Hypersonic Shock Wave Boundary Layer Interactions by Means of Infrared Technique. First European Symposium on Aerothermodynamics for Space Vehicles, ESTEC, Noordwijk, The Netherlands, May 28–30, 1991, ISBN 92-9092-114-5, pp. 159–164, ESA/ESTEC, Noordwijk, 1991.
13. A. Henckels, A.F. Kreins, F. Maurer: Application of Infrared Measurement Technique in Hypersonic Facilities. Proceedings of the 18th International Symposium on Shock Waves, held at Sendai, Japan, July 21–26, 1991, ed. by K. Takayama, ISBN 3-540-55686-9, Vol. 1, pp. 651–656, Springer, Berlin, 1992.
14. A.F. Kreins: Wärmestromverteilung und Strömungsfelduntersuchung in gestörten Hyperschall-Plattengrenzschichten. Dissertation an der RWTH Aachen, erschienen als DLR Forschungsbericht 94-03, Köln 1994.
15. U. Domröse: Ablösung einer hypersonischen, laminaren Plattengrenzschicht durch einen einfallenden Stoß. Dissertation an der RWTH Aachen. Shaker, Aachen, 1994.
16. U. Rist, S. Wagner: Direkte numerische Simulation der Stoß/Grenzschicht-Interaktion im Hyperschall. Finanzierungsantrag 2002-2003 zum DFG Sonderforschungsbereich 259 „Hochtemperaturprobleme rückkehrfähiger Raumtransportsysteme", Stuttgart 2001.
17. D. Schulte, A. Henckels, I. Schell: Boundary Layer Bleed in Hypersonic Inlets. Contribution to the 10th AG STAB/DGLR Symposium, Notes on Numerical Fluid Mechanics, Vol. 60, ed. by H. Körner and R. Hilbig, pp. 296–303. Vieweg & Sohn, Braunschweig, 1997.
18. D. Schulte: Beeinflussung viskoser Strömungseffekte in Hyperschall-Einläufen. Dissertation an der RWTH Aachen. Shaker, Aachen, 2001.
19. D. Schulte, A. Henckels, R. Neubacher: Manipulation of Shock/Boundary Layer Interactions in Hypersonic Inlets, ISABE-Paper 99-7038 presented on the 14$^{th}$ Int. Symp. on Airbreathing Engines, Florence, Italy, 1999.
20. D. Schulte, A. Henckels, R. Neubacher: Manipulation of Shock/Boundary Layer Interactions in Hypersonic Inlets. Journal of Propulsion and Power, Vol. 17, No. 3, pp. 585–590, American Institute of Aeronautics and Astronautics, 2001.

21. P. Gruhn, A. Henckels, S. Kirschstein: Flap contour optimization for highly integrated SERN nozzles, Aerospace Science and Technology, Vol. 4, pp. 555–565, 2000.
22. P. Gruhn, A. Henckels, G. Sieberger: Improvement of the SERN nozzle performance by aerodynamic flap design, Aerospace Science and Technology, Vol. 6, pp. 395–405, 2002.
23. G. Sieberger, A. Henckels, P. Gruhn: Experimentelle Untersuchung der Strömung im Heckbereich zukünftiger Raumflugzeuge, DLR IB-39113-2001A08, 2001.
24. P. Gruhn, A. Henckels: Experimentelle Untersuchung einer Düsenströmung für Hyperschallantriebe, Arbeits- und Ergebnisbericht Teilprojekt C8, SFB 253, 2001.
25. P. Gruhn, A. Henckels: Simulation of a SERN/Afterbody Flow Field Regarding Heated Nozzle Flow in the Hypersonic Facility H2K, Contribution to the 13th AG STAB/DGLR Symposium, Notes on Numerical Fluid Mechanics and Multidisciplinary Design, ed. by C. Breitsamter, B. Laschka, H.-J. Heinemann and R. Hilbig, ISBN 3-540-20258-7, Vol. 87, pp. 236–243, Springer, München, 2002.

## 5.7 Intake Flows in Airbreathing Engines for Supersonic and Hypersonic Transport

Birgit Ursula Reinartz, Joern van Keuk, Josef Ballmann *,
Carsten Herrmann, and Wolfgang Koschel

### Abstract

Main objective of this work is the numerical investigation of the intake flow of ELAC for supersonic and hypersonic inflow conditions. Because of the intake sidewalls the physical flow phenomena appearing in here are three-dimensional and very complex: The compression of the oncoming air is reached by a system of oblique shocks interacting with the wall boundary-layers, whereby massive separation can occur. Additionally, depending on the inflow conditions turbulence effects and – if necessary – laminar-turbulent transition have to be considered as well as real gas effects and chemical reactions. Consequently, the complete 3D Favre-averaged Navier-Stokes equations extended by species conservation equations are used as the physical model.

Starting from a finite-volume code with central discretization in space, in this case the well-known DLR FLOWer-Code, distinct extensions for internal flows at supersonic and hypersonic speed were necessary first. In this context several commonly used upwind methods for the convective part of the Navier-

* RWTH Aachen, Lehr- und Forschungsgebiet für Mechanik, Templergraben 64, 52062 Aachen; ballmann@lufmech.rwth-aachen.de

Stokes equations have been implemented. Additionally, detailed improvements of the turbulence modelling were performed. Finally, the code was extended for the simulation of real gas flows in local thermodynamical equilibrium and chemical non-equilibrium.

With the modified code extensive simulations for two- and three-dimensional supersonic and hypersonic flows have been carried out and the results have been compared with corresponding experimental data. The overall agreement regarding wall pressure distributions seems satisfactory and in several cases even the measured and calculated wall heat flux, which is the most important, but also a very sensitive quantity in hypersonic flows, are comparable.

### 5.7.1 Introduction

Future aerospace planes like the ELAC configuration developed in the framework of the Collaborative Research Centre 253 (SFB 253) should operate in a wide Mach number range ($0.2 \leq M \leq 9.0$) in altitudes up to 35 km at high Reynolds numbers ($Re \approx 10^8$). Since there exists a satisfactory amount of oxygen in these altitudes, an airbreathing propulsion system is possible, that is planned to work with combustion at subsonic speed. A critical issue for the overall performance of an airbreathing propulsion system is the aerodynamic design of the intake part. The primary purpose of the intake is to provide the engine with air at homogeneous high-pressure.

Thus, intake and diffusor for an airbreathing engine of a space transport vehicle play a significant role for the efficiency of the whole propulsion system. In the case of ELAC a combination of a turbo- and ramjet device should guarantee optimum thrust throughout the whole Mach number range. Objective is a high efficiency compression by deceleration of the oncoming supersonic or hypersonic air flow to subsonic speed with the aim of supplying the engine with as homogeneous a flow as possible. In the case of ELAC an intake with mixed outer and inner compression by several oblique shocks is planned. The final deceleration to subsonic flow is reached by a normal shock wave in the diffusor part. In close collaboration with the experimental project C2 of the SFB 253 the intake/diffusor flow of ELAC is investigated.

The physical phenomena appearing in such an intake/diffusor device are extremely complex. There are complicated interactions of shock waves that are again interacting with the boundary layers along the intake walls. These interactions can sometimes lead to massive separation forming extensive recirculation regions, in which the flow becomes unsteady.

For the numerical simulation of such intake/diffusor flows the complete 3D Favre-averaged Navier-Stokes equations have to be considered. This system of equations consists formally of a convective and a diffusive part as well as a source term, whereby each part has to be treated according to its mathematical character. The three main requirements concerning approximate solutions of the Navier-Stokes equations for the flows mentioned above are correct capturing of wave propagation phenomena, best possible turbulence modelling and – if necessary – adequate description of real gas effects and chemical reactions.

## 5.7.2 Physical Model

The Favre-averaged 3D Navier-Stokes equations with Fourier heat conduction form the basis for the numerical simulations presented in this work. The equations are used in conservation form and can be written for a three-dimensional control volume $V$ with boundary $\partial V$ as follows:

$$\int_V \frac{\partial \boldsymbol{U}}{\partial t} + \oint_{\partial V} \{(\boldsymbol{F} - \boldsymbol{F}_d)e_{n_x} + (\boldsymbol{G} - \boldsymbol{G}_d)e_{n_y} + (\boldsymbol{H} - \boldsymbol{H}_d)e_{n_z}\}dA = \int_V \boldsymbol{Q} dV \qquad (1)$$

$e_{n_x}$, $e_{n_y}$, $e_{n_z}$ are the components of the outwards directed unit normal $\boldsymbol{e}$ for $dA$. $\boldsymbol{U}$ is the vector of the unknown conservative variables and $\boldsymbol{F}$, $\boldsymbol{F}_d$, $\boldsymbol{G}$, $\boldsymbol{G}_d$, $\boldsymbol{H}$ and $\boldsymbol{H}_d$ the corresponding convective and diffusive flux functions. $\boldsymbol{Q}$ is the source term appearing for distinct turbulence models and in case of chemical non-equilibrium.

The flow problems investigated in here are characterized by complex geometries causing strong three-dimensional structures, massive separation and complicated turbulence effects. It is therefore necessary to use high quality turbulence models in order to account for as many physical effects as possible. Simple turbulence models like the Baldwin-Lomax [1] model or the standard $k$-$\omega$ model [34] often fail in connection with such flows. For better capturing the effect of compressibility there exist distinct corrections, that have been originally proposed by Coakley [4, 31]. The so-called Length-Scale Correction [31] defines an upper limit for the turbulent length scale and should prevent non-realistic high values for the wall heat flux in the case of shock/boundary-layer interactions. With the "Rapid-Compression Correction" [4] the production term in the $\omega$-equation is modified in order to capture the reduction of the turbulent length scale in the vicinity of shock waves. More recent proposals such as the Spalart-Allmaras model [24] or Menter's SST model [17] have been investigated, too. Finally, an explicit algebraic Reynolds-stress model after Wallin and Johannson [33], that includes nonlinear terms in the equation for the Reynolds-stress tensor, was tested. Since the FLOWer-Code was originally developed for the simulation of subsonic and transonic flows, it is working with the assumption of a thermally and calorically perfect gas. This assumption is surely justified for the flow problems FLOWer has been developed for, but not for high-speed and high-enthalpy flows, particularly not in the case of high temperatures and low densities. Although the thermal equation of state for a perfect gas may still be assumed, the calculation of enthalpy or pressure and the modelling of the transport coefficients have to be modified. One possibility is the assumption of a thermally perfect gas in local thermodynamical equilibrium, that represents one limiting case of the more general non-equilibrium approach to describe the excitation of internal degrees of freedom of particles as well as chemical reactions. This model implies the assumption of vanishing relaxation times so that chemical reactions take place instantaneously and the complete thermodynamic state is given by two independent state variables, e.g. $H=H(\rho,\in)$. One possibility to calculate state variables and gas mixture compound is to use tables based on statistical mechanics. Usually such tables contain the state variables and gas mixture compound as a function of internal energy and density in equally-spaced- and log-scale, respectively.

In the non-equilibrium case relaxation times for chemical reactions are no more negligible. This means that it is necessary to solve additional conservation equations for the chemical species, that include source terms. The forward and backward reaction rates in these source terms are usually formulated using an Arrhenius Ansatz and the equilibrium value from statistical mechanics [3]. Finally, for the calculation of the transport coefficients viscosity and thermal conductivity usually curve fits and different mixing rules are used (e.g. [2, 35]). Details for the implementation of local equilibrium and non-equilibrium chemistry into the FLOWer-Code can be found in [12, 28].

### 5.7.3 Numerical Method

For the numerical solution of the Navier-Stokes equations the solution domain is first subdivided into non-overlapping cells. For the geometrical description of these cells general coordinates $\xi$, $\eta$, and $\zeta$ are introduced and the Navier-Stokes equations are then formulated in integral form for each cell. Approximation of the surface integrals using the midpoint rule leads to the following expression of the Navier-Stokes equations in semi-discrete form:

$$\frac{d}{dt}(\boldsymbol{U}_{i,i,k}) = -\frac{1}{V_{i,j,k}}[\Delta_\xi(\hat{\boldsymbol{F}} - \hat{\boldsymbol{F}}_d) + \Delta_\eta(\hat{\boldsymbol{G}} - \hat{\boldsymbol{G}}_d) + \Delta_\zeta(\hat{\boldsymbol{H}} - \hat{\boldsymbol{H}}_d)] + \boldsymbol{Q}_{i,j,k} = \boldsymbol{Res}_{i,j,k} \quad (2)$$

In this equation the $\Delta_\lambda$, $\lambda=\xi,\eta,\zeta$ are the differences of the convective and diffusive flux vectors along the general coordinate directions.

As already mentioned above, the numerical method used to calculate approximate solutions of these equations for turbulent 3D supersonic and hypersonic perfect or real gas flows is based on the well-known DLR FLOWer-Code [13], which had to be extended for the simulation of supersonic and hypersonic flows.

FLOWer is formulated as a block-structured Finite-Volume scheme with optionally cell-vertex or cell-centered discretization. Time integration is performed either explicitly or implicitly depending on the particular flow problem. For steady calculations an explicit 5-stage Runge-Kutta scheme in connection with various convergence acceleration methods like local time stepping, multigrid and implicit residual smoothing is used for efficiency reasons. For unsteady computations an implicit discretization based on Jameson's "Dual-Time-Stepping" [10] is preferred. This method combines the implicit discretization with the convergence acceleration techniques for steady computations using an iteration parameter referred to as dual time. With this procedure storage and inversion of the system matrix, that is usually necessary in implicit methods, can be avoided. For the simulation of real gas flows with chemical reactions the so-called Strang Splitting method is used. This method is a pointwise semi-implicit scheme using two half-steps. The physical meaning of this procedure is that during one half-step of time integration an inert fluid is allowed to convect and diffuse, whereas during the second half-step a fluid at rest is allowed to react chemically. Thus, restrictions due to small chemical reaction time scales can be avoided.

The convective part of Eq. (1), the Euler-equations, are of hyperbolic type and therefore characterized by a directed propagation of information inside the solution domain. To account for this directed propagation of information in the discretized form as well, upwind methods are appropriate and a variety of commonly used upwind methods was implemented into the original FLOWer-Code in the framework of this research project. These are the flux-vector splittings of van Leer [29], AUSM [15] and several updates of this proposal like AUSMDV [32], AUSM+ [16], LDS [8] or MAPS [22] as well as the flux-difference splittings of Roe [21] and the simple HLLE [9] method.

All upwind schemes mentioned above are only of formally first order accuracy in space and consequently not applicable in Navier-Stokes solvers for reasons of consistency. It is therefore necessary to increase the formal order of accuracy in space up to at least second order. A commonly used method to construct second order schemes is the so-called MUSCL-extrapolation (*M*onotonic *U*pstream *S*cheme for *C*onservation *L*aws) [30]. This procedure means a piecewise linear reconstruction of the flow variables in the solution domain resulting in the calculation of left- and right-extrapolated values at the cell interfaces. To guarantee the TVD-property of the resulting schemes, so-called flux-limiters are commonly used [25]. These limiters control the extrapolation and, for example, switch off extrapolation in the vicinity of shock waves. Widely used limiters are the "minmod" and "superbee" of Roe, or the ones proposed by van Leer or van Albada.

Finally, for Roe's method there exists another variant for gaining a second order scheme, which was introduced by Harten and Yee, that is referred to as "modified flux approach" [36]. Harten and Yee used a Taylor expansion to construct a so-called "diffusive" formulation of the original Roe-scheme, where the truncation error of the first order scheme was appropriately approximated and compensated.

### 5.7.4 Results

#### 5.7.4.1 Turbulent 2D Supersonic Intake Flows with Internal Compression

In close collaboration with the experimental project C2 of the SFB 253 the intake-flow of ELAC is investigated in detail [20]. A two-dimensional mixed compression intake model with a subsequent isolator section is tested. All flow computations were performed with the compressible low Reynolds number $k$-$\omega$ model and the AUSM flux vector splitting in connection with the limiter functions of van Leer and van Albada. The flow field of an unthrottled supersonic intake for an isolator length of $l$=79.3 mm is presented in Fig. 5.7.1. Freestream conditions for this case are: $M_\infty$=2.41, $Re/m$=5.07×$10^7$ and $T_{t,0}$=305 K. The walls were assumed to be adiabatic regarding the comparatively long measurement times of ~20 minutes. The flow expands around the edge of the compression ramp corner, seen as red region in the Schlieren picture. Slightly downstream, the impingement of the lip shock on the ramp causes a small separation of the boundary layer. The corresponding separation and reattachment shocks are visible as blue lines. A further primary shock, which is caused by the cowl deflection angle, crosses the separation and reattachment shocks, is reflected at the ramp, and impinges at almost the same location as the reattachment shock on the cowl surface. Downstream, the repeatedly reflected shocks visibly weaken. Comparison of the Schlieren picture with the likewise presented computed Mach number contours shows an overall good agreement. The shock wave pattern, the separation, and the approximate boundary layer thickness of the Schlieren picture are also present in the simulation.

The surface pressure distributions, shown in Fig. 5.7.2, allow for a more quantitative comparison between the calculated and measured results. Here, a discrepancy in the ramp pressure distribution can be seen in the expansion region with subsequent separation. The computed separation appears smaller

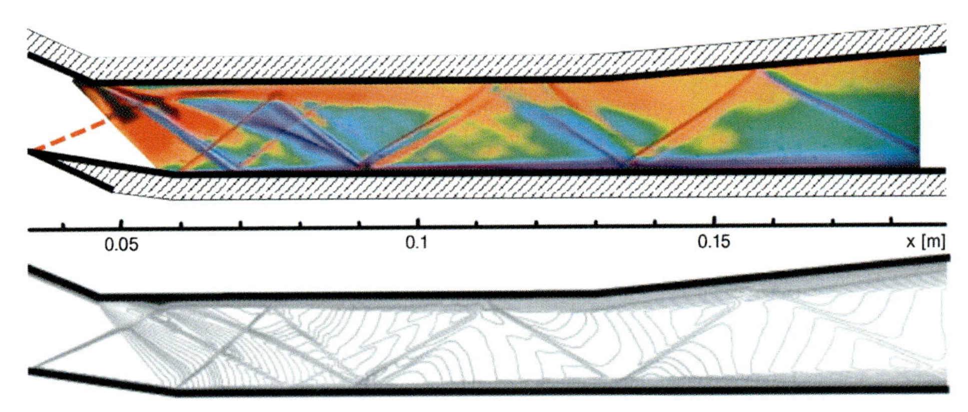


Figure 5.7.1: Comparison of an experimental colour Schlieren picture without throttling and corresponding Mach number contour lines from computation ($M_\infty$=2.41, $l$=79.3 mm).

than experimentally observed and, thus, the separation shock is weaker and impinges downstream of the measured location on the cowl surface. The reason for this discrepancy is probably a deficiency of the turbulence model to predict transitional behaviour in the flow. Due to the expansion, the boundary layer is laminarized, as will be discussed in more detail in the following part, and the turbulence model is not able to predict the changing state of the boundary layer correctly. On the other hand, the computed pressure distributions for the compression ramp as well as the cowl surface compare favourably with the experimentally obtained pressure measurements. The discrepancy between measurements and simulation in the diffusor section is due to the grid resolution in this section, which is reduced in favour of a better resolution in the intake and isolator. Overall, the agreement between simulation and experimental data is favourable.

During the intake experiments for another test case, a severe separation of the ramp boundary layer appeared that covered about one third of the isolator height (Fig. 5.7.3). Freestream conditions for this case are: $M_{\infty}=3.0$, $Re/m=4.87 \times 10^7$ and $T_{t,0}=290$ K. For an intake, a separation of this size is not tolerable due to the flow blockage and the resulting unsteadiness of the flow. Therefore, a wire was mounted on the ramp to increase the turbulence level of the boundary layer and to make it more resistant to adverse pressure gradients. This technique proved successful as is shown in Fig. 5.7.4. However, a numerical simulation was initiated to better understand the reasons and to visualize the flow pattern not visually accessible in the experiments. Therefore, the distur-

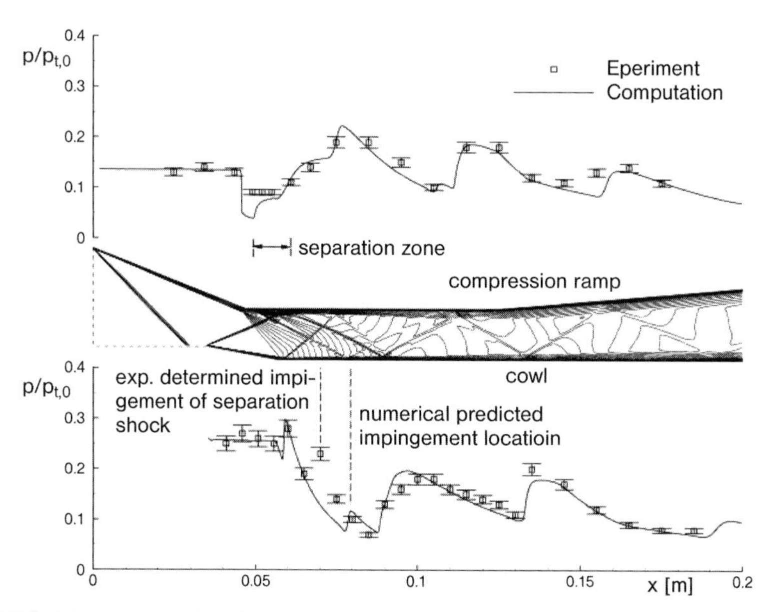


Figure 5.7.2: Experimental and computed Mach contours and surface pressure distribution without throttling ($M_{\infty}=2.41$, $l=79.3$ mm).

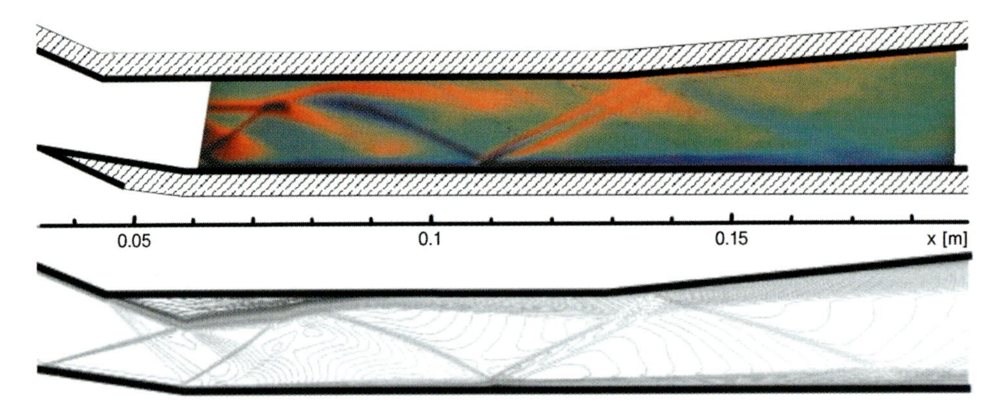


Figure 5.7.3: Large separation occurring after boundary layer is laminarized due to strong expansion. Colour Schlieren frame compared with computed contour lines of Mach number.

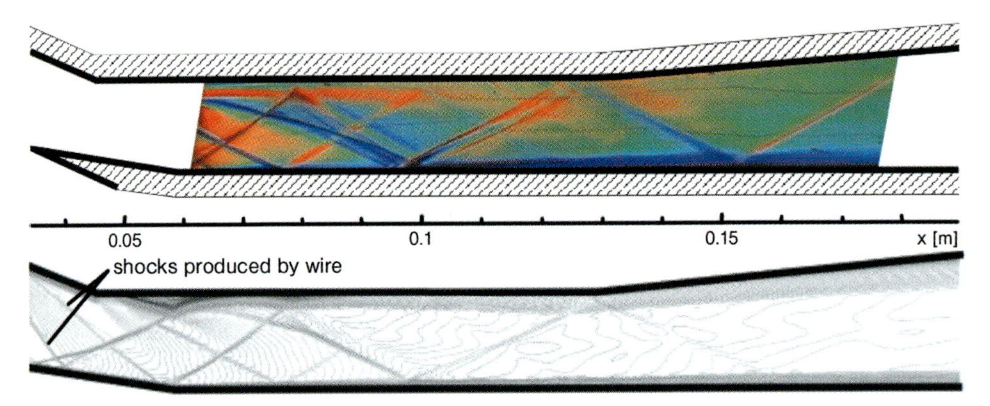


Figure 5.7.4: Size of separation is greatly reduced by insertion of a turbulence producing wire in the flow. Colour schlieren frame compared with computed lines of Mach number.

bance introduced by the wire was simulated with great care. Figure 5.7.5 shows the grid as well as the flow field in the vicinity of the wire. Prandtl-Meyer expansion theory yields that the expansion of the $M_\infty=3$ flow is 1.3 times stronger than the respective expansion at $M_\infty=2.41$. The subsequent computations of the intake flow with and without the wire showed that the expansion is indeed strong enough to change the state of the boundary layer flow. Even though, the turbulence model is not able to predict transition, the turbulence intensity reduced enough during the expansion that a separation of comparable size was determined for the wireless case (Fig. 5.7.3). In Fig. 5.7.4 the flow pattern obtained in the simulation including the wire is shown, yielding two additional induced shock waves (see also Fig. 5.7.5). Downstream of the wire, the boundary layer thickness is visibly increased and the vorticity lev-

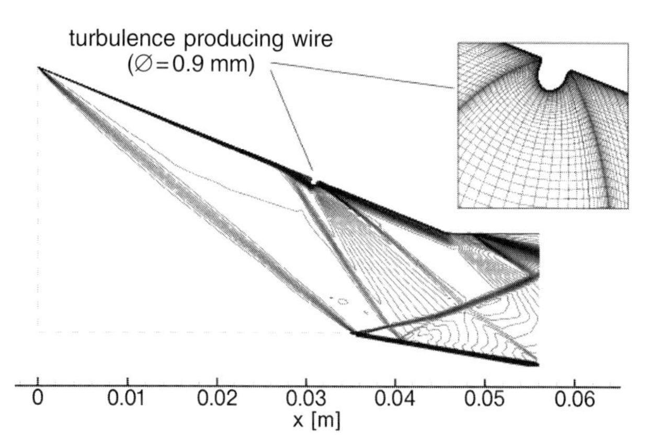


Figure 5.7.5: Flow field and grid detail in the vicinity of the mounted wire.

el is raised whereby the separation is greatly reduced. In Fig. 5.7.4, the concurrence of the shock wave locations between Schlieren picture and Mach number contours is very satisfactory. The computations fully explain the altered Schlieren wave pattern between the two experimental setups.

### 5.7.4.2 Laminar 3D Hypersonic Corner Flows

As first test cases for the simulation of three-dimensional flow problems with the modified FLOWer-Code laminar hypersonic corner flows were chosen. Such flows have been investigated extensively in literature both experimentally [11] and numerically [6]. Corresponding to the experiments by Hummel et al. [11] the following freestream conditions were chosen: $M_\infty = 12.3$, $Re/m = 5.0 \times 10^6$ and $T_\infty = 45.3$ K. The walls were assumed to be isothermal with $T_W = 300$ K.

Figures 5.7.6 and 5.7.7 show the computed Mach number distribution in comparison with the corresponding results of D'Ambrosio et al. [6] in the y-z-plane. The x-direction is the mean flow direction. There are two primary or wedge shocks interacting via the corner shock, whereby two reflected shock waves and contact discontinuities arise. The reflected shock waves impinge on the boundary layer causing a separation and additionally a small secondary separation region. For the result shown the AUSMDV method has been used in connection with the "minmod" limiter on a computational grid consisting of 100 gridpoints in each direction. The comparison with the results of D'Ambrosio et al. [6] shows a qualitatively nice agreement. This is particularly satisfying since the solutions have been achieved in differently extended solution domains on differently discretized grids with basically different methods (D'Ambrosio et al. [6] solve the parabolized Navier-Stokes equations with a space marching technique on grids with 120×120 gridpoints in each plane). Finally, in Figs. 5.7.8 and 5.7.9 a quantitative comparison between the computed results

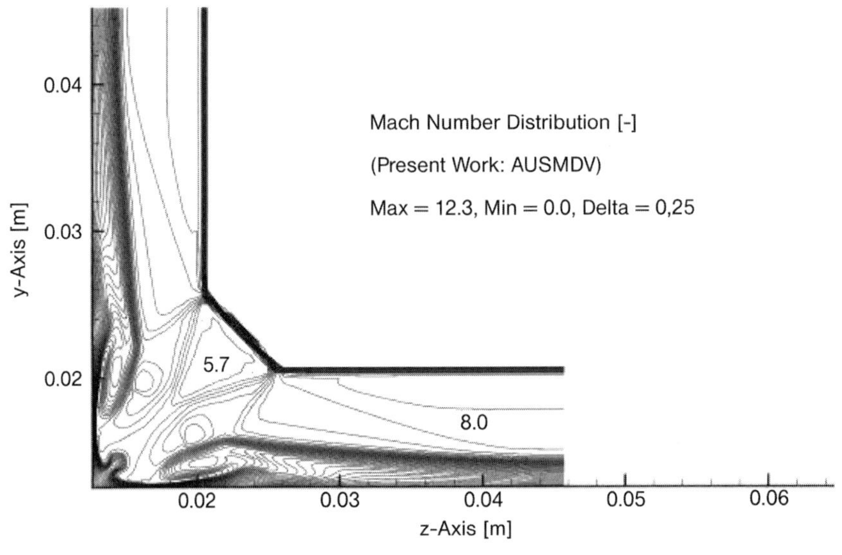


Figure 5.7.6: Computed results for Mach number distribution (present work), x=0.09 m.

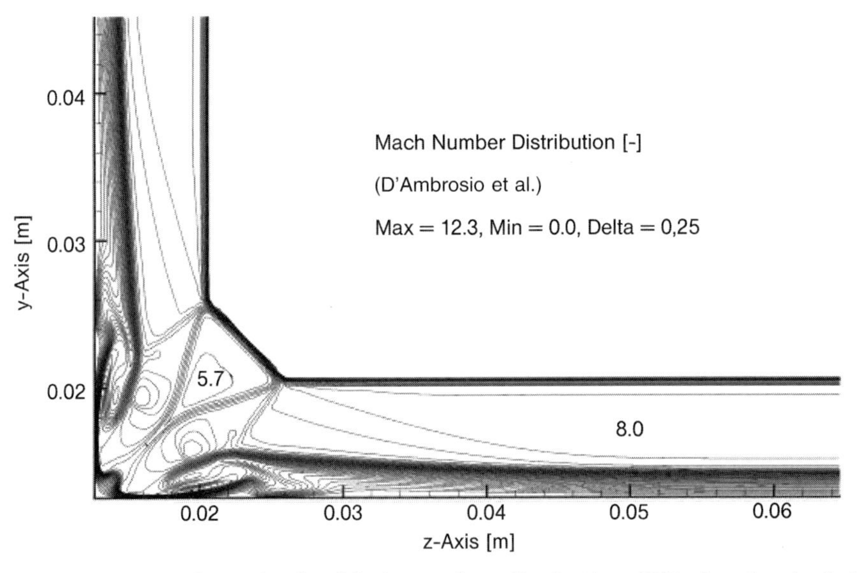


Figure 5.7.7: Computed results for Mach number distribution (D'Ambrosio et al. [6]), x=0.09 m.

and the experimental data of Hummel [11] regarding the wall pressure distribution is shown, whereby the region of secondary separation is enlarged. The overall agreement is again satisfactory.

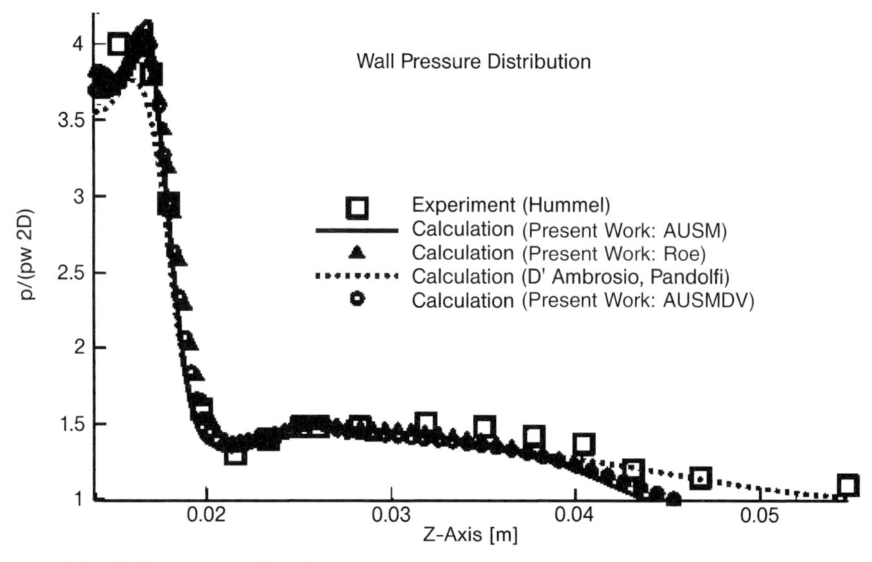


Figure 5.7.8: Wall pressure distribution (calculations/experiment).

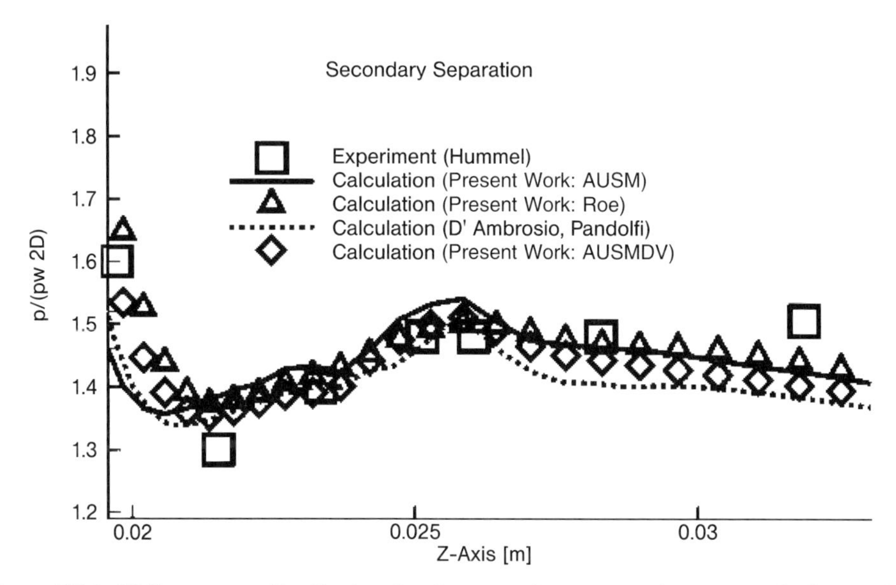


Figure 5.7.9: Wall pressure distribution for the secondary separation area (calculations/experiment).

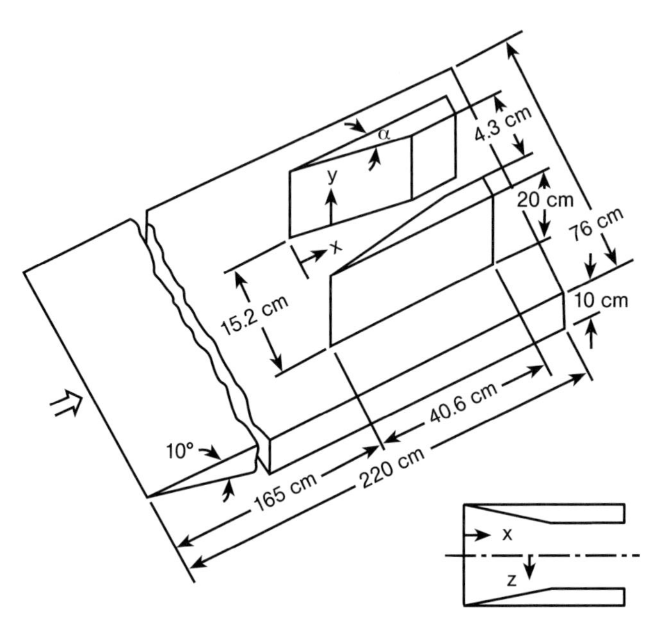


Figure 5.7.10: Flow geometry for crossing-shock interaction.

### 5.7.4.3 Turbulent 3D Hypersonic Flows through Symmetric/Asymmetric Double-Fin Configurations

As a preparation for the simulation of three-dimensional intake flows the flow through a symmetric double-fin configuration was investigated. Figure 5.7.10 shows the geometrical ratios of this flow problem. Corresponding to the experiments of Kussoy et al. [14] the following freestream conditions were chosen: $M_\infty = 8.3$, $Re/m = 5.3\times10^6$ and $T_\infty = 75.2$ K. All walls were assumed to be isothermal with $T_W = 300$ K. Two shocks are generated by the wedge-shaped sidewalls. These two shocks intersect downstream on the channel centreline creating a large interaction zone with complicated vortex systems. Additionally, the shocks interact with the boundary layer on the bottom causing primary and secondary separation regions. The computations were performed using the AUSMDV scheme in connection with the limiter of van Leer. Different turbulence models were tested. Figures 5.7.11 and 5.7.12 show comparisons with the experimental data regarding wall pressure distribution and wall heat flux in the symmetry plane. The overall agreement is completely satisfying, but it is worth mentioning that such a good agreement regarding the wall heat flux was rather the exception than the rule in the computations carried out.

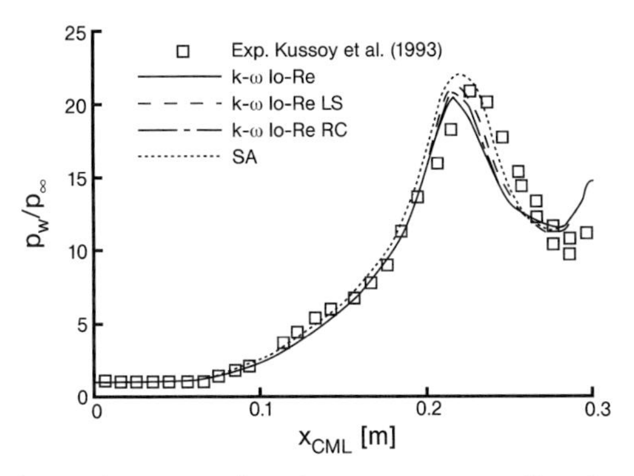


Figure 5.7.11: Comparison of computed surface pressure on the channel middle line with experimental data.

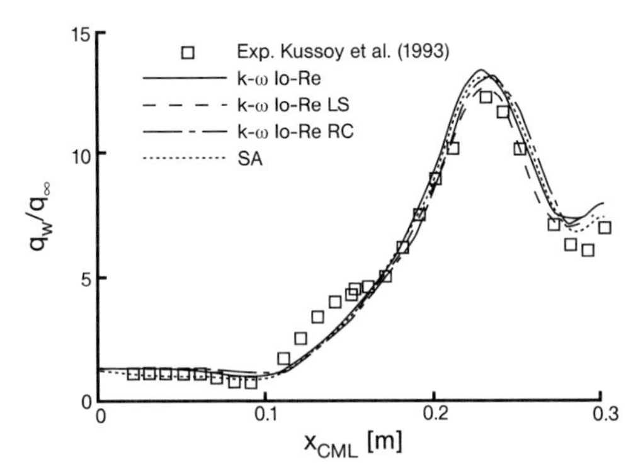


Figure 5.7.12: Comparison of computed heat transfer on the channel middle line with experimental data.

### 5.7.4.4 Laminar 2D Shock Interactions in Hypersonic Flows with Chemical Non-Equilibrium

Finally, first simulations of pure nitrogen real gas flows with chemical non-equilibrium were performed with the modified FLOWer-Code. Shock wave interactions in hypervelocity flows, that have been investigated experimentally at the Graduate Aeronautical Laboratories (California Institute of Technology) [23], serve as test cases. These interactions can cause extremely high surface pressure and particularly heat loads. A very severe case results from the impingement of a weak oblique shock wave on the bow shock ahead of a blunt body when so-called Edney-type shock interactions arise. Depending on the location of the impinging shock wave with respect to the body, Edney [7] observed and distinguished six different interaction regimes, referred to as Edney-type I–VI interactions. Figures 5.7.13–5.7.15 show an interferogram and corresponding computed density distributions as well as measured and calculated wall heat

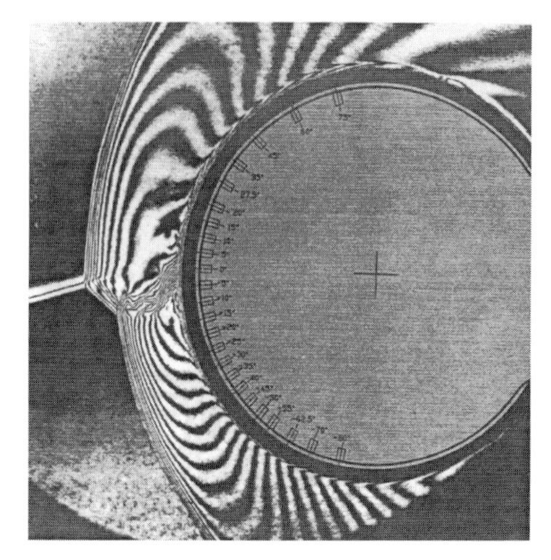

Figure 5.7.13: Interferogram for Type 4 shock interaction [23].

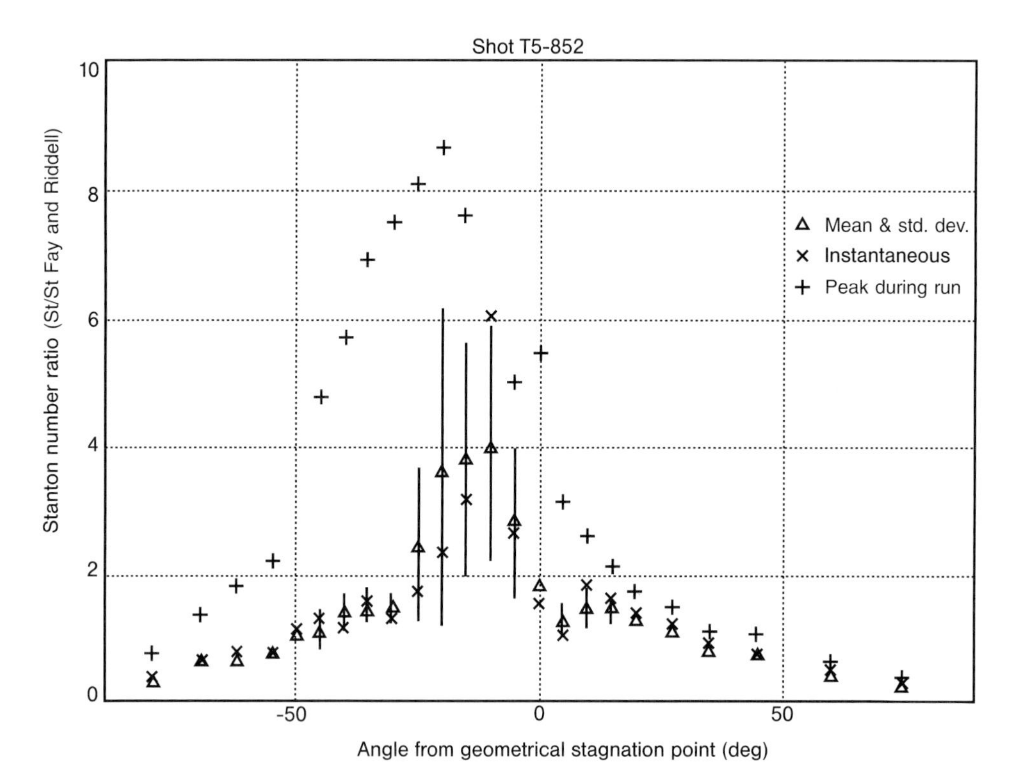


Figure 5.7.14: Measured heat flux for Type 4 shock interaction [23].

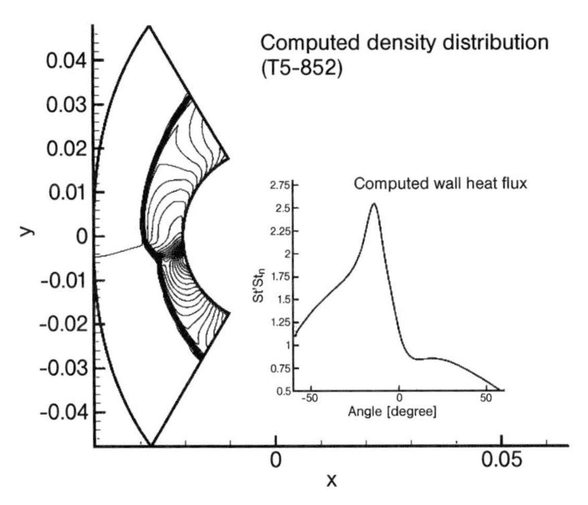


Figure 5.7.15: Computed density distribution and heat flux for Type 4 shock interaction [28].

fluxes for a Type 4 interaction. Corresponding to the experiments of Sanderson [23] the following freestream conditions were chosen: $M_\infty=6.29$, $Re/m=1.63\times10^6$, $T_\infty=1190$ K, $\rho_\infty=0.0155$ kg/m$^3$ and $x_{N\infty}=0.01$. The contour of the cylinder was assumed to be isothermal with $T_W$=300 K. This flow exhibits complex interactions between altogether five shocks and two contact discontinuities forming a jet that hits the body and causes the high thermal loads in this case. The overall agreement regarding the flow phenomena is satisfying, even though the numerical simulation almost completely fails in the resolution of the contact surface resulting from the interaction point. This may be due to the artificial viscosity needed for numerical stability which is introduced via the limiter functions in the connection with so-called shock sensors for the MUSCL-Extrapolation in the interaction regime. The agreement between measured and calculated wall heat fluxes is again acceptable, considering the enormous sensitivity of this quantity. This can be seen in the large error bars of the measurements near the region, where the jet hits the body.

Finally, clearly to see is the influence of the chemical non-equilibrium resulting in a distinct relaxation zone behind the bow shock, that cannot be captured using the local equilibrium assumption. Even not shown, a calculation with the perfect gas model results in a too big shock standoff distance by a factor of two. The same holds for the thermal loads on the contour.

### 5.7.5 Conclusions

In the framework of the research project C3 of the SFB 253 the DLR FLOWer-Code, that is originally developed for subsonic and transonic flows, has been extended step by step for turbulent super- and hypersonic optionally perfect or real gas flows. The extensions include implementation of different upwind methods, several improvements of the turbulence modelling and, finally, consideration of real gas effects and chemical reactions. In every step experimental results have been used for validation purposes showing an overall satisfying agreement so that a well validated and powerful tool for the simulation of real gas high speed flows has been made available.

## References

1. B. Baldwin, H. Lomax: Thin Layer Approximation and Algebraic Model for Separated Turbulent Flows, AIAA Paper 78-257, 1978.
2. F.G. Blottner, M. Johnsson, M. Ellis: Chemically Reacting Viscous Flow Program for Multi-Component Gas Mixtures, Report No. SC-RR 70-754, Sandia Laboratories, Albuquerque, New Mexico, 1971.
3. G.V. Candler: The Computation of Weakly-Ionized Hypersonic Flows in Thermo-Chemical Nonequilibrium, Ph.D. Thesis, Stanford University, 1988.
4. T.J. Coakley, P.G. Huang: Turbulence Modeling for High Speed Flows, AIAA Paper 92-0436, 1992.
5. T. Coratekin, J. van Keuk, J. Ballmann: On the Performance of Upwind Schemes and Turbulence Models in Hypersonic Flows, AIAA Journal, Vol. 42, No. 5, 2004, pp. 945–957.
6. D. d'Ambrosio, R. Marsillio, M. Pandolfi: Shock-Induced Separated Structures in Symmetric Corner Flows, AIAA Paper 95-2270, 1995.
7. B.E. Edney: Anomalous Heat Transfer and Pressure Distributions on Blunt Bodies at Hypersonic Speeds in the Presence of an Impinging Shock, FFA Report 115, 1968.
8. J.R. Edwards: A Low-Diffusion Flux-Splitting Scheme for Navier-Stokes Calculations, Computers & Fluids, Vol. 26, No. 6, pp. 635–659, 1997.
9. A. Harten, P.D. Lax, B. van Leer: On Upstream Differencing and Godunov-Type Schemes for Hyperbolic Conservation Laws, SIAM Review, Vol. 25, pp. 35–61, 1983.
10. A.J. Jameson: Time Dependent Calculations Using Multigrid, with Application to Unsteady Flows past Airfoils and Wings, AIAA Paper 91-1596, 1991.
11. K. Kipke, D. Hummel: Untersuchungen an längsangeströmten Eckenkonfigurationen im Hyperschallbereich. Teil I: Ecken zwischen ungepfeilten Keilen, Zeitschrift für Flugwissenschaften, Vol. 23, Heft 12, pp. 417–429, 1975.
12. A. Klomfass: Hyperschallströmungen im thermodynamischen Nichtgleichgewicht, PhD Thesis, RWTH Aachen, 1995.
13. N. Kroll, J. Raddatz, R. Heinrich et al.: FLOWer Version 116, Project MEGAFLOW, Institut für Entwurfsaerodynamik, Deutsches Zentrum für Luft- und Raumfahrt, 2000.
14. M.I. Kussoy, K.C. Horstman, C.C. Horstman: Hypersonic Crossing Shock-Wave/Turbulent Boundary-Layer Interactions, AIAA Journal, Vol. 31, No. 12, pp. 2197–2203, 1993.
15. M.S. Liou, C.J. Steffen: A New Flux Splitting Scheme, Journal of Computational Physics, Vol. 107, pp. 23–39, 1993.
16. M.S. Liou: A Sequel to AUSM: AUSM+, Journal of Computational Physics, Vol. 129, pp. 364–382, 1996.
17. F.R. Mentner: Two-Equation Eddy-Viscosity Turbulence Models for Engineering Applications, AIAA Journal, Vol. 32, No. 8, pp. 1598–1605, 1994.

18. B.U. Reinhartz, J. van Keuk, T. Coratekin, J. Ballmann: Computation of Wall Heat Fluxes in Hypersonic Inlet Flows, AIAA Paper 2002-0506, 2002.
19. B.U. Reinhartz, J. Ballmann: Details on the Computation of Hypersonic Inlet Flows, GAMM-Jahrestagung, Augsburg, 2002, PAMM, Vol. 3, Iss. 1, pp. 326–327.
20. B.U. Reinhartz, C.D. Herrmann, J. Ballmann, W.W. Koschel: Aerodynamic Performance Analysis of a Hypersonic Inlet Isolator using Computation and Experiment, AIAA Journal of Propulsion and Power, Vol. 19, No. 5, 2003, pp. 868–875.
21. P.L. Roe: Approximate Riemann Solvers, Parameter Vectors, and Difference Schemes, Journal of Computational Physics, Vol. 43, pp. 357–372, 1981.
22. C.-C. Rossow: A Simple Flux Splitting Scheme for Compressible Flows, Notes on Numerical Fluid Mechanics, Vol. 72, pp. 355–362, 1999.
23. S.R. Sanderson: Shock Wave Interaction in Hypervelocity Flow, PhD Thesis, Graduate Aeronautical Laboratories, California Institute of Technology, 1995.
24. P.R. Spalart, S.R. Allmaras: A One-Equation Turbulence Model for Aerodynamic Flows, AIAA Paper 92-0439, 1992.
25. P.K. Sweby: High Resolution Schemes Using Flux Limiters for Hyperbolic Conservation Laws, SIAM Journal on Numerical Analysis, Vol. 21, pp. 995–1011, 1984.
26. J. van Keuk, J. Ballmann, A. Schneider, W.W. Koschel: Numerical Simulation of Hypersonic Inlet Flows, AIAA Paper 98-1526, 1998.
27. J. van Keuk, J. Ballmann: Numerical Simulation of Symmetric Corner Flows in the Hypersonic Regime, Notes on Numerical Fluid Mechanics, 72, 250–257, 1999.
28. J. van Keuk, J. Ballmann, S.R. Sanderson, H.G. Hornung: Numerical Simulation of Experiments on Shock Wave Interactions in Hypervelocity Flows with Chemical Reactions, AIAA Paper 2003-0960, 2003.
29. B. van Leer: Flux Vector Splitting for the Euler Equations, Lecture Notes in Physics, Vol. 170, pp. 507–512, 1982.
30. B. van Leer: Towards the Ultimate Conservative Difference Scheme V. A Second-Order Sequel to Godunov's Method, Journal of Computational Physics, Vol. 32, pp. 101–136, 1979.
31. S.T. Vuong, T.J. Coakley: Modeling of Turbulence for Hypersonic Flows with and without Separation, AIAA Paper 87-0286, 1987.
32. Y. Wada, M.S. Liou: An Accurate and Robust Flux Splitting Scheme for Shock and Contact Discontinuities, SIAM Journal on Scientific Computing, Vol. 18, pp. 633 ff, 1997.
33. S. Wallin, A.V. Johansson: An Explicit Algebraic Reynolds Stress Model for Incompressible and Compressible Turbulent Flows, Journal of Fluid Mechanics, Vol. 403, pp. 89–132, 1999.
34. D.C. Wilcox: Turbulence Modeling for CFD, DCW Industries Inc., 1994.
35. C.R. Wilke: A Viscosity Equation for Gas Mixtures, Journal of Chemical Physics, Vol. 18, pp. 517–519, 1950.
36. H.C. Yee: Upwind and Symmetric Shock Capturing Schemes, NASA Technical Memorandum 89464, 1987.

# 6 Flight Mechanics and Control

## 6.1 Safety Improvement for Two-Stage-to-Orbit Vehicles by Appropriate Mission Abort Strategies

Michael Mayrhofer, Otto Wagner, and Gottfried Sachs *

### Nomenclature

|  |  |
|---|---|
| $C_D$ | drag coefficient |
| $C_L$ | lift coefficient |
| $D$ | drag |
| $f_\alpha$ | thrust factor for angle of attack dependency |
| $g$ | acceleration due to gravity |
| $h$ | altitude |
| $I_{sp}$ | specific impulse |
| $L$ | lift |
| $l$ | specific lift |
| $M$ | Mach number |
| $m$ | mass |
| $m_f$ | fuel mass |
| $n$ | load factor |
| $q$ | heat flux density |
| $\bar{q}$ | dynamic pressure |
| $r_E$ | radius of the Earth |
| $S$ | reference area |
| $T$ | thrust |
| $t$ | time |
| $V$ | speed |
| $\alpha$ | angle of attack |
| $\gamma$ | flight path angle |
| $\Delta$ | geocentric latitude |
| $\delta_T$ | throttle setting |
| $\varepsilon_T$ | thrust vector angle |
| $\Lambda$ | geographic longitude |
| $\mu$ | bank angle |
| $\sigma^*$ | specific fuel consumption at stoichiometric combustion |
| $\phi_f$ | equivalence ratio |
| $\chi$ | azimuth angle |
| $\omega_E$ | angular velocity of the Earth |

* Technische Universität München, Lehrstuhl für Flugmechanik und Flugregelung, Boltzmannstr. 15, 85748 Garching; sachs@lfm.mw.tum.de

*Basic Research and Technologies for Two-Stage-to-Orbit Vehicles*
DFG, Deutsche Forschungsgemeinschaft

ISBN: 3-527-27735-8

## Abstract

The improvement of the safety of two-stage-to-orbit vehicles is considered which consist of a carrier stage with airbreathing turbo/ram jet engines and a rocket propelled orbital stage. Formulating the nominal mission and the abort scenarios as an optimal control problem allows full exploitation of safety reserves. The shaping of the nominal mission significantly affects the prospective safety. Therefore, most relevant mission aborts are optimized together with the nominal mission treating them as a branched trajectory problem. With this method, the feasibility of the included mission aborts can be assured with minimum payload penalty. All other mission aborts can be separately treated, with the initial condition given by the state of the nominal trajectory at failure occurrence. A mission abort plan is set up for all emergency scenarios.


## 6.1.1 Introduction

There are world wide efforts concerning future space access vehicles for significantly reducing the cost of transportation. A promising concept is a two-stage-to-orbit vehicle consisting of a winged carrier powered by combined turbo/ram jet engines and a winged orbiter propelled by rocket engines (Fig. 6.1.1). This concept features inherent capabilities with regard to safety (gliding flight capability of both stages) and flexibility (launch window, orbit inclination, reentry, etc.).

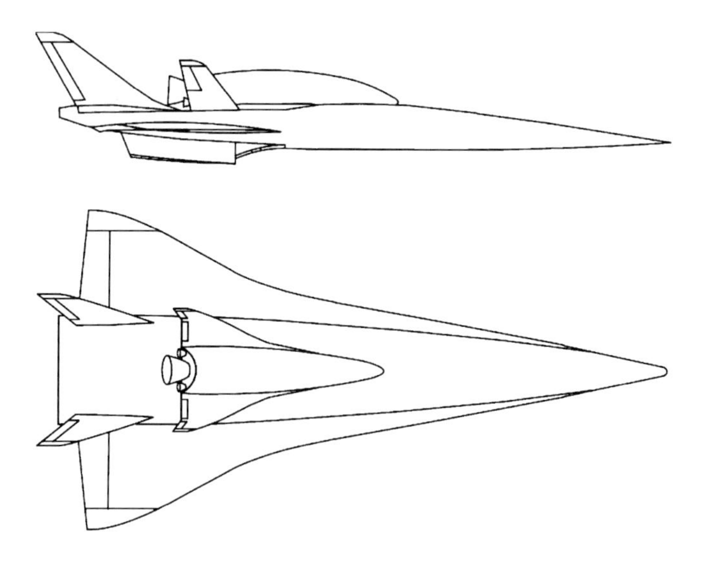

Figure 6.1.1: Two-stage-to-orbit vehicle.

Table 6.1.1: Failure classification.

| Failure | Consequence |
|---|---|
| Minor | Mission Continuation |
| Major | Intact Abort |
| Serious | Contingency Abort |
| Catastrophic | Loss of System |

Failures causing an abort may be grouped according to the consequences for the mission. A corresponding failure classification is given in Tab. 6.1.1.

According to recent investigations, 90% of all major system malfunctions are related to the propulsion system [1–3]. With regard to this failure type, intact abort scenarios are considered in the following.

An overview of intact mission abort scenarios is given in Fig. 6.1.2. There are intact mission aborts termed Abort Before Separation ABS for the complete flight system consisting of the combined carrier and orbital stages, Abort During Separation ADS of either the complete flight system or each stage alone and Abort After Separation AAS of the carrier stage. Mission aborts of the orbital stage are referenced to the ascent flight phase and the orbit, yielding Abort During Ascent ADA and Abort From Orbit AFO. The ADA mission abort type may be subdivided into Emergency Landing Site Landing ELSL, Abort Once Around AOA and Abort To Orbit ATO. The scenarios termed Direct Abort DA and Abort To Launch Site ATLS refer to mission aborts from the orbit.

The nominal mission is also shown in Fig. 6.1.2. It consists of the flight of the overall system to the separation location, the separation manoeuvre, the return of the carrier stage to the launch site and the ascent of the orbital stage to the target orbit. The target orbit of the orbital stage is circular at an altitude of 300 km with an inclination of 16.5°. After completing its operational task in the orbit, the orbital stage also returns to the launch site.

It is the purpose of this paper to consider the mission abort scenarios and to include them in the treatment of the nominal mission. Thus, a safety improved nominal trajectory can be determined which offers an enhancement for coping with possible emergency cases.

### 6.1.2 Dynamics Model of Two-Stage-to-Orbit Vehicle

For the trajectory optimization problem under consideration, a point mass model is used for describing the dynamics of the two-stage-to-orbit vehicle. The equations of motion read (Fig. 6.1.3):

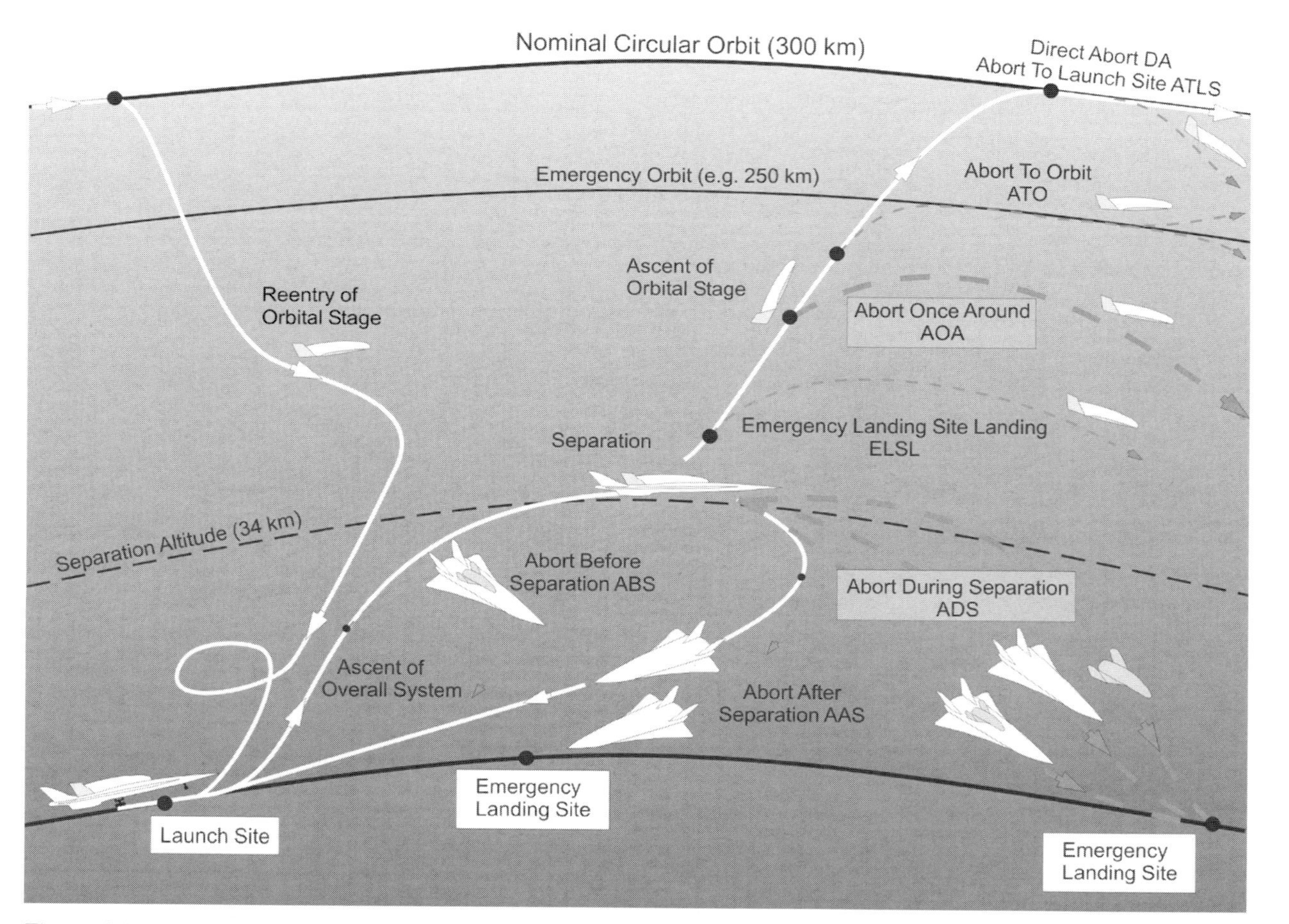


Figure 6.1.2: Mission abort scenarios.

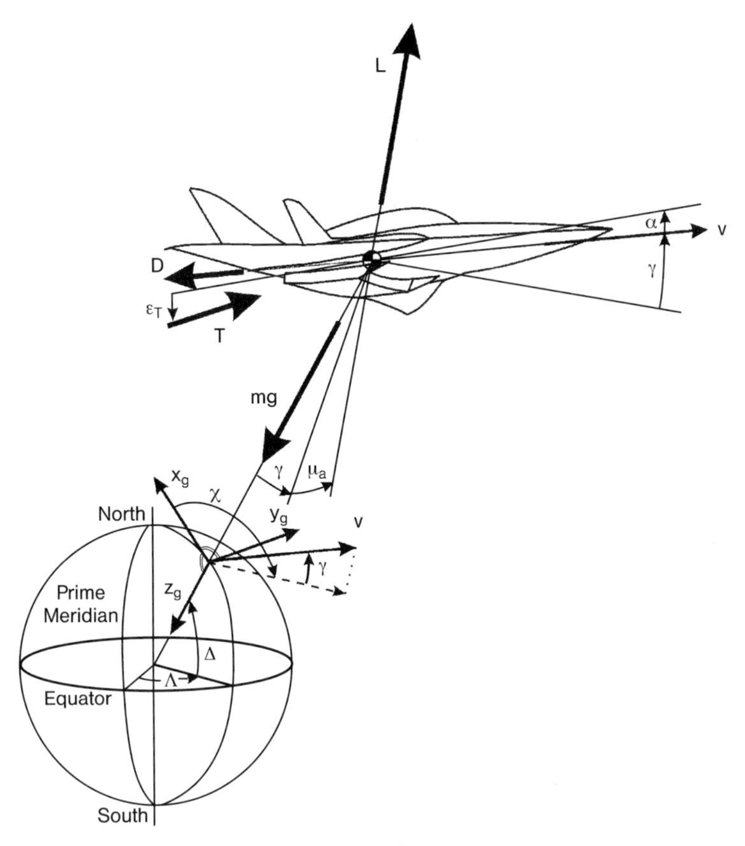


Figure 6.1.3: Forces on flight system and coordinate system.

$$\dot{V} = \frac{T \cos(a + \varepsilon_T) - D}{m} - g \sin\gamma + \omega_E^2 (r_E + h) \cos\Delta (\sin\gamma \cos\Delta - \cos\gamma \sin\Delta \cos\chi)$$

$$\dot{\gamma} = \frac{T \sin(a + \varepsilon_T) + L}{mV} \cos\mu_a + \left(\frac{V}{r_E + h} - \frac{g}{V}\right) \cos\gamma + 2\omega_E \cos\Delta \sin\chi$$

$$+ \frac{\omega_E^2 (r_E + h)}{V} \cos\Delta (\cos\gamma \cos\Delta - \sin\gamma \sin\Delta \cos\chi)$$

$$\dot{\chi} = \frac{T \sin(a + \varepsilon_T) + L}{mV \cos\gamma} \sin\mu_a + \frac{V \cos\gamma \sin\chi \tan\Delta}{r} - 2\,\omega_E (\tan\gamma \cos\Delta \cos\chi - \sin\Delta)$$

$$+ \frac{\omega_E^2 (r_E + h)}{V \cos\gamma} \sin\Delta \cos\Delta \sin\chi \qquad (1)$$

$$\dot{\Delta} = \frac{V \cos\gamma \cos\chi}{r_E + h}$$

$$\dot{\Lambda} = \frac{V \cos \gamma \sin \chi}{(r_E + h) \cos \Delta}$$

$$\dot{h} = V \sin \gamma$$

$$\dot{m} = -\dot{m}_f$$

These equations are used for the overall flight system prior to separation as well as for the carrier and orbital stages afterwards, applying the data sets corresponding to each configuration (mass, aerodynamics, propulsion system).

The aerodynamics model involves the lift and drag forces which can be expressed as

$$L = C_L \bar{q} S$$

$$D = C_D \bar{q} S \tag{2}$$

There are complex relationships describing the dependencies of the lift and drag coefficients on angle of attack $a$, Mach number $M$, and altitude $h$, yielding

$$C_L = C_L(a, M)$$

$$C_D = C_D(a, M, h) \tag{3}$$

Several sets were used for $C_L$ and $C_D$ to account for each flight system configuration (overall system, carrier and orbital stages) [4, 5].

The thrust and the fuel consumption model for the turbo/ram jet engines of the carrier stage are described as [6]

$$T = \delta_T T^* (M, h) f_a (M, a)$$

$$\dot{m}_f = \phi_f (\delta_T, M) \sigma^* (M, h) T(M, h, a) \tag{4}$$

The quantities $T^*$ and $\sigma^*$ denote thrust and fuel consumption at the stoichiometric operating condition. The factor $f_a$ describes the dependence of thrust on angle of attack. The transition from turbo to ram jet operation is modelled as a linear process in the Mach number range $3 \leq M \leq 3.5$.

Thrust and fuel consumption characteristics of the main rocket engine of the orbital stage are described by

$$T_R = \delta_{TR} T_{Rmax}$$

$$\dot{m}_{fR} = \frac{T_R}{g_0 \cdot I_{sp}} \tag{5}$$

Controls are the angle of attack $a$, the bank angle $\mu$, as well as the throttle settings $\delta_T$ *and* $\delta_{TR}$. The atmospheric model used in the optimization computations is based on [7].

Table 6.1.2: Constraints for overall system and carrier stage.

| | Minimum | Maximum |
|---|---|---|
| Angle of Attack [deg] | –1.5 | 20 |
| Throttle Setting (Turbo Jet) | 0 | 1 |
| Throttle Setting (Ram Jet) | 0 | *) |
| Dynamic Pressure [kPa] | 10 | 50 |
| Specific Lift [kN/m$^2$] | 0 | 4 |

*) The maximum throttle setting corresponds to the maximum equivalence ratio of 3.

Table 6.1.3: Constraints for orbital stage.

| | Minimum | Maximum |
|---|---|---|
| Angle of Attack [deg] | 0 | 45 |
| Throttle Setting | 0 | 1 |
| Dynamic Pressure [kPa] | 10 | 50 |
| Specific Lift [kN/m$^2$] | 0 | 8 |
| Stagnation Point Heat Flux [kW/m$^2$] | 0 | 375 |

Constraints have a significant effect on the performance of the flight system. To account for this effect, a realistic modeling has been applied. Tables 6.1.2 and 6.1.3 show the constraint values for the overall system and the carrier stage as well as for the orbital stage.

### 6.1.3 Optimization Problem

The goal of the trajectory optimization was to determine a nominal mission trajectory yielding the maximum orbital stage payload, including the capability of coping with all mission abort scenarios. Thus, the mission abort trajectories are coupled with the nominal trajectory, yielding a branched trajectory problem. For solving such problems, efficient optimization methods and computational procedures are required which are capable of coping with complex functional relationships including various kinds of constraints. The procedure used in the present paper is a parameterization optimization technique with the graphical environment GESOP [8, 9].

The initial conditions for the nominal mission refer to a flight condition shortly after take-off from the launch site. The initial conditions for a mission abort are given by the flight state at failure occurrence. The final conditions are due to the approach flight states either to the launch site or to a designated emergency landing site in case of special mission aborts.

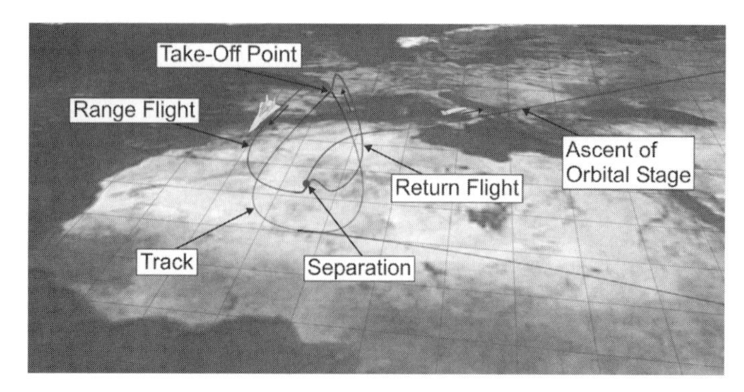


Figure 6.1.4: Reference mission trajectory.

The investigations are related to a reference mission for a Low Earth Orbit, as shown in Fig. 6.1.4. The carrier stage performs a range cruise to an optimal separation point. The latitude of the separation point corresponds to the target orbit inclination, in order to reduce the aerodynamic losses during the ascent of the orbital stage. The orbital stage is released during a highly dynamic flight manoeuvre at a Mach number of about 6.8. After stage separation the carrier returns to the launch site. The orbiter ascends to the target orbit and, after completing its orbital activities, reenters the Earth atmosphere and finally lands at the launch site.

### 6.1.4 Safety Improved Nominal Trajectory

Results on the safety improved nominal trajectory are presented in Fig. 6.1.5, showing state and control variables and other flight relevant quantities. Most significant trajectory parts are the accelerated climbing flight to the separation location, the highly dynamic separation manoeuvre and the return phase to the landing site. Several constraints become active at various sections of the trajectory.

In constructing the safety improved nominal trajectory, the separation conditions were treated as free. The Abort During Separation ADS and the Abort Once Around AOA which were implemented in the optimization process can be considered as dynamic constraints for the nominal trajectory, especially with regard to the separation conditions. The resulting optimal separation conditions for the safety improved nominal trajectory are presented in Tab. 6.1.4.

As particular aspect concerning the determination of a safety improved nominal mission was to keep the number of emergency landing sites as small as possible. In the development of the safety improved nominal trajectory, it turned out that there are two mission abort scenarios, both concerning the orbital stage, which have the most significant impact. These are the Abort During

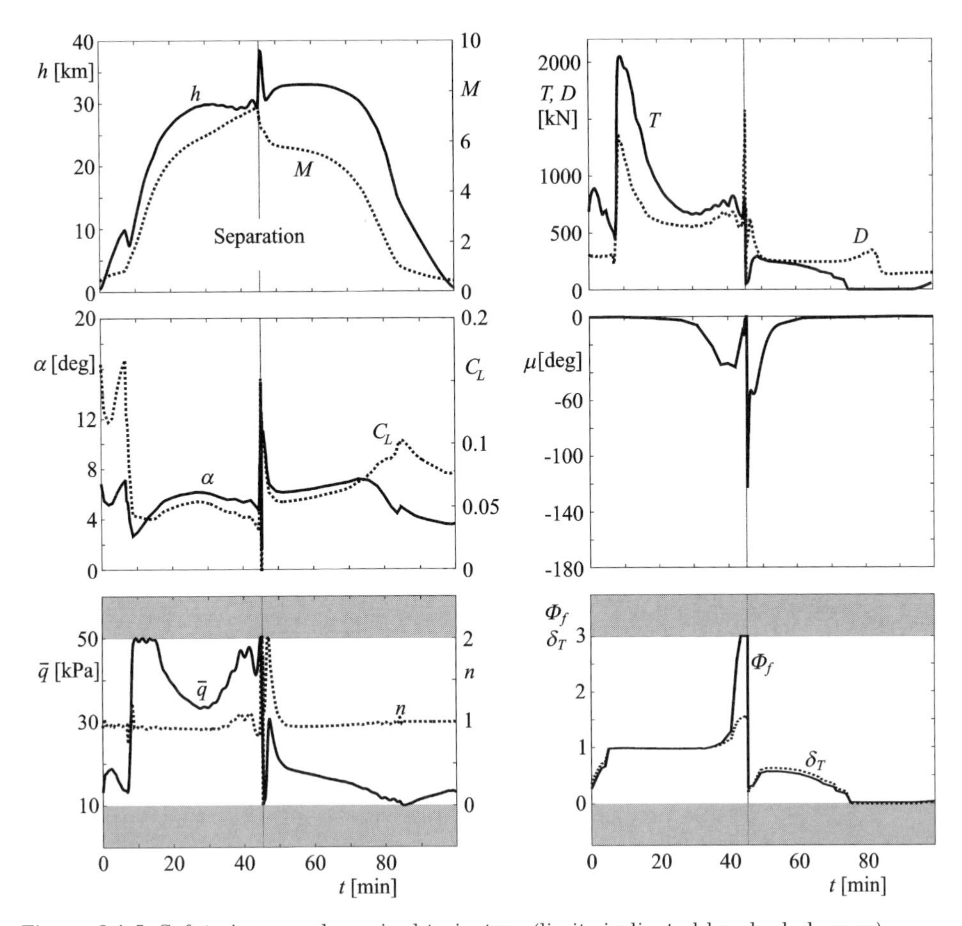


Figure 6.1.5: Safety improved nominal trajectory (limits indicated by shaded areas).

Table 6.1.4: Optimal separation conditions.

| Variable | |
|---|---|
| Altitude [km] | 33.9 |
| Speed [m/s] | 2189 ($M$=6.94) |
| Flight Path Angle [deg] | 8.17 |
| Azimuth Angle [deg] | 87.6 |
| Geocentric Latitude [deg] | 16.33 |
| Geographic Longitude [deg] | 12.28 |

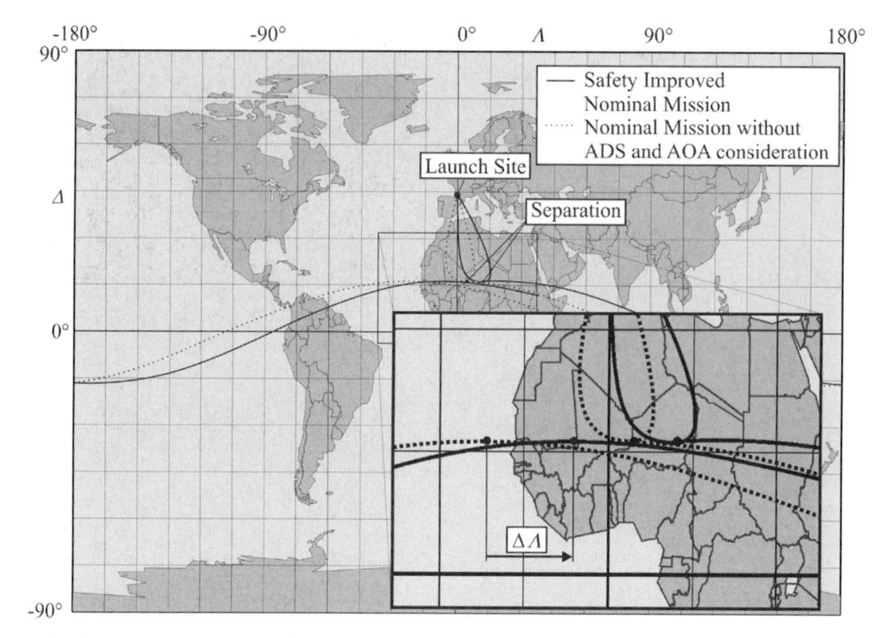


Figure 6.1.6: Trajectories for safety improved nominal mission and nominal mission without ADS and AOA consideration.

Separation ADS caused by an ignition failure of the main rocket engine and the Abort Once Around AOA after a re-ignition failure of the main rocket engine used for the on-orbit impulse.

The nominal mission trajectory resulting from an optimization without accounting for the ADS and AOA scenarios shows some differences. They primarily concern the separation flight condition of the carrier and orbital stages. A comparison of both cases is presented in Fig. 6.1.6. The flight path angle at separation is reduced in case of the safety improved trajectory so that the ADS mission abort can be performed without violating the constraint concerning dynamic pressure. Moreover, there is a shift of the separation point to the east, yielding a more favourable orbit plane for the AOA scenario.

## 6.1.5 Mission Aborts of Carrier Stage

The mission aborts of the carrier stage can be related to the flight phases in which an engine failure occurs. During outward flight, the engines operate at a significantly higher thrust level compared to the value at return flight. This contributes to increase the probability of an engine malfunction. For stage separation which is a highly dynamic flight manoeuvre, particular emphasis has to be placed on an all engine blow out.

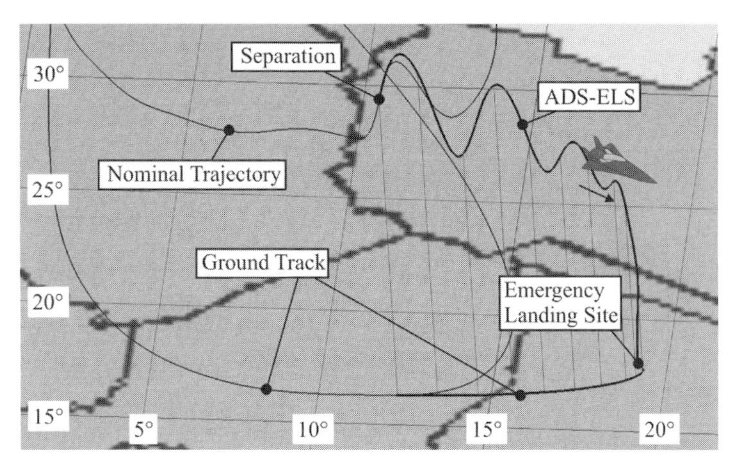


Figure 6.1.7: Optimal mission abort trajectory of flight system after an all engine blow out immediately before separation.

The most serious malfunction which is an all engine blow out just before separation with the orbiter positioned for release is considered in the following. Results are presented in Figs. 6.1.7 and 6.1.8 which show the trajectory and quantities describing the optimal mission abort of the carrier stage. In this case, it has to be taken into account that the initial conditions for a safe orbital stage ascent might not be achieved. As a worst case scenario, it is assumed that the orbital stage cannot be retracted. For terminating the mission, a gliding flight of the carrier stage with the orbital stage to an emergency landing site is considered.

From Fig. 6.1.8 it follows that the mission abort trajectory can be divided into two phases. The first phase corresponds to a deceleration flight within the dynamic pressure limits. The oxidizer of the orbital stage has to be dumped in order to reduce the weight of the overall system to an acceptable value for landing. The significantly lighter liquid hydrogen is not released due to safety reasons. There is a drag increase due to the non-operating engines ($T<0$ in Fig. 6.1.8). The second flight phase is a gliding flight close to or at minimum dynamic pressure, with finally attaining landing conditions.

To cope with a single engine failure at any point of the nominal flight, only two landing sites are required. One is the launch site and the other an emergency site appropriately chosen. This also holds for multiple engine failures in most cases.

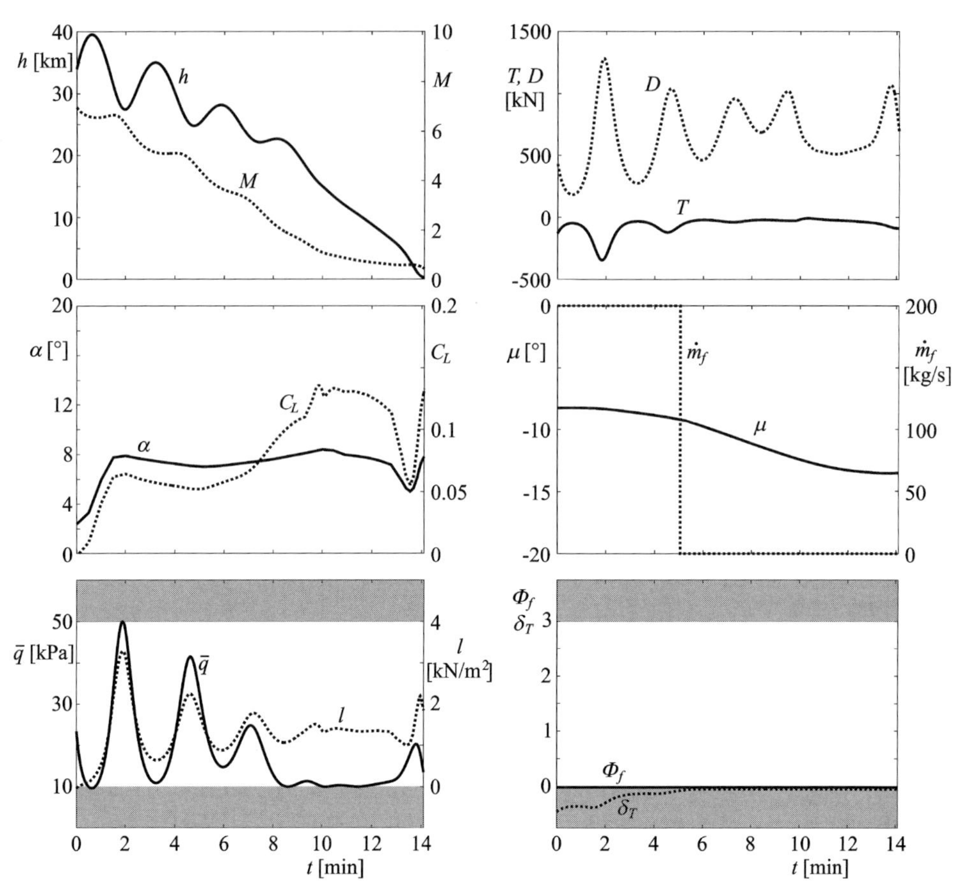


Figure 6.1.8: Optimal mission abort of overall system after an all engine blow out immediately before separation (limits indicated by shaded areas).

### 6.1.6 Mission Aborts of Orbital Stage

The mission abort types of the orbital stage are related to the ascent phase in which a failure of the main rocket engine occurs. For a failure at the beginning of the ascent, the only possibility for a safe mission abort is a gliding flight to an emergency landing site. Areas were determined in which emergency landings can be performed. The goal was to keep the number of emergency landing sites as small as possible. It turned out that only two landing sites are required in addition to the launch site. The two emergency landing sites cover engine failures in the first 88% of the initial boost phase. For the final 12% of the initial boost phase, an AOA mission abort to the launch site is possible.

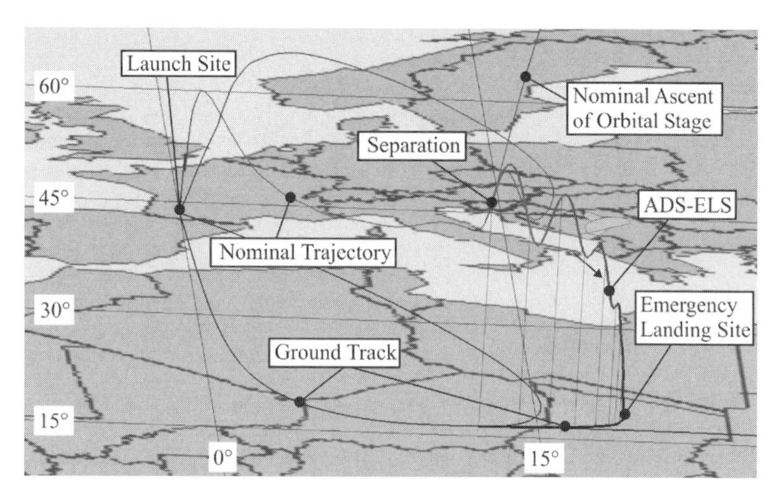


Figure 6.1.9: Optimal mission abort trajectory of orbital stage after main engine malfunction at latest failure for achieving the designated emergency landing site.

Results for two mission abort cases are presented in the following. The first one relates to a scenario requiring an emergency landing site while the launch site can be used for the second.

The first case concerns a mission abort of the orbital stage at the latest failure time for achieving the designated emergency landing site (ADA-ELS mission abort), Fig. 6.1.9 and Fig. 6.1.10. It basically consists of a decelerated gliding flight to the emergency landing site. In the final portion of the trajectory, the vehicle performs a turn. The constraints in dynamic pressure and stagnation point heat flux become active. It may be of interest to note that the fuel is jettisoned at high speed of the vehicle in the initial flight phase, contributing to an increase of the achievable range [10].

The second case concerns an AOA mission abort after a main engine reignition failure at the on-orbit-burn. Results are presented in Figs. 6.1.11 and 6.1.12. The orbital stage descends and reenters the atmosphere in a way similar to a nominal reentry. Finally, the orbiter lands at the launch site. The high crossrange reentry shows a small corridor. The maximum heat flux constraint becomes active at the beginning of the reentry and the minimum dynamic pressure limitation at an altitude of about 50 km.

For a main engine failure during the final 7% of the initial boost phase, it is possible for the orbital stage to reach an emergency orbit of at least 200 km altitude using the thrusters of the orbital manoeuvring system.

The flight performance of the orbital stage offers at least two or even more consecutive possibilities for returning from the orbit to the launch site within 24 hours. In case of a failure concerning the first de-orbit impulse, a further attempt can be made after the next orbit. In case of an immediate mission termination at any point, two additional emergency landing sites near the Equator are required.

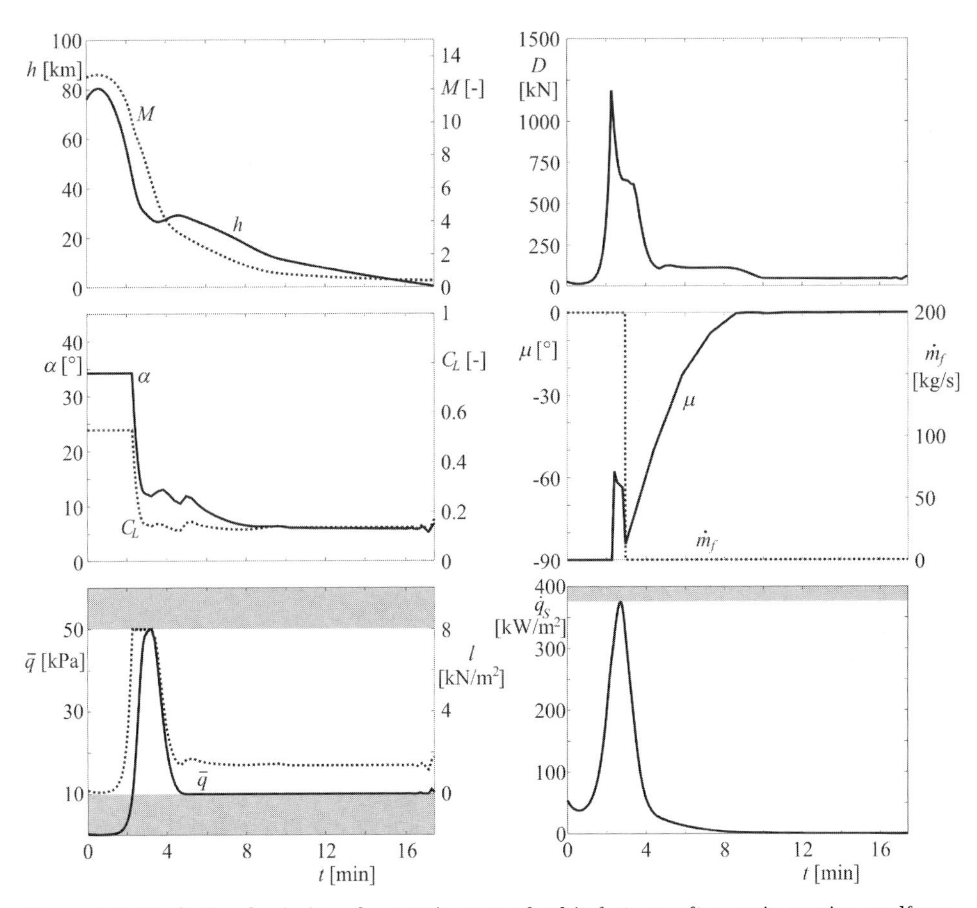


Figure 6.1.10: Optimal mission abort trajectory of orbital stage after main engine malfunction at latest failure for reaching the designated emergency landing site (limits indicated by shaded areas).

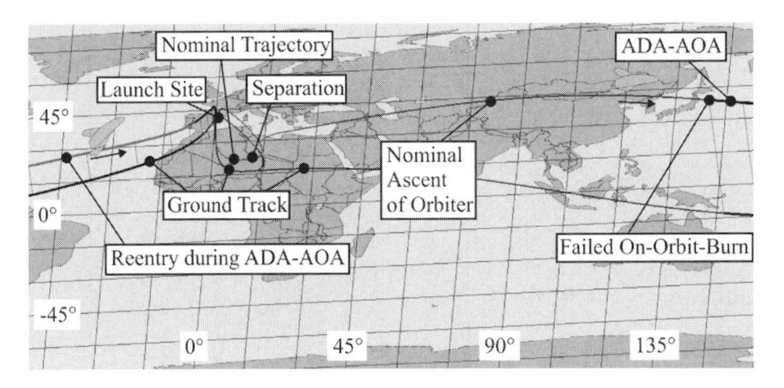


Figure 6.1.11: AOA mission abort trajectory of orbital stage after re-ignition failure of main engine.

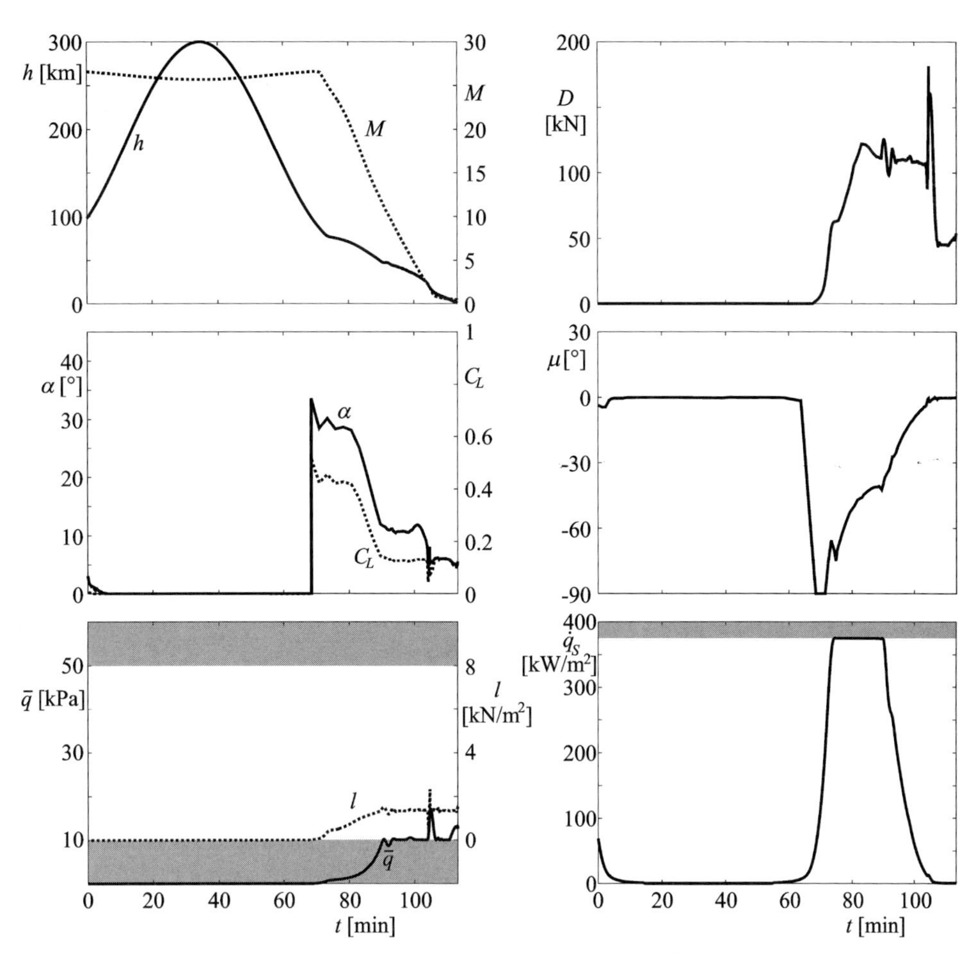


Figure 6.1.12: AOA mission abort of orbital stage after re-ignition failure of main engine (limits indicated by shaded areas).

### 6.1.7 Mission Abort Plan

Evaluating and combining the results from the mission safety improvement investigation, a complete mission abort plan for coping with emergency scenarios due to engine failures can be set up. Figure 6.1.13 presents an illustration of the mission abort plan. It shows the appropriate mission abort types for the combined vehicles as well as for the carrier and orbital stages after separation with reference to the flight time.

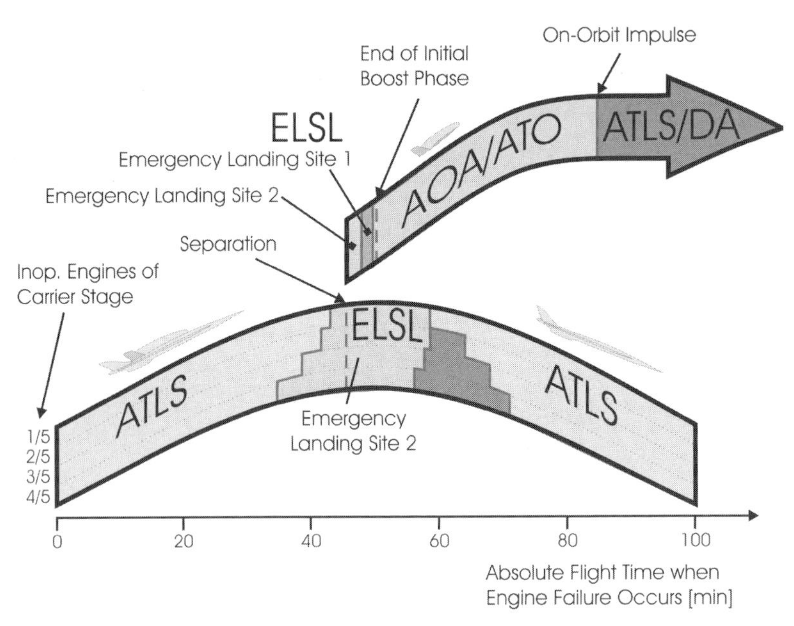


Figure 6.1.13: Mission abort plan.

## 6.1.8 Conclusions

Mission abort scenarios of a two-stage-to-orbit vehicle comprising a winged carrier stage with an airbreathing propulsion system and a rocket-propelled, winged orbital stage are considered. An evaluation of this consideration yields a safety improved nominal trajectory. Thus, a significant contribution to an improvement of overall flight safety can be achieved. With the safety improved nominal trajectory approach, only two emergency landing sites are required to cope with all mission abort scenarios. The results are based on the solution of a branched trajectories problem involving the most critical emergency scenarios.

## References

1. Allred, A. G., Sauvageau, D. R.: Crew Survival and Intact Abort Using Solid Rocket Boosters, Joint Propulsion Conference, Lake Buena Vista, FL, USA, July 1–3, 1996, AIAA 96-3156.
2. Gonzáles, P.: Influence of the Abort Capability in Reusable System Reliability. FESTIP-Overview, 9th International Space Planes and Hypersonic Systems and Technologies Conference, Norfolk, VA, USA, November 1–5, 1999, AIAA 99-4928.
3. Gonzáles, P.: Risk Management Procedures – Application of Technical Risk Assessment in FESTIP, AIAA Atmospheric Flight Mechanics Conference and Exhibit, Portland, OR, USA, August 9–11, 1999, AIAA 99-4253.

4. Kraus, W.: Aerodynamischer Basisdatensatz für die Sängerunterstufe. Messerschmidt-Bölkow-Blohm GmbH, Report-Nr. TN/HYP/44, 1988.
5. Cucinelli, G.: Sänger-Systemstudie – Aerodynamischer Datensatz HORUS 8-88, Messerschmidt-Bölkow-Blohm GmbH, Report-Nr. TN/HYP/71, 1989.
6. Dinkelmann, M.: Reduzierung der thermischen Belastung eines Hyperschallflugzeugs durch optimale Bahnsteuerung, PhD thesis, TU München, Institute of Flight Mechanics and Flight Control, 1997.
7. N.N., U.S. Standard Atmosphere 1976, Washington D.C., National Oceanic and Atmospheric Administration, National Aeronautics and Space Administration, United States Air Force, 1976.
8. Schnepper, K.: ALTOS (Advanced Launcher Trajectory Optimization Software), Software User Manual, PROMIS: Optimization Program, Institut für die Dynamik der Flugsysteme, DLR Oberpfaffenhofen, Issue 1: 20. 2. 1992.
9. N.N.: GESOP (Graphical Environment for Simulation and Optimization), Softwaresystem für die Bahnoptimierung, Institut für Robotik und Systemdynamik, DLR Oberpfaffenhofen, 1993.
10. Sachs, G., Mayrhofer, M., Wächter, M.: Optimal Control of Fuel Draining for Mission Abort Range Increase of an Orbital Stage. Workshop "Optimal Control" des Sonderforschungsbereichs 255 "Transatmosphärische Flugsysteme" und der Universität Greifswald, Greifswald, Oktober 2002, Tagungsband, ISBN 3-89791-316-X, TU München, S. 141–150, 2003.

# 6.2 Optimal Trajectories for Hypersonic Vehicles with Predefined Levels of Inherent Safety

Rainer Callies*

## Nomenclature

| | | | |
|---|---|---|---|
| $D$ | drag | $\alpha$ | angle of attack |
| $L$ | lift | $\mu$ | bank angle |
| $C_D$ | drag coefficient | $v$ | speed |
| $C_L$ | lift coefficient | $\gamma$ | flight path angle |
| $S$ | reference area | $\chi$ | azimuth angle |
| $T_{vac}$ | max. vacuum thrust | $h$ | altitude |
| $I_{sp,vac}$ | specific impulse in vacuum | $\Lambda$ | geographical latitude |
| $\bar{\varepsilon}$ | thrust angle | $\vartheta$ | geographical longitude |
| $\delta$ | throttle setting | $m$ | mass |
| $Ma$ | Mach number | $g_0$ | gravitational constant |
| $a$ | speed of sound | $R_0$ | mean radius of the Earth |
| $\rho$ | air density | $\omega$ | mean angular velocity of the Earth |
| $t$ | time | | |


## Abstract

Three-dimensional optimal flight trajectories including singular subarcs are computed for a single-stage, suborbital hypersonic demonstrator system. An optimal primary trajectory is only permissible if it offers a prescribed level of safety along the total flight path. As an example, the problem of mission abort is considered. The flight vehicle has to reach an emergency landing site from every point of the primary trajectory after a total failure of the aircraft engine. The mathematical model describing the motion of the rocket-powered vehicle is based on the three-dimensional equations of motion of a point mass with reference to the spherical rotating Earth. The safety condition is fulfilled pointwise; secondary optimal control problems describing mission abort are connected to an adaptively refined subset of points along the primary trajectory. The resulting problem is reformulated as a joint optimal control problem and transformed into a series of boundary value problems for systems of highly nonlinear differential equations. The numerical solution is by the advanced multiple shooting method JANUS. The demand for full safety in case of mission


* Technische Universität München, Zentrum Mathematik M2, Boltzmannstr. 3, 85748 Garching; *callies@ma.tum.de*

abort leads to significant deformations of the unperturbed primary trajectory, whereas the value of the objective function for the modified trajectory is only slightly reduced.

### 6.2.1 Introduction

The approach to inherent system safety presented here completely differs from feedback control, but it supports feedback control and makes it more efficient.

Optimal feedback control attempts to find a new optimal solution if there is a deviation from the nominal solution. Feedback control normally is a real time application. The controllability region of feedback control heavily depends on the nominal solution and is up to now computed a posteriori, i.e. after the nominal solution has been established. If the region is too small, the nominal solution is modified according to heuristic laws or by trial and error until the desired result is achieved – surely no optimal approach.

In contrast to that procedure, we try to find optimal nominal solutions that are so "good-natured" that later feedback control can be done very easily and the controllability region has a prescribed size. An optimal nominal solution is accepted only if it offers exactly the safety margin in every point of the trajectory necessary to fulfil the additional demands and criteria and thus guarantees a predefined level of safety along the total flight path. This leads to an extended problem of optimal control, which in general is not and does not need to be realtime-capable.

The results are constrained optimal solutions with a slightly higher cost functional (in case of a minimum problem), but with significantly higher inherent safety potentials. The region of controllability is now an important part of the control problem and no longer a somewhat accidental property of the solution.

As an example, the problem of mission abort is considered. Optimal flight trajectories for a hypersonic demonstrator system are investigated, on which maximum speed at a given altitude is achieved (primary problem). Moreover, the rocket-powered flight vehicle has to reach an emergency landing site from almost every point of the primary flight trajectory after a total failure of the aircraft engine. Every point of the primary trajectory is the starting point of a secondary control problem, the mission abort. The safety required (range at least to the next emergency landing site) is one of the constraints for a valid optimal solution of the primary problem.

### 6.2.2 Theoretical Background

#### 6.2.2.1 Classical Problem

In a more abstract way the core of the classical optimal control problem in flight dynamics can be stated as follows: Let us find a (primary) state function $x:[\tau_0, t_F] \to \mathbb{R}^n$ and a (primary) control function $u:[\tau_0, t_F] \to U \subset \mathbb{R}^m$, $U$ convex, which minimize the functional $I(u)$ (e.g. $I(u):=-m_{\text{payload}}$) subject to the conditions

$$\dot{x}(t) = \tilde{f}_i(t, x(t), u(t)) \text{ for } t \in [\tau_i, \tau_{i+1}[, \; \tau_{M+1} =: t_F, \; i = 0, 1 \ldots M \;, \tag{1}$$

$$0 = q_0(\tau_0, t_F, x(\tau_0), \; x(t_F)) \in \mathbb{R}^k, \; k \leq 2n \;, \tag{2}$$

$$0 = q_j(\tau_j, x(\tau_j^-), \; x(\tau_j^+)) \in \mathbb{R}^{l_j}, \; j = 1 \ldots M \;, \tag{3}$$

$$0 \leq C_i(t, x(t), \; u(t)) \in \mathbb{R}^{k_i}, \; k_i < m; \; q_i \text{ sufficiently smooth}; \; n, m, k, l_j, k_i, M \in \mathbb{N} \tag{4}$$

$\tau_0$ denotes the initial time, $t_F$ the final time and $\tau_i$ an intermediate time with interior point conditions; $\bar{U} := [\tau_i, \; \tau_{i+1}] \times \mathbb{R}^n \times \mathbb{R}^m$ and $\tilde{f}_i \in C^3\left(\bar{U}, \, \mathbb{R}^n\right)$, $C_i \in C^{N_2}\left(\bar{U}, \, \mathbb{R}^{k_i}\right)$ with $N_2 \in \mathbb{N}$ sufficiently large. In practical problems of optimal control, $n$ is approximately 10...50, the piecewise defined right hand sides $\tilde{f}_i(t, x, u)$ of the differential equations are highly nonlinear and in many cases so complicated, that they can only be set up by computer algebra programs (e.g. MAPLE) or implicitly formulated and evaluated by iterative algorithms. As usual we define $\tau_j^{\pm} := \lim_{\varepsilon \to 0, \varepsilon > 0} (\tau_j \pm \varepsilon)$.

#### 6.2.2.2 Related Boundary Value Problem

The above-defined problem of optimal control is transformed in a well-known manner (see e.g. [2, 10, 13]) into a multi-point boundary value problem. The transformation procedure is summarized in short for the simple case of

$$\dot{x} = \tilde{f}(t, x, u), \; t \in [\tau_0, t_F] \, .$$

Without additional interior point conditions, the following system of coupled nonlinear differential equations results

$$\dot{x} = \tilde{f}(t, x, u), \; \dot{\lambda} = -H_x\,(t, x, \lambda, u) \, .$$

$\lambda$ denotes the vector of the adjoint variables, $\tilde{f}$ the right-hand side of the system of the differential equations of motion and $H := \lambda^T \tilde{f}$ the Hamiltonian.

Constraints are coupled to the Hamiltonian of the unconstrained problem by Lagrangian multipliers (see e.g. [2, 11]) to define the augmented Hamiltonian. Today, this is possible only up to a certain complexity and dimension $k_i$ of the constraints, because only limited mathematical theory is available for the more difficult cases [9]. Modern approaches reformulate the problem into minimum coordinates to avoid the explicit treatment of complicated constraints $C(t, x, u) \geq 0$ [5, 8].

The controls are derived from

$$H_u(t, x, \lambda, u) = 0$$

and have to satisfy the Legendre-Clebsch condition: $H_{uu}(t, x, \lambda, u)$ pos. semidefinite.

If there exist nonlinear $(u_1)$ and linear controls $(u_2)$ in the system of ordinary differential equations

$$\dot{x} = f_1(t, x, u_1) + u_2 \cdot f_2(t, x, u_1),$$

then one gets in case of a linear control

$$u_2 = \begin{cases} u_{2,min} & \text{if } S > 0 \\ u_{2,max} & \text{if } S < 0 \end{cases}$$

with the *switching function* $S$ defined by $S := \lambda^T f_2$. The zeros of $S$ are called the *switching points*. For

$$S(t) \equiv 0 \ \ \forall \, t \in [t_1, t_2], \ \tau_0 \leq t_1 < t_2 \leq t_F,$$

singular control exists. Here the total time derivatives

$$S^{(j)} = \frac{d^j}{dt^j} S \equiv 0$$

have to be calculated, until for $j = q$ there is for the first time $\frac{\partial}{\partial u_2} S^{(q)} \neq 0$.

The control $u_2$ is then determined from $S^{(q)} \equiv 0$. It can be shown, that – if existing – $q$ is always even, i.e. $\exists \, p \in \mathbb{N} : q = 2p$; $p$ is called *order* of the singular control. For the optimal solution, the generalized Legendre-Clebsch condition has to be satisfied

$$(-1)^p \frac{\partial}{\partial u_2} \left[ \frac{d^q}{dt^q} \left( \frac{\partial}{\partial u_2} H \right) \right] \geq 0.$$

#### 6.2.2.3 Extended Problem (A)

In the safety-related problems under consideration in this paper, a (primary) state function $x : [\tau_0, t_F] \to \mathbb{R}^n$ and a (primary) control function $u : [\tau_0, t_F] \to$

$U \subset \mathbb{R}^m$, $U$ convex, have to be determined which minimize the objective function (Mayer problem)

$$J(x,u) := L(\tau_0, x(\tau_0), \ldots, \tau_M, x(\tau_M), t_F, x(t_F)) \tag{5}$$

subject to the conditions (1-4). The new and crucial element is added to this optimal control problem by the boundary value problems describing the safety demands (e.g. to reach an emergency landing site from every point of the optimal primary trajectory). With the (secondary) state functions $y(\cdot,t) : [t, \hat{t}_f(t)] \to \mathbb{R}^{\hat{n}}$ and the (secondary) control functions $w(\cdot,t) : [t, \hat{t}_f(t)] \to W(t) \subset \mathbb{R}^{\hat{m}}$, $W(t)$ convex, the secondary boundary value problems read as follows ($t$ plays the part of an additional parameter)

$$\frac{dy}{ds}(s,t) = g(s, y(s,t), w(s,t), t) \text{ with } s \in [t, t_f(t)], t \in [\tau_0, t_F], \tag{6}$$

$$y(t,t) = x(t), \tag{7}$$

$$0 = \hat{q}_0(t, t_F(t), y(t,t), y(t_F(t),t)) \in \mathbb{R}^{\hat{k}}, \ \hat{k} \le \hat{n},$$

$$0 = \hat{q}_j(\sigma_j(t), y(\sigma_j^-(t),t), y(\sigma_j^+(t),t), t) \in \mathbb{R}^{\hat{l}_j}, \ j = 1 \ldots \hat{M}, \ \sigma_j(t) < \sigma_{j+1}(t), \tag{8}$$

$$0 \le \hat{C}(s, y(s,t), w(s,t), t) \in \mathbb{R}^{\hat{p}}, \ \hat{p} < \hat{m}, \tag{9}$$

with $g, \hat{q}_0, \hat{q}_j, \hat{C}$ sufficiently smooth, $\hat{n}, \hat{m}, \hat{k}, \hat{l}_j, \hat{p} \in \mathbb{N}$ and $\sigma_j(t) \in ]t, t_F(t)[$, $j = 1 \ldots \hat{M}$. All secondary problems considered here have equal numbers of interior points and of interior point conditions for points with the same subscript. This is not essential and was chosen only to simplify the notation. Analogously to the primary problem, the formalism can be extended e.g. to piecewise defined right hand sides of the secondary problems or piecewise defined constraints. $\tau_0$ denotes the initial time and $t_F$ the final time for the primary trajectory $x(t)$, whereas $t$ is the initial and $t_F(t)$ the final time for the respective secondary trajectory $y(s,t)$, starting with $x(t)$ at time $t$.

**Lemma.** The extended problem (A) is numerically ill posed.

*Proof.* Let $(x^*(t), u^*(t))$ be a locally unique solution of the primary optimal control problem alone. Then for every $t$ the respective secondary problem in general is a nonlinear boundary value problem. From literature (see e.g. [17]) it is well known, that this problem may have no, one, several or even infinitely many solutions.

**Remark.** In case of no solution, the combined primary-secondary problem has to be investigated; a possible solution $(\bar{x}^*(t), \bar{u}^*(t))$ differs from $(x^*(t), u^*(t))$. In case of exactly one solution of the secondary problem, nothing has to be done,

$(x^*(t), u^*(t))$ is a solution of the combined problem too. Difficulties occur in case of several or infinitely many solutions.

**Example.** An optimal minimum-fuel trajectory is calculated for an experimental hypersonic vehicle from the launch site to the nominal landing site. In case of a partial engine failure at time $t \in ]\tau_0, t_F[$, an emergency landing site is available. If the aircraft can reach the emergency landing site with residual fuel left, there exist infinitely many (neighbouring) solutions of the mission abort problem starting at time $t$.

**Example.** The following example gives a mathematical insight into the problems discussed above. Consider the discrete case of a primary trajectory together with only one secondary trajectory starting at time $t_s$, both problems of the type defined above. After transformation to the joint time interval [0, 1], the differential equations for the state variables and the interior point conditions are

$$\dot{x} = f(x, u),\ \dot{y} = g(y, w);\ y(0) = x(t_s^-),\ x(t_s^+) = x(t_s^-);\ t_s \in ]0, 1[\,.$$

With the Hamiltonian $H := \lambda^T \dot{x} + \zeta^T \dot{y}$ ($\lambda, \zeta$ denote the respective adjoint variables) the problem of optimal control is transformed into a boundary value problem with

$$\dot{\lambda} = -H_x, \dot{\zeta} = -H_y$$

and with the additional boundary conditions

$$\zeta(0) + \lambda(t_s^+) - \lambda(t_s^-) = 0\,,\ \zeta(t_s^-) - \zeta(t_s^+) = 0\,.$$

The boundary value problem decomposes into two separate parts for $(x, \lambda)$ and $(y, \zeta)$, linked together only by some boundary and interior point conditions.

The objective function takes the minimum value for $\lambda(t_s^+) - \lambda(t_s^-) = 0$; this is the original primary problem without any additional constraint imposed by the secondary problem. In this case, every solution (if existing) of the secondary problem, that fulfils the boundary conditions, completes the solution of the total problem, no matter whether the differential equations for $\zeta$ are fulfilled or not. There is no influence of the selection of $w(t)$ on the objective function; the problem in general has no unique solution.

**Remark.** Uniqueness of the solution of a boundary value problem is essential for an efficient and accurate numerical solution. It will be crucial to omit secondary trajectories with the property not to modify the primary trajectory. What makes it difficult is, that this property cannot be predicted and may change in the course of the numerical solution procedure.

6.2.2.4 Extended Problem (B)

An elegant way to achieve at least local uniqueness of the solution of a secondary boundary value problem is to reformulate it into a secondary optimal control problem.

For $t \in [\tau_0, t_F]$ given, a (secondary) state function $y(\cdot, t) : [t, \hat{t}_f(t)] \to \mathbb{R}^{\hat{n}}$ and a (secondary) control function $w(\cdot, t) : [t, \hat{t}_f(t)] \to W(t) \subset \mathbb{R}^{\hat{m}}$, $W(t)$ convex, have to be determined, which minimize the functional

$$\hat{J}(y(\cdot, t),\ w(\cdot, t); t) := \hat{L}(t_f(t),\ y(t_f(t), t); t) \tag{10}$$

subject to the conditions (6-9) and $0 = \bar{q}_0(t, t_F(t), y(t, t), y(t_F(t), t)) \in \mathbb{R}^{\hat{k}}$, $\hat{k} < \hat{n}$.

The notation is identical with that for the extended problem (A). The only modifications are the definition of secondary objective functions and modified boundary conditions.

**Example.** An optimal trajectory is calculated for an experimental hypersonic vehicle from the launch site to the nominal landing site. In case of a partial engine failure at time $t \in ]\tau_0, t_F[$, an emergency landing site is available. Exactly that emergency (=secondary) trajectory has to be determined, on which the aircraft can reach the emergency landing site in minimum time.

**Remark.** For the minimum time objective function in connection with optimal flight trajectories, existence and uniqueness statements can be found in literature [1].

In the following we shall no longer consider the continuous set of starting times $t \in [\tau_0, t_F]$, but deal with a discrete subset $\{\hat{t}_1, \hat{t}_2, \ldots, \hat{t}_{\hat{p}}\}$, $\hat{t}_i \in [\tau_0, t_F]$, $i = 1, \ldots, \hat{p}$ and $\hat{t}_j < \hat{t}_{j+1}$, $j = 1, \ldots, \hat{p} - 1$. This model reduction is done with respect to the numerical calculations, the $\hat{t}_i \in \mathbb{R}$ and $\hat{p} \in \mathbb{N}$ have to be determined during the solution procedure.

To combine the primary and the $\hat{p}$ secondary optimal control problems into one extended problem, the objective functions (5), (10) are replaced by a new and joint objective function

$$\begin{aligned} J_{tot}(x, u, y_1, w_1, \ldots y_{\hat{p}}, w_{\hat{p}}) := L(\tau_0, x(\tau_0), \ldots, \tau_M, x(\tau_M), t_F, x(t_F)) + \\ + \varepsilon \left( \sum_{j=1}^{\hat{p}} \hat{L}_j(t_{Fj}, y_j(t_{Fj})) \right). \end{aligned} \tag{11}$$

To avoid cancellation of the individual parts of (11), we assume without loss of generality that

$$L(\tau_0, x(\tau_0), \ldots, \tau_M, x(\tau_M), t_F, x(t_F)) = |L(\tau_0, x(\tau_0), \ldots, \tau_M, x(\tau_M), t_F, x(t_F))|,$$

$$\hat{L}_j(t_{Fj}, y_j(t_{Fj})) = |\hat{L}_j(t_{Fj}, y_j(t_{Fj}))|, \ j = 1, \ldots \hat{p}.$$

The combined problem in its discretized version now can be stated as follows: A (primary) state function $x : [\tau_0, t_F] \to \mathbb{R}^n$, $\hat{p}$ (secondary) state functions $y_j : [\hat{t}_j, t_{Fj}] \to \mathbb{R}^{\hat{n}}$, a (primary) control function $u : [\tau_0, t_F] \to U \subset \mathbb{R}^m$, $U$ convex, and $\hat{p}$ (secondary) control functions $w_j : [\hat{t}_j, t_{Fj}] \to W_j \subset \mathbb{R}^{\hat{m}}$, $W_j$ convex, $j = 1, \ldots, \hat{p}$, have to be determined so that for a fixed $0 < \varepsilon \ll 1$ the joint objective function (11) is minimized subject to the conditions

$$\frac{dx}{dt}(t) = \tilde{f}_i(t, x(t), u(t)) \quad \text{for} \quad t \in [\tau_i, \tau_{i+1}[, \tau_{M+1} =: t_F \,, \; i = 0, 1 \ldots M \,, \tag{12}$$

$$\frac{dy_j}{ds}(s) = g_j(s, y_j(s), w_j(s)) \quad \text{for} \quad s \in [\,\hat{t}_j, t_{Fj}] \,, \; j = 1, \ldots, \hat{p} \,, \tag{13}$$

$$0 = q_0(\tau_0, t_F, x(\tau_0), x(t_F)) \in \mathbb{R}^k \,, \; k < 2n \,,$$

$$y_j(\hat{t}_j) = x(\hat{t}_j) \,, \tag{14}$$

$$0 = \bar{q}_{0j}\,(t_{Fj}, y_j(t_{Fj})) \in \mathbb{R}^{\hat{k}} \,, \; \hat{k} < \hat{n} \,, \tag{15}$$

$$0 = q_\mu(\tau_\mu, x(\tau_\mu^-), x(\tau_\mu^+)) \in \mathbb{R}^{l_\mu}, \mu = 1 \ldots M,$$

$$0 = \hat{q}_{\nu j}(\sigma_{\nu j}, y_j(\sigma_{\nu j}^-), y_j(\sigma_{\nu j}^+)) \in \mathbb{R}^{\hat{l}_\nu}, \nu = 1 \ldots \hat{M}, \; \sigma_{\nu j} < \sigma_{\nu+1,j}, \; \sigma_{\nu j} \in ]\,\hat{t}_j, t_{Fj}[,$$

$$0 \leq C_i(t, x(t), u(t)) \in \mathbb{R}^{k_i}, k_i < m,$$

$$0 \leq \hat{C}_j(s, y_j(s), w_j(s)) \in \mathbb{R}^\rho, \rho < \hat{m},$$

with $\tilde{f}_i, q_i, C_i, g_j, \bar{q}_{0j}, \hat{q}_{\nu j}, \hat{C}_j$ sufficiently smooth and $n, m, \hat{n}, \hat{m}, k, \hat{k}, l_\mu, \hat{l}_\nu, M, \hat{M}, k_i, \rho \in \mathbb{N}$. This problem is referred to as the extended problem (B). The restriction of (15) to $\bar{q}_{0j}(t_{Fj}, y_j(t_{Fj})) = 0$ instead of $\bar{q}_{0j}(\hat{t}_j, t_{Fj}, y_j(\hat{t}_j), y_j(t_{Fj})) = 0$ was chosen only to bring out the fundamental aspects of the linking between primary and secondary trajectories more clearly.

For $\hat{t}_j \neq \tau_i$, $j = 1 \ldots \hat{p}$, $i = 0, \ldots M + 1$, and thus $x(t)$ at least continuous in $\hat{t}_j$, the interior point condition at the junction point of the primary and the $j$-th secondary trajectory reads as

$$\lambda(\hat{t}_j^+) - \lambda(\hat{t}_j^-) + \hat{\lambda}_j(\hat{t}_j^+) = 0 \,. \tag{16}$$

$\lambda(t)$ denotes the adjoint variable for $x(t)$ and $\hat{\lambda}_j(s)$ denotes the adjoint variable for $y_j(s)$.

For $0 < \varepsilon$ every secondary trajectory makes its contribution to the joint objective function (11), no matter, whether this trajectory – starting for instance at $t = \hat{t}_s$ – is redundant or not. We call a secondary trajectory redundant, if there is no modification of the optimal primary trajectory in the extended problem (A) by adding the secondary trajectory starting at $t = \hat{t}_s$. If there are redundant secondary trajectories in the combined problem, then it becomes singularly perturbed for $\varepsilon \to 0$. For $\varepsilon = 0$, the combined problem is of the same type as the extended problem (A) with the same numerical difficulties caused by non-unique solutions in case of redundant secondary trajectories present.

(16) together with (11) offer a way to identify redundant secondary trajectories numerically. For $\varepsilon = 0$ and in case of a redundant secondary trajectory,

$$\lambda(\hat{t}_j^+) - \lambda(\hat{t}_j^-) = 0 \Rightarrow \hat{\lambda}_j(\hat{t}_j^+) = 0$$

holds, because for the optimal primary trajectory there is no jump in the adjoint variables at $t = \hat{t}_j$ with a secondary problem absent at that point.

$$\lambda(\hat{t}_j^+) \to \lambda(\hat{t}_j^-) \quad \text{for} \quad \varepsilon \to 0 \tag{17}$$

therefore is a theoretically well-founded indicator for a redundant secondary trajectory.

**Remark.** In the combined problem the accuracies of the single trajectories are not coupled. The secondary trajectories can be calculated with lower accuracy than the nominal trajectory. The adaptive selection of starting points for the secondary trajectories allows the safety constraint to be fulfilled with the precision necessary from an engineering point of view.

### 6.2.3 Numerical Method

For the solution, the extended optimal control problem (B) is transformed into a multi-point boundary value problem (MPBVP). The numerical treatment of the MPBVP is by the advanced version JANUS [4] of the multiple shooting method [3]. A detailed description is given in [4], only a short summary is presented here.

JANUS works with two types of discretisations, the *macro-discretisation*

$$t_0 < t_1 < \ldots < t_{N+1}$$

and the *micro-discretisation*

$$t_\nu =: t_{\nu,1} < t_{\nu,2} < \ldots < t_{\nu,K_\nu} := t_{\nu+1}, \nu = 0, \ldots, N$$

with $\{\tau_i, i = 0, \ldots, M+1\} \subset \{t_j\} \cup \{t_{\nu,\mu}\}$. In the original version of JANUS the solution of the MPBVP is equivalent to the solution of the following special system of nonlinear equations $(z := (Y_0^+, t_{N+1}, Y_1^+, t_1, \ldots, Y_N^+, t_N)^T)$

$$F(z) := \begin{pmatrix} r_0(t_0, t_{N+1}, Y_0^+, Y(t_{N+1}^-)) \\ r_1(t_1, Y_1^+, Y(t_1^-)) \\ r_2(t_2, Y_2^+, Y(t_2^-)) \\ \vdots \\ r_N(t_N, Y_N^+, Y(t_N^-)) \end{pmatrix} = 0 . \tag{18}$$

$Y(t_{\nu+1}^-) := Y(t_{\nu+1}; t_\nu, Y_\nu^+), \nu = 0, \ldots, N,$ is the solution at $t = t_{\nu+1}$ of the piecewise defined initial value problem

$$\begin{cases} \dot{Y} = f_{\nu,\mu}(t, Y),\ t \in [t_{\nu,\mu}, t_{\nu,\mu+1}],\ \mu = 1, \ldots, \kappa_\nu - 1\,, \\ r_{\nu,\mu}(t_{\nu,\mu}, Y(t_{\nu,\mu}^+), Y(t_{\nu,\mu}^-)) = 0, \quad \mu = 2, \ldots, \kappa_\nu - 1\,, \\ Y(t_\nu) = Y_\nu^+ \qquad \text{initial value.} \end{cases}$$

$f_{\nu,\mu} \in C^{N_3}(D_{\nu,\mu} \times \mathbb{R}^{\bar{n}}, \mathbb{R}^{\bar{n}}), r_{\nu,\mu} : \mathbb{R} \times \mathbb{R}^{\bar{n}} \times \mathbb{R}^{\bar{n}} \to \mathbb{R}^{d_{\nu,\mu}}$ and $r_0 : \mathbb{R}^2 \times \mathbb{R}^{\bar{n}} \times \mathbb{R}^{\bar{n}} \to \mathbb{R}^{d_0}$ with $\bar{n}, N, N_3, \kappa_\nu, d_{\nu,\mu}, d_0 \in \mathbb{N}$ and $D_{\nu,\mu} \supset [t_{\nu,\mu}, t_{\nu,\mu+1}]$ open; $Y(t_{\nu,\mu}^-) := \lim_{t \to t_{\nu,\mu}-0} Y(t; t_{\nu,\mu-1}, y_{\nu,\mu-1}^+) \in \mathbb{R}^{\bar{n}}$. For the micro-discretisation $\{t_{\nu,\mu}\}$ with $\nu=0, \ldots, N$, $\mu=2, \ldots, \kappa_\nu-1$ only so-called continuous design points (CDP) are permitted. On the other hand, CDPs may also be part of the macro-discretisation.

Def. 1 : $t_{\nu,\mu}$ is called CDP: $\Leftrightarrow Y(t_{\nu,\mu}^-) = Y(t_{\nu,\mu}^+)$ .

**Remark.** In optimal control problems for real life systems, there may be a huge number of CDPs. This is because important parts of the system model and dynamics are often determined from experimental data. The interpolation of these measurement data e.g. by multivariate Cardinal splines leads to a very large number of exceptional points with reduced differentiability of the right hand side of the system of differential equations (12), (13). Those points can be formulated as CDPs, the overall size of the nonlinear problem (18) is no longer increased.

The nonlinear system (18) is iteratively solved by a modified Newton method (iteration of the linearized system). The $\xi$-th iteration step reads $(DF(z^{(\xi)}) := \partial F(z)/\partial z|_{z=z^{(\xi)}})$:

$$z^{(\xi+1)} := z^{(\xi)} + \lambda \Delta F(z^{(\xi)})^{-1} F(z^{(\xi)})$$

with $\Delta F(z^{(\xi)}) \approx DF(z^{(\xi)})$. An iteration step is accepted, if at least one of the following tests is valid:

$$\|F(z^{(\xi+1)})\| \leq \|F(z^{(\xi)})\|$$
$$\|(\Delta F(z^{(\xi)}))^{-1} F(z^{(\xi+1)})\| \leq \|(\Delta F(z^{(\xi)}))^{-1} F(z^{(\xi)})\|$$

In addition we demand:

$$\|F(z^{(\xi+1)}\| \leq \|F(z^{(\xi-2)})\| \qquad \wedge \tag{19}$$

$$\|(\Delta F(z^{(\xi)}))^{-1} F(z^{(\xi)})\| \leq \|(\Delta F(z^{(\xi-2)}))^{-1} F(z^{(\xi-2)})\| \tag{20}$$

The conditions (19), (20) significantly stabilize and accelerate the iteration process. For the problem of mission abort in case of engine failure a straightforward approach is to transform the time intervals of all secondary trajectories to that of the primary trajectory:

$$[\hat{t}_j, t_{Fj}] \rightarrow [\tau_0, t_F],\ j = 1, \ldots, \hat{p}\,. \tag{21}$$

In that case the system of nonlinear equations is of the following modified type:

$$\tilde{F}(z) := \begin{pmatrix} \tilde{r}_0(\tau_0, t_F, Y_0^+, Y(t_F^-)) \\ \tilde{r}_1(\tau_0, t_F, Y_0^+, t_1, Y_1^+, Y(t_1^-), Y(t_F^-)) \\ \tilde{r}_2(\tau_0, t_F, Y_0^+, t_2, Y_2^+, Y(t_2^-), Y(t_F^-)) \\ \vdots \\ \tilde{r}_N(\tau_0, t_F, Y_0^+, t_M, Y_M^+, Y(t_M^-), Y(t_F^-)) \end{pmatrix} = 0 \tag{22}$$

The basic iteration scheme remains the same, but the structure of the iteration matrix changes. The matrix continues to have a block structure, but in a weakened form. An efficient direct solution method based on orthogonal Householder transformations for the large linearized system fully exploits the remaining structure. But there is a serious drawback with the formulation (22) derived from the transformation (21). Every new element $\bar{t}_\nu$ of the macro-discretisation necessary for only one (e.g. the primary) trajectory induces (in general trivial) interior point conditions for all the other trajectories. A complete and new condition

$$\bar{r}_\nu(\tau_0, t_F, Y_0^+, \bar{t}_\nu, \bar{Y}_\nu^+, Y(\bar{t}_\nu^-), Y(t_F^-)) = 0$$

has to be formulated and added to (22). This condition unnecessarily has a dimension of approximately $2(n + \hat{p} \cdot \hat{n})$ instead of a necessary dimension of e.g. $2n$, thus increasing the dimension of the iteration matrix $\Delta F(z^{(\xi)})$ by the respective number of rows and columns.

This difficulty can be overcome by the following time transformation replacing (21):

$$[\hat{t}_1, t_{F1}] \rightarrow [t_F, t_F + t_{F1}],\ [\hat{t}_j, t_{Fj}] \rightarrow [t_F + \sum_{k=1}^{j-1} t_{Fk}, t_F + \sum_{k=1}^{j} t_{Fk}]\,,\ j = 2, \ldots, \hat{p}\,. \tag{23}$$

The secondary trajectories are appended sequentially to the primary trajectory, instead of treating the primary and $\hat{p}$ secondary trajectories in parallel. The structure of the corresponding iteration matrix $\Delta F(z^{(\xi)})$ is given in Fig. 6.2.1.

This structure is well suited for blockwise storage. An efficient direct solution method for the large linearized system based on orthogonal Householder transformations of $\Delta F(z^{(\xi)})$ is successfully applied after an appropriate reordering of block columns and block rows.

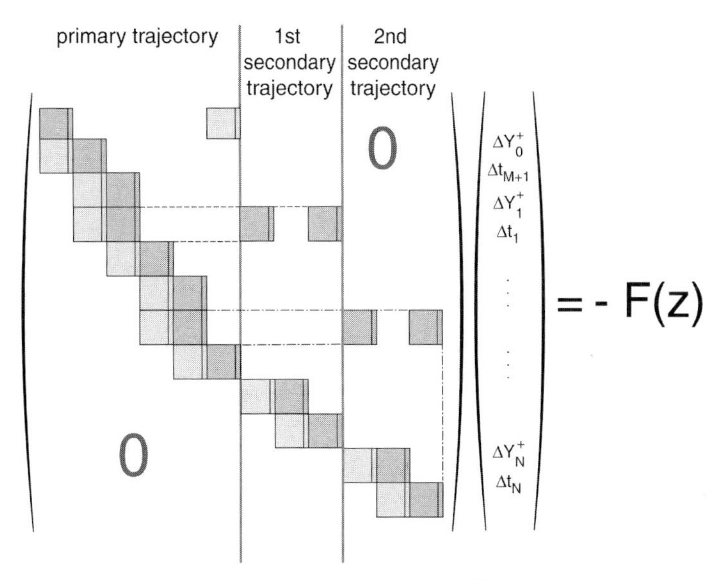


Figure 6.2.1: Basic structure of the iteration matrix $\Delta F(z^{(\xi)})$ schematically.

### 6.2.4 Model System

#### 6.2.4.1 Overview

A suborbital hypersonic demonstrator system (HYD) is investigated, the model is described in [6]. Attention is focused on the problems of optimal control; the technical data of the HYD are preliminary, but can easily be adjusted as soon as more accurate data are available.

The HYD is supposed to have an outer shape similar to that of the DC-XA (see Fig. 6.2.2). It has a launch mass of about $m_0 = 2.1 \cdot 10^4$ kg and a final mass of greater or equal $m_s = 7.4 \cdot 10^3$ kg; this includes the dry mass plus fuel for the landing phase.

#### 6.2.4.2 Thrust Model

Two main engines of type RL10-4N with multiple burn capability produce the thrust $T$. For these LOX/LH2 main engines a specific impulse in vacuum $I_{sp,vac} = 404$ s is estimated; vacuum thrust will be $T_{vac} = 311$ kN. All estimates are rather conservative. The available thrust as a function of the throttle setting $\delta$ is modelled by

$$T(h,\delta) = T_{vac} \cdot \delta \,.$$

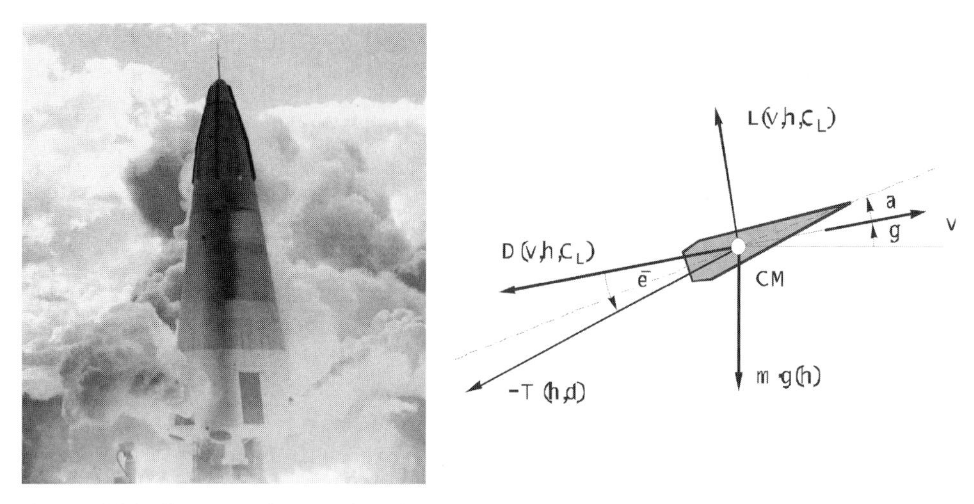


Figure 6.2.2: Computer image of and forces acting on the HYD. CM denotes the centre of mass.

### 6.2.4.3 Atmospheric and Aerodynamic Model

For the air density $\rho(h)$ the following model is valid:

$$\rho(h) = \rho_0 e^{-h/h_r} \left[\frac{\text{kg}}{\text{m}^3}\right]$$

with $\rho_0 = 1.225$ kg/m$^3$ and $h_r = 12760$ m. A more refined model can be found e.g. in [16].

For the lift $L(v, h, C_L)$ and the drag $D(v, h, C_L)$ a model is used with a quadratic polar:

$$D(v, h, C_L) = \frac{1}{2}\rho(h)v^2 SC_D = \frac{1}{2}\rho(h)v^2 S(C_{D0} + C_{D1}C_L^2)$$

$$L(v, h, C_L) = \frac{1}{2}\rho(h)v^2 SC_L$$

The reference area is $S = 33$ m$^2$; $C_{D0} = 0.05$ and $C_{D1} = 1/0.7$ are derived from [12, 15]. A more elaborate model has been developed in [5].

#### 6.2.4.4 Equations of Motion

The mathematical model of the HYD describes the motion of a point mass over a spherical rotating Earth [12]. Oblateness of the Earth and wind in the atmosphere are neglected. Calculations applying a non-rotating Earth show that the results differ less than 2.5%. Nevertheless, the complete model is used (also with respect to further research). The equations of motion read:

$$
\begin{aligned}
\dot{v} &= \{-D(v,h,C_L)/m - g(h)\sin\gamma\} + \omega^2 R\cos\Lambda(\sin\gamma\cos\Lambda - \cos\gamma\sin\chi\sin\Lambda) \\
&\quad + A\cdot T(h,\delta)\cos\bar{\varepsilon}/m \\
\dot{\gamma} &= \{L(v,h,C_L)\cos\mu/(mv) + (v/R - g(h)/v)\cos\gamma\} \\
&\quad + 2\omega\cos\chi\cos\Lambda + \omega^2 R\cos\Lambda(\sin\gamma\sin\chi\sin\Lambda + \cos\gamma\cos\Lambda)/v \\
&\quad + A\cdot T(h,\delta)\cdot\sin\bar{\varepsilon}\cos\mu/(mv) \\
\dot{\chi} &= L(v,h,C_L)\sin\mu/(mv\cos\gamma) - v\cos\gamma\cos\chi\tan\Lambda/R \\
&\quad + 2\omega(\sin\chi\cos\Lambda\tan\gamma - \sin\Lambda) - \omega^2 R\cos\Lambda\sin\Lambda\cos\chi/(v\cos\gamma) \\
&\quad + A\cdot T(h,\delta)\cdot\sin\mu\sin\bar{\varepsilon}/(mv\cos\gamma) \\
\dot{m} &= -A\cdot T_{vac}\cdot\delta/(g_0 I_{sp,vac}) \\
\dot{h} &= v\sin\gamma \\
\dot{\Lambda} &= v\cos\gamma\sin\chi/R \\
\dot{\vartheta} &= v\cos\gamma\cos\chi/(R\cdot\cos\Lambda)\ \text{(decoupled)}
\end{aligned}
\tag{24}
$$

with the abbreviations $g(h) := g_0 R_0^2/R^2$, $R := R_0 + h$; $A$ serves as a switch.

$$x := (v,\gamma,\chi,h,\Lambda,m,\vartheta) : [\tau_0, t_F] \to \mathbb{R}^7$$

is the state vector,

$$u := (C_L,\bar{\varepsilon},\delta,\mu) : [\tau_0, t_F] \to \mathbb{R}^4$$

denotes the control vector; $t$ is the independent variable (here: time). It should be mentioned, that $\bar{\varepsilon}$ is *not* defined as the thrust angle with respect to a body fixed axis of the rocket, but with respect to the velocity vector. The recalculation for the body fixed system is straightforward.

In this model $C_L$ is used as a control instead of $\alpha$. This avoids the additional evaluation of $C_L = C_L(\alpha, Ma) = C_L(\alpha, v, a(h))$ and its derivatives. A constant constraint $|\alpha| \le \alpha_0$ is transformed into a non-constant constraint $|C_L| \le C_{L0}(\alpha_0, v, h)$. In general this constraint is only active for a short period of time and this type of formulation partially decouples the problem; therefore the use of the control $C_L$ is preferable for the numerical treatment.

### 6.2.4.5 Primary Problem

The primary trajectory x($t$) (ascent and test trajectory) is the solution of the equations of motion

$$\dot{x} = \tilde{f}(t, x, u)\,,\ t \in [\tau_0, t_F]$$

with $\tilde{f}(t, \mathrm{x}, u) := (\dot{v}, \dot{\gamma}, \dot{\chi}, \dot{h}, \dot{\Lambda}, \dot{m}, \dot{\vartheta})^T|_{A=1}$ from (24).

The cost functional $I(u) := v(t_i)$ has to be maximized and $t_i \in ]\tau_0, t_F[$ has to be determined such that $h(t_i) = h_T$ ($h_T = 60\,000$ m prescribed) and $\gamma(t_i) = 0$.

The state variables at the boundaries x($\tau_0$) and x($t_F$) are totally fixed and prescribed:

$$\begin{aligned}
v(\tau_0) &= 80 \text{ m/s}\,, & v(t_F) &= 210 \text{ m/s}\,,\\
\gamma(\tau_0) &= 1.48\,, & \gamma(t_F) &= 0\,,\\
\chi(\tau_0) &= 1.69\,, & \chi(t_F) &= -1.29\,,\\
h(\tau_0) &= 1000 \text{ m}\,, & h(t_F) &= 1000 \text{ m}\,,\\
\Lambda(\tau_0) &= 0\,, & \Lambda(t_F) &= 0.0200\ v\,,\\
\theta(\tau_0) &= 0\,, & \theta(t_F) &= 0.0185\,,\\
m(\tau_0) &= 21\,061.9 \text{ kg}\,, & m(t_F) &= 7371.7 \text{ kg}
\end{aligned}$$

The control constraints are $\delta \in 0 \cup [0.12, 1]$, $|C_L| \le C_{L0}$, $|\bar{\varepsilon}| \le \varepsilon_0$ and $|\mu| \le \mu_0$; there is one state constraint: $v^2\rho \le S_0$ (dynamic pressure limit).

From the point of view of optimal control, the primary problem as a stand-alone problem is well defined and solvable [1]; some problems are introduced because the allowed domain $U$ for the control $\delta$ is not simply connected. This problem is efficiently treated by a special interior point condition derived from the minimum principle [4, 5].

For the problem under consideration here, one gets the following expressions for the transformed switching function $S$

$$S = -\sqrt{v^2\lambda_v^2 + \lambda_\gamma^2 + \frac{\lambda_\chi^2}{\cos^2\gamma}} - \sigma m v \lambda_m$$

and for the controls (apart from some special cases of minor practical interest)

$$\sin\mu = \frac{B \cdot \dfrac{\lambda_\chi}{\cos\gamma}}{\sqrt{\lambda_\gamma^2 + \dfrac{\lambda_\chi^2}{\cos^2\gamma}}} \qquad \sin\bar{\varepsilon} = -\frac{B \cdot \sqrt{\lambda_\gamma^2 + \dfrac{\lambda_\chi^2}{\cos^2\gamma}}}{\sqrt{v^2\lambda_v^2 + \lambda_\gamma^2 + \dfrac{\lambda_\chi^2}{\cos^2\gamma}}}$$

$$\cos\mu = \frac{B \cdot \lambda_\gamma}{\sqrt{\lambda_\gamma^2 + \dfrac{\lambda_\chi^2}{\cos^2\gamma}}} \qquad \cos\bar{\varepsilon} = -\frac{v\lambda_v}{\sqrt{v^2\lambda_v^2 + \lambda_\gamma^2 + \dfrac{\lambda_\chi^2}{\cos^2\gamma}}}$$

$$B := \operatorname{sign}(\lambda_\gamma) \qquad C_L = \frac{B}{2C_{D1}v\lambda_v}\sqrt{\lambda_\gamma^2 + \frac{\lambda_\chi^2}{\cos^2\gamma}}$$

It is well known (e.g. [7, 14]) that the optimal solution of problems of the type defined above may contain so-called singular controls. The singular control proves to be of order $p = 2$ for the present problem. Rather well understood in theory, the practical calculation and numerical programming of those controls is nevertheless very time-consuming: The formulation of the total time derivatives is straightforward, but tedious. By segmentation, this process can be accelerated. The equations are split up into several hierarchical parts (e.g. $C_L, D = D(\ldots, C_L), \dot{\lambda}_v = \dot{\lambda}_v(\ldots, D)$), which are then treated independently. The summation of the single contributions is done by the computer; the full expression for singular control has never to be evaluated. The calculation of $\delta$ by the solution of a nonlinear equation and the testing of the generalized Legendre-Clebsch condition are done numerically. By this technique, the effort for programming the complete singular control in the present case reduces to about 230 lines of FORTRAN code, compared with several thousands, as reported by other authors [7], who have calculated the derivatives for similar problems by help of MAPLE.

#### 6.2.4.6 Secondary Problem

The equations of motion for the $j$-th secondary trajectory $y_j(s)$, $j = 1, \ldots, \hat{p}$ (emergency landing) are very similar to the equations of motion of the primary trajectory; the only modification is $A = 0$ (no thrust available).

$y_j := (v_j, \gamma_j, \chi_j, h_j, \Lambda_j, m_j, \vartheta_j) : [\hat{t}_j, t_{Fj}] \to \mathbb{R}^7$ is the state vector, $w_j := (C_{Lj}, \mu_j) : [\hat{t}_j, t_{Fj}] \to \mathbb{R}^2$ denotes the control vector. $s$ is the independent variable (here: time) for trajectory no. $j$ that starts at the (primary) time $\hat{t}_j$ with the state vector $y_j(\hat{t}_j) = x(\hat{t}_j)$. For the equations of motion we get $\dot{y}_j(s) = (\dot{v}, \dot{\gamma}, \dot{\chi}, \dot{h}, \dot{\Lambda}, \dot{m}, \dot{\vartheta})_j^T|_{A=0}$ from (24).

The secondary objective functions

$$\hat{J}(w_j) = \hat{L}_j(t_{Fj}, y_j(t_{Fj})) := t_{Fj} - \hat{t}_j > 0$$

are to be minimized. If there exists a solution of the secondary problem $j$ for a given initial time $\hat{t}_j$, then there always exists a time minimal solution and the HYD is able to reach the emergency landing site. In this case the $j$-th secondary problem does not affect the primary trajectory. Other cost functionals are also possible in this context.

The boundary conditions at initial time $s = \hat{t}_j$ are $y_j(\hat{t}_j) = x(\hat{t}_j)$, at final time $t_{Fj}$ the emergency landing conditions are prescribed by

$$\begin{aligned} v_j &= 133\ \text{m/s}, & \Lambda_j &= 0.0095, \\ \gamma_j &= 0, & \theta_j &= 0.0120, \\ \chi_j &= \text{free}, & m_j &= m(\hat{t}_j), \\ h_j &= 1000\,\text{m}. \end{aligned} \tag{25}$$

Primary and secondary trajectories are coupled by the conditions at the respective initial times $\hat{t}_j$; the emergency landing site is the same for all secondary trajectories. The control constraints are $|C_{Lj}| \leq C_{L0}$ and $|\mu_j| \leq \mu_0$; dynamic pressure $v_j^2 \rho \leq S_0$ is the state constraint.

#### 6.2.4.7 Extended Problem (B)

The combined problem is formulated as an extended problem of type (B) with the joint objective function

$$J_{tot}(x, u, y_1, w_1, \ldots, y_{\hat{p}}, w_{\hat{p}}) := (4000 - v(t_i)) + \varepsilon \left( \sum_{j=1}^{\hat{p}} t_{Fj} - \hat{t}_j \right).$$

The $\kappa_j := t_{Fj} - \hat{t}_j$ are defined as additional state variables with the differential equations $\dot{\kappa}_j = 0$. The link condition $y_j(\hat{t}_j) = x(\hat{t}_j)$ connects the primary with a secondary trajectory; for that condition optimal control theory yields an interior point condition slightly different from (16) because of the relation $m_j(t_{Fj}) = m(\hat{t}_j)$ in (25):

$$\lambda_m(\hat{t}_j^+) + \hat{\lambda}_{m_j}(\hat{t}_j^+) - \lambda_m(\hat{t}_j^-) - \hat{\lambda}_{m_j}(t_{Fj}) = 0\,.$$

To detect redundant secondary trajectories according to (17), two otherwise identical optimal control problems with $\varepsilon_1 = 0.2$ and $\varepsilon_2 = 0.006$ are solved at each run with an overall relative precision of $10^{-9}$.

Not more than 8 secondary problems at a time are treated together with the primary problem. This is a dramatic reduction in the computational effort as compared with [6], where for a full combined problem about 50...100 secondary trajectories have to be considered. After the combined optimal control problem has been solved for $\varepsilon_1$ and $\varepsilon_2 \ll \varepsilon_1$, redundant trajectories are automatically omitted. Then a large number (about 300) of secondary optimal control problems are solved starting on adaptively determined points on the primary trajectory to identify those secondary trajectories which do not yet fulfil the safety conditions (i.e. the HYD cannot reach the emergency landing site when starting at those points). These trajectories are added to the (active) set of secondary trajectories; the so modified combined optimal control problem is solved again.

Computation time for one secondary trajectory is below 0.5 s and for a combined problem is in the order of 50 s on a Pentium IV (1.6 GHz) PC. Typically three to four times the selection of secondary problems is changed, i.e. nine combined problems have to be solved to get the desired result. Computation times are rather low, because very good estimates for a trajectory are available from the neighbouring trajectories and the preceding steps.

#### 6.2.4.8 Numerical Results

Figure 6.2.3 shows important results of the calculations. In Fig. 6.2.3 a the projection of the unperturbed primary trajectory $P_1$ of the HYD to the surface of the Earth is given (thick solid line) subject to the boundary conditions and constraints of the primary problem, but neglecting completely the safety considerations that lead to the secondary problems. Total flight time is 405.5 s, after 140.5 s the maximum speed of $v(t_i) = 2797.6$ m/s is achieved. Thin solid lines mark complete solutions of the secondary problem, that start with $x(\hat{t}_s)$ and reach the emergency landing site $E$; dashed lines indicate secondary trajectories that come as close as possible to $E$, but do not reach it.

Figure 6.2.3 b shows the projection of the *stabilized* primary trajectory $P_2$ (thick solid line) to the Earth's surface; $P_2$ is the solution of the combined problem. More than 95% of the secondary trajectories reach the landing site $E$. An exception has to be made only for a very short period after launch and immediately before landing; here a return to basis or a gliding down to the normal landing site is no problem. The notation of the secondary trajectories is as in Fig. 6.2.3 a. Additionally, the unperturbed trajectory $P_1$ (dashed-dotted line) is drawn for comparison purposes. For $P_2$ the total flight time is 351.4 s, after 159.0 s the maximum speed of $v(t_i) = 2772.4$ m/s is achieved. The dynamic

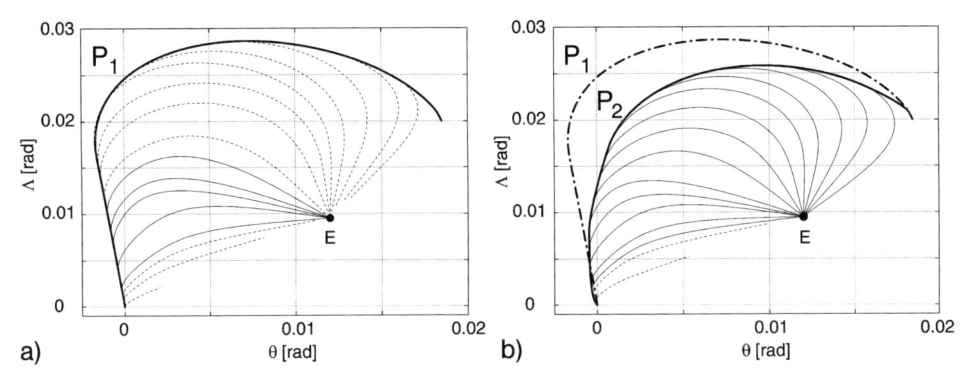


Figure 6.2.3: Projections of the unperturbed primary trajectory $P_1$ and the stabilized trajectory $P_2$ to the surface of the Earth. Thin solid lines mark secondary trajectories, that reach the emergency landing site $E$, dashed lines those secondary trajectories, that do not.

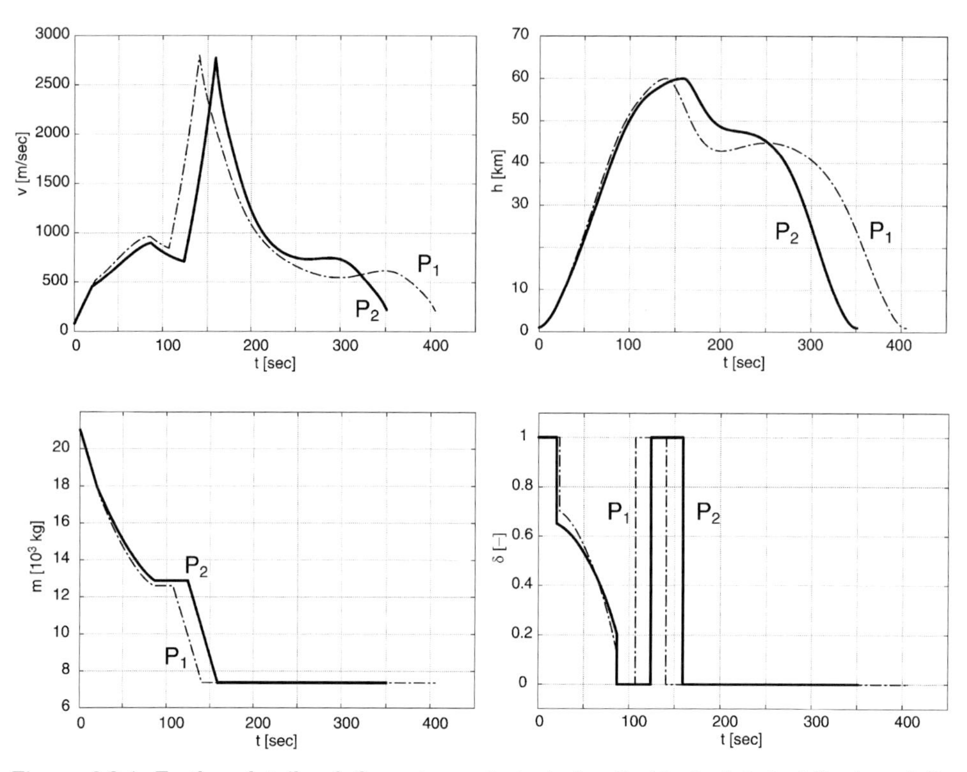


Figure 6.2.4: Further details of the primary trajectories $P_1$ (dashed-dotted line) and $P_2$ (solid line): Speed $v$, altitude $h$, mass $m$ and throttle $\delta$ setting are plotted as a function of flight time $t$ [6].

pressure constraint remains inactive. In Fig. 6.2.4 further details of the primary trajectories $P_1$ (dashed-dotted line) and $P_2$ (solid line) are given. Speed $v$, altitude $h$ and mass $m$ are plotted as a function of flight time $t$; moreover the throttle setting $\delta$ is shown.

### 6.2.5 Conclusion

The demand for extended inherent safety in case of mission abort leads to significant deformations of optimal flight trajectories. When handled in a mathematically correct way, maximum speed on the stabilized trajectory $P_2$ is only 25.2 m/s below that of the unperturbed trajectory $P_1$ for our example problem. The new approach significantly increases the safety potential of the system. Based on the control of the modified extended problem of type (B), an efficient numerical solution is possible even if redundant secondary trajectories are present. Moreover, those trajectories are automatically detected and omitted.

## Acknowledgement

The author is indebted to Prof. Dr. Dr h.c. R. Bulirsch who always encouraged and supported his work. This research has been supported by the DFG in the Collaborative Research Centre on Transatmospheric Vehicles (SFB 255). Special thanks are due to the speaker of the SFB 255, Prof. Dr. G. Sachs, who did an excellent job during all the years.

## References

1. L. Bittner: On the convexification of optimal control problems of flight dynamics, in Variational calculus, optimal control and applications, R. Bulirsch, W. H. Schmidt, K. Meier, and L. Bittner, eds., ISNM 124, Birkhäuser, Basel, 1998, pp. 3–14.
2. A. E. Bryson, Y.-C. Ho: Applied Optimal Control, Revised Printing, Hemisphere Publishing Corp., Washington D.C., 1975.
3. R. Bulirsch: Die Mehrzielmethode zur numerischen Lösung von nichtlinearen Randwertproblemen und Aufgaben der optimalen Steuerung, Report, Carl-Cranz-Gesellschaft e.V., Oberpfaffenhofen, 1971.
4. R. Callies: Entwurfsoptimierung und optimale Steuerung. Differential-algebraische Systeme, Mehrgitter-Mehrzielansätze und numerische Realisierung, Habilitationsschrift, Zentrum Mathematik, Technische Universität München, 2000.
5. R. Callies, R. Bulirsch: 3D Trajectory Optimization of a Single-Stage VTVL System, Paper AIAA-96-3903, San Diego, 1996.
6. R. Callies, G. Wimmer: Optimal Hypersonic Flight Trajectories with Full Safety in case of Mission Abort, Paper AIAA-2000-3996, Denver, 2000.
7. K. Chudej: Optimale Steuerung des Aufstiegs eines zweistufigen Hyperschall-Raumtransporters, PhD thesis, Mathematisches Inst., Technische Universität München, 1994.
8. K. Chudej: Effiziente und stabile Lösung eines reichweitenmaximalen Fluges mit Staudruckbegrenzung für ein Überschallflugzeug, in DGLR Jahrbuch 1997, Vol. 2, Bonn, 1997, pp. 1141–1147.
9. R. F. Hartl, S. P. Sethi, R. G. Vickson: A survey of the maximum principles for optimal control problems with constraints, SIAM Review, 37 (1995), pp. 181–218.
10. M. R. Hestenes: Calculus of Variations and Optimal Control Theory, Wiley, New York, 1966.
11. D. H. Jacobson, M. M. Lele, J. L. Speyer: New necessary conditions of optimality for control problems with state-variable inequality constraints, J. Math. Anal. Appl., 35 (1971), pp. 255–284.
12. A. Miele: Flight Mechanics, Vol. 1, Addison-Wesley, Reading, 1962.
13. H. J. Oberle: Numerische Behandlung singulärer Steuerungen mit der Mehrzielmethode am Beispiel der Klimatisierung von Sonnenhäusern, Habilitationsschrift, Technische Universität München, 1982.
14. H. J. Oberle: Numerical computation of singular functions in trajectory optimization problems, J. Guidance, Control, and Dynamics, 13 (1990), pp. 153–159.
15. F. J. Regan, S. M. Anandakrishnan: Dynamics of Atmospheric Re-Entry, AIAA Education Series, Washington, 1993.
16. H. Seywald, E. M. Cliff: A Feedback Control for the Advanced Launch System, Paper AIAA-91-2619-CP, New Orleans, 1991.
17. J. Stoer, R. Bulirsch: Numerische Mathematik 2, 3rd edn., Springer, Berlin, Heidelberg, New York, 1990.

## 6.3 Hypersonic Trajectory Optimization for Thermal Load Reduction

Michael Dinkelmann, Markus Wächter, and Gottfried Sachs *

### Nomenclature

| | |
|---|---|
| $C_D$ | drag coefficient |
| $C_L$ | lift coefficient |
| $C_P$ | heat capacity |
| $D$ | drag |
| $f_a$ | thrust factor for angle of attack dependency |
| $g$ | acceleration due to gravity |
| $h$ | altitude |
| $I$ | cost functional |
| $L$ | lift |
| $M$ | Mach number |
| $m$ | mass |
| $m_f$ | fuel mass |
| $n$ | load factor |
| $q$ | heat flux density |
| $\bar{q}$ | dynamic pressure |
| $r_E$ | radius of the Earth |
| $S$ | reference area |
| $s$ | range |
| $T$ | thrust |
| $T_i$ | temperature (at location $i$) |
| $t$ | time |
| $\vec{u}$ | control vector |
| $V$ | speed |
| $\vec{x}$ | state vector |
| $\alpha$ | angle of attack |
| $\Gamma$ | integrated heat flux |
| $\gamma$ | flight path angle |
| $\Delta$ | geocentric latitude |
| $\delta_T$ | throttle setting |
| $\varepsilon$ | emissivity |
| $\varepsilon_T$ | thrust vector angle |
| $\Lambda$ | geographic longitude |
| $\mu_a$ | bank angle |
| $\rho$ | atmospheric density |
| $\sigma$ | Boltzmann constant |
| $\sigma^*$ | specific fuel consumption at stoichiometric combustion |
| $\phi_f$ | equivalence ratio |
| $\chi$ | azimuth angle |
| $\omega_E$ | angular velocity of the Earth |

### Abstract

Thermal load reduction is considered for a hypersonic vehicle equipped with turbo/ram jet engines. Mathematical models for describing the unsteady heat input and the dynamics of the vehicle are presented. Two trajectory optimization problems are treated, one concerned with a range cruise and the other with a return-to-base flight. An efficient optimization technique is used constructing solutions. The results show that the thermal load can be significantly decreased.

* Technische Universität München, Lehrstuhl für Flugmechanik und Flugregelung, Boltzmannstr. 15, 85748 Garching; sachs@lfm.mw.tum.de

## 6.3.1 Introduction

In recent years, significant research efforts deal with new space transportation systems. There are various concepts for such transportation systems. A promising concept is concerned with a two-stage-to-orbit vehicle which employs an aerodynamic lifting capability and airbreathing propulsion.

The hot environment to which two-stage-to-orbit vehicles are exposed in hypersonic flight poses challenging problems. An insight is provided by Fig. 6.3.1 which shows temperatures which can be reached. For coping with these problems, complex thermal protection systems are necessary [1–3]. For thermal protection, insulating layers of adequate material and thickness are used.

The thermal load problem is investigated in this paper using trajectory optimization as a means to reduce the heat input into the vehicle. Particular emphasis is placed on a realistic modelling of the heat transfer from the hot outside environment into the vehicle. Since the performance requirements on hypersonic vehicles are high, fuel minimization is a primary goal in trajectory optimization. Accordingly, there is significant research in this field (e.g. [4–6]).

The research on thermal load problems shows significant results and important progress. This also concerns papers dealing with heat loads and input [7–15]. Problems of reentry vehicles related to unsteady heat flux are dealt with in [12, 13]. Unsteady temperature distributions are treated for trajectories which

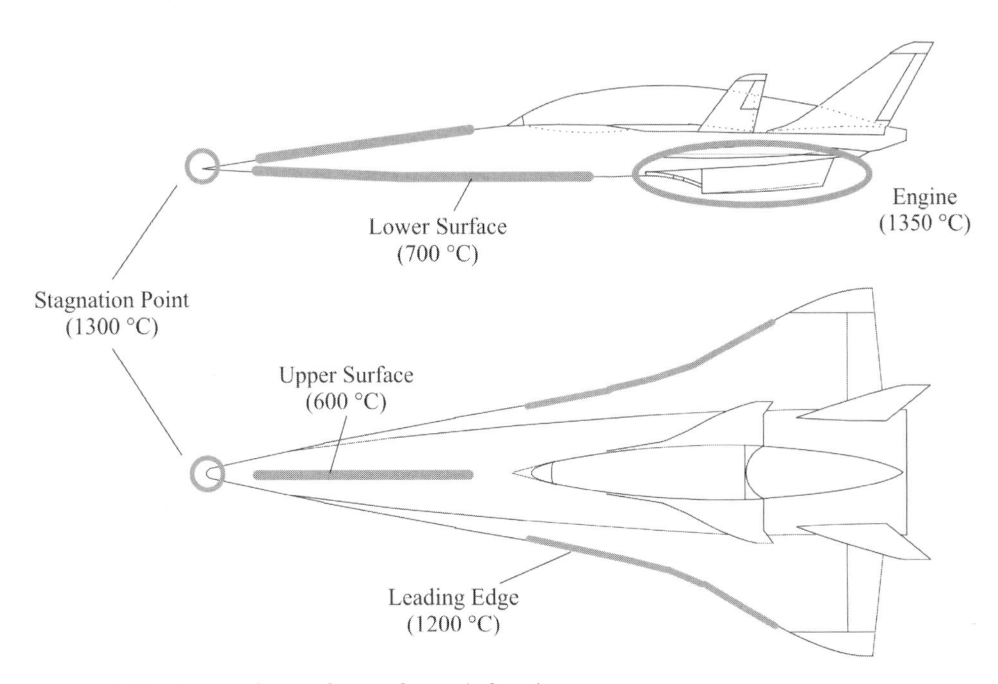


Figure 6.3.1: Areas of aero-thermodynamic heating.

have been determined in preceding calculations. The reduction of the temperature at various locations of a hypersonic vehicle on the basis of equilibrium conditions is subject of [9, 15]. Ref. [12] is concerned with unsteady temperature distributions in the wall of a reentry vehicle, using the energy-state method for flight path optimization.

It is the purpose of this paper to show the possibility for the reduction of the heat load by an appropriate trajectory control. For this purpose, an optimization technique is applied, with coupling of trajectory control and unsteady heat input in an integrated computational process.

Two trajectory control cases are considered. One relates to a range cruise which consists of a two-dimensional motion in the vertical plane. The other case is concerned with a three-dimensional trajectory. This is a return-to-base cruise problem of a two-stage-to-orbit vehicle, including a range requirement for releasing the orbital stage at a specified location.

## 6.3.2 Modelling of Vehicle Dynamics

For describing the dynamics of the vehicle, a point mass model is used. The equations of motion can be written as [16], Fig. 6.3.2:

$$
\begin{aligned}
\dot{V} &= \frac{T\cos(\alpha+\varepsilon_T)-D}{m} - g\sin\gamma + \omega_E^2(r_E+h)\cos\Delta(\sin\gamma\cos\Delta - \cos\gamma\sin\Delta\cos\chi) \\
\dot{\gamma} &= \frac{T\sin(\alpha+\varepsilon_T)+L}{mV}\cos\mu_a + \left(\frac{V}{r_E+h} - \frac{g}{V}\right)\cos\gamma + 2\omega_E\cos\Delta\sin\chi \\
&\quad + \frac{\omega_E^2(r_E+h)}{V}\cos\Delta(\cos\gamma\cos\Delta - \sin\gamma\sin\Delta\cos\chi) \\
\dot{\chi} &= \frac{T\sin(\alpha+\varepsilon_T)+L}{mV\cos\gamma}\sin\mu_a + \frac{V\cos\gamma\sin\chi\tan\Delta}{r_E+h} - 2\omega_E(\tan\gamma\cos\Delta\cos\chi - \sin\Delta) \\
&\quad + \frac{\omega_E^2(r_E+h)}{V\cos\gamma}\sin\Delta\cos\Delta\sin\chi \qquad (1) \\
\dot{\Delta} &= \frac{V\cos\gamma\cos\chi}{r_E+h} \\
\dot{\Lambda} &= \frac{V\cos\gamma\sin\chi}{(r_E+h)\cos\Delta} \\
\dot{h} &= V\sin\gamma \\
\dot{m} &= -\dot{m}_f
\end{aligned}
$$

where

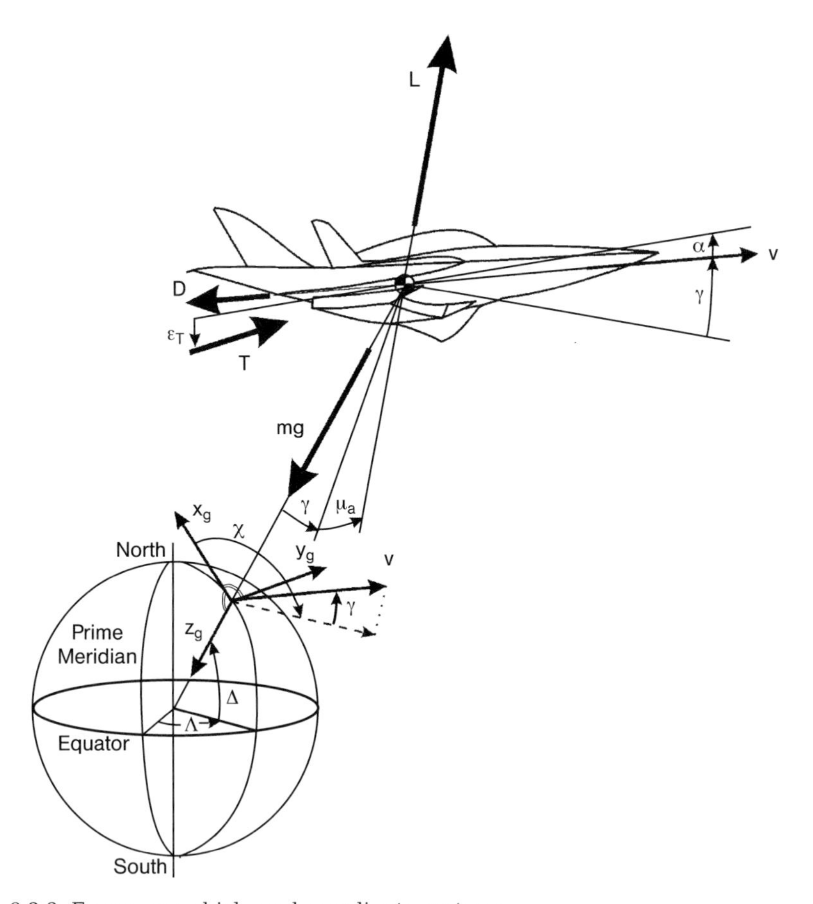


Figure 6.3.2: Forces on vehicle and coordinate systems.

$$g = g_0 \left( \frac{r_E}{r_E + h} \right)^2 \tag{2}$$

Complex mathematical models are used for describing the aerodynamics and powerplant characteristics. Emphasis is placed on realistic modellings.

The aerodynamics model involves the lift and drag forces which read

$$\begin{aligned} L &= C_L \bar{q} S \\ D &= C_D \bar{q} S \end{aligned} \tag{3}$$

There are complex relationships describing the dependencies of the lift and drag coefficients on angle of attack, $a$, Mach number, $M$, and altitude, $h$, yielding (Fig. 6.3.3)

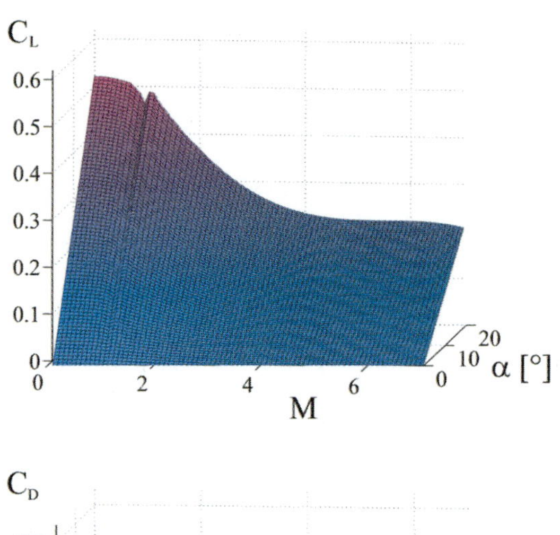


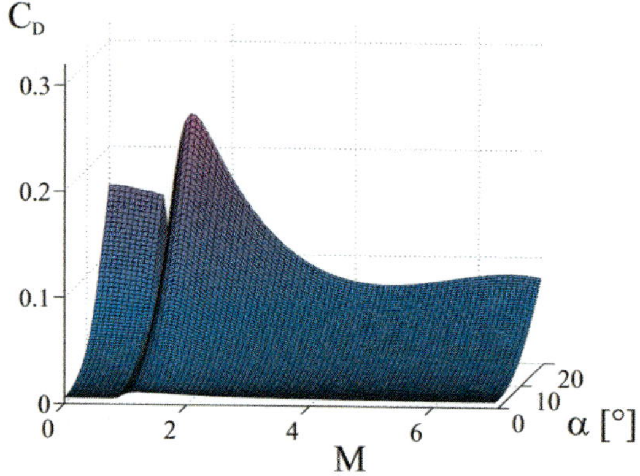


Figure 6.3.3: Models of lift and drag coefficients.

$$
\begin{aligned}
C_L &= C_L(M, h, \alpha) \\
C_D &= C_D(M, h, \alpha)
\end{aligned}
\tag{4}
$$

The thrust and the fuel consumption model for the turbo ram jet engines of the vehicle are described as

$$
\begin{aligned}
T &= \delta_T T^*(M, h) f_\alpha(M, h, \alpha) \\
\dot{m}_f &= \phi_f(M, h, \delta_T) \sigma^*(M, h) T
\end{aligned}
\tag{5}
$$

The quantities $T^*$ and $\sigma^*$ denote thrust and fuel consumption at the stoichiometric operating condition (Fig. 6.3.4). The engine thrust model accounts for the possibility of overfuelled combustion in the ram jet mode (Fig. 6.3.5). The factor $f_\alpha$ describes the dependence of thrust on angle of attack (Fig. 6.3.6). The transition from turbo to ram jet operation is modelled as a linear process in the Mach number range $3 \leq M \leq 3.5$.

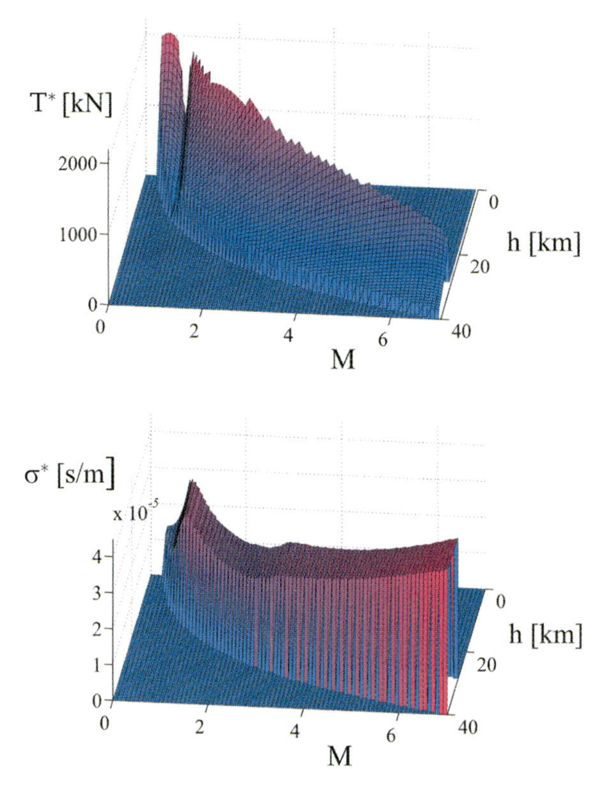


Figure 6.3.4: Models of reference thrust $T^*$ and thrust related fuel consumption $\sigma^*$.

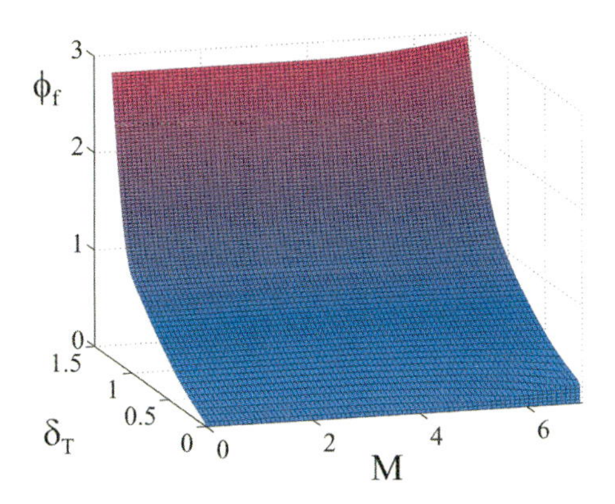


Figure 6.3.5: Relation beween equivalence ratio and thrust setting.

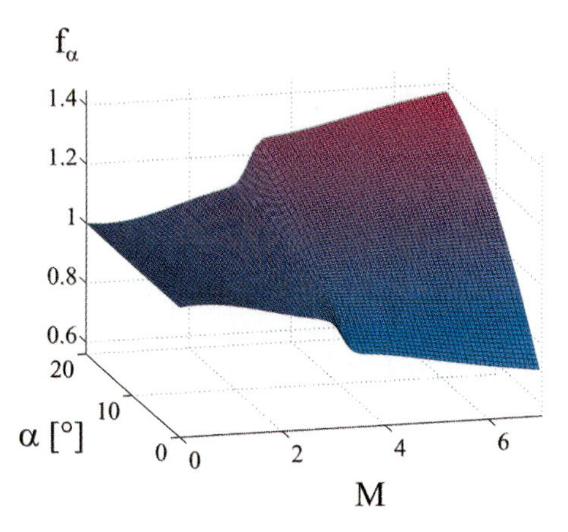


Figure 6.3.6: Effect of angle of attack on thrust of hypersonic vehicle.

### 6.3.3 Modelling of Heat Input

A complex mathematical model was developed to realistically describe the unsteady heat input from the hot outside through the thermal protection system into the vehicle. Two areas of the vehicle are considered. One is in front of the tanks and the other under the tanks (Fig. 6.3.7).

The wall structures of the thermal protection systems for both areas are shown in Figs. 6.3.8 and 6.3.9. The thermal protection systems consist of a set of layers, each of specific material and thickness. Models were developed for describing the heat flux through the layers. According to the build-up of the walls shown in Figs. 6.3.8 and 6.3.9, two models were considered, one showing 10 layers and the other 13 layers. The heat flux model is shown in Fig. 6.3.10. With regard to the two wall structures, the numbers for the layers are given by $n=10$ and $n=13$, respectively.

The heat flux from one layer to the next is described by a one-dimensional knot model. The heat flux in the first layer can be described by

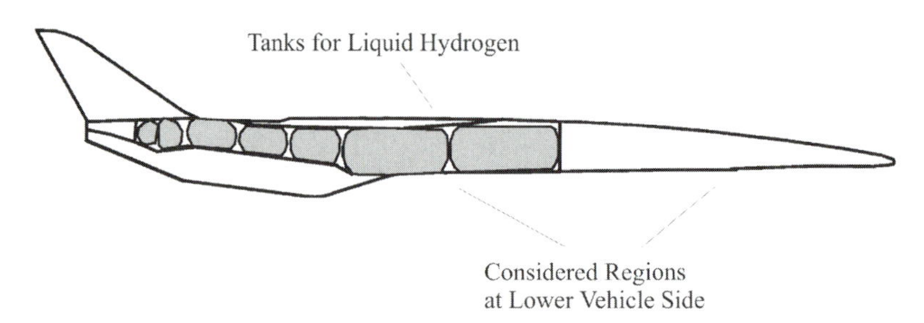


Figure 6.3.7: Hypersonic vehicle areas considered for heat input.

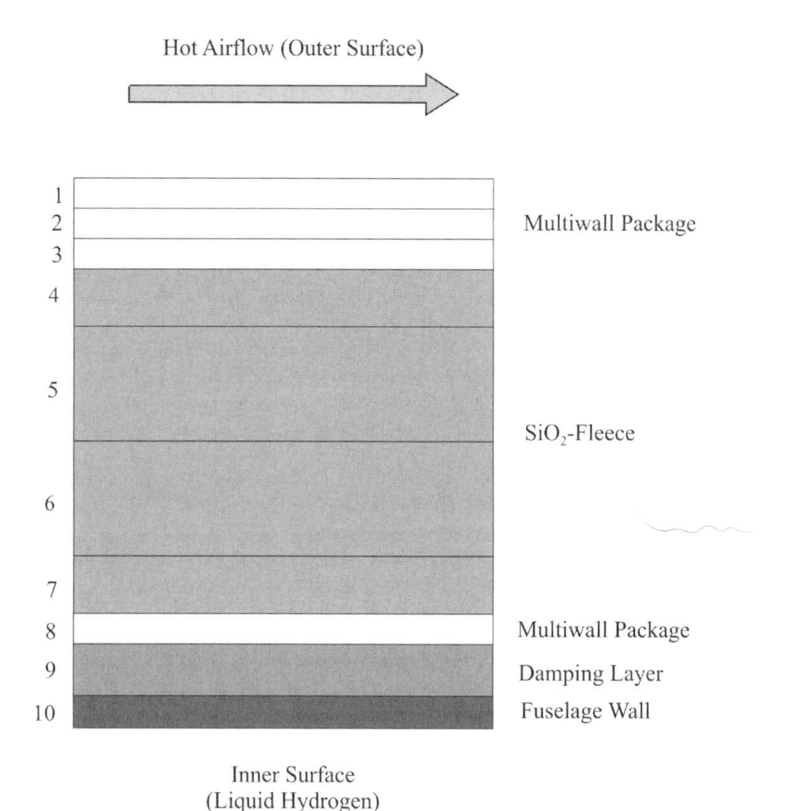


Figure 6.3.8: Heat protection in area in front of tanks.

$$q_1 = q_{air} - q_{radiation} = q_{air}(V, h, T_1; a) - \varepsilon\sigma(T_1^4 - T_\infty^4) \tag{6}$$

The temperature at the outer layer depends on the actual flight condition, yielding

$$T_1 = T_1(V, h; a) \tag{7}$$

The quantity

$$q_{air} = q_{air}(V, h, T_1; a) \tag{8}$$

denotes the heat flux into the vehicle

The heat fluxes for the remaining $n$–1 layers can be described by

$$q_i = C_i(T_i)(T_{i-1} - T_i) + \sigma\varepsilon(T_i)(T_{i-1}^4 - T_i^4) \tag{9}$$

With Eqs. (6) and (9), the following system of differential equations for the temperature $T_i$ of each element is obtained

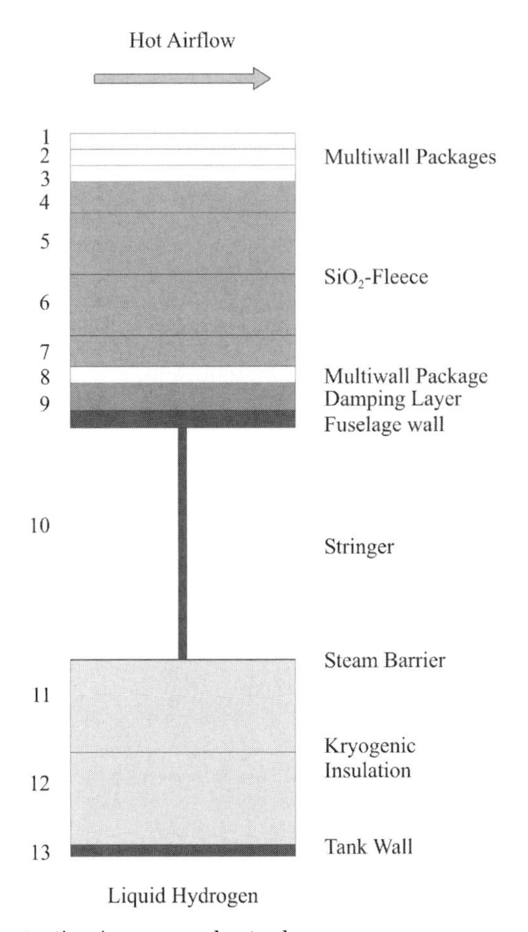


Figure 6.3.9: Heat protection in area under tanks.

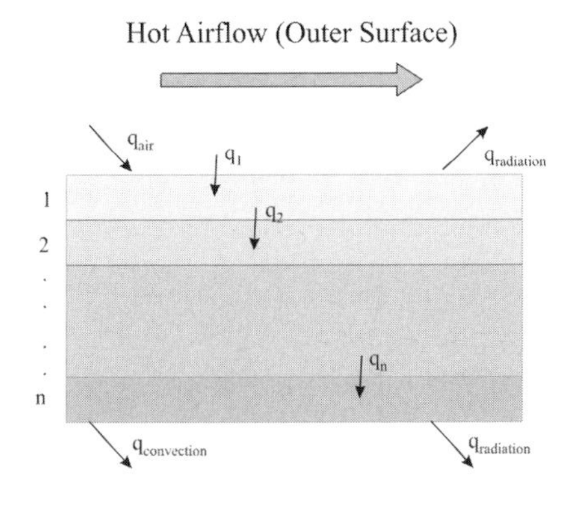


Figure 6.3.10: Heat flux model.

$$\dot{T}_i = \frac{q_i(T_{i-1}, T_i) - q_{i+1}(T_i, T_{i+1})}{C_{P,i}(T_{i-1}, T_i)}, \quad i = 1, ..., n-1 \tag{10}$$

For the inside, different temperature relations are considered. With regard to the area in front of the tanks, it is assumed that the temperature can change. Accordingly

$$\dot{T}_n = \frac{q_n(T_{n-1}, T_n) - \alpha_q(T_n - T_{IS}) - \varepsilon\sigma(T_n^4 - T_{IS}^4)}{C_{P,n}(T_{n-1}, T_n)} \tag{11}$$

For the tank area, the temperature is considered as constant, equal to the temperature of the fuel

$$\dot{T}_n = 0 \tag{12}$$

This is because there is a large amount of hydrogen fuel which shows a high specific heat capacity.

The overall heat flux through the inner layer can be expressed as

$$\dot{\Gamma}_n = q_n(T_{n-1}, T_n) \tag{13}$$

### 6.3.4 Optimization Problem

The optimal control problem is concerned with heat load reduction and fuel consumption minimization for two flight tasks. One is a range cruise consisting of a two-dimensional trajectory and the other is a return-to-base cruise showing a three-dimensional trajectory.

The problem can be formulated as to find a state function

$$\vec{x} = (V, \gamma, \chi, \Delta, \Lambda, h, m, T_i, \Gamma_n) : [t_0, t_f] \rightarrow R^{7+n+1} \tag{14}$$

and a control function

$$\vec{u} = (\alpha, \delta_T, \mu_a) : [t_0, t_f] \rightarrow R^3 \tag{15}$$

which minimize the cost functional

$$I[u] = -m|_{t_f} \rightarrow Min \tag{16}$$

subject to a set of differential equations $\dot{x}(t) = f(x(t), u(t))$, given by the Eqs. (1), (10)–(12) describing the dynamics and the unsteady heating of the vehicle, to state constraints

$$
\begin{aligned}
\bar{q}_{\min} &\leq \bar{q}(V,h) \leq \bar{q}_{\max} \\
n_{\min} &\leq n(V,h,m,\alpha) \leq n_{\max} \\
0 &\leq \Gamma_n(T_{n-1},T_n) \leq \Gamma_{n,\max}
\end{aligned} \tag{17}
$$

and to control constraints

$$
u_{i,\min} \leq u_i \leq u_{i,\max}, \quad i = 1,2,3 \tag{18}
$$

When the constraint concerning the heat flux, Eq. (17), becomes active the differential equations for the dynamics of the vehicle, Eq. (1), and for the heat input, Eqs. (10)–(12), are coupled.

The constraints for both problems are given in Tab. 6.3.1. The boundary conditions for the range cruise problem are shown in Tab. 6.3.2 and the boundary conditions for the return-to-base cruise problem in Tab. 6.3.3. In addition, the flight condition at the separation is presented in Tab. 6.3.3.

The procedure applied in this paper is a parameterization optimization technique [17].

Table 6.3.1: Constraints.

| | Minimum | Maximum |
|---|---|---|
| $\alpha$, deg | –1.5 | 20 |
| $\phi_f$ (ram) | $\phi_f(M, \delta_{T\min})$ | 3 |
| $\delta_T$ (turbo) | 0 | 1 |
| $\delta_T$ (ram) | 0 | $\delta_T(M, \phi_{f\max})$ |
| $\lvert\mu_a\rvert$, deg | free | varied |
| $\bar{q}$, kPa | 10 | 50 |
| $n$ | 0 | 2 |

Table 6.3.2: Conditions for range cruise.

| | Range Cruise | |
|---|---|---|
| | $t=0$ | $t=t_f$ |
| $h$, m | 500 | 500 |
| M | 0.44 | 0.44 |
| $\gamma$, deg | 0 | 0 |
| $\chi$, deg | 90 | 90 |
| $\Delta$, deg | 0 | 0 |
| $\Lambda$, deg | 0 | – |
| $m$, kg | 244 000 | – |
| $s$, km | 0 | 9000 |

Table 6.3.3: Conditions for return-to-base cruise.

| | Return-to-Base Cruise | | |
|---|---|---|---|
| | $t=0$ | Separation | $t=t_f$ |
| $h$, m | 500 | 33 800 | 500 |
| M | 0.44 | 6.8 | 0.44 |
| $\gamma$, deg | 3 | 8.7 | -3 |
| $\Delta$, deg | 43.5 | 16.5 | 43.5 |
| $\Lambda$, deg | 0 | – | 0 |
| $m$, kg | 340 000 | – | – |

## 6.3.5 Results

### 6.3.5.1 Range Cruise

As a reference, the minimum fuel range cruise without a heat input constraint is considered first. The optimal trajectory is presented in Fig. 6.3.11. A further insight is provided by Fig. 6.3.12 which shows the time histories of state and control variables.

A constraint on the heat input is considered next. Figure 6.3.13 presents the time histories of state and control variables. There is a significant decrease of the heat input, accompanied by a small fuel consumption penalty and a decrease of flight time.

Heat input characteristics of the reference case are illustrated in Fig. 6.3.14 which shows the temperatures in the layers of the thermal protection system. The temperatures of the outer layers basically follow the condition in the outside air, as given by the flight state. The temperatures in the inner part of the wall show a delayed behaviour.

The heat input for the area under the tanks is also presented in Fig. 6.3.14. There is a continuous increase of the heat input until the end of the flight. For comparison purposes, a steady-state model was also applied with re-

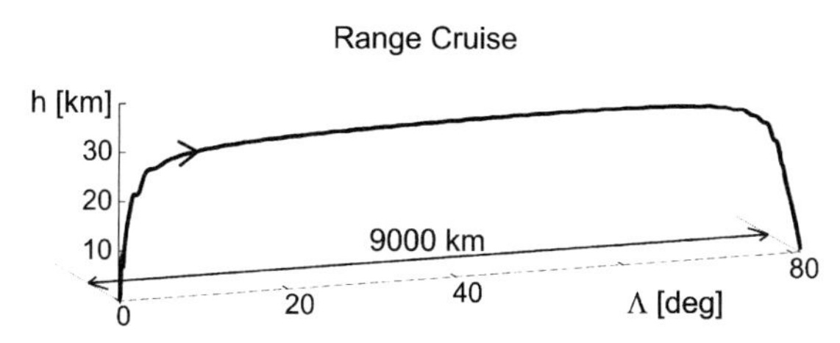


Figure 6.3.11: Minimum fuel range cruise trajectory without heat input constraint (9000 km range).

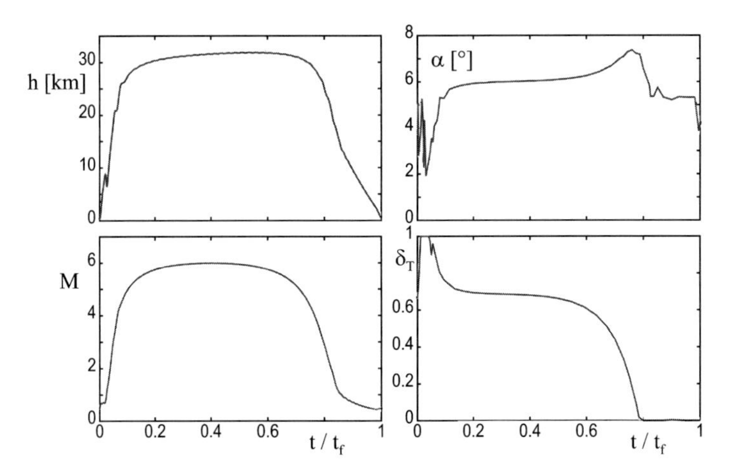


Figure 6.3.12: State and control variables of optimal 9000 km range cruise without heat input constraint ($\Gamma_n$=1214 kJ/m$^2$, $m_f$=62654 kg, $t_f$=113.3 min).

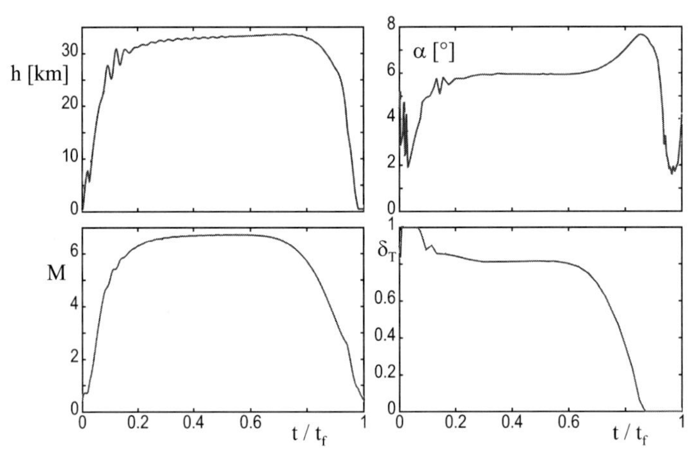


Figure 6.3.13: State and control variables of optimal 9000 km range cruise with heat input constraint ($\Gamma_n$ = 950 kJ/m$^2$, $m_f$=64459 kg, $t_f$=89.82 min).

gard to the heat input. The results presented in Fig. 6.3.14 show that there are significant differences between the heat inputs in the two model cases. This means that the unsteadiness of the heat flux plays an important role for the total heat input into the vehicle.

The temperature characteristics in the constrained heat input case are also shown in Fig. 6.3.14. There are significant differences with regard to the reference case. There is a higher temperature level in the constrained case. The increase of the temperature is reduced for the inner layers so that they show a similar characteristic as in the reference case. Consequently, the gradients of

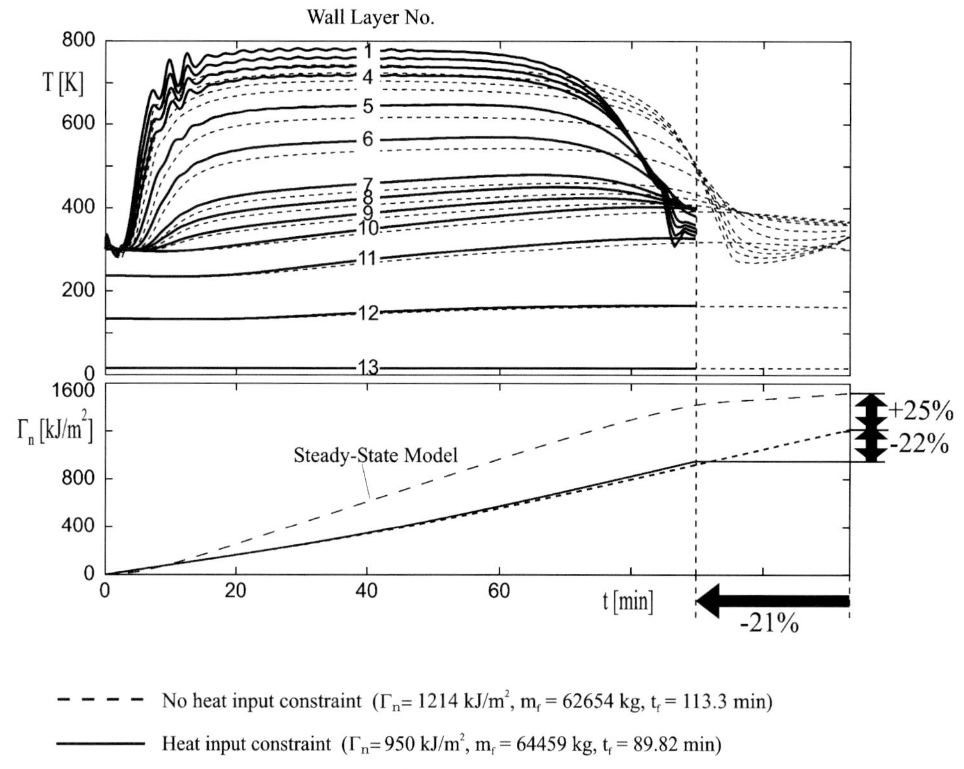


Figure 6.3.14: Temperatures and heat input for optimal 9000 km range cruise.

the heat inputs differ only a little. As a result, there is another mechanism for the significant heat input reduction in the constrained case. This mechanism is basically due to the reduction of the flight time. The overall reduction of the heat input amounts to 22%.

### 6.3.5.2 Return-to-Base Cruise

The return-to-base cruise without a heat input constraint is again used as a reference. Results for the minimum-fuel return-to-base cruise are given in Figs. 6.3.15 and 6.3.16. A perspective view on the trajectory is presented in Fig. 6.3.15. It reveals that the return-to-base cruise shows a three-dimensional trajectory where the turn constitutes a significant part of the overall flight. The time histories of state and control variables are presented in Fig. 6.3.16, providing a further insight into the characteristics of the return-to-base cruise. They show that the flight is comparatively unsteady. In particular, the separation ma-

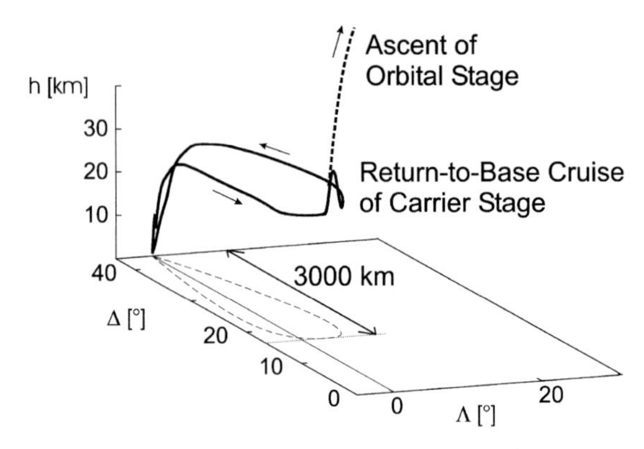


Figure 6.3.15: Optimal return-to-base cruise trajectory.

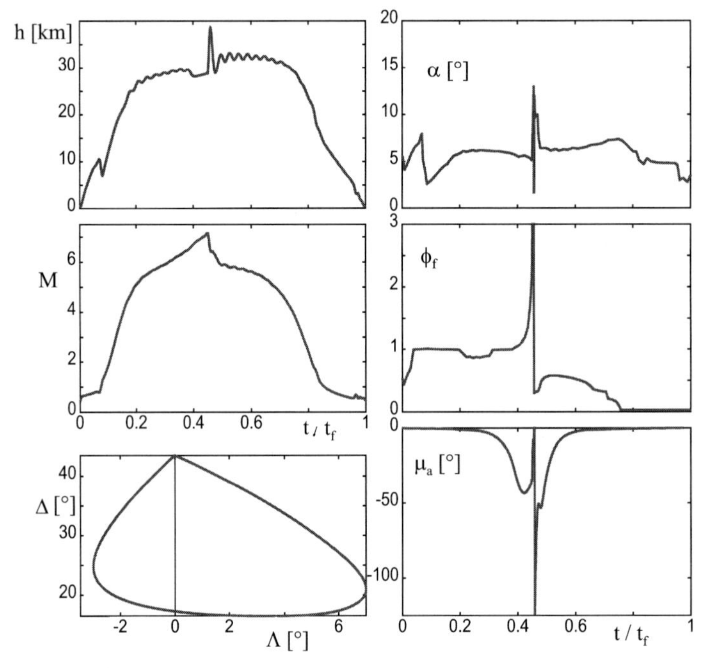


Figure 6.3.16: State and control variables of optimal return-to-base cruise without heat input constraint ($\Gamma_n = 1828$ kJ/m$^2$, $m_f = 85\,000$ kg, $t_f = 93.7$ min).

noeuvre which the flight system performs for releasing the orbital stage consists of a highly dynamic motion.

The time histories of state and control variables of the optimal return-to-base cruise with a heat input constraint are shown in Fig. 6.3.17. Here again, the flight is comparatively unsteady.

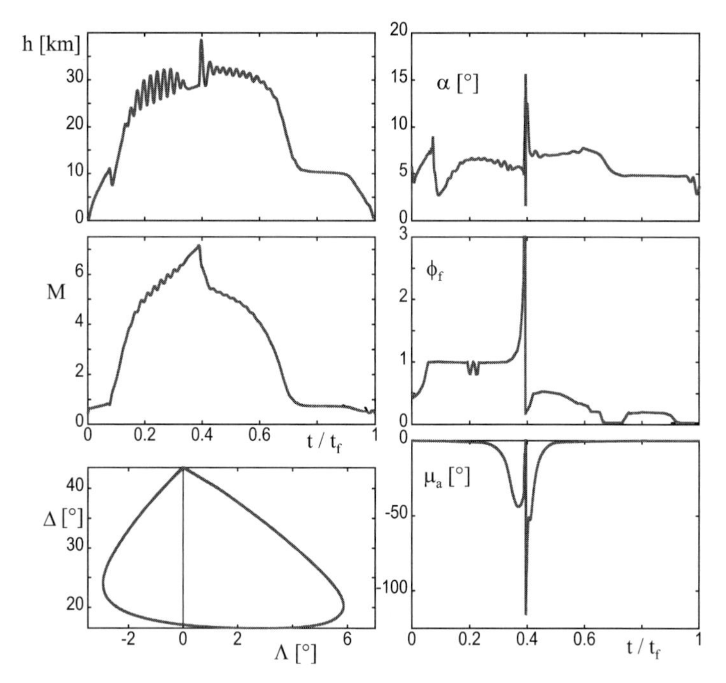


Figure 6.3.17: State and control variables of optimal return-to-base cruise with heat input constraint ($\Gamma_n$=1650 kJ/m$^2$, $m_f$=86 141 kg, $t_f$=115.3 min).

Results on the temperature and the heat input are presented in Fig. 6.3.18. There is a significant reduction of the overall heat input. It may be of interest to note that the reduced speed level during the return leg also contributes to decrease the heat load (Fig. 6.3.17). It is accompanied by an increase of overall flight time.

## 6.3.6 Conclusions

The heat input into a hypersonic vehicle can be reduced by an appropriate control of the trajectory. This possibility is considered for a two-dimensional range cruise and a three-dimensional return-to-base cruise problem. The hypersonic vehicle is equipped with a turbo/ram jet engines combination. A mathematical model for describing the heat flux through the thermal protection system of the vehicle is developed, with emphasis placed on accounting for unsteady heat transfer effects. For describing the aerothermodynamics and powerplant characteristics, complex modellings are applied. Solutions for the range cruise and return-to-base cruise problems are constructed using an efficient optimization technique. The results show that a significant heat input reduction with only a small fuel penalty can be achieved by an appropriate trajectory control.

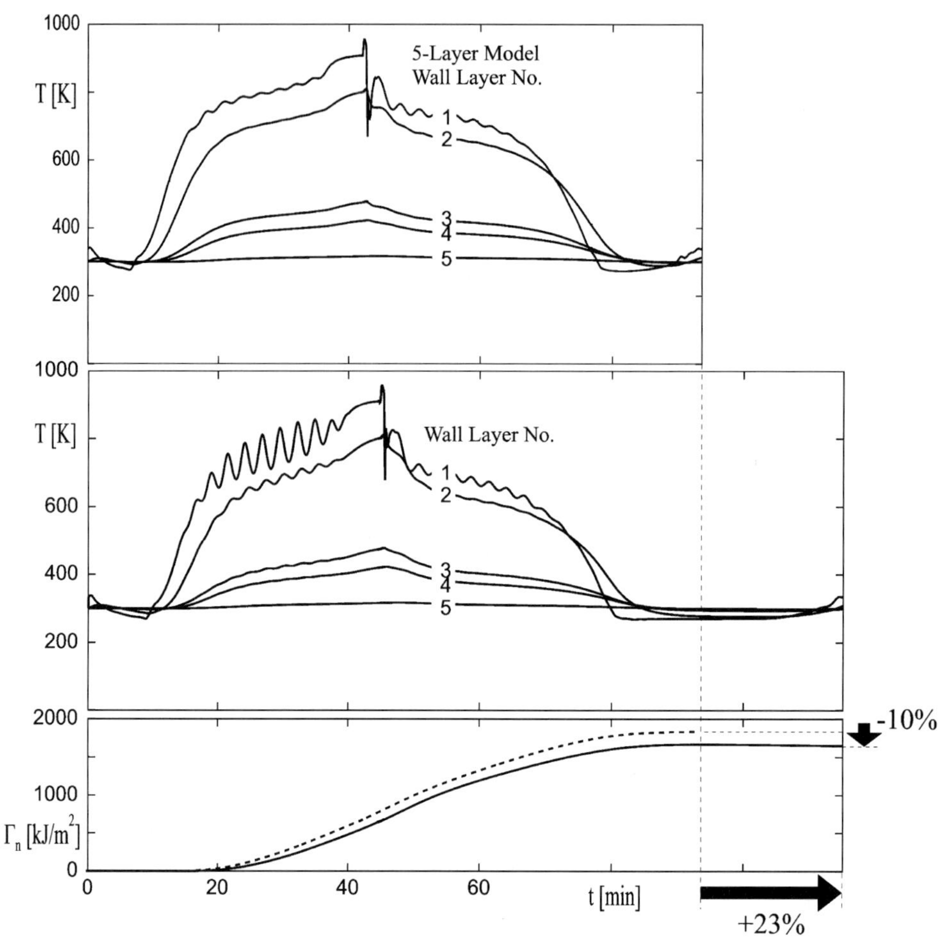


Figure 6.3.18: Temperature in the wall layers and heat input of optimal return-to-base cruise.

## References


1. Alwes, D.: TPS and Hot Structure Components for X-38 – A Status Report and Prospects on Future Transportation Systems. IAF 51st International Astronautical Congress, Rio de Janeiro, IAF-00-I.3.02, 2000.
2. Cowart, K., Olds, J.: Integrating Aeroheating and TPS Into Conceptual RLV Design. 9th International Space Planes and Hypersonic Systems and Technologies Conference, Norfolk, AIAA-99-4806, 1999.
3. Myers, D.E., Martin, C.J., Blosser, M.L.: Parametric Weight Comparison of Current and Proposed Thermal Protection Systems (TPS) Concepts. AIAA 33rd International Thermophysics Conference, Norfolk, AIAA-99-3459, 1999.
4. Bayer, R., Sachs, G.: Optimal Return-to-Base Cruise of Hypersonic Carrier Vehicles. Zeitschrift für Flugwissenschaften und Weltraumforschung, Vol. 19, No. 1, pp. 47–54, 1995.

5. Dewell, L. D., Speyer, J. L.: Fuel-Optimal Periodic Control and Regulation in Constrained Hypersonic Flight. Journal of Guidance, Control, and Dynamics, Vol. 20, No. 5, pp. 923–932, 1997.
6. Schmidt, D. K., Hermann, J. A.: Use of Energy-State Analysis on a Generic Airbreathing Hypersonic Vehicle. Journal of Guidance, Control, and Dynamics, Vol. 21, No. 1, pp. 71–76, 1998.
7. Bouquet, C., Vidal, J. P., et al.: Hot Structures and Thermal Protection Systems Investigations for Future RLV. IAF 51st International Astronautical Congress, Rio de Janeiro, IAF-00-I.3.01, 2000.
8. Glass, D. E., Merski, N. R., Glass, C. E.: Airframe Research and Technology for Hypersonic Airbreathing Vehicles. AIAA 11th International Space Planes and Hypersonic Systems and Technologies Conference, Orlèans, AIAA-2002-5137, 2002.
9. Zimmerman, C., Dukeman, G., Hanson, J.: Automated Method to Compute Orbital Reentry Trajectories with Heating Constraints. Journal of Guidance, Control, and Dynamics, Vol. 26, No. 4, pp. 523–529, 2003.
10. Scuderi, L. F., Orton, G. F.: Mach 10 Cruise/Space Access Vehicle Definition. AIAA 8th International Space Planes and Hypersonic Systems and Technologies Conference, Norfolk, AIAA-98-1584, 1998.
11. Klädtke, R., Obersteiner, M.: A European Roadmap for RLV's. IAF 52nd International Astronautical Congress, Toulouse, IAF-01-V.5.06, 2001.
12. Windhorst, R., Ardema, M. D., Bowles, J. V.: Minimum Heating Reentry Trajectories for Advanced Hypersonic Launch Vehicles. AIAA Guidance, Navigation, and Control Conference, New Orleans, LA, AIAA-97-3535, 1997.
13. Meese, E. A. Nørstrud, H.: Simulation of convective heat flux and heat penetration for a spacecraft at re-entry. Aerospace Science and Technology, Vol. 6, pp. 185–194, 2002.
14. Reich, G., Hinger, J., Huchler, M.: Thermal Protection Systems for Hypersonic Transport Vehicles. 20th International Conference on Environmental Systems, Williamsburg, Virginia, SAE TP 901306, 1990.
15. Ring, J. R.: Flight Trajectory Control for Thermal Abatement of Hypersonic Vehicles. AIAA 2nd International Aerospace Planes Conference, Orlando, 1990.
16. Vinh, N. X., Busemann A., Culp, R. D.: Hypersonic and Planetary Entry Flight Mechanics. Ann Arbor, The University of Michigan Press, 1980.
17. N. N. GESOP (Graphical Environment for Simulation and Optimization): Softwaresystem für die Bahnoptimierung. Institut für Robotik und Systemdynamik, DLR, Oberpfaffenhofen, 1993.

# 6.4 Flight Dynamics and Control Problems of Two-Stage-to-Orbit Vehicles

## 6.4.1 Flight Tests and Simulation Experiments for Hypersonic Long-Term Dynamics Flying Qualities

Robert Stich, Timothy H. Cox, and Gottfried Sachs *

Nomenclature

- $\boldsymbol{A}(s)$ coefficient matrix of homogeneous system
- $\boldsymbol{B}$ scaling matrix of control inputs
- $C_D$ drag coefficient
- $C_{D_V}$ $= \partial C_D/\partial(V/V_0)$
- $C_{D_\alpha}$ $= \partial C_D/\partial \alpha$
- $C_{D_\delta}$ $= \partial C_D/\partial \delta_e$
- $C_L$ lift coefficient
- $C_{L_V}$ $= \partial C_L/\partial(V/V_0)$
- $C_{L_\alpha}$ $= \partial C_L/\partial \alpha$
- $C_{L_\delta}$ $= \partial C_L/\partial \delta_e$
- $C_{m_h}$ $= (1/\rho_h)\partial C_m/\partial h$
- $C_{m_q}$ $= 2\partial C_m/\partial(q\bar{c}/V_0)$
- $C_{m_V}$ $= \partial C_m/\partial(V/V_0)$
- $C_{m_\alpha}$ $= \partial C_m/\partial \alpha$
- $C_{m_{\dot\alpha}}$ $= 2\partial C_m/\partial(\dot{\alpha}\bar{c}/V_0)$
- $C_{m_\delta}$ $= \partial C_m/\partial \delta_e$
- $\bar{c}$ mean aerodynamic chord
- $g$ acceleration due to gravity
- $h$ altitude
- $i_y$ radius of gyration
- $k_\rho$ factor, $k_\rho = -g/(\rho_h V_0^2)$
- $M$ Mach number
- $m$ mass
- $n_h$ thrust/altitude dependence, $n_h = (1/\rho_h)\partial(T/T_0)/\partial h$
- $n_V$ thrust/speed dependence, $n_V = (V_0/T_0)\partial T/\partial V$
- $S$ reference area
- $s$ Laplace operator
- $s_h$ height mode eigenvalue
- $T$ thrust, time constant
- $t$ time
- $\boldsymbol{u}$ control vector
- $V$ speed
- $\boldsymbol{x}$ state vector
- $z_T$ thrust lever arm
- $\alpha$ angle of attack
- $\gamma$ flight path angle
- $\Delta$ denoting a perturbation, e.g., $\Delta V$
- $\delta_a$ roll control
- $\delta_e$ pitch control
- $\delta_r$ yaw control
- $\delta_T$ throttle position
- $\zeta_p$ phugoid damping ratio
- $\zeta_{sp}$ short period damping ratio
- $\mu$ relative mass parameter, $\mu = 2m/(\rho S\bar{c})$
- $\rho$ air density
- $\rho_h$ density gradient, $\rho_h = -(1/\rho_0)\mathrm{d}\rho/\mathrm{d}h$
- $\sigma$ real part
- $\tau$ reference time, $\tau = \mu\bar{c}/V_0$
- $\phi$ roll angle
- $\chi_a$ azimuth angle
- $\omega$ imaginary part
- $\omega_p$ undamped natural phugoid frequency
- $\omega_{sp}$ undamped natural short period frequency

* Technische Universität München, Lehrstuhl für Flugmechanik und Flugregelung, Boltzmannstr. 15, 85748 Garching; sachs@lfm.mw.tum.de

Abstract

In hypersonic flight, there are unique aircraft dynamics and flying qualities properties which substantially differ from the characteristics of vehicles in the conventional speed regime. Longitudinal stability and control problems are considered, with particular emphasis put on long-term dynamics. This concerns inherent dynamics properties in regard to the modes of motion as well as flight path control. Results from flight tests and simulation experiments are presented.

### 6.4.1.1 Introduction

New concepts are considered for enhancing performance and improving productivity of space transportation systems [1–4]. A promising concept features a two-stage-to-orbit vehicle consisting of a carrier and an orbital stage. The carrier stage is equipped with a lifting wing and an airbreathing propulsion system so that it is capable of an airborne flight at hypersonic speed.

The hypersonic flight regime poses challenging problems in many areas. This also concerns the dynamics of the vehicle and its flying qualities. There are substantial differences when compared to characteristics of conventional aircraft. Accordingly, unique stability and control characteristics exist in hypersonic flight. This is due to several reasons. One is the high kinetic energy level which vehicles have when flying at hypersonic speed [5]. This substantially changes the interrelation between speed and altitude variations in a disturbed state. Another reason concerns the aerodynamic forces and moments which show specific properties in the hypersonic regime. A further issue is the strong interaction between vehicle dynamics and the engine [6]. The engine-related forces and moments are significantly larger in hypersonic flight than in the conventional speed regime.

The differences between the hypersonic and the conventional speed regimes also concern the long-term dynamics of aerospace planes. As an example, an additional mode of motion termed height mode exists which is not known in the conventional speed regime. Other specific characteristics of long-term dynamics in hypersonic flight are related to the control and response behaviour of aerospace planes. There are substantial effects concerning both control and stability [4].

This paper reports on results from a flight test and simulator experiment programme dealing with the addressed vehicle dynamics and flying qualities issues. The programme is part of a joint research effort of the Institute of Flight Mechanics and Flight Control of the Technische Universität München and the NASA Dryden Flight Research Center within the framework of the research at the Sonderforschungsbereich 255 "Transatmosphärische Flugsysteme" of the Deutsche Forschungsgemeinschaft. The research aircraft and flight simulation facilities of the NASA Dryden Flight Research Center provide excellent and unique opportunities for conducting flying qualities research for aerospace planes.

#### 6.4.1.2 Hypersonic Flight Dynamics

For a rather general treatment of the problems in mind, analytical expressions are introduced which can be derived from the equations describing the dynamics of the vehicle and its response characteristics. For this purpose, reference is made to the linearized equations of motion for longitudinal flight at hypersonic speed:

$$\boldsymbol{A}(s)\boldsymbol{x}(s) = \boldsymbol{B}\,\boldsymbol{u}(s) \tag{1}$$

where

$$\boldsymbol{x}(s) = [\Delta V/V_0, \Delta a, \Delta h/(\mu\bar{c})]^T$$

$$\boldsymbol{u}(s) = [\delta_e, \delta_T]^T$$

$$\boldsymbol{A}(s) = \begin{bmatrix} s\tau + \left(2 + \frac{C_{D_V}}{C_D} - n_V\right)C_D & C_{D_a} & s\tau^2 g/V_0^2 + \mu\bar{c}\rho_h C_D(1 - n_h) \\ 2C_L + C_{L_V} & \begin{matrix} C_{L_a} + C_D \\ -(s\tau)^2(i_y/\bar{c})^2 \\ + s\tau(C_{m_q} + C_{m_{\dot{a}}}) + \mu C_{m_a} \end{matrix} & \begin{matrix} -(s\tau)^2 + \mu\bar{c}\rho_h C_L \\ -(s\tau)^3(i_y/\bar{c})^2 + (s\tau)^2 C_{m_q} \\ + \mu^2\bar{c}\rho_h C_{m_h} \end{matrix} \\ \mu C_{m_V} & & \end{bmatrix}$$

$$\boldsymbol{B} = -\begin{bmatrix} C_{D_\delta} & -C_D \\ C_{L_\delta} & 0 \\ \mu C_{m_\delta} & \mu(z_T/\bar{c})C_D \end{bmatrix}$$

The dynamic system described by Eq. (1) accounts for the effects of altitude dependent forces by introducing the atmospheric density gradient $\rho_h$ which exerts a significant influence in hypersonic flight but can be ignored in the conventional speed regime [7]. At larger hypersonic speeds at about $M \approx 10$ and above, the curvature and rotation of the Earth are also of influence on the dynamics of the vehicle. However, this is not included in the following.

First, the eigenvalues are addressed. Only those characteristics will be considered which are unique for hypersonic flight. Due to the density gradient effect $\rho_h$, there are three modes of motion (as compared with two modes in the conventional speed regime):

- short period mode,
- phugoid,
- height mode.

The short period mode describes the short-term dynamics of the aircraft with inherent stability and shows similar characteristics as conventional vehicles when considering the effects of aerodynamic stability derivatives and c.g. positions. The following expressions apply for vehicles with inherent stability

$$\omega_{sp}^2 \approx -\left(\frac{V_0}{i_y}\right)^2 \frac{C_{L_\alpha}}{\mu}\left(\frac{C_{m_\alpha}}{C_{L_\alpha}} + \frac{C_{m_q}}{\mu}\right), \quad 2\zeta_{sp}\omega_{sp} \approx -\frac{V_0}{\mu\bar{c}}\left[C_{L_\alpha} - \left(\frac{\bar{c}}{i_y}\right)^2 (C_{m_q} + C_{m_{\dot\alpha}})\right] \tag{2}$$

The phugoid shows unique characteristics in hypersonic flight. It may be described by

$$\omega_p^2 \approx -g\rho_h(1+2k_\rho), \quad 2\zeta_p\omega_p \approx (1 - n_h + k_\rho n_V)\frac{C_D}{C_L}\frac{g}{V_0} \tag{3}$$

where

$$k_\rho = -\frac{g/\rho_h}{V_0^2} \tag{4}$$

The quantity $k_\rho$ is introduced as a measure for weighting the speed effect $V_0$ on phugoid dynamics vs. the density gradient effect $\rho_h$.

Fig. 6.4.1.1 shows that the relation

$$k_\rho \ll 1 \tag{5}$$

holds for the whole hypersonic regime, at all altitudes. The relation $k_\rho \ll 1$ basically means that the speed effect is small when compared with the density gradient effect. This is a general property of hypersonic flight, independent of altitude. It is due to the high kinetic energy and the fact that the density gradient shows little change with altitude, Fig. 6.4.1.2.

Because of $k_\rho \ll 1$, the relations Eq. (2) reduce to $(n_h \approx 1)$

$$\omega_p^2 \approx -g\rho_h, \quad \zeta_p \approx 0 \tag{6}$$

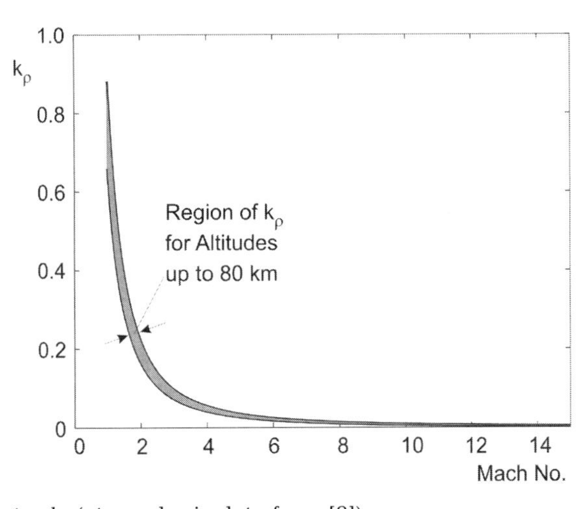


Figure 6.4.1.1: Factor $k_\rho$ (atmospheric data from [8]).

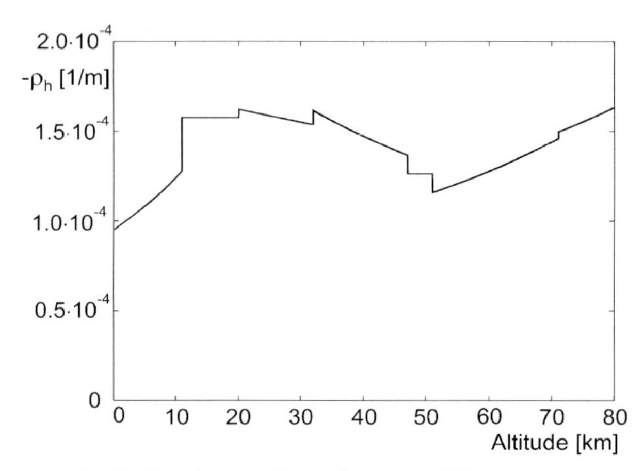


Figure 6.4.1.2: Atmospheric density gradient (data from [8]).

From Eq. (6) and the $\rho_h$ properties shown in Fig. 6.4.1.2 it follows that the phugoid frequency is rather constant in hypersonic flight, practically independent of altitude.

The height mode which is not existent in the conventional speed regime or can be neglected is due to the effects of the atmospheric density and speed-of-sound gradients on aircraft dynamics. This low frequency eigenvalue may be expressed as

$$s_h \approx (n_V - 2n_h) \frac{C_D}{C_L} \frac{g}{V_0} \tag{7}$$

#### 6.4.1.3 Research Aircraft and Flight Simulator

The test facilities used at the NASA Dryden Flight Research Center are a SR-71 aircraft, an operational SR-71 simulator and a dedicated flight simulator for hypersonic vehicles.

The SR-71 aircraft (Fig. 6.4.1.3) is a vehicle capable of sustained supersonic cruise at a Mach number greater than three and at altitudes higher than 70 000 ft. This vehicle is, for sustained flight, the highest flying, fastest aircraft in the world. Thus, it provides unique testing possibilities which no other aircraft can offer.

The addressed capability of the SR-71 for a flight at a Mach number greater than three provides an excellent possibility to conduct flight tests on hypersonic flying qualities issues and to address hypersonic simulation validity. The reason for this application is that there are dynamics similarities between the high supersonic and the hypersonic regions.

Figure 6.4.1.3: NASA Research Aircraft SR-71 (the aircraft shown in the photo was used in the flight tests).

The SR-71 aircraft used in the experimental programme is especially equipped with a measurement and instrumentation system for flight testing at Mach 3 flight conditions and beyond. It provides a test bed for a variety of experiments relevant for aerospace planes. Among these are tests on dynamic problems in hypersonic flight such as control and flying qualities issues, the near ultraviolet spectrometer and other research topics, and even potentially the separation of two-stage systems [9].

In addition to the aircraft used for flight tests, an operational SR-71 flight simulator was available in the experimental programme. It was used in preparation for flight testing as well as for conducting separate simulator experiments on the flying qualities issues addressed above.

The other flight simulator used in the experiments features a fixed base cockpit equipped with adequate instrumentation devices and computers for simulating hypersonic vehicle dynamics, Fig. 6.4.1.4. Control force characteristics are generated by electrically driven control loaders. The cockpit instrumentation is based on a configuration used for the Space Shuttle. It employs a central field and a simulated head up display. The simulated head up display which was especially developed for the flying qualities experiments in mind features indicator elements appropriate for hypersonic flight [10].

For simulating the aerodynamics and powerplant characteristics of an aerospace plane, the six-degree-of-freedom model GHAME (Generic Hypersonic Aerodynamic Model Example) was used [11]. It comprises a complex mathematical model for describing the aerodynamic characteristics of a single-stage hypersonic flight system for a Mach number range up to $M = 25$. The powerplant system is modelled as a combination of turbo, ram and scram jet engines.

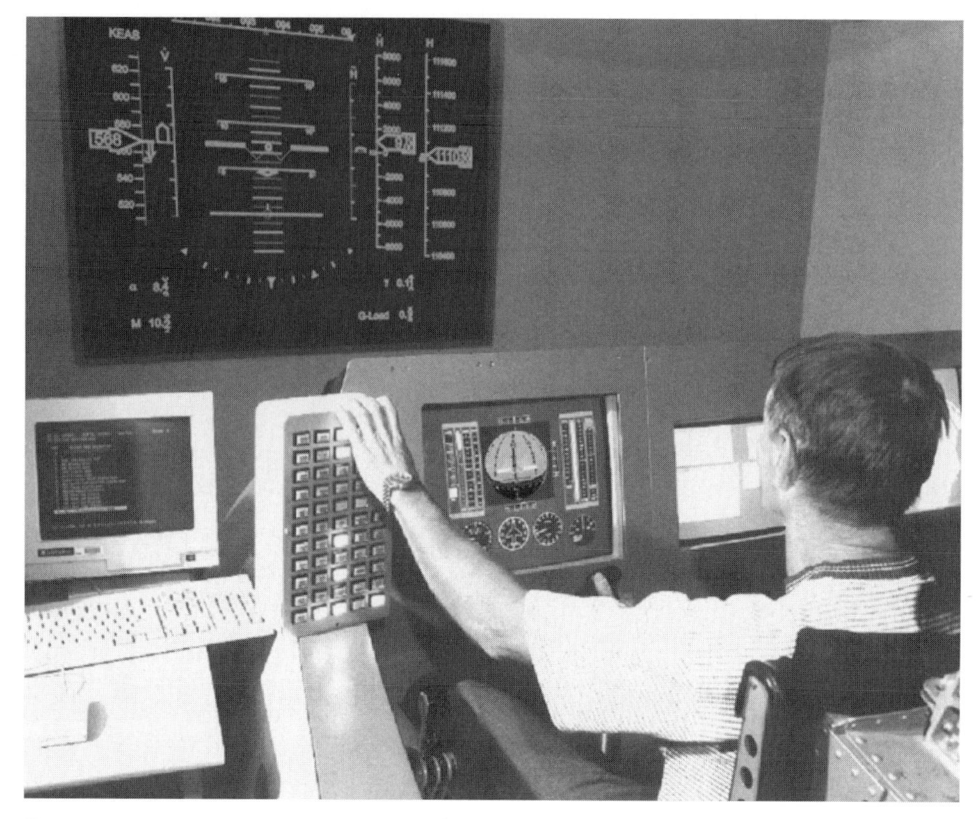


Figure 6.4.1.4: Cockpit of hypersonic flight simulator.

The basic dynamics characteristics of the simulated aerospace plane correspond to Level 1 conditions when applying flying qualities requirements as in [12]. Two types of flight control systems may be used for longitudinal dynamics. One comprises feedback of pitch rate to the pitch control surface. It represents the basic structure of a pitch damper. The other control system is of the type rate command / attitude hold.

#### 6.4.1.4 Results

Results concerning the long-term dynamics eigenvalues are presented in Fig. 6.4.1.5, for the YF-12 [13] and for the GHAME configuration [10]. Basically, the relations given by Eqs. (3) and (6), respectively, are confirmed by the flight test and computation data presented in Fig. 6.4.1.5. There is only a small change in the frequency of the phugoid when comparing the super- and hypersonic cases. Further, the damping of the phugoid is close to zero for all flight

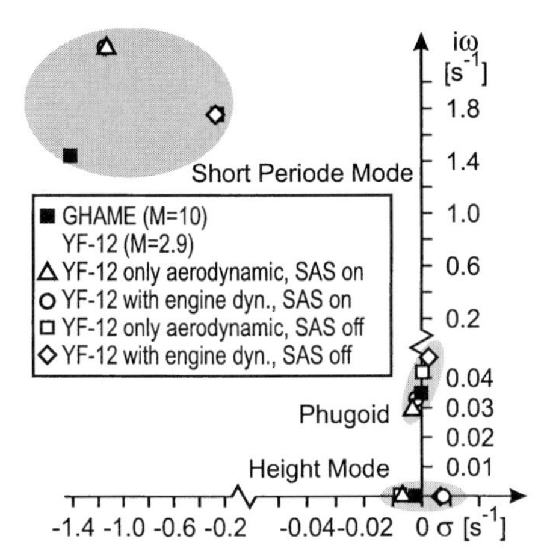


Figure 6.4.1.5: Phugoid and height mode characteristics in super- and hypersonic flight (YF-12 data from [14] and GHAME data from [15]).

conditions considered. The height mode eigenvalue shows only small values, indicating a slow aperiodic mode of motion. The fact that the long-term eigenvalues of the super- and the hypersonic aircraft compare well with each other supports the above argument that there is a dynamics similarity between the related speed regions.

Three manoeuvres which are considered as relevant for hypersonic cruise vehicles (e.g. [14]) have been selected:

- horizontal turn,
- vertical plane altitude change,
- ascent turn.

The horizontal turn manoeuvre proved to be a demanding task [15]. The pilot has to perform a heading change of 12° at constant equivalent airspeed and constant altitude (554.8 kt and 110000 ft, corresponding to $M = 10$), followed by a flight at the final heading for 1 minute. For some runs the pilot had to account for an initial vertical speed of 20 ft/s, others started with a density perturbation right at the beginning. After each flight the pilot answered a questionnaire concerning longitudinal handling qualities and gave a Cooper-Harper Rating [16] for longitudinal long-term dynamics. Table 6.4.1.1 shows the refer-

Table 6.4.1.1: Controlled quantities and performance values for horizontal turn manoeuvre.

| Controlled Quantities | Adequate Performance | Desired Performance |
|---|---|---|
| Target Heading | ±1° | ±0.5° |
| Target Altitude | ±600 ft | ±300 ft |
| Trim Speed (EAS) | ±10 kt | ±5 kt |

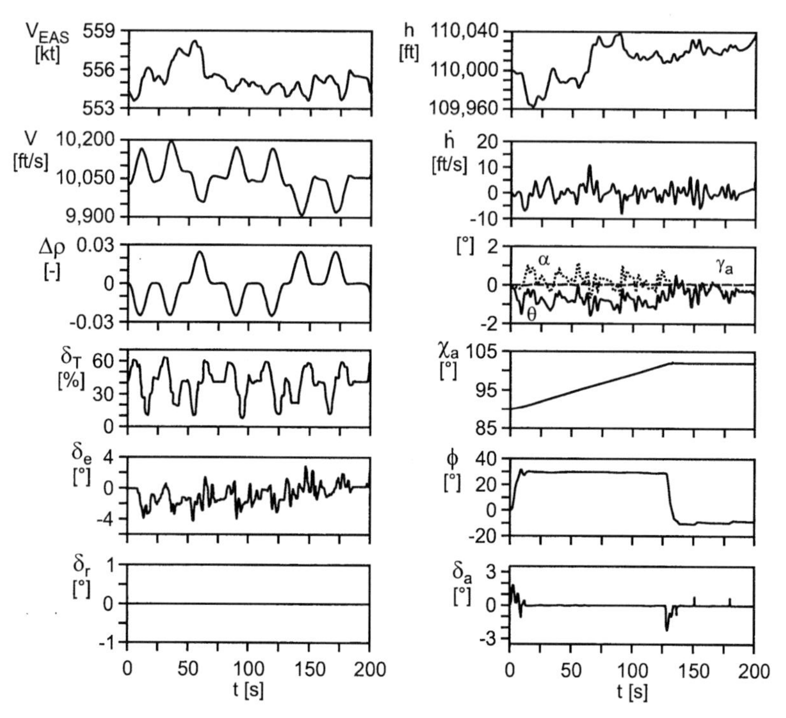


Figure 6.4.1.6: State and control variables for a horizontal turn manoeuvre (GHAME model, $M=10$, $h=110000$ ft).

ence values for adequate and desired pilot performance for the horizontal turn task with respect to Cooper-Harper Ratings. Results typical for the horizontal turn manoeuvre experiments are shown in Fig. 6.4.1.6.

The horizontal turn was used as a flight manoeuvre to investigate the effects of aircraft properties on flying qualities, like changed stability characteristics of the long-term modes. Results for nominal stability characteristics from flight tests and simulation experiments are presented in Fig. 6.4.1.7. This Fig. shows pilot ratings obtained in flight tests with the SR-71 and in GHAME simulator experiments. The values from the flight tests compare well with those from the simulator experiments, thus again confirming that there is a dynamics similarity.

Another issue of concern is a flight path control problem in hypersonic flight. It relates to the correlation between flight path and pitch attitude changes following a pitch control input. For conventional aircraft, there is usually a well balanced correspondence between path and attitude responses which may be qualified as path-attitude consonance or path-attitude coupling. It concerns the lag of the flight path angle for a step change in pitch attitude. In hypersonic flight, there can be a large flight path lag of the order of 25 seconds or more, yielding a characteristic which may be qualified as path-attitude decoupling.

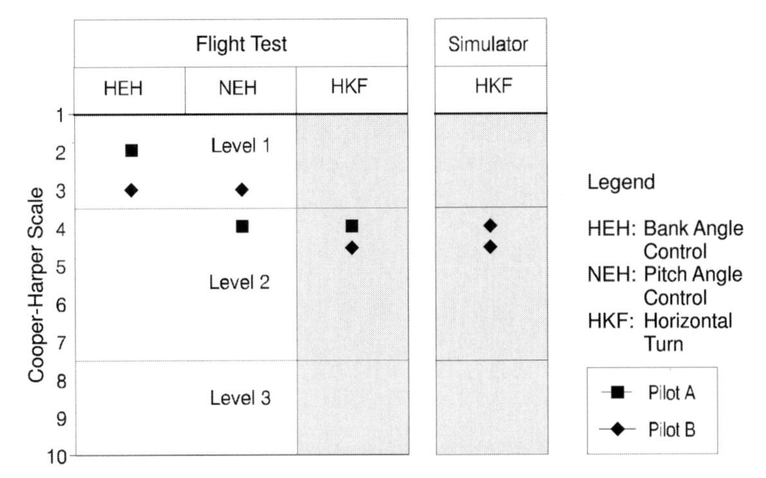


Figure 6.4.1.7: Results on pilot ratings for flight tests with SR-71 and simulator experiments.

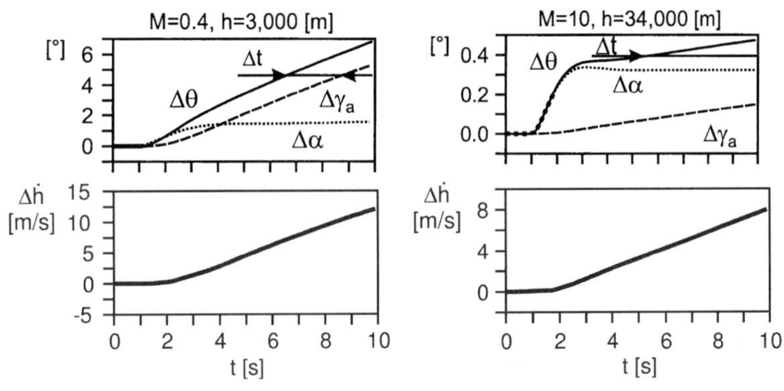


Figure 6.4.1.8: Responses to step pitch control inputs in subsonic flight ($M$=0.4, $h$=3400 m) and in hypersonic flight (GHAME model, $M$=10, $h$=37400 m).

The path-attitude decoupling problem is illustrated in Fig. 6.4.1.8. In the subsonic case the time lag between pitch attitude and flight path is small, i.e. the pilot can use both quantities synonymously. By contrast, the time lag in the hypersonic case is much larger. Furthermore, Fig. 6.4.1.8 shows that the size of the flight path angle in relation to the pitch attitude is significantly decreased in the hypersonic case.

For describing the time lag, the numerator zero $1/T_{\theta 2}$ of the pitch attitude to pitch control transfer function is used. The development of this quantity in the sub-, super- and hypersonic regimes is illustrated in Fig. 6.4.1.9 for a variety of aircraft. From the results presented in Fig. 6.4.1.9 it follows that the large values $T_{\theta 2}$ are a general characteristic of super- and hypersonic flight.

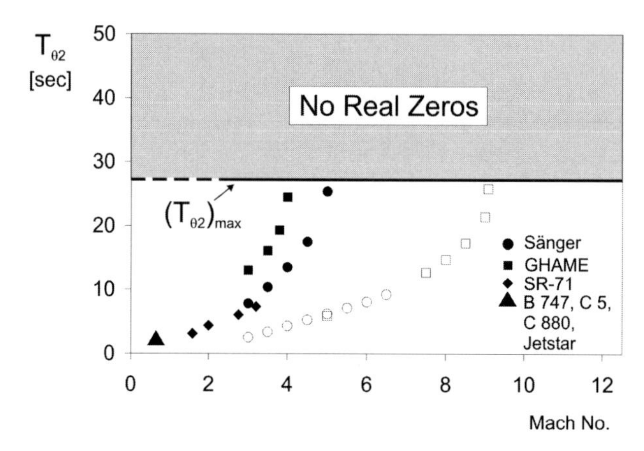


Figure 6.4.1.9: Development of $T_{\theta 2}$ in the hypersonic region; symbols filled: flight at maximum lift-to-drag ratio; symbols open: high dynamic pressure flight ($q_{max}$ = 50 kPa).

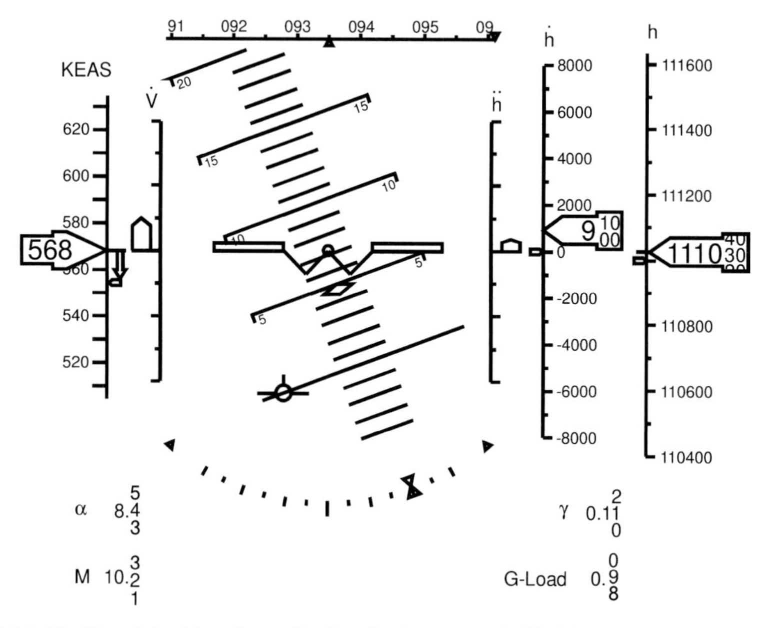


Figure 6.4.1.10: Simulated head up display for hypersonic flight.

The path-attitude decoupling problem was subject of dedicated simulations experiments. A simulated head up display was used which was customized for hypersonic flight (Fig. 6.4.1.10). In the centre of the head up display, there is a high resolution pitch ladder which moves with respect to a fixed airplane symbol (⌐W¬). On the pitch ladder a diamond shaped symbol (◇) moves up and down and indicates vertical speed and vertical acceleration. This

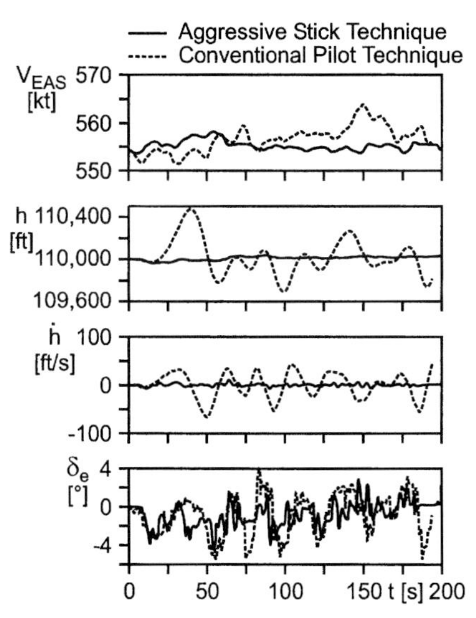


Figure 6.4.1.11: Comparison of time histories for horizontal turn with conventional and aggressive stick control technique.

indicator is based on experience with the YF-12 and SR-71 and proved to be important for manual control. Referenced to linear scales, vertical acceleration, vertical speed, altitude, airspeed and longitudinal acceleration are indicated.

Using vertical speed indication for flight path control yields better results with lower pilot workload. One test pilot called this procedure aggressive stick technique. Results are presented in Fig. 6.4.1.11 which provides a comparison of the conventional control and the aggressive stick techniques. It was possible to cope with large path-attitude lag values, given a precise vertical speed indicator without time lag [17].

#### 6.4.1.5 Conclusions

Aircraft dynamics and flying qualities properties unique for hypersonic flight are considered, with particular emphasis put on long-term dynamics. The characteristics of the phugoid and the height mode are described. The flight path control problem due to path-attitude decoupling is addressed. Results from flight tests and simulation experiments are presented.

References to Section 6.4.1


1. Covault, C.: 'Global Presence' Objective Drives Hypersonic Research. Aviation Week & Space Technology, Vol. 150, No. 14, pp. 54–55, 1999.
2. Grallert, H.: Synthesis of a FESTIP Air-Breathing TSTO Space Transportation System. Paper of 9th International Space Planes and Hypersonic Systems and Technologies Conference, Norfolk, AIAA-99-4884, 1999.
3. Taniguchi, H.: Concept Study on Winged Vehicle. Paper of 9th International Space Planes and Hypersonic Systems and Technologies Conference, Norfolk, AIAA-99-4804, 1999.
4. Morring, F. Jr.: NASA Drawing New Launch Road Map. Aviation Week & Space Technology, Vol. 158, No. 4, 2003.
5. Stich, R., Sachs, G.: Unconventional Stability and Control Characteristics in Hypersonic Flight. UNCONF '97 "Unconventional Flight Analysis", Book 1, Selected Paper of the First International Conference on Unconventional Flight, Published by the Department of Aircraft and Ships and R-Group Ltd, Budapest, ISBN 963 420 699 9, pp. 17–33, 1999.
6. McRuer, D.: Design and Modelling Issues for Integrated Airframe/Propulsion Control of Hypersonic Flight Vehicles. Proceedings of the American Control Conference, Boston, Mass., pp. 729–735, 1991.
7. Sachs, G.: Längsstabilität im Überschall- und Hyperschallflug. Zeitschrift für Flugwissenschaften, Vol. 24, No. 6, pp. 310–329, 1976.
8. Committee for Extension to the Standard Atmosphere, U. S. Standard Atmosphere, 1976, U.S. Government Printing Office, Washington, DC, 1976.
9. Lockheed SR-71, Supersonic/Hypersonic Research Facility. Researcher's Handbook, Vol. I Executive Summary, Vol. II Technical Description, 1992.
10. Stich, R., Sachs, G., Cox, T.: New Longitudinal Handling Qualities Criterion for Unstable Hypersonic Vehicles. Proceedings of the AIAA Atmospheric Flight Mechanics Conference and Exhibit, 11.–14. August 2003, Austin, Texas, ISBN 1-56347-638-X, AIAA 2003-5309, pp. 1–11, 2003.
11. Bowers, A., Noffz, G., Gonda, M., Iliff, K.: A Generic Hypersonic Aerodynamic Model Example (GHAME). Edwards: NASA Dryden Flight Research Facility, 1989.
12. N.N.: MIL-HDBK-1797 – Flying Qualities of Piloted Aircraft, 1997.
13. Powers, B. G.: Phugoid Characteristics of a YF-12 Airplane With Variable-Geometry Inlets Obtained in Flight Tests at a Mach Number of 2.9. NASA Technical Paper 1107, 1977.
14. Berry, D.: National Aerospace Plane Flying Qualities Task Definition Study. NASA Dryden Flight Research Centre, Edwards. NASA-Technical-Memorandum 100452, 1988.
15. Sachs, G., Knoll, A., Stich, R., Cox, T.: Simulations- und Flugversuche über Flugeigenschaften von Hyperschall-Flugzeugen. Zeitschrift für Flugwissenschaften und Weltraumforschung, Vol. 19, No. 1, pp. 61–70, 1995.
16. Harper, R., Cooper, G.: Handling Qualities and Pilot Evaluation. Journal of Guidance, Control, and Dynamics, Vol. 9, No. 5, pp. 515–529, 1986.
17. Stich, R., Sachs, G., Cox, T.: Path-Attitude Inconsonance in High Speed Flight and Related Path Control Issues. Proceedings of the 22nd Congress of Aeronautical Sciences, Harrogate, 27–31 August 2000, Les Mureaux Cedex: International Council of the Aeronautical Sciences, 2000, ICAS-2000-4.10.4.

### 6.4.2 Wind Tunnel Tests for Modelling the Separation Dynamics of a Two-Stage-to-Orbit Vehicle

Christian Zähringer and Gottfried Sachs *

#### Nomenclature

$C_l$ rolling moment coefficient, $C_l = 2L/(\rho V^2 Ss)$
$C_m$ pitching moment coefficient, $C_m = 2M/(\rho V^2 Sl_\mu)$
$C_n$ yawing moment coefficient, $C_n = 2N/(\rho V^2 Ss)$
$C_X$ longitudinal force coefficient, $C_X = 2X/(\rho V^2 S)$
$C_Y$ side force coefficient, $C_Y = 2Y/(\rho V^2 S)$
$C_Z$ vertical force coefficient, $C_Z = 2Z(\rho V^2 S)$
$h$ Vertical distance between orbital and carrier stages
$l_{EOS}$ reference length
$M_\infty$ Mach number
$\alpha$ angle of attack
$\beta$ sideslip angle
$\Delta$ denoting a perturbation, e.g., $\Delta\alpha$
$\phi$ bank angle

#### Abstract

Results from wind tunnel tests concerning the separation of a two-stage-to-orbit vehicle are presented. Force and moment characteristics of the orbital stage show significant aerodynamic interference effects between the orbital and the carrier stages. This concerns changes in magnitude as well as dependencies which are solely due to the interference.

#### 6.4.2.1 Introduction

The separation manoeuvre of two-stage-to-orbit vehicles poses challenging problems in the field of flight mechanics [1–3]. An important goal is to achieve a safe separation manoeuvre. For dealing with the flight mechanics problem, an aerodynamics model is required which accounts for the complex flow phenomena during separation. In particular, the aerodynamic interference effects

* Technische Universität München, Lehrstuhl für Flugmechanik und Flugregelung, Boltzmannstr. 15, 85748 Garching; sachs@lfm.mw.tum.de

due to the proximity of the two stages are of concern [4]. For this purpose, wind tunnel tests were conducted to determine the aerodynamic forces and moments during separation in hypersonic flight.

The wind tunnel tests on stage separation are a joint research effort of the Institute of Fluid Mechanics and the Institute of Flight Mechanics and Flight Control of the Technische Universität München as well as the Collaborative Research Centre 253 "Fundamentals of Space Plane Design" at RWTH Aachen. The wind tunnel tests were conducted at the Institute of Theoretical and Applied Mechanics of the Russian Academy of Sciences, Siberian Branch, Novosibirsk, Russia.

#### 6.4.2.2 Test Facility and Wind Tunnel Models

As test facility, the T-313 wind tunnel of the Institute of Theoretical and Applied Mechanics was used. This is a blowdown wind tunnel which operates on dried air compressed in large gas holders. The facility is composed of input pipelines with pressure adapter, plenum chamber, nozzle block, pressure chamber with test section, supersonic diffuser, two ejectors, and output pipeline, from which the air is fed to a noise damping chamber before it is released into the atmosphere. The size of the rectangular test section is $0.6\,\mathrm{m}\times 0.6\,\mathrm{m}\times 2\,\mathrm{m}$. The test section Mach number can be varied between $M_\infty = 2\text{–}6$ by means of a plane contoured nozzle with inserts of different throat sizes.

Carrier and orbital stage models of 1/150 scale are used, based on the hypersonic ELAC1C/EOS vehicles of the Collaborative Research Centre 253 "Fundamentals of Space Plane Design" at RWTH Aachen, Fig. 6.4.2.1. ELAC1C represents the carrier stage and EOS the orbital stage. The carrier stage ELAC1C has a delta wing planform with a leading-edge sweep of 75°, an elliptical forebody and afterbody, the latter with integrated winglets. The EOS vehicle is a wing-body configuration with a centre-line fin. The configuration consists of an elliptical nose and forebody, a centre body with half cylindrical upper side, wing strakes of elliptical cross sections, and a highly swept main wing and fin showing standard NACA profiles.

The stage separation is simulated in a quasi-steady manner by mounting the orbital stage at different vertical positions $h/l_{EOS}$ as well as at different angles of attack $\Delta\alpha$, sideslip $\Delta\beta$ and bank $\Delta\Phi$ relative to the carrier stage. The corresponding coordinate systems ($x_{ELAC1C}$, $y_{ELAC1C}$, $z_{ELAC1C}$) and ($x_{EOS}$, $y_{EOS}$, $z_{EOS}$) are shown in Fig. 6.4.2.1.

The separation test bench is presented in Fig. 6.4.2.2. Main elements are referred to numbers used below. The carrier stage model is mounted by means of a tail sting. An automatic four-component mechanical balance is employed to measure the loads acting on the carrier stage model ELAC1C. This balance is fitted with an arm system to decompose the total forces and moments of the balance coordinate system into the aerodynamic components lift, drag, pitching moment and rolling moment.

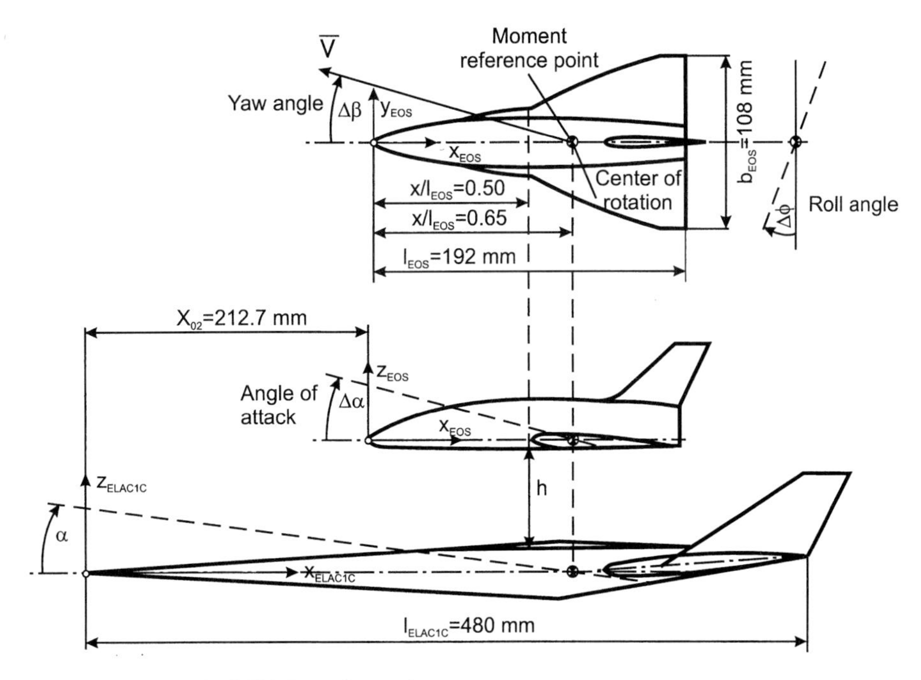


Figure 6.4.2.1: ELAC1C/EOS configuration.

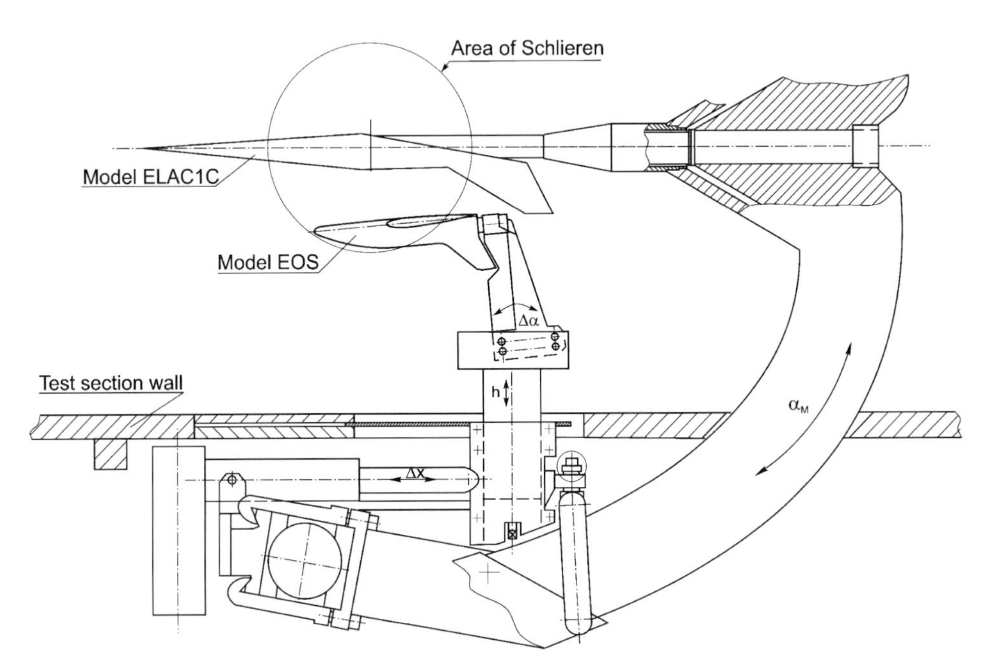


Figure 6.4.2.2: Separation test bench.

The orbital stage model is attached to a pylon. An internal six-component strain-gage balance is used for determining the loads acting on the orbital stage model. The construction of this balance is made in form of a tail sting measuring all components of the aerodynamic forces (drag, side force, lift) and moments (rolling moment, pitching moment, yawing moment) in the model-fixed coordinate system.

#### 6.4.2.3 Results

In the following, results from the wind tunnel tests on aerodynamic interference effects with regard to the orbital stage are presented. Reference is made to the complete report about the test programme [5]. Two groups of aerodynamic interference effects are considered. One group relates to force and moment dependencies which are changed due to the interference. The other group concerns force and moment dependencies which are usually not existent but introduced by aerodynamic interference.

*Longitudinal Characteristics*

The force and moment coefficients for the longitudinal motion belong to the first group addressed. Results are presented in Figs. 6.4.2.3–6.4.2.5. Basically, the forces in the longitudinal and normal directions as well as the pitching moment are significantly influenced by the aerodynamic interference. The longitudinal force coefficient shows an increase with the reduction of the distance be-

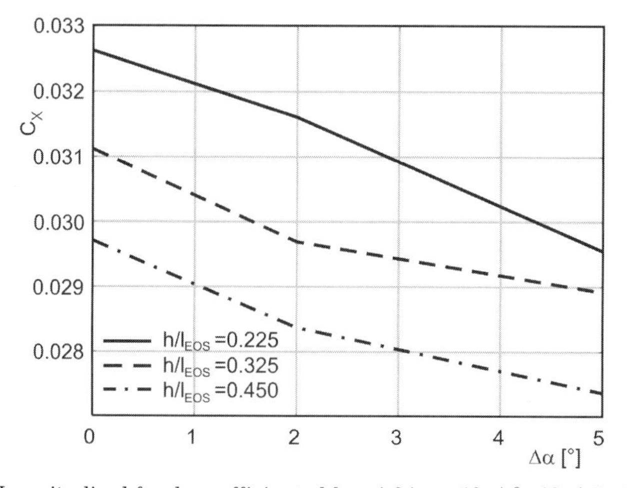


Figure 6.4.2.3: Longitudinal forch coefficient; $M_\infty=4.04$, $\alpha=1°$, $\Delta\beta=0°$, $\Delta\Phi=0°$.

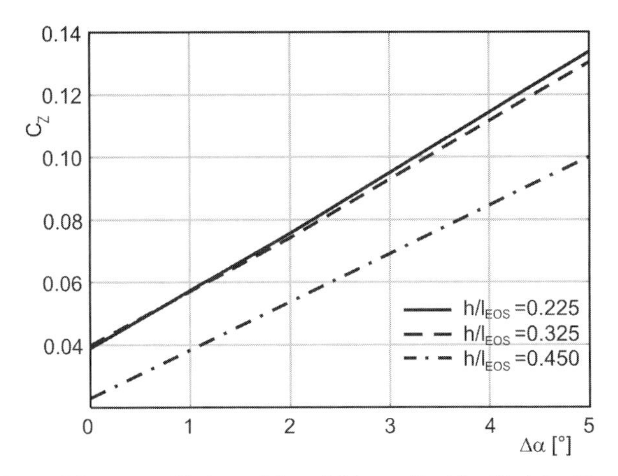


Figure 6.4.2.4: Normal force coefficient; $M_\infty = 4.04$, $\alpha = 1°$, $\Delta\beta = 0°$, $\Delta\Phi = 0°$.

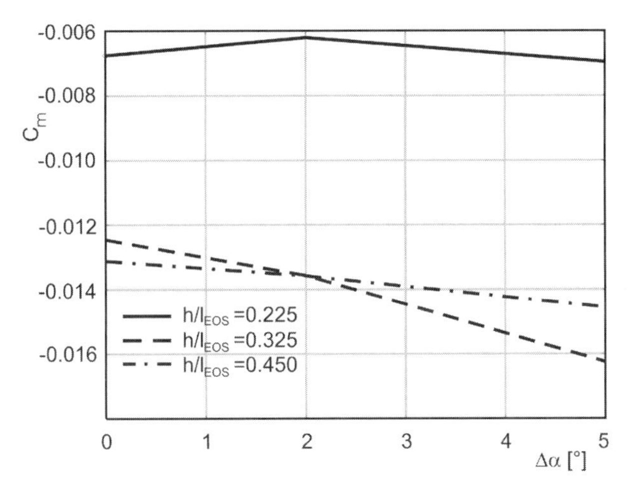


Figure 6.4.2.5: Pitching moment coefficient; $M_\infty = 4.04$, $\alpha = 1°$, $\Delta\beta = 0°$, $\Delta\Phi = 0°$.

tween the orbital and carrier stages (Fig. 6.4.2.3). This holds for the entire range considered for the relative angle of attack.

The vertical force coefficient shows a significant increase due to aerodynamic interference, Fig. 6.4.2.4. This effect stays at a constant level. The increase in the vertical force is advantageous with regard to flight mechanics. This is because it supports the separation process by leading to a faster moving away of the orbital stage. The advantage is particularly relevant for small angles of attack. The point in mind is addressed in Fig. 6.4.2.6 which shows two flight conditions of the orbital stage at the beginning of the separation. With regard to flight safety, any motion towards the carrier stage is not acceptable. This particularly concerns the rear of the orbital stage in case of a rotation for

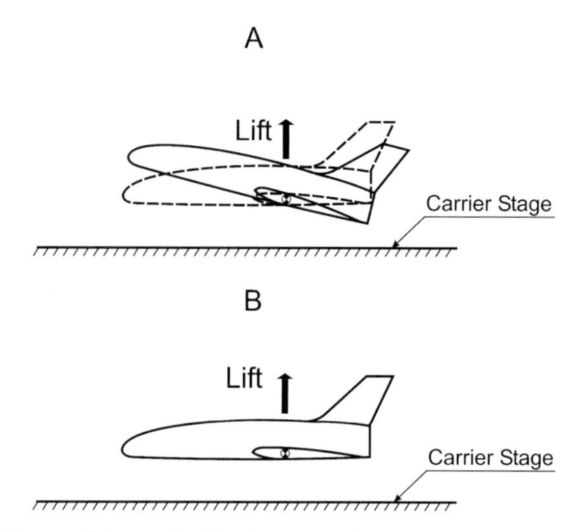


Figure 6.4.2.6: Flight conditions of orbital stage at begin of separation.

increasing the angle of attack, as shown in part A of Fig. 6.4.2.6. For avoiding such a situation, the availability of a vertical force without the necessity of rotation is supportive (part B of Fig. 6.4.2.6).

The pitching moment coefficient is shown in Fig. 6.4.2.5. There is a change in positive direction due to aerodynamic interference.

*Lateral-Directional Characteristics*

With regard to lateral-directional characteristics, there are dependencies which usually do not exist but are introduced by aerodynamic interference effects. Results are given in Figs. 6.4.2.7 and 6.4.2.8 which present the rolling and the yawing moment coefficients due to banking of the orbital stage relative to the carrier stage.

The aerodynamic interference causes a rolling moment which increases in magnitude with the bank angle (Fig. 6.4.2.7). Further inspection reveals that it acts in a direction opposite to the bank angle. This is an advantageous effect from a flight mechanics standpoint. It means that a restoring moment is produced when there is a bank angle disturbance between the orbital and the carrier stages. The described flight condition is illustrated in Fig. 6.4.2.9. The restoring moment effect is of interest with regard to the separation manoeuvre because of the proximity of the orbital and the carrier stages. The effect of the relative bank angle on the yawing moment is shown in Fig. 6.4.2.8. It also increases in size with the bank angle.

Other effects concern changes in the magnitude of lateral-directional force and moment characteristics due to the interference. Results are presented in

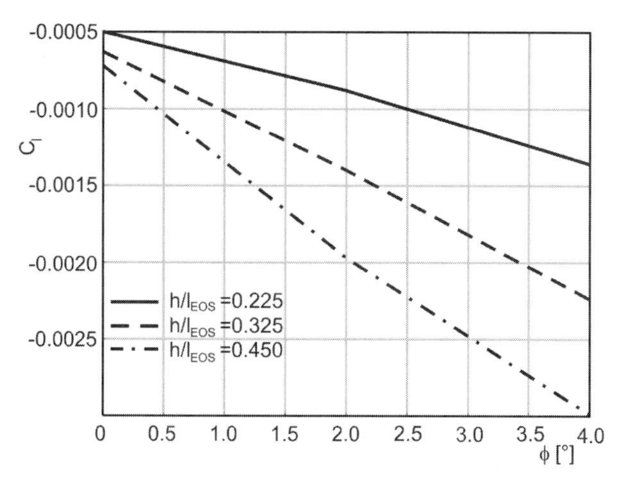


Figure 6.4.2.7: Rolling moment coefficient; $M_\infty = 4.04$, $\alpha = 1°$, $\Delta\alpha = 5°$, $\Delta\beta = 0°$.

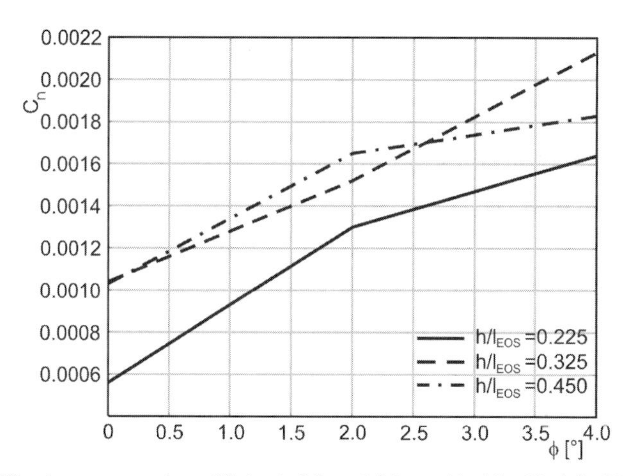


Figure 6.4.2.8: Yawing moment coefficient; $M_\infty = 4.04$, $\alpha = 1°$, $\Delta\beta = 0°$, $\Delta\beta = 0°$.

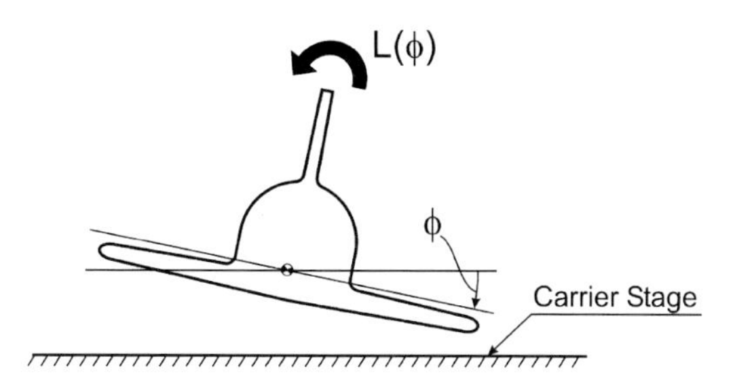


Figure 6.4.2.9: Roll moment due to banking.

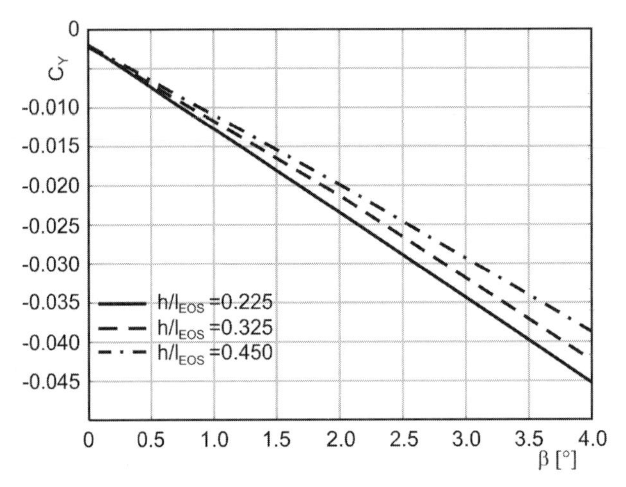


Figure 6.4.2.10: Side force coefficient; $M_\infty = 4.04$.

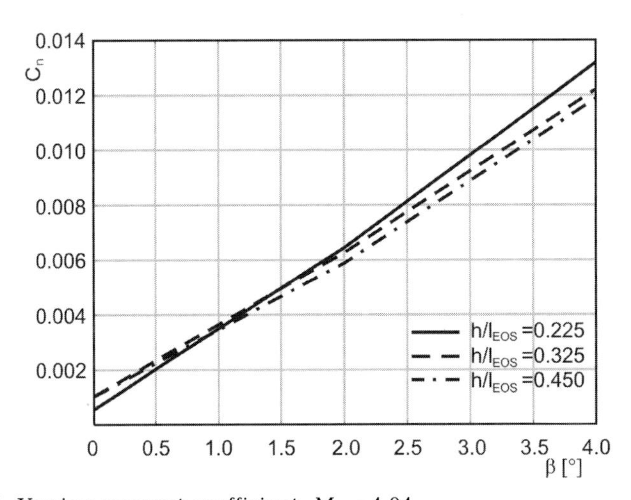


Figure 6.4.2.11: Yawing moment coefficient; $M_\infty = 4.04$.

Figs. 6.4.2.10 and 6.4.2.11 on the dependencies of the side force and the yawing moment on the sideslip angle. The amount of the side force is reduced for smaller distances between the orbital and the carrier stages (Fig. 6.4.2.10). The yawing moment shows a similar behaviour (Fig. 6.4.2.11).

### 6.4.2.4 Conclusions

Wind tunnel test results from a joint research programme of the Institute of Fluid Dynamics and the Institute of Flight Mechanics and Flight Control of the Technische Universität München on stage separation of two-stage-to-orbit vehicles are presented. The wind tunnel tests were conducted in the Institute of Theoretical and Applied Mechanics of the Russian Academy of Sciences, Siberian Branch, Novosibirsk. Models of the ELAC1C/EOS (1/150 scale) of the Collaborative Research Centre 253 "Fundamentals of Space Plane Design" at RWTH Aachen were used. Force and moment characteristics show significant interference effects. This concerns changes in magnitude as well as dependencies which are solely due to aerodynamic interference and otherwise not existent.

## References to Section 6.4.2


1. Schoder, W.: Untersuchung zur Steuerung und Regelung eines Hyperschall-Flugsystems beim Separationsmanöver. Dissertation, Technische Universität München, 1995.
2. Reubush, D., Martin, J., Robinson, J., Bose, D., Strovers, B.: Hyper-X Stage Separation – Simulation Development and Results. AIAA Paper 2001-1802, AIAA 10th International Space Planes and Hypersonic Systems and Technologies Conference, Kyoto, Japan, Apr. 2001.
3. Cvrlje, T., Breitsamter, C., Heller, M., Sachs, G.: Lateral Unsteady Aerodynamics and Dynamic Stability Effects in Hypersonic Flight. Proceedings of the International Conference on Methods of Aerophysical Research, Part I, (Novosibirsk, Russia), June 1998, pp. 44–54.
4. Cvrlje, T., Laschka, B: Simulation of an Orbiter Separation Manoeuvre at Hypersonic Speed. AIAA Paper 2001-1850, AIAA 10th International Space Planes and Hypersonic Systems and Technologies Conference, Kyoto, Japan, Apr. 2001.
5. Kharitonov, A., Brodetsky, M., Vasenyov, L., Adamov, N., Breitsamter, C., Heller, M.: Investigation of Aerodynamic Characteristics of the Models of a Two-Stage Aerospace System During Separation. Final Report, Institute of Theoretical and Applied Mechanics of the Russian Academy of Sciences, Siberian Division, and Institute of Fluid Dynamics and Institute of Flight Mechanics and Flight Control, Technische Universität München, Nov. 2000.

# 7 High-Temperature Materials and Hot Structures

## 7.1 Ceramic Matrix Composites – the Key Materials for Re-Entry from Space to Earth

Martin Frieß *, Walter Krenkel, Richard Kochendörfer, Rüdiger Brandt, Günther Neuer, and Hans-Peter Maier

### 7.1.1 Introduction and Overview

Based on more than 20 years of experience with polymer and metal matrix composites DLR decided to enter the field of ceramic matrix composites (CMCs) in the end of the eighties. To tackle very high temperature applications above 1200 °C – like re-entry or propulsion components – only carbon fibres were feasible as reinforcement. The main criteria in selecting a suitable manufacturing process were low cost aspects and a fast process to realize short component delivery times as well as high learning rates to enter not only the space but also the terrestrial market. A thorough screening of all processes known at that time lead to the conclusion, that a liquid silicon infiltration process provided the highest potential to meet these requirements.

It was well known that a reaction of silicon (Si) and carbon (C) to silicon carbide (SiC) cannot be avoided as soon as liquid Si contacts C. However, over the years it became feasible to localize and limit these contact areas and design the microstructure within the material according to the requirements. Here the activities within SFB 259 had a strong impact on the material development. Starting with one type of C-fibre and one carbonaceous matrix system (XP60), over the years almost all C-fibres with different sizings and more than ten matrix systems were evaluated, keeping in mind the main criteria: to use only commercially available fibres without coatings and to avoid any reinfiltration steps. Due to the fact that a variation of the manufacturing parameters, like

* DLR, Institut für Bauweisen und Konstruktionsforschung, Pfaffenwaldring 38–40, 70550 Stuttgart; martin.friess@dlr.de

*Basic Research and Technologies for Two-Stage-to-Orbit Vehicles*
DFG, Deutsche Forschungsgemeinschaft

ISBN: 3-527-27735-8

temperature, time, pressure, atmosphere, etc., have a strong impact on the microstructure and consequently on the properties of the material, a large variation of material qualities was feasible and investigated. Thus a thorough understanding of how to tailor the microstructure could be achieved. To furnish the partner institutes with a constant material quality for investigations of the thermomechanical and thermophysical properties a standard material was defined in each investigation period sponsored by DFG.

Finally the know-how to tailor the microstructure allowed on the one hand the manufacture of graded materials with different properties of the core and the surface material, respectively. On the other hand it allowed a process integrated in-situ joining technique with identical properties of the joint as well as in the basic material.

Concerning structural aspects it was clear that C-fibre based CMCs are rather multiphase materials which are sensitive to oxidation and can hardly be protected against high temperature cycling conditions in oxidative atmosphere. As a consequence the designer has to realize concepts to live with limited life time structures. For multi mission re-entry vehicles a replacement of the high temperature stable skin panels must be possible from the outside of the vehicle which requires a fastening system based on CMCs as well. This design philosophy was consequently realized and validated by ground tests in heated test chambers and plasma channels as well as in real re-entry flight tests.

## 7.1.2 Liquid Silicon Infiltration: Process Development

Ceramic matrix composites (CMCs) are comprised of carbon or ceramic fibres embedded in a ceramic matrix, combining the superior properties of ceramics such as low specific weight, chemical resistance, extreme high temperature stability and thermal shock resistance with a high damage tolerance. Especially for applications in aerospace under re-entry conditions, materials with such properties are required and used. On the other hand CMCs find more and more entry to non-aerospace areas.

For the manufacture of non-oxide CMCs three different processes are currently used: The CVI- (chemical vapour infiltration), the LPI- (liquid polymer infiltration) or PIP- (polymer infiltration and pyrolysis) as well as the LSI-process (liquid silicon infiltration). The LSI-process, developed at DLR in Stuttgart, is regarded as the most promising route for industrial products, especially if cost aspects are considered, as short processing times can be achieved and only cheap raw materials – carbon fibres, carbonaceous resins and silicon granulate (semiconductor quality) – are necessary [1].

The synthesis of the ceramic matrix of CMCs manufactured via the LSI-process is carried out by liquid phase infiltration of molten silicon into porous carbon fibre preforms, followed by a chemical reaction to silicon carbide. The end products of the three-step process are C/C-SiC materials, consisting of load-bearing carbon fibres (C-fibres) and amorphous carbon, residual silicon

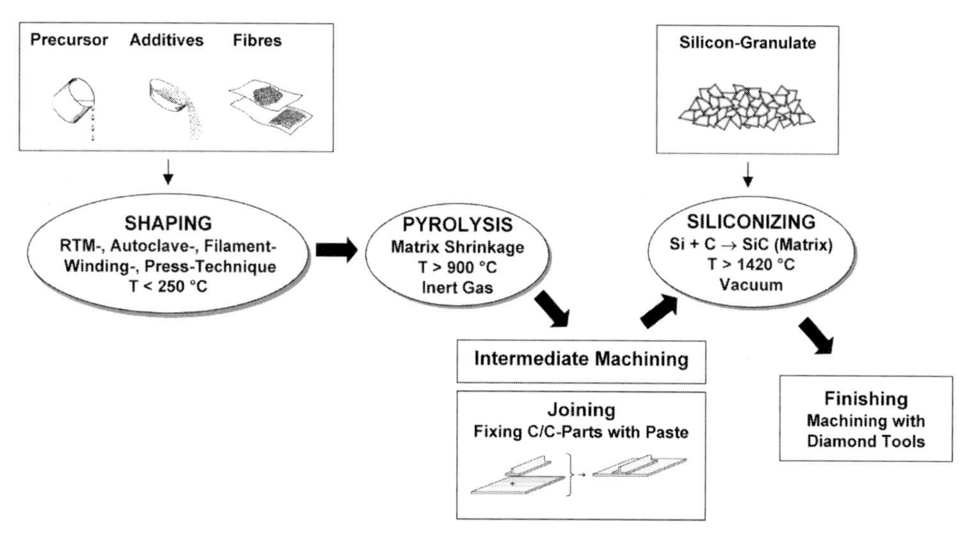


Figure 7.1.1: Overview of the fundamental processing steps for the manufacture of C/C-SiC components.

(Si), and crystalline silicon carbide (SiC) as the most important matrix components. Figure 7.1.1 gives an overview of the fundamental LSI processing steps.

C/C-SiC composites were originally developed for high temperature and short term applications such as aircraft engine parts or thermal protection for spacecraft. They show a considerably low open porosity (less than 5%), a low density (about 2 g/cm$^3$), and a SiC ceramic matrix content of at least 20 weight percent. To follow a consequent low cost approach, common carbon fibres without any additional surface coating are mostly used and lead to a moderate but in many cases sufficient strength level and fracture toughness. Based on carbon fibres, which are normally applied in carbon fibre reinforced plastics (CFRP), tensile strengths of 190 MPa and failure strains of up to 0.35% are achievable [1]. One further essential advantage of the LSI-process lies in the aspect, that the C/C-SiC microstructures and properties can be intentionally adjusted within a broad range by variation of the process parameters and by varying the dimensionality of the fibre preform.

Meanwhile, C/C-SiC materials are an interesting alternative to conventional construction materials in various applications of mechanical and automotive engineering. They show very good wear properties, extreme temperature stability and thermal shock resistance, a very low thermal expansion and a high mass-specific mechanical stability [2].

## 7.1.3 Microstructural Design of C/C-SiC Composites

The microstructure and phase composition of C/C-SiC composites can be tailored to specific requirements by modifying the fibre surfaces or by varying the process parameters and fibre preform architecture. Rovings, woven fabrics, braids etc. of different carbon fibres and modifications – high tenacity (HT), intermediate modulus (IM) and high modulus (HM) – without any additional surface coatings are used as reinforcements for C/C-SiC materials.

### 7.1.3.1 C/C-SiC Composites Derived from As-Received Carbon Fibres

Commercially available carbon fibres normally show a high amount of active surface groups (OH, COH, COOH, etc.) to increase the adhesion with the polymer matrix via chemical bonding. The use of as-received fibres in the CFRP preform of the LSI-process therefore leads to strong fibre/matrix bonding (FMB) which can be measured in terms of high interlaminar shear strengths (ILSS). Typical values for CFRP composites based on fabric-reinforcements lie in the range of 40–50 MPa.

After curing, these composites are postcured for a complete polymerization of the matrix. Bidirectionally reinforced CFRP composites with fibre contents of 60 Vol.% typically show densities of 1.49 $g/cm^3$ and an open porosity of less than 1%. Subsequently, the CFRP composites are pyrolysed under inert atmosphere ($N_2$) at temperatures between 900 °C and 1650 °C to convert the polymer matrix to amorphous carbon.

Thermo-optical analysis of the pyrolysis step shows that first fibre/matrix debondings occur beyond 505 °C [3]. At lower temperatures, the fibre bundles (i.e. filaments plus matrix) are submitted to tensile stresses parallel to the fibre axis due to the thermal mismatch between polymer and carbon fibre. Perpendicular to the fibres, the bundles are submitted to compression stresses. Dilatometer experiments revealed that at about 505 °C this stress state changes, and tensile stresses in matrix and fibre occur perpendicular to the fibre axis whilst the fibres come under compression. With increasing pyrolysis temperature the stresses locally exceed the tensile strength of the matrix resulting in a relaxation by cracking. In case of high fibre/matrix bonding this procedure repeats several times for each fibre bundle. As a result, segmentation of all bundles occur and lead to a translaminar crack pattern with dense C/C segments consisting of about 300–500 individual fibres. These segments are hardly accessible to silicon but the crack pattern can act as a communicating channel system with high capillary forces for liquids of low viscosity.

The subsequent infiltration of liquid silicon at temperatures above 1410 °C rapidly fills the cracks within some minutes. The dense carbon matrix inside the fibre bundles shields off the highly reactive silicon. A layer of silicon carbide is formed around the bundles and only a distinct amount of load carrying

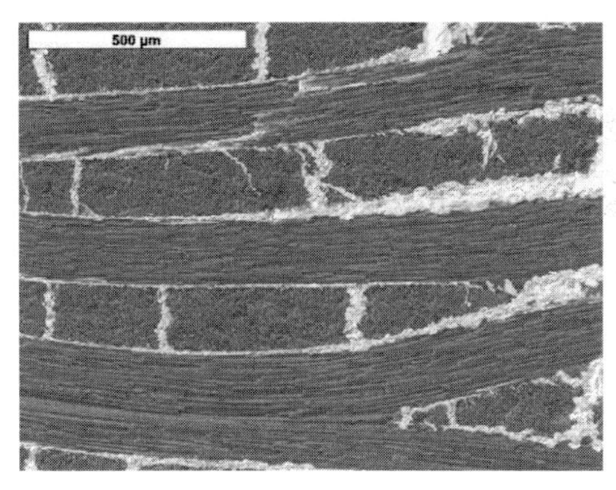


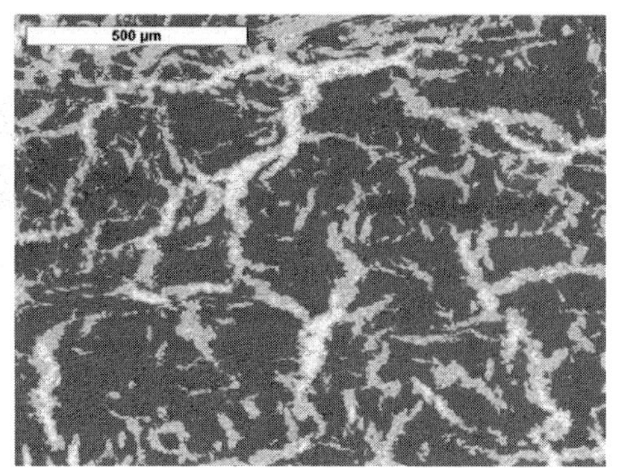


Figure 7.1.2: SEM micrographs of long (woven fabric) and short (40 mm) fibre reinforced C/C-SiC composites.

carbon fibres are converted and damaged. Fig. 7.1.2 shows two representatives of these C/C-SiC composites derived from as-received carbon fibres.

#### 7.1.3.2 C/C-SiC Composites Derived from Thermally Pre-Treated Carbon Fibres

A thermal treatment of the carbon fibres prior to their implementation into the polymer matrix de-activates the fibre surface and reduces the FMB in the polymer stage, by governing the formation of the microcrack pattern during pyrolysis. Exposing the fibres in inert atmosphere at temperatures beyond 600 °C the functional surface groups of the fibres are removed to a certain degree, resulting in reduced bonding forces in the fibre-matrix interface. Applying pre-treatment temperatures of up to 900 °C, moderate ILSS values of up to 30 MPa are measured in a 2D CFRP laminate.

During pyrolysis, these bondings are strong enough to form a crack pattern similar to that of high FMB, but the fibre bundles are additionally interspersed with small cracks. This effect leads to a more relaxed stress state within the bundles with a higher amount of sliding areas. However, more carbon fibres come in contact with silicon and react to SiC.

By increasing the pre-treatment temperature beyond 900 °C the FMB decreases further and ILSS values for the CFRP composite are typically below 20 MPa. During pyrolysis, these weak bondings lead to a shrinkage of the matrix away from the fibres. In contrast to the previous cases, no geometrical reduction in the laminate's thickness occurs and the individual fibres remain in their original position. The C/C composite is totally microcracked but not delaminated, resulting in a microstructure without any segmentation into fibre segments.

During siliconizing all cracks are filled and much more silicon is absorbed, resulting in a higher density of about 2.3 g/cm$^3$. As there is no protection of the

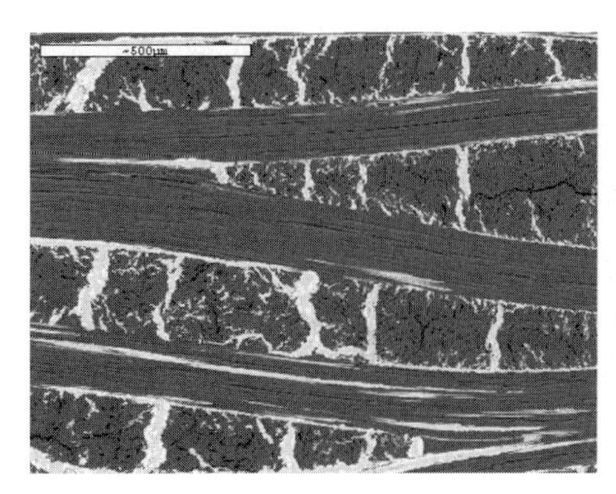

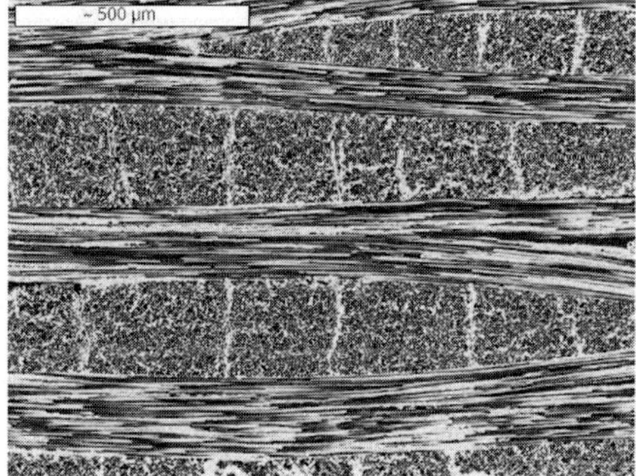


Figure 7.1.3: SEM micrographs of C/C-SiC composites, reinforced with thermally pre-treated woven fabrics (left: treatment temperature below 900 °C, right: beyond 900 °C).

fibres, the liquid silicon can react with most of the filaments. This reduces the load-capability of the fibres considerably and the fracture behaviour of the composite becomes more similar to that of monolithic ceramics. Figure 7.1.3 depicts two C/C-SiC microstructures, derived from thermally pre-treated woven fabrics.

#### 7.1.3.3 Graded C/C-SiC Composites

The different microstructures, derived from distinct thermal pre-treatments of the fibre reinforcements, lead to various structural materials which can be used for different applications. As the LSI-processing conditions of all these composites can be kept constant, fabrics of different pre-treatments can also be combined within one laminate. As a result, microstructures with gradients can be manufactured easily, showing graded properties over the thickness of the laminate [4].

One example of a graded C/C-SiC composite is shown in Fig. 7.1.4, used for friction pads. The symmetrical lay-up of the composite consists of 26 fabric layers which are exposed to four different temperatures (600 to 1100 °C) before stacking-up the laminate. By thermal pre-treatments, the FMBs of the composite are reduced gradually in transverse direction resulting in a continuous increase of silicon up-take respectively SiC formation from the centre to the surfaces. Due to the higher coefficient of thermal expansion of the outer SiC-rich layers, transverse cracks occur when cooling down from the sintering temperature. They stop when reaching the first intact in-plane fibres and lead to a relaxation of the composite. The 2D-laminate has a density of 2.2 g/cm$^3$, a residual content of free silicon of less than 5 weight percent, and shows a 4-pt-bending strength of about 80 MPa.

Figure 7.1.5 shows the ground surface of such a graded C/C-SiC composite. The original woven structure of the fibre preform and the microcrack pat-

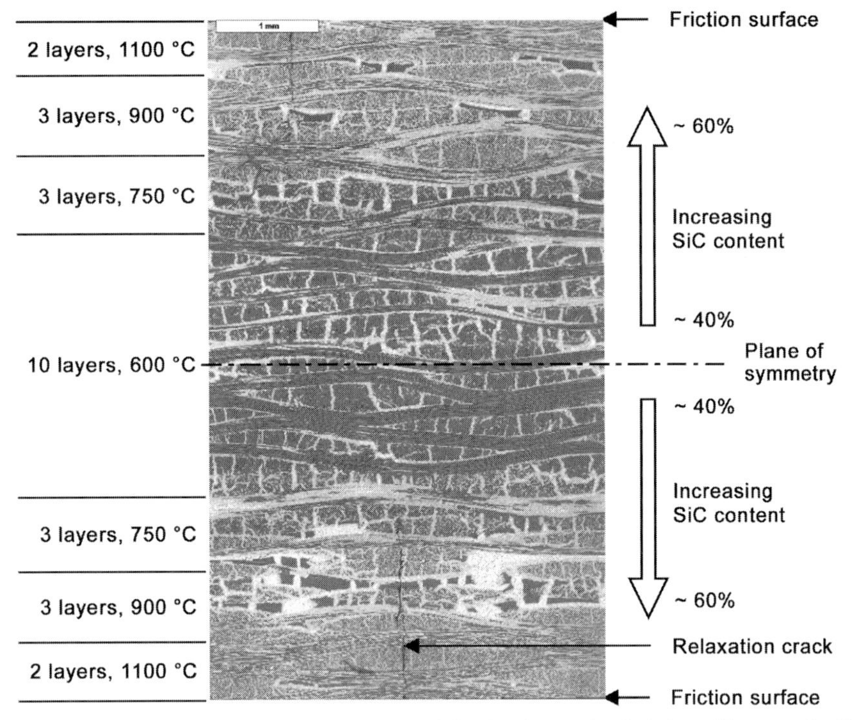


Figure 7.1.4: Graded C/C-SiC composite used for brake pads (bright: silicon and silicon carbide, dark: carbon fibres).

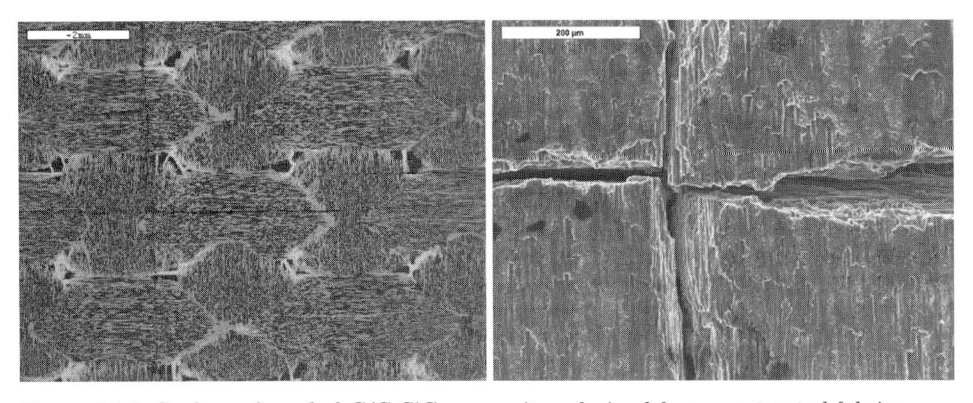

Figure 7.1.5: Surface of graded C/C-SiC composites, derived from pre-treated fabrics.

tern can be seen clearly. The cracks have a width of 17–19 μm and are running in-plane in a rectangular order with a distance of 1–2 mm from each other. The crack depth from the surface into the laminate's thickness lies between 1.2 and 1.8 mm. The stress distribution as well as the crack pattern was modelled by FEM which is in good accordance with the experiments (see section 7.3.3).

If an even finer grading (1100, 1000, 900, 825, 750, 675 and 600 °C, one layer each, symmetrical lay-up) is chosen, the mechanical properties can be significantly improved up to a 3-pt bending strength of 156 MPa. These experimental results agree with a theoretical calculation of the stress distribution within the graded C/C-SiC composite. Applying classical laminate theory the tensile stress in the different pre-treated layers was calculated and compared to their tensile strengths. These strength values were determined with composites of the respective layers. The coefficient of thermal expansion (20–1500 °C, reference temperature 20 °C) was measured by high temperature dilatometry, whereas Young's modulus and Poisson's ratio (at room temperature) were determined in tensile testing. For calculation according to the super imposition principle, the following formulas were used:

$$\sigma_i = (a_{\text{total}} - a_i)\frac{E_i}{1 - \nu_i}(T_{\text{RT}} - T_{\text{Silicon}})$$

$$a_{\text{total}} = \frac{a_i E_i}{\sum_{i=1}^{n} E_i}$$

The siliconizing temperature $T_{\text{Silicon}}$ was determined to be 1650 °C. The results are shown in Tab. 7.1.1.

When the calculated stresses in the layers are compared with the measured or estimated tensile strengths, it is obvious that at least the layers subjected to a fibre pretreatment (FP) at 1100 °C and 1000 °C will be cracked. In addition, the calculated tensile stress in the next layer (900 °C) almost reaches the tensile strength of this layer, imposing that cracks are to be expected in this layer, too. Considering the crack depths in this material as well as in the sam-

Table 7.1.1: Comparison of the calculated stresses (>0: tensile, <0: compression) in the layers to the measured tensile strengths of the layers.

| Temperature of FP [°C] | Calculated stress in the layer [MPa] | Tensile strength [MPa] | Young's modulus [GPa] | Coefficient of thermal expansion ** $[10^{-6}\ K^{-1}]$ | SiC-content [mass-%] |
|---|---|---|---|---|---|
| 1100 | 114 | 19.4 | 110.9 | 3.8 | 60.9 |
| 1000 | 62 | 24.9* | 88.75* | 3.6* | 56.3* |
| 900 | 25 | 30.4 | 66.60 | 3.4 | 50.7 |
| 825 | –12 | 61.4* | 64.50* | 3.05* | 47.2* |
| 750 | –47 | 92.4 | 62.40 | 2.7 | 43.6 |
| 675 | –65 | 93.3* | 55.65* | 2.45* | 42.3* |
| 600 | –77 | 94.1 | 48.90 | 2.2 | 40.9 |

* estimated values by averaging,
** 20–1500 °C, reference temperature 20 °C.

Figure 7.1.6: SEM micrograph of a graded C/C-SiC composite, derived from pre-cured fibre sizings.

ple chosen for brake disk application a pretty good accordance with the calculated crack depth by means of classical laminate theory was achieved, and thus emphasises this assumption.

Generally, these relaxation cracks reduce the overall strength level of the composite, but they maintain the structural integrity of such graded composites. The achieved strength values, although less than those derived from the uncracked core material, are still sufficient to use the composites in brake pads. By applying a finer grading it is feasible to further improve the strength values and thus expand the fields of applications, where structural materials in combination with functional properties such as high wear and corrosion resistance are of essential importance.

One other possibility to form graded C/C-SiC composites lies in the use of as-received sized carbon fibres whose sizings (mostly an epoxy resin) are precured prior to the matrix impregnation. Varying the pre-curing temperature from RT to 200 °C, the sizing agent cures in different stages and leads to different fibre/matrix bondings between the sizing and the matrix resin. Gradients can be achieved by laying-up different pre-cured fabrics symmetrically to the laminate's plane of symmetry, impregnating them with the matrix resin and curing, pyrolysing and siliconizing the laminate via the LSI-process.

Figure 7.1.6 shows a 2D C/C-SiC microstructure with a graded lay-up based on HT-fibres and a phenolic resin, where the outer four layers have been pre-cured. The low FMBs of these layers lead to a high silicon uptake and therefore a high silicon carbide content, resulting in a mean density of 2.0 g/ $cm^3$, an open porosity of 3.6%, and a 3 pt-bending strength of 100 MPa of the C/C-SiC composite. The combination of SiC-rich surfaces with quasi-ductile cores predestines such graded C/C-SiC composites for components where a high abrasive as well as a high thermal shock resistance are mandatory.

#### 7.1.3.4 C/C-SiC Composites Derived from Graphitized C/C

Pyrolysis of the polymer matrix is generally performed at 900 °C in nitrogen atmosphere. This temperature has been chosen as a compromise between processing time and char yield of the polymer. Nevertheless, the resulting matrix still contains some residual hydrogen which can be removed at temperatures above 1300 °C. Final heat treatments of C/C materials up to 2700 °C normally do not graphitize the amorphous carbon, but show a big influence on the shrinkage of the matrix, i.e. quite different C/C composites concerning microstructure, density, strength level and ceramic content can be achieved [5].

Carbon/carbon composites with different fibre reinforcements (HTA, T800, M40) have been pyrolysed at 900 °C and subsequently exposed to temperatures, ranging from 1650 to 2700 °C, prior to their siliconizing. The heat treatment at 1650 and 1950 °C was performed in a high temperature graphite furnace which was conductively heated (2 K/min) to the graphitization temperature in nitrogen atmosphere under reduced pressure (ca. 0.1 mbar). Ultra-high temperature treatments ($T$=2400 and 2700 °C) were performed in a graphite furnace which was inductively heated (10 K/min) to the graphitization temperature in argon atmosphere under reduced pressure (ca. 300 mbar). After anneal-

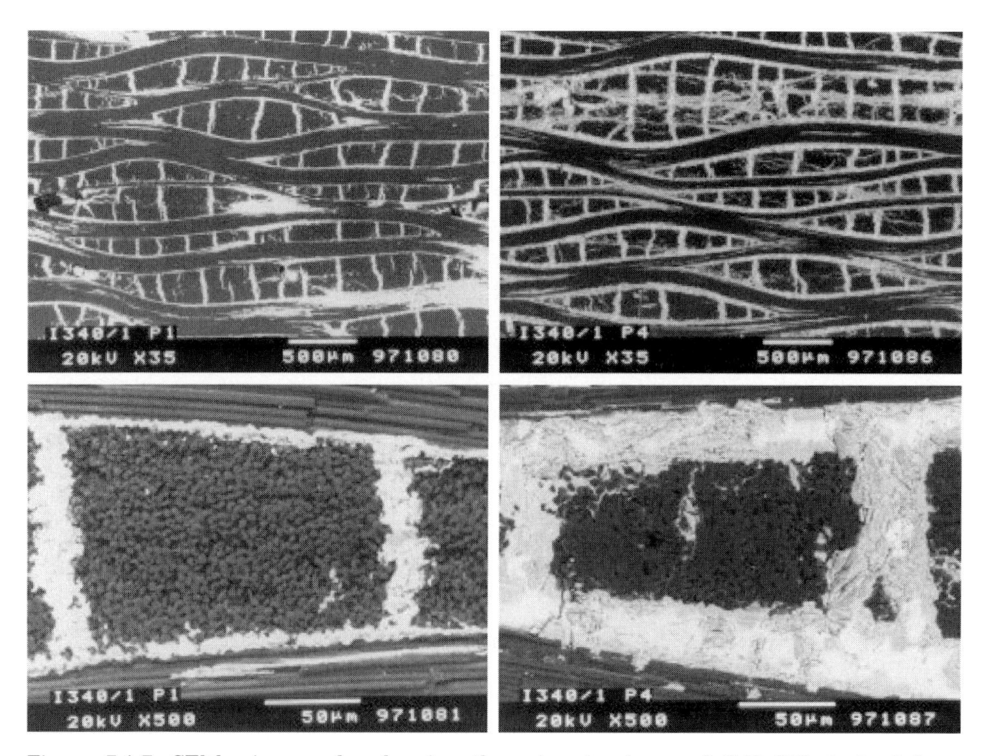


Figure 7.1.7: SEM micrographs showing the microstructures of C/C-SiC derived from C/C (T800/XP-60) after heat treated at 1650 °C (left) and 2700 °C (right).

ing at this temperature for 1 h, the composites were cooled to ambient temperature.

The porosity of the carbon/carbon composites increases considerably after the graphitization step due to the weight loss of the precursor derived matrix and a small (up to 1%) increase of the composite's length. However, the C/C composites show a similar microcrack pattern as before – except a distinct increase in medium crack spacing. They can be siliconized as easily as carbon/carbon composites which are only pyrolysed at 900 °C, although the silicon uptake after the high temperature treatment is much higher.

Figure 7.1.7 shows the microstructures of C/C-SiC derived from heat treatments at 1650 °C and 2700 °C. With increasing temperature the SiC areas around the C/C segments become thicker (30–40 μm) and a certain amount of carbon fibres additionally convert to silicon carbide during siliconizing.

The higher Si-consumption influences the mechanical properties essentially. The mechanical properties, i.e. flexural and tensile strengths are lower compared to C/C-SiC composites which are pyrolysed at 900 °C. Summarizing, these kinds of composites are not candidates for structural materials, but show some attractive properties as functional materials of high thermal or electrical conductivity (see section 7.1.5).

## 7.1.4 Macroscopic Design Aspects

### 7.1.4.1 Dimensional Stability

The macroscopical changes of C/C-SiC components during their manufacture are determined by the fibre/matrix bonding forces, the preform architecture and the volumetric change of the precursor, mainly during the pyrolysis step. As the carbon fibres show different coefficients of thermal expansion in longitudinal and radial direction, the shrinkage of the C/C-SiC composite mainly depends on the fibre orientation.

Theoretically, short fibre reinforced C/C-SiC components with a random fibre orientation show a more or less isotropic shrinkage during the pyrolysis step. Actually, short fibre reinforced C/C-SiC composites are manufactured by an axial pressing of compounds, consisting of chopped fibre rovings and precursors. The densifying and curing of the compound in hydraulic presses with pressures between 5 and 10 MPa and temperatures of up to 250 °C lead to a distinct in-plane orientation of the short fibres. Therefore, short fibre reinforced C/C-SiC components normally show an anisotropic behaviour which dominates the shrinkage rates during pyrolysis.

In case of bi-directional reinforcements (for example fabrics) this anisotropic shrinkage is much more pronounced and macroscopic dimensional changes of such 2D C/C-SiC components occur only transverse to the fabric layers due to the shrinkage impediment of the continuous fibres.

During a 900 °C pyrolysis, short fibre reinforced CFRP compounds, based on high FMB, HT fibres and a phenolic precursor, shrink by approx. 1% in the in-plane direction and about 5% in the transverse direction [6]. In comparison, a 2D-laminate, also based on high FMB, HT fibres and a highly aromatic precursor (XP60), leads to an unchanged geometry in length and width after pyrolysis, but shrinks in the laminate thickness by approximately 4.4% [6]. A pyrolysis step at 1650 °C under vacuum results in an additional volumetric contraction of 1.74%. Here, the main geometrical change also occurs in the direction of the 2D-laminate thickness. Minor changes (–0.25%) occur in breadth and length.

In total, the anisotropic characteristics of 2D-laminates with high fibre volume contents of more than 50% result in dimensional changes over the whole pyrolysis of less than –7%. Although high mass changes occur during the final siliconizing step with silicon uptakes of more than 40%, the chemical reactions do not cause macroscopical changes in shape and geometry between the C/C-SiC and the previous C/C material state.

The anisotropic irreversible changes in thickness during the pyrolysis step lead to changes in the shape of curved 2D-laminates. In case of angled plates, these geometrical distortions can be calculated from the original angle $\alpha$ and the ratio $D$, which is the quotient of the change of thickness $\Delta d$ during pyrolysis and the original thickness of the laminate $d$. These correlations for a first approximation of this spring forward effect are shown in Fig. 7.1.8.

For example, a transverse shrinkage during pyrolysis of –4.5% reduces a rectangular angle by 4.3° to 85.7°. Vice versa, to get an angle of 90° in the C/C-SiC composite the initial CFRP component must show an angle $\alpha$ of 85.95°.

These dimensional changes normally do not lead to delaminations of plates or non-closed profiles due to the unconstrained shrinkage of the composite. However, the transverse shrinkage of the matrix in shells with a closed contour like tubes is hindered by the geometrical stability of the wound fibres which result in additional matrix stresses in radial direction.

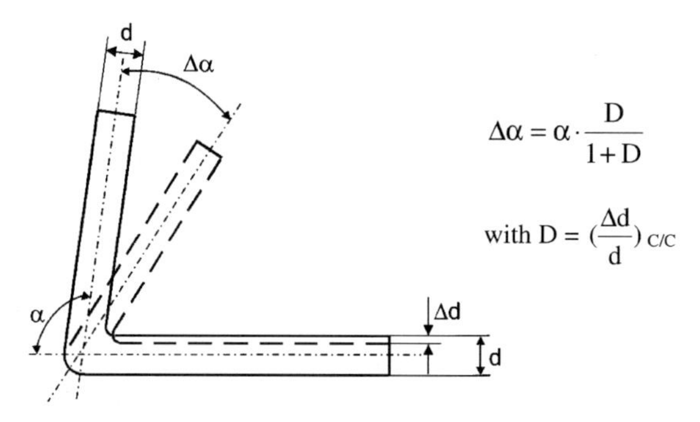


Figure 7.1.8: Spring forward effect during pyrolysis of orthotropic reinforced 2D-CFRP laminates due to irreversible changes only in transverse direction.

Due to the high integrity of the fibre reinforcement the changes of diameter amount normally to a total of less than 1%. This is the equivalent of a reduction in wall thickness of 3% and 4% for thin-walled tubes with a ±45° and 0°/90° fibre orientation, respectively [7]. In general, the higher the amount of fibre reinforcement in circumferential direction and the higher the tube's wall thickness, the higher the radial shrinkage impediments which cause local delaminations. Wound C/C-SiC tubes are therefore limited in their size if a high tightness and a delamination-free microstructure are required [8].

#### 7.1.4.2 Modular Construction by *In-Situ* Joining

Due to the fact, that most of the dimensional changes within C/C-SiC composites are complete after the pyrolysis step, the manufacture of complex components is possible through modular construction. Different parts can be manufactured and pyrolysed separately and no traditional fastening techniques like metallic bolts or ceramic adhesives are necessary. The C/C components can be machined into their final shape and joined together by shape or force locking the individual parts. In order to enrich the gaps between the parts to be joined with carbon and to fix them during siliconizing, a temporary bonding material is suitable.

Thus, a paste has been developed which consists of a phenolic resin and a graphite powder as a solid filler. The two materials are mixed and applied to the surface of the C/C parts to be joined. To achieve a good wettability and to fill even small gaps, resin shares of 80–85% within the mixture and small powder sizes of about 4 μm are necessary. Wider gaps can be additionally filled with carbon felts or fabrics [9, 10].

In a first step, the porous C/C parts are fixed together with the paste. During the subsequent infiltration of silicon, the reactive metal is caused to flow into the gaps, reacts with the carbon material and converts to SiC. The resulting joint strength is in the same range as the interlaminar shear strengths of

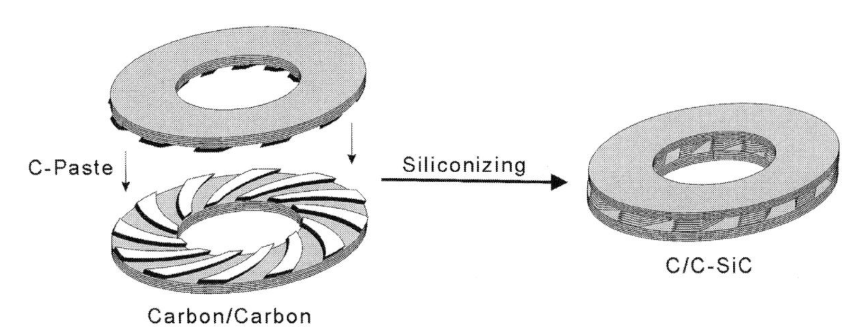


Figure 7.1.9: Manufacture of internally ventilated brake disks by *in-situ* joining of two pre-fabricated C/C half-disks.

the C/C-SiC composites. Components of more than hundred individual parts have been assembled successfully within one siliconizing step.

From an engineering point of view, this in-situ joining method by reaction bonding is one of the key features of the LSI process which reduces the costs and weight of C/C-SiC products. Fig. 7.1.9 shows schematically the manufacture of internally ventilated brake disks for passenger cars, which are produced in series by joining two symmetrical half-discs prior to siliconizing [11].

## 7.1.5 Thermophysical Characterization of C/C-SiC

### 7.1.5.1 Methods to Measure Thermophysical Properties

Measurements of the specific heat capacity have been carried out at IKE (Institut für Kernenergetik, University of Stuttgart) by means of differential scanning calorimetry on a DSC2 of Perkin Elmer at temperatures up to 730 °C. Thermal conductivity can be measured at IKE either directly by means of the longitudinal comparative technique in the range up to 800 °C or indirectly up to 2000 °C by measuring the thermal diffusivity by heating with an intensity modulated light beam of a xenon lamp [13]. In the latter case the thermal conductivity was calculated from measured thermal diffusivity, specific heat capacity and density of the samples. The spectral and the total emissivity have been measured at IKE by means of the radiation comparison technique in the temperature range between 800 °C and 1600 °C on disc shaped samples of 15 mm diameter and 3 mm to 5 mm thickness [14]. Two apparatuses are available, one with electron beam heating for measurements in vacuum [15] and the second with induction heating for measurements in air or in a gas atmosphere [16].

### 7.1.5.2 Materials and Specimen Preparation

In order to investigate the influence of process parameters on the thermophysical properties of C/C-SiC [17], samples of various sizes were cut from C/C-SiC-plates which were manufactured according to the LSI-process being described elsewhere in more detail [12]. Therefore, three C-precursors, two phenolic resins JK25 and JK27 as well as the highly aromatic resin XP60, and three pan-based C-fibres (HTA from Tenax, T800 and M40 from Toray) as well as a pitch-based C-fibre (NGF from Nippon Graphite Fibre) were used. The fibre diameters ranged from 5 μm for T800-fibres to 6.5 μm for M40-fibres and 7 μm for HTA- as well as NGF-fibres. All parameter variations together with the individual sample designations are listed in Tab. 7.1.2.

At first CFRP plates were processed from XP60 via RTM and from the phenolic resins via autoclave technique. Then, the CFRP plates with a fibre vol-

Table 7.1.2: Process parameters, physical properties and chemical compositions of C/C-SiC specimens. The structure of the sample designation is: fibre/precursor+pyrolysis temperature.

| Specimen | H/X9 | H*/X9 | T/X9 | T*/X9 | M/X9 | N/X9 | H/X16 | H/J16 | H/K16 |
|---|---|---|---|---|---|---|---|---|---|
| Fibre | HTA | HTA[1] | T800 | T800[2] | M40 | NGF | HTA | HTA | HTA |
| Precursor | XP60 | XP60 | XP60 | XP60 | XP60 | XP60 | XP60 | JK25 | JK27 |
| Pyrolysis temperature [°C] | 900 | 900 | 900 | 900 | 900 | 900 | 1650 | 1650 | 1650 |
| Porosity [%] | 3.5 | 2.8 | 3.0 | 3.7 | 2.3 | 6.7 | 3.4 | 2.2 | 2.0 |
| Density [g/cm$^3$] | 1.90 | 2.31 | 1.88 | 1.92 | 2.05 | 2.03 | 1.96 | 2.05 | 1.85 |
| Si [mass-%] | 6.1 | 2.2 | 8.7 | 4.7 | 5.1 | 1.4 | 8.3 | 5.5 | 2.3 |
| C [mass-%] | 60.9 | 36.9 | 63.5 | 64.1 | 65.5 | 59.2 | 60.7 | 50.5 | 67.5 |
| SiC [mass-%] | 33.0 | 60.9 | 27.8 | 31.2 | 29.4 | 39.4 | 31.0 | 44.0 | 30.2 |

[1] HTA-fibre pretreated at 1100 °C, [2] T800-fibre pretreated at 600 °C.

ume content of roughly 60% (56–63%) were pyrolysed to C/C at 900 °C or optionally at 1650 °C. Some specimens were annealed above 2000 °C in an inductively heated graphite furnace in Ar-atmosphere for 1 h. In the last step the C/C plates were infiltrated with liquid silicon at 1650 °C to form C/C-SiC. From the C/C-SiC plates, small samples were cut with a diamond blade for the determination of porosity and density (Archimedes method) as well as for microstructural characterization in SEM. The determination of Si, C and SiC in the samples was obtained by gravimetric analysis. Therefore, Si was removed by dissolving in a mixture of hydrofluoric and nitric acid (20/80) at 40 °C for 48 h, whereas the C-content was obtained by burning off C at 700 °C for 20 h in air yielding residual SiC.

A typical microstructure of C/C-SiC derived from CFRP with high fibre matrix bonding (H/X9) is shown in Fig. 7.1.10 (left). Dense C/C segments (dark), which are not accessible by liquid Si during siliconizing, are surrounded by a SiC-matrix containing some residual Si. In contrast Fig. 7.1.10 (right) shows the microstructure of C/C-SiC when it is obtained from a CFRP with low fibre matrix bonding (H*/X9). In this case, due to many small cracks in C/C state, liquid Si can enter these cracks and attack all fibres during siliconizing. Consequently a large amount of C-fibres and C-matrix is converted to SiC resulting in a significantly increased SiC-content (60.9% compared to 33.0% for H/X9). All other samples based on the precursor XP60 and fibres without fibre pretreatment revealed a microstructure similar to that shown in Fig. 7.1.2 (left) due to high fibre matrix bonding. The microstructure based on the two types of phenolic resins is shown in Fig. 7.1.11. Whereas the sample H/J16 (JK25-type) has an increased SiC-content (44%) and a microstructure showing C/C segments with incorporated SiC-areas, the H/K16 sample is similar to the XP60-type in microstructure as well as in SiC-content. The annealing of C/C specimens (graphitization) prior to siliconizing does not change the microstructure of

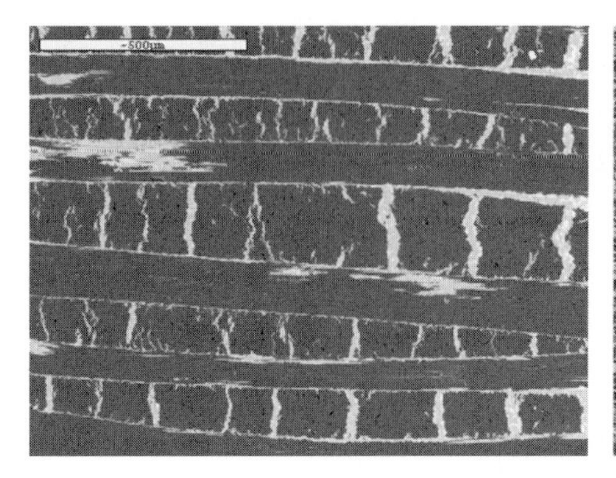
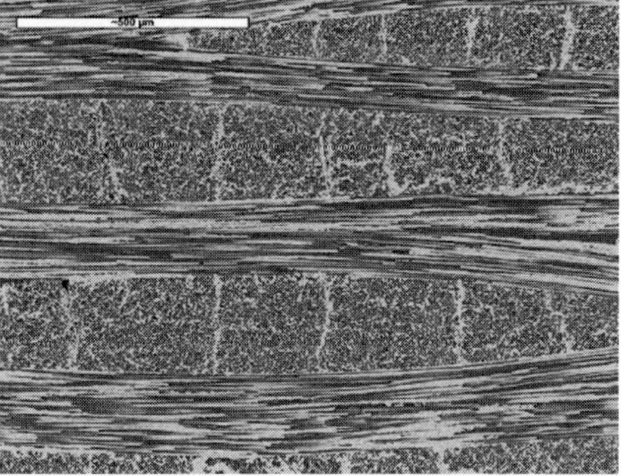

Figure 7.1.10: SEM-micrographs of C/C-SiC with HTA-fibres, pyrolysed at 1650 °C; left: high fibre matrix bonding (H/X9), right: low fibre matrix bonding (H*/X9).

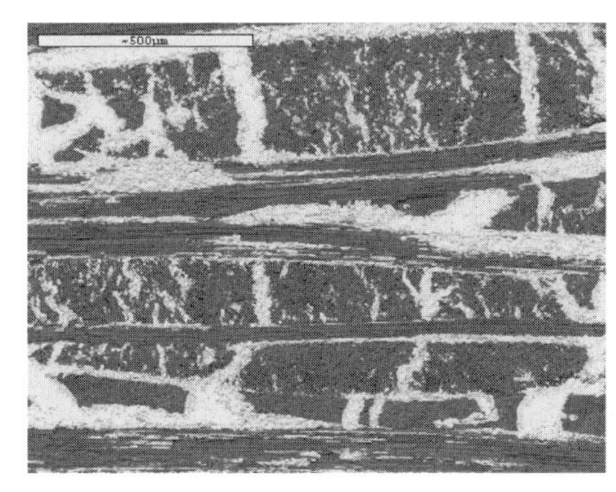
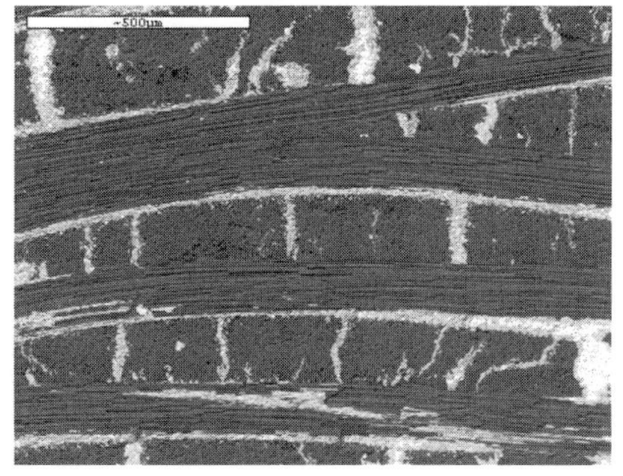

Figure 7.1.11: SEM-micrograph of C/C-SiC with HTA-fibres, pyrolysed at 1650 °C; left: JK25-type, right: JK27-type.

C/C-SiC samples in principle. However, the SiC- and Si-content is increased due to the formation of larger spacing between the C/C segments.

#### 7.1.5.3 Specific Heat Capacity

Considering the specific heat capacity, there are two groups of materials: one with low C- and high Si- as well as high SiC-content and the second with a low Si- and SiC- and a high C-content. For comparison the measured curves are plotted in Fig. 7.1.12 together with literature values for C, SiC, and Si, as well as the calculated theoretical values assuming mean chemical compositions similar to these of the two types. It is evident that Si reduces heat capacity values whereas increasing C-content shifts them upwards.

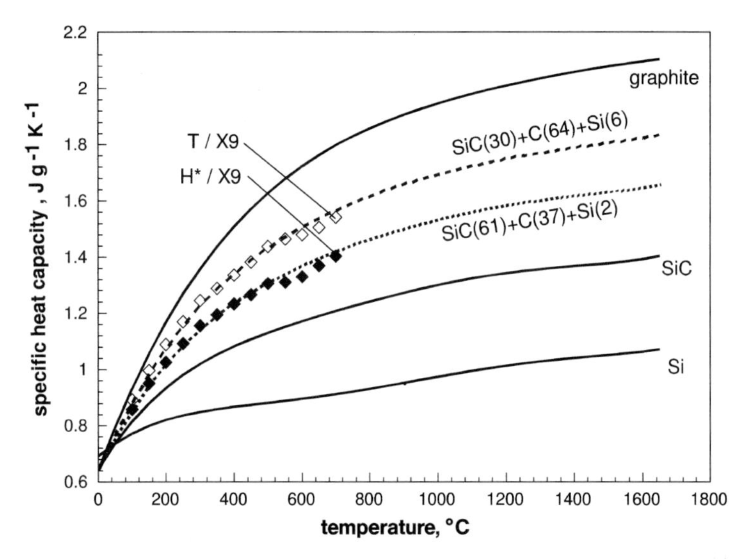


Figure 7.1.12: Specific heat capacity of C/C-SiC with different fibres and annealed samples, respectively. For comparison the values of graphite, SiC and Si are also plotted.

### 7.1.5.4 Thermal Conductivity

The influence of the fibre type on the thermal conductivity of the composite is shown in Fig. 7.1.13. Four different composites have been tested, all of them with about 60 vol-% of fibres, the same precursor (XP60) and pyrolysed at

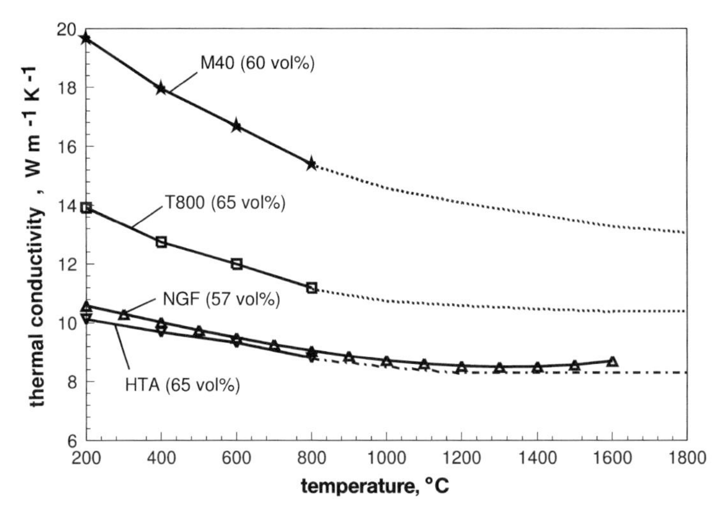


Figure 7.1.13: Influence of C-fibre type on the thermal conductivity (in transversal direction to the fibre planes) of C/C-SiC composites with precursor XP60, pyrolysed at 900 °C.

900 °C. Only the type of fibres (HTA, T800, NGF, and M40) has been varied. Due to the low fabrication temperature of the HTA- and NGF-fibres (below 1300 °C) the conductivity of these C/C-SiC composites is similarly low, whereas composites of the intermediate modulus fibre T800 and especially the high modulus fibre M40 show significantly higher values of about 40% and 100% at 200 °C, respectively. As it is expected for ceramics and carbon, the thermal conductivities of all four composites decrease with increasing temperature. Because of the pronounced orientation of the fibres in one plane there is a strong anisotropy of the thermal conductivity parallel and transversal to main fibre orientation. Thermal conductivity parallel to the fibre planes (not shown here) are by a factor 2 to 3 higher compared to the transversal direction as it is indicated by the anisotropy factor in Fig. 7.1.14.

The type of precursor also influences the thermal conductivity of the composite, as it can be seen in Fig. 7.1.15. Here the conductivity of three composites with different precursors (XP60, JK27, and JK25) but the same type of fibres (HTA, 60 vol-%) and pyrolysed at 1650 °C, is plotted as function of temperature. The composite with the precursor XP60 shows the lowest conductivity, the use of precursor JK27 causes an increase of about 18% at 200 °C, and the precursor JK25 improves the thermal conductivity by about 64%. This is mainly due to a change in microstructure (Fig. 7.1.11). The precursor JK25 leads to C/C-SiC with a higher SiC-content and thus an increase in thermal conductivity.

Further improvement in thermal conductivity of C/C-SiC can be achieved by annealing the C/C at higher temperatures, as it is shown in Fig. 7.1.16. Four different composites (different fibers and different precursors) were annealed in the C/C state between 900 °C and 2800 °C. For the both composites with low

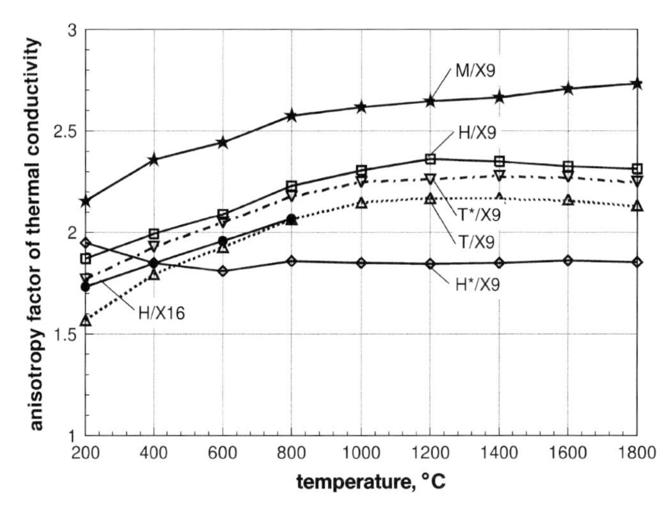


Figure 7.1.14: Anisotropy factor of thermal conductivity of C/C-SiC composites with different fibres and precursor XP60.

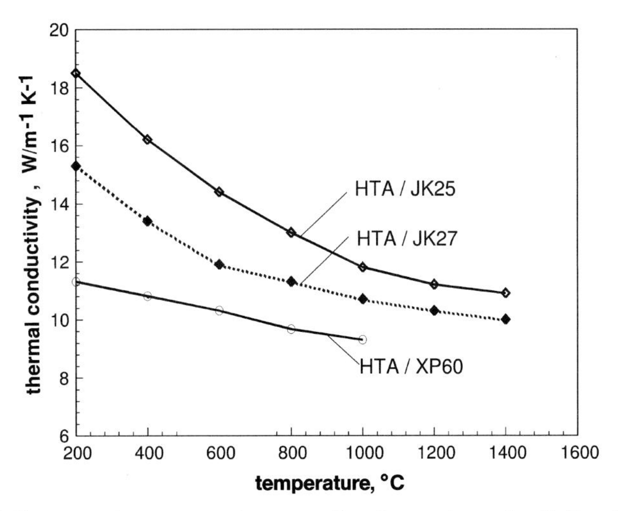


Figure 7.1.15: Influence of precursor type on the thermal conductivity of C/C-SiC with HTA-fibres (60 vol-%), pyrolysed at 1650 °C.

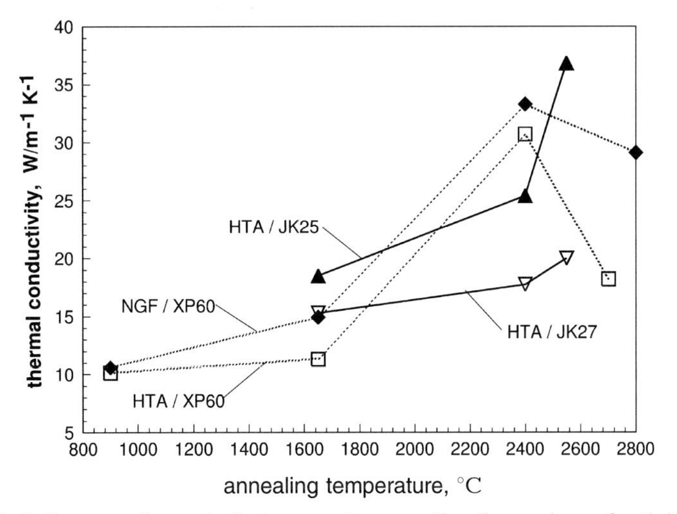


Figure 7.1.16: Influence of pyrolysis temperature on the thermal conductivity of various C/C-SiC composites at 200 °C.

thermal conductivity (HTA/XP60 and NGF/XP60), which were originally pyrolysed at 900 °C, the enhancement in the pyrolysis temperature to 1650 °C caused an increase in thermal conductivity by about 12 and 40% at 200 °C.

However, an enhancement in the annealing temperature to 2400 °C improved the conductivity enormously (about 200%). Further increase of annealing temperature to 2700 °C or 2800 °C caused a drop in thermal conductivity. According to these observations the influence of annealing temperature on the two other composites (HTA/JK25 and HTA/JK27) was tested only in the range from 1650 °C to 2550 °C. For both these composites the rise in annealing temperature from 1650 °C to 2400 °C increases the conductivity only by about 37% and 16%, respectively, but a further increase in annealing temperature (2550 °C) causes further improvement in thermal conductivity. Thus, compared to XP60, the two phenolic precursors JK25 and JK27 reveal a different trend in thermal conductivity versus annealing temperature.

#### 7.1.5.5 Spectral and Total Emissivity

The emissivity is much less influenced by fibre type and treatment as we have seen for the thermal conductivity. Here mainly such effects are important that result in changes directly at the surface. Due to the crystal structure of the fibres with a high electrical conductivity in longitudinal direction the spectral behaviour of the composite material with fabric plane parallel to the surface is like that of metals, as shown by Neuer [14]. If the spectral emissivity measured at various temperatures is plotted versus wavelength (Fig. 7.1.17), a metallic character can be observed. This is shown by a so-called X-point where the spectral emissivity is independent of temperature. This effect cannot be observed on SiC-SiC composites. Obviously, the C-fibers are responsible for the metallic character due to their low electric resistance at room temperature of about 15 μΩm.

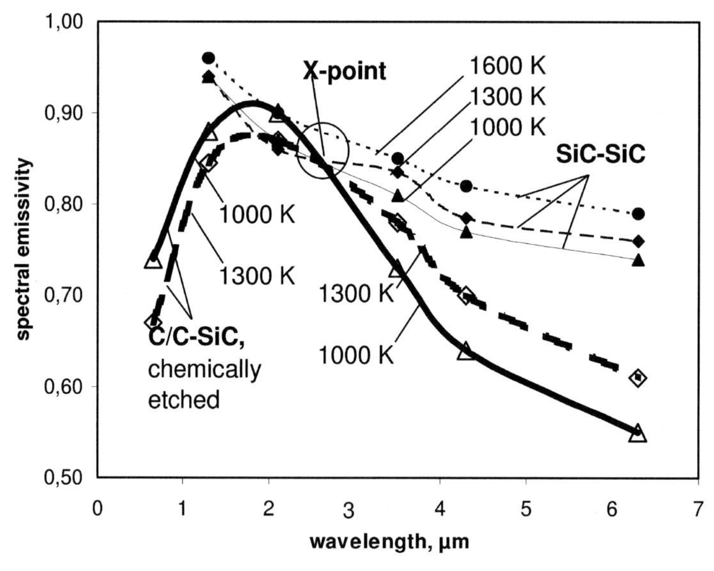


Figure 7.1.17: Spectral normal emissivity of C/C-SiC (HTA/XP60), chemically etched, in comparison with a SiC-SiC.

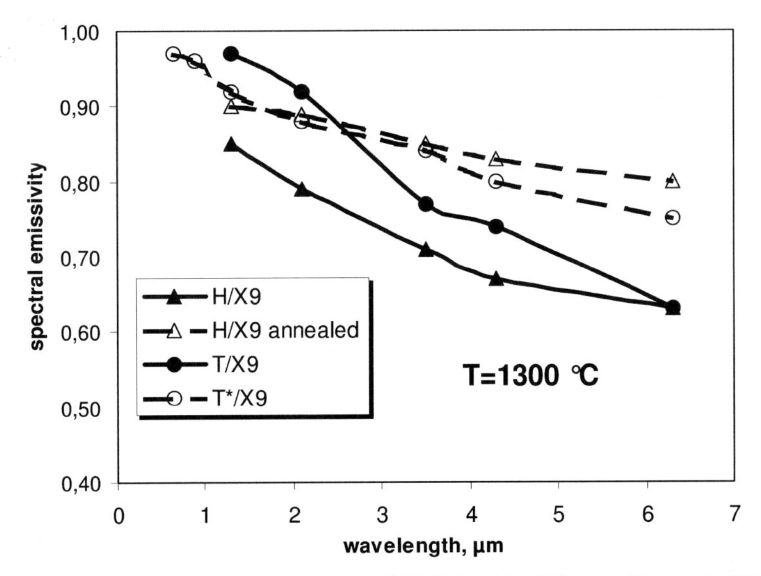


Figure 7.1.18: Spectral normal emissivity of C/C-SiC with different fibers at 1300 °C. The sample with HTA-fibre was measured before and after annealing.

However, after annealing the samples at temperatures above 1450 °C the crystal structure of the fibres is changed resulting in a remarkable increase of the spectral emissivity at long wavelengths. This is valid for HTA-fibres, as Fig. 7.1.18 shows, but also at the samples with T800-fibres the slope of the spectral emissivity is reduced by pretreatment of the fibres. In the case of the total emissivity (Fig. 7.1.19) the influence of annealing is more pronounced at samples with HTA-fibres which correspond to the difference between the spectral emissivity curves of both samples being larger in the untreated case than after annealing. Annealing of C/C-SiC samples also leads to a reduction of resi-

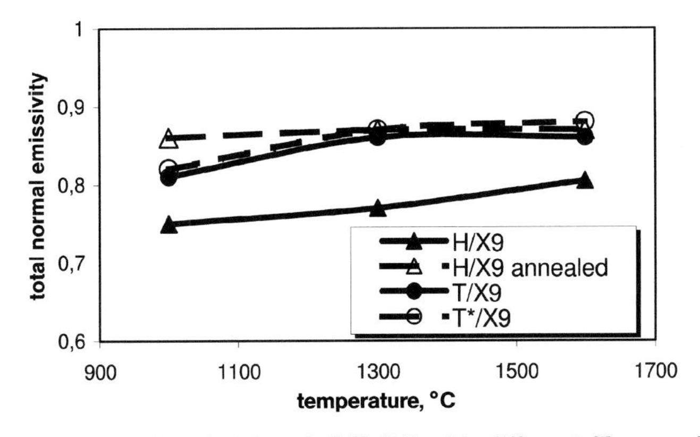


Figure 7.1.19: Total normal emissivity of C/C-SiC with different fibers and various pretreatments of fibres (T800) or samples (HTA) respectively.

dual free silicon in the material, but earlier published investigations [18] have shown, that desiliconizing by chemical etching much less influences the spectral behaviour of the emissivity. So it must be assumed that changes in emissivity by thermal treatment are caused by changing of fibre structure.

## 7.1.6 Thermomechanical Characterization of C/C-SiC

### 7.1.6.1 Failure Mechanism of C/C-SiC Materials

Due to the different coefficients of thermal expansion of fibre and matrix, residual stresses are generated during the cooling down from siliconizing temperature to room temperature when manufacturing C/C-SiC materials. The residual stresses lead to matrix cracks along the fibres. Consequently, in the unloaded state a C/C-SiC specimen already has a certain degree of initial micro cracks (small damages) in the matrix. In the literature, the failure course is described as follows: Just prior to fracture initiation, long cracks develop and increasingly affect the limit surface detachment of the fibre from the matrix. In case of crack propagation perpendicular to the fibres, cracks branch at each fibre and run over the so-called dependent length along the fibres. The fibres will be pulled out of the matrix during additional deformation of the composite. This process will be influenced both by the friction coefficient of the fibre/matrix interface as well as the mismatch of the thermal expansion coefficient of fibre and matrix in radial direction. When the limit of the fracture strain is exceeded, fibre failure occurs and the end of the fibres will be pulled out of the matrix [19]. In cases of high strength fibres and low fibre/matrix bonding, failure occurs due to multi-crack formation in the matrix with subsequent fracture of fibres at higher stress. Depending on the strength of the fibre and fibre/matrix bonding, various failure mechanisms take place.

Crack model investigations, performed at MPA (Materialprüfungsanstalt, University of Stuttgart), on unloaded and variably tensile loaded specimens showed a significant increase in the number – and not in length – of micro cracks in the range of $<100\,\mu m$ in the matrix. This finding is supported by further investigations showing changes in electrical conductivity and in stiffness as well as fractographic appearance in SEM [20].

### 7.1.6.2 Influence of the Temperature on the Stress-Strain Behaviour

In order to investigate the effect of test temperature on the fracture behaviour of C/C-SiC, material type P2 (T800, JK60), tensile tests were carried out between room temperature and $1600\,^{\circ}C$ on a series of flat specimens. In Fig. 7.1.20 the microstructure of the C/C-SiC test specimen is shown. The test

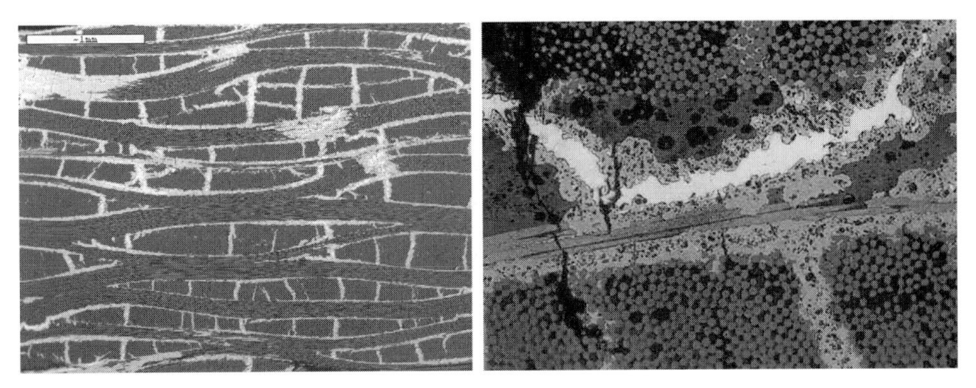

Figure 7.1.20: SEM micrograph of C/C-SiC specimen (material type P2, left) and a magnified site showing free silicon (white) in SiC-matrix (right).

at room temperature was performed in air whereas all high temperature tests were operated under vacuum conditions [21].

The experimental results are presented in the following figures showing each characteristic value versus test temperature, such as tensile strength (Fig. 7.1.21), fracture strain (Fig. 7.1.22) and E-modulus (Fig. 7.1.23).

No significant dependence of the tensile strength on the test temperature could be found for test temperatures up to 1350 °C. All measured values range in a scatter band between 122 and 148 MPa. However, when the test temperature is increased to 1400 °C and higher (up to 1600 °C), a significant decrease in

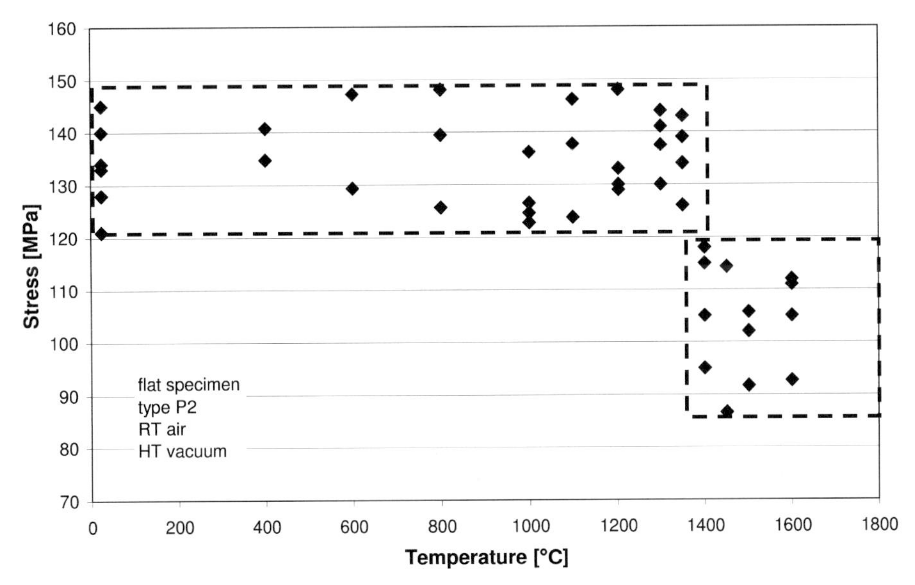


Figure 7.1.21: Progression of the tensile strength of C/C-SiC specimen (material type P2, flat specimen) versus test temperature.

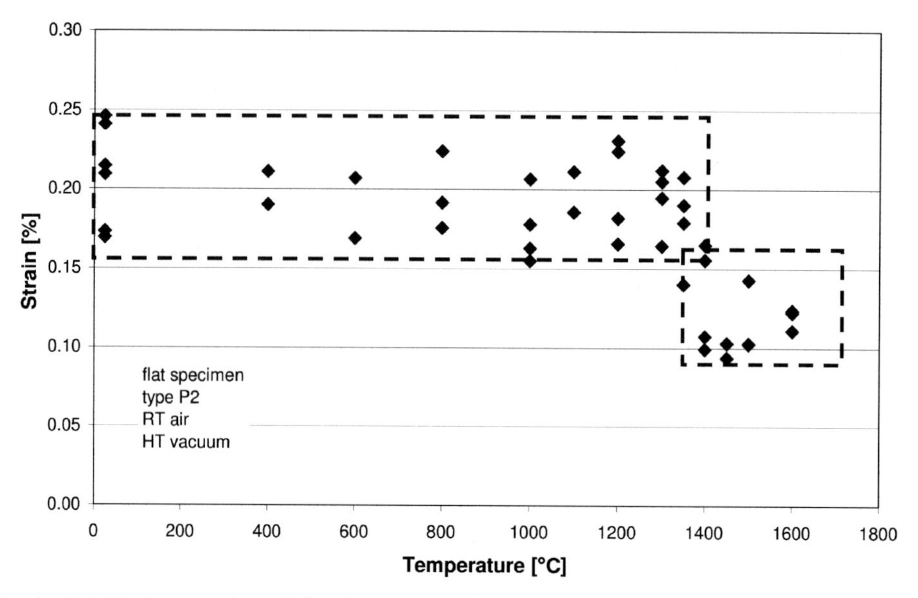


Figure 7.1.22: Progression of the fracture strain of C/C-SiC specimen (material type P2, flat specimen) versus test temperature.

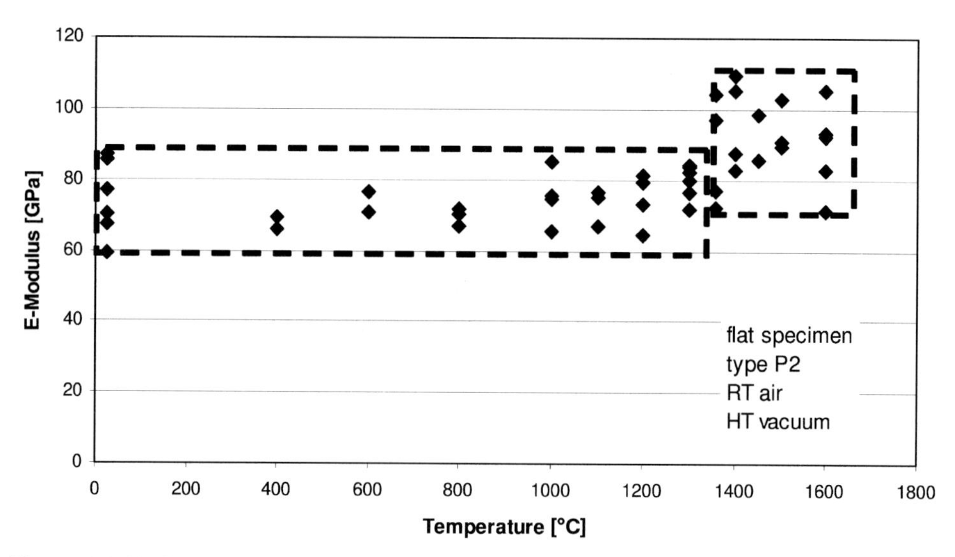


Figure 7.1.23: Progression of the E-modulus of C/C-SiC specimen (material type P2, flat specimen) versus test temperature.

tensile strength was measured. This strength level remains constant up to a test temperature of 1600 °C in a limited scatter band. The scattering of the strength characteristics within the scatter band is due to the individually distributed defects within the material. The progression of fracture strain versus test temperature shows a similar behaviour but without a distinct transition between the scatter bands. In contrast, the E-modulus versus test temperature remains almost constant up to 1350 °C within the scatter band (Fig. 7.1.23). However, when the test temperature is raised to 1400 °C and higher (up to 1600 °C), the E-modulus is slightly increased.

One possible explanation of the discrete drop in the tensile strength (Fig. 7.1.21) could be found in the presence of free silicon, which upon solidification during cooling in the siliconizing step shows a volumetric expansion (anomalous behaviour like water). Consequently, a change from liquid into solid state will result in stresses (compression) within the silicon and stresses (mainly tensile) in the ambient SiC-matrix (Fig. 7.1.24) causing additional micro cracks in the matrix.

The local residual stresses reduce again if the test temperature is above the melting point of the silicon and some micro cracks close again. In addition, open micro cracks may be filled again with liquid silicon which reacts with the carbon fibres to form SiC causing cavities and thus cause a reduction in strength.

Figure 7.1.25 (left) shows the fracture surface of a specimen tested at 1400 °C in vacuum. The (free) silicon melted during the test performance (the melting point of Si in the specimen may be lowered due to impurities). After fracture of the specimen re-crystallisation occurred on the fracture surface showing typical structures with deposited large crystals. Tests performed below 1400 °C did not cause superficial fusion with re-solidification of free silicon on the fracture surface (Fig. 7.1.25, right).

If specimens were annealed at temperatures above the melting point of Si (30 min at 1500 °C and 1600 °C, respectively) and subsequently tested below the Si-melting point (at 1200 °C and 1300 °C, respectively), the strength behaviour of these specimens was similar to non-aged specimens. The obtained

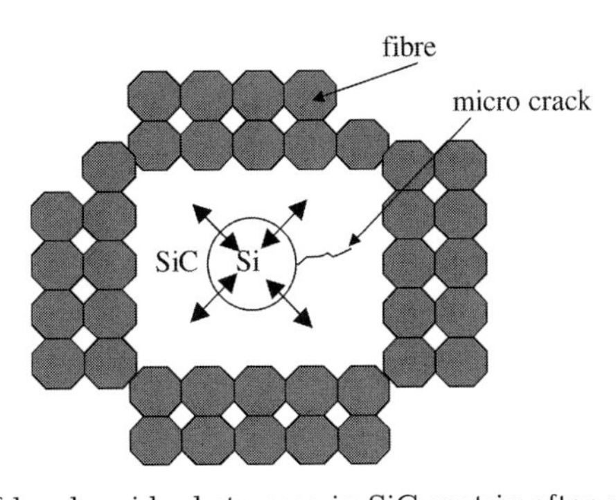


Figure 7.1.24: Model of local residual stresses in SiC-matrix after solidification of silicon.

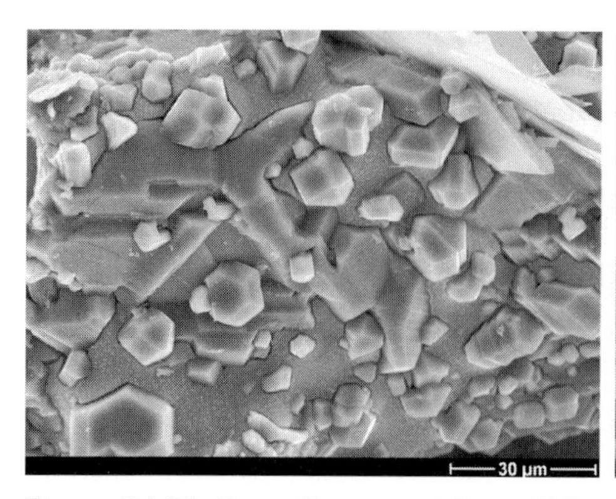


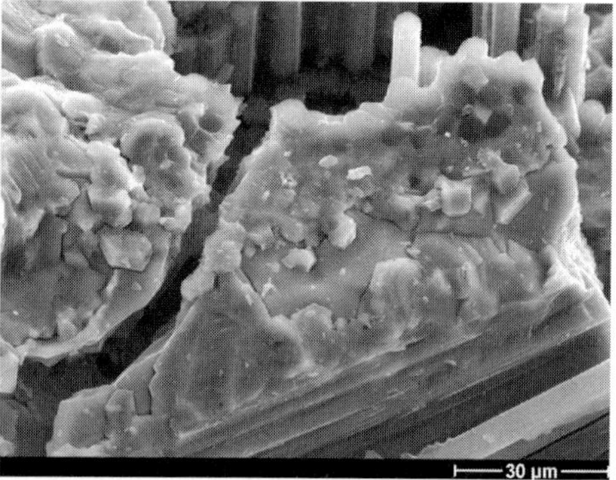


Figure 7.1.25: Free silicon crystals on SiC-matrix, testing temperature 1400 °C (left) and fracture surface, test temperature lower 1400 °C (right).

strength values lay within the scatter band (Fig. 7.1.21). After this ageing small pure silicon crystals appeared on the fracture surface.

It can be concluded that a temporary ageing at temperatures above 1400 °C has none or only little influence on the strength behaviour of C/C-SiC. Therefore, the dominating temperature at loading is of importance to the strength of the material. Extending the duration of ageing considerably will cause a kind of reverse siliconizing processes (as for example by diffusion of the molten silicon out of the specimen) which may influence the strength. In summary, it can be concluded from the results that the change of the silicon from liquid into solid as well as the attack of liquid Si on the fibres will have a negative influence on the local stress situation in connection with the development of cavities. Such a high decrease in strength, however, may not be expected since the Si-content is typically in the range of 1 to 5 wt-%.

However, other explanations based on different stress levels in other C/C-SiC materials are feasible, too. In Fig. 7.1.26 the tensile strength versus temperature of another C/C-SiC material (type A) is shown in comparison to C/C-SiC, material type P2. This material does not show such a drop in tensile strength at high temperature. It is interesting to note that at high temperature both materials reveal a similar level in tensile strength. Therefore, further investigations with different types of C/C-SiC materials (influence of process parameters, fibre and matrix type, fibre/matrix bonding, etc.) are necessary in order to understand these observations in this kind of key materials designed and evaluated (not only) for re-entry from space to earth.

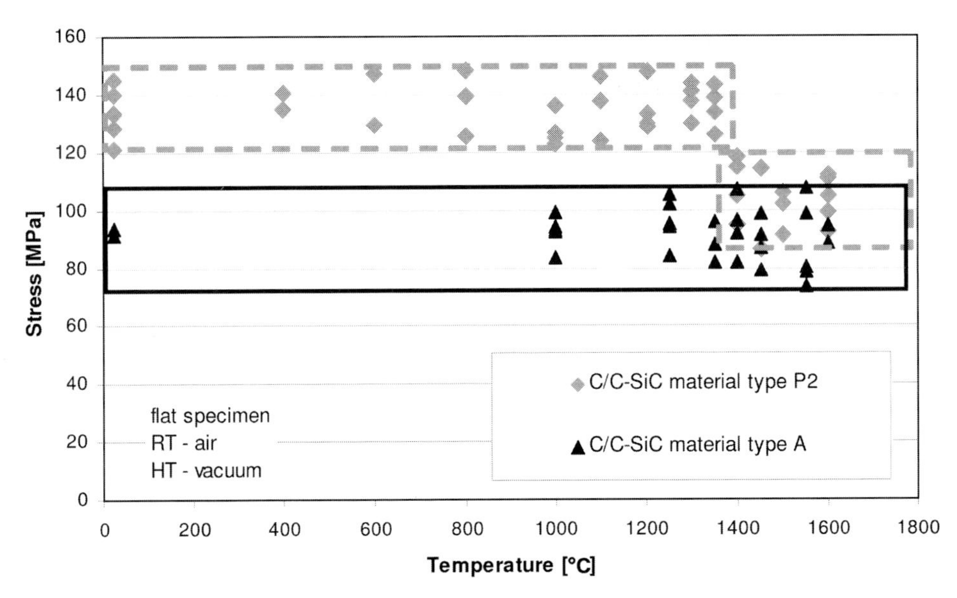


Figure 7.1.26: Progression of the tensile strength of two different C/C-SiC materials (material type P2 and type A, flat specimen) versus test temperature.

## Acknowledgement

Joachim Fabig, Frank Gern, Nicole Lützenburger, Thomas Reimer, Peter Schanz, Thomas Ullmann, Peter Winkelmann, Eckehard Schreiber, Grazyna Jaroma-Weiland, Joachim Kuhnle, Karl Maile, Franz Josef Arendts, Werner Geiwiz, Andreas Theuer.

## References

1. W. Krenkel: Cost Effective Processing of CMC Composites by Melt Infiltration (LSI-Process), Ceramic Engineering and Science Proceedings, Volume 22, Issue 3, pp. 443–454, 2001.
2. W. Krenkel: Design of Ceramic Brake Pads and Disks, Ceramic Engineering and Science Proceedings, Volume 23, Issue 3, p. 319, 2002.
3. J. Schulte-Fischedick, M. Frieß, W. Krenkel, R. Kochendörfer, M. König: Crack Microstructure during the Carbonization of Carbon Fibre Reinforced Plastics to Carbon/Carbon Composites, Proceedings of the 12$^{th}$ Int. Conference on Composite Materials (ICCM-12), Paris, France, July 5–7, 1999.
4. M. Frieß, R. Renz, W. Krenkel: Graded Ceramic Matrix Composites by LSI-Processing, Proceedings of the 10$^{th}$ International Congress and 3$^{rd}$ Forum on New Materials, CIMTEC, Florence, Italy, July 14–18, 2002.
5. M. Frieß, W. Krenkel: Influence of Graphitization Temperature on the Properties of C/C-SiC manufactured via LSI-Processing, Proceedings of Topical Symposium V

(Ed.: P. Vincenzini) of the 9th International Conference on Modern Materials and Technologies, CIMTEC, Florence, Italy, June 14–19, p. 127, 1998.
6. B. Heidenreich, R. Renz, W. Krenkel: Short Fibre Reinforced CMC Materials for High Performance Brakes, in High Temperature Ceramic Matrix Composites (Eds.: W. Krenkel, R. Naslain, H. Schneider), WILEY-VCH, Weinheim, Germany, p. 809, 2001.
7. W. Krenkel, N. Lützenburger: Near Net Shape Manufacture of CMC Components, Proceedings of the 12th International Conference on Composite Materials (ICCM-12), Paris, France, July 5–7, 1999.
8. J. Schmidt, M. Scheiffele, W. Krenkel: Engineering of CMC Tubular Components, in High Temperature Ceramic Matrix Composites (Eds.: W. Krenkel, R. Naslain, H. Schneider), WILEY-VCH, Weinheim, Germany, p. 826, 2001.
9. W. Krenkel, T. Henke, N. Mason: In-Situ Joined CMC Components, in Key Engineering Materials, Proceedings of the International Conference on Ceramic and Metal Matrix Composites CMMC 96, Part 1, Trans Tech Publications Ltd., Zurich, Switzerland, p. 313, 1997.
10. W. Krenkel, T. Henke: Modular Design of CMC Structures by Reaction Bonding of SiC, in Joining of Advanced and Specialty Materials II (Eds.: M. Singh, J.E. Indacochea, J.N. DuPont, K. Ikeuchi, J. Martinez-Fernandez), ASM Conference Proceedings from Materials Solutions 99, Cincinnati, USA, 1999.
11. US Patent 6 086 814, 2000.
12. W. Krenkel: Entwicklung eines kostengünstigen Verfahrens zur Herstellung von Bauteilen aus keramischen Verbundwerkstoffen, PhD-thesis, University of Stuttgart, 2000.
13. R. Brandt, G. Neuer: Messung von Wärmetransporteigenschaften mit Hilfe eines instationären Verfahrens, Brennstoff-Wärme-Kraft, 33, 108–112, 1981.
14. G. Neuer: Spectral and total emissivity measurement of highly emitting materials, Int. J. Thermophys., 16, 257–265, 1995.
15. G. Neuer, G. Jaroma-Weiland: Spectral and total emissivity measurement of high temperature materials, Int. J. Thermophys., 19, 917–929, 1998.
16. G. Neuer, R. Kochendörfer, F. Gern: High temperature behaviour of the spectral and total emissivity of CMC materials, High Temp. High Press., 27/28, 183–189, 1995/1996.
17. R. Brandt, M. Frieß, G. Neuer: Thermal conductivity, specific heat capacity, and emissivity of ceramic matrix composites at high temperatures, High Temp. High Press., 35, 169–177, 2003.
18. G. Neuer, P. Pohlmann, E. Schreiber: in: VDI-Berichte 1379, VDI-Verlag, Düsseldorf, 173–178, 1998.
19. A. G. Evans: Perspective on the Development of High-Toughness Ceramics, J. Am. Ceram. Soc. 73 [2], 187–206, 1990.
20. K. Maile, A. Udoh: Ermittlung des Zusammenhangs Fehlerverhalten – Beanspruchung – ZfP-Größe in faserverstärkter Keramik C/C-SiC, report SFB 381 (1997–1999), University of Stuttgart, pp. 71–100.
21. H.-P. Maier: Thermomechanisches Verhalten von C/C-SiC, Materialkennwerte-Ermittlungen, report SFB 259 (1999–2001), University of Stuttgart, pp. 177–210.

## 7.2 Behaviour of Reusable Heat Shield Materials under Re-Entry Conditions

Fritz Aldinger, Monika Auweter-Kurtz, Markus Fertig *,
Georg Herdrich, Kurt Hirsch, Peter Lindner, Dirk Matusch,
Günther Neuer, Uwe Schumacher, and Michael Winter

For their re-entry into the atmosphere of the earth reusable space transportation vehicles have to rely on safe and economic heat shield materials, which are optimized with respect to high temperature application, to low total weight as well as to resistance against plasma-chemical erosion in the upper atmosphere. Very high demands are made on these heat shield materials: They not only have to be chosen with respect to low specific weight and high mechanical strength, their temperature stability and their resistance against chemical erosion have to be extraordinarily high. Since efficient radiation cooling is important, these materials, moreover, should be qualified by high total emissivity.

During the re-entry process the plasma, which is formed in front of the heat protection materials, is in a non-equilibrium state. The material erosion is dominated by plasma chemical reactions, by the passive-active transition for SiC based materials and by the catalytic efficiency, by which the recombination of atomic species from the gas phase is increased associated with the release of formation enthalpy at the material surface. For the selection of the appropriate materials and their protection layers detailed modelling of the catalysis and of the erosion was performed, and numerous investigations of the material behaviour at the transition from passive to active oxidation (and vice-versa), of the catalytic efficiency and of the heat flux distributions were necessary. Moreover, for the detailed erosion studies specific in-situ and ex-situ diagnostics were developed and applied. Due to the dramatic increase of the temperature and of the erosion of all materials in question above the high temperature threshold (which depends on the pressure), the re-entry trajectories have to be optimized, that the appearance of active oxidation is avoided.

Multiple use of the heat shield based on the composite material C/C-SiC (carbon fibre reinforced silicon carbide structure material) is successfully achieved by specific oxidation protection layers, which change their composition during the plasma exposure. A variety of protection layers on the basis of oxides, nitrides, carbides, metals, and amorphous covalent ceramics were developed and investigated in detail in different plasma wind tunnels (PWK). An improved protection is achieved by an increase of the activation energy of the erosive reactions. A double layer combined with a glazing process, moreover, led to protection of the carbon composites as well, resulting in strong reduction

* Universität Stuttgart, Institut für Raumfahrtsysteme, Pfaffenwaldring 31,
70550 Stuttgart; fertig@irs.uni-stuttgart.de

of the erosion rate. Only materials with high total emissivity have been considered, since they increased the radiation cooling and hence resulted in the reduction of the surface temperature.

### 7.2.1 Principles and Modelling of Heterogeneous Reactions

One of the most important phenomena concerning surface heat loads of re-entry vehicles is catalysis. In principle, catalysis increases the reaction rates. However, forward as well as backward reaction rates are speeded up by the same factor. Hence, catalysis has no influence on chemical equilibrium. In chemical nonequilibrium in contrast catalysis has a significant influence on composition and therefore on energy distribution among the different degrees of freedom.

#### 7.2.1.1 Heterogeneous Catalysis

Catalysis at the heat shield becomes an important issue if air is in chemical nonequilibrium close to the surface. The degree of chemical nonequilibrium at the surface mainly depends on freestream pressure, flight velocity and vehicle geometry [20]. In principle, the degree of dissociation of air increases with rising temperature. Since surface temperature is much lower than air temperature outside of the boundary layer, dissociation degree of oxygen and nitrogen would be lower close to the surface in chemical equilibrium. If air is in chemical nonequilibrium close to the surface, mainly atomic oxygen but also atomic nitrogen is present at the heat shield. In case of atom recombination at the surface, dissociation enthalpy is released, hereby increasing surface heat load. For most re-entry vehicles, effects due to surface catalysis are negligible at altitudes above about 100 km and below about 40 km. At high altitudes, reaction rates are very low due to the low density. Therefore, the amount of dissociation behind the shock wave is low, i.e. chemical reactions within the air can be neglected. At low altitudes, air is always in chemical equilibrium due to higher density. Hence, any speed-up of reaction rates does not lead to differences in vehicle loads. Within the intermediate altitude range the boundary layer is chemically frozen if atoms do not recombine at the surface. Therefore, dissociation degrees close to the surface becomes high and catalytic recombination at the surface may increase heat load by up to a factor of three for typical re-entry vehicles and flight paths [7]. Within this intermediate altitude range where catalysis becomes important, maximum heat loads along the trajectory occur. The associated conditions called peak heating typically arise within the altitude range from 80 km to 60 km. Up to now, heat shield design is based on the fully catalytic design assumption i.e. all atoms recombine at the surface which is believed to be the worst case. However, this worst case assumption is justified only in the vicinity of the stagnation point. Downstream from the stagnation point partial catalytic heat loads may become higher if catalysis increases along the surface depending on surface temperature [9]. Therefore, it is necessary to determine catalytic phenomena at the surface accurately in order to optimize heat shield mass and reliability. Hence, detailed experimental and numerical investigations of the reactive processes at the surface are required.

The catalyst itself does not appear within the chemical balance equation, i.e. after the reaction the catalyst reaches again its initial state. Several operating principles for catalytic reactions are possible. In chemical industry, the most important operating principle is the decrease of activation energy for the balance reaction. If for example activation energy of a reaction is higher than dissociation energy of the starting substances, the desired product is not formed without catalyst. However, this is not the case for recombination of oxygen and nitrogen during re-entry, since no activation energy is required [30]. Another possibility for speed-up is an increase in collision probability. As already mentioned, nonequilibrium effects during re-entry depend on density. If density is low, collisions between gas species are rare. Moreover, the collision of two atoms does not immediately form a stable molecule. Actually, a so-called activated complex is formed with enough energy for dissociation. Hence, another collision is necessary in order to release the excess energy hereby stabilizing the molecule. Although this is possible at the surface, effectiveness of this process is not sufficient in order to explain high recombination probability. Catalytic surfaces usually allow for adsorption, i.e. sticking of at least one of the starting substances [37]. Generally, two kinds of adsorption processes are distinguished. While adsorption energy for physisorption is low (of the order of $20\ \mathrm{kJ\ mol^{-1}}$) since only Van der Waals forces are involved, in case of chemisorption a valence bond forms between surface and adsorbate such that adsorption energy is of the order of $200\ \mathrm{kJ\ mol^{-1}}$. Usually, density of adsorbates is of the order of density within a fluid [37]. Hence, collision probability between atoms in the gas and adsorbed atoms is higher by several orders of magnitude as compared to collisions between gaseous species alone. Moreover, adsorbed atoms already have less energy than gaseous atoms so that the deactivation of activated complex is not necessary. With adsorbed atoms again two recombination mechanisms may be distinguished [5, 8]. If a gaseous atom recombines with an adsorbed one the reaction mechanism is called Eley-Rideal type. Activation energies for Eley-Rideal type mechanism are typically about 5% to 10% the adsorption energy. In the second recombination mechanism of Langmuir-Hinshelwood type two adsorbed atoms recombine with each other. The rate determining step of this reaction is typically the energy required to mobilize an adsorbed atom hereby allowing the atom to move along the surface. The mobilizing energy is usually considered to be about half of the adsorption energy.

Catalytic efficiencies in terms of recombination coefficients as defined in section 4.5 for atomic oxygen recombination on Silicon Dioxide ($SiO_2$) and Silicon Carbide (SiC) are shown in Fig. 7.2.1. Atomic nitrogen recombination coefficients are plotted in Fig. 7.2.2. The figures show results from the catalysis models described in [8] together with experimental data from Stewart [38]. Apparently, the results for $SiO_2$ are not accurate for $T^{-1}$ higher than $0.001\ \mathrm{K^{-1}}$,

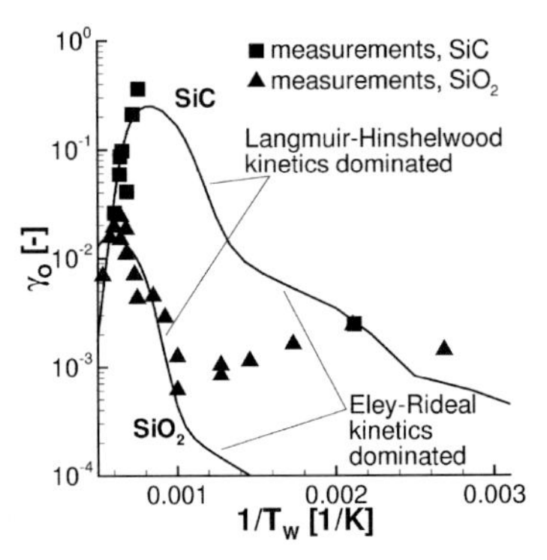


Figure 7.2.1: Recombination coefficients for atomic oxygen on SiC and $SiO_2$. Measurements are from Stewart [38].

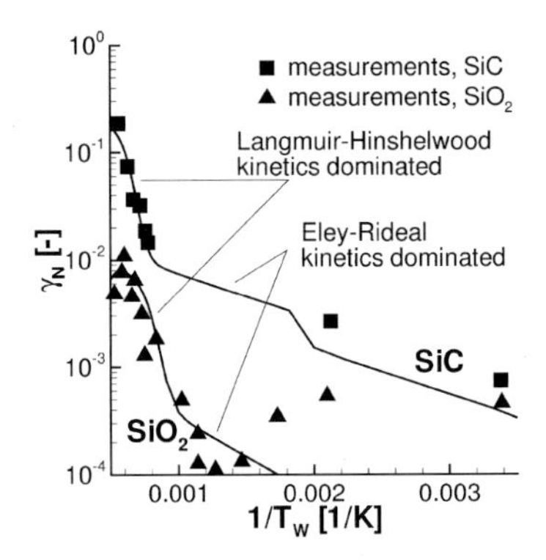


Figure 7.2.2: Recombination coefficients for atomic nitrogen on SiC and $SiO_2$. Measurements are from Stewart [38].

i.e. for temperatures about 1000 K. This inaccuracy arises since physisorption, which is assumed to be dominant for catalysis below 1000 K [24] was neglected based on the assumption that lower SiC surface temperatures are unimportant for typical re-entry missions. Due to the lower activation energy Eley-Rideal recombination dominates at low temperature. Associated with lower activation energy is the smaller gradient as shown in the figures. Hence, Langmuir-Hinshelwood type reactions dominate at high temperatures. The decrease of recombination coefficients at the upper temperature limit depends on thermal desorption of adsorbates. Depending on adsorption energy, adsorbed atoms are released into gas-phase with increasing temperature.

### 7.2.1.2 Redox Reactions Including Active and Passive Oxidation

If only catalytic reactions would apply at the surface, infinite flights without exchange of the heat shield were possible. Unfortunately, additional reactions at the surface generate gaseous species such that mass and thickness of the heat shield decrease. The recession speed mainly depends on the resistivity of the heat shield with regard to oxidation. In the past, carbon based materials have been used for thermally highly loaded surface areas e.g. for the US-Shuttle Orbiter. Although carbon is stable up to temperatures of about 3000 K, surface protection is required since carbon can be easily oxidized. Although the high-temperature heat shield of the US-Shuttle Orbiter made of Reinforced Carbon Carbon (RCC) has a protective SiC layer, multiple use upper temperature limit is 1900 K. The maximum use temperature limit of the newly developed C/C-SiC, C/SiC and SiC/SiC heat shield materials is some 100 K higher. If no additional protection is applied the surface of all of these composite materials consists of SiC. Three major constituents arise if SiC is oxidized in air at high temperature.

$$\mathrm{SiC} + \frac{3}{2}\mathrm{O_2} \rightleftharpoons \mathrm{SiO_2} + \mathrm{CO} \quad \Delta H^0 = -948.8\,\frac{\mathrm{kJ}}{\mathrm{mol}} \tag{1}$$

$$\mathrm{SiC} + \mathrm{O_2} \rightleftharpoons \mathrm{SiO} + \mathrm{CO} \quad \Delta H^0 = -138.5\,\frac{\mathrm{kJ}}{\mathrm{mol}} \tag{2}$$

As can be seen from the negative sign of the enthalpy difference both oxidizing reactions are exothermal i.e. energy is released so that gas and surface are heated. The major difference between the reactions is that $SiO_2$ is solid (or liquid if temperature is above 2000 K) while SiO is gaseous. In both reactions gaseous CO is formed. Reaction (1) is usually associated with the term passive oxidation. If only this reaction applies one carbon atom from the surface is replaced by two oxygen atoms. Therefore, a mass increase of 20 kg $mol^{-1}$ follows. The $SiO_2$ formed at the surface results in a solid quartz layer at the surface. This layer acts as a diffusion barrier for oxygen such that oxidation decreases with time. Within the reaction (2), which is associated with the term active oxidation, only gaseous products are formed such that a mass loss of 40 kg $mol^{-1}$ arises. In order to determine whether active or passive oxidation is present equilibrium considerations are an initial step. Hilfer investigated several reactions and reaction models for SiC in air [14]. His finding was that the model of Wagner [45] fits best to experimental results. Following Hilfer and Wagner, transition from active to passive oxidation is determined by the equilibrium reaction between SiO and $SiO_2$, i.e.

$$\mathrm{SiO_2} \rightleftharpoons \mathrm{SiO} + \frac{1}{2}\mathrm{O_2} \quad \Delta H^0 = 819.3\,\frac{\mathrm{kJ}}{\mathrm{mol}} \tag{3}$$

It must be noted, that similar reactions apply for the oxidation of the surface due to atomic oxygen and nitrogen [14].

A problem arises concerning the definition of the transition from active to passive oxidation. One possible definition is the mass change of a sample. In this case, increasing mass would indicate passive oxidation while decreasing mass indicates active oxidation. This definition has the disadvantage that mass increase reduces with time since the increasing $SiO_2$ layer thickness hampers oxygen access to the underlying SiC. Without gas exchange, stationary conditions are characterized by constant mass of the sample. In a gas flow, however, stationary conditions are always combined with a decrease of sample mass since gaseous reaction products are transported away by the flow. Hence, this definition is unsuitable in order to characterize oxidation regime in flows.

Other definitions rely on the thickness of the $SiO_2$ layer in the broadest sense. Passive oxidation is associated with $SiO_2$ layer formation, active oxidation with formation of SiO without generation of a $SiO_2$ layer. Discrepancies arise concerning the definition of the transition from active to passive and vice versa. The transitional process from passive to active oxidation involves a phase where the $SiO_2$ layer is removed. This process consumes 819.3 kJ $mol^{-1}$ of energy according to reaction (3) such that surface is cooled. Simultaneously, SiC removal probability increases depending on decreasing protection and reaches its maximum with complete removal of the $SiO_2$ layer (cf. Fig. 7.2.3 and Fig. 7.2.4). Without quartz layer only reaction (2) remains. Hence, missing of additional cooling and species formed due to reaction (3) together with the exposure of the bare SiC surface to the hot gas lead to a sudden change of surface heat load and near wall gas composition. In case of a gas flow, where the conditions at the edge of the boundary layer are nearly unaffected by the change of surface kinetics, a rapid surface temperature increase follows which in turn leads to a rapid increase of surface erosion rate. The inverse transition from active to passive requires lower surface temperature than the transition from passive to active since erosion of SiC tends to decrease oxygen partial pressure below the value during transition from passive to active oxidation (cf. Fig. 7.2.4). Hence, a kind of hysteresis arises due to the dependence of transition on temperature and oxygen partial pressure and the coupling of these values with surface reactions. Figure 7.2.3 shows isobars for SiC re-

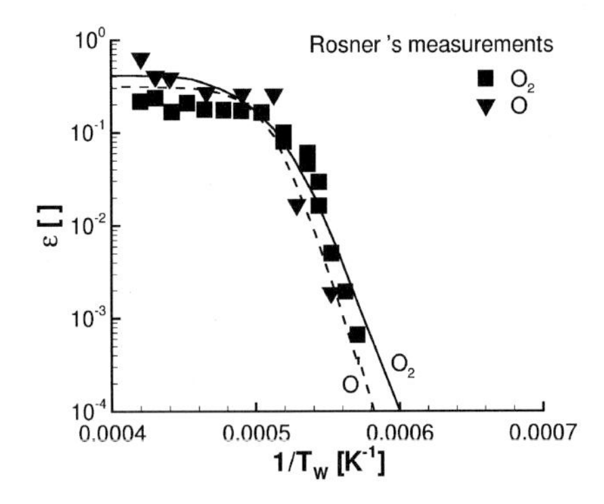


Figure 7.2.3: Isobars for SiC removal probability for atomic oxygen at 1.33 Pa and molecular oxygen at 4 Pa. Measurements are from Rosner and Allendorf [35].

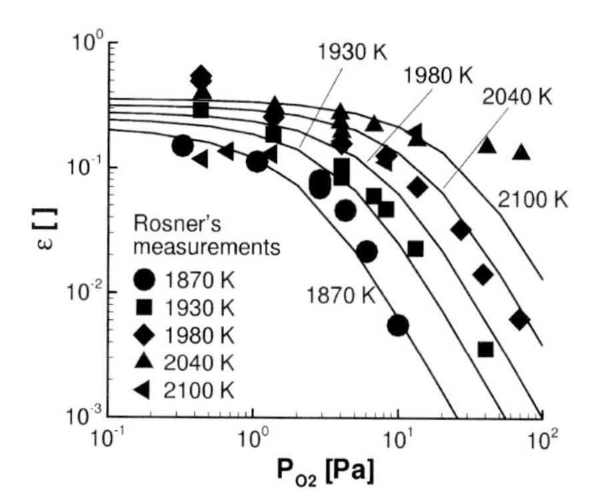


Figure 7.2.4: SiC removal probability isotherms for molecular oxygen. Measurements are from Rosner and Allendorf [35].

moval probability, i.e. oxidation rate divided by oxygen impingement rate, for oxidation of SiC by atomic oxygen at 1.33 Pa and molecular oxygen at 4 Pa comparing measurements of Rosner and Allendorf [35] and the surface reaction model developed within SFB 259 [8]. As can be seen from the figure the low pressure isobars are reproduced very well for temperatures below 2250 K. The isotherms plotted in Fig. 7.2.4 show differences comparing surface model [8] and measurements [35]. Differences may arise due to the phase change of the quartz layer from solid to liquid. However, no final statement concerning differences is possible so far.

#### 7.2.1.3 Surface Reaction Model Applied to MIRKA Re-Entry Flow

Within the framework of the ASTRA programme, the first successful Western European re-entry experiment named MIRKA has been reinvestigated by applying the surface reaction models described in [8] to three flow conditions along the trajectory as defined in [7]. MIRKA had a spherical shape with a diameter of 1 m.

Figure 7.2.5 shows a plot of computed surface heat flux distributions versus the angle with regard to role axis. Multiple surface models were applied to the flow at $t = 1301$ s when peak heating conditions arose in about 60 km of altitude. The plot shows that the heat flux computed with the fully catalytic design assumption is about three times higher compared to the assumption of a non catalytic surface. Between these curves, which mark the upper and lower limits for the catalytic surface heat flux in the stagnation area, computed heat fluxes determined with the $SiO_2$ and the SiC catalysis models are up to 1.25 MW $m^{-2}$ and 1.5 MW $m^{-2}$, respectively. Numerical investigations were performed in order to obtain whether the conditions allow for passive to active transition and vice versa. For the study of passive to active transition an initial SiC simulation

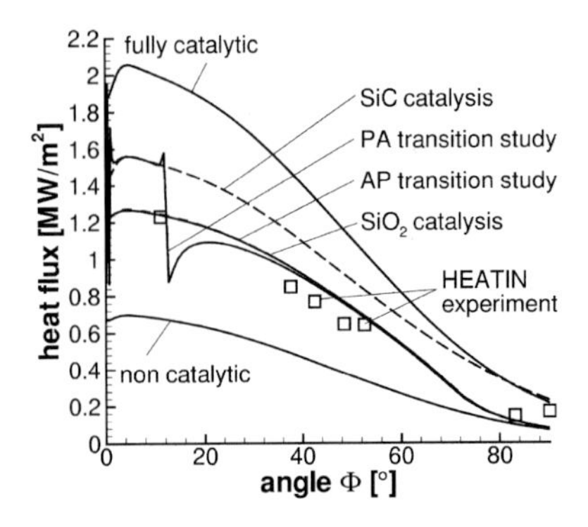


Figure 7.2.5: Several gas-surface interaction models for MIRKA peak heating conditions at $t = 1301$ s in about 60 km of altitude in comparison to HEATIN experiment.

with a quartz layer of 1 mm thickness on top in radiation equilibrium was generated. Allowing for active oxidation, only little changes in surface heat flux were observed. The $SiO_2$ layer remained intact all over the surface, although thickness increases downstream from stagnation point. Since no higher heat loads arose during flight, transition from passive to active oxidation is unlikely. Therefore, this solution is assumed to represent reaction kinetics of the MIRKA during flight most accurately. Comparing the computed heat fluxes with the data from the HEATIN experiment shows very good agreement.

The following study focused on active to passive transition. Initially, a simulation without $SiO_2$ layer in radiation equilibrium was carried out. In a second step, the formation of a quartz layer was allowed hereby fixing surface temperature to the values found within the initial step. Formation of a $SiO_2$ layer was observed downstream of $\Theta \approx 15^\circ$. In a final step a radiation equilibrium boundary condition was applied. As can be seen from Fig. 7.2.5 transition from active to passive oxidation was found close to the stagnation point and downstream of $\Theta \approx 12^\circ$. This result indicates that MIRKA was already within the hysteresis curve of the transitional area. It must be noted, that heat flux computed with active oxidation equals heat flux obtained with the SiC catalysis model. A similar finding was made for the interrelation between passive oxidation simulation and $SiO_2$ catalysis model. From the computed data it follows, that the difference in heat flux between active and passive oxidation mode is determined by higher nitrogen recombination probability on SiC.

Similar comparative simulations were performed for the MIRKA flow conditions at about 70 km and 46 km of altitude. For both flow conditions no active oxidation mode was found.

## 7.2.2 Characterization of High-Temperature Oxidation and Catalytic Behaviour of TPS Materials

The oxidation regimes and the catalytic efficiency of materials under investigation are strongly related. This relation is mainly built via the heat flux to the heat shield material which, as shown above, is influenced by the catalytic efficiency and significantly affects the oxidation behaviour via the surface temperature.

### 7.2.2.1 Experimentally Observed Influence of Catalytic Efficiency

The heat flux experienced by a material under a specified plasma condition is an important piece of information for statements on the catalytic behaviour of such a material. In the past, oxides of copper were used as reference material for calorimetric measurements (see section 4.7.2) due to their high catalytic efficiency. If a water-cooled probe is used, the convective heat flux due to the temperature difference between plasma and surface should remain largely unchanged. Higher catalytic efficiency then yields higher recombination rates and therefore higher heat fluxes. Figure 7.2.6 shows the heat flux reduction relative to highly catalytic constantan for different materials investigated by Stöckle [41] measured with a water-cooled probe with Gardon Gage sensors. The figure shows that with different materials decisive heat flux reductions under the

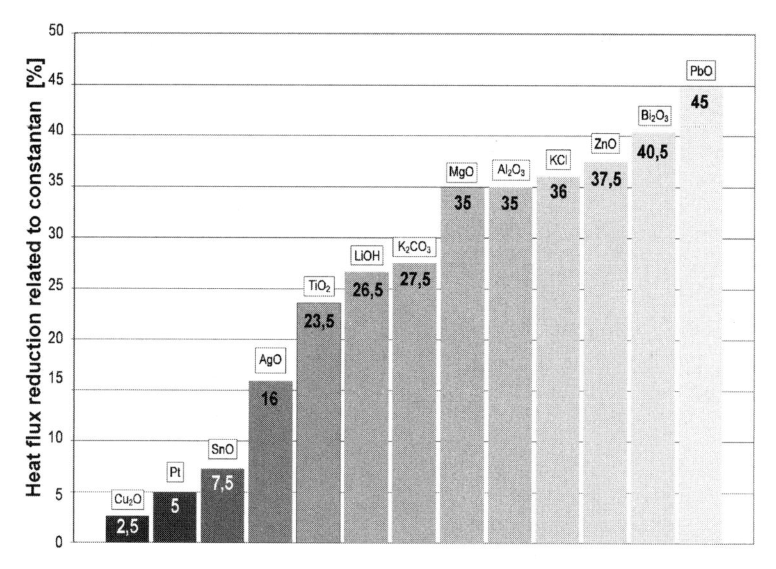


Figure 7.2.6: Relative heat flux reduction in relation to constantan reduction using Gardon Gage based heat flux measurements with a water-cooled probe.

same plasma conditions are observed. For lead oxide with rather low catalytic efficiency, a heat flux reduction of more than 40% is measured in comparison with copper.

These investigations offer initial information on the significance of catalytic efficiency. Subsequently the need for TPS materials with low catalytic efficiencies with regard to the recombination of atomic oxygen and nitrogen becomes evident. By using materials with known catalytic efficiency for a flight experiment, for example, information on the entry plasma composition can be obtained depending on the measured heat-flux difference. However, such an investigation requires detailed calibrations using plasma wind tunnels (PWKs). In addition, surface catalytic efficiency is a function of surface temperature such that new measurement techniques have to be applied to determine the real wall temperature dependent heat fluxes.

For the investigation of catalytic effects on hot surfaces heat-flux measurements with the pyrometric double probe as described in section 4.7.2 were performed with PWK2-RD5 (compare section 4.7). The total heat flux and front-side temperatures at maximum heat flux and in the stationary state are depicted in Fig. 7.2.7 versus the time difference to the point when the probe started moving into the plasma.

Different heat-flux reductions from the maximum to the steady state value are seen for the diverse materials. Since the effects which are responsible for this decrease can not be quantified yet as explained in section 4.7.2, a comparison between the heat fluxes to the different materials is given for both values in Tab. 7.2.1.

Although the quantitative differences in convective heat flux and structural losses are not yet known general tendencies for the different catalytic effects

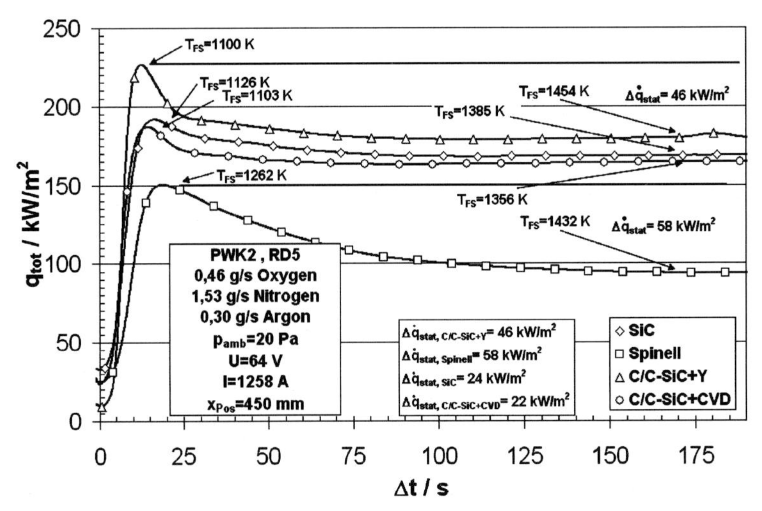


Figure 7.2.7: Heat-flux measurements on different materials using catalytic pyrometer probe.

Table 7.2.1: Pyrometrically measured heat fluxes on different materials.

| | $T_{Wall,\ stat}$ [K] | $\dot{q}_{tot,stat}$ [kW/m$^2$] | $\frac{\dot{q}_{tot,stat}}{\dot{q}_{tot,SiC}}$ | $T_{Wall,\ max}$ [K] | $\dot{q}_{tot,max}$ [kW/m$^2$] | $\frac{\dot{q}_{tot,stat}}{\dot{q}_{tot,SiC}}$ |
|---|---|---|---|---|---|---|
| SiC | 1390 | 166 | 1 | 1126 | 192 | 1 |
| CVD SiC C/C | 1360 | 164 | 0.99 | 1103 | 189 | 0.98 |
| Yt-C/C | 1450 | 180 | 1.08 | 1100 | 230 | 1.20 |
| Synthetic Spinel | 1430 | 94 | 0.55 | 1262 | 150 | 0.78 |

can be seen: The catalytic efficiency of Yttrium coated carbon composite material (Y–C/C) is higher than the catalytic efficiency of SSiC and carbon composite material with a SiC coating (C/C–CVD–SiC). This effect is stronger at the heat flux maximum at lower temperatures. Furthermore, the catalytic efficiency of SSiC is higher than the catalytic efficiency of Spinel ($MgAl_2O_4$). This effect is stronger in the steady state at higher temperatures. Therefore, within the cooperation with the Japanese Aerospace Agency NASDA where a full scale mock-up model of the HYFLEX nose structure was tested in the SCIROCCO plasma wind tunnel, IRS contributed catalytic efficiency sensors using SSiC and synthetic Spinel ($MgAl_2O_4$) material [34].

Based on such measurements, particularly the catalytic efficiency probe measurements in PWK3 using pure oxygen and nitrogen, it is planned to determine the catalytic recombination coefficients for SSiC and synthetic Spinel. For this purpose, it is planned to apply the theory of Goulard (see section 4.6.1) with the support of copper heat flux measurements, total pressure measurements and enthalpy measurements. Presently the obtained data are being evaluated [34].

#### 7.2.2.2 Oxidation Behaviour

Oxidation regimes of Si based heat shield materials are investigated using the plasma wind tunnels at IRS [2]. Here, various investigations [10, 14] have shown that the so-called transition from passive to active oxidation is accompanied by a sudden increase in wall temperature by several hundreds of Kelvin which in turn leads to a significant increase of the specific mass loss rate. Therefore, the transition can become mission critical for an atmospheric entry vehicle and must be avoided. Concerning the X-38, maximum temperature levels in the stagnation point region are quite high. A comparison with the oxidation regimes shows that a transition could appear, see Fig. 7.2.8.

Other investigations depict flight experiments where the transition seems to have occurred [10]. Within the investigations the passive–active transition (PAT) has been investigated such that a data base could be obtained and incorporated into present entry trajectory designs. Within the ASTRA programme,

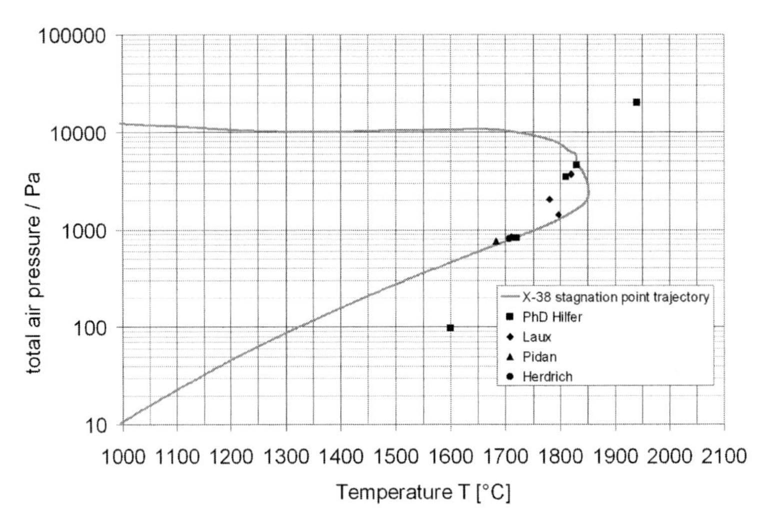


Figure 7.2.8: Passive to active transition (PAT) using plasma source RD5 together with X-38 stagnation point trajectory.

IRS is investigating basic SSiC materials and C/C-SiC (DLR-S and MAN-T) using pure oxygen, pure nitrogen and air plasmas to identify the involved reaction schemes. This is of great importance because previous investigations imply that a decisive share of the energy seen by the thermal protection system during the active oxidation could be contributed by nitrogen reactions that are ongoing within the oxidation regime [10]. Test campaigns in oxygen atmospheres have been performed using the inductively heated plasma wind tunnel PWK3. This facility offers the advantage of being able to be operated even with pure reactive gases such as oxygen [12]. The investigation of the erosion behaviour during the active oxidation showed major differences concerning both the temperature increase and the erosion rate. In PWK3-IPG3 oxygen plasmas, specific mass loss rates for SSiC material samples between 100 and 170 $kg/(m^2\,h)$ were measured. In air plasmas using PWK2-RD5 mass loss rates between 30 and 60 $kg/(m^2\,h)$ were detected. Simultaneously, the temperature jumps in the RD5 air plasmas are up to 400 K while the $\Delta T$ in the inductively driven PWK3 with oxygen plasmas was about 200 K for all oxygen conditions that have been investigated [26]. These data have to be treated carefully due to the different facilities and the corresponding flow conditions. However, the correspondence with the modelling results in section 7.2.2 is fairly good as they predict a higher mass loss rate in pure oxygen plasmas due to the increased erosion of the material sample. In addition, a smaller temperature increase is predicted during active oxidation in oxygen plasmas. This has also been confirmed by the measurements.

The transition from passive to active oxidation can be detected with different diagnostic systems as shown in Fig. 7.2.9. For better visibility, the offset of the time axis was chosen that way that each transition point is found at 0 s. For

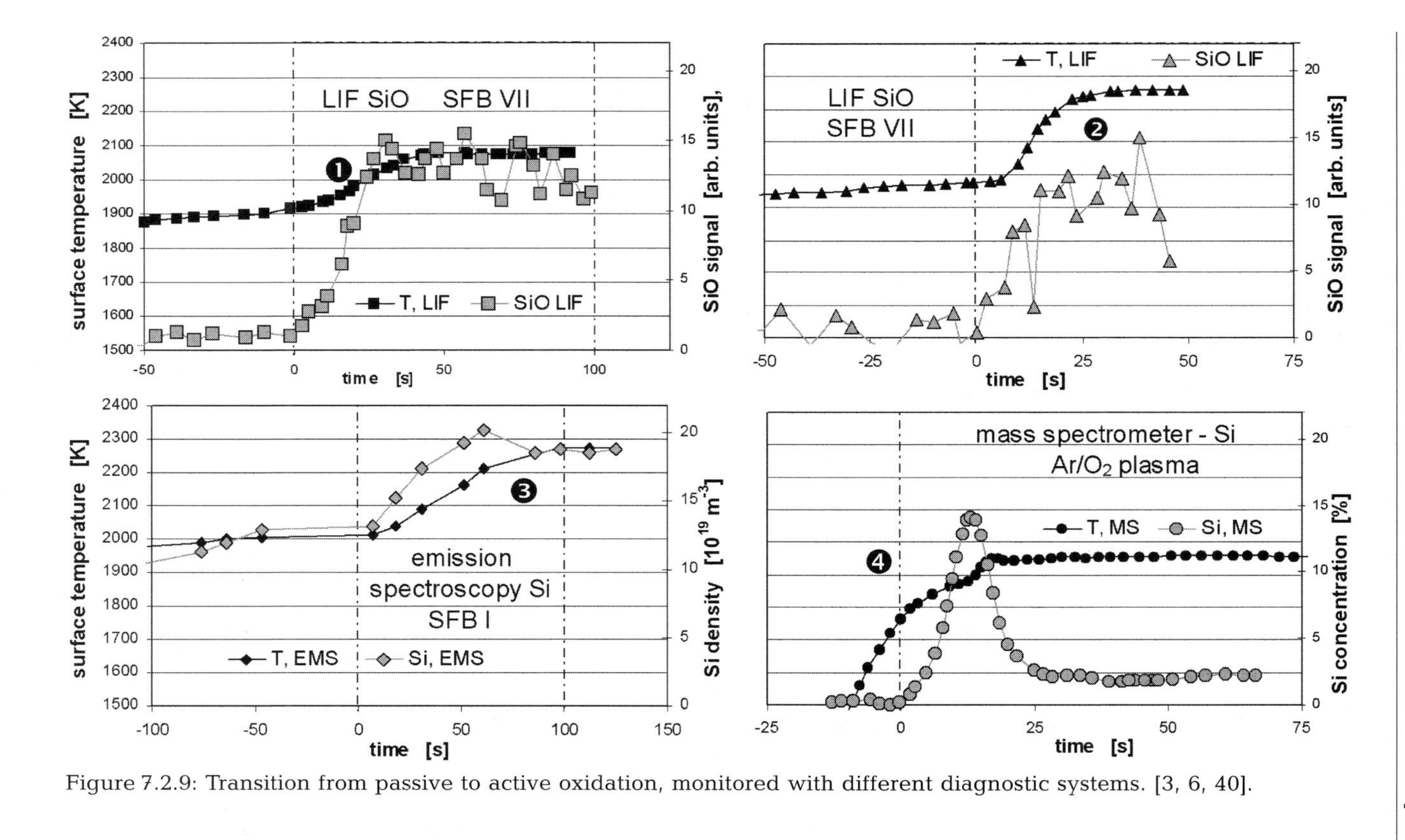


Figure 7.2.9: Transition from passive to active oxidation, monitored with different diagnostic systems. [3, 6, 40].

each method the measured quantity as well as the surface temperature is shown. As already described in section 4.7.3, Silicon Oxide (SiO) can only be detected in the case of active oxidation or in the transition regime and can be monitored with planar LIF [6].

Two data sets for the plasma condition of the test case SFB VII (graph 2) and a lower ambient pressure of 110 Pa (graph 1) are presented. The signals show a steep increase of the SiO signal at a discrete time which correlates very well with the temperature jump of the material sample. Similar behaviour can be found in the Silicon (Si) density determined from emission spectroscopic measurements for the test condition SFB I (graph 3) [40].

In addition, mass spectrometer investigations in which the plasma was extracted through a small hole in the material sample show this steep increase of the Si mass number (graph 4) [3]. After some seconds a maximum is reached and the Si signal decreases to an almost constant value. This behaviour might indicate the degradation of the $SiO_2$ layer while the almost constant signal after some 30 s represents the regime of active oxidation. These investigations were performed in an argon–oxygen plasma. This same tendency, i.e. the sinking signal of the erosion products after a discrete time, may be seen also in the LIF data. Unfortunately, the data acquisition was stopped too early to make a clear statement. Further measurements will investigate this phenomenon in more detail to verify whether the $SiO_2$ layer is already lost at the time of the temperature jump or not.

In Fig. 7.2.10 the transition from passive to active oxidation of an SSiC material sample ($t < 1300$ s) is shown. This was achieved by moving the material

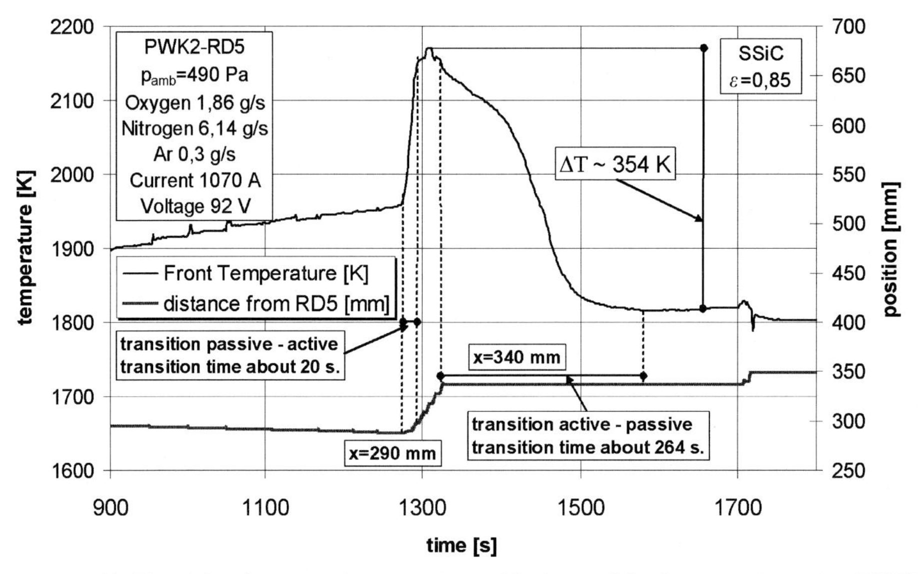


Figure 7.2.10: Transition from passive to active oxidation and back to passive using PWK2 RD5.

sample closer and closer to the plasma generator until the transition arose. This can be seen by the sudden increase of the wall temperature. Then the sample was moved back quickly into a zone where usually passive oxidation appears. However, it took about 264 s until the wall temperature reached the passive oxidation value. In fact, such a hysteresis has to be considered carefully, as similar behaviour during flight, i.e. that the active oxidation is kept alive for several minutes, despite regions of lower heat loads being reached, would be mission critical.

### 7.2.3 Developments and Investigations of Protection Layers for Reusable Heat Shield Materials

A great variety of different heat shield materials for protection against oxidation and plasma chemical erosion during the re-entry phase of reusable space transportation vehicles was developed and investigated in detail by simulation experiments applying the plasma jet of a plasma wind tunnel PWT (see section 7.2.4.7). In this chapter the different methods of protection layer production, the diagnostics and the verification of the simulation experiments in the plasma wind tunnel as well as the success and the progress obtained in protecting reusable heat shield materials are presented.

#### 7.2.3.1 Production and Characteristics of Protection Layers

The production of these layers in most cases needs specific conditions and treatments: Layers produced by PCVD (plasma or physical chemical vapour deposition) like SiC, $SiO_2$ and $SiN_x$ [31, 32] ask for large vacuum reactors. Plasma sprayed layers of $Y_2SiO_5$ [44] and pure metal layers like Ti and carbides like TiC [25] need reduced pressure, clean room conditions and heat treatment. Amorphous Si–C–N and Si–B–C–N ceramic layers were derived applying the polymer-pyrolysis method by the lamination with metal organic precursors in Argon atmosphere followed by thermal treatment [11]. The lamination of the C/C and C/C-SiC substrates is reached by the dip coating procedure, the immersion of the substrate into a slurry of precursor solved in toluol and a solid filler material, followed by drying and heat treatment in Ar at 1050 °C for conversion into a pyrolysed ceramic layer of Si–B–C–N [11]. The pyrolysis step is due to the thermal composition of silicon and boron containing organic precursors to an amorphous covalently bound layer during the heat treatment in Ar atmosphere, which leads to gaseous species formation like $CH_4$ and $H_2$. Repeating the dip-coating and pyrolysis processes several times results in a coating thickness of as high as 60 to 100 μm.

A simple and cheep coating method was developed for $TiO_2$, $HfO_2$ and $ZrO_2$: $TiO_2$ powder of size > 1 μm, for example, was sprayed or painted onto the substrate together with some organic polymers and a volatile solvent [33] and successfully tested in the plasma jet [16, 17]. Due to the volatile solvent this coating method results in nearly perfect surface infiltration into materials like C/C-SiC and SiC or Si(B)CN.

#### 7.2.3.2 Diagnostics for the Tests of the Protection Layers in the Plasma Wind Tunnel

The analysis of the interaction region between the plasma and the heat shield target is performed by several different spectroscopic diagnostic and laser scattering systems [19, 21, 36, 43] to determine the plasma parameters. The characteristics of the space and time dependent erosion products and the *in-situ* erosion rates are deduced from absorption and emission spectroscopy [1, 13, 16–18, 22, 39, 43]. The experimental investigations are supported by numerical simulation codes, using the atomic data from the ADAS (Atomic Data Analysis Structure) system [42], which allow to calculate the spatial distributions of electron density and temperature in the boundary region [27]. The long-term reproducibility of the plasma parameters is registered by monitor systems using selected line and band emissions.

The *in-situ* analysis of the target surface and of the qualification of the protection layers during the experiment was obtained from visible and near infrared

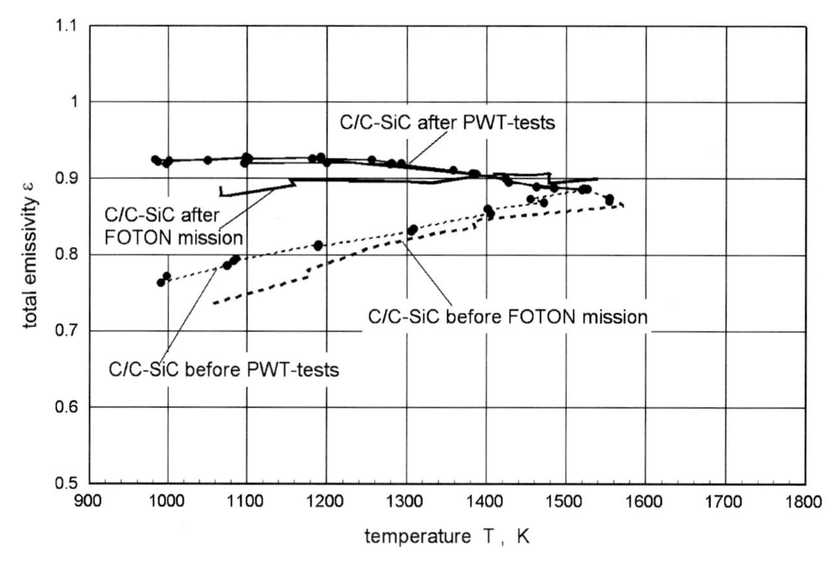


Figure 7.2.11: The agreement of material emissivity changes before and after plasma wind tunnel (PWT) experiments with those before and after a space mission with the Russion FOTON capsule verifies the simulation results obtained from material tests in the plasma wind tunnel.

imaging and mid-infrared detection systems. A 2-d CCD-camera with spatial and temporal resolutions of 0.2 mm and 0.3 s recorded macroscopic surface effects like local overheating, rapid temperature rise due to the transition from passive to active oxidation and its propagation as well as local failures in the protection layer [1, 18]. The formation of a $SiO_2$ self protection layer on C/C-SiC was detected by the Fourier transform infrared spectrometer (FTIR) [18]. Comparative measurements of the relative emissivities with respect to uncoated segments are applied to register the temporal behaviour of the surface structure in the plasma jet [15].

The simulation of the material behaviour in space by laboratory experiments in the plasma wind tunnel (PWT) was impressively verified by comparing the emissivities of C/C-SiC before and after the PWT tests with those before and after a space mission with the Russian FOTON capsule [28, 29]. Under comparable conditions the material changes in both cases lead to similar emissivities, as given in Fig. 7.2.11.

### 7.2.3.3 Protection Material Tests and Results

Material tests for protection layers of $TiO_2$, $SiO_2$, $ZrO_2$, Si–B–C–N or CVD-SiC were performed in the plasma wind tunnel. The qualification of all heat shield protection layers was obtained from comparative plasma tests in the plasma wind tunnel (PWT) for different exposure times within the region of passive oxidation [13, 14, 17], which is given by the region between the low and the high temperature threshold as given in [17].

The erosion rates for surface temperatures of about 2000 K and exposure times up to 9000 s are presented in Fig. 7.2.12 [17, 27].

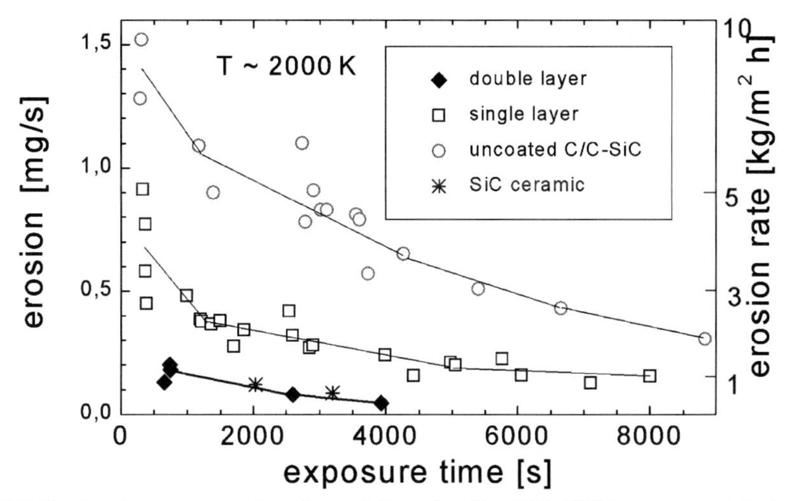


Figure 7.2.12: Erosion rates as function of time for the C/C-SiC base material, for single protection coatings and for double layers in comparison with SiC sinter ceramics.

For unprotected C/C-SiC base material the erosion rate is continuously higher compared to protected material. The decreasing slope is due to the formation of a self protection layer of $SiO_2$ with increasing exposure time within the region of passive oxidation. Single protection layers like $TiO_2$, $SiO_2$, $ZrO_2$, Si–B–C–N, CVD-SiC, TiC or Ti, nearly independent from the materials used, generally show a significantly reduced erosion rate.

During these experiments two different kinds of protection mechanisms could be distinguished: Covalently bound materials like CVD-SiC or Si–B–C–N work like solid barrier layers, they prevent oxygen diffusion and do not react with the carbon or SiC of the substrate. The surface of these materials is covered only by a thin $SiO_2$ self protection layer during the experiment. Ionicly bound materials like $TiO_2$ or $SiO_2$ form a melt and by this way cover the surface of the substrate, they reduce the oxygen and carbon diffusion and are able to heal cracks or small delaminations, but they do react with the substrate.

Additionally, $TiO_2$ exhibit an extremely high thermal emission coefficient at high temperatures. For temperatures above 1700 K the $TiO_2$ coatings reach an emissivity as high as 0.9, which is responsible for the excellent radiation cooling. *Ex-situ* measurements of the emissivity were developed by Neuer [28] for heat shield materials before and after space missions [29] showing no major difference in emissivity between tests in the plasma jet and in real space missions. Comparative *in-situ* measurements of the relative emissivity between coated and uncoated parts of the target [15] are applied to register the temporal behaviour of the surface layers during plasma erosion.

Protection layers of combined materials, so-called double layers, have been derived by using a covalently bound Si–B–C–N or SiC layer covered by a second ionicly bound $TiO_2$ layer. After a formation time these double layers result in erosion rates as low as about 0.1 $kg/m^2\,h$ at 2000 K and 1500 Pa. These values are comparable to those of bulk SiC ceramics and are one order of magnitude lower than those for unprotected C/C-SiC [18]. This high protection quality is accomplished by the addition of the two protection effects of covalent and ionic shield materials. The strong glazing effect enforced by the $TiO_2$ coating closes all cracks and holes in the SiC or Si–B–C–N layer and the SiC or Si–B–C–N layer works like a barrier, mostly chemically inert against the substrate and the titanium oxide. Additionally, molten $TiO_2$ increases the radiation cooling by high emissivity and possesses a surpassing adhesion to covalently bound materials and no smearing effect during experiments exposed under 45° angle to the plasma jet. After completion of the plasma jet experiments with the double layers, the SiC or Si–B–C–N layer is still present, and the cracks caused by the plasma treatment are closed by a separated $SiO_2$-$TiO_2$ glass [15]. In Fig. 7.2.13 the strong glazing effect is demonstrated in a microscope image of a SiC layer with a $TiO_2$ coating on top after an exposure of $<2000$ s at temperatures up to 2000 K and stagnation pressures of 1500 Pa in an air like plasma. The $TiO_2$-$SiO_2$ glass, however, crumbles from the surface, but it can be renewed by the cheep and simple procedure described before.

The physics behind this effective high temperature protection of the C/C–SiC basis material by additional layers is the increase of the effective binding

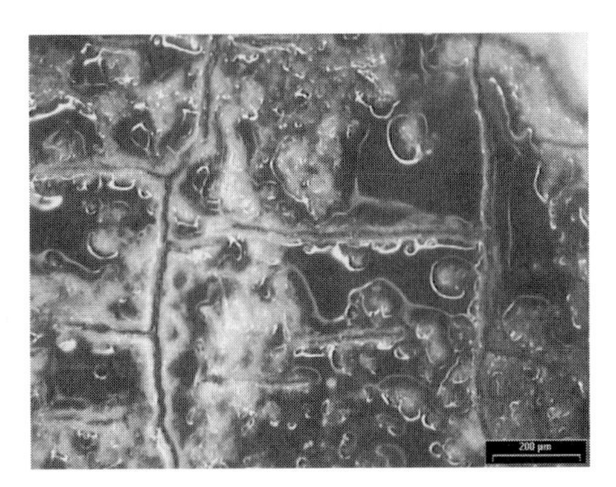

Figure 7.2.13: Glazing effect on a double layer consisting of SiC and $TiO_2$ on top.

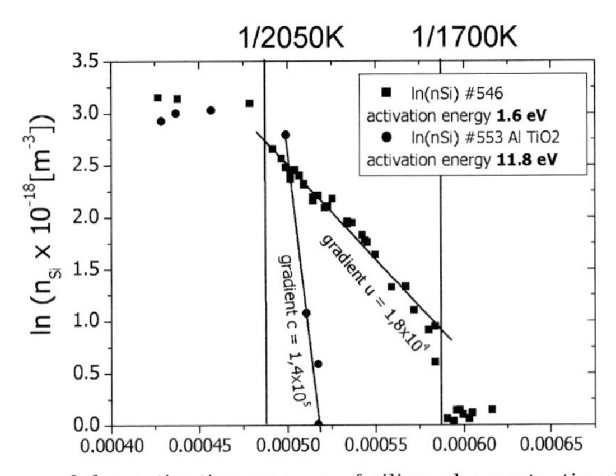


Figure 7.2.14: Increase of the activation energy of silicon by protection layers.

energy for Si and C typically by nearly one order of magnitude. This increase is shown in the Arrhenius plot in Fig. 7.20.14 for Silicon. In this case the activation energy is enhanced from 1.6 eV to 11.8 eV in case of the $TiO_2$ protection layer [27]. The Arrhenius plot is based on the *in-situ* recorded erosion rate of Si measured by absorption spectroscopy during a scan of the surface temperature from below the low temperature threshold across the region of passive oxidation to above the high temperature threshold.

Most of the protection coatings change their composition during the exposure in an air like plasma. For the exploration of the erosion characteristics and mechanisms the composites and the surface structure were analysed applying depth profiles and EDX. For uncoated C/C-SiC materials only the SiO ($SiO_2$) resonances and SiC were found with EDX and *in-situ* FTIR diagnostic as well [18]. The $SiO_2$ formation within the region of passive oxidation improves the resistance of pure C/C-SiC against oxygen attack. $TiO_2$ coated surfaces were investigated by XPS. The resonances of TiO ($TiO_2$) were found even after a very long plasma exposure of 5000 s. Moreover, the Ti resonance lines can be observed in the *in-situ* spectra during the whole experiment. This demonstrates the advantage of the thin (the thickness is about 20 μm) coatings of oxides of the transition element oxides like $TiO_2$ as a single layer or on top of a barrier layer consisting of SiC or Si(B)CN.

For economic re-coating with respect to reusability of the heat shield components the investigations presented here show, that the methods of paint coating or open air spraying of the $TiO_2$ powder with a volatile solvent in normal atmosphere seem to be unbeatable and easily to carry out. Since the October 2002 FOTON capsule mission failed, the verification of the achieved progress in a re-entry space mission with the competition of different double layers, however, still has to be performed. This experiment has to be repeated in the near future.

## Acknowledgements

Ingo Altmann, Jennifer Baer-Engel, Gerd Bauer, Kurt Behringer, Joachim Bill, Bruno Bitzenberger, Rüdiger Brandt, Martin Bross, Patrizia Dabalá, Andreas Daiß, Markus Feigl, Hans-Heiner Frühauf, Axel Greiner, Horst Hailer, Hermann Hald, Manfred Hartling, Detlef Heimann, Gerhard Hilfer, Jörg Hübner, Farid Infed, Herbert Jentschke, Stefan Jonas, Sebastian Kanne, Mathias Kayser, Oliver Knab, Stefan Laure, Thorsten Laux, Stefan Löhle, Thomas Ludwig, Michael Niethammer, Ferdinand Olawsky, Sergej Pidan, Bernhard Roth, Detlef Schinköth, Hellmuth Schmidt, Joachim Schneider, Karl-Heinz Schneider, Eberhard Schöll, Edgar Schreiber, Karl Schwörer, Ronnie Stirn, Thomas Stöckle, Abdelnasser Tawfik, Gerd Volk, Sigrid Wagner, Thomas Wegmann and Manfred Wiese as well as all staff members of the Institute for Space Systems and the Institute of Plasma Research and many students are gratefully acknowledged by the authors for their contributions, either in form of scientific works or by assistance for the development, manufacturing and operation of the experimental set-ups and facilities.

## References

1. Altmann, I., Bauer, G., Hirsch, K., Jentschke, H., Klenge, S., Roth, B., Schinköth, D. Schumacher, U.: In situ diagnostics of the interaction region between a nitrogen-oxygen plasma jet and hot C/C-SiC ceramic materials, High Temp.-High Press. 32, 573–579, 2000.
2. Auweter-Kurtz, M., Wegmann, T.: Overview of IRS Plasma Torch Facilities, RTO-EN-8. AC/323(AVT)TP/23, RTO Educational Notes 8, April 2000; von Karman Institute for Fluid Dynamics, RTO AVT/VKI Special Course on Measurement Techniques for High Enthalpy Plasma Flows, Belgium, October 1999.
3. Dabalá, P.: Massenspektrometrische Untersuchungen zum Erosionsverhalten von Hitzeschutzmaterialien für Wiedereintrittskörper, PhD Thesis, Institute of Space Systems, Universität Stuttgart, 1998.

4. Daiß, A., Frühauf, H.-H., Messerschmid, E.W.: Chemical Reactions and Thermal Nonequilibrium on Silica Surfaces, Molecular Physics and Hypersonic Flows, ed. by Capitelli, M., NATO ASI Series, Kluwer Academic Publishers, Dordrecht, Boston, London, May 1995.
5. Daiß, A., Frühauf, H.-H., Messerschmid, E.W.: Modeling of Catalytic Reactions on Silica Surfaces with Consideration of Slip Effects, Journal of Thermophysics and Heat Transfer, Vol. 11, No. 3, July 1997.
6. Feigl, M., Auweter-Kurtz, M.: Investigation of SiO production in front of Si-based material surfaces to determine the transition from passive to active oxidation using planar laser-induced fluorescence. 35th AIAA Thermophysics Conference, Anaheim, CA, USA, 2001.
7. Fertig, M., Frühauf, H.-H.: Detailed Computation of the Aerothermodynamic Loads of the MIRKA Capsule, Proceedings of the Third European Symposium on Aerothermodynamics for Space Vehicles, pp. 703–710, ESA, ESTEC, Noordwijk, The Netherlands, November 1998.
8. Fertig, M., Frühauf, H.-H., Auweter-Kurtz, M.: Modelling of Reactive Processes at SiC Surfaces in Rarefied Nonequilibrium Airflows, AIAA-Paper 2002-3102, 8th AIAA Joint Thermophysics and Heat Transfer Conference, St. Louis, Missouri, USA, 2002.
9. Fertig, M., Frühauf, H.-H.: Reliable Prediction of Aerothermal Loads at TPS-Surfaces of Reusable Space Vehicles, Proceedings 12th European Aerospace Conference, Paris, November 29–December 1, 1999.
10. Hald, H.: Faserkeramiken für heisse Strukturen von Wiedereintrittsfahrzeugen – Simulation, Test und Vergleich mit experimentellen Flugdaten, PhD Thesis, Institute of Space Systems, University of Stuttgart, 2002.
11. Heimann, D.: Oxidationsschutzschichten für kohlefaserverstärkte Verbundwerkstoffe durch Polymer-Pyrolyse, Dissertation, Universität Stuttgart, 1996.
12. Herdrich, G., Auweter-Kurtz, M., Kurtz, H., Laux, T., Winter, M.: Operational Behavior of Inductively Heated Plasma Source IPG3 for Entry Simulations, Journal of Thermophysics and Heat Transfer, Vol. 16, No. 3, July–September 2002.
13. Hilfer, G., Auweter-Kurtz, M.: Experimental and theoretical investigations of oxidation behaviour of thermal protection material under oxygen attack, High Temperatures – High Pressures 27/28, 1995/6, 435–448.
14. Hilfer, G.: Experimentelle und theoretische Beiträge zur Plasma-Wand-Wechselwirkung keramischer Hitzeschutzmaterialien unter Wiedereintrittsbedingungen, Shaker, Aachen, ISBN 3-8265-7118-5, 2000.
15. Hirsch, K., Altmann, I., Bauer, G., Brandt, R., Lindner, P., Neuer, G., Roth, B., Schneider, J., Stirn, R., Schumacher, U.: In situ determination of the emissivity and of the erosion rate of heat shield materials in re-entry simulation experiments, 16th ECTP2002: European Conference on Thermophysical Properties, 1.–4. 09. 2002, Imperial College, London (National Physical Laboratory), accepted for publication in High Temperature – High Pressure, Vol. 34 (2003), SFB-report 2000/52.
16. Hirsch, K., Altmann, I., Bauer, G., Kollermann, T., Roth, B., Schinköth, D., Schneider, J., Schumacher, U., Tawfik, A.: Development and investigation of erosion protection layers for heat shields of re-entry space vehicles, ECAMPVII, 7th European Conference on Atomic and Molecular Physics, Berlin 2.–6. April, P12.13, Ed. Europ. Phys. Soc., 2001.
17. Hirsch, K., Altmann, I., Bauer, G., Roth, B., Schinköth, D., Schneider, J., Schumacher, U., Tawfik, A.: 2nd International Symposium on Atmospheric Re-entry Vehicles and Systems, Arcachon/France, March 26–29, Proceedings of the Association Aeronautique et Astronautique de France, SFB-report 2003/62, 2001.
18. Hirsch, K., Roth, B., Altmann, I., Barth, K-L, Jentschke, H., Lunk, A., Schumacher, U.: Plasma-induced silica-like protection layer formation on C/C-SiC heat-shield materials for re-entry vehicles, High Temp. – High Press. 31, 1999, 455–465.
19. Hirsch, K., Volk, G.: Thomson scattering with a gated intensified CCD camera using a frequency doubled periodically pulsed YAG laser, Rev. Sci. Instr. 66, 5369 (1995).

20. Ingeri, G.R.: Nonequilibrium Hypersonic Stagnation Flow with Arbitrary Surface Catalycity Including Low Reynolds Number Effects. International Journal of Heat and Mass Transfer, Vol. 9, No. 8, pp. 755–772, 1966.
21. Jentschke, H., Bauer, G., Behringer, K., Hirsch, K.: Spectroscopical investigation of an air-like plasma jet, 5th Europ. Conf. on Atomic and Molecular Phys. Vol. 19a, Part II, 613–616, ed. by R.C. Thompson (Mulhouse, European Physical Society), 1995.
22. Jentschke, H., Hirsch, K., Klenge, S., Schumacher, U.: High resolution emission and absorption spectroscopy for erosion product analysis in boundary plasmas, Rev. Sci. Instrum. 70, 336–339, 1999.
23. Jentschke, H., Schumacher, U., Hirsch, K.: Studies of silicon erosion in plasma-target interaction from optical emission and absorption spectroscopy, Contrib. to Plasma Phys. 38, 501-2, 1998.
24. Kim, Y.C., Boudart, M.: Recombination of O, N and H Atoms on Silica: Kinetics and Mechanism, Langmuir, Vol. 7, pp. 2999–3005, 1991.
25. Laure, S.: Private communication.
26. Laux, T.: Untersuchungen zur Hochtemperaturoxidation von Siliziumkarbid in Plasmaströmungen, submitted for PhD Thesis, Institute of Space Systems, Universität Stuttgart, 2003.
27. Lindner, P., Hirsch, K., Roth, B., Schneider, J., Stirn, R., Schumacher, U.: In situ determination of the erosion rate of heat shield materials in re-entry simulation experiments, 3rd International Symposium on Atmospheric Re-entry Vehicles and Systems, Arcachon/France, March 24–28, Proc. of the Association Aeronautique et Astronautique de France, SFB-report 2003/63.
28. Neuer, G., Fiessler, L., Groll, M., Schreiber, E.: Temperature, its Measurement and Control in Science and Industry, 6, ed. by J.F. Schooley, American Institute of Physics, New York 1992, 787–790.
29. Neuer, G., Jaroma-Weiland, G.: Spectral and total emissivity of high temperature materials, Int. J. Thermophys. 19, 917–921, 1998.
30. Park, C.: Nonequilibrium Hypersonic Aerothermodynamics, John Wiley & Sons, New York, 1989.
31. Patent 1999: Deutsches Patentamt Nr. 199 28 173.4; Hirsch, K., Kaiser, M., Greiner, A., Bill, J., Aldinger, F.: Beschichtung zur Verminderung der Erosion an thermisch hochbelasteten Oberflächen aus faserverstärkter Keramik und Verfahren zu deren Herstellung.
32. Patent 2000a: Europäisches Patentamt München PCT/EP 00/05555, Hirsch, K., Kaiser, M., Greiner, A., Bill, J., Aldinger, F.: Verwendung einer Beschichtung mit faserverstärkter Keramik und deren Herstellung.
33. Patent 2000b: Deutsches Patentamt München, Nr. 100 48 764.5, Hirsch, K., Roth, B., Altmann, I., Schinköth, D., Kochendörfer, R., Krenkel, W., Lützenburger, N.: Bauteil, das eine Schutzschicht gegen Erosion durch thermische Belastung aufweist und Verfahren zur Herstellung einer Schutzschicht auf einem Bauteil.
34. Pidan, S., Auweter-Kurtz, M., Herdrich, G., Laux, T., Fertig, M.: Experimentelle Erprobung eines katalyzitätsbasierten Sensorsystems zur Bestimmung der Wärmestromdichte auf dem HYFLEX-Nasen-Mockup im Plasmawindkanal SCIROCCO, DGLR Jahrestagung 2003, München, November 2003.
35. Rosner, D.E., Allendorf, H.D.: High Temperature Kinetics of the Oxidation and Nitridation of Pyrolytic Silicon Carbide in Dissociated Gases, Journal of Physical Chemistry, Vol. 74, No. 9, pp. 1829–1839, 1970.
36. Schinköth, D.: Laser-Streu-Diagnostik im Vergleich mit Emissionsspektroskopie an einem Freistrahlplasma, Dissertation, Universität Stuttgart, 2001.
37. Smorjai, G.A.: Principles of Surface Chemistry, Prentice-Hall, Inc., Englewood Cliffs, New Jersey, 1972.
38. Stewart, D.A.: Determination of Surface Catalytic Efficiency for Thermal Protection Materials – Room Temperature to Their Upper Use Limit, AIAA-Paper 96-1863, 31st Thermophysics Conference, New Orleans, LA, 1996.

39. Stirn, R., Altmann, I., Hirsch, K., Lindner, P., Roth, B., Schumacher, U.: Bestimmung der räumlichen Verteilung der Plasmaparameter in der Wechselwirkungszone zwischen einem Luftplasma und Hitzeschutzmaterialien, Verhandl. DPG (VI), 37, 5/2002.
40. Stirn, R., Hirsch, K., Lindner, P., Roth, B., Schneider, J., Schumacher, U.: Investigation of the erosion of coated and uncoated heat shield materials by a plasma jet, Course on low temperature plasma physics and applications (CLTPP-7), Sept. 8–17, 2002, Bad Honnef (Germany).
41. Stöckle, Th.: Experimentelle und numerische Untersuchung der Oberflächenkatalyzität metallischer und keramischer Werkstoffe in Hochenthalpieströmungen, PhD Thesis, Institute of Space Systems, Universität Stuttgart, 2000.
42. Summers, H.P., Wood, L.: Atomic Data and Analysis Structure (ADAS), User Manual JET-R(94)06.
43. Tawfik, A., Quell, S., Hirsch, K., Roth, B., Schneider, J., Schumacher, U.: Time correlated measurements of spectra from process and beam plasmas in the range from 240 nm to 1000 nm, ECAMPVII, 7th European Conf. on Atomic and Molecular Physics, Berlin, 2.–6. April 2001, paper P12.15.
44. Ullmann, T., Schmücker, M., Hald, H., Henne, R., Schneider, H.: Plasma Sprayed Yttrium Silicates for Oxidation/Erosion Protection of C/C–SiC Components, pp. 147–153, Proc. 4th European Workshop, "Hot Structures and Thermal Protection Systems for Space Vehicles", Palermo, Italy, 26–29 November 2002 (ESA SP-521, April 2003).
45. Wagner, C.: Passivity during the Oxidation of Silicon at Elevated Temperatures, Journal of Applied Physics, Vol. 29, No. 9, pp. 1295–1297, September 1958.

## 7.3 Design and Evaluation of Fibre Ceramic Structures

Bernd-Helmut Kröplin, Richard Kochendörfer, Thomas Reimer *,
Thomas Ullmann, Ralf Kornmann, Roger Schäfer,
and Thomas Wallmersperger

### 7.3.1 Introduction

Space applications present extreme challenges to the efficiency and the lightweight design of structures as typical flight conditions during re-entry are approximately 20 minutes of very high heat loads with 1800 K to 2100 K in the stagnation point of a winged vehicle and even more than 2300 K in the case of ballistic return flights of capsules. Therefore, the selected materials play a decisive role. With the pure payload share of present space transportation systems (STS) being as low as 0.5–2%, a need for further mass reduction in space transportation systems is evident. Mass savings, however, have often to be paid for with additional cost, because expensive materials with better performance have to be used or manufacturing or maintenance requires a bigger effort. So, the incorporation of new materials like ceramic matrix composites (CMC) has to be accompanied by new design concepts to reduce costs.

* DLR, Institut für Bauweisen und Konstruktionsforschung, Pfaffenwaldring 38–40, 70569 Stuttgart; thomas.reimer@dlr.de

In addition to the targeted mass reduction, the focus of new developments is also on issues to enhance operations, durability or non destructive test and inspection methods and to enable better performance through improved design tools. This means for instance a better understanding of the effects of the fluid-structure interaction during the hypersonic flight regime and the capability for analysing these effects on local structural features like gaps between panels.

In the severe re-entry environment, significant improvements in the performance of space vehicles are expected by the use of ceramic materials, and especially by the use of fibre-ceramics or CMC which may also be used as gradient type materials. Besides the expectation of improved overall system performance, the use of ceramics in certain applications is in fact a key technology for the realization of advanced concepts, especially with regard to future reusable launch vehicles (RLV). In the field of high temperature materials for instance, there is no alternative to ceramic matrix composites, which are the key technology for lightweight ultra-high temperature thermal protection systems and hot structures. In addition to that and on the background of the above mentioned issues especially CMC systems also offer a potential for cost reduction compared to sensitive tile systems like they are used on the Space Shuttle as one of the present STS.

The main goal of the work carried out in the Collaborative Research Centre 259 was to design, analyse, manufacture and verify a structure representative for the hot outer skin of an STS. The verification should be carried out in a ground test facility with the intention of a flight opportunity.

There have been a number of technology programmes in which CMC materials were characterized under re-entry relevant environments, with the development of new concepts for TPS designs and hot primary structures and test campaigns of representative structural components. The most important programmes in this context were the European HERMES project [1], the German national Hypersonic-Technology Programme (HTP), the "Hot Structures Programme", studies for return capsules (e.g. COLIBRI [2]) and flight experiments on re-entry capsules like FOTON [3], EXPRESS [4] and MIRKA. These research activities were continued by the German participation in the "Manned Space Transportation Programme" (MSTP) and the "Future Space Transportation Investigation Programme" (FESTIP) [5] by ESA, and by the German TETRA [6] programme. Since DLR was actively participating in almost all of these, there are strong influences and cross-links to the work done in the Collaborative Research Centre 259.

#### 7.3.1.1 Concept Design and Manufacturing Studies

The vehicle that was initially chosen as a baseline for the realization of the CMC component was COLIBRI, a research capsule designed for ballistic re-entry. The capsule should be equipped with a segmented rigid-surface TPS. Since the front section of the conical part of the capsule experiences loads comparable to typical concepts of Reusable Launch Vehicles (RLV), one of the front section surface segments (see Fig. 7.3.1) was then selected as a typical cut-out for realization within the Collaborative Research Centre.

In parallel to the selection of a reference vehicle, generic requirements for the overall TPS as well as for the structural components were developed as a result of system investigations.

System requirements:
- a segmented TPS needs to pay special attention to the gap regions between panels;
- the segmentation scheme of panels and insulation should be different;
- assembly of the fibre-ceramic panels shall be carried out from the outside of a vehicle with no access from the vehicle interior;
- replacement of single panels shall be possible;
- panels shall be inspectable with non-destructive test methods;
- different thermal expansion values of cold structure and hot surface have to be compensated;
- the TPS concept shall be fail-safe.

Component requirements:
- design aware of manufacturing needs;
- stresses due to anisotropic shrink during manufacturing are to be minimized;
- minimum mass with small wall thicknesses;
- temperature gradients over thickness shall lead to low values of thermally induced stresses;
- panels shall be stiff with low deformation of surfaces relevant for sealing purposes.

During the early phases of the project, a joining element was developed which can be described as a ceramic rivet. It is an essential feature of the design concepts that were elaborated since it is the basis for panel assembly from the vehicle outside and can also be disassembled easily. In the final phases of the programme it was accompanied by a threaded version shown in Fig. 7.3.2 which has the advantage to be much smaller and so generates more design flexibility.

The compensation of thermal expansion mismatch between cold and hot structures can be executed via slip joints or elastic deformation of stand-offs. The solution based on elasticity was preferred over slipping joints since it avoids problems due to micro movement and component wear in a vibration

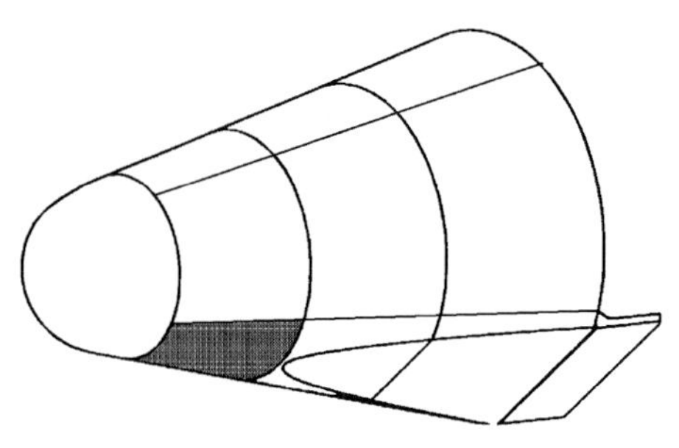

Figure 7.3.1: The COLIBRI capsule with TPS segmentation (in grey the selected segment).

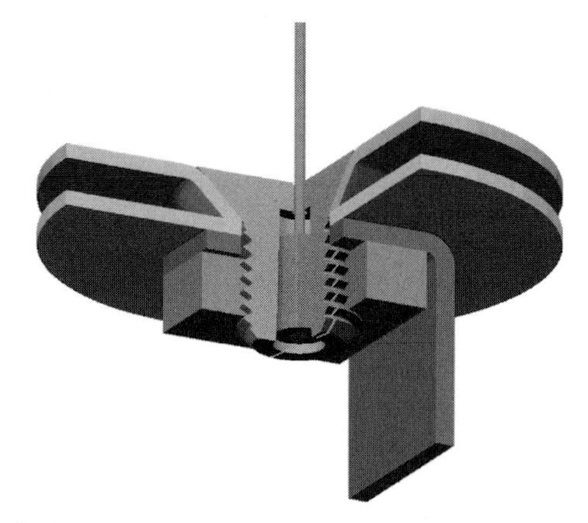

Figure 7.3.2: Ceramic fastener.

environment. Deformation in general due to thermal expansion of the structures and surface pressure is also a critical issue with regard to the interface regions between the surface panels. There are gaps between the panels which creates the need for a sealing system. In the case of deformed structures the gaps may widen or steps may come up which has to be handled by the system. A thorough investigation of such effects was carried out in an arc jet environment.

The fail-safe requirement essentially means that the loss of one panel should not lead to a complete failure of the TPS with subsequent failure of the cold structure. This is reached by using a different segmentation scheme for the outer hot skin panels of the vehicle surface than for the internal flexible insulation mats underneath the surface panels. So, in case of a panel loss, there is still the insulation mat which is also qualified for re-entry on the vehicle surface up to certain temperature and pressure values.

#### 7.3.1.2 Manufacturing

In parallel to the design studies manufacturing studies were carried out in order to establish the most efficient way for the fabrication of the curved and stiffened panel.

Manufacturing problems in the C/C-SiC fabrication are usually a result of the very radical alteration of the material properties during the manufacturing process. Characteristic values like strength, modulus, thermal expansion coefficient, density or porosity and – above all – the wall thickness change very much and in the case of integral components with complex geometries this may lead to failure due to excessive stresses as a result of suppressed deformations.

Therefore, a solution was looked for to reduce these problems. It was decided to separate the fabrication of the panel shell and the stiffener completely to avoid these problems (see Fig. 7.3.3). Shell and stiffener were separately fabricated as CFRP components and only after pyrolysis they were joined and siliconized into one component; a processing technique which is described as *in-situ* joining [7].

The final component with dimensions of 420×450 mm was thoroughly examined to learn more for a further development of the processing and joining technology. The most important findings were the following:

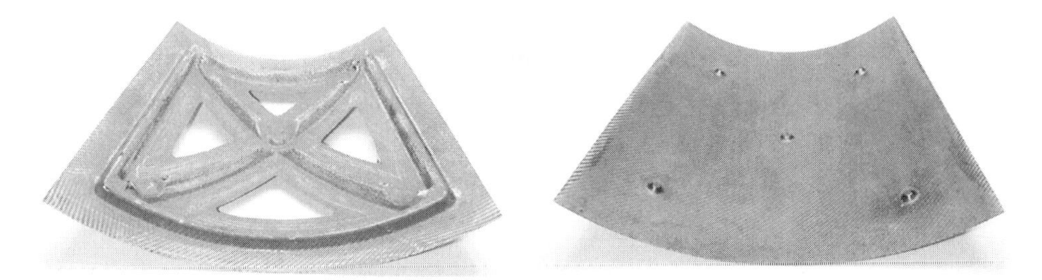

Figure 7.3.3: Manufacturing of the COLIBRI panel.

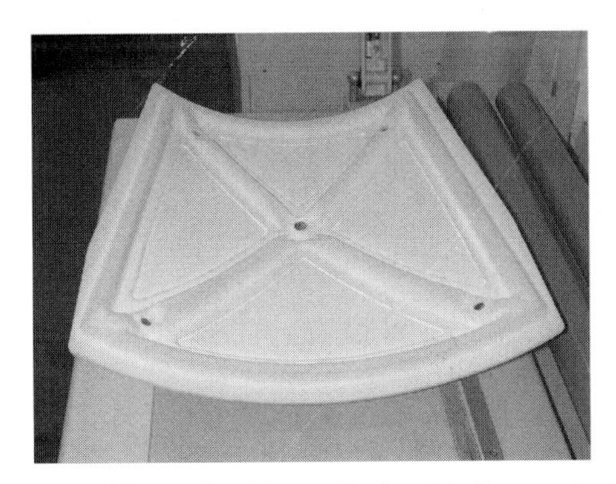

Figure 7.3.4: Finished COLIBRI panel with applied oxidation protection coating.

The interface area between the stiffener and the shell was not fully siliconized, but there were visible flaws at the edge region of the stiffener profile. The reasons were found mainly in the manual preparation of the joined components that led to an uneven distribution of joining adhesive and varying gap conditions, plus an insufficient pressure during joining. In the manufacturing of a flight experiment which will also be described, these issues were resolved with a much better result.

After completion of the panel, a high-temperature test campaign was carried out that will also be described. After that the panel was coated with a new type of oxidation protection coating (see Fig. 7.3.4) that was developed in parallel to the manufacturing investigations.

#### 7.3.1.3 Test

The panel complete with z-standoffs and central post was assembled into DLR's high-temperature test facility Maxitherm to subject it to thermal tests.

Since the heat load was applied onto the convex side of the panel, the upper, concave side of the panel was not covered with insulation to have unobstructed visibility of the panel for the **H**igh **T**emperature **G**rating **M**ethod (HTGM) measurements.

Due to the lack of insulation on the concave side of the panel, a significant share of the applied heat load was radiated again into the test chamber and did not contribute to heating up of the panel, which means that peak temperatures were at around 1320 K compared to 1900 K that can be reached with insulated components [8]. The heat load was applied as thermal shock with the sudden opening of a thermal shutter to produce thermal gradients as large as

Figure 7.3.5: View on the panel during high-temperature test.

possible. A view on the concave side of the panel during high-temperature testing is presented in Fig. 7.3.5.

#### 7.3.1.4 Plasma Sprayed Yttrium Silicates for Oxidation Protection of C/C-SiC Panels

Due to oxidation sensitivity of carbon fibres above 670 K, the use of CMC structures for reusable launch vehicles (RLVs) is limited under atmospheric re-entry conditions. Therefore, oxidation protection coatings are inevitable to improve reusability and reliability of structural components based on carbon fibres. Currently used coating systems consist of several layers – so-called multi-layers – with different materials, each proposed to perform a specific function within the protecting system [9, 10]. Advanced multi-layers have to proof adequate oxidation resistance as well as some self healing properties at temperatures up to 1870 K. Furthermore the application process of the coating itself has to support both, applicability on large structural parts with partly complex shaped geometry as well as repair procedures of damaged surface areas which eventually becomes necessary after the first flights.

In order to improve both oxidation protection and erosion resistance, a two-layer coating system has been developed for the C/C-SiC composite material. As depicted in Fig. 7.3.6a, the bonding layer consists of SiC applied by chemical vapour deposition (CVD) also preventing outward diffusion of carbon from the C/C-SiC substrate. The outer erosion protection layer is based on yttrium silicates ($Y_2SiO_5$ and $Y_2Si_2O_7$) with inlays of amorphous silica and is deposited by using the low pressure plasma spraying technique (LPPS). The silicate layer (thickness: ca. 120 μm) is interspersed by angular – sometimes even

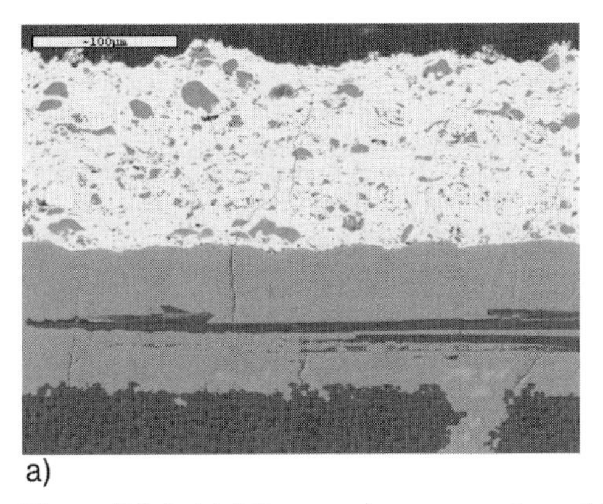

a)

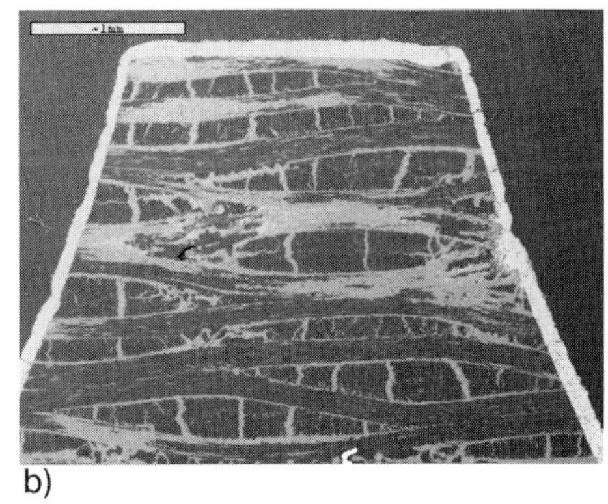

b)

Figure 7.3.6: (a) Microscopic cross section of the coating system with low pressure plasma sprayed yttrium silicate (top) and CVD-SiC (bottom); SEM image, white bar = 100 μm. (b) C/C-SiC sample with an irregular shaped surface covered by a plasma sprayed yttrium silicate coating; SEM image, white bar = 1 mm.

spherical shaped – particles of amorphous silica. These $SiO_2$ grains are residuals from the spray powder. Due to the high transmission of ultra-violet radiation of pure silica, the particles do not melt inside the plasma jet within that short time (approximately $10^{-3}$ s). Thus the partly viscous silica particles were projected to the substrate by the plasma jet and embedded in between the solidified yttrium silicate droplets during the coating build-up. The surrounding matrix of yttrium silicate consists of numerous flat and irregular shaped bodies which are mechanically linked to each other. In literature this is usually described as a "pancake"-like microstructure which sometimes even appears like microscopic lamellae in a cross section.

Thermo-gravimetric oxidation tests were performed in flowing air at 1770 K with C/C-SiC specimens provided with an all around coating. As depicted in Fig. 7.3.7, the uncoated CMC material shows a relative mass loss of approximately 60% as the carbon fibres and the carbon matrix burn out completely within 80 minutes whereas with the coated material there is nearly no weight change nor any visible material degradation for at least 170 minutes. Further qualification tests for thermal shock resistance between room temperature and 1870 K (mean thermal gradient: 20 K/s) were performed in a solar furnace which is operated by DLR in Cologne. Beside a very good adhesion between the yttrium silicate layer, the CVD-SiC bonding layer and the C/C-SiC substrate, the coating's surface did not show any visible degradation, except the colour which turned from grey to white – just as the original colour of the spray powder. Yttrium silicate coated samples were also tested inside the plasma wind tunnels (PWK-1 and -2) which are operated by the Institute of Space Systems (IRS, University of Stuttgart). Different re-entry relevant test cycles between 1623 and 1923 K for up to 1380 s (stagnation pressure: 4.1 hPa) have

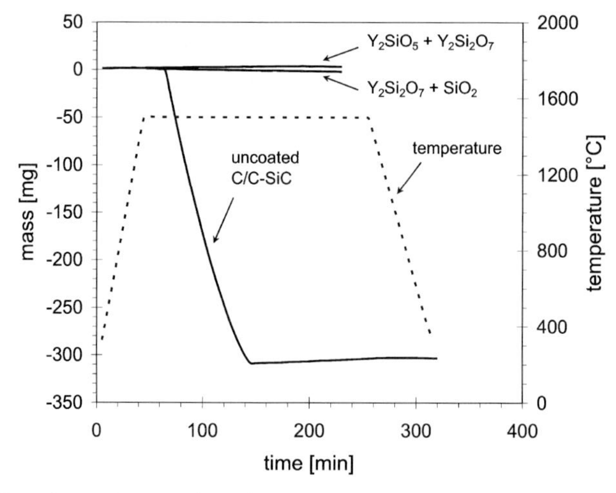


Figure 7.3.7: Mass loss of coated and uncoated C/C-SiC specimens in flowing air at 1773 K.

shown very high erosion resistance of the coating system. The specific mass loss rates were less than 0.01 kg/m$^2$h for 1623 K and less than 0.10 kg/m$^2$h for 1923 K, respectively [11, 12].

In order to adapt the coating technique for large scaled and complex shaped structures by using off-angle plasma spraying, first efforts with samples of certain geometries were successfully performed as shown in Fig. 7.3.6 b. In addition further experience was achieved by optimizing the automatic plasma torch guidance which is necessary to scale-up the LPPS coating processing for applications on large panel like C/C-SiC structures. Finally a C/C-SiC panel of approximately 40×41 cm was coated by low pressure plasma spraying of yttrium silicate. The panel is shown from its rear side in Fig. 7.3.4.

In autumn 2001 the coating system was applied on two designated areas on a C/C-SiC panel (Fig. 7.3.9) as part of the FOTON-M1 experiment which was supposed to fly on a real re-entry mission with an unmanned Russian capsule. For the application of the yttrium silicate coating, it was necessary to cover the panel with a ceramic mask which was mounted on top and allowed a proper coating build-up in these designated areas whereas the rest of the panel's surface was protected during the plasma spray process [13].

#### 7.3.1.5 Flight Experiment

After the successful tests with the COLIBRI panel in the ground facility, the next logical step was to look for a flight opportunity to take the system to the real test with a full re-entry. For the flight experiment with a fully ceramic TPS design, a co-operation was started with CSDB Progress in Russia on the FOTON-M1 mission, slated for launch on October, 15 in 2002.

The FOTON missions are microgravity research flights of roughly 2 weeks with a return capsule of 2.3 m diameter with ablative heat shield in which the experiment then had to be integrated.

The time between the agreement and the flight date was very short with less than a year and a big challenge for the involved team. The experiment design, which originally had aimed towards a carrier with a flat surface, had to be adapted to the spherical shape of the FOTON capsule and all the lessons learned from the demonstrator panel in the ground test as well as from other programmes that were completed in the meantime (e.g. TETRA/X-38) had to be incorporated.

Since re-entry flight opportunities are rare when the experiment has to fly as passenger, the carrier may not provide the optimum trajectory for the intended environment. In this case the FOTON capsule re-enters on a typical ballistic trajectory with short duration and high peak loads in the stagnation point, leading to temperatures of up to 2600 K and high dynamic pressures. This is rather high compared to the environment usually expected for an RLV, which is the conceptual basis for the experiment. Therefore, the experiment location on the capsule surface was selected on the equator of the vehicle at almost 90°

away from the flight direction where temperatures were expected to be in the range of 1900–2000 K.

The primary goals for the experiment were:

- full re-entry;
- system test with CMC panels, CMC attachments and fasteners and a sealing system;
- test and comparison of different oxidation protection coatings.

The experiment design according to the FOTON boundary conditions were:

- circular experiment with two semi-circular panels of 300 mm diameter;
- close-out ring of 340 mm diameter;
- total diameter of the aluminium carrier structure of 385 mm;
- experiment thickness was 23.5 mm;
- 3 load introduction components per panel;
- one stiff central post and two flexible z-stand-offs;
- seal system underneath the gaps between the panel edges;
- C/C-SiC fasteners.

The design (Fig. 7.3.8) of the flight-experiment panels accounted for the manufacturing experience from the COLIBRI panel that was tested in the ground facility. There were again flexible stand-offs plus one central post per panel. The *in-situ* joining technique was further developed and in many details improved. In contrast to the COLIBRI panel, the stiffeners that were joined to the shells were made as a number of individual parts. Much more attention was given to the preparation of the joining areas, both on the stiffeners and on the panels which was machined, and also the joining itself was carried out in a very controlled fashion with a joining rig.

There were 9 thermocouples integrated at characteristic locations, as well as one accelerometer that was placed in the capsule internal in the data acquisition box. A CIMT (crystal indicators of maximum temperatures) peak temperature measurement was added to the experiment that is based on the altera-

Figure 7.3.8: TPS experiment design for the FOTON-M1 mission.

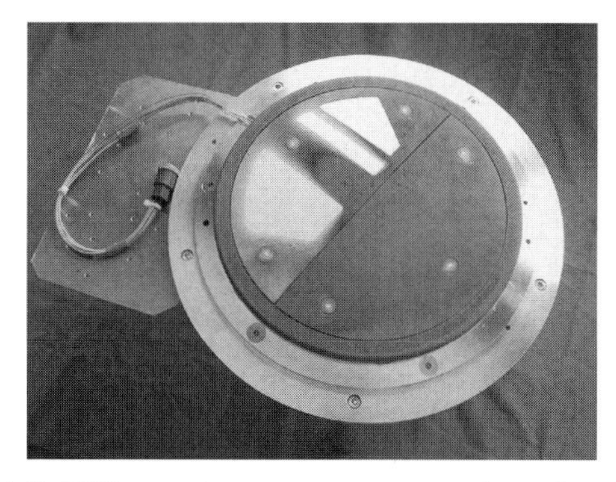

Figure 7.3.9: FOTON-M1 TPS experiment on transport rig with white areas of oxidation protection coatings.

tion of physical properties in special crystals subjected to a temperature load. After flight the peak temperature can be determined.

The manufacturing and assembly of the experiment took place at DLR in Stuttgart. The assembled experiment (Fig. 7.3.9) was then shipped to Samara in Russia and was integrated on the surface of the FOTON capsule.

The capsule was then transferred to the launch site in Plesetsk where the assembly to the launcher, a Soyuz rocket took place.

Launch was on October, 15 in 2002, as planned. Unfortunately, a malfunction in one of the four liquid-fuel boosters of the Soyuz launcher after roughly 20 seconds led to destruction of the rocket, the capsule and all the experiments aboard.

The experiment is scheduled for a re-flight on the FOTON-M2 mission which is now scheduled for launch in spring 2005. Pointing out the positive, that will give the opportunity to carry out a pre-flight test campaign using the HTGM technique that was developed in the meantime for use in arc jets and to make a comparison between ground and flight test which was not possible before FOTON-M1.

## 7.3.2 Measuring Model Deflections by Thermo-Mechanical Loads in a Plasma Wind Tunnel

### 7.3.2.1 Overview

In cooperation with the DLR in the framework of the IMENS (**I**ntegrated **M**ultidisciplinary d**E**sig**N** of Hot **S**tructures) project and the Collaborative Research Centre 259, an extensive measuring campaign has been performed, measuring the deformations of a model in DLR's arc heated facility L3K in Cologne.

Aim of the cooperation was to sustain the displacement measurements of the built-in sensor in the tested model by an additional measuring method and also developing the measuring method based on the HTGM (**H**igh **T**emperature **G**rating **M**ethod) itself in the high-enthalpy flow in a plasma wind tunnel.

The need for accurate measurement of the thermomechanical behaviour of materials at high temperatures has led to the development of manifold measurement techniques, including extensometry, strain gauges, moiré and speckle interferometry. Often these techniques can only be used for small laboratory-size specimens and within a limited temperature range.

When considering the more expensive and critical testing of larger parts of specimens or entire components, a suitable method for investigating such objects is needed. Therefore, at the **I**nstitute for **S**tatics and **D**ynamics of Aerospace Structures (ISD) an optical method on the basis of the stereo vision technique has been adapted to fulfil the requirements of high temperature measurements. The method can be applied to object sizes of about $10\times10\,mm^2$ up to $0.5\times0.5\,m^2$.

The HTGM is based on photogrammetry, using digital image stereo pairs. The method uses a full-field surface scanning technique and calculates the absolute spatial coordinates (3-D) of a facet mesh consisting of rectangular pixel-areas. To correlate the deformed image to the undeformed reference images, each image is divided into small subsets. The discrete matrix of the pixel gray level values in each subset form a unique "fingerprint" identification within the image. A computer programme correlates the subset pixel gray level matrices in the deformed and undeformed images and gives the displacements of the centre point of each facet. Comparable to a finite-element procedure, the subsets can be processed simultaneously to provide a continuous displacement field, which leads to a higher accuracy than with conventional feature-based systems.

From these data one gets the displacement vectors, the local strain values and the contour difference, if the object is loaded. If the marked object points are concentrated like they are in a grating, this technique is well-known in experimental mechanics as grating method. The results of such measurements are suitable for comparison with finite element calculations or for phenomenological investigations of materials, for example.

In the previous periods of the collaborative research centre, investigations of structures and specimens made of fibre ceramic materials were conducted from ambient temperatures up to 1773 K. The results showed, that the digital image correlation technique remained fully capable within this temperature range, providing accurate measurements of the displacements and strains of the investigated objects [14, 15].

In a further step of this development the measuring method should be used to measure the behaviour of a model in a high-enthalpy flow. As an optical non-contact method, it is applicable in such an environment without affecting neither the hypersonic flow field nor the investigated object. In the following the application and results of the method for measuring the deformations of the IMENS-model in a high-enthalpy flow are shown.

Background of the campaign is, that the thermo-mechanical behaviour of highly loaded components of space vehicles, for instance nose caps, thermal protection shields or control surfaces, in high-enthalpy flow fields can be predicted with larger constraints only. Numerical simulations are normally performed for flow and structure independently, not regarding the complete effects of the flow-structure interaction. Therefore, the design of such components requires high safety margins which increase the mass and subsequently the costs of space craft.

A challenging task for a more reliable design of hot structures is the coupling of flow solver codes with codes that provide the thermal and mechanical response of the structure. In addition, it is essential to validate the numerical results by well-suited experiments, using sophisticated measurement techniques, before coupled simulations become common design tools.

#### 7.3.2.2 Model Design

Long-duration high-enthalpy facilities which are normally used for ground testing of gas/surface-interaction phenomena in high-temperature flow fields are not capable of simulating the flight environment of a complete space vehicle configuration [16]. In the IMENS project, a generic model (see Fig. 7.3.7) of a re-entry vehicle nosecap region was developed which enabled a strong thermo-mechanical coupling during investigation in the high-enthalpy flow [17].

The nosecap model ($295 \times 194 \times 60\,mm^3$) is a simplified model of a T-gap section of a real nosecap assembly. It basically consists of thermal protection material parts made of DLR Stuttgart's C/C-SiC and an $Al_2O_3$ insulation inside, in order to avoid any heat transfer to the water-cooled base plate and the model holder. The gap width can be changed by moving the rear plates and by varying their heights to adjust different step configurations. The generic model was designed with special focus on maximum deformation during the test. For the thermo-mechanical deformation measurement a sensor was integrated inside the model, located 42 mm upstream from the T-gap. The aerodynamic pressure was measured on two locations of the model's surface with a specially designed pressure-port system. Moreover, the surface temperature was mea-

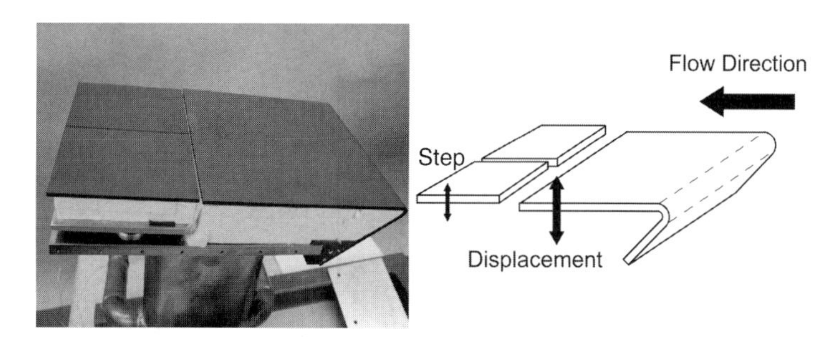


Figure 7.3.10: IMENS model and principle.

Table 7.3.1: Flow conditions during the tests.

| | |
|---|---|
| Reservoir pressure [hPa] | 4550 |
| Reservoir temperature [K] | 5400 |
| Total enthalpy [MJ/kg] | 11.0 |
| Mass flow rate [g/s] | 142 |
| Free stream Mach number | 7.6 |
| Free stream pressure [hPa] | 0.50 |
| Free stream temperature [K] | 491 |
| Flow velocity [m/s] | 3730 |
| $N_2$ mass fraction | 0.757 |
| $O_2$ mass fraction | 0.012 |
| NO mass fraction | 0.018 |
| N mass fraction | $<10^{-6}$ |
| O mass fraction | 0.213 |

sured with 5 pyrometers, also an infrared camera was applied. In addition, 20 thermocouples were placed inside the model.

Due to the orthotropic thermal expansion of the C/C–SIC material the deformation of the curved nosecap plate is mainly influenced by its tendency to increase the radius of curvature. Additionally, the aerodynamic pressure deforms the plate in opposite direction compared to the thermally driven deformation.

For the test campaign, conducted in DLR's arc-heated facility L3K, air was used as the working gas which was accelerated to hypersonic velocities. The angle of attack was 20 and 30 degrees, respectively. The flow conditions can be taken from Tab. 7.3.1.

#### 7.3.2.3 Adaptation of the HTGM to the L3K Facility

In a pre-test with a dummy-model representing the contour of the original nosecap model used in the subsequent tests, the conditions in the wind tunnel were investigated concerning the deformation measurement requirements using the HTGM. Special issues were:

- perspective, field of view and distances of the investigated surfaces relative to the stereoscopic pair of cameras;
- diffraction effects due to flow and shock waves;
- durability of the marking material for producing a grating;
- contrast of the grating;
- possible interference of the digital cameras with electromagnetic disturbances of the arc-heater.

Due to the fact, that the test chamber of the L3K facility was not originally designed for conducting stereoscopic measurements on the investigated models, the base distance referred to the object distance was not optimal but sufficient for the robust working grating method. The convergence angle between the two cameras was found to be between 10 to 20 degrees, dependent of the investigated surface. An optimal angle normally is in the range of 30 to 45 degrees for normal operation. Maximum spatial accuracy is achieved with a convergence angle of 60 degrees. This counts for the theoretical value of the stereoscopic principle. In practice the matching process of finding corresponding pixel-facets in the images influences also the accuracy. As an effect of the smaller convergence angle the perspective distortion of the stereoscopic pair is reduced and the matching process of corresponding image facets is more precise. On account of this, both effects are compensating each other up to a certain extent.

Additionally, the depth of view is covering the whole field of view reproducing sharp images on the complete investigated surface. Both these aspects led to the conclusion that the expected accuracy is in the range of the requirements.

The diffraction problem was considered as neglectable because of the very low total pressures in the plasma flow. It was decided to perform additional measurements taking images through the plasma flow to the wall of the measuring chamber. By switching on and off the plasma flow, a possible influence of the plasma should be detected by observed displacements of the rigid wall.

One of the most important aspects was to see if a grating with sufficient contrast could be realized on the model's surface and if the marks were stable against erosion by the plasma flow.

The flow parameters during the conducted tests are depicted in Tab. 7.3.1. The L3K facility is operated using high-enthalpy air so under the given test conditions the contained atomic oxygen could have been capable to corrode the surface as well as the marking material. It was shown that the marking material of $Al_2O_3$ adhered very well to the model surface and was not affected by the influence of the plasma flow. During the subsequent measurements, in fact, it could be shown that the grating could be exposed to the plasma flow over numerous experiments without visible degradation.

In previous measurements on high temperature specimen and components, the self radiation of the hot objects was used to expose the measuring images taken at temperatures above $1073\,^{\circ}$K. At temperatures lower than 1073 K, the objects were incidently lit by an external light source. This was necessary because of the effect that the contrast of the grating is established using materials with different emissivity values. The thermal radiation leads to the problem that areas with high emissivity and also absorption values occur dark during incident lighting. When the temperature of the objects rises, these parts of the surface radiate more intensively, which leads to the result that these parts now seem brighter. In the images taken of the heated objects, the reflected light by the incident illumination and the radiated light by the object itself are extinguishing the former contrast between the different regions. By switching off the external light source and only using the heat radiation of the object, this problem can be overcome.

The first tests in the Plasma Wind Tunnel have shown, that the conditions there are different. The arc-heater of the plasma source has a power rating of 6 MW. The tests were performed at a reservoir temperature of 4500 K and a pressure of 4550 hPa. At such conditions there is a radiation of light through the nozzle of the arc heater. The reflection of this light from the surface of the model proved to be stronger than the heat radiation of the model's surface in the visible spectrum range, and the former observed effect of extinguishing contrasts as a result of the prescribed radiation problem was not existent.

The maximum temperatures in the observed regions on the model surface were expected not to exceed 1473 K as shown in Fig. 7.3.11. So the maximum emission of the model is at infrared wavelength whereas the higher temperature of the arc radiation produces a spectrum at a shorter wavelength. By cutting off the light spectrum above 525 nm via an edge filter, the image quality respectively the contrast of the grating could be satisfactorily achieved during the complete course of temperature during the execution of a test in the L3K.

Since the plasma flow is accelerated through a nozzle, the arc radiation is a shaped beam pointing in axial flow direction from the nozzle throat to the investigated model. To illuminate shaded parts of the measuring regions, an additional halogen lamp was used. The final test setup is shown in Fig. 7.3.12.

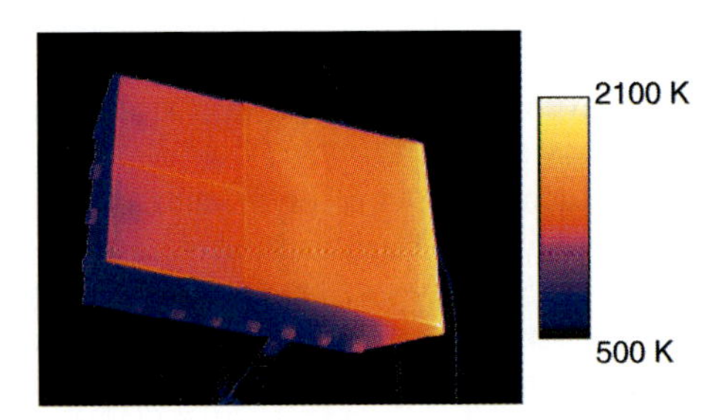


Figure 7.3.11: IR-image of the IMENS model.

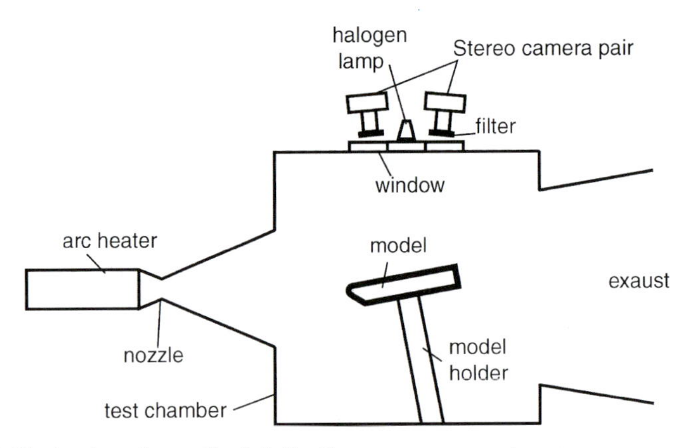


Figure 7.3.12: Test set-up for optical deflection measurement.

Electromagnetical interferences of the L3K facility with the digital cameras Kodak DCS 420 could not be detected.

#### 7.3.2.4 Results

A systematic overview of the behaviour of the test set-up has been gained in a series of measurements. In the subsequent discussion the results of two specific measurements are presented. Figure 7.3.13 shows a measuring image taken from the left rear side of the model near the T-gap section. The white marks on the surface are the necessary grating, whereas the small crosses represent the centres of the evaluated image facets, giving the displacement field as shown in Fig. 7.3.14. The dimension of the grated surface is about $120 \times 90\ mm^2$. The image was taken during plasma operation shortly before the plasma is switched off. This means, that the temperatures in the measuring zone are as high as 1473 K.

A comparison of the results measured by the High Temperature Grating Method with the built-in mechanical sensor of the model is shown in Fig. 7.3.15. The diagram shows the results of two different measurements with the same test configurations and plasma parameters. The sensor is located 42 mm upstream of the transversal gap on the symmetry axis of the model. The measuring results of both methods show a good conformity in qualitative and quantitative values. The difference between the tactile mechanical result and the optical result can be explained by the thermal expansion of the measuring pin of the sensor. Calculations have shown, that approximately 0.2 mm expansion of the pin has to be added to the measured values of the mechanical sensor.

At the test start with ambient temperatures of the model and measuring pin the results correspond with each other. With increasing test time, the model and the built-in sensor pin heats up, which leads to an expansion of the pin. Due to the fact that the pin is located at the back surface of the model, the positive displacements will be measured too low. The complete run of the curve

Figure 7.3.13: Measuring image of the HTGM.

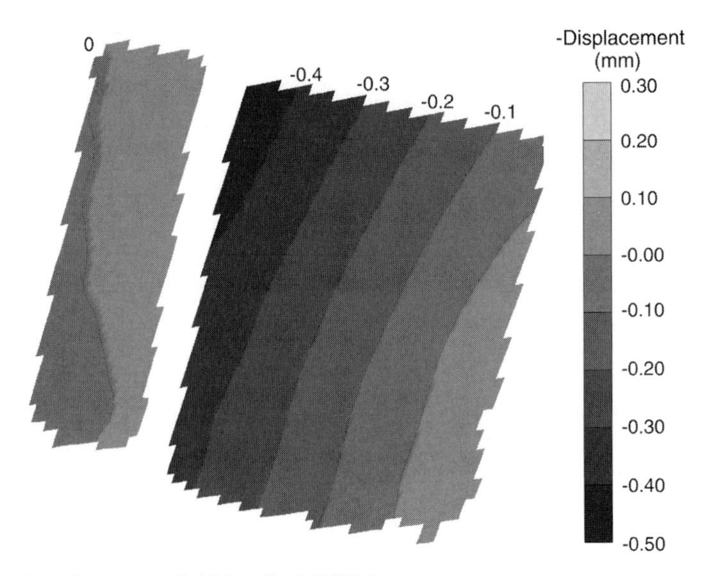


Figure 7.3.14: Displacement field by the HTGM.

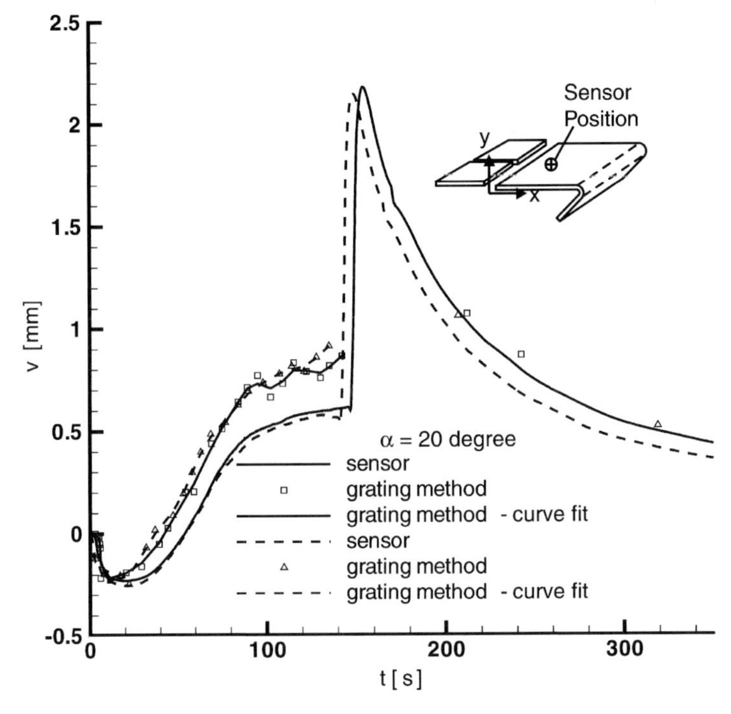


Figure 7.3.15: Comparison of the measured Y-displacement at the sensor position of the HTGM and the model's built-in sensor.

can be explained in the following manner: Directly after exposing the model to the hypersonic flow, the pressure loads will bend the C/C-SiC-plate down to negative Y-deflections. On account of the heating of the orthotropic C/C-SiC material, the V-shaped nose of the model increases its radius of curvature, bending the plain surface against the pressure-induced deflection to positive values. After switching off the plasma flow, the pressure vanishes causing the plate to bounce to the maximum peak of the curve. In the following cool-down period, the deflection relaxes to its initial state.

As a further result, made possible by the full field HTGM measuring method, a warpage of the flat plate symmetric to the middle axis of the model, was detected in the displacement field of Fig. 7.3.14. This warpage disappears after switching off the plasma flow, which leads to the conclusion, that the pressure forces are responsible for this effect.

All these solutions were obtained by using the rear segment of the plates behind the transversal gap as a reference surface for the HTGM method. This means that the reference coordinate system of the HTGM is fixed to the rear plate by using the stochastic grating pattern of the grating as basis for the reference system. This procedure is necessary in order to subtract the rigid body motions of the complete model due to movements of the model holder or the plasma wind tunnel from the measuring results of the ground fixed camera system of the HTGM. The underlying assumption of this procedure is, that the rear plate itself behaves like a rigid body and is not subjected to any distortion.

To prove this assumption in two subsequent measurements, the plate surfaces and the sidewall of the model have been chosen for investigation.

The investigated model section including the sidewall is shown in Fig. 7.3.16. It shows the right hand side (lower part) at the gap position of the model, also shown in Fig. 7.3.10 without the sidewall. The sidewall of the model itself was made of C/C-SiC too and was screwed to the water-cooled base plate by the use of fibre ceramic screws made of C/C-SiC. All these features should result in very low deflections of the sidewall during test, which leads to the conclusion to choose the sidewall as a reference surface for the HTGM measurement.

Figure 7.3.16: Field of view including the sidewall.

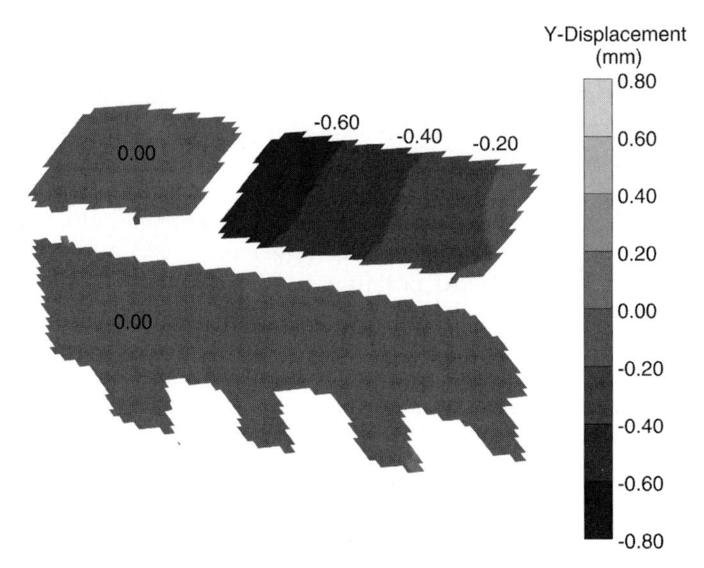


Figure 7.3.17: Displacement field of Fig. 7.3.16.

Figure 7.3.17 shows the result of the displacement field of the HTGM for $t = 20$ s after starting the high-enthalpy flow.

In Fig. 7.3.18, the displacement results of the built-in sensor and the HTGM at the X-position of the sensor 42 mm upstream the transversal gap, are presented. As in the results shown before, the coincidence between the tactile

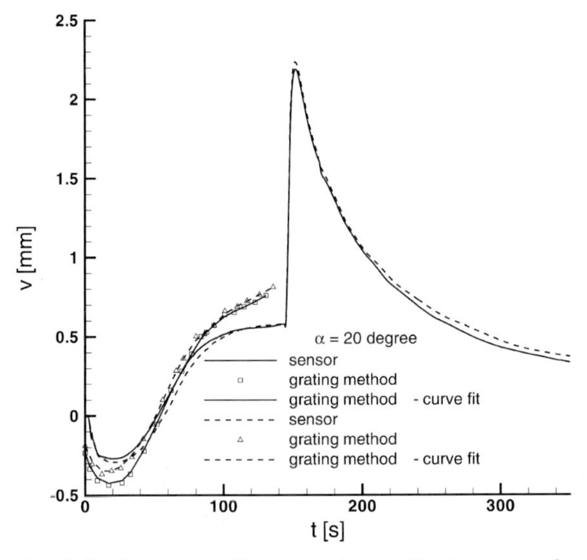


Figure 7.3.18: Results of displacement 42 mm upstream the transversal gap.

and the optical method is good. Differences are observed for negative displacements at $t<30$ s and positive displacements for $t>80$ s. These differences can be explained for the negative displacement values by the measured warpage of the model's surface. Due to the fact, that the built-in sensor is located on the model's symmetry axis, whereas the optically obtained values are, in this case, measured near the sidewall of the model, the warpage at this position leads to increased negative deflections as observed. For positive displacements, the warpage and the thermal expansion of the measuring pin of the built-in sensor are overlaid. Compared with the results in Fig. 7.3.15, the complete displacement curve is linear transformed by about 0.15 mm, also obtained from the warpage results in the displacement field in Fig. 7.3.14. As expected, the displacement between the sidewall and the rear plate was small and under 0.05 mm for $t<80$ s. For the following period a linear slope up to 0.15 mm was noticed.

## 7.3.3 Material Description of Fibre Ceramics

### 7.3.3.1 Phenomena in C/C-SiC Materials

To identify the phenomena occurring in C/C-SiC materials, uniaxial tests applying multiple unloading-reloading cycles are used. Experimental results of cyclic tension tests on a planar specimen are plotted in Fig. 7.3.19.

Unloading beyond a certain limit stress reveals inelastic strains. Increasing the load leads to a decrease of the unloading and reloading modulus; this indi-

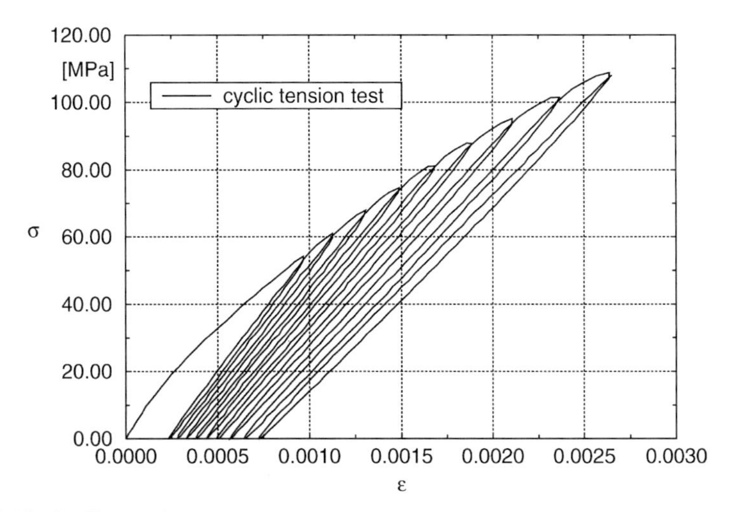


Figure 7.3.19: Cyclic tension test.

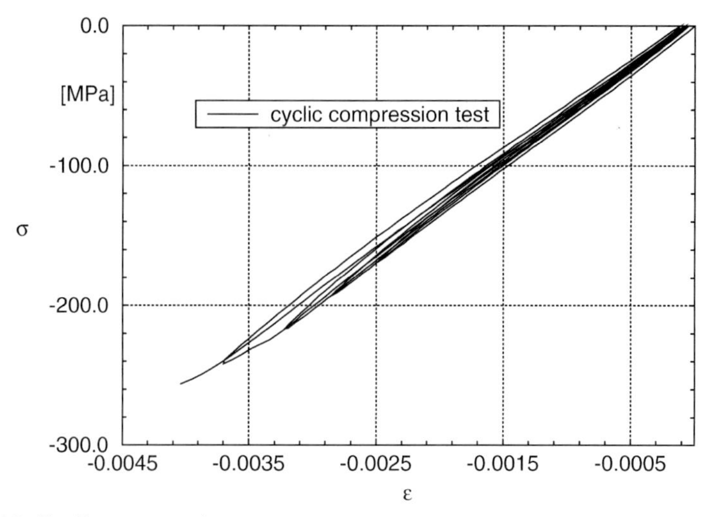


Figure 7.3.20: Cyclic compression test.

cates an internal damage of the material. The hysteresis of the unloading and reloading curves is very small and can thus be neglected. The same phenomena can be observed in torsion tests on tubular specimen.

In Fig. 7.3.20, the stress-strain curve obtained by a cyclic compression test on a planar specimen. The material behaves almost linear elastic up to fracture. Neither inelastic strain nor significant damage is observed. The failure stress in compression is approx. twice the failure stress in tension.

To describe the phenomena occurring in the material behaviour, in the following we present two different models: First a macroscopic (phenomenological) model which describes the global behaviour of the material, second a micromechanically based phenomenological model for fibre reinforced ceramics which describes the phenomena occurring in the fibres and the matrix in a smaller length scale (see Fig. 7.3.21).

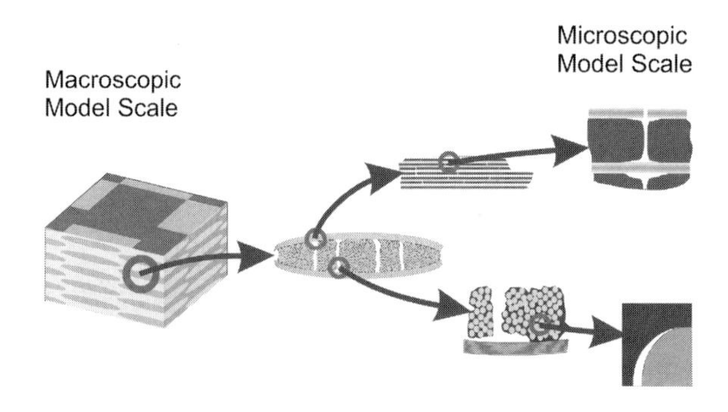


Figure 7.3.21: Macroscopic and microscopic model scales.

### 7.3.3.2 Phenomenological Model

In this section a model based on the macroscopic behaviour is presented. It includes

- elasticity,
- inelasticity,
- damage,
- fracture.

Each of these phenomena can be described by one constitutive model. It is assumed, that the stress state determines the state of the material and thus the constitutive models being active. Therefore, a complex multi-axial stress state exists in a material point. Interaction relations are used to transfer a multi-axial stress state in a scalar equivalent stress. These relations can also be interpreted as surfaces in a six-dimensional stress space. All stress states on a surface are characterized by the same material behaviour. In Fig. 7.3.22, a two-dimensional intersection of the limit surface model for C/C-SiC is depicted.

In the following the different constitutive models describing the behaviour of fibre reinforced ceramics are given.

*Elastic Model*

The elastic behaviour of materials with a crystal structure is caused by a reversible change of the interatomic distances of the crystal lattice. On the macroscopic level, elastic material behaviour can be characterized by a reversible thermodynamical process. For a thermodynamic state completely determined

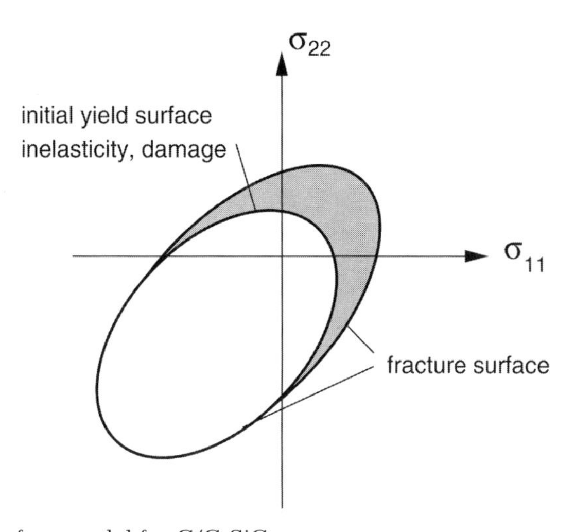


Fig 7.3.22: Limit surface model for C/C-SiC.

by the strain $\varepsilon$ or the stress $\sigma$ state, the Gibbs free energy $g$ describing the linear elastic behaviour is given by

$$g = \frac{1}{2}\sigma^{ij} F_{ijkl} \sigma^{kl} \tag{1}$$

where $\boldsymbol{F}$ is the flexibility tensor. Splitting the fourth-order tensor $\boldsymbol{F}$ in a scalar elasticity tensor $E$ and a fourth-order elastic mapping tensor $\boldsymbol{J}^e$ leads to

$$\boldsymbol{F} = \frac{1}{E}\boldsymbol{J}^e \tag{2}$$

The strain tensor $\boldsymbol{\varepsilon}^e$ which stems from the first and second law of thermodynamics can be formulated by

$$\boldsymbol{\varepsilon}^e = \rho \frac{\partial g}{\partial \boldsymbol{\sigma}} = \frac{1}{E}\boldsymbol{J}^e : \boldsymbol{\sigma} \tag{3}$$

*Inelastic Model*

In general, the inelastic strains of metals are – on the macroscopic level – caused by the migration of crystal defects such as dislocations. On the micromechanical level, the inelastic strains of fibre reinforced ceramics are caused by formation and growth of micro-cracks and the relative displacement of fibres and matrix. In the macroscopic model, the formulation is given by an elastoplastic formulation. If a theory of infinitesimal strains is assumed, the total strain tensor can be split in elastic and inelastic parts.

$$\dot{\boldsymbol{\varepsilon}} = \dot{\boldsymbol{\varepsilon}}^e + \dot{\boldsymbol{\varepsilon}}^i \tag{4}$$

where the elastic strain rate can be derived from the elastic strain and the inelastic strain rate.

*Damage*

Destruction of interatomic bonds causes micro-defects like micro-cracks and micro-voids. This phenomenon is called damage. On the macroscopic level, a degradation of the stiffness of the material, caused by the reduction of the effective material area by the micro-defects, is observed.

In this model, a phenomenological approach is chosen to account for the damage. The material parameters of the elastic model are functions of internal damage variables. For isotropic damage, only one scalar damage variable has to be used.

The formation of micro-cracks, which is a micromechanical damage mechanism, is the underlying principle of the relative displacement of fibres and matrix that causes the inelastic strain. Thus, the damage as well as the inelastic phenomena are a result of the same micromechanical process.

*Fracture*

Fracture denotes the formation of macroscopic failure zones in the material. The failure mechanism depends on the load applied to the material: Tension and shear loads cause failure by initiation and growth of macroscopic cracks; multiaxial compression can cause failure due to crushing of the material.

In this paper, a smeared approach is applied. The fracture of the material is simulated by softening constitutive relations. If the fracture model is active, the strain rate consists of an elastic and a fracture part.

$$\dot{\varepsilon} = \dot{\varepsilon}^e + \dot{\varepsilon}^c \qquad (5)$$

More details on the fracture model can be found in Fink et al. [18].

#### 7.3.3.3 Micromechanically Based Phenomenological Model

As C/C-SiC can be subdivided in two components: the first one consists of longitudinal fibres embedded in the C-matrix, while the second component consists of the transverse fibre bundles in the SiC-matrix.

Upon loading, failure of the composite begins in the second component due to rupture of the bridging fibres, leading to a growth of the processing-induced cracks.

These cracks can extend in the longitudinal fibre bundles. With increasing loads, stochastic fibre failure takes place in the first component. This leads to a rupture of the specimen.

After complete unloading, large inelastic strains can be detected for relative low tensile stresses due to interlocking of the surfaces of the shrinkage cracks upon unloading.

In the micromechanical model, this behaviour is modelled by formulating the different effects in the fibres and the matrix for various load cases.

*Stochastic Fibre Failure*

The stochastic fibre model (see Fig 7.3.23) is formulated by a constitutive law of a longitudinal ply under in-axis tension. It is based on the model by Hild et al. [20] and Phoenix and Raj [21].

Note, in this model, it is assumed that the stress in a broken fibre is built up linearly through shear stress transfer within a recovery length.

$$L_f = R\sigma_L^{l,i}/(2\tau) \quad (6)$$

When unloading the ply, the shear stress $\tau$ is reversed, beginning at the plane of the fibre crack [22].

*Transverse Cracking*

Another problem is the crack initiation in the matrix in fibre ceramics when tension is applied. In literature, often a shear-lag model is used. The stress distribution in the transverse layers resulting from the actual crack distribution determines the initiation of cracks. If the stress value is larger than a limit, a new crack is formed. Note, details for shear damage modelling can be found in Weigel et al. [19].

*Compression Failure*

Compression failure due to microbuckling of single fibres or fibre bundles has been modelled e.g. by Gupta et al. [22]. In addition to the failure caused by large compressive or tensile strains in the buckling of the fibre bundles, a loss of stability due to debonding of the fibre bundles is considered.

*Parameter Identification and Numerical Simulation*

For applying this micromechanical model, we have to determine many material parameters. Due to the difficulties to find all these micromechanical parameters through direct experiments, we have tried to determine these parameters by pure tension, compression and torsion experiments by an identification procedure using the least squares method.

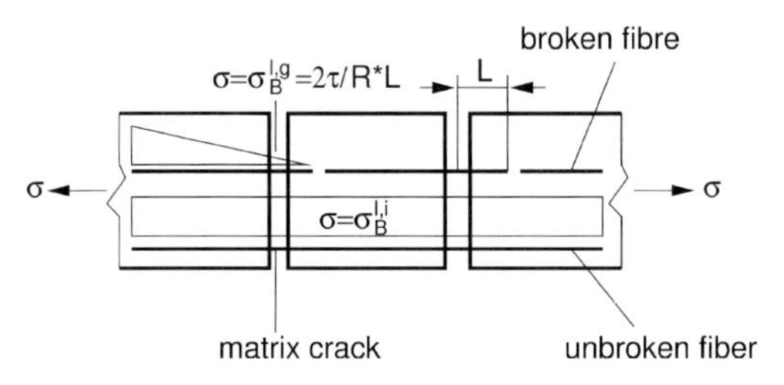


Figure 7.3.23: Stochastic fibre failure.

Although, only two states of each curve have been used for parameter identification, the model is able to reproduce the complete stress-strain behaviour with good accuracy. In Fig. 7.3.24, simulation results using the models given above and the determined parameters have been performed. We can find a very promising correlation between theory and experiments.

### 7.3.3.4 Functionally Graded Materials

Reusable re-entry vehicles request strict requirements to resist thermo-mechanical loads for a high number of missions. A main problem is to minimize the degradation of the thermal protective shield during re-entry. Due to large temperature differences and different thermal expansion coefficients of the protective coating and the base material, high stress gradients at the interface are obtained. To solve this problem, functionally graded materials (FGM) have been developed. In these materials, the composition of the components is dependent

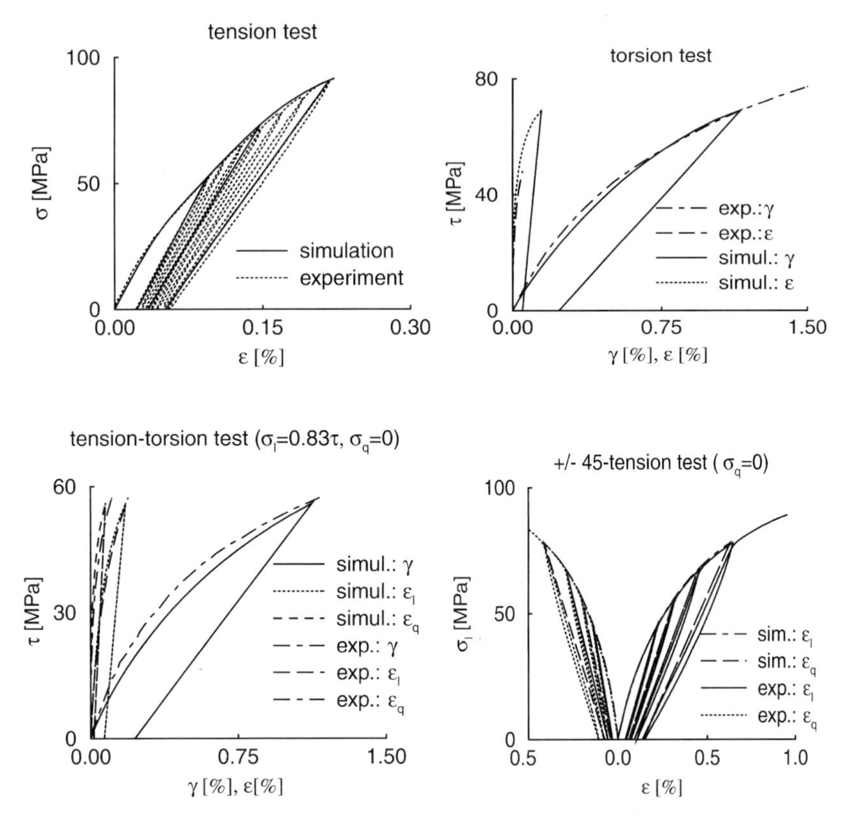


Figure 7.3.24: Comparison between numerical and experimental stress-strain curves for various test cases.

of the position over the thickness. Thus, the stress-concentration in the FGFM is reduced and therefore the adhesion between the protective layer and the base material is improved.

In the following, a FGM probe consisting of four different layers stacked symmetrically to the horizontal axis is investigated. For further details see Section 7.1. In Fig. 7.3.25, the model of this FGM layer composition is plotted. The material parameters for each layer are given in Tab. 7.3.2.

The FGM is stress-free at the manufacturing temperature of approx. 1800–1900 K. In the numerical simulation, the temperature of the probe is reduced from the manufacturing temperature to the room temperature. The probe is supported by a minimal bearing. In the first test case, the simulation has been computed for an undamaged probe, in the second test case the same probe with one single crack perpendicular to the surface has been investigated.

In Fig. 7.3.26, the result for the first test case is depicted. The probe geometry is contracted in the 1-direction, i.e. parallel to the layers. The contraction of the top and bottom layers is larger than of the inner-layer due to the demand of constant strain at the layer-interfaces. The in-plane stress $\sigma_{11}$ is nearly constant in each layer and changes stepwise at the interfaces.

In Fig. 7.3.27, the stress distribution $\sigma_{11}$ of the probe with crack is plotted.

The undamaged areas show the same behaviour as in the first test case. Due to the temperature reduction the crack extension increases. In the region near the crack a stress concentration can be determined. The stress concentration especially in horizontal direction induces high interlaminar strains which

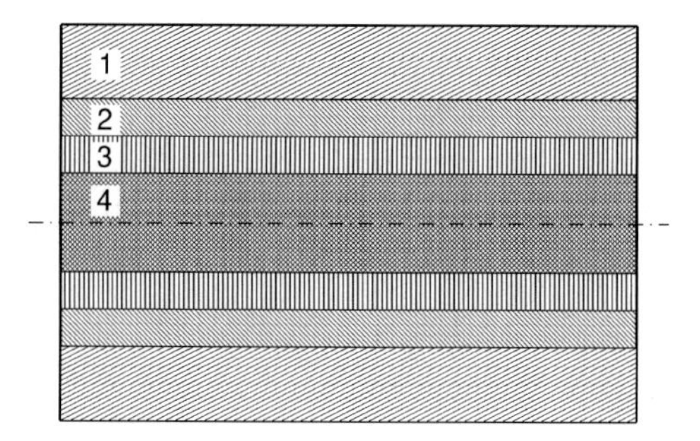


Figure 7.3.25: Model of the FGM material.

Table 7.3.2: Characteristics of the different layers in the FGM.

| Layer | Layer thickness [mm] | Elastic modulus [GPa] | Thermal expansion factor [1/K] |
|---|---|---|---|
| 1 | 1.2 | 110.9 e-3 | 3.8 e-6 |
| 2 | 0.6 | 66.6 e-3 | 3.4 e-6 |
| 3 | 0.6 | 62.4 e-3 | 2.7 e-6 |
| 4 | 1.0 | 48.9 e-3 | 2.2 e-6 |

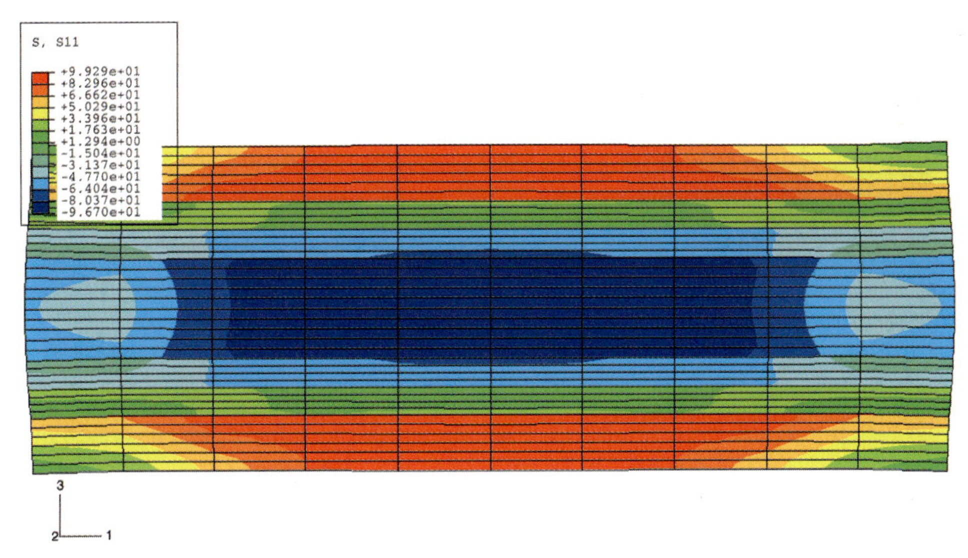


Figure 7.3.26: Numerical $\sigma_{11}$-stress distribution of the undamaged probe.

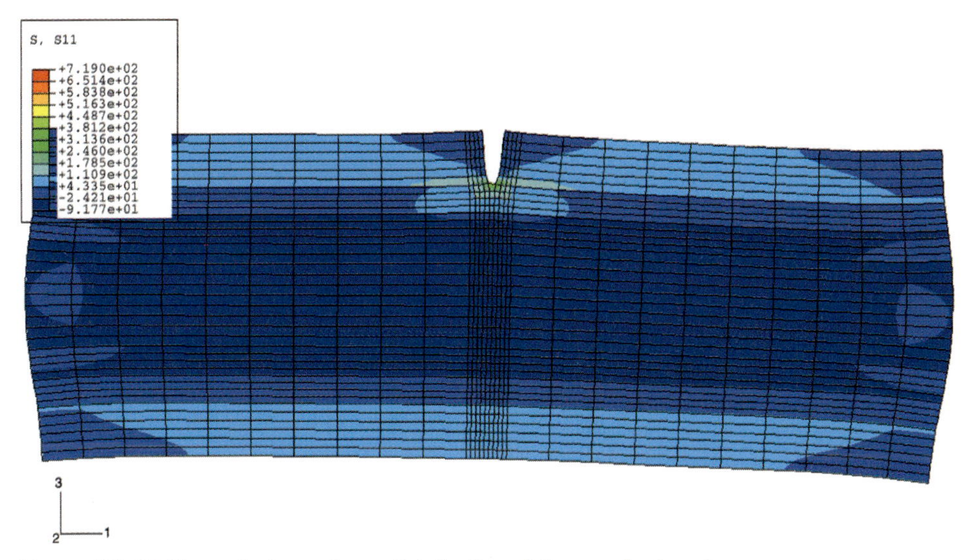


Figure 7.3.27: Numerical $\sigma_{11}$-stress distribution of the cracked probe.

may lead to additional horizontal cracks which can be confirmed in the experimental investigations.

For the investigation of typical crack patterns of the material, a test has been conducted by the use of the high temperature grating method [14]. In the experiment, the FGM-specimen as described before, was heated from room-temperature up to 673 K. The resulting strain pattern indicates the cracks in

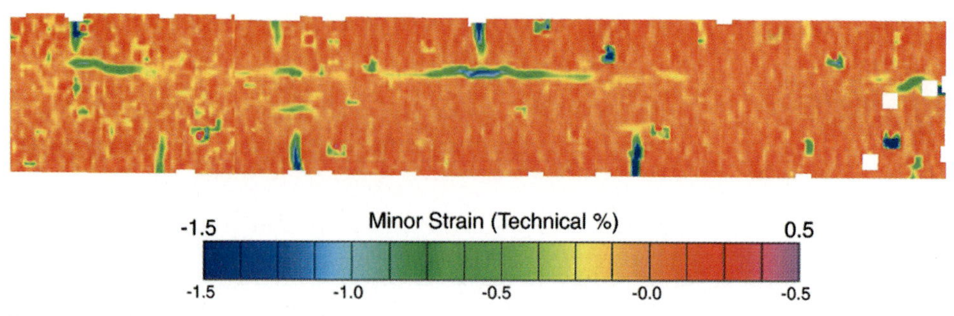


Figure 7.3.28: Experimentally measured minor strain distribution.

the surface of the probe. In Fig. 7.3.28, cracks perpendicular to the layer surface and horizontal cracks parallel to the boundaries of the different layers are detected.

In the numerical simulation, a formation of a large horizontal stress-gradient in the region near the crack tip confirms the experimental results.

### 7.3.4 Conclusions

Over a period of more than ten years, high-temperature structural components made from fibre ceramics – a new generation of materials – were developed from scratch as reported in 7.3.1. The development phase cumulated in a sophisticated flight experiment incorporating all the valuable experience gained through a great number of manufacturing prototypes and ground tests. Although the flight was not successful, it is justified to say that the technology of a CMC TPS has reached a high level of maturity which probably can not be enhanced much further by ground tests alone. There will be a re-flight of the experiment that will be the beginning of a new phase of the CMC TPS development in which the technology has to prove its performance under real re-entry conditions, followed by other flights on different vehicles.

The results in section 7.3.2 show, that the **H**igh **T**emperature **G**rating **M**ethod is applicable under the conditions prevailing in the long duration high-enthalpy facility L3K. As a non-contact full-field optical measurement method, the principle offers significant advantages over conventional techniques. Without using probes, which would disturb the hypersonic flow field around the investigated object, displacements and deflections, even of complex structures, can be measured with high accuracy. The comprehensive information about the whole investigated area leads to a better understanding of component behaviour.

In section 7.3.3, two models have been presented to describe the phenomena occurring in fibre ceramic materials: a phenomenological macroscopic model and a phenomenologically based microscopic model. Both show a very good correlation compared to the experimental results. Finally, experimental and numerical results for FGM materials – which are very suitable to resist for example high thermomechanical loads – are compared. For this cases also a good qualitative agreement can be observed.

## References


1. C. Petiau: The Challenge of the Hermes TPS Shingles, Proc. Int. Conf.: Spacecraft Structures and Mechanical Testing, Noordwijk, Netherlands, 24–26 April 1991.
2. U. Schöttle, J. Burkhardt: Konzeptstudie auftriebserzeugender Rückkehrkörper COLIBRI, Universität Stuttgart, IRS, IRS-95-P7, 1995.
3. H. Hald, M. von Bradke, K. Maile, H. Schneider, R. Pfleger: From Plasma Channel to Real Reentry Testing, $44^{th}$ Int. Astronautical Congress, 1993, Graz, Austria.
4. H. Hald, P. Winkelmann: Post Mission Analysis of the Heat Shield Experiment CETEX for the EXPRESS Capsule, 48th Int. Astronautical Congress, 1997, Turin, Italy.
5. H. Hald, D. Petersen, T. Reimer, F. Rühle, P. Winkelmann, H. Weihs: Development of a CMC-based TPS for two Representative Specimens of Cryogenic Tank RLVs, AIAA 8th Int. Space Planes and Hypersonic Systems and Technologies Conference, 1998, Norfolk, Virginia, USA.
6. H. Weihs, H. Hald, T. Reimer, I. Fischer: Development of a CMC Nose Cap for X-38, Int. Astron. Congress, IAF-01-I.3.01, 2001, Toulouse, France.
7. W. Krenkel, T. Henke, N. Mason: *In-situ* Joined CMC Components, CMMC 9, 1996, San Sebastian, Spain.
8. SFB 259, Hochtemperaturprobleme rückkehrfähiger Raumtransportsysteme , Arbeits- und Ergebnisbericht 2001, Universität Stuttgart, 2001, Stuttgart.
9. J.R. Strife, J.E. Sheehan: Ceramic Coatings for Carbon-Carbon Composites, Am. Ceram. Soc. Bull. (1988), 369–374.
10. M.E. Westwood et al.: Review – Oxidation Protection for Carbon Fibre Composites, J. Mater. Sci., Vol. 31 (1996), 1389–1397.
11. T. Laux, T. Ullmann, M. Auweter-Kurtz, H. Hald, A. Kurz: Investigation of Thermal Protection Materials along an X-38 Re-entry Trajectory by Plasma Wind Tunnel Simulations, Proc. $2^{nd}$ Int. Symp. on Atmospheric Re-entry Vehicles and Systems, March 2001, Arcachon, France, available on CD.
12. T. Ullmann, M. Schmücker, H. Hald, R. Henne, H. Schneider: Low Pressure Plasma Sprayed Yttrium Silicates for Oxidation Protection of C/C-SiC Composite Structures, Proc. of $7^{th}$ Japan Int. SAMPE Symp. and Exhib., 13–16 Nov. 2001, Tokyo, Japan, ISBN 4-9900028-7-3, 325–328.
13. T. Ullmann, M. Schmücker, H. Hald, R. Henne, H. Schneider: Plasma Sprayed Yttrium Silicates for Oxidation/Erosion Protection of C/C-SiC Components, Proc. of $4^{th}$ Europ. Workshop on Hot Structures and Thermal Protection Systems for Space Vehicles, 26–29 Nov. 2002, Palermo, Italy, ESA SP-521, 147–153.
14. R. Kornmann, K. Eberle, M. Dogigli: Measurement of thermal expansion of ceramic matrix composite tubes at temperatures up to 1400 °C, High Temperatures-High Pressures, 2000, volume 32, 621–628.
15. R. Kornmann, B. Kröplin: Objektrasterverfahren zur räumlichen Verformungsmessung an Hochtemperaturbauteilen, tm – Technisches Messen 67 (2000), Oldenbourg Verlag, 267–273.
16. A. Gülhan, B. Esser: Arc-Heated Facilities as a Tool to Study Aerothermodynamic Problems of Re-entry Vehicles, in: F. K. Lu, D. E. Marren: Advanced Hypersonic Test Facilities, Progress in Astronautics and Aeronautics, Vol. 198, 375–403, AIAA, 2002.
17. R. Schäfer, A. Mack, B. Esser, A. Gülhan: Fluid-structure interaction on a generic model of a re-entry vehicle nosecap. In: $5^{th}$ International Congress on Thermal Stresses, Blacksburg, Virginia, 2003.

18. A. Fink, D. Dinkler, B. Kröplin: A phenomenological material model for fiber reinforced ceramics, Z. Flugwiss. Weltraumforsch. 19, 197–204 (1995), Springer Verlag.
19. N. Weigel, B.-H. Kröplin, D. Dinkler: Micromechanical modeling of damage and failure mechanisms in C/C–SiC, Computational Materials Science 16, 126–137 (1999).
20. F, Hild, A. Burr, F. A. Leckie: Fiber Breakage and Fiber Pull-Out of Fiber-Reinforced Ceramic-Matrix Composites. European Journal of Mechanics A, 13, 731–749 (1994).
21. S. L. Phoenix, R. Raj: Scaling in Fracture Probabilities for Brittle Matrix Fiber Composite. Acta metall. Mater., 40, 2813–2828 (1992).
22. V. Gupta, K. Anand, M. Kryska: Failure Mechanics of Laminated Carbon-Carbon Composites – 1. Under Uniaxial Compression. Acta metall. Mater., 42, 781–795 (1994).

# 8 Cooperation with Industry and Research Establishments, Participation in National and International Research Programmes

Dieter Jacob, Gottfried Sachs, and Siegfried Wagner

From the start of the three Collaborative Research Centres, a successful cooperation with industry and non-university research establishments developed. This cooperation concerns a great number of joint activities for which a description is given in the following paragraphs.

In the preparation phase, industry expressed great interest in the Collaborative Research Centres and their basic research and declared to provide a continuous support. The support implied that computational methods and data were exchanged on an adequate level. This was very helpful because the Collaborative Research Centres got access to important material, e.g. experimental results from comprehensive wind tunnel tests on stage separation at hypersonic speed were made available. The theoretical insight into hypersonics, the experimental know-how, the algorithms and software tools developed within the Collaborative Research Centres on the other hand enabled scientists to support both German and European technology programmes for future space transportation systems.

There were many cooperative efforts between the Collaborative Research Centres and the industry and non-university research establishments on a working level. This concerns the individual cooperation between experts as well as the participation in working groups established by the Collaborative Research Centres. The working groups were concerned with subjects such as configuration, numerical methods, engines, etc.

Members of the Collaborative Research Centres were active in the review process of the Hypersonic Technology Programme of the Ministry for Research and Technology of the Federal Government of Germany. Furthermore, representatives of the Collaborative Research Centres were appointed members of the Hypersonic Technology Programme Committee of this Ministry.

A most important activity concerns the participation of the Collaborative Research Centres in national and international research programmes. The participation also implied a close cooperation with industry companies and non-university research establishments involved in these programmes. The participation of the Collaborative Research Centres includes the following programmes:

*Basic Research and Technologies for Two-Stage-to-Orbit Vehicles*
DFG, Deutsche Forschungsgemeinschaft

ISBN: 3-527-27735-8

- German TETRA-Programme (Technologien für zukünftige Raumtransportsysteme);
- German ASTRA-Programme (Ausgewählte Technologien für zukünftige Raumtransport-Anwendungen);
- European FESTIP-Programme (Future European Space Transportation Investigations Programme).

The Collaborative Research Centres performed demanding research projects in their respective fields, such as aerothermodynamics, flight mechanics and guidance, engines, structures, etc. With this involvement, the experience and knowledge resulting from the basic research of the work conducted at the Collaborative Research Centres could be transferred to the Research Programmes at a more applied level. For instance, there was a close collaboration in the field of thermal protection materials with the French companies Dassault, SEP and Aerospatiale. In pursuit of a more evolutionary re-entry technology development, Europe subsequently (i.e. early 1990s) embarked on ballistic capsule missions and later on lifting re-entry missions. In cooperation with space companies (e.g. Astrium, Kayser-Threde) scientists of the Collaborative Research Centres contributed to the capsule missions EXPRESS (cooperation with Russia and Japan), MIRKA (cooperation with Russia), IRDT-1/2 (cooperation with Russia) by development of flight experiments (CETEX, PYREX, HEATIN, FIPEX). There was also a collaboration with the Japanese space agency NASDA, with the Japanese companies Kawasaki Heavy Industries (KHI), Mitsubishi Heavy Industries (MHI), and Nissan. The cooperation with industry and non-university research establishments also implied joined teams, thus contributing to the efficiency of the research work.

The participation of the Collaborative Research Centres in the European programme FESTIP was extended to the development of the demonstrator vehicle X-38 in 1998. This programme was initiated by the American and European space agencies to prepare and develop the Crew Return Vehicle (CRV). Significant support was provided to Germany's (DLR) TETRA programme "Technologies for Future Space Transportation Systems" and ESA's "Applied Reentry Technology Programme" (ARTP). The development of advanced non-equilibrium Navier–Stokes codes was strongly supported in joint research with Moscow State University in Russia, Interprofessional Aerothermochemistry Research Complex (CORIA) in France, with University of Minnesota, Scientific Research and Technology, Inc., and NASA, USA, with University of Bari and with Centre for Advanced Studies, Research and Development in Sardinia, both Italy. In addition, meta-computing was applied in cooperation with Japan Atomic Energy Research Institute (JAERI), Tsukuba Advanced Computing Centre (TACC), both Japan, Manchester Computing Centre (MCC), UK, National Centre for High-Performance Computing (NCHC), Taiwan, Sandia National Laboratories, USA, and Korea Institute of Science and Technology (KISTI) in Korea.

Mission and vehicle conditions of the space transportation system HOPPER within the German ASTRA programme were investigated. The diagnostic system for determining temperature distribution and thermal flow as well as a miniature spectrometer for verifying chemical models are foreseen to be used on the ESA demonstrator EXPERT.

In 1999, a cooperation between the Interprofessional Aerothermochemistry Research Complex (CORIA), France, and an Institute of the Collaborative Research Centres was initiated. Working fields and related groups of scientists were identified for future cooperation. One of these fields was the comparison of the inductively driven plasma wind tunnel facilities. In addition, code development and validation were strongly supported in international cooperation. A successful cooperation between ONERA, Départment Mécanique du Solide et de l'Endomagement (DMSE), and several research groups at the University of Stuttgart started in 1999 in the field of materials properties and component tests. In cooperation with Laboratoire National d'Essais (LNE) comparative measurements of the thermal diffusivity and conductivity of a ceramic high temperature material were performed, providing good agreement in the test results.

During the last two years the activities of the Collaborative Research Centres were especially supported by the Ministry for Science, Research and Arts of Baden-Württemberg, by the Universität Stuttgart, and by MAN-Technologie within the research project "The development of protective layers for re-usable space transportation systems". This development of protective layers for fiber ceramic thermal protection materials for space transportation vehicles plays an important role worldwide because these materials are only then re-usable when the layers prevent atomic oxygen and atomic nitrogen from attacking the base material. Therefore, this project was very important for the overall research work. It was planned to test the new protective layers on a return flight of the Russian Foton capsule and to compare the results with the investigations in the ground test facilities. However, the mission failed due to a malfunction of the carrier rocket. The cooperation with MAN-Technologie dealt with plasma tests of ceramic thermal protection system material samples.

In Europe, within the framework of the ESA programme AURORA, technologies are currently developed for un-crewed and crewed interplanetary missions. A large part of the know-how regarding interplanetary return missions can be offered by the Collaborative Research Centres to these missions.

The experiences gained within the Collaborative Research Centres led to the formation of several companies. One of these companies, Dr. Laure Plasmatechnologie, develops plasma sources for coating processes in particular for solar-thermal applications and high resistant surface requirements.

Currently, the European FLPP-Programme (Future Launcher Preparatory Programme) is discussed in the preparation of which the Collaborative Research Centres are taking part.

# 9 Conclusions and Perspectives

Dieter Jacob, Gottfried Sachs, and Siegfried Wagner

The innovative idea to initiate three Collaborative Research Centres at different universities, namely in Aachen, Munich and Stuttgart, that perform research work on a common subject and by a well-defined work sharing has been concluded successfully. The aim of the investigation has been to study basic and technological problems that evolve during the design of re-usable space transportation systems, specifically of two-stage configurations.

Several themes were agreed upon by the chairmen of the three centres. They included overall design of the configuration with ability to take-off and land horizontally on airports with a proper runway. The second theme was aerodynamics that covered low speed flow, supersonic and especially hypersonic flow. The third theme – flight mechanics – foresaw the fields of stability, control and flying qualities, particularly at hypersonic speeds and for two-stage configurations. A further central problem has been the complete propulsion system that consisted of an airbreathing engine for the first stage and a rocket propulsion for the second stage. The fifth theme treated lightweight structures and new materials that were able to withstand temperatures of the order of 2000 Kelvin and beyond. In close relation to this problem was the design and testing of hot structures. In order to be able to handle these problems, proper methods and technologies had to be created.

In the field of aerodynamics proper numerical simulation procedures as well as wind tunnel and measuring techniques had to be developed. The investigations included experimental studies of the Reynolds number influence at low speeds, the development of numerical methods at supersonic and hypersonic speeds, and experimental studies of the outer flow and the internal flow through the propulsion system at a broad range of Mach numbers. For instance, further direct numerical simulation (DNS) techniques for viscous flows had to be established in order to study the laminar-turbulent transition of boundary layers, and the separation of the second stage from the first stage was studied in wind tunnels and by numerical simulations.

In flight mechanics, trajectory optimization was a research subject. Goals were performance optimization and exploitation of the safety potential which two-stage-to-orbit vehicles offer. Another research subject concerns stability, control and flying qualities related to the unique dynamics characteristics of hy-

*Basic Research and Technologies for Two-Stage-to-Orbit Vehicles*
DFG, Deutsche Forschungsgemeinschaft

ISBN: 3-527-27735-8

personic aircraft. Furthermore, the separation manoeuvre of the two-stage system was investigated, with the objective of maximizing safety and performance.

A further subject of investigations was the re-entry phase of the second stage. It was studied by numerical simulations. New theoretical models were developed to represent material properties. The latter were also investigated in plasma wind tunnel tests and were used to validate the theoretical models.

A major critical part of the configuration was the complete propulsion system. Besides studying the boundary layer on the lower side of the fore-body, the inlet with its shock wave structure was a major issue of investigation. The hybrid configuration consisted of a jet engine with an additional extra ramjet/scramjet engine. Thus combustion – especially supersonic combustion – was thoroughly studied by numerical simulation and in special combustion test stands. A further point of interest was the performance of the nozzle which influences not only the delivered thrust but also the stability and control of the vehicle.

Because of high temperatures during the ascent phase and the re-entry phase, new materials and protective layers of the structures were developed. This included new gradient materials and two-layer coating layers to prevent atomic oxygen and atomic nitrogen from attacking the base material. In addition, so-called hot structures had to be designed which not only can accept the high temperature but can also carry the high loads.

The cooperation between the Collaborative Research Centres that included DLR and Max-Planck-Institut let to a large network of research engineers and scientists that planned and carried out common tests and numerical simulations using different measuring techniques and different simulation procedures, respectively. The expertise that was collected during the common research work let to a cooperation not only with German industry but also with research groups and industry in Russia, France, Italy, Japan, Korea, Taiwan, UK and the USA.

Despite the fact that a lot of new insights into complex mechanisms have been gained and a lot of progress has been achieved, a variety of problems still remains to be solved. The current European concept for future space transportation systems foresees a Two-Stage Transportation System (TSTO) with a reusable rocket-driven first stage (RLV, re-usable launching vehicle) in order to reduce current costs of space transportation. The universities in this stage of development can contribute to the optimization of rocket nozzles with efficient external and internal aerodynamics in the wide range of ambient pressure and temperature. There are many aerodynamic interference and flight stability problems and control laws to be solved. Wind tunnel tests and numerical simulations should be performed for all these new concepts and the corresponding measuring techniques and simulation tools should be developed. The universities are prepared to perform these tasks.

# 10 Appendix

## 10.1 Publications

Abstiens, R., Fühling, S., Vetlutsky, V.N.: Boundary-Layer Measurements on the ELAC 1 Configuration at Re=$4\times10^7$. GAMM Annual Meeting, Bremen, April 1998 ZAMM Volume 79 (1999), Suppl. 3, pp. 927–928.
Abstiens, R., Fühling, W., Schröder, W.: Boundary-Layer Measurements on the ELAC Configuration at Re=$20\times10^6$. ICTAM 2000, 20th Int. Congress of Theoretical and Applied Mechanics, Chicago, USA, 2000.
Albus, J., Miermeister, M., Friese, H., Öry, H.: Stress and stability analysis of shells with elliptical cross-section. ZFW, Vol. 20, Nr. 2, 1996.
Albus, J., Öry, H.: Comparison of different liquid hydrogen tank integration concepts for the ELAC-I research configuration. ZFW, Vol. 17 (1993), pp. 149–156.
Aldinger, F., Weinmann, M., Bill, J.: Precursor-Derived Si-B-C-N Ceramics. Pure Appl. Chem., 70(2), p. 439, 1998.
Allgayer, U., Friedrich, R.: Scalar PDF Approach Applied to Turbulent Combustion in a Compressible Shear Flow. Submitted to Flow, Turbulence and Combustion, 2001.
Allgayer, U., Friedrich, R.: Turbulente Vermischung in einer Überschallgrenzschicht mit Stickstoffeinblasung. Seminar des Sonderforschungsbereichs 255, München, 12. Dez. 1996, Technische Universität München, Tagungsband, pp. 9–16, 1997.
Allgayer, U., Lechner, R.: Turbulence and Combustion Modelling. – In: Schaller, Ch., Brehm, M. (Eds.): Overview of Research Projects on the Cray Y-MP at the Leibniz-Rechenzentrum München, LRZ-Bericht Nr. 9601, 1996.
Althaus, W.: Linear stability analysis of the near wake of a flat plate. ZFW, Vol. 19, Nr. 6, 1995.
Altmann, I., Bauer, G., Hirsch, K., Jentschke, H., Klenge, S., Roth, B., Schinköth, D., Schumacher, U.: In situ Diagnostics of the Interaction Region between a Nitrogen-Oxygen Plasma Jet and hot C/C-SiC Ceramic Materials. High Temperatures – High Pressures 32, 573–579, 2000.
Altmann, I., Quell, S., Hirsch, K., Roth, B., Schneider, J., Schumacher, U.: In Situ Determination of Erosion Rates by Application of Emission and Absorption Spectroscopy. ECAMP VII, 7th Europ. Conf. on Atomic and Molecular Physics, Berlin 2.–6. April 2001, pp. 12–14, 2001.
Andres, Y.A., Zimmermann, F., Schöttle, U.M.: Optimization and Control of the Early Deployment Phase during a Tether-Assisted Deorbit Maneuver. Proceedings of the 22nd International Symposium on Space Technology and Science, Morioka, Japan, 2000.
Aoki, R., Heyduck, J., Klinge, B., Kraft, H.: Tensile Test of C/C-SiC Specimens. European Conf. on Composites Testing and Standardisation, Hamburg, Sept. 13–15, 1994, Cambridge/UK: Woodhead Pub. (ed.: Hogg, P.I. et al.), PP. 217–227, 1994.

*Basic Research and Technologies for Two-Stage-to-Orbit Vehicles*
DFG, Deutsche Forschungsgemeinschaft

ISBN: 3-527-27735-8

Aoki, R.M., Busse, G., Eberle, K., Hänsel, C., Schanz, P., Wu, D.: NDI Evaluation of Local Oxidised C/C-SiC Specimens. INSIGHT-Journal of the British Institute of Non-Destructive Testing 40, 10, pp. 706–711, 1998.

Aoki, R.M., Schanz, P.: NDI Assessment of CMC (C/C-SiC) Manufactouring. INSIGHT-Journal of the British Institute of Non-Destructive Testing 41, 3, pp. 173–175, 1999.

Aoki, R.M.: Non-Destructive Inspection of Fibre Reinforced Ceramics. 1st Joint Belgian-Hellenic Conf. on Non Destructive Testing; Patras/GR, May 22–23, 1995, in: "Non Destructive Testing", Edited by Van Hemelrijk, D. and Anastassopoulos, A., pp. 137–142, Balkema Pub., Rotterdam/NL, 1996.

Aoki, R.M.: Phenomenological Investigations on C/C-SiC Specimens with NDT Methods. Proc 4th Int. Conf. on High Temperature Ceramic Matrix Composites (HT-CMC4), München, Okt. 1–3, 1998, High Temperature Ceramic Matrix Composites, pp. 517–523, Wiley-VCH, Weinheim, 2001.

Arbeiter, P.L., Neuer, G.: Pulse relaxation method, a new way for investigating anisotropic homogeneous high temperature composites. High Temp. – High Press. 27/28, 599–610, 1995/96.

Ardey, N., Mayinger, F.: Optical Diagnostics on the Structure of Premixed Turbulent Hydrogen-Air-Flames. German-Russian Workshop on Problems of Reaction Kinetics in High-Temperature Systems, Munich, 10.–13. Oktober, 1993.

Ardey, N., Mayinger, F.: Survey of Hydrogen Combustion Research at the Intstitute A for Thermodynamics of the TU Munich 2nd German-Canadian Workshop on Hydrogen, October 1993, Pinawa, Canada.

Ardey, N., Durst, B., Mayinger, F.: Influence of Transport Phenomena on the Structure of Premixed Hydrogen-Air-Flames, Proc. of the 3rd. Int. Workshop on Hydrogen Combustion in Reactor Safety Research, München, Sept. 1994.

Ardey, N., Mayinger, F., Durst, B.: Influence of Transport Phenomena on the Structure of Lean Premixed Hydrogen-Air Flames. American Nuclear Society Transactions, Vol. 73, TANSAO 73 1–522, 1995.

Ardey, N., Mayinger, F.: Aerosol Resuspension by Highly Transient Containment Flow – Insight by Means of Laser-Optical Methods. Journal for Nuclear Engineering Energy Systems and Radiation, Vol. 63, 1998.

Ardey, N., Mayinger, F.: Flame Acceleration by Turbulent Promotion and Jet Ignition. Submitted for Journal of Progress in Energy & Combustion Science, 1998.

Ardey, N., Mayinger, F.: Highly Turbulent Hydrogen Flames/Explosions in Partially Obstructed Confinements. Proceedings of the 1st Trabzon Int. Energy and Environment Symp., Karadeniz Techn. Univ., Trabzon, Türkei, 29.–31. Juli, 1996, pp. 676–692. And Proceedings of the 8th Canadian Hydrogen Workshop, Canadian Hydrogen Association, Toronto, Kanada, 27.–29. Juni, 1997.

Ardey, N., Strube, G., Mayinger, F.: Complementary use of Optical Flow Diagnostics in Combustion Research. Proc. EUROMECH 309, Göttingen, 28. Sept.–1. Okt. 1993.

Arendts, F.J., Theuer, A., Maile, K., Kuhnle, J.: Mechanical Behaviour of Different Sized C/SiC-Tubes under Multiaxial Loading and Temperatures up to 1600°C. Silicates Industrial – Journal of the Belgian Ceramic Society, 1995.

Arendts, F.J., Theurer, A., Maile, K., Kuhnle, J., Neuer, G., Brandt, R.: Thermomechanical and Thermophysical Properties of Liquid Siliconized C/C-SiC. ZFW 19, 189–196, 1995.

Arendts, F.J.: Aktuelle Entwicklungen in der Strukturtechnik. Zeitschrift für Flugwissenschaften und Weltraumforschung ZFW 16, pp. 231–246, Springer Verlag, 1992.

Arning, R., Alles, W.: Eine Low-Cost-Methode zur Bestimmung flugmechanischer Eigenschaften von Raumflugzeugen im Freiflugexperiment. Vortrag zur DGLR Jahrestagung, Hamburg, September 2001, DGLR-2001-100, DGLR-Jahrbuch.

Arning, R.: Bestimmung der aerodynamisch-flugmechanischen Parameter der Unterstufe eines Raumtransportersystems mit Hilfe frei fliegender Modelle. DGLR-Jahrestagung, Leipzig, September 2000, DGLR-Jahrbuch, DGLR-JT2000-056.

Auweter-Kurtz, M., Bauer, G., Behringer, K., Dabalà, P., Habiger, H., Hirsch, K., Jentschke, H., Kurtz, H., Laure, S., Stöckle, T., Volk, G.: Plasma Diagnostics within the Plasma Wind

Tunnel PWK. Zeitschrift für Flugwissenschaften und Weltraumforschung, Juni 1995, Vol. 19, Nr. 3, pp. 166–179, 1995.
Auweter-Kurtz, M., Feigl, M., Winter, M.: Diagnostic Tools for Plasma Wind Tunnels and Reentry Vehicles at the IRS, RTO-EN-8. AC/323(AVT)TP/23, RTO Educational Notes 8, April 2000; von Karman Institute for Fluid Dynamics, RTO AVT/VKI Special Course on Measurement Techniques for High Enthalpy Plasma Flows, Belgium, 1999.
Auweter-Kurtz, M., Fertig, M., Herdrich, G., Laux, T., Schöttle, U., Wegmann, Th., Winter, M.: Entry Experiments at IRS – In-flight Measurement during Atmospheric Entries. 53rd Int. Astronautical Congress, Houston, TX, USA, Oct. 2002, accepted for the Space Technology Journal (ST), Vol. 23, Issue 4, pp. 217–234, August 2003.
Auweter-Kurtz, M., Habiger, H., Kurtz, H., Laure, S., Röck, W, Tubanos, N.: Die Plasmawindkanäle PWK-IRS – Werkzeuge zur Untersuchung von Hitzeschutzmaterialien für Rückkehrfahrzeuge. Zeitschrift für Flugwissenschaften und Weltraumforschung, Springer Verlag, Februar 1993, Vol. 17, No. 1, pp. 1–15, 1993.
Auweter-Kurtz, M., Hald, H., Koppenwallner, G., Speckmann, H.-D.: German Experiments Developed for Reentry Missions. Acta Astronautica, Vol. 38, No. 1, pp. 47–61, 1996.
Auweter-Kurtz, M., Hilfer, G., Habiger, H., Yamawaki, K., Yoshinaka, T., Speckmann, H.-D.: Investigation of Oxidation Protected C/C Heat Shield Material in Different Plasma Wind Tunnels. Acta Astronautica, Vol. 45, No. 2, pp. 93–108, 1999.
Auweter-Kurtz, M., Laure, S.: Plasma Generators for Re-Entry Simulation. Journal of Propulsion and Power, Vol. 12, No. 6, pp. 1053–1061, November–Dezember, 1996.
Auweter-Kurtz, M., Wegmann, T.: Overview of IRS Plasma Torch Facilities, RTO-EN-8. AC/323(AVT)TP/23, RTO Educational Notes 8, April 2000; von Karman Institute for Fluid Dynamics, RTO AVT/VKI Special Course on Measurement Techniques for High Enthalpy Plasma Flows, Belgium, 1999.
Auweter-Kurtz, M.: Lichtbogenantriebe für Weltraumaufgaben. B.G. Teubner Verlag, Stuttgart, 1992.
Auweter-Kurtz, M.: Plasma Source Development for the Qualification of Heat Shield Materials of Atmospheric Entry Vehicles at IRS. Angenommen zur Veröffentlichung in der Zeitschrift Vacuum 2001, 2001.
Auweter-Kurtz, M.: Plasma Thruster Development Program at the IRS. Acta Astronautica, Vol. 32, No. 5, pp. 337–391, 1994.
Awakowicz, P., Behringer, K.: Population Densities and Rate Coefficients for Electron Impact Excitation in Singly Ionized Oxygen. Plasma Phys. Contr. Fusion 37, 551, 1995.
Backstein, S., Staufenbiel, R.: Some experimental results on spiral vortex breakdown. ZFW, Vol. 19, Nr. 6, 1995.
Ballmann, J., Sanaknaki, H.: A multi-dimensional time-marching method of characteristics for viscous and heat-conducting flows. Archive of Applied Mechanics 70 (2000), Springer Verlag 2000, pp. 65–80.
Bareis, B., Braig, W., Rahn, M., Schöttle, U.M.: Performance optimization of turboramjet engines propelling a two-stage space transportation system. ZFW 19, 236–245, 1995.
Bauer, A., Ludäscher, M., Rick, H., Sachs, G.: Simulation of Engine/Aircraft Dynamic Behaviour for Hypersonic Flight Vehicles. International Council of the Aeronautical Sciences, ICAS-94-6.7.2, Anaheim, 1994.
Bauer, A., Ludäscher, M., Rick, H.: Leistungsrechnung und Simulation des Betriebsverhaltens von Antriebssystemen für Hyperschallflugzeuge, DGLR-Jahrestagung, 93-03-079, Göttingen, 1993.
Beauvais R., Strube, G., Mayinger, F.: Hydrogen combustion – safety hazards inhydrogen systems. Proc. of the Int. Conf. "ATHENS", Athens, Greece, June 3–6, 1991. Eds.: D.A. Kouremenos et al. Athens: Greg. Foundas, pp. 407–417, 1991.
Beauvais R., Strube, G., Mayinger, F.: Turbulent flame acceleration – mechanisms and significance of safety considerations. Proc. of the 9th World Hydrogen Energy Conference, Paris, France, June 22–25, 1992, Eds.: T.N. Veziroglu et al. Paris, M.C.I. 1992, Vol. 2, pp. 1093–1102. Int. J. Hydrogen Energy, Vol. 19/8, pp. 701–708, 1994.

Behringer, K., Summers H.P., Denne B., Forrest, M., Stamp, M.: Spectroscopic Determination of Impurity Influx form Localized Surfaces. Plasma Phys. and Contr. Fusion 31, 2059, 1989.
Behringer, K.: Measurement of CH4/CD4 Fluxes and of Chemical Carbon Erosion from DH/CD Molecular Band Emission. J. Nucl. Mater. 176 & 177, 606, 1990.
Behringer, K.: Diagnostic and Modelling of ECRH Microwave Discharges. Review Article, Plasma Phys. And Contr. Fusion 33, 997, 1991.
Beisel, T., Gabriel, E., Resch, M.: An Extension to MPI for Distributed Computing on MPPs. In: Marian Bubak, Jack Dongarra, Jerzy Wasniewski (Eds), 'Recent Advances in Parallel Virtual Machine and Message Passing Interface', Lecture Notes in Computer Science, pp. 75–83, Springer, 1997.
Berkmann, P., Pesch, H.J.: Abort landing of an airplane under different windshear conditions: An optimal control problem with a third-order state constraint and a multifarious switching structure. J. of Optim. Theory and Appl. (1995).
Bestek, H., Eissler, W.: Direct numerical simulation of transition in Mach 4.8 boundary layers at flight conditions. In: Rodi, W., Bergeles, G. (eds.): Engineering Turbulence Modelling and Experiments 3, 611–620, Elsevier Sci. B.V., 1996.
Beylich, A.E.: Performance of a Hypersonic Twin Nozzle System. AIAA Journal, Vol. 34, No. 5, pp. 953–960, 1996.
Beylich, A.E., Zeutzius, M.: Performance and control of asymmetric nozzle jets. Proceedings of the Asian-Pacific Conference on Aerospace Technology and Science, International Academic Publishers, Beijing 100 010, China, pp. 917–922, 1994.
Bill, J., Aldinger, F.: Progress in Material Synthesis. Z. Metallkd. 87, pp. 827, 1996.
Bill, J., Heimann, D.: Polymer-Derived Ceramic Coatings on C/C-SiC Composites. J. Europ. Ceram. Soc. 16, p. 1115, 1996.
Bissinger, N.C., Koschel, W.W., Sacher, P.W., Walther, R.: SCRAM-JET Investigation within the German Hypersonics Technology Program. Scramjet Propulsion, edited by Curran, E.T. and Murthy, S.N.B. published by AIAA, Progress in Astronautics and Aeronautics, Reston, Virginia 2001, Vol. 189, pp. 119–158.
Bittner, L.: Dynamic and multistage versions of matrix models. GAMM-Tagung 1997, Bremen, 6.–9. April 1998.
Bittner, L.: Existence of optimal control for multistage optimal control problems. Workshop des Sonderforschungsbereichs 255 "Optimalsteuerungsprobleme von Hyperschall-Flugsystemen", Greifswald, 23.–24. Okt. 1997, Technische Universität München, 1998.
Bittner, L.: Konvexifizierung von Optimalsteuerproblemen. Seminar des Sonderforschungsbereichs 255, München, 12. Dez. 1996, Technische Universität München, Tagungsband, pp. 95–108, 1997.
Bittner, L.: On the Convexification of Optimal Control Problems for Flight Dynamics. ISNM 124, Birkhäuser Verlag, Basel, 1998, pp. 3–14.
Bleilebens, M., Kumar, S., Olivier, H.: Entwicklung der heißen Modelltechnik zur Untersuchung hypersonischer Strömungen. DGLR Jahrestagung Berlin, 1999, Band III, pp. 1381–1392.
Bleilebens, M., Olivier, H.: Experimental Investigations on Transition of Hypersonic Ramp Flows. Third European Symposium on Aerothermodynamics for Space Vehicles, Nov. 24–26, 1998, ESTEC, Nordwijk, The Netherlands, ESA-426, pp. 205–211.
Bleilebens, M., Olivier, H.: Surface Temperature Effects on Shock-Wave Boundary-Layer Interaction on Ramp Flows. Proceedings of 12$^{th}$ DGLR-Fachsymposium der AG-STAB, Univ. Stuttgart, Nov. 2000, to be published in Notes on Numerical Fluid Mechanics, Vieweg Verlag.
Bleilebens, M., Olivier, H.: Surface Temperature Effects on Shock-Wave Boundary-Layer Interaction of Ramp Flows. In: Wagner, S., Rist, U., Heinemann, J., Hilbig, R. (Eds.): New Results in Numerical and Experimental Fluid Mechanics III (Notes on Numerical Fluid Mechanics, Vol. 77), Springer Berlin Heidelberg 2002, pp. 161–168.
Boeing, H., Jeltsch, R.: On the numerical entropy and explicit difference schemes for conservation laws. Institut für Geometrie und Praktische Mathematik, RWTH Aachen, Bericht Nr. 48, 1987.

Boie, C., Auweter-Kurtz, M., Kaeppeler, H.J., Sleziona, P.C.: Numerical Simulation of MPD Thrusters on Adaptive Unstructured Mesh. Editors: S. Wagner et. al.: Computational Fluid Dynamics 1994. John Wiley & Sons Ltd. 1994, pp. 296–300, 1994.

Brandstetter, A., Rocci Denis, S., Kau, H.-P., Rist, D.: Experimental Investigation of Supersonic Combustion with Strut Injection. AIAA Paper 2002-5242, 2002.

Brandstetter, A., Rocci Denis, S., Kau, H.-P., Rist, D.: Flame Stabilization in supersonic Combustion. AIAA Paper 2002-5224, 2002.

Brandstetter, A., Rocci Denis, S., Kau, H.-P.: Stability of a Supersonic Model Combustion Chamber. Seminar des Sonderforschungsbereichs 255, 12. Dezember 2002, Tagungsband, ISBN-3-89791-334-8, S. 75–87, 2004.

Brandstetter, A.: Auslegung und Konstruktion einer Modellbrennkammer zur Untersuchung der Wasserstoffverbrennung in Überschallverbrennungen. DGLR Jahrestagung 1999, Berlin.

Brandt, R., Neuer, G.: Determination of thermal conductivity of carbon fibres perpendicular to their axis. High Temp. – High Press., 27/28, 267–272, 1995/96.

Breitner, M.H., Koslik, B., von Stryk, O., Pesch, H.J.: Optimal Control of Investment, Level of Employment and Stockkeeping. In: Bachem, A., Derigs, U., Jünger, M., Schrader, R. (Eds.), Operations Research 1993, Physica-Verlag Heidelberg, pp. 60–63, 1994.

Breitner, M.H.: Construction of the optimal feedback controller for constrained optimal control problems with unknown disturbances. In: Computational optimal control. Eds.: R. Bulirsch et al. Basel: Birkhäuser, 1994, pp. 147–162. (ISNM 115)

Breitner, M.H.: Real-time capable approximation of optimal strategies in complex differential games. Proc. of the sixth inter. sympos. on dynamic games and applications, St-Jovite, 1994. Eds.: M. Breton et. al., Montreal: GERAD, École des Hautes Études Commerciales, pp. 370–374, 1994.

Breitner, M.H., Koslik, B., von Stryk, O., Pesch, H.J.: Iterative design of economic models via simulation, optimization and modeling. Proc. of the 1st Mathmod Vienna (IMACS Sympos. on Mathematical Modelling 1994), Wien, 1994. Eds.: I. Troch et al. Wien: Technische Universität, Vol. 5, pp. 816–819, 1994.

Breitner, M.H., Pesch, H.J.: Construction of the Optimal Feedback Controller for Constrained Optimal Control Problems with Unknown Disturbances. Proc. of the Fifth Inter. Symp. on Dynamic Games and Applications, Grimentz, July 15–17, 1992. Ed.: The International Society of Dynamic Games, Genf: Université de Genève, Département d'économie commerciale et industrielle, pp. 77–84, 1992.

Breitner, M.H., Pesch, H.J.: Reentry trajectory optimization under atmospheric uncertainty as a differential game. In: Advances in dynamic games and applications. Eds.: T. Başar et al. Boston: Birkhäuser, pp. 70–88, 1994. (Annals of the Inter. Society of Dynamic Games; 1)

Breitner, M., Grimm, W., Pesch, H.J.: Barrier Trajectories of a Realistic Missile/Target Pursuit-Evasion Game. – In: Differential Games – Developments in Modelling and Computation, Proc. of the 4th Inter. Symp. on Differential Games and Applications, Helsinki, Aug. 9–10, 1990. Eds.: R. P. Hämäläinen et al. Berlin: Springer, 1991, pp. 48–57. (Lect. Notes in Contr. and Inf. Sci., 156), 1991.

Breitner, M., Pesch, H.J., Grimm, W.: Complex Differential Games of Pursuit-Evasion Type with State Constraints, Part 1: Necessary Conditions for Optimal Open-loop Strategies. J. of Optimization Theory and Applications 78, No. 3, pp. 419–441, 1993.

Breitner, M., Pesch, H.J., Grimm, W.: Complex Differential Games of Pursuit-Evasion Type with State Constraints, Part 2: Numerical Computation of Optimal Open-loop Strategies. J. of Optimization Theory and Applications 78, No. 3, pp. 443–463, 1993.

Breitner, M.H., Heim, A.: Robust Optimal Control of a Reentering Space Shuttle. DGLR-Jahrestagung 1995, Bonn, 26.– 29. Sept. 1995, DGLR-Jahrbuch II 1995, pp. 583–592.

Breitner, M.H., Rettig, U., Stryk, O. von: Robust optimal control with large neural networks emulated on the neuro-computer board SYNAPSE-PC. Proceedings of the 15th IMACS World Congress 1997 on Scientific Computation, Modelling and Applied Mathematics, Berlin, 24.–29. Aug. 1997.

Breitsamter, C., Cvrlje, T., Heller, M., Sachs, G.: Instationäre Aerodynamik und dynamische Stabilität von Hyperschall-Fluggeräten. DGLR-Jahrbuch 1998, pp. 89–102.
Breitsamter, C., Cvrlje, T., Laschka, B., Heller, M., Sachs, G.: Lateral-Directional Coupling and Unsteady Aerodynamic Effects of Hypersonic Vehicles. AIAA Journal of Spacecraft and Rockets, Vol. 38, No. 2, März–April 2001, pp. 159–167.
Breitsamter, C., Laschka, B., Zähringer, C., Sachs, G.: Wind Tunnel Tests for Separation Dynamics Modeling of a Two-Stage Hypersonic Vehicle. AIAA 10th International Space Planes and Hypersonic Systems and Technologies Conference Proceedings, Kyoto, Japan, 24.–27. April 2001, AIAA Paper 2001-1811, 2001.
Brun, Chr., Friedrich, R., da Silva, C.B., Métais, O.: A new mixed model based on the velocity structure function. Proceedings of the EUROMECH Colloquium 412, R. Friedrich, W. Rodi (Eds.), Kluwer Academic, 2002.
Brun, Chr., Friedrich, R.: Modeling the test sgs tensor Tij: An issue in the dynamic approach. Physics of Fluids, Vol.13, pp. 2373–2385, 2001.
Brüning, G., Hafer, X., Sachs, G.: Flugleistungen. Springer Verlag, Berlin, 3. Auflage, 1993.
Buchauer, O., Hiltmann, P., Kiehl, M.: Sensitivity Analysis of Initial Value Problems with Applications to Shooting Techniques. Numer. Math. 67, pp. 157–159, 1994.
Bulirsch, R., Bank, R., Merten, K.: Mathematical Modelling and Simulation of Electrical Circuits and Semiconducter-Devices. Proc. of a Conf. in Oberwolfach. Birkhäuser, Basel, (ISNM 93), 1990.
Bulirsch, R., Breitner, M.: Wilhelm Martin Kutta. In: 125 Jahre Technische Universität München. München: Tech. Univ. München, 1993, pp. 98–99 und in: DMV-Mitteilungen 2, pp. 7–8, 1994.
Bulirsch, R., Callies, R.: Mathematik und Hochtechnologie. Naturw. Rdsch., Vol. 49, Nr. 4, pp. 127–135, 1996.
Bulirsch, R., Callies, R.: Optimal Trajectories for a Multiple Rendezvouz Mission to Asteroids. Acta Astronautica 26, pp. 587–597, 1992.
Bulirsch, R., Callies, R.: Optimal Trajectories for a Multiple Rendezvous Mission to Asteroids. Proc. of the 42nd IAF-Congress, Montreal, Oct. 5–11, 1991. IAF Paper No. 91-342, 1991.
Bulirsch, R., Callies, R.: Optimal Trajectories for an Ion Driven Spacecraft from Earth to the Planetoid Vesta. Proc. of the AIAA Guidance, Navigation and Control Conference, New Orleans, Aug. 12–14, 1991. AIAA Paper No. 91-2683, 1991.
Bulirsch, R., Callies, R.: Optimale Flugbahn einer Raumsonde zum Planetoiden Vesta. DGLR-Jahrestagung, Friedrichshafen; DGLR-90-147, 1.–4. Okt. 1990, DGLR-Jb. II, 1990, pp. 895–904.
Bulirsch, R., Chudej, K., Reinsch, K.-D.: Optimal Ascent and Staging of a Two-Stage Space Vehicle System. DGLR-Jahrestagung, Friedrichshafen; DGLR-90-137, 1.–4. Okt. 1990, DGLR-Jb. I, 1990, pp. 243–249.
Bulirsch, R., Chudej, K.: Ascent Optimization of an Airbreathing Space Vehicle. Proc. of the AIAA Guidance, Navigation and Control Conference, New Orleans, Aug. 12–14, 1991. Washington, pp. 520–528, 1991.
Bulirsch, R., Chudej, K.: Combined optimization of trajectory and stage separation of a hypersonic two-stage space vehicle. Z. f. Flugwiss. Weltraumforsch. Vol. 19, Nr. 1, 1995, pp. 55–60.
Bulirsch, R., Chudej, K.: Effiziente und stabile Lösung eines reichweitenmaximalen Fluges mit Staudruckbeschränkung für ein Überschallflugzeug. DGLR-Jahrestagung 1997, München, 14.–17. Okt. 1997, DGLR-Jahrbuch II 1997, pp. 1141–1147.
Bulirsch, R., Chudej, K.: Guidance and trajectory optimization under state constraints – applied to a Sänger-type vehicle. In: D.B. DeBra, E. Gottzein (Eds.): Automatic Control in Aerospace 1992. Selected Papers from the 12th IFAC-Symposium, Ottobrunn, Germany, 7–11 September 1992. IFAC Symposia Series 1993, No. 12. Pergamon, Oxford, UK, pp. 483–488, 1993.
Bulirsch, R., Chudej, K.: Staging and ascent optimization of a dual-stage space transporter. Z. Flugwiss. Weltraumforsch. 16, pp. 143–151, 1992.

Bulirsch, R., Hardt, M: Virtual Reality – Symbiosis of Science and Art. In: Walter Hruby (ed.): Digital (R)Evolution in Radiology, Springer Verlag, Wien New York, 2001.

Bulirsch, R., Montrone, F., Pesch H. J.: Optimal Control in Abort Landing of a Passenger Aircraft. In: P. Bernhard und H. Bourdache-Siguerdidjane (Hrsg.), Proc. of the 8th IFAC Workshop on Control Applications of Nonlinear Programming and Optimization, Paris, 7.–9. 6. 1989. – Oxford: IFAC Publications, 1992.

Bulirsch, R., Montrone, F., Pesch, H. J.: Abort Landing in the Presence of a Windshear as a Minimax Optimal Control Problem, Part 1: Necessary Conditions. J. of Optim. Theory and Appl. 70 Nr. 1, pp. 1–23, 1991.

Bulirsch, R., Montrone, F., Pesch, H. J.: Abort Landing in the Presence of a Windshear as a Minimax Optimal Control Problem, Part 2: Multiple Shooting and Homotopy. J. of Optim. Theory and Appl. 70 Nr. 2, pp. 221–252, 1991.

Bulirsch, R., Nerz, E., Pesch, H. J., von Stryk, O.: Combining Direct and Indirect Methods in Nonlinear Optimal Control: Range Maximization of a Hang Glider. Published in: R. Bulirsch, A. Miele, J. Stoer, K. H. Well (Eds.): Optimal Control, Proc. of the Conf. in Optimal Control and Variational Calculus, Oberwol., 1991.

Bulirsch, R., von Stryk, O.: Direct and Indirect Methods for Trajectory Optimization. Published in: Annals of Operations Research.

Bulirsch, R.: Alfred Pringsheim der Mathematiker. In: Alfred Pringsheim, Hans Thoma, Thomas Mann. Eine Münchner Konstellation. H.-W. Kruft. München: Verlag der Bayerischen Akademie der Wissenschaften, pp. 25–34, 1993.

Bulirsch, R.: Mathematik und Informatik – Vom Nutzen der Formeln. – In: Informatik und Mathematik. Ed.: M. Broy. Berlin: Springer Verlag, pp. 3–27, 1991.

Bulirsch, R.: Mathematik und Raumfahrt – In: Mitteilungen d. Alexander-von-Humboldt-Stiftung 53, pp. 11–26, 1989.

Bulirsch, R., Nerz, E., Pesch, H. J., von Stryk, O.: Combining Direct and Indirect Methods in Nonlinear Optimal Control: Range Maximization of a Hang Glider. In: Optimal Control, Calculus of Variations, Optimal Control Theory and Numerical Methods. Eds.: R. Bulirsch et al. Basel: Birkhäuser, 1993, pp. 273–288 (ISNM 111)

Buss, M., von Stryk, O., Bulirsch, R., Schmidt, G.: Towards hybrid optimal control. Automatisierungstechnik, Vol. 48, Issue 09, pp. 448–459, 2000.

Butz, T., von Stryk, O.: Parallel Parameter Estimation for Vehicle Dynamics Simulations. Proceedings of the 14th Supercomputer Conference, Mannheim, 10.–12. Juni 1999.

Butzek, S. Schmidt, W. H.: Relaxation gaps in optimal control processes with state constraints. ISNM 124, Birkhäuser Verlag, Basel, pp. 21–29, 1998.

Butzek, S., Schmidt, W. H.: Konstruktion von Näherungslösungen für Steuerprobleme mit Hilfe von Lösungen relaxierter Probleme. SFB-Report, Nr. 26, Greifswald und München, 1996.

Callies, M., Callies, R.: Hochgenaue Konturliniengenerierung GAMM-Jahrestagung 2000, Göttingen, 2.–7. April 2000.

Callies, R., Bulirsch, R.: 3-D trajectory optimization of a single-stage VTVL system. Proceedings of the AIAA Guidance, Navigation and Control Conference, San Diego, USA, 29.– 31. Juli 1996, AIAA-96-3903, 1996.

Callies, R., Bulirsch, R.: Wissenschaftliches Rechnen, Mathematik und optimale Steuerungen. Hrsg.: Bungartz, H.-J., Zenger, Ch. NNFM Notes on Numerical Fluid Mechanics, Friedr. Vieweg & Sohn Verlagsgesellschaft, Braunschweig 2002.

Callies, R., Wimmer, G.: Effiziente Behandlung von Design-Kennlinien in Optimalsteuerungsproblemen. Seminar des Sonderforschungsbereichs 255, München, 3. Februar 2000, Tagungsband, ISBN 3-89791-175-2, Hieronymus Verlag, München, pp. 117–128, 2001.

Callies, R., Wimmer, G.: Optimal Hypersonic Flight Trajectories with Full Safety in case of Mission Abort. Proc. of the AIAA Atmospheric Flight Mechanics Conference 2000, AIAA-Paper 2000-3996, Denver, 14.–17. August 2000.

Callies, R., Wimmer, G.: Optimierung von Testflugbahnen. ZAMM 79, S3, pp. 929–931, 1999.

Callies, R., Wimmer, G.: Stabilisierte Hyperschallflugbahnen. ZAMM 81, S 3, pp. 769–770, 2001.
Callies, R.: A robotic satellite with simplified design. In: Computational optimal control. Eds.: R. Bulirsch et al. Basel: Birkhäuser, 1994. pp. 281–290. (ISNM 115)
Callies, R.: Advanced Integration Techniques for the Numerical Solution of Optimal Control Problems. In: Optimalsteuerungsprobleme in der Luft- und Raumfahrt. Hrsg.: G. Sachs, SFB 255. ISBN 3-89791-147-7, Hieronymus Verlag, München, pp. 13–26, 2000.
Callies, R.: Antriebsoptimierung bei Raumsonden. Z. angew. Math. Mech. 74 (1994) 6, pp. T589–591.
Callies, R.: Design optimization of high performance satellites. In: Numerical simulation in science and engineering. Eds.: M. Griebel et al. Braunschweig: Vieweg, 1994, pp. 19–30. (NNFM 18)
Callies, R.: Efficient Treatment of Experimental Data in Integrated Design Optimization. In: Advances in Computational Engineering & Sciences, Hrsg.: S. N. Atluri, F. W. Brust, Tech Science Press, Palmdale/CA, pp. 1188–1193, 2000.
Callies, R.: Multidimensional Stepsize Control. ZAMM 81, S3, pp. 743–744, 2001.
Callies, R.: Multi-Model Approach in Trajectory Optimization. Workshop Optimal Control, Greifswald 2002, Tagungsband ISBN 3-89791-316-X, Hieronymus Verlag, München, p. 27–38, 2003.
Callies, R.: Optimal Design of a Mission to Neptune. In: Optimal Control, Calculus of Variations, Optimal Control Theory and Numerical Methods. Eds.: R. Bulirsch et al. Basel: Birkhäuser, 1993, pp. 341–349. (ISNM 111)
Callies, R.: Optimal design of a small venus mission. Proc. of the 44th Congress of the International Astronautical Federation, Graz, Österreich, Oct 16–22, 1993. Paper No. IAF-93-A.6.53.
Callies, R.: Optimal Design of Multi-Stage Satellites subject to Life Analysis. Proc. of the AIAA Guidance, Navigation and Control Conference 2000, AIAA-Paper 2000-4561, Denver, 14.–17. August 2000.
Callies, R.: Optimaler Satellitenentwurf unter Berücksichtigung von Lebensdaueraspekten. Proc. des Workshops Optimalsteuerungsprobleme von Hyperschall-Flugsystemen, Greifswald, 14.–15. Nov. 1994, pp. 63–72.
Callies, R.: Small satellites for deep space operation – a challenge to optimal control. – In: Neunzert, H. (Ed): Progress in Industrial Mathematics at ECMI 94, John Wiley, Chichester, pp. 13–24, 1996.
Chucholowski, C., Vögel, M., von Stryk, O., Wolter, T.-M.: Real time simulation and online control for virtual test drives of cars. In: H.-J. Bungartz, F. Durst, Chr. Zenger (Hrsg.): High Performance Scientific and Engineering Computing. Lecture Notes. In: Computational Science and Engineering 8, Springer Verlag, pp. 157–166, 1999.
Chudej, K., Günther, M.: A global state space approach for the efficient numerical solution of state-constrained trajectory optimization problems. Journal of Optimization Theory and Applications, 103, 1, pp. 75–93, 1999.
Chudej, K.: Accelerating Multiple Shooting for State-Constrained Trajectory Optimization Problems. – In: Schmidt, W. H., Heier, K., Bittner, L., Bulirsch, R. (Eds.): Variational Calculus, Optimal Control and Applications. ISNM 124, Basel: Birkhäuser, 1998, pp. 197–206.
Chudej, K.: Effiziente indirekte Lösung zustandsbeschränkter Flugbahnoptimierungsaufgaben mit Minimalkoordinaten. Proceedings des Workshops des Sonderforschungsbereichs 255 "Optimalsteuerungsprobleme von Hyperschall-Flugsystemen", Greifswald, 23.–24. Okt. 1997, Technische Universität München, 1998, pp. 115–123.
Chudej, K.: Extended necessary conditions for challenging state-constrained optimal control problems in aerospace engineering. Z. ang. Math. Mech., Vol. 76, Nr. 3, pp. 1–4, 1996.
Chudej, K.: Konvergenzbeschleunigung der Mehrzielmethode für Flugbahnoptimierungsaufgaben. Zeitschrift für Angewandte Mathematik und Mechanik, 78, Suppl. 3, pp. 879–880, 1998.
Chudej, K.: Numerical computation of optimal ascent trajectories with a dynamic pressure limit. – In: Neunzert, H. (Ed): Progress in Industrial Mathematics at ECMI 94, John Wiley, Chichester, pp. 25–31, 1996.

Chudej, K.: Numerische Berechnung staudruckbeschränkter optimaler Flugbahnen mit der Mehrzielmethode. Proc. des Workshops Optimalsteuerungsprobleme von Hyperschall-Flugsystemen, Greifswald, 14.–15. Nov. 1994, pp. 73–79.
Chudej, K.: Optimal ascent of a hypersonic space vehicle. – In: Optimal Control, Calculus of Variations, Optimal Control Theory and Numerical Methods. Eds.: R. Bulirsch et.al. Basel: Birkhäuser, 1993, pp. 317–326. (ISNM 111)
Chudej, K.: Optimale Steuerung und Stufung eines zweistufigen Raumtransporters. – In: Z. angew. Math. Mech. 71 6, pp. T700–T703, 1991.
Chudej, K.: Optimization of the stage separation and the flight path of a future launch vehicle. In: J. Henry, J.-P. Yvon (Eds.): System Modelling and Optimization. Springer, London (1994) 491–500. (LNCIS 197)
Chudej, K.: Realistic Modelled Optimal Control Problems in Aerospace Engineering – A Challenge to the Necessary Optimality Conditions. Mathematical Modelling of Systems, Vol. 2, Nr. 4, pp. 252–261, 1996.
Chudej, K.: Verallgemeinerte notwendige Bedingungen für zustandsbeschränkte Optimalsteueraufgaben mit stückweise definierten Modellfunktionen. Z. Angew. Math. Mech. 75, SII, pp. 587–588, 1995.
Chudej, K.: Zulässige Schaltstrukturen für Flugbahnoptimierungsprobleme mit Ungleichungsbeschränkungen. Seminar des Sonderforschungsbereichs 255, München, 12. Dez. 1996, Technische Universität München, Tagungsband, pp. 119–129, 1997.
Chudej, K., Bulirsch, R.: Numerical solution of a simultaneous staging and trajectory optimization problem of a hypersonic space vehicle. AIAA/DGLR Fifth International Aerospace Planes and Hypersonics Technologies Conference, Munich, Nov. 30–Dec. 3, 1993, Paper No. AIAA-93-5130, 1993.
Coratekin, T., van Keuk, J., Ballmann, J.: On the performance of upwind schemes and turbulence models in hypersonic flows. AIAA Paper 01-1752AIAA, 10$^{th}$ International Space Planes and Hypersonic Systems and Technologies Conference, Kyoto, Japan, April 2001.
Coratekin, T., van Keuk, J., Ballmann, J.: On the Performance of Upwind Schemes and Turbulence Models in Hypersonic Flows. AIAA Journal, reviewed and accepted, 2003.
Cox, T.H., Sachs, G., Knoll, A., Stich, R.: A Flying Qualities Study of Longitudinal Long-Term Dynamics of Hypersonic Planes. Proceedings of the AIAA 6th International Aerospace Planes and Hypersonics Technologies Conference, Chattanooga, USA, 3.–7. April 1995, AIAA Paper 95-6150, 1995.
Cox, T.H., Sachs, G., Knoll, A., Stich, R.: A Flying Qualities Study of Longitudinal Long-Term Dynamics of Hypersonic Planes. NASA Technical Memorandum 104308, 1995.
Cvrlje T., Decker, K.: Unsteady Aerodynamics of Hypersonic Vehicles. Annual GAMM Meeting, Göttingen, 2.–7. April, 2000.
Cvrlje, T., Breitsamter, C. Heller, M., Sachs, G.: Instationäre Aerodynamik und Dynamische Stabilität von Hyperschall-Fluggeräten. Eingereicht für den Deutschen Luft- und Raumfahrtkongreß/DGLR-Jahrestagung, Bremen, 5.–8. Okt. 1998.
Cvrlje, T., Breitsamter, C., Heller, M., Sachs, G.: Lateral Unsteady Aerodynamics and Dynamic Stability Effects in Hypersonic Flight. Proceedings of the International Conference on Methods of Aerophysical Research ICMAR '98 (Part I), 1998, pp. 44–54.
Cvrlje, T., Breitsamter, C., Laschka, B., Heller, M., Sachs, G.: Unsteady and Coupling Aerodynamic Effects on the Lateral Motion in Hypersonic Flight. 9th International Space Planes and Hypersonic Systems and Technologies Conference, Norfolk (VA), USA, 1.–5. Nov. 1999, AIAA Paper 99-4832, 1999.
Cvrlje, T., Breitsamter, C., Laschka, B.: Numerical simulation of lateral aerodynamics of an orbital stage at stage separation flow conditions. Journal of Aerospace Science and Technology, Vol. 4, No. 3, pp. 157–171, 2000.
Cvrlje, T., Breitsamter, C., Weishäupl, C., Laschka, B.: Euler and Navier-Stokes Simulations of Two-Stage Hypersonic Longitudinal Motions. AIAA Journal of Spacecraft and Rockets, Vol. 37, No. 2, März–April 2000, pp. 242–251, 2000.
Cvrlje, T., Laschka, B.: Simulation of an Orbiter Separation Maneuver at Hypersonic Speed. 10th International Space Planes and Hypersonic Systems and Technologies Conference, Kyoto, Japan, 24.–27. April 2001, AIAA Paper 2001-1850, 2001.

Cvrlje, T.: Separationsbewegung im Hyperschall: Gittergenerierung. Seminar des Sonderforschungsbereichs 255, München, 12. Dez. 1996, Technische Universität München, Tagungsband, pp. 1–8, 1997.

Cvrlje, T.: Simulation of an orbiter separation maneuver at hypersonic speed. Festschrift zur Emeritierung von Professor Dr.-Ing. Boris Laschka, Herbert Utz Verlag München, Sept. 2002, pp. III-2-1–III-2-11, 2002.

Cvrlje, T.: Unsteady Separation of a Two-Stage Hypersonic Vehicle. 30th Fluid Dynamics Conference & Exhibit, Norfolk (VA), USA, 28. Juni–1. Juli 1999, AIAA Paper 99-3412, 1999.

Cvrlje, T.: Unsteady Separation of an Idealized Two-Stage Hypersonic Vehicle. Vortrag auf der GAMM-Jahrestagung 1997 in Regensburg. – In: Krause, E. (Ed.): Gasdynamic Problems of Space Transportation Systems, Special Issue, Eigenverlag, Aerodynamisches Intitut, RWTH Aachen, 1997.

Cvrlje, T.: Unsteady Separation of an Idealized Two-Stage Hypersonic Vehicle. Vortrag auf der GAMM-Jahrestagung 1998 in Bremen. – In: Krause, E. (Ed.): Gasdynamic Problems of Space Transportation Systems, Special Issue, Eigenverlag, Aerodynamisches Institut, RWTH Aachen, 1998.

da Costa, O., Kriegel, M., Wagner, O., Sachs, G.: Missionsanalyse von Wiedereintritts-Fluggeräten mit degradierten Kontrollflächen. Seminar des Sonderforschungsbereichs 255, 12. Dezember 2002, TU München, ISBN 3-89791-334-8, S. 119–132, 2004.

Daiß, A., Frühauf, H.-H., Messerschmid, E.W.: Chemical Reactions and Thermal Nonequilibrium on Silica Surfaces. Proceedings of NATO Advanced Study Institute, Molecular Physics and Hypersonic Flows, Maratea, May 21–June 3, 1995, NATO ASI Series C: Mathematical and Physical Sciences. Vol. 482, Kluwer, 1996.

Daiß, A., Frühauf, H.-H., Messerschmid, E.W.: Modelling of Catalytic Reactions on Silica Surfaces with Consideration of Slip Effects. AIAA Journal of Thermophysics and Heat Transfer, Vol. 11, No. 3, July–September, 1997.

Decker, F., Neuwerth, G., Staufenbiel, R.: Low-speed aerodynamics of the hypersonic research configuration ELAC I. ZFW 17 (1993), pp. 99–107.

Decker, K., Cvrlje, T.: Unsteady Aerodynamics of a Hypersonic Vehicle During a Separation Phase. 4th Asian Computational Fluid Dynamics Conference, Mianyang, China, 18.–22. September 2000.

Decker, K., Laschka, B.: Unsteady Aerodynamics of a Hypersonic Vehicle During a Separation Phase. 10th International Space Planes and Hypersonic Systems and Technologies Conference, Kyoto, Japan, 24.–27. April 2001, AIAA Paper 2001-1852, 2001.

Denk, G.: A new numerical method for the integration of highly oscillatory second-order ordinary differential equations. Applied Numerical Mathematics 13, pp. 57–67, 1993.

Dewell, L., Speyer, J., Dinkelmann, M., Sachs, G.: Fuel-Optimal Periodic Solutions to Hypersonic Cruise with Active Engine Cooling. Proceedings of the AIAA Guidance, Navigation and Control Conference, New Orleans, USA, 11.–13. Aug. 1997, AIAA-97-3478, pp. 234–242, 1997.

Dieterle, L., Kompenhans, J., Peiter, U., Prengel, K.: Flow investigations on a large delta wing using LSI and PIV. 8$^{th}$ International Symposium on Flow Visualization, Sorrento (NA), Italy, Sept. 1–4, 1998, Conference Proceedings, pp. 204.1–204.11.

Dieterle, L., Peiter, U.: ELAC-1: Experimental investigation using vortex structures using PIV; High Reynolds number wind tunnel tests with an ELAC-Model in the DNW. GAMM Annual Meeting, Bremen, 1998, ZAMM Volume 79 (1999), Suppl. 3, pp. 931–932.

Dinkelmann, M., Sachs, G.: Optimale Bahnsteuerung zur Reduzierung der thermischen Belastung im Hyperschallflug. Zeitschrift für Angewandte Mathematik und Mechanik, Vol. 79, Supplement 3, pp. 933–934, 1999.

Dinkelmann, M., Sachs, G.: Trajectory Optimization for Reducing Unsteady Heat Input of Aerospace Craft. Proceedings of the AIAA Atmospheric Flight Mechanics Conference, Denver, USA, pp. 295–305, 2000.

Dinkelmann, M., Wächter, M., Sachs, G.: Flugbahn-Optimalsteuerung für ein Hyperschall-Flugsystem zur Verringerung des instationären Wärmetransfers, GAMM Jahrestagung, Göttingen, 2.–7. April 2000.

Dinkelmann, M., Wächter, M., Sachs, G.: Modelling and Simulation of Unsteady Heat Input for Trajectory Optimization of Aerospace Vehicles. Proceedings of the 3rd MATHMOD IMACS Symposium on Mathematical Modelling, TU Wien, 2.–4. Februar 2000, Band 1, pp. 81–85, 2000.
Dinkelmann, M., Wächter, M., Sachs, G.: Modelling and simulation of unsteady heat transfer for aerospacecraft trajectory optimization. IMACS Journal Mathematics and Computers in Simulation, Vol. 53, pp. 389–394, 2000.
Dinkelmann, M., Wächter, M., Sachs, G.: Modelling of Heat Transfer and Vehicle Dynamics for Thermal Load Flight Optimization. In: Mathematical and Computer Modelling of Dynamical Systems, Vol. 8, pp. 237–257, September 2002.
Dinkelmann, M., Wächter, M., Sachs, G.: Reducing Heat Loads with Optimal Flight Path Control for a Hypersonic Vehicle. Workshop des SFB 255, Greifswald, 7.–8. Oktober 1999, Tagungsband ISBN 3-89791-147-7, TU München, pp. 105–113, 2000.
Dinkelmann, M., Wächter, M., Sachs, G.: Unsteady Heat Input Modeling for Trajectory Optimization of Aerospace Vehicles. GAMM-Jahrestagung 2000, Göttingen, 2.–7. April 2000.
Dinkelmann, M., Wächter, M., Sachs, G.: Unsteady Heat Input Modelling for Trajectory Optimization of Aerospace Vehicles. IAF-00-A.4.09, 51. International Astronautical Congress, Rio de Janeiro, 2.–6. Oktober 2000.
Domröse, U., Krause, E., Meinke, M.: Numerical simulation of laminar hypersonic shock-boundary layer interaction. ZFW, Vol. 20, Nr. 2, 1996.
Dürr, J., Lamparter, H.-P., Bill, J., Steeb, S., Aldinger, F.: An X-ray and neutron scattering investigation of precursor derived Si24C43N33 ceramics. Journal of Non-Crystalline Solids 234, p. 155, 1998.
Edwards, G., Levick, A., Neuer, G., Schreiber, E., u. a.: Laser absorption radiation thermometry and industrial temperature measurement. J. F. Dubbelmam, M. J. de Groot (Hrsg.) TEMPMECO 99, ISBN-90-9013359-3, pp. 613–618, 1999.
Eich, E., Mehlhorn, R., Sachs, G.: Stabilization of Numerical Solutions of Boundary Value Problems Exploiting Invariants. Sonderforschungsbereich 255 "Transatmosphärische Flugsysteme", Report Nr. 16, pp. 1–28, 1993.
Eich, E., Mehlhorn, R., Sachs, G.: Stabilization of Numerical Solutions of Boundary Value Problems Exploiting Invariants. AIAA Guidance, Navigation and Control Conference Proceedings, Scottsdale, AZ, Aug. 1–3, pp. 338–348, 1994.
Eickermann, Th., Henrichs, J., Resch, M., Stoy, R., Voelpel, R.: Metacomputing in gigabit environments: Networks, tools and applications. Parallel Computing 24, pp. 1847–1872, 1998.
Eissler, W., Bestek, H.: Direct numerical simulation of transition in Mach 4.8 boundary layers at flight conditions. In: Rodi, W., Bergeles, G. (eds): Engineeering Turbulence Modelling and Experiments 3, 611–620, Elsevier Science B.V. 1996.
Eissler, W., Bestek, H.: Numerical simulations of initial transition in Mach 4.8 boundary layers at wind tunnel and flight conditions. ZFW 19/3, 228–235, 1995.
Eissler, W., Bestek, H.: Spatial numerical simulations of linear and weakly nonlinear wave instabilities in supersonic boundary layers. Theoret. Comp. Fluid Dynamics 8, 219–235, 1996.
Eissler, W., Bestek, H.: Spatial Numerical Simulations of Nonlinear Transition Phenomena in Supersonic Boundary Layers. Kral, L.D., Zang, T.A. (eds.): Transitional and Turbulent Compressible Flows. ASME-FED-Vol. 151, 77–92, 1993.
Eissler, W., Bestek, H.: Wall-temperature effects on transition in supersonic boundary layers investigated by direct numerical simulations. In: Henkes, R.A.W.M., Van Ingen, J.L. (eds.): Transitional Boundary Layers in Aeronautics, 459–467, North-Holland, 1996.
El-Askary, W. A., Schröder, W., Meinke, M.: LES of Compressible Wall-Bounded Flows. $16^{th}$ AIAA Computational Fluid Dynamics Conference, AIAA 2003-3554, Orlando, USA, 2003.
Engler, V., Coors, D., Jacob, D.: Sensitivity based design optimization of a space transportation system. 3rd European Symposium on Aerothermodynamics for Space Vehicles, ESTEC, Noordwijk, The Netherlands, 24.–26. November, 1998, ESA SP 426, pp. 49–57.

Engler, V., Coors, D., Jacob, D.: Optimization of a space transportation system including design sensitivities. AIAA Paper 98-1553, 8$^{th}$ International Space Planes and Hypersonic Systems and Technologies Conference, Norfolk, Virginia, USA, April 27–30, 1998, Proceedings pp. 252–261.
Engler, V., Coors, D., Jacob, D.: Optimization of a space transportation system including design sensitivities. Journal of Spacecraft and Rockets, Vol. 35, No. 6, Nov.–Dec. 1998, pp. 785–791.
Engler, V.: Optimierung der Unterstufenauslegung eines zweistufigen, horizontal startenden Raumtransportsystems. Tagungsband II der DGLR-Jahrestagung 1994 in Erlangen.
Esch, Th., Bauer, A., Rick, H.: Simulation and Nozzle/Afterbody Integration of Hypersonic Propulsion Systems. Zeitschrift für Flugwissenschaften und Weltraumforschung, ZFW, Band 19, Nr. 1, 1995.
Esch, Th., Giehrl, M.: Numerical Analysis of Nozzle and Afterbody Flow of Hypersonic Transport Systems. ASME, 94-GT-391, Den Haag, 1994.
Esch, Th., Hollmeier, S., Rick, H.: Design and Off-Design Simulation of Highly Integrated Hypersonic Propulsion Systems. Proceedings of the AGARD/PEP 86th Symposium on Advanced Aero-Engine Concepts and Controls, Seattle, USA, Sept. 1995.
Fabig, J., Krenkel, W.: Principles and New Aspects in LSI-Processing, Advanced Structural Fiber Composites. Proceedings of Topical Symposium V – "Advanced Structural Fiber Composites" of the Forum on New Materials of the 9th CIMTEC-World Ceramics Congress and Forum on New Materials, Florence, Italy, June 14–19, 1998/ed. by P. Vincenzini, Faenza: Techna, 1999 XIV, pp. 141–148, 1999.
Fabre, D., Laurent Jacquin, L., Sesterhenn, J.: Linear interaction of a cylindrical entropy spot with a shock. Physics of Fluids, Vol. 13, No. 8, pp. 2403–2422, 2001.
Fares, E., Schröder, W.: A Differential Equation to Approximate the Wall Distance. Int. Journ. Numer. Meth. Fluids, Vol. 39, pp. 743–762, 2002.
Fares, E., Schröder, W.: Numerical Simulation of Wake Flows. ECCOMAS CFD 2001, Swansea, Sep. 2001.
Fasel, H., Thumm, A., Bestek, H.: Direct Numerical Simulation of Transition Phenomena in Supersonic Boundary Layers: Oblique Breakdown. In Kral, L.D., Zang, T.A. (eds.): Transitional and Turbulent Compressible Flows. ASME-FED-Vol.151, 77–92, 1993.
Fasoulas, S., Auweter-Kurtz, M., Habiger, H. A.: Experimental Investigation of a Nitrogen High-Enthalpy Flow. AIAA Journal of Thermophysics and Heat Transfer, Vol. 8, No. 1, pp. 48–58, January–March, 1994.
Fasoulas, S., Sleziona, P. C., Auweter-Kurtz, M., Habiger, H., Laure, S. H., Schönemann, A. T.: Characterization of a Nitrogen Flow Within a Plasma Wind Tunnel. AIAA Journal of Thermophysics and Heat Transfer, Juli–September 1995, Vol. 9, No. 3, pp. 422–431, 1995.
Fedorova, N. N., Gaisbauer, U., Kharlamova, Y. V., Knauss, H., Wagner, S.: Numerical and Experimental Investigations of Shock Wave/Turbulent Boundary Layer Interaction on a Double Ramp Configuration, Annual Scientific Conference. GAMM, Augsburg, 2002.
Ferrière, A., Rodriget, G. P., Mason, N. P., Nedele, M. R.: High temperature measurements of thermophysical properties of DLR C/C-SiC material in the CNRS solar furnace. High Temperatures – High Pressures, Vol. 27/28, p. 307–312, 1996.
Fertig, M., Dohr, A., Frühauf, H.-H.: Transport Coefficients for High Temperature Nonequilibrium Air Flows. AIAA Journal of Thermophysics and Heat Transfer, Vol. 15, No. 2, 2001.
Fertig, M., Frühauf, H.-H., Auweter-Kurtz, M.: Modelling of Reactive Processes at SiC-Surfaces in Rarefied Nonequilibrium Airflows. Journal of Thermophysics and Heat Transfer, accepted for publication 2003.
Fey, M, Jarausch, H., Jeltsch, R., Karmann, P.: On the Interaction of Euler and ODE Solver when Computing Reactive Gas Flow. Institut für Geometrie und Praktische Mathematik, RWTH Aachen, Bericht Nr. 55, 1989.
Fey, M., Jeltsch, R., Müller, S.: Stagnation point computations of nonequilibrium inviscid blunt body flow. Computers Fluids, Vol. 22, No. 4/5, pp. 501–515, 1993.

Fezer, A., Kloker, M.: DNS of point-source-induced transition in a hypersonic cone boundary layer. Part I: Linear and weakly nonlinear regime. Part II: Dynamical structures and breakdown. Submitted to J. Fluid Mech., 2003.
Fezer, A., Kloker, M.: Spatial direct numerical simulation of transition phenomena in supersonic boundary layers. In: Fasel, H., Saric, W. (eds.): Laminar-Turbulent Transition 1999, 415–420, Springer Verlag, 2000.
Fezer, A., Kloker, M.: Transition processes in Mach 6.8 boundary layers at varying temperature conditions investigated by spatial numerical simulation. In: Nitsche, W., Heinemann, H.-J., Hilbig, R. (eds.): New Results in Numerical and Experimental Fluid Dynamics II. Notes on Num. Fluid Mech. 72, 138–145, Vieweg-Verlag, 1999.
Fezer, A., Kloker, M.: DNS of transition mechanisms at Mach 6.8 – flat plate vs. sharp cone. In: Zeitoun, D.E., Periaux, J., Desideri, J.A., Marini, M. (eds.): West East High Speed Flow Fields 2002, 434–441, CIMNE, 2003.
Fink, A., Dinkler, D., Kröplin, B.: A Phenomenological Material Model for Fibre Reinforced Ceramics. ZFW 19, 197–204, 1995.
Fischer, J., Neuer, G., Schreiber, E., Thomas, R.: Specification of Radiation Thermometers and Metrological Characterisation of a New Transfer-Standard. In: Proceedings TEMPMEKO 2001, eds. B. Fellmuth, J. Seidel, G. Scholz, VDE Verlag GmbH, 801–806, 2002.
Fischer, M., Magens, E., Weisgerber, H., Winandy, A., Cordes, S.: CARS Temperature Measurement on an Air Breathing RAM jet Model. AIAA Paper 98-0961, $36^{th}$ Aerospace Sciences Meeting & Exhibit, January 12–15, Reno, NV, 1998.
Fischer, M., Magens, E., Weisgerber, H., Winandy, A., Cordes, S.: Coherent Anti-Stokes Raman Scattering Temperature Measurements on an Air-Breathing Ramjet Model. AIAA Journal, Vol. 37, No. 6, June 1999, pp. 744–750.
Fox, U., Rick, W., Koschel, W.: Computation of hypersonic high-temperature nozzle flow. ZFW, Vol. 17 (1993), pp. 139–148.
Foysi, H., Sarkar, S., Friedrich, R.: Compressibility effects and turbulence scalings in supersonic channel flow. Journal of Fluid Mechanics, Vol. 509, pp. 207–216, 2004.
Friedrich, R., Bertolotti, F.P.: Compressibility effects due to turbulent fluctuations. Applied Scientific Research, Vol. 57, pp. 165–194, 1997.
Friedrich, R., Hannappel, R., Hauser, Th.: Stoß-Wellen- und Stoß-Turbulenz-Wechselwirkung. – In: Turbulente Strömungen in Forschung und Praxis. A. Leder (Hrsg.), Verlag Shaker, Aachen, pp. 131–144, 1993.
Friedrich, R., Hannappel, R.: On the interaction of wave-like disturbances with shocks – two idealizations of the shock/turbulence interaction problem. – In: Acta Mechanica [Suppl.], pp. 69–77, 1994.
Friedrich, R., Hüttl, T.J., Manhart, M., Wagner, C.: Direct numerical simulation of incompressible turbulent flows. Computers & Fluids, Vol. 30, No. 5, pp. 555–580, 2001.
Friedrich, R., Rodi, W. (Eds.): Advances in LES of Complex Flows. Vol. 65 of Fluid Mechanics and its Applications, Kluwer Academic Publishers, 2002.
Friedrich, R.: Compressible turbulence. SPACE COURSE 1993, TU München, Oct. 11–22, pp. 15-1 to 15-45.
Friedrich, R.: Modelling of Turbulence in Compressible Flows. Transition, Turbulence and Combustion Modelling. Edited by Hanifi A., Alfredsson P.H., Johansson A.V. and Henningson D.S., Kluwer Academic Publishers (ERCOFTAC Series) Dordrecht Boston London, pp. 243–348, 1999.
Friedrich, R.: Turbulence modelling of compressible flows. ESA-Bericht. Projekt-Nr. 11018/94, WP 1300, pp. 1–49, 1995.
Frieß, M., Krenkel, W., Brandt, R., Neuer, G.: Influence of Process Parameters on the Thermophysical Properties of C/C-SiC. Proc 4th Int. Conf. on High Temperature Ceramic Matrix Composites (HT-CMC 4), München, Okt. 1–3, 2001, High Temperature Ceramic Matrix Composites, pp. 328–333, Wiley-VCH/Weinheim, 2001.
Frieß, M., Krenkel, W., Nestler, K., Marx, G.: CVD-Coating on Fabric Sheets in Combination with the LSI-Process. Proc 4th Int. Conf. on High Temperature Ceramic Matrix Composites (HT-CMC 4), München, Okt. 1–3, 2001, High Temperature Ceramic Matrix Composites, pp. 199–204, Wiley-VCH/Weinheim, 2001.

Frieß, M., Krenkel, W.: Flüssigsilicierung von graphitierten C/C-Vorkörpern. Tagungsband Verbundwerkstoffe und Werkstoffverbunde, Hamburg, 1999, Wiley-VCH, Editor: K. Schulte, K.U. Kainer, pp. 442–447, 1999.

Frieß, M., Krenkel, W.: Influence of the Graphitization Temperature on the Properties of C/C-SiC Manufactured via the LSI-Processing, Advanced structural fiber composites: proceedings of Topical Symposium V – "Advanced Structural Fiber Composites" of the Forum on New Materials of the 9th CIMTEC-World Ceramics Congress and Forum on New Materials, Florence, Italy, June 14–19, 1998/ed. by P. Vincenzini, Faenza: Techna, 1999 XIV, pp. 127–134, 1999.

Frieß, M., Renz, R., Krenkel, W.: Graded Ceramic Matrix Composites by LSI-Processing. In: Advanced Inorganic Structural Fiber Composites IV, (ED.: P. Vincenzini, C. Badini), 2003, pp. 141–148, 2003.

Frieß, M., Stantschev, G., Kochendörfer, R.: Langfaserverstärkte keramische Verbundwerkstoffe im LPI- und LSI-Verfahren. Tagungsband Verbundwerkstoffe und Werkstoffverbunde Chemnitz 2001, pp. 334–346, Wiley-VCH/Weinheim, 2001.

Fritz, J., Kröner, M., Sattelmayer, T.: Flashback in a Swirl Burner with Cylindrical Premixing Zone. The 46th ASME International Gas Turbine & Aeroengine Technical Congress, ASME Paper 2001GT-0054, New Orleans, Lousiana, USA, 2001.

Fritz, J., Kröner, M., Sattelmayer, T.: Simultaner Einsatz optischer Messmethoden zur Untersuchung der Flammenausbreitung in verdrallten Rohrströmungen. 8. Fachtagung "Lasermessmethoden in der Strömungsmesstechnik", Tagungsbericht GALA, Freising, 2000.

Fritz, J., Kröner, M., Sattelmayer, T.: Simultaner Einsatz optischer Messmethoden zur Untersuchung instationärer Drallflammen. Gaswärme International, Vol. 50, Nr. 9, pp. 411–415, 2001.

Frühauf, H.-H., Fertig, M., Kanne, S.: Validation of the Enhanced URANUS Nonequilibrium Navier-Stokes Code. AIAA Journal of Spacecraft and Rockets, Vol. 37, No. 2, March – April, 2000.

Frühauf, H.-H., Fertig, M., Olawsky, F., Bönisch, T.: Upwind Relaxation Algorithm for Reentry Nonequilibrium Flows. High Performance Computing in Science and Engineering 99, Springer, Berlin, pp. 365–378, 2000.

Frühauf, H.-H., Fertig, M., Olawsky, F., Infed, F., Bönisch, T.: Upwind Relaxation Algorithm for Reentry Nonequilibrium Flows. High Performance Computing in Science and Engineering 2000, Springer, Berlin, pp. 440–445, 2001.

Frühauf, H.-H., Knab, O., Daiss, A., Gerlinger, U.: The URANUS Code – An Advanced Tool for Reentry Nonequilibrium Flow Simulations. Zeitschrift für Flugwissenschaften und Weltraumforschung, Vol. 19, No. 3, Springer, June, 1995.

Frühauf, H.-H.: Computation of High Temperature Nonequilibrium Flows. Proceedings of NATO Advanced Study Institute, Molecular Physics and Hypersonic Flows, Maratea, May 21–June 3, 1995, NATO ASI Series C: Mathematical and Physical Sciences, Vol. 482, Kluwer, 1996.

Gabler W., Mayinger, F., Hannappel, R.: Experimental and Numerical Investigation of the flow field in a supersonic combustor with backward facing steps, Z. Flugwiss., pp. 29–33, 1995.

Gabler, W., Allgayer, U., Esch, Th., Haindl, H., Hantschk, C., Hollmeier, S., Hönig, R.: Systemuntersuchungen zu Scramjet-Antrieben, Arbeitsgruppe Grenzschicht/Verbrennung – SFB 255. DGLR-Jahrestagung 1995, Bonn, Sept. 1995, DGLR-JT95-121, 1995.

Gabler, W., Gleis, S., Hauser, T., Heindl, H., Hönig, R., Reindl, U., Reisinger, D., Friedrich, R., Kappler, G., Mayinger, F., Rick, H., Schmitt-Thomas, G., Vortmeyer, D.: Systemuntersuchungen zu SCRAMJET-Antrieben, eingereicht bei ZFW, 6/95.

Gabler, W., Haibel, M., Mayinger, F.: Dynamic Structures and Mixing Processes in Sub- and Supersonic Hydrogen Air Flames in Combustion Chambers with Cascades of Rearward Facing Steps, Proc. of the 19th ICAS Congress, Anaheim USA 1994.

Gabler, W., Haibel, M., Mayinger, F.: Mixing Process in Reacting and Non Reacting Supersonic Flows. 5th European Turbulence Conference in Siena July 5–8, 1994.

Gabler, W., Mayinger, F.: Non-intursive Laser-based Optical Measurement Techniques for Reacting and Non-reacting Suspersonic Flows. Proceedings of the 33rd Aircraft Symp., Hiroshima, Japan, 8.–11. Nov., 1995. Soc. of Aeronautics and Space Sciences (Ed.), pp. 369–374, 1995.

Gafert, J., Behringer, K., Coster, D., Dorn, C., Hirsch, K., Niethammer, M., Schumacher, U.: First experimental determination of ion flow Velocities and temperatures in the ASDEX Upgrade divertor with high resolution spectroscopy. Plasma phys. Control Fusion 39, 1981–1995, 1997.

Gaisbauer, U., Knauss, H., Wagner, S., Kharlamova, Y.V., Fedorova, N.N.: Shock/Turbulent Boundary Layer Interaction on a Double Ramp Configuration – Experiments and Computation. XI. International Conference on the Methods of Aerophysical Research ICMAR, Novosibirsk, Russland, 2002.

Gaisbauer, U., Knauss, H., Weiss, J., Wagner, S.: Measurement Techniques for the Detection of Low Disturbance and Transition Localization in a Short Duration Wind Tunnel. X. International Conference on the Methods of Aerophysical Research ICMAR, Novosibirsk, Russland, 2000.

Gaisbauer, U., Knauss, H., Weiss, J., Wagner, S.: The Meaning of Disturbance Fields in Transition Experiments and their Detection in The Test Section Flow of a Short Duration Wind Tunnel. Notes on Numerical Fluid Mechanics, Vieweg Series, 2001.

Garbey, M., Hess, M., Resch, M., Tromeur-Dervout, D.: Numerical Algorithms and software tools for efficient Meta-Computing. Parallel CFD 2000, International Parallel Computational Fluid Dynamics 2000 Conference at Trondheim/Norway, May 22–25, 2000.

Gerlinger, P., Aigner, M.: Assumed PDF Modeling with Detailed Chemistry. In: High Performance Computing in Science and Engineering 01, Krause, Jäger (eds.), pp. 318–327, Springer Verlag, 2001.

Gerlinger, P., Aigner, M.: Assumed PDF Modeling with Detailed Chemistry. In: High Performance Computing in Science and Engineering, Springer Verlag, 2002.

Gerlinger, P., Algermissen, J., Brüggemann, D.: Matrix Dissipation for Central Difference Schemes with Combustion. AIAA Journal, 33, pp. 1865–1870, 1995.

Gerlinger, P., Algermissen, J., Brüggemann, D.: Numerical Simulation of Mixing for Turbulent Slot Injection. AIAA Journal, 34, pp. 73–78, 1996.

Gerlinger, P., Brellochs, F., Algermissen, J., Brüggemann, D.: Mixing and Combustion Phenomena in Supersonic Flows Studied by CFD and Laserdiagnostics. ZFW 19, pp. 246–252, 1995.

Gerlinger, P., Brüggemann, D., Algermissen, J.: Numerical Simulation of Supersonic Mixing and Combustion. Twenty-Fifth Symposium (International) on Combustion/The Combustion Institute, pp. 21–27, 1994.

Gerlinger, P., Brüggemann, D.: An Implicit Multigrid Scheme for the Compressible Navier-Stokes Equations with Low-Reynolds-Number Turbulence Closure. Journal of Fluids Engineering, 120, pp. 257–262, 1998.

Gerlinger, P., Brüggemann, D.: Multigrid Convergence Acceleration for Turbulent Supersonic Flow. International Journal for Numerical Methods in Fluids, 24, pp. 1019–1035, 1997.

Gerlinger, P., Brüggemann, D.: Multigrid Convergence Acceleration for Non-Reactive and Reactive Flows. In: High Performance Computing in Science and Engineering '99, Krause, Jäger (Eds.), pp. 344–353, Springer Verlag, 1999.

Gerlinger, P., Brüggemann, D.: Numerical Investigation of Hydrogen Strut Injection into Supersonic Air Flows. Journal of Propulsion and Power, 16, pp. 22–28, 2000.

Gerlinger, P., Möbus, H., Brüggemann, D.: An Implicit Multigrid Method for Turbulent Combustion. Journal of Computational Physics, 167, pp. 247–276, 2001.

Gerlinger, P., Noll, B., Aigner, M.: Assumed PDF Modelling and PDF Structure Investigation Using Finite-Rate Chemistry. Angenommen zur Veröffentlichung in: Progress in Computational Fluid Dynamics, 2003.

Gerlinger, P., Stoll, P., Brüggemann, D.: An Implicit Multigrid Method for the Simulation of Chemically Reacting Flows. Journal of Computational Physics, 146, pp. 322–345, 1998.

Gerlinger, P., Stoll, P., Brüggemann, D.: Robust Implicit Multigrid Method for the Simulation of Turbulent Supersonic Mixing. AIAA Journal, 37, pp. 766–768, 1999.
Gerlinger, P.: Investigation of an Assumed PDF Approach for Finite-Rate Chemistry. Combustion Science and Technology, 175, pp. 841–872, 2003.
Gern, F., Kochendörfer, R., Neuer, G.: High Temperature Behaviour of the Spectral and Total Emmissivity of CMC-Materials. High Temperatures – High Pressures, 1995/1996, Vol. 27/28, p. 183–189, 1996.
Gern, F., Kochendörfer, R.: Liquid Silicon Infiltration: Description of Infiltration Dynamics and Silicon Carbide Formation. Composites: Part A, Applied Science and Manufacturing, Vol. 28/4, p. 355–364, 1997.
Gern, F., Kochendörfer, R.: Microstructural Approach to Infiltration Behaviour of Liquid Siliconized Carbon/Carbon. SAMPE Journal of Advanced Materials, Oktober 1996, Vol. 28/1, 1996.
Geurts, J.B., Friedrich, R., Métais, O. (Eds.): Direct and Large-Eddy Simulation IV. ERCOFTAC SERIES, Vol. 8, Kluwer Academic Publishers, Dordrecht, 2001.
Gleis, S., Rau, W., Vortmeyer, D.: Beruhigung von Verbrennungsschwingungen im Prozeßgaserhitzer einer Claus-Abgasreinigungsanlage – Eine Fallstudie. VDI-Berichte Nr. 922, pp. 337–347, 1991.
Gleis, S., Vortmeyer, D., Rau, W.: Experimental Investigations on the Transition from Stable to Unstable Combustion by means of Active Instability Control. AGARD CP Nr. 479, pp. 22-1–22-7, 1990.
Glocker, M., Vögel, M., von Stryk, O.: Trajectory optimization of a shuttle mounted robot. In: Proc. Workshop on Optimal Control in Hypersonic Flight, Greifswald, 7.–8. Oktober 1999 Tagungsband ISBN 3-89791-147-7, TU München, pp. 71–82, 2000.
Glößner, C., Olivier, H.: Aerothermodynamics of a winged reentry vehicle in hypersonic flow. 4th European Symposium on Aerothermodynamics for Space Vehicles, Capua, October 15–18, 2001.
Glößner, C., Olivier, H.: Experimente im Hyperschall an der ELAC-Oberstufe EOS. DGLR-Jahrestagung, Leipzig, September 2000, DGLR-Jahrbuch, Band III, DGLR-JT2000-206.
Glößner, C., Olivier, H.: Aerothermodynamics of a Winged Re-entry Vehicle in Hypersonic Flow. Proceedings of the 4th European Symposium on Aerothermodynamics for Space Vehicles, Capua/Italy, 15.–18. October 2001, ESA SP487, pp. 233–240.
Glößner, C., Olivier, H.: Force and Moment Coefficients of the HOPPER/PHOENIX-Configuration in Hypersonic Flow. Proceedings of the 3rd Int. Symposium on Atmospheric Re-entry Vehicles and Systems, Arcachon/France, 24.–27. March 2003.
Glößner, C., Olivier, H.: Force and moment measurements on the HOPPER configuration for high enthalpy conditions. 24th International Symposium on Shock Waves (ISSW 24), Beijing/China, scheduled for 20.–25. July 2003 (postponed).
Glößner, C., Olivier, H.: Kraft- und Momentenmessung im Hyperschall an der ELAC-Oberstufe EOS. Deutscher Luft- und Raumfahrtkongress, Hamburg, 17.–20. September 2001, DGLR-Jahrbuch 2001, Bd. 3, veröffentlicht als CD-ROM.
Gogel, T., Auweter-Kurtz, M., Gölz, T., Messerschmid, E., Schrade, H., Slezonia, C.: Numerical Study of High Enthalpy Flow in a Plasma Wind Tunnel. Computer Methods in Applied Mechanics and Engineering, 89, 1–3, pp. 425–434, 1991.
Gogel, T.H., Auweter-Kurtz, M., Gölz, T.M,. Messerschmid, E.W., Schrade, H.O., Sleziona, P.C.: Numerical Study of High Enthalpy Flow in a Plasma Wind Tunnel. Journal of Computer Methods in Applied Mechanics and Engineering, 1991, Vol. 89, No. 1–3, pp. 425–434, 1991.
Gogel, T.H., Dupuis, M., Messerschmid, E.: Radiation Transport Calculation in High Enthalpy Environments for 2D-Axisymmetric Geometries. Journal of Thermophysics and Heat Transfer, Vol. 8, No. 4, pp. 744–750, Oct.–Dec., 1994.
Gogel, T.H., Messerschmid, E.: Coupled Flow- and Radiation Field Computations in Hypersonic Flow. Computational Fluid Dynamics 94, Proceedings of the Second European Fluid Dynamics Conference, September 1994, Stuttgart, pp. 744–752, John Wiley and Sons, Chinchester, 1994.

Gogel, T.H., Sedghinasab, A., Keefer, D.R.: Radiation Calculation in Cylindric Arc Columns Using a Monte Carlo Method. Journal of Quantitative Spectroscopy and Radiative Transfer, Vol. 52, No. 2, pp. 179–194, Pergamon Press, 1994.
Gomez Garcia, J., Albus, J., Reimerdes, H.-G.: Mass-Estimation of Shell Structures with Elliptical Cross-Section under Combined Loadings. AIAA Paper 98-1594, 8th International Space Planes and Hypersonic Systems and Technologies Conference, Norfolk, Virginia, USA, April 27–30, 1998.
Gräsel, J., Beylich, A.E.: Passive Control for the Off-Design Performance of a Sigle Expansion Ramp Nozzle. AIAA Paper 98-1602, 8th International Space Planes and Hypersonic Systems and Technologies Conference, Norfolk, Virginia, USA, April 27–30, 1998.
Greza, H.: Automatische Neugenerierung lösungsadaptierter unstrukturierter Rechennetze zur numerischen Strömungssimulation nach der Methode der Finiten Elemente. Deutscher Luft- und Raumfahrtkongreß 1991, DGLR-Jahrestagung, 10.–13. Sept. 1991, Berlin, Jahrbuch der DGLR, 1991.
Gröbner, J., Kolitsch, U., Seifert, H.J., Fries, S.G., Lukas, H.L., Aldinger, F.: Re-Assessment of the Y-O Binary System. Z. Metallkd. 87, p. 88, 1996.
Grönig, H., Olivier, H.: Experimental Hypersonic Flow Research in Europe. JSME Int. Journal, Series B, Vol. 41, No. 2, pp. 397–407.
Grotowsky, I.M.G., Ballmann, J.: A numerical algorithm for calculating flows in hypersonic inlets. ZFW, Vol. 20, Nr. 2, 1996.
Grotowsky, I.M.G., Ballmann, J.: Efficient time integration of Navier-Stokes equations. Computers & Fluids 28, 1999, pp. 243–263.
Grotowsky, I.M.G., Ballmann, J.: Numerical investigation of hypersonic step-flows. Shock Waves (2000) 10, Springer Verlag 2000, pp. 57–72.
Gruhn, P., Henckels, A., Kirschstein, S.: Flap contour optimization for highly integrated SERN nozzles. Aerospace Science Technology, Vol. 4, No. 8, December 2000, pp. 555–565.
Gruhn, P., Henckels, A., Sieberger, G.: Improvement of the SERN nozzle performance by aerodynamic flap design. "Aerospace Science and Technology", Vol. 5, No. 6, pp. 395–405, Elsevier, Paris, 2002.
Gruhn, P., Henckels, A.: Simulation of a SERN/Afterbody Flow Field regarding heated Nozzle Flow in Hypersonic Facility H2K. 13. DGLR-Fach-Symposium der AG STAB, publiziert in der Reihe des Springer Verlags "Notes on Numerical Fluid Mechanics", publication planned for 2003.
Grünig, C., Avrashkov, V., Mayinger, F.: Fuel Injection into a Supersonic Airflow by Means of Pylons. Journal of Propulsion and Power, Vol. 16, No. 1, Jan.–Feb. 2000, pp. 29–34, 2000.
Grünig, C., Avrashkov, V., Mayinger, F.: Influence of Shock Waves on Mixing Processes in Supersonic Hydrogen-Air Flames. INTAS Work Report, Ref.No. INTAS-94-0079, 1996.
Grünig, C., Avrashkov, V., Mayinger, F.: Self-Ignition and Supersonic Reaction of Pylon-Injected Hydrogen Fuel. Journal of Propulsion and Power, Vol. 16, No. 1, Jan.–Feb. 2000, pp. 35–40, 2000.
Grünig, C., Avrashkov, V., Mayinger, F.: Shock-Induced Supersonic Combustion of Pylon-Injected Hydrogen. Accepted for AIAA Journal of Propulsion and Power, 1998.
Grünig, C., Mayinger, F.: Experimental Investigation of Supersonic Flame Stabilisation based in Fuel Self-Ignition. Chemical Engineering and Technology, Vol. 23, pp. 909–918, 2000.
Grünig, C., Mayinger, F.: Experimentelle Untersuchung stoß-induzierter H2- und Kerosin/H2-Überschallflammen in Scramjet-Brennkammern. Tagungsband DGLR-Jahrestagung 1998, Bremen, 5.–8. Oktober 1998.
Grünig, C., Mayinger, F.: Experimentelle Untersuchungen zur Flammenstabilisierung durch Selbstzündung in Überschallbrennkammern. Thermodynamik-Kolloquium des GVC/VDI-GET, Leipzig, 5.–6. Oktober 1998.
Grünig, C., Mayinger, F.: Induktion, Stabilisierung und Optimierung von H2-Überschallflammen durch Verdichtungsstöße. Seminar des Sonderforschungsbereichs 255, München, 12. Dez. 1996, Technische Universität München, Tagungsband, pp. 83–94, 1997.

Grünig, C., Mayinger, F.: Supersonic Combustion of Kerosene/H2-Mixtures in a Model Scramjet Combustor. Combustion Science and Technology, Vol. 146, pp. 1–22, 1999.
Grünig, C., Mayinger, F.: Überschallverbrennung von Kerosin/Wasserstoffgemischen. 2. Workshop "Hyperschallantriebe" der SFB's 253, 255, 259 und der DLR, Stuttgart, 22. November 1997.
Grünig, C., Mayinger, F.: Untersuchung der physikalischen Mechanismen zur Stabilisierung von Kerosin/Wasserstoff-Überschallflammen – "Dual Fuel Concept". 1. Workshop "Hyperschallantriebe" der SFB's 253, 255, 259 und der DLR, München, Mai 1997.
Grünig, C., Mayinger, F.: Zerstäubung von Kerosin-Strahlen als Voraussetzung zur Verbrennung in einem Überschallluftstrom. Conference Proceedings, 4th Workshop Spray '98, Essen, 13.–14. Okt. 1998.
Haarmann, T., Koschel, W.: Computation of Wall Heat Fluxes in Cryogenic H2/O2 Rocket Combustion Chambers. AIAA 2002-3693, 2002.
Haarmann, T., Koschel, W.: Numerical Simulation of Head Loads in a Cryogenic H2/O2 Rocket Combustion Chamber. PAMM Proceedings in Applied Mathematics and Mechanics, 2(1), pp. 360–361, 2002.
Haber, J., Heim, A, Alefeld, M.: A new 3D Graphics Library: Concepts, Implementation, and Examples. – In: Hege, H.-C. et al. (Eds.): Visualization and Mathematics, Springer, Berlin, pp. 211–225, 1996.
Habermann, M., Olivier, H., Grönig, H.: Operation of a High Performance Detonation Driver in Upstream Propagation Mode for a Hypersonic Shock Tunnel. In: Ball, G.J., Hillier, R., Roberts, G.T. (eds.) Proceedings of the 22nd International Symposium on Shock Waves, London, July 1999, Vol. I, pp. 447–452.
Habermann, M., Olivier, H.: Experimental Studies in a Detonation Driven Shock Tube at Elevated Pressures. Symposium on Shock Waves, Saitama, Japan, March 1998, pp. 551–554.
Habermann, M., Olivier, H.: Upgrade of the Aachen Shock Tunnel with a Detonation Driver. International Conference on the Methods of Aerophysical Research (ICMAR 2000), Novosibirsk-Tomsk, Rußland, July 2000, Part I, pp. 116–121.
Habiger, H. A., Auweter-Kurtz, M.: Investigation of a High Enthalpy Air Plasma Flow with Electrostatic Probes. Journal of Thermophysics and Heat Transfer, Vol. 12, No. 2, pp. 198–205, April–June 1998.
Hafer, X., Sachs, G.: Flugmechanik – Moderne Flugzeugsentwurfs- und Steuerungskonzepte, Springer Verlag, Berlin, 3. Auflage, 1993.
Haibel, M,. Mayinger, F., Strube, G.: Application of Non-Intrusive Diagnostic Methods to Sub- and Supersonic Hydrogen-Air-Flames; 3rd. Int. Symposium on Special Topics in Chemical Propulsion, Scheveningen, pp. 109–112, 1993.
Haibel, M., Mayinger F.: The Effect of Turbulent Structures on the Development of Mixing and Combustion Processes in Sub- and Supersonic Hydrogen Flames. Int. J. Heat Mass Transfer, Vol. 37, Suppl. 1, pp. 249–253, 1994.
Haibel, M., Mayinger, F., Strube, G.: High Speed Hydrogen Combustion Phenomena. IUTAM-Symposium on Aerothermodynamics in Combustors, Abstract Volume, Taipei, R.O.C., II-30, 1991.
Haibel, M., Mayinger, F.: Dynamik stationärer Wasserstoff-Luft-Flammen in sub- und supersonisch durchströmten Brennräumen – turbulente Gemischbildung, Flammenstabilisierung und Flammenstruktur. – In: Proc. des Thermodynamik-Kolloquiums '93 des VDI-GET, Dresden, Sept. 20–21, 1993. Hrsg.: VDI (Düsseldorf 1993).
Haibel, M., Mayinger, F.: Experimental Investigation of the Mixing Process and the Flame Stabilisation in Sub- and Supersonic Hydrogen-Air-Flames. Proc. of the 10th Int. Heat Transfer Conference, Brighton UK Aug. 14–18 1994, Vol. 3, pp. 63–64, 1994.
Haibel, M., Mayinger, F.: Fundamental Combustion Technology. 2nd Space Course on Low Earth Orbit Transportation 1993, Munich, FRG, pp. 26.1–26.21, 1993.
Haibel, M., Mayinger, F.: The effect of turbulent structures on the development of mixing and combustion processes in sub- and supersonic H2 flames. – In: Int. Journal of Heat and Mass Transfer, Vol. 37., 241–253, 1994.

Haibel, M., Mayinger, F.: Turbulenzgestützte Gemischbildung und Stabilisierung von sub- und supersonischen Wasserstoff-Luft-Flammen. – In: Jahrbuch 1992, 1, der Deutschen Gesellschaft für Luft- und Raumfahrt e.V. (DGLR), Bonn.
Haibel, M., Strube, G., Mayinger, F.: High speed combustion phenomena. – In: IUTAM-Symp. on Aerothermodynamics in Combustors, Taipei, June 3–5, 1991. Eds.: R.S. Lee et al. Taipei: Nat. Univ., II, pp. 30–32, 1991.
Haidinger, F.A., Friedrich R.: Computation of shock wave/turbulent boundary layer interactions using a two-equation model with compressibility corrections. – In: Advances in Turbulence IV. Appl. Sci. Research 51, pp. 501–504, 1993.
Haidinger, F.A., Friedrich R.: Numerical simulation of strong shock/turbulent boundary layer interactions using a Reynolds stress model. Z. Flugwiss. Weltraumforsch. 19, pp. 10–18, 1995.
Haidinger, F.A., Friedrich, R.: Computation of shock wave/turbulent boundary-layer intersections using a Reynolds stress model. Proc. 9th Symp. "Turbulent Shear Flows", Kyoto, Japan, August 16–18, pp. 4-3-1 to 4-3-6, 1993.
Hald, H.: CMC Materials from Plasma Channel Tests to Real Capsule Reentry. High Temp. Chem. Processes 3, pp. 153–165, 1994.
Hänel, D., Henze, A., Krause, E.: Supersonic and hypersonic flow computations for the research configuration ELAC I and comparison to experimental data. ZFW, Vol. 17 (1993), pp. 90–98.
Hannappel, R., Friedrich, R.: Direct numerical simulation of a Mach 2 shock interaction with isotropic turbulence. Applied Scientific Research 54, pp. 205–221, 1995.
Hannappel, R., Friedrich, R.: DNS of a M=2 shock intracting with isotropic turbulence. – In: Direct and Large Eddy Simulation I. Selected papers from the first ERCOFTAC Workshop, Guildford, March 28.30, 1994. Eds P.R. Voke, L. Kleiser, J.P. Chollet, Doordrecht, Kluwer Academics, pp. 359–373, 1994.
Hannappel, R., Friedrich, R.: Interaction of isotropic turbulence with a normal shock wave. Advances in Turbulence IV, Ed. F.T.M. Nieuwstadt, Doordrecht: Kluwer Academic Publisher, pp. 507–512, ISBN 0-7923-2282-7, 1993.
Hannappel, R., Hauser, Th., Friedrich, R.: A comparison of ENO and TVD schemes for the computation of shock-turbulence interaction. J. Comp. Phys., Vol. 121, pp. 176–184, 1995.
Haug, R., Heimann, D., Bill, J., Aldinger, F.: Keramische Oxidationsschutzschichten aus polymeren Vorstufen. In: Verbundwerkstoffe und Werkstoffverbunde, Ziegler G. (Ed.); DGM Informationsgesellschaft Verlag, Oberursel, pp. 429, 1996.
Hauser, Th., Friedrich, R.: Erfahrungen bei der Parallelisierung des Navier-Stokes Lösers COMPRES. Seminar des Sonderforschungsbereichs 255, München, 12. Dez. 1996, Technische Universität München, Tagungsband, pp. 37–42, 1997.
Hauser, Th., Friedrich, R.: Hypersonic flow around a blunt cone under the influence of external wave-like perturbations. Vortrag auf der GAMM-Jahrestagung 1997 in Regensburg, 24.–27. 3. 1997. – In: Krause, E. (Ed.): Gasdynamic Problems of Space Transportation Systems, Special Issue, Eigenverlag, Aerodynamisches Institut, RWTH Aachen, pp. 29–30, 1997.
Hauser, Th., Friedrich, R.: On the effect of freestream perturbations in hypersonic flow around a blunt cone. Paper accepted for ECCOMAS 98, Athen. – To appear in: Proceedings of the 4th ECCOMAS Comp. Fluid Dyn. Conf., J. Wiley & Sons.
Hauser, Th., Hannappel, R., Friedrich, R.: Testing High Order Shock-Capturing Schemes in 2d Super- and Hypersonic Flows. Aerothermodynamics for Space Vehicles, ESA SP-318, 1991.
Heim, A., von Stryk, O.: Trajectory optimization of industrial robots with application to computer-aided robotics and robot controllers. Optimization 47, pp. 407–420, 2000.
Heimann, D., Bill, J., Aldinger, F., Schanz, P., Gern, F.H., Krenkel, W.: Development of Oxidation Protected Carbon/Carbon. Z. Flugwiss. und Weltraumforschung, ZFW 19, Heft 3, 1995.
Heimann, D., Bill, J., Aldinger, F.: Keramische Schutzschichten aus polymeren Vorstufen und deren Anwendung als Oxidationsschutz für CFC-Materialien. In: Werkstoff und

Verfahrenstechnik, (Ziegler G., Cherdron H., Hermel W., Hirsch J., Kolaska H., Ed.), Symposium 6, Werkstoffwoche '96, 28.–31. 5. 1996, Stuttgart, DGM-Informationsgesellschaft Verlag, p. 817, 1997.
Heimann, D., Wagner, T., Bill, J., Aldinger, F., Lange, F. F.: Epitaxial Growth of C/C-SiC Thin Films on a 6H-SiC Substrate Using the Solution Precursor Method. J. Mat. Res. 12, p. 3099, 1997.
Heinold, B.: Von optimalen Prozessen mit diskreten Phasenbeschränkungen zu solchen mit kontinuierlichen Phasenbeschränkungen, Workshop Universität Greifswald 14./15. November 1994, Tagungsband, pp. 97–101, 1994.
Heinz, S. (2003): A Model for the Reduction of the Turbulent Energy Redistribution by Compressibility, Phys. Fluids.
Heinz, S. (2003): On Fokker-Planck Equations for Turbulent Reacting Flows. Part 1. Probability Density Function for Reynolds-averaged Navier-Stokes Equations, Flow, Turb. Combust.
Heinz, S. (2003): On Fokker-Planck Equations for Turbulent Reacting Flows. Part 2. Filter Density Function for Large Eddy Simulation, Flow, Turb. Combust.
Heinz, S. and Roekaerts, D.: PDF Modelling of Turbulent Reacting Flows: Reynolds and Damköhler Number Effects. Turbulence and Shear Flow-1. Edited by S. Banarjee and J. K. Eaton, Begell House Inc., New York Wallingford, pp. 339–344, 1999.
Heinz, S., Friedrich, R.: Scalar PDF Transport Equations for Subsonic and Supersonic Combustion. Combustion Theory and Modelling, 2001.
Heinz, S., Friedrich, R.: PDF/FDF – Methoden zur Prognose turbulenter Überschallverbrennung. Seminar des Sonderforschungsbereichs 255, 12. Dezember 2002, TU München, ISBN 3-89791-334-8, S. 37–46, 2004.
Heinz, S., Roekaerts, D.: Reynolds Number Effects on Mixing and Reaction in a Turbulent Pipe Flow. Chem. Eng. Sci. 56, pp. 3197–3210, 2001.
Heinz, S.: Advanced Methods to Compute Multiphase Turbulent Reacting Flows. Z. Angew. Math. Mech. 81, pp. 471–472, 2001.
Heinz, S.: On the Kolmogorov Constant in Stochastic Turbulence Models. Phys. Fluids. 14, 4095–4098, 2002.
Heinz, S.: Statistical Mechanics of Turbulent Flows. Springer Verlag, Berlin, Heidelberg, New York, Tokyo, 2003.
Heiser, W., Olejak, D., Reisinger, D., Wagner, S.: Determination of particle size by means of a relaxation-length method at TWM. DANTEC Information, Nr. 11, 1992.
Heitmeier, F., Schmitt-Thomas, Kh. G., Dietl, U.: Hypersonic Propulsion Challenges in Design and Materials. Space Course 1993, TU München, Oct. 11–22 1993.
Heller, M., Holzapfel, F., Sachs, G.: Phygoide der Seitenbewegung bei Hyperschall-Flugsystemen. Seminar des Sonderforschungsbereichs 255, TU München, 3. Februar 2000, Tagungsband, pp. 129–144, ISBN 3-89791-175-2, 2001.
Heller, M., Holzapfel, F., Sachs, G.: Robust Lateral Control of Hypersonic Vehicles. AIAA Atmospheric Flight Mechanics Conference Proceedings, Denver, USA, 14.–17. Aug. 2000, AIAA Paper 2000-4248, 2000.
Heller, M., Holzapfel, F., Sachs, G.: Robuste Regelung der Seitenbewegung von Hyperschall-Flugzeugen. DGLR-Jahrestagung, Leipzig, JT2000-061, 2000.
Heller, M., Holzapfel, F., Sachs, G.: Unique Lateral Coupling Problems of Hypersonic Aircraft. Proceedings of the 2nd International Conference on Unconventional Flight, Balatonfüred 14.–16. Juni 2000, publication planned for 2001.
Heller, M., Sachs, G., Löfberg, J., Helmersson, A.: Robustes Steuerungs- und Regelungskonzept für die Seitenbewegung eines Hyperschallflugzeugs. Zeitschrift für Angewandte Mathematik und Mechanik, Vol. 79, Supplement 3, pp. 941–942, 1999.
Heller, M., Sachs, G., Löfberg, J., Helmersson, A.: Robustes Steuerungs- und Regelungskonzept für die Seitenbewegung eines Hyperschallflugzeugs. GAMM-Jahrestagung 1998, Bremen, 6.–9. April 1998, – In: Krause, E. (Ed.): Gasdynamic Problems of Space Transportation Systems, Special Issue, Eigenverlag, Aerodynamisches Institut RWTH Aachen, 1998.
Heller, M., Sachs, G., Ståhl-Gunnarson, K., Frank, H., Rylander, D.: Flight Dynamics and Robust Control of a Hypersonic Test Vehicle with Ramjet Propulsion. Proceedings of

the AIAA 8th International Space Planes and Hypersonic Systems and Technology Conference, Norfolk, USA, 17.–30. April 1998, pp. 126–136, 1998.
Heller, M., Sachs, G.: Robustes Steuerungs- und Regelungskonzept für die Seitenbewegung eines Hyperschallflugzeugs. Workshop "Optimalsteuerungsprobleme von Hyperschall-Flugsystemen" des Sonderforschungsbereichs 255, Greifswald, 23.–24. Okt. 1997, Tagungsband, Technische Universität München, pp. 51–62, 1998.
Heller, M.: Stabilitäts- und Regelungsprobleme im Hyperschallflug. Seminar des Sonderforschungsbereichs 255, München, 12. Dez. 1996, Technische Universität München, Tagungsband, pp. 141–153, 1997.
Henckels, A., Kreins, A.F., Maurer, F.: Experimental investigations of hypersonic shock-boundary layer interaction. ZFW, Vol. 17 (1993), pp. 116–124.
Henckels, A., Maurer, F., Olivier, H., Grönig, H.: Fast temperature measurement by infrared line scanning in a hypersonic shock tunnel. Experiments in Fluids, Vol. 9, pp. 298–300, 1990.
Henze, A., Houtman, E.M., Jacobs, M., Vetlutzky, V.N.: Comparison between experimental and numerical heat flux data for supersonic flow around ELAC 1. ZFW, Vol. 20, Nr. 2, 1996.
Henze, A., Schröder, W., Bleilebens, M., Olivier, H.: Numerical and Experimental Investigations on the Influence of Thermal Boundary Conditions on Shock Boundary-Layer Interaction. Computational Fluid Dynamics Journal, 12(2):46, July 2003.
Henze, A., Schröder, W., Meinke, M.: Computation of Flows around Space Configurations. LNCSE, High performance scientific and engineering computing, FORTWIHR Conference, Erlangen, 12.–14. März 2001, Springer 2002, pp. 131–138.
Henze, A., Schröder, W., Meinke, M.: Numerical Analysis of the Supersonic Flow around Reusable Space Transportation Systems. 4$^{th}$ European Symposium on Aerothermodynamics for Space Vehicles, Capua, Italien, 2001, pp. 191–197.
Henze, A., Schröder, W.: On the Influence of Thermal Boundary Conditions on Shock-Boundary-Layer Interaction. DGLR Jahrestagung, Leipzig, September 2000, DGLR-Jahrbuch, DGLR-JT2000-175.
Henze, A.: Flow Simulation of the Hypersonic Research Configuration ELAC 1. GAMM Annual Meeting, Bremen, April 1998, ZAMM Vol. 79 (1999), Suppl. 3, pp. 943–944.
Henze, A.: Hypersonic flow computations for the reusable transport system HOPPER. 11$^{th}$ AIAA/AAAF International Conference, AIAA 2002-5194, Orleans, Frankreich, 2002.
Herdrich, G., Auweter-Kurtz, M., Endlich, P.: Mars Entry Simulation using the Inductively Heated Plasma Generator IPG4. Angenommen als Engineering Note for the Journal of Spacecrafts and Rockets Vol. 40, No. 5, 2003.
Herdrich, G., Auweter-Kurtz, M., Habiger, H.: Pyrometric Temperature Measurements on Thermal Protection Systems. ZAMM 79 (1999) Suppl. 3, pp. 945–946, 2003.
Herdrich, G., Auweter-Kurtz, M., Kurtz, H.: A New Inductively Heated Plasma Source for Re-Entry Simulations. Journal of Thermophysics and Heat Transfer, Vol. 14, No. 2, April–June 2000, pp. 244–249, 2003.
Herdrich, G., Auweter-Kurtz, M., Kurtz, H.L., Laux, T., Winter, M.: Operational Behavior of the Inductively Heated Plasma Source IPG3 for Re-entry Simulations. Journal of Thermophysics and Heat Transfer, Vol. 16, No. 3, pp. 440–449, July–September 2002.
Hesse, M., Reinartz, B.U., Ballmann, J.: Inviscid Flow Computation for the Shuttle-Like Configuration PHOENIX. in: Notes on Numerical Fluid Mechanics, Ed.: Chr. Breitsamter, B. Laschka, H.-J. Heinemann, R. Hilbig, Springer, Berlin Heidelberg New York, 2003.
Hirsch, K., Altmann, I., Bauer, G., Brandt, R., Lindner, P., Neuer, G., Roth, B., Schneider, J., Stirn, R., Schumacher, U.: In situ determination of the emissivity and of the erosion rate of heat shield materials in re-entry simulation experiments. 16th ECTP2002: European Conference on Thermophysical Properties, 1–4. 09.2002, Imperial College, London (National Physical Laboratory), (www.ectp.npl.co.uk). accepted for publication in High Temperature – High Pressure, vol 34, 2003.
Hirsch, K., Altmann, I., Bauer, G., Kollermann, T., Roth, B., Schinköth, D., Schneider, J., Schumacher, U., Tawfik, A.: Development and investigation of erosion protection layers

for heat shields of re-entry space vehicles. ECAMPVII, 7th European Conference on Atomic and Molecular Physics, Berlin 2.–6. April 2001, paper P12.13, 2001.
Hirsch, K., Altmann, I., Bauer, G., Roth, B., Schinköth, D., Schneider, J., Schumacher, U., Tawfik, A.: Heat Shield Materials with and without Oxidation Protection Layer – High Temperature Emergency Behaviour. 2nd International Symposium on Atmospheric Reentry Vehicles and Systems, Arcachon/France, March 26.–29. 2001, Proc of the Association Aeronautique and Astronautique de France, paper 14_9_P, 2001.
Hirsch, K., Bross, M., Jentschke, H., Schumacher, U.: High Resolution Emission and Absorption Spectroscopy for Temperature and Density Determination of Erosion Products in Plasma-Target Interaction. Plasma 97, International Symposium on Plasma Research and Application, Jarnoltowek (near Opole), Poland, 12. June 1997.
Hirsch, K., Roth, B., Altmann, I., Barth, K-L., Jentschke, H., Lunk, A., Schumacher, U.: Plasma induced SiO2 like Protection Layer Formation on C/C-SiC Heat Shield Materials for Re-entry Vehicles. High Temperatures – High Pressures 31, 455–465, 1999.
Hirsch, K., Volk, G.: Thomson Scattering with a Gated Intensified CCD Camera Using a Frequency Doubled Periodically Pulsed YAG Laser. Rev. Sci. Instrum. 66, No. 11, 5369–5370, 1995.
Hollmeier, S., Esch, Th., Herrmann, O., Rick, H.: Propulsion/Aircraft Integration and Dynamic Performance of Hypersonic TSTO Vehicles. Proceedings of the 12th International Symposium on Airbreathing Engines, ISABE, Melbourne, Australien, Sept. 1995.
Hollmeier, S., Kopp, S., Fiola, R., Rick, H.: Simulationsmodelle zur Regelung von Antriebssystemen für Hochgeschwindigkeitsflugzeuge. DGLR-Jahrestagung 1997, München, Okt. 1997, DGLR-JT97-081, 1997.
Hollmeier, S., Kopp, S., Herrmann, O., Rick, H.: Propulsion/Aircraft Integration of Hypersonic Vehicles enabling Active Thrust Vectoring and Realtime Simulation of Dynamic Engine Behavior. Eingereicht zur Veröffentlichung im International Journal of Turbo- and Jet Engines, Mai 1998.
Hollmeier, S., Kopp, S., Rick, H.: FSSC-12 Installed Turbojet Engine Performance. WP.-No. 4010-C, FESTIP Final Report, Lehrstuhl für Flugantriebe, TU München, 1999.
Hollmeier, S., Kopp, S.: Statusbericht zur Echtzeitsimulation von Hyperschallantrieben mittels Leistungssyntheserechnung. 1. Workshop Hyperschallantriebe der Sonderforschungsbereiche 253, 255, 259 und der DLR, München, 15. Mai 1997.
Hollmeier, S., Rick, H., Wagner, O.: Zur Echtzeitsimulation von Hyperschall-Kombinationsantrieben im Flugsimulator. DGLR-Jahrestagung 1995, Bonn, Sept. 1995, DGLR-JT95-100, 1995.
Holzapfel, F., Heller, M., Zähringer, C., Sachs, G.: Generic Flight Dynamics Modeling Environment for Control System Design and Analysis of Hypersonic Vehicles. AIAA 10th International Space Planes and Hypersonic Systems and Technologies Conference Proceedings, Kyoto, Japan, 24.–27. April 2001.
Holzapfel, F., Mayrhofer, M., Dinkelmann, M., Sachs, G.: Flugmechanische Untersuchung des Dual-Fuel-Konzepts für ein Hyperschall-Flugsystem, Workshop des Sonderforschungsbereichs 255, Greifswald, 23.–24. Oktober 1997, Tagungsband, TU München, pp. 73–79, 1998.
Holzapfel, F., Zähringer, C., Sachs, G., Laschka, B.: Computergestützte Entwicklungsumgebung zur Automatisierung flugdynamischer Analyse- und Regelungsaufgaben. DGLR-JT2001-187, Deutscher Luft- und Raumfahrtkongress, Hamburg, 17.–22. Sept. 2001.
Holzapfel, F., Zähringer, C., Sachs, G.: Conception of an Efficient Computational Environment for Aircraft Dynamics and Control Research. 7th Mini Conference on Vehicle System Dynamics, Identification and Anomalies, Budapest, Ungarn, 2000.
Holzapfel, F., Zähringer, C., Sachs, G.: PC Based Configurable Aircraft Simulation Environment for Dynamics and Control Analysis. AIAA Modeling and Simulation Technologies Conference 2001, Montreal, Quebec, Kanada, 2001.
Holzapfel, F.: Linearisierung des Flugzeugverhaltens für große seitliche Schwerpunktverschiebung. DGLR-Jahrestagung, Leipzig, JT2000-167, 2000.
Hönig, R., Theisen, D., Kappler, G., Andresen, P.: Experimental Investigation of a Scramjet Model Combustor Employing Injection Through a Swept Ramp Using Laser-Induced

Fluorescence with Tunable Excimer Lasers, submitted to: IUTAM Symposium "Combustion in Supersonic Flows", 2.–6. Oktober 1995, Poitier, France, 1995.

Hornik, A.: Belastung der Atmosphäre durch ARIANE X und Sänger. Luft- und Raumfahrt, DGLR, Jahrg. 12, 7/8 91.

Hornik, A.: Pollution of the Atmosphere by Space Launchers, Acta Astronautica.

Hornik, A.: Struktur – Growthfaktor, Luft- und Raumfahrt, DGLR.

Hornung, M., Lentz, S., Friedrich, W., Staudacher, W.: Integration und Auswirkungen der Ergebnisse der Teilprojekte des Sonderforschungsbereiches 255 auf den Gesamtentwurf eines Transatmosphärischen Raumtransportsystems (HTSM). Seminar des Sonderforschungsbereichs 255, TU München, 3. Februar 2000, Tagungsband, pp. 161–172, ISBN 3-89791-175-2, 2001.

Hornung, M., Staudacher, W.: Design of an Airbreathing Upper Stage for a Reusable Two-Stage-to-Orbit, AIAA International Air & Space Symposium and Exposition "The Next 100 Years" 14.–17. 07. 2003, Dayton/Ohio, 2003.

Hornung, M., Staudacher, W.: Generation of a Pre-Design Dataset for an Airbreathing Upper Stage of a Reusable Space Transportation System, Notes on Numerical Fluid Mechanics, Vol. 87, Springer-Verlag, 2003.

Hornung, M., Staudacher, W.: Impact of an Airbreathing Upper Stage on a Reusable Two-Stage-to-Orbit, RTO/AVT Symposium, 6.–10. 10. 2003, Warschau, Polen, 2003.

Hornung, M., Staudacher, W.: Verwendung eines luftatmenden Antriebes in der Oberstufe eines zweistufigen Raumtransportsystems, DGLR Jahrestagung 2002 Stuttgart 23.09–26.09, 2002.

Hörschler, I.: Entwicklung von Gitteroptimierungsmethoden für die ELAC-1 Konfiguration. DGLR-Jahrestagung 1998, Bremen, 5.–8. Oktober 1998, Sammelband pp. 137–146.

Höss, B., Fottner, L.: Experimental Setup, Measurement and Analysis of the Onset of Compressor Flow Instabilities in an Aeroengine. Proceedings of the 17th ICIASF, Monterey, USA, 29. Sept.–2. Okt. 1997.

Höss, B., Leinhos, D., Fottner L.: Stall Inception in a Compressor System of a Turbofan Engine. ASME Journal of Turbomachinery, Vol. 122, 1998, pp. 32–44.

Höss, B.: Untersuchungen zum Einfluß von Eintrittsstörungen auf das dynamische Leistungsverhalten von Turbostrahltriebwerken unter besonderer Berücksichtigung instabiler Verdichterströmungen. Institut für Strahlantriebe, UniBw München, Abschlußbericht DFG-Fo 136/4, 1998.

Hu, J., Fottner, L.: Calculations of Effects of Rotating Inlet Distortion on Flow Instabilities in Compression Systems. Proceedings of the ASME 1995, ASME 95–GT–196, 1995.

Hu, J., Fottner, L.: Numerical Simulation of Active Supression of Rotating Stall in Axial Compression Systems. Journal of Thermal Science, Vol. 5, Nr. 4, 1996, pp. 231–242.

Hüttl, T.J., Friedrich, R.: Direct numerical simulation of turbulent flows in curved and helically coiled pipes. Computers & Fluids, Vol. 30, No. 5, pp. 591–606, 2001.

Iglseder, H., Arens-Fischer, W., Keller, H.U., Arnold, G., Callies, R., Fick, M., Glaßmeier, K.H., Hirsch, H., Hoffmann, M., Rath, H.J., Kührt, E., Lorenz, E., Thomas, N., Wäsch, R.: INEO – Imaging of Near Earth Objects. In: Proc. of the World Space Congress 1992, XXIX. COSPAR Plenary Meeting, Washington, D.C., Aug. 28–Sept. 5, 1992. AIAA, Washington, 1992. Paper No. 1-M.1.03.

Infed, F., Fertig, M., Olawsky, F., Auweter-Kurtz, M.: Simulation of Hypersonic Flows in Thermo Chemical Nonequilibrium Around the Re-entry Vehicle X-38 with URANUS Code. 3rd International Symposium Atmospheric Reentry Vehicles and Systems, Arcachon, France, 24–27 March 2003.

Infed, F., Olawsky, F., Auweter-Kurtz, M.: Stationary Coupling of 3D Hypersonic Nonequilibrium Flows and TPS Structure with URANUS. AIAA-Paper 2003-3984, 16th AIAA Computational Fluid Dynamics Conference, Orlando, Florida, USA, June 23–26, 2003.

Infed, F., Olawsky, F., Frühauf, H.-H.: X-38 Oberflächentemperaturunsicherheiten, DGLR-Jahrbuch, 2001.

Jacob, D., Ballmann, J.: Forschungs- und Entwicklungstrends bei Luft- und Raumtransportsystemen. Erschienen in "Horizonte: Die RWTH Aachen an der Schwelle des 21. Jahrhunderts", Springer Verlag 1999, pp. 473–481.

Jacob, D., Neuwerth, G., Peiter, U.: High Reynolds number wind tunnel tests with an ELAC-Model in the DNW. GAMM Annual Meeting, Bremen, April 1998, ZAMM Vol. 79 (1999), Suppl. 3, pp. 951–954.

Jacob, D.: Research on the Hypersonic Space Plane Configuration ELAC at the RWTH Aachen. Proceedings of the 12$^{th}$ European Aerospace Conference, Paris, 29.11.–1.12.99, (CEAS-Conference), Paper 11–59.

Jalowiecki, A., Bill, J., Aldinger, F., Mayer, J.: Interface Characterization of Nanosized B-Doped Si3N4/SiC Ceramics. Composites Part A, 27A, pp. 717, 1996.

Janovsky, R., Staufenbiel, R., Rüggeberg, T., Koschel, W.: Flugleistungssimulation und Bahnoptimierung horizontal startender Raumtransportsysteme. DGLR-Jahrestagung, Friedrichshafen, 1.–4. Okt. 1990, Jahrbuch III der DGLR, pp. 1421–1432, 1990.

Janovsky, R., Staufenbiel, R.: Design Optimization of an Airbreathing Aerospaceplane. in: Oertel, H., Koerner, H.: "Orbital transport, technical, meteorological, and chemical aspects", Springer Verlag, 1993, pp. 149–163.

Jänsch, C., Kraft, D., Schnepper, K., Well, K.H.: Survey Paper on Trajectory Optimization Methods. Technical Report TR1, ESTEC Noordwijk, 1989.

Jänsch, C., Kraft, D., Well, K.H.: Comparative Study of Nonlinear Programming Codes for Trajectory Optimization. Technical Note TN1, ESTEC Noordwijk, 1989.

Jänsch, C., Markl, A.: Trajectory Optimization and Guidance for a Hermes-Type Reentry Vehicle. AIAA Guidance Navigation & Control Conference, New Orleans, AIAA Proc. 91-2659-CP, Vol.1/3, pp. 543–553, 12.–14. 8. 1991.

Jänsch, C., Schnepper, K., Well, K.H.: Trajectory Optimization of a Transatmospheric Vehicle. Invited Paper: American Control Conference, Boston. ACC FA4, Vol. 3/3, pp. 2232–2237, 26.–28. 6. 1991.

Jaroma-Weiland, G., Neuer, G., Tischendorf, M.: PC-Programme to standardize the description of measurements results of thermophysical properties. Int. J. Thermophys. 17, 233–238, 1996.

Jaroma-Weiland, G., Neuer, G.: Database-Related Standardization to describe Experimental Results on the Thermophysical Properties of Solids. In: Thermodynamic Modeling and Materials Data Engineering. Hrsg.: J.-P. Caliste, A. Truyol und J. H. Westbrook. Springer Verlag, Berlin, Heidelberg, 349–354, 1998.

Jentschke, H., Bauer, G., Behringer, K., Hirsch, K.: Spectroscopical Investigation of an Air-Like Plasma Jet. 5th Europ. Conf. On Atomic and Molecular Phys., ECAMP, Vol. 19a, Part II, 613–616, 1995.

Jentschke, H., Hirsch, K., Klenge, S., Schumacher, U.: High resolution emission and absorption spectroscopy for erosion product analysis in boundary plasmas. Rev. Sci. Instrum. 70, 336–339, 1999.

Jentschke, H., Schumacher, U., Hirsch, K.: Studies of Silicon Erosion in Plasma-Target Interaction from Optical Emission and Absorption Spectroscopy. Contrib. to Plasma Phys. 38, 501–512, 1998.

Jessen, C., Vetter, M, Grönig, H.: Experimental studies in the Aachen hypersonic shock tunnel. ZFW, Vol. 17 (1993), pp. 73–81.

Jiang, L., Breitsamter, C., Laschka, B.: Numerical simulation of the stage separation flow physics on an idealized two stage transport vehicle. Deutscher Luft- und Raumfahrtkongress 2003, München, 17.–20. Nov. 2003.

Jiang, L., Zähringer, C., Breitsamter, C., Sachs, G.: Experimentelle und Numerische Simulationen zur Konfiguration "EOS" – Ebene Platte. Seminar des Sonderforschungsbereiches 255, 12. Dezember 2002, München, Tagungsband ISBN 3-89791-334-8, TU München, S. 21–36, 2004.

Jordan, M., Ardey, N., Gerlach, C., Mayinger, F.: Quenching Effects at Jet Ignition of Lean Hydrogen- and Methane-Air Mixtures. Proceedings of the 10th International Symposium on Transport Phenomena in Thermal Science and Process Engineering, Kyoto, Japan, 30. Nov.–3. Dez. 1997, 1997, pp. 19–24.

Jordan, M., Ardey, N., Mayinger, F. et al.: Improved Modelling of Turbulent Hydrogen Combustion and Catalytic Recombination for Hydrogen Mitigation. Proceedings of the

FISA – 97 Symposium on EU Research on Severe Accidents, Luxembourg, 17.–19. Nov. 1997, European Commission, Directorate-General XII, 1997.
Jordan, M., Ardey, N., Mayinger, F., Carcassi, M.N., Heitsch, M., Martin-Fuertes, F., Monti, R.: Influence of Turbulence on the Deflagrative Flame Propagation in Lean Premixed Hydrogen Air Mixtures. Proceedings of the FISA – 97 Symposium on EU Research on Severe Accidents, Luxembourg, 17.–19. Nov. 1997, European Commission, Directorate-General XII, 1997, pp. 324–333.
Jordan, M., Tauscher, R., Mayinger, F.: New Challenges in Thermo-Fluiddynamic Research by Advanced Optical Methods. Proceedings of the 15th UIT National Heat Transfer Conference, Turin, Italien, 1997, Edizioni ETS, 1997, pp. 79–100 (Int. Journal of Heat and Technology, Vol. 15, Nr. 1, 1997, pp. 43–54).
Jumel, J., Lepoutre, F., Frieß, M., Krenkel, W., Neuer, G.: Microscopic Thermal Characterization of C/C-SiC. Proc 4th Int. Conf. on High Temperature Ceramic Matrix Composites (HT-CMC 4), München, Okt. 1–3, 2001, High Temperature Ceramic Matrix Composites, pp. 120–126, Wiley-VCH/Weinheim, 2001.
Juranek N., Strube G., Mayinger F.: Structure and burning velocity of premixed, turbulent hydrogen-air-flames. Proc. ICHMT Int. Symp. (Energy Systems and Environmental Effects) Cancun, Mexico, Aug. 22–25, 1993. Eds.: J.C. de Gortari et al. (Mexico 1993) 481–486.
Kanne, S., Frühauf, H.-H., Messerschmid, E.W.: Thermochemical Relaxation Through Collisions and Radiation. AIAA Journal of Thermophysics and Heat Transfer, Vol. 14, No.4, 2000.
Kasagi, N., Eaton, J.K., Friedrich, R., Humprey, J.A.C., Leschziner, M.A., Miyauchi, T. (Eds.): 3rd International Symposium on Turbulence and Shear Flow Phenomena, Sendai, Japan, 25–27 June, 2003, Proceedings, 2003.
Kasal, P., Gerlinger, P., Walther, R., von Wolfersdorf, J., Weigand, B.: Supersonic Combustion: Fundamental Investigations of Aerothermodynamic Key Problems. AIAA/AAAF 11th International Space Planes and Hypersonic Systems and Technology Conference, AIAA-2002-5119, Sept. 29–Oct. 4, 2002 Orleans, France, 2002.
Kau, H.P., Rist, D., Brandstetter, A., Rocci Denis, S.: Experimental Investigation of Supersonic Combustion with Strut Injection. Journal of Propulsion and Power, 2001.
Kiehl, M., Mehlhorn, R., Schumann, M.: Parallel Multiple Shooting for Optimal Control Problems under NX. J. Optimization Methods and Software, Vol. 4, 1995, pp. 259–271.
Kiehl, M., von Stryk, O.: Generalized Necessary Conditions for Optimal Control Problems of Bolza Type: Theory and Application. Z. Ang. Math. Mech. Vol. 74, Nr. 6, 1994, pp. T591–T593.
Kiehl, M., Zenger, Chr.: An Improved Starting of the G-B-S Method for the Solution of Ordinary Differential Equations. Computing, 41 131–136, 1989.
Kiehl, M.: A parallel waveformrelaxation and iterated deferred correction method for the solution of IVPs. Z. Ang. Math. Mech., Vol. 75, Nr. 2, 1995, pp. 689–690.
Kiehl, M.: A Vector Implementation of an ODE Code for Multi-Point-Boundary Value Problems. – In: Parallel Computing 17, pp. 347–352, 1990.
Kiehl, M.: Circuit simulation an application for parallel ODE solver? In: Mathematical modelling and simulation of electrical circuits and semiconductor devices. Eds.: R. E. Bank et al. Basel: Birkhäuser, 1994, pp. 61–71 (ISNM 117).
Kiehl, M.: Increasing the Vector Length for Matrix Multiplication with Reduced Memory Access. – In: Mathematical Modelling and Simulation of Electrical Circuit and Semiconductor Devices. Eds.: R.E. Bank et al. Basel: Birkhäuser, pp. 101–108. (ISNM 93), 1990.
Kiehl, M.: Parallel multiple shooting for the solution of initial value problems. Par. Comp. 20 (1993) pp. 275–295.
Kiehl, M.: Parallel one-step methods with minimal parallel stages. J. Appl. Num. Math., Vol. 17, 1995, pp. 397–409.
Kiehl, M.: Partioning methods for the simulation of fast reactions. – In: Günther, M., Ostermann, A. (Eds.): Numerical treatment of ordinary and algebro-differential equations, Preprint Nr. 1934, Fachbereich Mathematik, TH Darmstadt, 1997, pp. 33–36.

Kiehl, M.: Partitioning in reaction kinetics. – In: Keil, F. et al. (Eds.): Scientific computing in chemical engineering, Springer, Berlin, 1996, pp. 122–128.
Kiehl, M.: Partitioning methods for the simulation of fast reactions. ZAMM, 78 (3), pp. 967–970, 1998.
Kiehl, M.: Sensitivity Analysis of Stiff and Non-stiff Initial-value Problems. – In: Schmidt, W.H., Heier, K., Bittner, L., Bulirsch, R. (Eds.): Variational Calculus, Optimal Control and Applications, ISNM 124, Birkhäuser, Basel, 1998, pp. 143–152.
Kiehl, M.: Smoothing the Function of the Multiple-Shooting Equation. – In: Appl. Math. Optim. 24 171–181, 1991.
Kiehl, M.: Vectorized Multiple Shooting with Reduced Number of Integrations. – In: Z. Angew. Math. Mech. 70 6, T603–T604, 1990.
Kiehl, M.: Vectorizing the Multiple-Shooting Method for the Solution of Boundary-Value Problems and Optimal-Control Problems. Proc. of the 2nd Int. Conf. on Vector and Parallel Computing Issues in Applied Research and Development, Tromsö, Norwegen, 6.–10.6.1988, Eds.: J. Dongarra et al. London: Ellis Horwood Limited, pp. 179–188, 1989.
Kiehl, M., von Stryk, O.: Real-Time Optimization of a Hydroelectric Power Plant. Computing, (1992), 49, pp. 171–191.
Kloker, M., Konzelmann, U., Fasel, H.: Outflow boundary conditions for spatial Navier-Stokes simulations of transition boundary layers. AIAA-Journal 31, 620–628, 1993.
Kloker, M., Stemmer, C.: Three-dimensional steady disturbance modes in the Blasius boundary layer – a DNS study. In Wagner, S., Kloker, M., Rist, U. (eds.): Recent Results in Laminar-Turbulent Transition. NNFM 86, 91–110, Springer 2003.
Kloker, M.: A robust high-resolution split-type compact finite-difference scheme for spatial direct numerical simulation of boundary-layer transition. Applied Sci. Research 59, No. 4, 353–377, Kluwer Acad. Publ., 1998.
Kloker, M.: A rubost high-resolution split-type compact FD-scheme for spatial direct numerical simulation of boundary-layer transition. Applied Scientific Research 59 (4), 353–377, 1998.
Kloker, M.: DNS of transitional boundary-layer flows at sub- and hypersonic speeds. DGLR-JT2002-017, DGLR, Bonn, 2002.
Klomfaß, A., Müller, S., Ballmann, J.: Modelling of Transport Phenomena for the Hypersonic Stagnation Point Heat Transfer Problem. AIAA-Paper 93-5047, 1993.
Klutchnikov, I., Ballmann, J.: Direct Numerical Simulation of Hypersonic Flow over a Cavity. PAMM – Proceedings in Applied Mathematics and Mechanics, Vol. 2, Issue 1, WILEY-VCH, Weinheim Germany, 2003, pp. 350–351.
Klutchnikov, I., Ballmann, J.: Direct Numerical Simulation of Supersonic Ramp and Step Flow. in: C. Liu, S. Sakell, T. Beutner (Eds.), DNS/LES Progress and Challenges, Proceedings of the 3rd AFOSR Int. Conference on DNS/LES, Greyden Press, Columbus, 2001, pp. 479–486.
Klutchnikov, I., Ballmann, J.: Direkte numerische Simulation turbulenter kompressibler Strömungen in Kanälen unterschiedlicher Konfiguration. ZAMM Volume 81 (2001), pp. 925–926.
Klutchnikov, I., Ballmann, J.: DNS of Turbulent Compressible Fluid Flow with a High Order Difference Method. AIAA Paper 2001–2542.
Klutchnikov, I., Ballmann, J.: DNS with a High-Order Difference Method of Cavity Flows at Hypersonic Speed. reviewed and submitted to "Shock Waves", Springer Verlag, 2003.
Knab, O., Frühauf, H.-H., Messerschmid, E.: Theory and Validation of the Physically Consistent Coupled Vibration-Chemistry-Vibration Model. AIAA Journal of Thermophysics and Heat Transfer, Vol. 9, No. 2, April–June, 1995.
Knauss, H., Gaisbauer, U., Schlaich, F., Wagner, S.: Complementary Measurement Techniques for the Detection of Wall Conditions in Supersonic Flow. Experiments in Fluids, 30/5, 2001 pp. 597–601, 2001.
Knauss, H., Gaisbauer, U., Wagner, S., Buntin, D., Maslov, A. A., Smorodsky, B., Betz, J.: Calibration Experiments of a New Active Fast Response Heat Flux Sensor to Measure Total Temperature Fluctuations – Part I: Introduction to the problem. XI. International

Conference on the Methods of Aerophysical Research ICMAR, Novosibirsk, Russland, 2002.

Knauss, H., Gaisbauer, U., Wagner, S., Buntin, D., Maslov, A. A., Smorodsky, B., Betz, J.: Calibration Experiments of a New Active Fast Response Heat Flux Sensor to Measure Total Temperature Fluctuations – Part II: Preliminary measures for a performance of comparative experimental and theoretical heat flux determination with an ALTP. XI. International Conference on the Methods of Aerophysical Research ICMAR, Novosibirsk, Russland, 2002.

Knauss, H., Gaisbauer, U., Wagner, S., Buntin, D., Maslov, A. A., Smorodsky, B., Betz, J.: Calibration Experiments of a New Active Fast Response Heat Flux Sensor to Measure Total Temperature Fluctuations – Part III: Heat flux density determination in a short duration wind tunnel. XI. International Conference on the Methods of Aerophysical Research ICMAR, Novosibirsk, Russland, 2002.

Knoll, A., Lesch, K., Möller, H., Sachs, G., Wagner, O.: Flight Simulator. 2nd Space Course on Low Earth Orbit Transportation, Sonderforschungsbereich 255 "Transatmosphärische Flugsysteme", Technische Universität München, Proceedings, Band 2, 1993, pp. E-1–E-5.

Kochendörfer, R., Lützenburger, N.: Applications of CMCs Made via the Liquid Silicon Infiltration Technique. Proc. 4th Int. Conf. on High Temperature Ceramic Matrix Composites (HT-CMC 4), München, Okt. 1–3, 2001, High Temperature Ceramic Matrix Composites, pp. 277–287, Wiley-VCH/Weinheim, 2001.

Kochendörfer, R.: Ceramic Matrix Composites – from Space to Earth: The Move from Prototype to Serial Production. International Conference on Advanced Ceramics and Composites, Cocoa Beach, Florida, USA, January 21–26, 2001, Ceram. Eng. Sci. Proc. 22 (3), pp. 11–22, 2001.

Kochendörfer, R.: Ermittlung der Werkstoffqualität über ZfP-Methoden an Faserkeramik-Materialien (C/C-SiC) mit und ohne Schäden. SFB & Industrie-Kolloquium "Schäden in Werkstoffen; Ursachen – Erkennung – Vorhersage", DLR-Stuttgart, 20.–22. Oktober 1999, Universität Stuttgart, Sonderforschungsbereich 381, 1999.

Kochendörfer, R.: Extremely High Temperatures Stable Composites, Int. Conf on Wood and Wood Fiber Composites. Stuttgart, 13.–15. April 2000, FMPA Stuttgart, Proc. of Int. Conf. on Wood and Wood Fiber Composites, pp. 453–462, 2000.

Kochendörfer, R.: Low Cost Processing of C/C-SiC Composites by means of Liquid Silicon. Proc 3rd Int. Conf. on High Temperature Ceramic Matrix Composites (HT-CMC 3), Osaka/Japan, Sept. 6–9, 1998, The Ceramic Society of Japan. High Temperature Ceramic Matrix Composites III, pp. 451–456, Trans Tech Pub. LTD/CH-GE-UK-USA, 1998.

Kolitsch, U., Ijevskii, V. A., Seifert, H. J., Wiedmann, I., Aldinger, F.: Formation and General Characterisation of a Previously Unknown Ytterbium Silicate (A-Type Yb2SiO5). J. Mater. Sci., p. 6135, 1997.

König, M., Krüger, R., Rinderknecht, S.: Finite element analysis of delamination growth in a multidirectional composite ENF specimen. Composite Structures: Theory and Practice, ASTMSTP 1383, 345–365, 2000.

Konrath, R., Schröder, W.: Telecentric Lenses for Imaging in Particle-Image Velocimetry: A New Stereoscopic Approach. Experiments in Fluids, Vol. 33, pp. 703–708, 2002.

Kopp, S., Hollmeier, S., Rick, H., Herrmann, O.: Airbreathing Hypersonic Propulsion System Integration within FESTIP FSSC-12. 9th International Space Planes and Hypersonic Systems and Technologies Conference, Norfolk, USA, AIAA 99-4813, 1999.

Kopp, S., Hollmeier, S., Rick, H., Herrmann, O.: Hypersonic Airbreathing Propulsion System and Aircraft Integration within FESTIP. 7th European Propulsion Forum – Aspects of Engine/Airframe Integration, Pau, France, 10.–12. März 1999.

Kopp, S., Hollmeier, S., Rick, H.: Zur Leistungssynthese von Hyperschallantrieben. Seminar des Sonderforschungsbereichs 255, München, 12. Dez. 1996, Technische Universität München, Tagungsband, pp. 43–56, 1997.

Kopp, S., Hollmeier, S.: Simulation des Betriebsverhaltens von Antrieben für Hochleistungsflugzeuge. Forum der Luft- und Raumfahrttechnik, Technische Universität München, 27. Nov. 1997.

Kopp, S.: Simulationsmodelle zur Regelung von Antriebssystemen für Hochgeschwindigkeitsflugzeuge. 2. Workshop Hyperschallantriebe der Sonderforschungsbereiche 253, 255, 259 und der DLR, Stuttgart, 20. Nov. 1997.

Körber, S., Ballmann, J.: Mechanisms and acoustics of Blade-Vortex-Interactions. ZFW, Vol. 19, Nr. 6, 1995.

Kornmann, R., Eberle, K., Dogigli, M.: Measurement of Ceramic matrix composite tubes at temperatures up to 1400°C. Journal of High Temperatures – High Pressures, volume 32, pp. 621–628, 2000.

Kornmann, R., Kröplin, B.: Objektrasterverfahren zur räumlichen Verformungsmessung an Hochtemperaturbauteilen. tm-Technisches Messen 67, Oldenbourg Verlag, pp. 267–273, 2000.

Koschel, W., Rick, W., Bikker, S.: Application of Finite Element Method to Hypersonic Nozzle Flow Computation. 77th AGARD PEP Symposium on CFD Techniques for Propulsion Applications, 27.–31. May 1991, San Antonio, AGARD CP-510, pp. 33-1/33-14, 1991.

Koschel, W., Rick, W.: Design Considerations for Nozzles of Hypersonic Airbreathing Propulsion. Third International Airspace Planes Conference, 3.–5. Dec. 1991, Orlando, Florida, AIAA-91-5019, 11 p., 1991.

Koschel, W.: Airbreathing Propulsion Technology. Space Course Aachen 1991, Feb. 18–Mar. 8, Aachen, Proceedings Vol. I, pp. 16-1/16-30, 1991.

Koschel, W. W., Link, Th., Fox, U.: Computation of Hypersonic Propulsion System Expansion Flows Considering Hydrogen/Air High Temperature Effects. AIAA Paper 98-1601, 8th International Space Planes and Hypersonic Systems and Technologies Conference, Norfolk, Virginia, USA, April 27–30, 1998.

Koschel, W. W.: Basic Hypersonic Propulsion Research in German Universities. AIAA Paper 98-1634, 8th International Space Planes and Hypersonic Systems and Technologies Conference, Norfolk, Virginia, USA, April 27–30, 1998.

Koschel, W. W.: Design and Analysis of Scramjet Components. VKI LS 1998-01, High Speed Propulsion, ISSN0377-8312, Sept. 1998.

Koslik, B., Breitner, M. H.: An Optimal Control Problem in Economics with Four Linear Controls. Journal of Optimization Theory and Applications, Vol. 94, Nr. 3, 1997, pp. 619–634.

Kostyukova, O., Schmidt, W. H.: Convexification of a simplified model of airplane dynamics. Workshop des SFB 255, Greifswald, 23./24. Oktober 1997, Tagungsband, TU München, pp. 91–98, 1998.

Krapez, J.-C., Spagnolo, L., Frieß, M., Maier, H.-P., Neuer, G.: Profile inversion of principal diffusivities through the use of a spatially modulated heating and a Fourier analysis. Journée d'Etudes de la SFT, Paris, 2003.

Krause, E., Abstiens, R., Fühling, S., Vetlutsky, V. N.: Boundary Layer Investigations on a Model of the ELAC 1 Configuration at High Reynolds Numbers in the DNW. European Journal of Mechanics-B/Fluids, Vol. 19 (2000), pp. 745–764.

Krause, E., Limberg, W., Kharitonov, A. M., Brodetzky, M. D., Henze, A.: An experimental investigation of the ELAC 1 configuration at supersonic speeds. Exp. in Fluids, Vol. 26 (1999), pp. 423–436, Springer Verlag 1999.

Kreichgauer, O.: Systemtechnische Methodik zur Sicherheitsanalyse von Luftverkehrsszenarien. – In: Deutscher Luft- und Raumfahrtkongress, DGLR 92-03-091, Bremen, 29. Sept. – 2. Okt. 1992.

Kreim, H., Kugelmann, B., Pesch, H. J., Breitner, M.: Minimizing the maximum heating of a re-entering space shuttle: an optimal control problem with multiple control constraints. Optimal Control Applications & Methods, Vol. 17, 1996, pp. 45–69.

Kreiner, A., Kopp, S., Herrmann, O., Hollmeier, S., Rick, H.: Integration von Kombinationsantrieben in Hyperschallflugzeuge der FESTIP-Studie. Seminar des Sonderforschungsbereichs 255, München, 3. Februar 2000, Technische Universität München, Tagungsband, pp. 33–46, ISBN 3-89791-175-2, 1999.

Kreiner, A., Kopp, S., Preiss, A., Erhard, W., Rick, H.: Advances in Turbo-Engine Real-Time Simulation for Modern Control System Development. Proceedings of the 22nd In-

ternational Congress of Aeronatical Sciences, Harrogate, U.K., 27 August–1. September 2000.
Kreiner, A., Lietzau, K., Gabler, R., Rick, H.: Modellbasierte Regelungskonzepte für Turbo-Luftstrahl-Triebwerke. DGLR-Jahrestagung, Leipzig, 2000.
Kreins, A.F., Henckels, A., Maurer, F.: Experimental studies of hypersonis shock induced boundary layer separation. ZFW, Vol. 20, Nr. 2, 1996.
Krenkel, W., Lützenburger, N.: Near Net Shape Manufacture of CMC Components. Proceedings of the 12th Int. Conference on Composite Materials (ICCM-12), Paris, 5.–7., 1999.
Krenkel, W., Nedele, M.R.: Novel Concept of a High Temperature Heat Exchanger, Innovative materials in advanced energy technologies: proceedings of Topical Symposium VII – "Innovative Materials in Advanced Energy Technologies" of the Forum on New Materials of the 9th CIMTEC-World Ceramics Congress and Forum on New Materials, Florence, Italy, June 14–19, 1998/ed. by P. Vincenzini Faenza: Techna, 1999 XVIII, pp. 289–296, 1999.
Krenkel, W., Renz, R.: C/C-SiC Components for High Performance Application. ECCM-8, Neapel, 3.–6.Juni 1998, Proc. of ECCM 8, Woodhead Pub., (ed. Visconti, C.) Vol 4, pp. 23–29, 1998.
Krenkel, W.: Cost Effective Processing of CMC Composites by Melt Infiltration (LSI-Process). International Conference on Advanced Ceramics and Composites, Cocoa Beach, Florida, USA, January 21–26, 2001, Ceram. Eng. Sci. Proc. 22 (3), pp. 443–454, 2001.
Krenkel, W.: Faserverstärkte Keramiken für Bremsenanwendungen. Materialwissenschaft und Werkstofftechnik 31, 655–660, 2000.
Kröplin, B., Fink, A., Weihe, S.: Heiße tragende Strukturen – Bauweisen und Werkstoffe. DGLR-Jahrbuch, Göttingen, 1994.
Kugelmann, B., Pesch, H.J.: Real-Time Computation of Feedback Controls with Applications in Aerospace Engineering. Proc. of the AIAA Guidance, Navigation and Control Conference, New Orleans, Aug. 12–14, 1991. Washington, pp. 537–542, 1991.
Kugelmann, B., Pesch, H.J.: Serielle und parallele Algorithmen zur Korrektur optimaler Flugbahnen in Echtzeit. DGLR-Jahrestag., Friedrichshafen; DGLR-90-136, 1.–4. Okt. 1990, DGLR-Jb. 1990, I, pp. 233–241.
Kugelmann, B., Pesch, H.J.: Controllability Investigations of a Two-Stage-to-Orbit Vehicle. In: Optimal Control, Calculus of Variations, Optimal Control Theory and Numerical Methods. Eds.: R. Bulirsch et al. Basel: Birkhäuser, 1993, pp. 327–339. (ISNM 111) [SFB 3, 1991]
Kugelmann, B., Pesch, H.J.: A New General Guidance Method in Constrained Optimal Control, Part 1: The Numerical Method, Part 2: Application to Space Shuttle Guidance. Jota, Vol. 67, 1990, pp. 421–435, pp. 437–446.
Kugelmann, B.: Färbealgorithmen zur Darstellung der Erdkugel. – In: H. Jürgens, D. Saupe (Hrsg.), Visualisierung in Mathematik und Naturwissenschaften, Bremer Computergraphik-Tage 1988, Bremen, 4.–6. Juli 1988, Berlin: Springer Verlag, 75–86, 1989.
Kugelmann, B.: Feedback Control in Space Dynamics. Proc. of the Int. Symp. Mécanique Spatiale – Space Dynamics, Toulouse, Nov. 6–10, 1989. Centre National d'Etudes Spatiales. Toulouse: Cepadues Edition, pp. 737–747, 1990.
Kugelmann, B.: Minimizing the Noise of an Aircraft during Landing Approach – In: Schmidt, W.H., Heier, K., Bittner, L., Bulirsch, R. (Eds.): Variational Calculus, Optimal Control and Applications. ISNM 124, Birkhäuser, Basel, 1998, pp. 271–280.
Kugelmann, B.: Optimal Guidance of Dynamic Systems. Proc. of the 14th IFIP Conf. on System Modelling and Optimization, Leipzig, July 3–7, 1989. Eds.: H.-J. Sebastian et. al. Berlin: Springer, pp. 332–341. (Lect. notes in contr. and inf. sci., 143), 1990.
Kugelmann, B.: Parallel computation in air traffic guidance. – In: Neunzert, H. (Ed): Progress in Industrial Mathematics at ECMI 94, John Wiley, Chichester, 1996, pp. 51–58.
Kugelmann, B.: Parallel computation of feedback controls in aerospace engineering. Z. Ang. Math. Mech., Vol. 76, Nr. 3, 1996, pp. 9–12.

Kugelmann, B.: Performance of a feedback method with respect to changes in the air-density during the ascent of a two-stage to orbit vehicle. – In: Bulirsch, R. et al. (Eds.): Computational Optimal Control, Birkhäuser, Basel, ISNM 115, 1994, pp. 329–338.

Kussmaul, K., Föhl, J., Maile, K.: Qualifizierung von Keramik für hochbeanspruchte Bauteile. Ingenieur-Werkstoffe 3, pp. 64 ff., 1991.

Lachner, R., Breitner, M.: Steuerung von Luft- und Raumfahrzeugen mit Differentialspielen. Proc. des Workshops Optimalsteuerungsprobleme von Hyperschall-Flugsystemen, Greifswald, 14.–15. Nov. 1994, pp. 35–44.

Lachner, R., Breitner, M. H., Pesch, H. J.: Three-dimensional air combat: Numerical solution of a complex differential game. Proc. of the Sixth Inter. Sympos. on Dynamic Games and Applications, St-Jovite, 1994. Eds.: M. Breton et al. Montreal: GERAD, école des Hautes Études Commerciales, 1994, pp. 287–296.

Lachner, R., Hönig, R., Hoßfeld, H.-Ch., Magrini, G. L., Kappler, G.: One-dimensional model of the chemically reacting flow in a supersonic combustion chamber, ZFW, Bd. 19, pp. 34–40, 1995.

Lachner, R., Breitner, M. H., Pesch H. J.: Optimal Strategies of a Complex Pursuit-Evasion Game. In: Fritsch, R., Toepell, M., Proceedings of the 2. Gauß-Symposium in München, August 2–8. Ludwig-Maximillians-Universität München, 1993; J. of Computing and Information 4, pp. 87–110, 1994.

Lang, N., Limberg, W.: Construction of three-dimensional flow structure out of two-dimensional steady flow field velocity measurements. Exp. in Fluids, Vol. 27 (1999), pp. 351–358, Springer Verlag 1999.

Lang, N., Schröder, W.: Experimentelle Untersuchung der Leeseitenwirbel in Unterschallströmungen: Strömungssichtbarmachung und Geschwindigkeitsmessung mir vapour-Screen und PIV. GALA e.V., Lasermethoden in der Strömungsmeßtechnik, München-Freising, September 2000, Shaker-Verlag Aachen.

Lang, N., Schröder, W.: Experimentelle Untersuchung der Hyperschallkonfiguration ELAC mit berührungslosen Meßverfahren. DGLR-Jahrestagung, Leipzig, September 2000, DGLR-Jahrbuch, DGLR-JT2000-208.

Laux, T., Killinger, A., Auweter-Kurtz, M., Gadow, R., Wilhelmi, H.: Functionally Graded Ceramic Materials for High Temperature Applications for Space Planes. In: Materials Science Forum, Vol. 308–311, Hrsg. W. A. Kayser, Trans Tech Publications Inc., Uetikon-Zürich, Schweiz, 1999.

Laux, T., Ullmann, T., Auweter-Kurtz, M., Hald, H., Kurz, A.: Investigation of thermal protection materials along the X-38 re-entry trajectory by plasma wind tunnel simulations. Atmospheric Reentry Vehicles and Systems, Proc. of the 6th Int. Conf. on protection of Materials and Structures from the LEO, Arcachon, France, 2001.

Le Ribault, C., Friedrich, R.: Investigation of transport equations for turbulent heat fluxes in compressible flows. Int. J. Heat Mass Transfer, Vol. 40, 1997, pp. 2721–2738.

Lechner, R., Sesterhenn, J., Friedrich, R.: Effects of compressibility and fluid properties in turbulent supersonic channel flow. – Proc. 2nd Int. Symp. Turbulence and Shear Flow Phenomena, Stockholm, 27–29 June 2001, 2001.

Lechner, R., Sesterhenn, J., Friedrich, R.: Turbulent supersonic channel flow. Journal of Turbulence, Vol. 2, pp. 001, 2001.

Leinhos, D., Höss. B., Fottner, L.: Rotating Stall Inception with Inlet Distortion in the Low Pressure Compressor of a Turbofan Engine. Proceedings of the 19th International Symp. Aircraft Integrated Monitoring Systems, Garmisch Partenkirchen, 1998.

Leinhos, D. C., Schmid, N. R., Fottner, L.: The Influence of Transient Inlet Distortions on the Instability Inception of a Low Pressure Compressor in a Turbofan Engine. ASME Journal of Turbomachinery, Vol. 123, 2000, pp. 1–8.

Lentz, S., Hornung, M., Staudacher, W.: Konzeptuntersuchungen zum HTSM. Seminar des Sonderforschungsbereichs 255, 12. Dezember 2002, TU München, ISBN 3-89791-334-8, S. 133–149, 2004.

Lentz, S., Staudacher, W.: Application of an Air Collection and Enrichment System aboard a TSTO Space Transport, Notes on Numerical Fluid Mechanics, Vol. 87, Springer Verlag, 2003.

Lentz, S., Staudacher, W.: Entwurf eines wiederverwendbaren zweistufigen Raumtransportsystems mit einer Anlage zur Sauerstoffgewinnung während des Fluges, DGLR Jahrestagung 2002 Stuttgart 23.09.–26.09.2002, DGLR-JT2002-180, 2002.

Lentz, S., Staudacher, W.: Reusable TSTO Launch vehicles using in-flight LOX-Collection, AIAA 2003-PP8095, International Air & Space Symposium and Exposition "The Next 100 Years" 14.–17. 07. 2003, Dayton/Ohio, 2003.

Lentz, S., Staudacher, W.: Reusable TSTO Space Transport Systems using in-flight LOX-Collection, RTO/AVT Symposium, 06.–10. 10. 2003, Warschau, Polen, 2003.

Lenzner, S., Auweter-Kurtz, M., Heiermann, J., Sleziona, P. C.: Energy Partitions in Inductively Heated Plasma Sources for Reentry Simulations. Journal of Thermophysics and Heat Transfer, Vol. 14, No. 3, July–September 2000, pp. 388–395, 2003.

Leyland, P., Favre, J., Merazzi, S., Henze, A., Olivier, H.: ELAC: A High Speed Transportation System. Experiments, Calculations and Visualisation. West East High Speed Flow Fields, CIMNE, Barcelona, Spanien, 2002.

Liang, J.-J., Navrotsky, A., Ludwig, T., Seifert, H. J., Aldinger, F.: Enthalpy of Formation of Rare Earth Silicates Y2SiO5 and Yb2SiO5 and N-containing silicate Y10(SiO4)6)N2. J. Mater. Res., 14(4) p. 1181, 1999.

Lietzau, K., Kreiner, A.: Model Based Control for Jet Engines. ASME Turbo Expo, New Orleans, 2001.

Limberg, W., Stromberg, A.: Pressure measurements at supersonic speeds on the research configuration ELAC I. ZFW, Vol. 17 (1993), pp. 82–89.

Link, T., Koschel, W. W.: Computation of a Nonequilibrium Expansion Flow in a SERN-type Nozzle. AIAA Journal of Propulsion and Power, Vol. 17, No. 5, September–October 2001.

Link, T., Koschel, W. W.: Computation of the Expansion Flow of a Hypersonic Propulsion System. AIAA Paper 01-1866, 10$^{th}$ International Space Planes and Hypersonic Systems and Technologies Conference, Kyoto, Japan, April 2001.

Link, T., Koschel, W. W.: Computation of the Two-Dimensional Flow in SERN Nozzles. Proceedings des 12. DGLR-Fachsymposium der AG-STAB, Univ. Stuttgart, Nov. 2000, reviewed and accepted to be published in Notes on Numerical Fluid Mechanics, Vieweg Verlag.

Loesener, O., Auweter-Kurtz, M., Hartling, M., Messerschmid, E.: Linear Pyrometer for Investigations of Thermal Protection Systems. Journal of Thermophysics and Heat Transfer, Vol. 7, No. 1, pp. 82–87, Jan.–Mar, 1993.

Loesener, O., Messerschmid, E.: Qualifikation eines Pyrometers für Raumfahrtanwendungen. Zeitschrift für Flugwissenschaften und Weltraumforschung, ZFW 19, Heft 3, 1995.

Loesener, O., Neuer, G.: A new far-infrared pyrometer for radiation temperature measurement on semitransparent and absorbing materials in an arc heated wind tunnel. Measurement 14, 125–134, 1994.

Löfberg, J.: Robust $H_\infty$- and $\mu$-Synthesis Control for Lateral Dynamics of a Hypersonic Test Vehicle. Master Thesis, Universität Linköping, 1998.

Lösch, M., Hillesheimer, M., Hornik, A., Krysta, D.: ARIANE X – logische Schritte zu einem wiederverwendbaren Trägersystem. DGLR-Jahrbuch Band III, pp. 1405–1417, 1990.

Ludwig, A., Quested, P., Neuer, G.: How to find thermophysical material property data for casting simulations. Advanced Engineering Materials 3, no.1–2, pp. 11–14, 2001.

Lyubar, A., Sander, T., Sattelmayer, T.: Einfluss des Vorheizerausbrands auf die Flammenstabilisierung in einer Überschallbrennkammer, Seminar des Sonderforschungsbereichs 255, 2002, TU München, ISBN 3-89791-334-8, S. 47–59, 2004.

Lyubar, A., Sander, T., Sattelmayer, T.: Novel Two Stage Injector for Hydrogen Flame Stabilization in Supersonic Airflow, 11th AIAA/AAAF International Conference, Space Planes and Hypersonic Systems and Technologies, 2002, Orléans, France.

Lyubar, A., Sander, T., Sattelmayer, T.: Numerical Investigation of Fuel Mixing, Ignition and Flame Stabilization by a Strut Injector in a Scramjet Combustor, Proceedings of the 11th International Conference on Methods of Aerophysical Research (ICMAR), Volume 2, pp. 122–127, 3.–7. July 2002, Akademgorodok, Novosibirsk – Denisova Cave, Altai region, Russia.

Lyubar, A., Sattelmayer, T.: Numerische Voruntersuchung zu einer reagierenden Überschallströmung in einer Modellbrennkammer. Seminar des Sonderforschungsbereichs 255, München, 3. Februar 2000, Tagungsband ISBN 3-89791-175-2, TU München, pp. 65–72, 2001.

Maier H.-P., Maile K.: Deformation and Failure Behaviour of thermomechanically loaded C/C-SiC. High Temperatures – High Pressures, 1999, volume 31, pages 337–345, 1999.

Maier, H.-P., Maile K.: Beschreibung des Festigkeits- und Versagensverhaltens von Faserkeramik (C/C-SiC) unter komplexer mechanischer und korrosiver Beanspruchung im Bereich T >1650°K. Werkstoffwoche München 1998; Editor: Heinrich, Günter, Ziegler, Hermel, Riedel; Band VII; Symposium 9 Keramik, Seite 601–606, WILEY-VCH Verlag GmbH, Weinheim; ISBN 3-527-29944-0, 2003.

Manhart, M., Friedrich, R., Deng, G., Piquet, J.: Direct versus statistical simulation of accelerated/retarded and separating/reattaching turbulent boundary layers. – In: Notes on Numerical Fluid Mechanics and Multidisciplinary Design, Vol. 82, pp. 244–260, 2003.

Manhart, M., Friedrich, R.: DNS of a turbulent boundary layer with separation. Int. J. Heat and Fluid Flow, Vol. 23, pp. 572–581, 2002.

Mathew, J., Lechner, R., Foysi, H., Sesterhenn, J., Friedrich, R.: An explicit filtering method for LES of compressible flows. Phys. Fluids, Vol. 15, pp. 2279–2289, 2003.

Maurer, H., Pesch, H.J.: Solution Differentiability for Nonlinear Parametric Control Problems. SIAM J. Control and Optimization, 32, 6, pp. 1542–1554, 1994.

Maurer, H., Pesch, H.J.: Solution differentiability for parametric nonlinear control problems with control-state constraints. Control and cybernetics 23 (1994) 1/2, pp. 201–227.

Maurer, H., Pesch, H.J.: Solution differentiability for parametric nonlinear control problems with inequality constraints. In: System modelling and optimization. Eds.: J. Henry et al. London: Springer, 1994, pp. 437–446. (LNCIS 197).

Mayinger F.: Image-forming optical techniques in heat transfer: Revival by computer-aided data processing. – In: Journal of Heat Transfer, vol. 115 ( Nov. 1993), 824–834.

Mayinger, F. (Ed.): Optical Measurement Techniques; Springer, New York, 1994.

Mayinger, F., Gabler, W., Hönig, R., Kappler, G.: Spectroscopic Techniques for Ram-Combustors; 2nd. Space Course on Low Earth Orbit Transportation 1993, Munich, FRG , pp. 20.1–20.60, 1993.

Mayinger, F., Jordan, M., Ardey, N.: Slow Turbulent Combustion and Jet Formation in a Tube. Proceedings of the 10th International Symposium on Transport Phenomena, 30. Nov.–3. Dez., 1997, Kyoto, Japan, 1997.

Mayinger, F.: Modern electronics in image-processing and physical modelling – A new challenge for optical techniques. Invited keynote lecture, Proc. of the 10th. Int. Heat Transfer Conference, Brighton, UK., Aug. 14–18, 1994. Eds.: G. F. Hewitt et al. Vol. 1, p. 61–79.

Mayrhofer, M., Sachs, G.: Reichweitenerhöhung durch Luftbetankung eines Hyperschall-Flugsystems – Dual-Fuel-Konzept. Zeitschrift für Angewandte Mathematik und Mechanik, Vol. 79, Supplement 3, 1999, pp. 965–966.

Mayrhofer, M. Sachs, G.: Steigerung der Missionssicherheit von Hyperschall-Flugsystemen durch Optimalsteuerung von Notflugbahnen. DGLR-Jahrbuch 1999, Band II, 1999, pp. 1101–1108.

Mayrhofer, M., Dinkelmann, M., Sachs, G.: Optimal Three-Dimensional Range Cruise of a Dual-Fuel Hypersonic Vehicle. Proceedings of the 21st ICAS Congress, Melbourne, Australien, 13.–18. September, 1998, ICAS 98-1,3,1, 1998.

Mayrhofer, M., Sachs, G.: A Contribution to Mission Safety for a Two-Stage Hypersonic Vehicle. AIAA 9th International Space Planes and Hypersonic Systems and Technologies Conference, AIAA 99-4886, Norfolk, VA, USA, November 1999.

Mayrhofer, M., Sachs, G.: Abort Trajectory Range Increase by Optimal Fuel Draining. Workshop des Sonderforschungsbereichs 255 "Transatmosphärische Flugsysteme" und der Universität Greifswald, Greifswald, 7.–8. Oktober 1999, Tagungsband, TU München, ISBN 3-89791-147-7, 2000, pp. 115–125.

Mayrhofer, M., Sachs, G.: Aspekte der Flugbahn-Optimierung für zweistufige Raumtransportsysteme. Seminar des SFB 255, München, 2002, Tagungsband ISBN 3-89791-334-8, TU München, S. 89–102, 2004.
Mayrhofer, M., Sachs, G.: Mission Safety Concept for a Two-Stage Space Transportation Vehicle. Proceedings of the 10th International Space Planes and Hypersonic Systems and Technologies Conference, Kyoto, Japan, 24.–27. April 2001, AIAA Paper Nr. 2001-1789, 2001.
Mayrhofer, M., Sachs, G.: Mission Safety Improvement: A Key Factor for the Success of Next Generation Launchers. AIAA/AAAF 11th International Space Planes and Hypersonic Systems and Technologies Conference, AIAA 2002-5255, Orleans, France, October 2002.
Mayrhofer, M., Sachs, G.: Notflugbahnen eines zweistufigen Hyperschall-Flugsystems ausgehend vom Trennmanöver. Seminar des Sonderforschungsbereichs 255, München, 12. Dez. 1996, Tagungsband, Technische Universität München, 1997, pp. 109–118.
Mayrhofer, M., Sachs, G.: Notflugbahnen für einen Triebwerksausfall beim Orbitalstufen-Aufstieg. Seminar des Sonderforschungsbereichs 255, München, 3. Februar 2000, Tagungsband ISBN 3-3-89791-175-2, TU München, 2001, pp. 103–115.
Mayrhofer, M., Sachs, G.: Optimal Abort Trajectory Control for a Winged Orbital Stage. AIAA Atmospheric Flight Mechanics Conference Proceedings, Boston, USA, 10.–12. Aug. 1998, AIAA-98-4150, 1998.
Mayrhofer, M., Sachs, G.: Optimal Mission Aborts of a Winged Orbital Stage. Journal of the Brazilian Society of Mechanical Sciences, Vol. XXI (Special Issue: Proceedings of the 14th International Symposium on Space Flight Dynamics – ISSFD XIV, 1999), pp. 250–259.
Mayrhofer, M., Sachs, G.: Optimal Three-Dimensional Range Cruise of a Dual-Fuel Hypersonic Vehicle. 21st Congress of the International Council of the Aeronautical Sciences, ICAS 98-1.9.1, Melbourne, Australien, September 1998.
Mayrhofer, M., Sachs, G.: Optimale Notflugbahnen der Orbitalstufe nach dem Trennmanöver. Workshop "Optimalsteuerungsprobleme von Hyperschall-Flugsystemen" des Sonderforschungsbereichs 255, Greifswald, 23.–24. Okt. 1997, Tagungsband, Technische Universität München, 1998, pp. 63–72.
Mayrhofer, M., Sachs, G.: Orbital Stage Abort Trajectories after Separation from Carrier. – In: Proceedings of the 12th International Symposium on Space Flight Dynamics, Darmstadt, 2.– 6. Juni 1997, pp. 463–469.
Mayrhofer, M., Sachs, G.: Range Increase in Gliding Flight by Optimal Jettisoning Control. Proceedings of the 2nd International Conference Conference on Unconventional Flight, Juni 2000.
Mayrhofer, M., Sachs, G.: Reichweitenerhöhung durch Luftbetankung eines Hyperschall-Flugsystems – Dual-Fuel-Konzept. Vortrag auf der GAMM-Jahrestagung 1998, Bremen, 6.–9. April 1998.
Mayrhofer, M., Sachs, G.: Safety Issues for a Carrier Stage of a Lifting Two-Stage Space Transportation System. AIAA Atmospheric Flight Mechanics Conference, AIAA 2001-4068, Montreal, Canada, August 2001.
Mayrhofer, M., Wächter, M., Sachs, G.: Branched Trajectory Optimization Problems of Mission Abort Scenarios. Workshop Optimal Control, University of Greifswald, Germany, ISBN 3-89791-316-X, Hieronymus Verlag München, pp. 61–70, October 2002.
Mayrhofer, M., Wächter, M., Sachs, G.: Flight Safety Improvement: A Basic Issue for the Success of Future Launchers. 4th Asian Control Conference, Singapore, ISBN 981-04-6440-1, Published by ASCC, September 2002.
Mayrhofer, M.: Notflugbahnen eines zweistufigen Hyperschall-Flugsystem. DGLR-Jahrestagung 1997, München, 14.– 17. Okt. 1997, DGLR-Jahrbuch II, 1997, pp. 1081–1088.
Mehlhorn, R., Dinkelmann, M., Sachs, G.: Application of Automatic Differentiation to Optimal Control Problems. Control Applications of Optimization, International Series of Numerical Mathematics, Band 115, Birkhäuser, Basel, Boston, Stuttgart, pp. 255–267, 1994.
Mehlhorn, R., Schumann, M., Kiehl, M.: Parallelisierung der Mehrzielmethode und Implementierung auf einem iPSC-Hypercube mit Anwendungen in der Flugbahnoptimierung. Z. Ang. Math. Mech., Vol. 75, Nr. 2, 1995, pp. 599–600.

Mehlhorn, R.: Integrierte Arbeitsumgebung zur numerischen Berechnung von Problemen der optimalen Steuerung. Herbert Utz Verlag, München, 1996.
Messerschmid, E., Fasoulas, S., Gogel, T.H., Grau, T.: Numerical Modeling of Plasma Wind Tunnel Flows. Zeitschrift für Flugwissenschaften und Weltraumforschung, ZFW 19, Heft 3, 158–165, Springer Verlag Berlin, 1995.
Messerschmid, E., Weigand, A.: How to Enhance Safety for Future Space Transportation Systems. In: Space Technology, Vol. 13, No. 3, pp. 329–347, 1993.
Messerschmid, E.: A European Perspective: International Cooperation for Future Space Science Missions. Space Times, Vol. 34, No. 3, 1995.
Messerschmid, E.: Support of the University of Stuttgart for Space Science and Technology. Special Issues des Journal of the British Interplanetary Society, London, Vol. 47, pp. 414–426, 1994.
Messerschmid, E.: The Role of man in Space, Keynote Speech 9th IAA Man in Space Symposium, June 17–21, Köln, Zeitschrift für Flugwissenschaften und Weltraumforschung ZFW 19, 1–7, 1992.
Miesbach, S., Pesch, H.J.: Symplectic Phase Flow Approximation for the Numerical Integration of Canonical Systems. Numer. Math. 61 (1992), pp. 501–521.
Möbus, H., Gerlinger, P., Brüggemann, D.: Comparison of Eularian and Lagrangian Monte Carlo PDF Methods for Turbulent Diffusion Flames. Combustion and Flame, 124, pp. 519–534, 2001.
Möbus, H., Gerlinger, P., Brüggemann, D.: Efficient Methods for Particle Temperature Calculation in Monte Carlo PDF. Proceedings of the 4th ECCOMAS Conference, 1, pp. 163–168, John Wiley & Sons, 1998.
Möbus, H., Gerlinger, P., Brüggemann, D.: Scalar and Joint Scalar-Velocity-Frequency Monte-Carlo PDF Simulation of Supersonic. Combustion and Flame, 132, pp. 3–24, 2003.
Moelyadi, M.A., Breitsamter, C., Laschka, B.: Aerodynamic Investigation on ASTRA System Concept 2. Seminar des Sonderforschungsbereichs 255, 12. Dezember 2002, TU München, ISBN 3-89791-334-8, S. 1-20, 2004.
Moravszki, C., Sachs, G.: Hypersonic simulation tests for altitude control with 3D-tunnel/predictor display. Proceedings of the 3rd UNCONF, Budapest, Sept. 2001.
Moravszki, Cs., Sachs, G.: Predictive Head Up Display at High Speed for Altitude Control. 7th Mini Conference on Vehicle System Dynamics, Identification and Anomalies VSDIA2000, Budapest, Ungarn, 6.–8. November 2000.
Morikawa, K., Grönig, H.: Formation and structure of vortex systems around a translating and oscillating airfoil. ZFW, Vol. 19, Nr. 6, 1995.
Moser, L.K., Hindelang, F.J.: Application of diode laser spectroscopy on the measurement of boundary layer-induced temperature changes in shock tubes. Proceedings of the 3rd International Symposium on Monitoring of Gaseous Pollutants by Tunable Diode Lasers, Freiburg, Germany, 1991.
Moser, L.K., Hindelang, F.J.: Boundary layer effect on thermal NO decomposition behind incident shock waves. Shock Waves, accepted for publication.
Moser, L.K., Hindelang, F.J.: Shock-tube study of the vibrational relaxation of nitric oxide. Proceedings of the 17th International Symposium on Shock Waves and Shock Tubes, Bethlehem PA, USA, 1989.
Moser, L.K., Hindelang, F.J.: Vibrational relaxation of NO behind shock waves. Experiments in Fluids, 7, 67, 1989.
Müller J., Olejak D., Reisinger D., and Wagner S.: An investigation on the quality of the flow in the empty test section of TWM, Part II. 79th Semi-Annual Meeting of the Supersonic Tunnel Association, Arlington, TX, USA, Mar. 1993.
Müller, C.A., Ballmann, J.: Flow computation for the hypersonic configuration ELAC I at low speeds and large incidence. ZFW, Vol. 17 (1993), pp. 108–115.
Müller, J., Mümmler, R., Staudacher, W.: Comparison of some Measurement Techniques for Shock Induced Boundary Layer Separation. Aerospace Science and Technology, pp. 383–395, Volume 5, Issue 6, 2001.

Müller, J., Mümmler, R., Staudacher, W.: Separated Shock Boundary Layer Interaction. 92nd Semiannual STAI Meeting, Huntsville, Alabama, USA, 3.–5. Oktober 1999.

Müller, J., Mümmler, R., Staudacher, W.: Shock Wave/Boundary Layer Interaction. 22nd International Congress of Aeronautical Sciences, Harrogate, United Kingdom, 28. August–1. September 2000, ICA0351.1, 2000.

Müller, J., Mümmler, R., Staudacher, W.: Stoßinduzierte Ablösungen an Klappen im Überschall. Jahrestagung der DGLR, Bremen, 5.–8. Oktober, 1998.

Müller, J., Mümmler, R.: The New Control and Data Acquisition System for the Trisonic Windtunnel Munich. 93rd STAI Meeting, Sunnyvale, CA, USA, 30. April–2. Mai 2000.

Müller, J., Olejak, D., Staudacher, W.: Different Techniques to Enhance the Accuracy of LDA-Measurements. Proceedings of the 87th semi-annual meeting of the Supersonic Tunnel Association, Centre Paul Langevin, Aussois, Frankreich, Mai 1997.

Müller, J., Olejak, D., Staudacher, W.: Nonintrusive Untersuchungen an supersonischen Rezirkulationsgebieten. Seminar des Sonderforschungsbereichs 255, München, 12. Dez. 1996, Tagungsband, Technische Universität München, 1997, pp. 17–36.

Müller, J.E., Fottner, L.: Numerical and Experimental Investigation of the Flow Field inside the Intake Duct of a Combined Cycle Engine for Hypersonic Flight. Proceedings of the Symposium on Computational Fluid Dynamics in Aeropropulsion 1995, ASME Int. Mech. Eng. Congress & Exp., San Francisco, USA, 12.–17. Nov. 1995, AD – Vol. 49, 1995, pp. 157–168.

Müller, P.N., Reinsch, K.-D., Bulirsch, R.: On the complete solution of $\varepsilon y''=y^3$. J. of optim. theory and appl. 80 (1994) 2, pp. 367–372.

Müller, S., Dickopp, C., Ballmann, J., Jeltsch, R.: Computation of viscous hypersonic nonequilibrium blunt body flow. ZFW, Vol. 17 (1993), pp. 125–130.

Mümmler, R., Müller, J., Staudacher, W.: Non-Intrusive Investigation of Separated Flow at a 24 Degree Flat Plate/Ramp Konfiguration. 90th STAI-Meeting, Ottawa, Canada, 4.–6. Oktober 1998.

Neubacher, R., Henckels, A., Gawehn, T.M.: Experimental Investigation of a Hypersonic Inlet for the TSTO-Configuration ELAC. Proceedings of the 12[th] AG STAB/DGLR Symp., Stuttgart 2000, Notes on Numerical Fluid Mechanics, Springer Verlag, pp. 129–136.

Neuer, G., Brandt, R., Maglic, K.D. u.a.: Thermal diffusivity of the candidate standard reference material cordierite. High Temp. – High Press., 31, pp. 517–524, 1999.

Neuer, G., Fiessler, L., Groll, M., Schreiber, E.: Critical analysis of the different methods of multiwavelength pyrometry. In: Temperature, its Measurement and Control in Science and Industry, Vol. 6, Hrsg.: J.F. Schooley, American Institute of Physics, New York, p. 787–790, 1992.

Neuer, G., Jaroma-Weiland, G.: Spectral and total emissivity of high temperature materials. Int. J. Thermophys. 19, 917–929, 1998.

Neuer, G., Kochendörfer, R., Gern, F.: High temperature behaviour of the spectral and total emissivity of CMC-materials. High Temp. – High Press. 27/28 (1995/96), 183–189, 1995/1996.

Neuer, G., Schreiber, E.: New Techniques in Radiation Thermometry to eliminate the influence of Emissivity and Absorption in the optical path. In: E.W.P. Hahne et al. Hrsg. Proc. 3rd European Thermal Science Conference 2000, Edicioni ETS, Pisa, Italy, pp. 691–695, 2000.

Neuer, G.: Specific problems of temperature measurements and calibration in the range 1000 °C to 2500 °C. In: Ultra High Temperature Mechanical Testing, Hrsg. R.D. Lohr und M. Steen, Woodhead Publishing Limited, Cambridge, England, 81–96, 1995.

Neuer, G.: Spectral and total emissivity measurement of highly emitting materials. Int. J. Thermophys. 16, 257–265, 1995.

Neuer, G.: Total normal and spectral emittance of refractory materials for high temperature ovens. Thermochimica Acta, 218, 211–219, 1993.

Neuwerth, G., Peiter, U., Decker, F., Jacob, D.: Reynolds number effects on the low speed aerodynamics of the hypersonic configuration ELAC-1. AIAA Paper 98-1578, 8[th] Inter-

national Space Planes and Hypersonic Systems and Technologies Conference, Norfolk, Virginia, USA, April 27–30, 1998, Proceedings pp. 376–390.
Neuwerth, G., Peiter, U., Decker, F., Jacob, D.: Reynolds number effects on the low – speed aerodynamics of a hypersonic configuration. Journal of Spacecraft and Rockets, Vol. 36, No. 2, March–April 1999, pp. 265–272.
Niestroy, O., Gallus, H.: Two- and Three-Dimensional Flow Simulation in Hypersonic Inlet Diffusors Using an Explicit Time Stepping Scheme. AIAA-93-5044, AIAA/DGLR Fifth International Aerospace Planes and Hypersonics Technologies Conference, 30.11.–3.12.1993, Munich, Germany.
Nockemann, M.: Investigations of tip vortices with the triple hot-wire anemometry. ZFW, Vol. 19, Nr. 6, 1995.
O'Hare, P.A.G., Tomaszkiewicz, I., Seifert, H.J.: The standard molar enthalpies of formation of $\alpha$-$Si_3N_4$ and $\beta$-$Si_3N_4$ by combustion calorimetry in fluorine, and the enthalpy of the $\alpha$ to $\beta$ transition at the temperature 298.15 K, J. Mater. Res., 12, p. 3201, 1997.
Ohmenhäuser, F., Weihe, S., Kröplin, B.: Algorithmic Implements of a Generalized Cohesive Crack Model. Comp. Materials Science, 16, 294–306, 1999.
Olawsky, F., Infed, F., Auweter-Kurtz, M.: Preconditioned Newton-Method for Computing Supersonic and Hypersonic Nonequilibrium Flows. AIAA-Paper 2003-3702, 16th AIAA Computational Fluid Dynamics Conference, Orlando, Florida, USA, June 23–26, 2003.
Olejak D., Heiser W., Reisinger D., Wagner S.: Laser Doppler Anemometry and determination of particle size by a relaxation-length method at TWM. DANTEC Information 11, Skovlundem, Denmark, June 1992.
Olejak D., Müller J., Reisinger D., Wagner S.: An Investigation on the Quality of the Flow in the Empty Test Section of TWM, Part I. 78th Semi-Annual Meeting of the Supersonic Tunnel Association, Melville, Long Island, USA, Oct. 1992.
Olejak D., Müller J., Reisinger D., Wagner S.: Recent Advances of LDA-measurements in Shock Wave/Turbulent Boundary Layer Interactions. 81th Semi-Annual Meeting of the Supersonic Tunnel Association, Buffalo, NY, USA, Apr. 1994.
Olejak D., Müller J., Reisinger D., Staudacher W.: Nonintrusive Investigation in Separated Supersonic Flows. 83th Semi-Annual Meeting of the Supersonic Tunnel Association, Washington D.C., USA, April. 1995.
Olejak, D., Heiser, W., Reisinger, D., Wagner, S.: The Application of Laser Doppler Anemometry in a Trisonic Windtunnel. LASER ANEMOMETRY, Advances and Applications, edited by A. Dybbs, B. Ghorashi, Vol. 1, pp. 217–228, 1991.
Olivier, H., Habermann, M., Bleilebens, M.: Use of Shock Tunnels for Hypersonic Propulsion Testing. AIAA Paper 99-2447, 35$^{th}$ Joint Propulsion Conference, Los Angeles, USA, June 20–24, 1999.
Olivier, H., Schulte-Rödding, J.-H., Grönig, H., Vetlutzky, V.N.: Measurements with the ELAC 1 configuration at Mach 7. ZFW, Vol. 20, Nr. 2, 1996.
Olivier, H., Vetter, M., Grönig, H.: High Entalphy testing in the Aachen Shock Tunnel TH2. Proc. of the First European Symp. Aerodynamics for Space Vehicles, ESTEC, Netherlands, 1991.
Olivier, H.: A theoretical model for the shock stand-off distance in frozen and equilibrium flows. Journal of Fluid Mechanics (2000), Vol. 413, pp. 345–353.
Olivier, H.: Prediction of the shock stand-off distance in reacting flows. in: Ball, G. J., Hillier, R., Roberts, G. T. (eds.), Proceedings of the 22$^{nd}$ International Symposium on Shock Waves, London, 1999, Vol. II, pp. 1649–1652, 2000.
Pagella, A., Rist, U., Wagner, S.: Numerical investigations of a transitional flat-plate boundary layer with impinging shock waves. Proc. CEAS Aerospace Aerodynamics Research Conference, June 10–12, 2002, Cambridge, UK, Royal Aeronautical Society, London, CD-ROM 269, 2002.
Pagella, A., Rist, U., Wagner, S.: Numerical investigations of small-amplitude disturbances in a laminar boundary-layer with impinging shock waves. In: Wagner, S., Rist, U., Heinemann, H.-J., Hilbig, R. (eds.): New Results in Numerical and Experimental Fluid Dynamics III. Notes on Num. Fluid Mech. 77, reviewed articles of the 12. STAB/DGLR-Symposium, Stuttgart, Nov. 15–17, 2000, pp. 146–153, Springer, 2001.

Pagella, A., Rist, U., Wagner, S.: Numerical Investigations of Small-Amplitude Disturbances in a Laminar Boundary Layer with Impinging Shock Waves. Notes on Numerical Fluid Mechanics, Vol. 77, 'New Results in Numerical and Experimental Fluid Mechanics III', S. Wagner, U. Rist, J. Heinemann, R. Hilbig (eds.), pp. 153–160, Springer, 2002.

Pagella, A., Rist, U., Wagner, S.: Numerical Investigations of Small-Amplitude Disturbances in a Boundary Layer with Impinging Shock Wave at Ma=4.8. Phys. Fluids 14, 2088–2101 July, 2002

Pagella, A., Rist, U., Wagner, S.: Numerical Investigations of Small-Amplitude Disturbances in a Boundary Layer with Impinging Shock Wave at Ma=4.8: Compression Corner vs. Impinging Shock Wave, submitted to Phys. Fluids, 2003.

Pagella, A., Rist, U., Wagner, S.: Numerical Investigations of Transitional Flat-Plate Boundary Layers with Impinging Shock Waves – Non-Linear Cases. CEAS TRA Aerodynamics Research Conference, June 10–12, 2002 Cambridge, United Kingdom, 2002.

Pagella, A., Rist, U., Wagner, S.: Numerical Investigations of Transitional Flat-Plate Boundary Layers with Impinging Shock Waves. In: Zeitoun, D. E., Periaux, J., Desideri, J. A., Marini, M. (eds.): West East High Speed Flow Fields 2002. Proc. W.E. H.S.F.F. conference, Marseille, France, April 22–26, 2002, CIMNE (Barcelona, Spain), pp. 377–384, 2003.

Peiter, U., Neuwerth, G., Jacob, D.: Aerodynamik der Hyperschallkonfiguration ELAC 1 im Langsamflugbereich. Deutscher Luft- und Raumfahrtkongress 2000, DGLR-Jahrbuch 2000, Paper DGLR-JT 2000-020.

Pesch, H. J.: Optimal Control Problems under Disturbances. Proc. of the 14th IFIP Conf. on System Modelling and Optimization, Leipzig, July 3–7, 1989. Eds.: H.-J. Sebastian et al. Berlin: Springer, pp. 377–386 (Lect. notes in contr. and inf. sci., 143), 1990.

Pesch, H. J.: The Accessory Minimum Problem and its Importance for the Numerical Computation of Closed-Loop Controls. Proc. of the 29th IEEE Conf. on Decision and Control, Honolulu, Dec. 5–7, 1990. Ed.: IEEE Control System Society. New York: Publishing Services IEEE, Vol. 2, pp. 952–953, 1990.

Pesch, H. J., Bulirsch, R.: The Maximum Principle, Bellman's Equation and Carathéodory's Work. J. of optim. theory and appl. 80 (1994) 2, pp. 203–229.

Pesch, H. J.: Offline and Online Computation of Optimal Trajectories in the Aerospace Field. In: Applied mathematics in aerospace science and engineering. Eds.: A. Miele et al. New York: Plenum Press, 1994, pp. 165–219.

Pesch, H. J.: A practical guide to the solution of real-life optimal control problems. – Control and cybernetics 23 (1994) 1/2, pp. 7–60.

Pesch, H. J.: Solving optimal control and pursuit-evasion game problems of high complexity. In: Computational optimal control. Eds.: R. Bulirsch et al. Basel: Birkhäuser, 1994, pp. 43–64. (ISNM 115)

Pesch, H. J., Gabler, I., Miesbach, S., Breitner, M. H.: Synthesis of optimal strategies for differential games by neural networks. Proc. of the Sixth Inter. Sympos. on Dynamic Games and Applications, St-Jovite, 1994. Eds.: M. Breton et al. Montreal: GERAD, école des Hautes études Commerciales, 1994, pp. 45–60.

Pesch, H. J., Heim, A., von Stryk, O., Schäffler, H., Scheuer, K.: Parameteridentifikation, Bahnoptimierung und Echtzeitsteuerung von Robotern in der industriellen Anwendung. – In: Hoffmann, K.-H., Jäger, W., Lohmann, Th., Schunck, H. (Eds.): Mathematik – Schlüsseltechnologie für die Zukunft, Springer, Berlin, 1996, pp. 539–550.

Pesch, H. J., Rentrop, P.: Numerical Solution of Asymptotic Two-Point Boundary Value Problems with Application to the Swirling Flow over a Plane Disk. – In: V. Boffi und H. Neunzert (Hrsg.), Proc. of the Third German-Italian Symp. Applications of Mathematics in Industry and Technology, Siena, 18.–22. 6. 1988. – Stuttgart: Teubner, 327–338, 1989.

Pesch, H. J.: A Survey of Certain Methods for the Guidance of Space Vehicles, Z. Angew. Math. Mech. 70, 6, T 745–747, 1990.

Pesch, H. J.: Implementation Concepts for ODE Solvers with Comments on Extrapolation Methods, Stepsize Control Algorithms, Order Selection Techniques, and Dense Output, Oberpfaffenhofen: Carl-Cranz Gesellschaft (Course C 1.02: Numerical Integration of

Ordinary Differential-Algebraic Equations in Mechanical System Simulation), 1990, p. 38, 2nd revised and extended edition (Course DR 1.02: Numerical Integration Methods for the Simulation of Constrained and Unconstrained Mechanical Systems), 1991.
Pesch, H.J.: Introduction to the Theory and Numerical Treatment of Optimal Control Problems, Oberpfaffenhofen: Carl-Cranz Gesellschaft (Course DR 3.10: Optimierungsverfahren – Software und praktische Anwendungen), 1991.
Pesch, H.J.: Off-line and On-line Computation of Optimal Trajectories in the Aerospace Field, In: A. Miele, A. Salvetti (Hrsg.): Applied Mathematics in the Aerospace Science and Engineering – New York: Plenum Press (Mathematical Concepts and Methods in Science and Engineering; 44), pp. 165–220, 1994.
Pesch, H.J.: Optimal and Nearly Optimal Guidance by Multiple Shooting. – In: Centre National d'Etudes Spatiales (Hrsg.), Proc. of the Inter. Symp. Mécanique Spatiale – Space Dynamics, Toulouse, 6.–10.11.1989. – Toulouse: Cepadues Editions, 761–771, 1990.
Pesch, H.J.: Optimal Re-Entry Guidance under Control and State Constraints. In: P. Bernhard und H. Bourdache-Siguerdidjane (Hrsg.), Proc. of the 8th IFAC Workshop on Control Applications of Nonlinear Programming and Optimization, Paris, 7.–9.6.1989. – Oxford: IFAC Publications, 1992.
Pesch, H.J.: Real-Time Computation of Feedback Controls for Constrained Optimal Control Problems, Part 1: Neighboring Extremals, Optimal Control Applications & Methods 10, No. 2, 129–145, 1989.
Pesch, H.J.: Real-Time Computation of Feedback Controls for Constrained Optimal Control Problems, Part 2: A Correction Method Based on Multiple Shooting, Optimal Control Applications Methods 10, No. 2, 147–171, 1989.
Pfaffenzeller, G., Rentrop, P., Schmidt, W.: Robot control based on neural networks. Proc. of ECMI VI, Limerick 1991, ed. F. Hodnett, Teubner-Kluwer, 1992, 247–250.
Pickles, S., Costen, F, Brooke, J., Gabriel, E., Müller, M., Resch, M., Ord, S.: The problems and solutions of the metacomputing experiment in SC99. HPCN'2000, Amsterdam/The Netherlands, May 10–12, 2000.
Pickles, S.M., Brooke, J.M., Costen, F.C., Garbriel, T., Müller, M., Resch, M., Ord, S.M.: Metacomputing across intercontinental networks. Future Generation Computer Systems (17) 8, pp. 911–918, 2001.
Rahn, M., Schöttle, U.M., Messerschmid, E.: Impact of mission requirements and constraints on conceptual launch vehicle design. Aerospace Science and Technology, no. 6, 391–401, 1999.
Rahn, M., Schöttle, U.M.: Decomposition Algorithm for Performance Optimization of a Launch Vehicle. Journal of Spacecraft and Rockets, Vol. 33, No. 2, March–April, pp. 214–221, 1996.
Raible, T., Jacob, D.: Multidisciplinary space plane design and optimization using a genetic algorithm. CEAS Conference on Multidisciplinary Aircraft Design and Optimization, 25.–26. June 2001, Köln, Proceedings pp. 139–153.
Raible, T., Jacob, D.: Sensitivity Based Optimization of Two-Stage-To-Orbit Space Planes with Lifting Body and Waverider Lower Stages. AIAA-2003-6955, 12$^{th}$ International Space Planes and Hypersonic Systems and Technologies Conference, Norfolk, Virginia, USA, 15–19 December 2003.
Rausch, O., Rochholz, H., Laschka, B.: Source Term Control of Elliptically Generated Block-Structured Grids. ICAS-Conference Proceedings, Bejing, China, 1992.
Reinartz, B.U., Ballmann, J.: Details on the Computation of Hypersonic Inlet Flows. PAMM – Proceedings in Applied Mathematics and Mechanics, Vol. 2, Issue 1, WILEY-VCH, Weinheim Germany, 2003, pp. 326–327.
Reinartz, B.U., Ballmann, J.: Numerical Simulation of Turbulent Flows Inside a Hypersonic Inlet. In: Notes on Numerical Fluid Mechanics, Ed.: Chr. Breitsamter, B. Laschka, H.-J. Heinemann, R. Hilbig, Springer, Berlin Heidelberg New York, 2003.
Reinartz, B.U., Herrmann, C.D., Ballmann, J., Koschel, W.W.: Analysis of Hypersonic Inlet Flows with Internal Compression. AIAA Paper 2002-5230.

Reinartz, B.U., Herrmann, C.D., Ballmann, J., Koschel, W.W.: Aerodynamic Performance Analysis of a Hypersonic Inlet Isolator Using Computation and Experiment. Journal of Propulsion and Power, 2003.

Reinartz, B.U., Hesse, M., Ballmann, J.: Numerical Investigation of the Shuttle-Like Configuration Phoenix. In: Krause, E., Jäger, W. (Eds.), High Performance Computing in Science and Engineering '02, Springer Verlag Berlin, Germany, 2002, pp. 379–390.

Reinartz, B.U., Koschel, W.W.: Thermal Analysis of Fluid-Structural Interaction in High-Speed Engine Flows. AIAA Journal of Propulsion and Power, Vol. 17, No. 6, pp. 1339–1346, 2001.

Reinartz, B.U., van Keuk, J., Coratekin, T., Ballmann, J.: Computation of Wall Heat Fluxes in Hypersonic Inlet Flows. AIAA Paper 2002-0506.

Reinsch, K.-D.: Interpolation by pieces of Euler's elastica. – In: Le Mehaute, A., Rabut, C., Schumaker, L.L. (Eds.): Curves and Surfaces with Applications in CAGD, Vanderbilt University Press, Nashville, USA, 1997, pp. 379–386.

Reisinger, D., Müller, J., Olejak, D., Staudacher, W.: Effect of Mach Number and Reynolds Number on Flap Efficiencies in the supersonic Regime. Proceedings of the 85th semi-annual meeting of the Supersonic Tunnel Association, Atlanta, USA, April 1996.

Reisinger, D.: Ablösung an der Rampe im Überschall und Auswirkungen auf die Klappenwirksamkeit. Tagungsband des 7. STAB-Workshop, Göttingen, Nov. 1995.

Rentrop, P., Roche, M., Steinebach, G.: The Application of Rosenbrock-Wanner Type Methods with Stepsize Control in Differential-Algebraic Equations. – In: Numer. Math. 55 pp. 545–563, 1989.

Rentrop, P., Wever, U.: Parametrization for Curve Interpolation in Technical Applications. Proc. of the 14th IFIP Conf. on System Modelling and Optimization, Leipzig, July 3–7, 1989. Eds.: H.-J. Sebastian et al. Berlin: Springer, 1990, pp. 575–582. (Lect. Notes in Contr. and Inf. Sci., 143), 1990.

Rentrop, P., Wever, U.: Theory and Applications of the Exponential Spline. Submitted to: Numer. Math. 1992.

Resch, M., Rantzau, D., Stoy, R.: Metacomputing Experience in an Transatlantic Wide Area Application Testbe. Future Generation Computer Systems (15) 5–6, pp. 807–816, 1999.

Resch, M., Babovsky, H.: Workstation Clustering: A powerful tool for numerical simulation in flight sciences and space research, J. Flight Sciences and Space Research 19, pp. 253–258, 1995.

Riedel, R., Bill, J., Kienzle, A.: Boron-Modified Inorganic Polymers – Precursors for the Synthesis of Multicomponent Ceramics. Applied Organometallic Chemistry, 10, p. 241, 1996.

Riedel, R., Kienzle, A., Dressler, W., Ruwisch, L., Bill, J., Aldinger, F.: A Novel Silicoboron Carbonitride Ceramic with Ultra-High Thermal Stability. Nature 382, p. 797, 1996.

Rocci Denis, S., Brandstetter, A., Kau, H.-P.: Transition between Ramjet and Scramjet Modes in a Combustor for an Air-Breathing Launcher, AIAC Paper 2003-054.

Rochholz, H., Huber, Th., Matyas, F.: Unsteady airloads during the separation of an idealized two-stage hypersonic vehicle. Zeitschrift für Flugwissenschaften und Weltraumforschung, Vol. 19, Nr. 1, 1995, pp. 2–9.

Rochholz, H.: Three-Dimensional Euler-Calculations on the Inria-Delta-Wing, AIAA-93-5049, AIAA/DGLR fifth Int. Aerospace Planes and Hypersonics Technologies Conference, Munich, 1993.

Rochholz, H., Huber, Th., Matyas, F.: Unsteady Airloads during Separation of a two-stage hypersonic vehicles, ZFW Sonderheft, 2 1995.

Rochholz, H., Matyas, F., Laschka, B.: Aerodynamics during the Separation Phase of two-staged Hypersonic Vehicles, United Nations/Indonesia Regional Conference on Space Science and Technology, Bandung, Indonesia, 1993.

Roos, E., Maier, H.-P.: Deformation and failure behaviour of thermomechanically loaded C/C-SiC; High temperature fibre composite materials, Editor V.K. Srivastava; Department of mechanical engineering, Institute of technology, Banaras Hindu University, Varanasi; Indo-German Workshop on High Temperature Fibre Composite Materials; Sept. 11–15. 2000; pages 1–10; ISBN 81-7764-067-4, 2000.

Roubícek, T., Schmidt, W. H.: Existence of solutions of certain nonconvex optimal control problems governed by nonlinear integral equations. Optimization, 1997, Vol. 42, pp. 91–108.
Ruppe, H. O.: Design Considerations for Future Space Launchers. Acta Astronautica, Vol. 29, No. 9, pp. 705–722, Sept. 1993.
Ruppe, H. O.: Zur Optimierung von Stufenraketen. – In: Der Himmel hat viele Gesichter: Winfried Petri zum 80. Geburtstag, Institut für Geschichte der Naturwissenschaften, München, 1994.
Ruppe, H. O., Hornik, A., Lösch, M.: Analysis of Future Space Transportation Systems (AZURA). 2nd Space Course on Low Earth Orbit Transportations, Paper No. 41, München, October 1993.
Ruppe, H. O., Hornik, A., Lösch, M., Simon, M.: Flexible Ariane-Trägerbaureihe aus Derivaten der cryogenen Stufen der Ariane 4/5. Deutscher Luft- und Raumfahrtkongress, DGLR 93-03-011, Göttingen, 28. Sept.–1. Okt. 1993.
Ruppe, H. O., Hornik, A., Lösch, M., Sabath, D.: A Method for Evaluation of Space Launchers. Zeitschrift für Flugwissenschaft und Weltraumforschung (ZFW), Band 19, Heft 1, pp. 71–79, Februar 1995.
Sachs, G., Knoll, A., Stich, R., Cox, T.: Hypersonic Simulator Experiments for Long-Term Dynamics Flying Qualities. AIAA/DGLR 5th International Aerospace Planes and Hypersonics Technologies Conference, München, 30. November–3. Dezember, 1993, AIAA Paper Nr. 93-5088, 1993.
Sachs, G., Bayer, R., Drexler, J.: Optimal Trajectories for Sänger-Type Vehicles. AGARD CP 489, pp. 21-1–21-12, 1990.
Sachs, G., Bayer, R., Drexler, J.: Optimum Ascent Performance of Winged Two-Stage Vehicles. ESA SP-293, pp. 271–278, 1989.
Sachs, G., Bayer, R., Lederer, R., Schaber, R.: Hypersonic Flight Performance Improvements by Overfueled Ramjet Combustion. AIAA Paper 91-5074, 1991.
Sachs, G., Bayer, R., Lederer, R., Schaber, R.: Improvement of aerospace plane performance by overfueled ramjets combustion. Zeitschrift für Flugwissenschaften und Weltraumforschung, Band 17, Nr. 1, pp. 25–32, 1993.
Sachs, G., Bayer, R., Lederer, R., Schaber, R.: Hypersonic Flight Performance Improvements by Overfueled Ramjet Combustion: AIAA Third International Aerospace Planes Conference, Orlando, FL, 3.–5. Dez., 1991, AIAA Paper No. 91-5074, pp. 1–8.
Sachs, G., Bayer, R.: Optimal Direct Ascent with Carrier Return-to-Base. Journal of Guidance, Control, and Dynamics, Vol. 18, Nr. 5, 1995, pp. 1206–1207.
Sachs, G., Bayer, R.: Optimal Return-to-Base Cruise of Hypersonic Carrier Vehicles. Zeitschrift für Flugwissenschaften und Weltraumforschung, Vol. 19, Nr. 1, 1995, S. 47–54. Sonderheft für den Sonderforschungsbereich 255 "Transatmosphärische Flugsysteme".
Sachs, G., Dinkelmann, M.: Heat Input Reduction by Optimal Trajectory Control. AIAA Guidance, Navigation and Control Conference, San Diego, USA, 29.–31. Juli 1996, AIAA-96-3905, 1996.
Sachs, G., Dinkelmann, M.: Optimal Three-Dimensional Cruise and Ascent of a Two-Stage Hypersonic Vehicle. 46th IAF Congress, Oslo, 1995.
Sachs, G., Dinkelmann, M.: Optimale Bahnsteuerung zur Reduzierung des Treibstoffverlustes von Hyperschall-Antrieben. Zeitschrift für Flugwissenschaften und Weltraumforschung, Vol. 20, Nr. 6, 1996, pp. 257–264.
Sachs, G., Dinkelmann, M.: Optimization of Three-Dimensional Range and Ascent Trajectories of a Two-Stage Hypersonic Vehicle. Space Technology, Vol. 16, Nr. 5/6, 1996, pp. 365–373.
Sachs, G., Dinkelmann, M.: Optimum three-dimensional hypersonic cruise for stage separation. Third International Congress on Industrial and Applied Mathematics, Hamburg, 3.–7. Juli, 1995.
Sachs, G., Dinkelmann, M.: Reduction of Coolant Fuel Losses in Hypersonic Flight by Optimal Trajectory Control. Journal of Guidance, Control, and Dynamics, Vol. 19, Nr. 6, 1996, pp. 1278–1284.

Sachs, G., Dinkelmann, M.: Reduzierung der thermischen Belastung im Hyperschallflug durch optimale Bahnsteuerung. DGLR-Jahrestagung 1996, Dresden, 24.–27. 9. 1996, DGLR-Jahrbuch III 1996, pp. 1583–1589.
Sachs, G., Dinkelmann, M.: Trajectory Optimization for Reducing Coolant Fuel Losses of Aerospace Planes. Proceedings of the AIAA Guidance, Navigation and Control Conference, Baltimore, USA, 7.–10. Aug. 1995, pp. 1806–1812.
Sachs, G., Dinkelmann, M.: Treibstoffoptimaler räumlicher Reichweitenflug zweistufiger Hyperschall-Flugsysteme. Tagungsband des Workshops "Optimalsteuerungsprobleme von Hyperschall-Flugsystemen", Sonderforschungsbereich 255, "Transatmosphärische Flugsysteme", pp. 81–90, 1994.
Sachs, G., Dobler, K., Theunissen, E.: Verbesserung der Flugführung durch perspektivisches Flugbahn-Display. Deutscher Luft- und Raumfahrtkongreß 1998, Bremen, 5.–8. Okt. 1998.
Sachs, G., Dobler, K.: Synthetic Vision Flight Tests for Low Level Guidance in Poor Visibility. Proceedings of the AIAA Guidance, Navigation and Control Conference, San Diego, USA, 29.–31. Juli 1996, AIAA Paper 96-3744, 1996.
Sachs, G., Drexler, J., Schoder, W.: Optimal Separation and Ascent of Lifting Upper Stages. Lecture Notes in Engineering, SpringerVerlag, pp. 325–331, 1993.
Sachs, G., Drexler, J., Schoder, W.: Optimal separation and ascent of lifting upper stages. Orbital Transport – Technical, Meteorological and Chemical Aspects. Eds.: H. Ortel jr., H. Körner. Third Aerospace Symposium, Braunschweig 26.–28. August 1991, Berlin Heidelberg New York: Springer Verlag, pp. 325–331, 1993.
Sachs, G., Drexler, J., Stich, R.: Ascent Optimization and Control Problems of Wingless Upper Stages. IAF-90-296, 1990.
Sachs, G., Drexler, J., Stich, R.: Optimal Control of Ascent Trajectories and Rotary Dynamics for Wingless Orbital Stages. Acta Astronautica, Band 25, Nr. 8/9, pp. 463–471, 1991.
Sachs, G., Drexler, J.: Optimal Ascent of a Horus/Sänger Type Space Vehicle. AIAA/AAS Astrodynamics Conference Proceedings, 1988.
Sachs, G., Heller, M., Wahlberg, L.: Robust Control of a Hypersonic Experimental Vehicle with Ramjet Engines. Proceedings of the AIAA Guidance, Navigation and Control Conference, San Diego, USA, 29.–31. Juli 1996, AIAA Paper 96-3728, 1996.
Sachs, G., Heller, M.: Robuster Regler für einen Hyperschall-Erprobungsträger. Submitted to: Jahrbuch der DGLR, 1995.
Sachs, G., Knoll, A., Stich, R., Cox, T.H.: Simulations- und Flugversuche über Flugeigenschaften von Hyperschall-Flugzeugen. Zeitschrift für Flugwissenschaften und Weltraumforschung, Vol. 19, Nr. 1, 1995, pp. 61–70, Sonderheft für den Sonderforschungsbereich 255 "Transatmosphärische Flugsysteme".
Sachs, G., Knoll, A., Stich, R., Cox, T.: Flugeigenschaftsuntersuchungen mit dem Hyperschall-Flugsimulator der NASA. DGLR-Jahrestagung 1993, Göttingen, 28 September–1 Oktober 1993, DGLR-Jahrbuch 1993 I, pp. 77–83, 1993.
Sachs, G., Knoll, A., Stich, R., Cox, T.: Simulator and Flight Tests on Aerospace Plane Long-Period Control and Flying Qualities. AIAA Atmospheric Flight Mechanics Conference Proceedings, Scottsdale, AZ, 1–3 August, 1994, Proceedings, pp. 410–418, 1994.
Sachs, G., Mayrhofer, M., Wächter, M.: Optimal Control of Fuel Draining for Mission Abort Range Increase of an Orbital Stage. Workshop Optimal Control, University of Greifswald, Germany, ISBN 3-89791-316-X, Hieronymus Verlag München, pp. 141–150, October 2002.
Sachs, G., Mayrhofer, M.: Range Increase in Gliding Flight by Optimal Jettison Control. 2nd International Conference on Unconventional Flight, Balatonfüred, Ungarn, Juni 2000.
Sachs, G., Mayrhofer, M.: Reichweitensteigerung bei Hyperschall-Notflugbahnen durch Optimalsteuerung des Treibstoff-Ablaßvorgangs. GAMM-Jahrestagung 2000, Göttingen, 2.–7. April 2000.
Sachs, G., Mehlhorn, R., Dinkelmann, M.: Efficient Convexification of Flight Path Optimization Problems. ISNM International Series of Numerical Mathematics, Band 124, Birkhäuser Verlag, Basel, 1998, pp. 321–330.

Sachs, G., Mehlhorn, R., Dinkelmann, M.: Flight Path Optimization Problems Related To Vehicle Fixed Thrust Direction. Proceedings of the AIAA Guidance, Navigation and Control Conference, San Diego, USA, 29.– 31. Juli 1996, AIAA-96-3869, 1996.

Sachs, G., Mehlhorn, R.: Flight Path Optimization Problems Related to Vehicle Fixed Thrust Direction. Tagungsbericht 4/1996 "Optimalsteuerung und Variationsrechnung – Optimal Control", Oberwolfach, 21.–27. Jan. 1996, Mathematisches Forschungsinstitut Oberwolfach, 1997.

Sachs, G., Mehlhorn, R.: Improvement of Endurance Performance by Periodic Optimal Control of Variable Camber. AGARD-CP-547, pp. 8-1–8-7, 1993.

Sachs, G., Möller, H.: Visualization of the Separation Maneuver of Hypersonic Vehicles. 2nd Space Course on Low Earth Orbit Transportation, Sonderforschungsbereich 255 "Transatmosphärische Flugsysteme", Technische Universität München, Proceedings, Band 2, pp. D2-1–D2-5, 1993.

Sachs, G., Moravszki, C.: Entwurf und Erprobung eines Tunnel-Prädiktordisplays für den Hochgeschwindigkeitsflug. Deutscher Luft- und Raumfahrtkongress, DGLR 2002-200, Stuttgart, Sept. 2002.

Sachs, G., Moravszki, C.: Simulation experiments for hypersonic trajectory control using a predictive flight path display. Proceedings of the AIAA Guidance, Navigation and Control Conference, AIAA Paper 2002-4696, Monterey, Aug. 2002.

Sachs, G., Schoder, W., Drexler, J.: Optimal Separation and Ascent of Lifting Upper Stages. In: Lecture Notes in Engineering, Springer Verlag, 1992.

Sachs, G., Schoder, W., Heller, M., Sacher, P.: Robust Control Concept for a Hypersonic Test Vehicle. Proceedings of the AIAA 6th International Aerospace Planes and Hypersonics Technologies Conference, Chattanooga, USA, 3.– 7. April 1995, AIAA Paper 95-6061, 1995.

Sachs, G., Schoder, W., Kraus W.: Separating of Lifting Vehicles at Hypersonic Speed. – Wind Tunnel Tests and Flight Dynamics Simulation. – AGARD CP "Space Systems Design and Development Testing", pp. 5-1–5-2, 1994.

Sachs, G., Schoder, W., Kraus, W.: Separation of Lifting Vehicles at Hypersonic Speed – Wind Tunnel Tests and Flight Dynamics Simulation. AGARD FVPS Symposium, Cannes, 3–5 Oktober, 1994, AGARD CP 561, pp. 5-1–5-2, 1995.

Sachs, G., Schoder, W., Sacher, P.: Robust Control Concept for a Hypersonic Test Vehicle. 6th International Aerospace Planes and Hypersonics Technologies Conference, Chattanooga, 3–7 April, 1995, AIAA Paper No. 95-6061, 1995.

Sachs, G., Schoder, W.: Control Problems of Separating Lifting Vehicles at Hypersonic Speed. Proceedings of the 2nd Mini Conference on Vehicle System Dynamics, Identification and Anomalies, Technical University of Budapest, pp. 185–187, 1992.

Sachs, G., Schoder, W.: Control Problems of Separating Lifting Vehicles at Hypersonic Speed. Proceedings of the 2nd Mini Conference on Vehicle Dynamics, Identification and Anomalies, Technical University of Budapest, pp. 185–197, 1992.

Sachs, G., Schoder, W.: Ein robustes Steuerungs- und Regelungskonzept für das Separationsmanöver eines zweistufigen Hyperschall-Flugsystems. DGLR Jahrbuch 1992 III, pp. 1465–1472, 1992.

Sachs, G., Schoder, W.: Flugmechanische Probleme beim Trennvorgang auftriebserzeugender, zweistufiger Hyperschall-Flugsysteme. DGLR-Jahrbuch 1990 I, pp. 215–222, 1990.

Sachs, G., Schoder, W.: Flugmechanische Probleme beim Trennvorgang auftriebserzeugender, zweistufiger Hyperschall-Flugsysteme. DGLR-Jahrestagung 1990, Friedrichshafen, 1–4 Oktober, 1990, DGLR-Jahrbuch 1990 I, pp. 215–222, 1990.

Sachs, G., Schoder, W.: Optimal Separation of Lifting Vehicles in Hypersonic Flight. AIAA Guidance, Navigation and Control Conference Proceedings, pp. 529–536, 1991.

Sachs, G., Schoder, W.: Optimization Problems of Separation and Ascent of Hypersonic Two-Stage Lifting Vehicles. Kolloquium "Bahnoptimierung und Lenkung von transatmosphärischen Flugsystemen", DLR und SFB 255, Oberpfaffenhofen, 3. 6. 1991.

Sachs, G., Schoder, W.: Robust Control of the Separation of Hypersonic Lifting Vehicles. AIAA 4th International Aerospace Planes Conference, Orlando, FL, 1.–4. Dez. 1992, AIAA Paper 92-5013, 1992.

Sachs, G., Stich, R.: Steuerungs- und Regelungsprobleme im Hyperschallflug, DGLR-Jahrestagung 1991, Berlin, 10–13 September, 1991, DGLR-Jahrbuch 1991 I, pp. 103–112, 1991.

Sachs, G., Wagner, O., Drexler, J.: Optimale Aufstiegsbahnen für die Sänger-Oberstufe. DGLR-Jahrbuch, 1988.

Sachs, G.: Control and Optimization of Transatmospheric Vehicles. Review of the German Hypersonic Research and Technology Programme, BMFT, Bonn, 16.–17. 4. 1991, Tagungsband, pp. 51–66, 1992.

Sachs, G.: Control and Optimization Problems of Transatmospheric Vehicles. Review of the German Hypersonic Research and Technology Programme, BMFT, Bonn, 16.–17. April, 1991, Tagungsband, 1992, pp. 51–66.

Sachs, G.: Deficiencies of Long-Term Dynamics Requirements and New Perspectives. AIAA Atmospheric Flight Mechanics Conference, Boston, 14–16 August, 1989, Proceedings, pp. 394–403, 1989.

Sachs, G.: Dynamic Systems Visualization Applied to Flight Mechanics Problems. Notes on Numerical Fluid Mechanics, Vieweg, Braunschweig, Band 48, pp. 157–172, 1994.

Sachs, G.: Effect of Thrust/Speed Dependence on Long-Period Dynamics in Supersonic Flight. Journal of Guidance, Control, and Dynamics, Band 13, Nr. 6, pp. 1163–1166, 1990.

Sachs, G.: Flugeigenschaftskriterien und Langzeit-Dynamik im Überschall- und Hyperschallflug. Zeitschrift für Flugwissenschaften und Weltraumforschung, Band 15, Heft 4, 1991.

Sachs, G.: Flugeigenschaftsprobleme der Langzeit-Dynamik im Über- und Hyperschall. DGLR-Jahrestagung 1989, Hamburg, 2–4 Oktober, 1989, DGLR-Jahrbuch 1989, pp. 461–469, 1989.

Sachs, G.: Flying Qualities Problems of Aerospace Craft. AIAA Atmospheric Flight Mechanics Conference, Portland, 20–22 August, 1989, Proceedings, pp. 43–53, 1990.

Sachs, G.: Increase and Limit of $T\theta 2$ in Super- and Hypersonic Flight. Journal of Guidance, Control, and Dynamics, Band 22, Nr. 1, 1999, pp. 181–183.

Sachs, G.: Longitudinal Dynamics and Control Problems in Hypersonic Flight. Blue Book "Space Planes", SVS "Leonardo da Vinci", Delft University of Technology, Delft, Niederlande, 1995, pp. 249–275.

Sachs, G.: Novel Stability and Control Problems of Aerospace Planes. 12th World Congress, IFAC, Sydney, Australien, 18–23 Juli 1993, Proceedings, Band 3, pp. 121–124, 1993.

Sachs, G.: Optimal Ascent Trajectories of Two-Stage Transatmospheric Vehicles. Space Course Aachen, Forum Weltraumforschung, RWTH Aachen, pp. 22-1–22-21, 1991.

Sachs, G.: Optimale Aufstiegsbahnen zukünftiger Raumtransportsysteme mit Rückkehrflug der Trägerstufe. – In: Festschrift 65. Geburtstag Prof. Ruppe, Lehrstuhl für Raumfahrttechnik, Technische Universität München, pp. 61–75, 1994.

Sachs, G.: Path-Attitude Decoupling and Flying Qualities Implications in Hypersonic Flight. Aerospace Science and Technology, Band 2, 1998, Nr. 1, pp. 49–59.

Sachs, G.: Reply by Author to Robert F. Stengel. Journal of Guidance, Control, and Dynamics, submitted 1992.

Sachs, G.: Stability and Control Problems in Hypersonic Flight. 2nd Space Course on Low Earth Orbit Transportation, München, 11–22 Oktober, 1993, Proceedings, pp. 35-1–35-19, 1993.

Sachs, G.: Thrust/Speed Effects on Aerospace Plane Long-Term Dynamics. Journal of Guidance Control, and Dynamics, submitted 1992.

Sachs, G., Heller, M.: Robuster Regler für einen Hyperschall-Erprobungsträger. DGLR-Jahrestagung, Bonn, 26.– 29. Sept. 1995, DGLR-Jahrbuch 1995, pp. 1031–1039.

Sander, T., Lyubar, A., Sattelmayer, T.: Gasdynamically Optimized Two Stage Injectors for Flame Stabilization in Supersonic Flows with Low Losses, eingereicht bei AIAA Journal.

Sander, T., Lyubar, A., Sattelmayer, T.: Neuartiger Wasserstoffinjektor zur Flammenstabilisierung in einer Überschallströmung, Seminar des Sonderforschungsbereichs 255, 2002, TU München, ISBN 3-89791-334-8, S. 61–73, 2004.

Sander, T., Lyubar, A., Sattelmayer, T.: Raman Scattering to the Flowfield of a Scramjet Combustor, Proceedings of the 11th International Conference on Methods of Aerophysical Research (ICMAR), Volume 2, pp. 143–147, 3.–7. July 2002, Akademgorodok, Novosibirsk – Denisova Cave, Altai region, Russia.

Sander, T., Lyubar, A., Shafranovsky, E. A., Sattelmayer, T.: Influence of the Burnout in Vitiators on Ignition Limits in Scramjet Combustors, to be published in AIAA Journal 01/04.

Sander, T., Sattelmayer, T.: Ramanspektroskopie an einer reagierenden Überschallströmung und reaktionskinetische Untersuchung eines Wasserstoff-Vorheizers. Seminar des Sonderforschungsbereichs 255, München, 3. Februar 2000, Tagungsband ISBN 3-89791-175-2, TU München, 2001, pp. 77–88.

Sarkar, S., Foysi, H., Friedrich, R. (2003): On the turbulence structure in compressible turbulent isothermal channel flow. – To appear in: Direct and Large-Eddy Simulation V, R. Friedrich, B. Geurts, O. Métais (Eds.), Kluwer Academic Publishers, 2004.

Scheidler, S., Fottner, L.: Das Dynamische Betriebsverhalten eines Turbostrahltriebwerkes bei Einlaufstörungen. DGLR-Jahrestagung 2001, Hamburg, JT-024, 2001.

Schlamp, S., Rösgen, T., Kasal, P., Weigand, B.: Experimental Considerations for Laser-Induced Thermal Acoustics (LITA) in Compressible Turbulent Flows. AIAA 33rd Fluid Dynamics Conference and Exhibit, AIAA 2003-8370, June 23–26, 2003, Orlando, FL, USA, 2003.

Schmid, N., Fottner, L.: Beeinflussung des Strömungsfeldes im Einlaufkanal eines Hyperschall-Kombinationstriebwerks mittels Rampenabsenkung. 1. Workshop Hyperschallantriebe der Sonderforschungsbereiche 253, 255, 259 und der DLR, München, 15. Mai, 1997.

Schmid, N., Fottner, L.: Leistungsverhalten des Turboteils von Kombinationsantrieben unter dem Einfluß von Eintrittsstörungen. 2. Workshop Hyperschallantriebe der Sonderforschungsbereiche 253, 255, 259 und des DLR, Stuttgart, 22. Nov., 1997.

Schmid, N., Fottner, L.: Numerische und Experimentelle Optimierung des Strömungsfeldes im Einlaufkanal eines Hyperschall-Kombinationstriebwerks. Seminar des Sonderforschungsbereichs 255, München, 12. Dez. 1996, Technische Universität München, 1997, pp. 69–82.

Schmid, N., Hildebrandt, T., Fottner, L.: Numerical Investigation of the Flow Field Inside the Intake-Diffusor of a Combined Cycle Engine for Hypersonic Flight. Proceedings of the ECCOMAS 98, Athen, Griechenland, 7.–8. Sept., 1998.

Schmid, N. R., Leinhos, D. C., Fottner, L.: Performance of a Turbofan Engine with Inlet Distortions from the Inlet Diffuser of a Combined Cycle Engine for Hypersonic Flight, Int. Society for Airbreathing Engines, 5.–10. September 1999, Florence, 99-7076, XIV ISABE, 1999.

Schmid, N. R., Leinhos, D. C., Fottner, L.: Steady Performance Measurements of a Turbofan Engine with Inlet Distortions Containing Co- and Counter-Rotating Swirl from an Inlet Diffuser for Hypersonic Flight. ASME 2000-GT-0011, 2000.

Schmidt, W. H., Heier, K., Bittner, L., Bulirsch, R. (Eds.): Variational Calculus, Optimal Control and Applications. ISNM 124, Birkhäuser, Basel, 1998.

Schmidt, W. H.: An existence theorem for a special control problem. Proceedings of the 3rd Int. Symp. on "Methods and Models in Automation and Robotics", Miedzyzdroje 96, 1, Szczecin, 1996, pp. 263–266.

Schmidt, W. H.: Volterra Integral Processes with delay – necessary optimality conditions. Published in: Corduneanu "Volterra equations and applications", Gordon and Breach Sciences Publishers, 1998.

Schmitt-Thomas, K.G., Dietl, U., Haindl, H.: New Developments in Thermal Barrier Coatings (TBCs) for Gas Turbine Use. Industrial Ceramics, Vol. 16, Nr. 3, 1996, pp. 195–198.

Schmitt-Thomas, K.G., Dietl, U., Haindl, H.: New developments in Thermal Barrier Coatings (TBCs) for Gas Turbine Application. 8th Cimtec, Florenz, 1994.

Schmitt-Thomas, K.G., Dietl, U.: Diffusionssperrschichten aus Al2O3 zur Beeinflussung der Haftschichtoxidation von MCrAlY-Legierungen. Werkstoffe und Korrosion, 43, pp. 492–495, 1992.

Schmitt-Thomas, K.G., Dietl, U.: Neue Wärmedämmschichten für den Gasturbinenbau. Posterpräsentation BDLI Hamburg, Tagungsband, Werkstofftag, 1991.

Schmitt-Thomas, K.G., Dietl, U.: Thermal barrier coatings with improved oxidation resistance. Surface and Coating Technology, 68/69, pp. 113–115, 1994.

Schmitt-Thomas, K.G., Dietl, U.: Verbesserung der Haftschichtoxidation bei keramischen Wärmedämmschichtsystemen. VDI Berichte Nr. 1021, 1993.

Schmitt-Thomas, K.G., Gregory, J.K., Haindl, H., Hertter, M.: Verfahrens- und Werkstofftechnische Optimierung von keramischen Wärmedämmschicht (WDS)-Systemen. Seminar des Sonderforschungsbereichs 255, München, 12. Dez. 1996, Technische Universität München, 1997, pp. 57–67.

Schmitt-Thomas, K.G., Haindl, H., Fu, D.: Modifications of Thermal Barrier Coatings (TBCs). Surface and Coatings Technology, Nr. 94/95, 1997, pp. 149–154.

Schmitt–Thomas, K.G., Haindl, H., Fu, D.: Modifications of Thermal Barrier Coatings (TBCs). Proceedings of the ICMCTF97 Conference, San Diego, 21.–25. April 1997.

Schmitt-Thomas, K.G., Steppe, P., Dietl, U.: Phasenanalysen an plasmagespritzten Wärmedämmschichten. Fortschritte in der Metallographie, Dr. R. Rieder-Verlag, Stuttgart, Band 22, pp. 639–647, 1991.

Schmitt-Thomas, K.G., Dietl, U., Haindl, H.: Thermal barrier coatings for airbreathing combustion systems. Zeitschrift für Flugwissenschaft und Weltraumforschung (ZFW), Vol. 19, Nr. 1, 1995, pp. 41–46.

Schmitz, E., Henze, A., Schröder, W.: Investigation of the Flow Field of the TSTO Concept at Supersonic Mach Numbers. Proceedings of the 12$^{th}$ European Conference, Paris, 29.11.–1.12.99, (CEAS-Conference), Paper 11–58.

Schmitz, E.: Simulation of the Nozzle Flow with a Cold Jet on a Space Transportation System. GAMM Annual Meeting, Bremen, April 1998, ZAMM Volume 79 (1999), Suppl. 3, pp. 967–968.

Schmitz, E.: Simulation of the nozzle with a cold jet on a space transportation system. Proceedings of the GAMM Annual Meeting, GAMM 98, Bremen, 1998.

Schneider, A., Koschel, W.W.: Detailed analysis of a mixed compression hypersonic intake. 14$^{th}$ International Symposium on Air Breathing Engines, Florence, Italy, Sept. 1999, ISABE Paper No, 99-7036.

Schoder, W., Rochholz, H.: Separation of Two-Stage Hypersonic Vehicles – In: Blue Book "Space Planes", Society of Aerospace Students SVS "Leonardo da Vinci", Delft University of Technology, Delft, Niederlande, pp. 276–315, 1994.

Schoder, W., Rochholz, H.: Separation of Two-Stage Hypersonic Vehicles. 2nd Space Course on Low Earth Orbit Transportation, Sonderforschungsbereich 255, Band 2, pp. 36-1–36-22, 1993.

Schönemann, A., Auweter-Kurtz, M.: Mass Spectrometric Investigation of High Enthalpy Plasma Flows. Journal for Thermophysics and Heat Transfer, Vol. 9, Nr. 4, pp. 620–628, Oktober–Dezember, 1995.

Schönemann, A.T., Auweter-Kurtz, M., Habiger, H.A., Sleziona, P.C., Stöckle, T.: Analysis of the Argon Additive Influence on a Nitrogen Arcjet Flow. Journal of Thermophysics and Heat Transfer, Juli–September 1994, Vol. 8, No. 3, pp. 466–472, 1994.

Schöttle, U.M., Bregman, E.R., Hillesheimer, M., Inatani, Y.: Conceptual study of a small semiballistic reentry experiment vehicle. Z. Flugwiss. Weltraumforsch. 15, 362–372, 1991.

Schöttle, U.M., Jahn, G., Bregman, E.R.: Design Studies of a Class of Aeroassisted Orbit Transfer Vehicles. Proceedings of the Twentieth International Symposium on Space Technology and Science, Volume II, Gifu, Japan, pp. 876–883, 1996.

Schöttle, U.M., Messerschmid, E.: Evaluation of a Hydrazine Arcjet Propulsion System for Spacecraft Geosynchronous Orbit Transfer and Station Keeping. Proceedings of the 18th International Symposium on Space Technology and Science, Kagoshima, Japan, 1992.

Schöttle, U.M.: Consideration on Propulsion Systems and Flight Trajectories for Launch Vehicles. Proceedings of the Twentieth International Symposium on Space Technology and Science, Volume I, Gifu, Japan, 1996, pp. 19–28, 1996.

Schreiber, E., Neuer, G.: Considerations for the Realisation of the Temperature Scale above 2000 °C by means of Radiation Thermometry. In: Proceedings TEMPMEKO 2001, eds. B. Fellmuth, J. Seidel, G. Scholz, VDE Verlag GmbH, 143–148, 2002.

Schreiber, E., Neuer, G.: The laser absorption pyrometer for simultaneous measurement of surface temperature and emissivity. In: TEMPMECO '96 – 6th Int. Symposium on Temperature and Thermal Measurements in Industry and Science, Ed. P. Marcarino, Published by Leverotto & Bella, Torino, p. 365–370, 1997.

Schreiber, E.: Fast determination of pyrometer interference blocking. In: Temperature, its Measurement and Control in Science and Industry, Vol. 6, Hrsg. J.F. Schooley, American Institute of Physics, New York, pp. 791–795, 1992.

Schröder, W., Henze A.: Numerical Analysis of Flows over Space Transportation Systems. $10^{th}$ International Conf. on Meth. of Aerophys. Research, July 2000, Novosibirsk, Russia, Invited Paper, Proceedings Part I, pp. 177–182.

Schröder, W., Henze, A.: Numerical Analysis of Flows over Space Transportation Systems. $10^{th}$ Int. Conf. on Meth. Of Astrophys. Res., Juli 2000, Novosibirsk, Russland.

Schröder, W., Meinke, M., Henze, A.: First numerical results on the European Hypersonic Vehicle HOPPER. ICAS Conference, Toronto, Sept. 2002.

Schröder, W., Henze, A.: Numerical Solutions on the hypersonic configuration ELAC-EOS. $4^{th}$ International Symposium on Aerothermodynamics, 2001, Capua.

Schulte, D., Heckels, A., Neubacher, R.: Manipulation of Shock/Boundary Layer Interaction in Hypersonics Inlets. AIAA Journ. of Propulsion and Power, Vol. 17, No. 3, pp. 585–590, May–June 2001.

Schulte, D., Henckels, A., Neubacher, R.: Manipulation of Shock/Boundary Layer Interactions in Hypersonic Inlets. $14^{th}$ International Symposium on Airbreathing Engines, Florence, Sept. 5–10, 1999, IASBE-Paper 99-7038.

Schulte, D., Henckels, A., Neubacher, R.: Manipulation of Shock/Boundary Layer Interactions in Hypersonic Inlets. AIAA Journal of Propulsion and Power, Vol. 17, No. 3, pp. 585–590, 2001.

Schulte, D., Henckels, A., Wepler, U.: Reduction of Shock Induced Boundary Layer Separation in Hypersonic Inlets Using Bleed. Aerospace Science and Technology, Vol. 2, No. 4, 1998, pp. 231–239.

Schulte-Fischedick, J., Zern, A., Mayer, J., Rühle, M., Frieß, M., Krenkel, W., Kochendörfer, R.: The Morphology of C/C-SiC Composites. Mat. Sci. Eng., A332, pp. 146–152, 2002.

Schulte-Rödding, J.-H., Olivier, H.: Experimental Investigations on Hypersonic Inlet Flows. AIAA Paper 98-1528AIAA, $8^{th}$ International Space Planes and Hypersonic Systems and Technologies Conference, Norfolk, Virginia, USA, April 27–30, 1998.

Schulte-Rödding, J.-H., Olivier, H.: Intake Flow Study of Hypersonic Vehicles in the Shock Tunnel TH 2. GAMM Annual Meeting, Bremen, April 1998, ZAMM Volume 79 (1999), Suppl. 3, pp. 973–974.

Schumann, M., Lamperts, S., Stellner, G., Kiehl, M., Mehlhorn, R., Ludwig, T., Bode, A.: Parallel multiple shooting on paragon XP/S, iPSC/860, and clusters of workstations. Aims, porting efforts and performance evaluation. Proc. of the INTEL Supercomputing User Group, Annual North American Users Conference. San Diego, Ca., 1994, 90–97.

Schwer, A.G., Schöttle, U.M., Messerschmid, E.: Operational Impacts and Environmental Effects on Low-Thrust Transfer-Missions of Telecommunication Satellites. Acta Astronautica Vol. 38, Nos. 4–8, pp. 561–575, 1996.

Seifert, H.J., Aldinger, F.: Applied Phase Studies. Z. Metallkd. 87, p. 841, 1996.

Seifert, H.J., Lukas, H.L., Aldinger, F.: Development of Si-B-C-N Ceramics Supported by Constitution and Thermochemistry. Ber. Bunsenges, Phys. Chem., 102 (9) p. 1309, 1998.

Seifert, H.J., Peng, J., Golczewski, J., Aldinger, F.: Phase Equilibria of Precursor-derived Si-B-C-N Ceramics. Appl. Organometal. Chem. 15, p. 794, 2001.

Seifert, H.J., Peng, J., Lukas, H.L., Aldinger, F.: Phase Equilibria and Thermal Analysis of Si-C-N Ceramics. J. of Alloys and Compounds, 320, p. 251, 2001.

Seitz, J., Bill, J., Egger, N., Aldinger, F.: Structural Investigations of Si/C/N-Ceramics from Polysilazane Precursors by Means of Nuclear Magnetic Resonance. J. Europ. Ceram. Soc., 16, p. 885, 1996.

Sesterhenn, J.: A characteristic-type formulation of the Navier–Stokes equations for high order upwind schemes. Computers & Fluids, 30(1), 2001, pp. 37–67.

Siepenkötter, N., Kasberg, W., Alles, W.: Approximated Nonlinear Stability Analysis of the Dynamics of Flexible Aircraft; Multidisciplinary Research on Modulation of Flow and Fluid-Structure Interaction at Airplane Wings. Numerical Notes on Fluid Mechanics, Springer Verlag, 2001.

Sleziona, P.C., Auweter-Kurtz, M., Boie, C., Heiermann, J., Lenzner, S.: Numerical Code for Magneto-Plasma Flows. In: Hyperbolic Problems: Theory, Numerics, Applications, Seventh International Conference in Zürich, Vol. II, Hrsg. M. Fey, R. Jeltsch, International Series of Numerical Mathematics Vol. 130, 1999, Birkhäuser Verlag, Basel, pp. 905–914, 1999.

Sleziona, P.C., Auweter-Kurtz, M., Gogel, T.H., Gölz, T.M., Messerschmid, E.W. Schrade, H.O.: Non-Equilibrium Flow in an Arc Heated Wind Tunnel. Veröffentlicht in Desideri et al. (Hrsg.): "Hypersonic Flow for Reentry Problems. Test Cases: Experiment and Computations", Springer Verlag, Berlin New York, 1991.

Sleziona, P.C., Auweter-Kurtz, M., Heiermann, J., Kaeppeler, H.J., Lenzner, S.: Numerical Simulation of the Flow in Magneto-Plasma Accelerators. Systems Analysis-Modelling-Simulation Journal (SAMS), Vol. 34, 1999, pp. 137–151, 2003.

Sleziona, P.C., Auweter-Kurtz, M., Schrade, H.O.: Computation of MPD Flows and Comparison with Experimental Results. International Journal of Numerical Methods in Engineering, Vol. 34, pp. 759–771, 1992.

Sleziona, P.C., Auweter-Kurtz, M., Schrade, H.O.: Numerical Model for MPD Thruster Calculation. Editors: Wagner, S. et al.: Computational Fluid Dynamics '94, John Wiley and Sons, Ltd, pp. 790–795, 1994.

Spagnolo, L., Krapez, J.-C., Friess, M., Maier, H.-P., Neuer, G.: Flash thermography with a periodic mask: profile evaluation of the principal diffusivities for the control of composite materials. Thermosense XXV, X.P. Maldague, E. Cramer, Eds., Proc. of SPIE, vol. 5073, pp. 392–400, 2003.

Staudacher, W.: Entwurfsproblematik luftatmender Raumtransportsysteme. – In: Dinkler, D., Messerschmid, W. (Eds.): 3rd Space Course on Low Earth Transportation, Sonderforschungsbereich 259, Universität Stuttgart, 1995, pp. 65–97.

Staufenbiel, R., Coors, D, Janovsky, R.: AZURA Endbericht: Flugleistungssimulation und Sensitivitätsanalyse horizontal startender Raumtransportsysteme. Bericht Nr. B-6200-2, Auftraggeber IABG, 8.1.90.

Staufenbiel, R.: Second stage trajectories of air-breathing spaceplanes. Journal of Spacecraft and Rockets, Vol. 27, No. 6, pp. 618–622, Nov.–Dec. 1990.

Staufenbiel, R.: Transport Efficiency of Space Planes with Airbreathing Phases. Journal of Spacecraft and Rockets, Vol. 29, No. 4, pp. 460–465, Jul.–Aug. 1992.

Steinebach, D.A., Kühl, W., Gallus, H.E.: Design Aspects of the Propulsion System for Aerospace Planes. AIAA/DGLR Fifth International Aerospace Planes and Hypersonics Technologies Conference, 30.11.–3. 12. 1993, Munich, Germany, AIAA 93-5127, C1-Nr. 7, 1994.

Steinebach, D.A., Franke, M., Gallus, H.E.: Integration des Antriebssystems in einen Raumtransporter vom Typ ELAC-I. Deutscher Luft- und Raumfahrt-Kongreß 1994, 4.–7. Oktober in Erlangen, DGLR 94-C3-51, C1-Nr. 8, 1994.

Steinebach, D.A., Kühl, W., Gallus, H.E.: Performance analysis of a turbofan as a part of an airbreathing propulsion system for space shuttles. ZFW, Vol. 17 (1993), pp. 131–138.

Steinhoff, J., Mersch, T., Decker, F.: Computation of Incompressible Flow over Delta Wings Using Vorticity Confinement. AIAA-Paper 94-0646, 32nd Aerospace Sciences Meeting & Exhibits, Reno, Nv., January 10–13, 1994.
Stemmer, C., Kloker, M., Wagner, S.: Navier-Stokes Simulation of Harmonic Point Disturbances in an Airfoil Boundary Layer. AIAA-Journal 38, No. 8, 1369–1376, 2000.
Stemmer, C., Kloker, M.: Interference of Wave Trains with Varying Phase Relations in a Decelerated Two-Dimensional Boundary Layer. In: Rist, U., Wagner, S., Heinemann, H.-J., Hilbig, R. (eds.): New Results in Numerical and Experimental Fluid Dynamics III. NNFM 77, 239–246, Springer, 2001.
Stemmer, C., Kloker, M.: Later Stages of Transition of an Airfoil Boundary Layer Excited by a Harmonic Point Source. In: Fasel, H., Saric, W. (eds.): Laminar-Turbulent Transition, 499–504, Springer 2000.
Stengel, R. F.: Comments on "Effect of Thrust/Speed Dependence on Long-Period Dynamics in Supersonic Flight". Journal of Guidance, Control, and Dynamics, submitted 1992.
Stich, R., Sachs, G., Cox, T.: Bahn-Lage-Entkopplung im Hochgeschwindigkeitsflug. Seminar des Sonderforschungsbereichs 255, TU München, 3. Februar 2000, Tagungsband ISBN 3-89791-175-2, TU München, 2001, pp. 145–160.
Stich, R., Sachs, G., Cox, T.: Path-Attitude Decoupling Problems of Aerospace Craft. Paper of the 9th International Space Planes and Hypersonic Systems and Technologies Conference, Norfolk, USA, 1.–5. Nov. 1999, AIAA-99-4847, 1999.
Stich, R., Sachs, G., Cox, T.: Path-Attitude Inconsonance in High Speed Flight and Related Path Control Issues. Proceedings of the 22nd International Congress of Aeronautical Sciences, Harrogate, U.K., 27 August–1. September 2000, ICAS-2000-4.10.4., 2000.
Stich, R., Sachs, G., Cox, T. H.: Simulation Tests for Investigating Flying Qualites of Aerospace Planes. Proceedings of the AIAA 8th International Space Planes and Hypersonic Systems and Technology Conference, Norfolk, USA, 27.–30. April 1998, pp. 115–125.
Stich, R., Sachs, G.: Simulator- und Flugversuche zu Flugeigenschaftsproblemen im Hyperschallflug. DGLR-Jahrestagung, München, 14.– 17. Okt. 1997, DGLR-Jahrbuch 1997, pp. 475–484.
Stich, R., Sachs, G.: Simulator- und Flugversuche zur Langzeitdynamik im Hyperschall. Seminar des Sonderforschungsbereichs 255, München, 12. Dez. 1996, Technische Universität München, 1997, pp. 131–140.
Stich, R., Sachs, G.: Unconventional Stability and Control Characteristics in Hypersonic Flight. Proceedings of the 1st International Conference on Unconventional Flight, Budapest, Ungarn, 13.–15. Okt. 1997.
Stirn, R., Altmann, I., Hirsch, K., Lindner, P., Roth, B., Schumacher, U.: Bestimmung der räumlichen Verteilung der Plasmaparameter in der Wechselwirkungszone zwischen einem Luftplasma und Hitzeschutzmaterialien". Verhandl. DPG (VI), 37, 5/2002.
Stoer, J., Bulirsch, R.: Numerische Mathematik 2, 3. verb. Auflage, Springer-Lehrbuch, Berlin, 1990.
Stoll, P., Gerlinger, P., Brüggemann, D.: Implicit Preconditioning Method for Turbulent Reacting Flows. Proceedings of the 4th ECCOMAS Conference, 1, pp. 205–212, John Wiley & Sons, 1998.
Störkmann, V., Olivier, H., Grönig, H., Paris, S.: Force Measurements on ELAC 1a and on a Reentry Capsule in Short Duration Hypersonic Facilities. 21st International Symposium on Shock Waves, Great Keppel Island, Australia, July 20–25, 1997, Paper 63.
Störkmann, V., Olivier, H., Grönig, H.: Force Measurements in Hypersonic Impulse Facilities. AIAA Journal, Vol. 36, No. 3, March 1998, pp. 342–348.
Stragies, R., Chudej, K., Bulirsch, R.: Effiziente Lösung einer realistischen Optimalsteuerungsaufgabe durch Transformation des Randwertproblems. – In: Günther, M., Ostermann, A. (Eds.): Numerical treatment of ordinary and algebro-differential equations, Preprint Nr. 1934, Fachbereich Mathematik, TH Darmstadt, 1997, pp. 27–28. Submitted to: Z. Ang. Math. Mech.
Stromberg, A., Decker, F.: Aachener Hyperschallkonfiguration ELAC I im Hochdruckkanal der DLR. Luft- und Raumfahrt (DGLR), Heft 6, 1993.

Stromberg, A., Henze, A., Limberg, W., Krause, E.: Investigation of vortex structures on delta wings. ZFW, Vol. 20, Nr. 2, 1996.

Stromberg, A., Limberg, W.: Measurements of pressures, forces and moments at subsonic and supersonic speeds on ELAC 1. ZFW, Vol. 20, Nr. 2, 1996.

Stromberg, A., Schmitz, E.: Windkanalwaage zum Messen aerodynamischer Kräfte an der Hyperschall-Konfiguration ELAC 1. MTB 31 (1995), Heft 1.

Swamy, V., Seifert, H.J., Aldinger, F.: Thermodynamic properties of Y2O3 phases and the yttrium-oxygen phase diagram. J. Alloys Comp., 269, p. 201, 1998.

Tamme, R., Krenkel, W., Hasert, U.-F.: Entwicklung und Einsatzmöglichkeiten keramischer Hochtemperaturwärmetauscher aus C/C-SiC Verbundwerkstoffen. VDI-Bericht Nr. 1385, pp. 369–382, 1998.

Tawfik, A., Quell, S., Hirsch, K., Roth, B., Schneider, J., Schumacher, U.: Time Correlated Measurements of spectra from process and beam plasmas in the range from 240 nm to 1000 nm. ECAMPVII, 7th European Conference on Atomic and Molecular Physics, Berlin 2.–6. April 2001, paper P12.15, 2001.

Tetlow, M.R., Schöttle, U.M., Schneider, G.M.: Comparison of Glideback and Flyback Boosters. Journal of Spacecraft and Rockets, Vol. 38, No. 5, pp. 752–758, 2001.

Ting, C.: Visualization of the Supersonic Flow Past EOS. GAMM Conference, Göttingen, Germany, 2000, ZAMM 2001.

Ting, C.-C., Henze, A., Schröder, W.: Heat Transfer Measurements in Supersonic Flow using the Liquid Crystal Display Technique. $11^{th}$ AIAA/AAAF International Conference, AIAA 2002-5200, Orleans, Frankreich, 2002.

Ting, C.-C., Henze, A.: Heat Transfer Measurements and Flow Visualization in Supersonic Flow. STAB-Tagung, München, CD, 2002.

Tremblay, F., Manhart, M., Friedrich, R.: LES of flow around a circular cylinder at a subcritical Reynolds number with cartesian grids. Proceedings of the EUROMECH Colloquium 412, R. Friedrich, W. Rodi (Eds.), Kluwer Academic, 2001.

Tscharnuter, D.: Determining the Controllability Region for the Re-Entry of an Apollo-Type Spacecraft. – In: Schmidt, W.H., Heier, K., Bittner, L., Bulirsch, R. (Eds.): Variational Calculus, Optimal Control and Applications, ISNM 124, Birkhäuser, Basel, 1998, pp. 333–342.

Ullmann, T., Schmücker, M., Hald, H., Henne, R., Schneider, H.: Low Pressure Plasma Sprayed Yttrium-Silicates for Oxidation protection of C/C-SiC Composite Structures. 7th Japan International Sampe Symposium & Exhibition, Proc. of the JISSE-7 Tokyo, Japan, 2001.

Ullmann, T., Schmücker, M., Hald, H., Henne, R., Schneider, H.: Oxidation protection of C/SiC composites based on plasma-sprayed Y-silicates. 8th Symp. of Materials in a Space Environment, Proc. of the 5th Inf. Conf. on Protection of Materials and Structures from the LEO, Arcachon, France, 2000.

Ullmann, T., Schmücker, M., Hald, H., Henne, R., Schneider, H.: Yttrium-Silicates for Oxidation Protection of C/C-SiC Composites. Proc 4th Int. Conf. on High Temperature Ceramic Matrix Composites (HT-CMC 4), München, Okt. 1–3, 2001, High Temperature Ceramic Matrix Composites, pp. 230–235, Wiley-VCH/Weinheim, 2001.

van Keuk, J., Ballmann, J.: Numerical Simulation of Laminar Symmetric Corner Flows in the Hypersonic Regime. NNFM 72, 1999, pp. 250–257.

van Keuk, J., Ballmann, J., Sanderson, S.R., Hornung, H.G.: Numerical Simulation of Experiments on Shock Wave Interactions in Hypervelocity Flows with Chemical Reactions. AIAA Paper 2003-6393, $41^{st}$ Aerospace Sciences Meeting and Exhibit, Reno, Nevada, 2003.

van Keuk, J., Ballmann, J.: Assessment Tests for Upwind Methods for Multi-Dimensional Supersonic Flow Problems. Proceedings of the Second International Symposium on Finite Volumes for Complex Applications-Problems and Perspectives, Duisburg, Juli 19–22, 1999.

van Keuk, J., Ballmann, J.: Hysteresis Phenomena in Shock Wave Reflections for Steady Wedge Flows. PAMM – Proceedings in Applied Mathematics and Mechanics, Vol. 1, Issue 1, WILEY-VCH, Weinheim Germany, 2002, pp. 272–273.

van Keuk, J., Ballmann, J.: Numerical Flow Simulations in the Hypersonic Regime. GAMM Annual Meeting, Bremen, April 1998, ZAMM Volume 79 (1999), Suppl. 3, pp. 959–960.
van Keuk, J., Ballmann, J.: Numerical Simulation of 3D Supersonic Inlet Flows. GAMM Annual Meeting, Metz, France, April 1999, ZAMM 80 (2000), pp. 645–646.
van Keuk, J., Ballmann, J.: Numerical Simulation of Inviscid Shock Interactions on Double-Wedges. GAMM Annual Meeting, Göttingen, 2000, ZAMM 81 (2001), pp. 457–458.
van Keuk, J., Ballmann, J., Schneider, A., Koschel, W.W.: Numerical Simulation of Hypersonic Inlet Flows. AIAA Paper 98-1526, 8$^{th}$ International Space Planes and Hypersonic Systems and Technologies Conference, Norfolk, Virginia, USA, April 27–30, 1998.
Vinckier, A., Höld, R., Mundt, Ch., Wagner, S.: Ein Viscous-Shock-Layer-Verfahren für Hyperschallströmungen. ZFW, 18, pp. 203–211, 1994.
Vinckier, A., Wagner, S.: A Flux Filtering Scheme Applied to the Navier-Stokes-Equation. Proc. Second European Fluid Dynamics Conference, Stuttgart, Sept. 1994, Vol. I (Wagner, S., Hirschel, E.H., Periaux, J., Piva, R., Eds), John Wiley 1994, pp. 258–264, 1994.
von Stryk, O., Glocker, M.: Decomposition of mixed-integer optimal control problems using branch and bound and sparse direct collocation. In: S. Engell, S. Kowalewski, J. Zaytoon (eds.): Proc. ADPM 2000 – The 4th International Conference on Automation of Mixed Processes: Hybrid Dynamic Systems, Dortmund, 18.–19. September (Aachen: Shaker, 2000), pp. 99–104.
von Stryk, O., Vögel, M.: A guidance scheme for full car dynamics simulations. Z. Angew. Math. Mech. 79, Suppl. 2, 1999, pp. 363–364.
von Stryk, O.: Ein direktes Verfahren zur Bahnoptimierung von Luft- und Raumfahrzeugen unter Berücksichtigung von Beschränkungen. – In: Z. Angew. Math. Mech. 71 6, T 705–T 706, 1991.
von Stryk, O.: Numerical Solution of Optimal Control Problems by Direct Collocation. Published in: R. Bulirsch, A. Miele, J. Stoer, K.H. Well (Eds.): Optimal Control, Proc. of the Conf. in Optimal Control and Variational Calculus, Oberwolfach, 1991.
von Stryk, O.: Numerical solution of optimal control problems by direct collocation. In: Optimal Control, Calculus of Variations, Optimal Control Theory and Numerical Methods. Eds.: R. Bulirsch et al. Basel: Birkhäuser, 1993, pp. 129–143. (ISNM 111)
von Stryk, O.: Numerische Lösung optimaler Steuerungsprobleme: Diskretisierung, Parameteroptimierung und Berechnung der adjungierten Variablen. Fortschritt-Berichte VDI, Reihe 8, Nr. 441, VDI-Verlag, Düsseldorf, 1995, p. 158.
von Stryk, O.: Optimal control of multibody systems in minimal coordinates, Z. Angew. Math. Mech. 78, Suppl. 3, 1998, pp. 1117–1120.
von Stryk, O.: Optimal control of multibody systems in minimal coordinates. – In: Günther, M., Ostermann, A. (Eds.): Numerical treatment of ordinary and algebro-differential equations, Preprint Nr. 1934, Fachbereich Mathematik, TH Darmstadt, 1997, pp. 19–22. Submitted to: Z. Ang. Math. Mech.
von Stryk, O.: Optimization of dynamic systems in industrial applications. In: Proc. 2nd European Congress on Intelligent Techniques and Soft Computing (EUFIT). Ed.: H.J. Zimmermann. Aachen: 1994, pp. 347–351.
von Stryk, O., Bulirsch, R.: Direct and Indirect Methods for Trajectory Optimization. Annals of Oper. Res. 37 (1992) pp. 357–373.
von Stryk, O., Schlemmer, M.: Optimal control of the industrial robot Manutec r3. In: Computational optimal control. Eds.: R. Bulirsch et al. Basel: Birkhäuser, 1994, pp. 367–382. (ISNM 115)
Wächter, M., Chudej, K.: Efficient hybrid solution of a realistic trajectory optimization problem with singular arcs. Report des Sonderforschungsbereichs 255, Nr. 35, Technische Universität München, 1998.
Wächter, M., Mayrhofer, M., Sachs, G.: Reichweitenerhöhung durch Treibstoffablassen beim Notabstieg einer geflügelten Orbitalstufe. 20. Norddeutsches Kolloquium über Angewandte Analysis und Numerik, 4.–5. Juni, Ernst-Moritz-Arndt-Universität Greifswald, Tagungsband (Kurzfassungen, 1999), p. 5.

Wächter, M., Sachs, G.: Heat Input Reduction in Hypersonic Flight by Trajectory Optimization with Multipoint Constraints, eingereicht bei: Asian Journal of Control, 2003.
Wächter, M., Mayrhofer, M. Sachs, G.,: Verminderung der instationären Aufheizung im Hyperschallflug durch Mehrpunkt-Beschränkungen. Tagungsband zum SFB 255 – Seminar, TU München, München, Germany, ISBN 3-89791-334-8, Dezember 2002, S. 103–118, 2004.
Wächter, M., Sachs, G.: Hypersonic Heat Transfer Reduction by Trajectory Optimization with Multipoint Constraints. Proceedings of the 47th Asian Control Conference, 25.–27. September 2002, Singapur, ISBN 981-04-6440-1, pp. 869–874, 2002.
Wächter, M., Sachs, G.: Unsteady Heat Load Reduction for a Hypersonic Vehicle with a Multipoint Approach. Workshop des Sonderforschungsbereichs 255 "Transatmosphärische Flugsysteme und der Universität Greifswald, Greifswald, 1.–3. Oktober 2002, Tagungsband, ISBN 3-89791-316-X, TU München, pp. 15–26, 2003.
Wächter, M., Sachs, G.: Unsteady Heat Load Simulation for Hypersonic Cruise Optimization. High-Performance Scientific and Engineering Computing, Lecture Notes in Computational Science and Engineering, Springer Verlag, Heidelberg, Vol. 21, pp. 325–332, 2002.
Wagner, C., Hüttl, T.J., Friedrich, R.: Low-Reynolds number effects derived from numerical simulations of turbulent pipe flow. Computers & Fluids, Vol. 30, No. 5, pp. 581–590, 2001.
Wagner, S., Hirschel, E.H., Periaux, J., Piva, R.: Computational Fluid Dynamics '94. Proceedings of the Second European Fluid Dynamics Conference, Stuttgart, Sept. 1994, 1029 pages, Vol. I John Wiley, 1994.
Wagner, S., Hirschel, E.H., Periaux, J.: Computational Fluid Dynamics '94. Proceedings of the Second European Fluid Dynamics Conference, Stuttgart, Sept. 1994, 267 pages, Vol. II John Wiley, 1994.
Wagner, S., Michl, Th.: An Efficient Modular Parallel 3D-Navier-Stokes Solver. Proc. Second European Fluid Dynamics Conference, Stuttgart, Sept. 1994, Vol. I (Wagner, S., Hirschel, E.H., Periaux, J., Piva, R., Eds), John Wiley 1994, pp. 449–454, 1994.
Wagner, S., Seifert, H.J., Aldinger, F.: High-Temperature Reactions of C/C-SiC Composites with Precursor-Derived Ceramic Coatings. Mat. Manu. Proc., 17 (5) 619–635, 2002.
Wagner, S., Ullmann, T., Seifert, H.-J., Aldinger, R., Schmücker, M., Schneider, H.: Yttrium Silicate Coatings: Thermodynamic Assessment and High Temperature Investigations. J. Am. Ceram. Soc., 2001.
Wagner, S.: Direct Numerical Simulation of Laminar-Turbulent Transition. Invited Lecture at the 3rd ECCOMAS Europ. Fluid Dynamics Conference, Paris, 1996. In: Computational Methods in Applied Sciences 1996 (J.-A. Desideri, C. Hirsch, E. Ouate, M. Pandolfi, J. Periaux, E. Stein (Eds.)). John Wiley 1996, 70–90, 1996.
Wahlberg, L.: Stability Analysis of a Hypersonic Test Vehicle Based on Trim Curves. Seminar des Sonderforschungsbereichs 255, München, 12. Dez. 1996, Technische Universität München, 1997, pp. 155–160.
Wang, B.L., Habermann, M., Lenartz, M., Olivier, H., Grönig, H.: Detonation formation in $H_2$–$O_2$/He/Ar mixtures at elevation initial pressures. Shock Waves, Vol. 10, pp. 295–300, 2000.
Weigel, N., Kröplin, B., Dinkler, D.: Micromechanical Modeling of Damage and Failure Mechanisms in C/C-SiC. Comp. Materials Science, 16, 120–132, 1999.
Weimer, M., Hofhaus, J., Althaus, W.: Simulation of vortex breakdown. ZFW, Vol. 19, Nr. 6, 1995.
Weiner, R., Arnold, M., Rentrop, P., Strehmel, K.: Partitioning Strategies in Runge-Kutta Type Methods. Report M-9102, Mathematisches Institut, TU München, 1991. Submitted to: IMA J. Numer. Anal. 1992.
Weiner, R., Arnold, M., Rentrop, P., Strehmel, K.: Partitioning strategies in Runge-Kutta type methods. IMA J. Numer. Anal. 13, 1993, pp. 303–319.
Weinmann, M., Kamphowe, T.W., Schuhmacher, J., Müller, K., Aldinger, F.: Design of Polymeric Si-B-C-N Ceramic Precursors for the Application in Fiber-Reinforced Composite Materials. Chem. Mater., 12, p. 2112, 2000.

Weisgerber, H., Fischer, M., Magens, E., Beversdorff, M., Link, T., Koschel, W.W.: Experimental and numerical analysis of an expansion flow in chemical and thermal nonequilibrium. ISABE-99-7134, $14^{th}$ International Symposium on Air Breathing Engines, Florence, Italy, 1999.
Weisgerber, H., Fischer, M., Magens, E., Winandy, A., Foerster, W., Beversforff, M.: Experimental analysis of the flow of exhaust gas in a hypersonic nozzle. AIAA Paper 98-1600, $8^{th}$ International Space Planes and Hypersonic Systems and Technologies Conference, Norfolk, Virginia, USA, April 27–30, 1998.
Weiss, J., Knauss, H., Wagner, S., Kosinov, A.D. (ITAM): Constant Temperature Hot Wire Measurements in a Short Duration Supersonic Wind Tunnel. Paper 2646 in Aerodynamics Research Conference 2001, April 9–10, 2001, Royal Aeronautical Society, London, U.K., Special Edition, Vol. 105, No. 1050, pp. 435–441, August, 2001.
Weiss, J., Knauss, H., Wagner, S.: Method for the Determination of Frequency Response and Signal to Noise Ratio for Constant Temperature Hot Wire Anemometers. Review of Scientific Instruments, Vol. 72, No. 3, March, 2001.
Well, K.H.: ARIANE V Ascent Trajectory Optimization with a First-Stage Splash-Down Constraint. 8th Ifac Workshop on Control Applications of Nonlinear Programming and Optimization, Paris, 7.–9.6.1989. Pergamon Press, Oxford, Vol.2, June 1990.
Werner, K.-D.: The Evolution of Discontinuities in Solutions of Inhomogeneous Scalar Hyperbolic Conservation Laws in Several Space Dimensions. Institut für Geometrie und Praktische Mathematik, RWTH Aachen, Bericht Nr. 53, 1988.
Wittel, F., Kun, F., Kröplin, B., Herrmann, H.: A Study of Transverse Ply Cracking Using a Discrete Element Method. Comp. Materials Science, im Druck, 2003.
Yamaleev, N.K., Ballmann, J.: Iterative Space-Marching Method for Compressible Sub-, Trans- and Supersonic Flows. AIAA Journ., Vol. 38, No. 2, Febr. 2000, pp. 225–233.
Yamaleev, N.K., Ballmann, J.: Space-Marching Method for Calculating Steady Supersonic Flows on a Grid Adapted to the Solution. Journ. of Computational Physics, Vol. 146, 1998, pp. 436–463.
Zähringer, C., Breitsamter, C., Sachs, G., Laschka, B.: Windkanalversuche zum Trennvorgang eines zweistufigen Hyperschall-Flugsystems. DGLR-Jahrbuch 2001, ISSN 0070-4083, Band I, pp. 159–165, 2001.
Zähringer, C., Heller, M., Sachs, G.: Lateral Separation Dynamics and Stability of a Two-Stage Hypersonic Vehicle. 12th AIAA Space Planes and Hypersonic Systems and Technologies Conference, 15.–19. Dezember 2003, Norfolk, Virginia, USA, AIAA-2003-7080, 2003.
Zeutzius, M., Beylich, A.E.: Experimental investigation of asymmetric nozzles for advanced hypersonic space planes – structure of nozzle jets and thrust vector control. ZFW, Vol. 17 (1993), pp. 311–322.
Zhong, C., Reimerdes, H.-G.: A Higher-Order Theorie for Stability Analysis of Cylindrical and Conical Sandwich Shells with Flexible Cores. Fifth International Conference on Sandwich Construction, Zürich, 2000, Vol. I, pp. 113–127.
Zimmermann, F., Calise, A.J.: Numerical Optimization Study of Aeroassisted Orbital Transfer. Journal of Guidance, Control, and Dynamics, Vol. 21, No. 1, Jan.-Febr., pp. 127–133, 1998.

## 10.2 Dissertations

Albus, J.: Analytische und semi-analytische Berechnungsmethoden zur Auslegung zylindrischer Schalenstrukturen beliebiger Querschnittsform. RWTH Aachen, 1996.
Alefeld, M.: Visualisierung dreidimensionaler Szenen. Fakultät für Mathematik, Technische Universität München, 1998.
Allgayer, U.: Turbulente Verbrennung in kompressiblen Scherschichten. Fachgebiet Strömungsmechanik, Technische Universität München, 1999.
Arning, R.: Untersuchung flugmechanischer Eigenschaften eines Raumflugzeugs mit Hilfe frei fliegender Modelle. RWTH Aachen, 2001.
Bareis, B.: Systematische Untersuchung und Optimierung der Auslegungsparameter luftatmender Kombinationstriebwerke für eine Hyperschallbeschleunigungsmission, Universität Stuttgart, ILA, 1997.
Bauer, A.: Betriebsverhalten luftatmender Kombinationsantriebe für den Hyperschallflug unter besonderer Berücksichtigung der Triebwerksintegration. Lehrstuhl für Flugantriebe, Technische Universität München, 1994.
Bayer, R.: Optimierung von Flugbahnen der Unterstufe eines luftatmenden Raumtransporters. Lehrstuhl für Flugmechanik und Flugregelung, Technische Universität München, 1993.
Beauvais, R.: Brennverhalten vorgemischter, turbulenter Wasserstoff-Luft-Flammen in einem Explosionsrohr. Lehrstuhl A für Thermodynamik, Technische Universität München, 1994.
Bleilebens, M.: Einfluss der Wandtemperatur auf die Stoß-Grenzschicht Wechselwirkung an einer Rampe im Hyperschall. RWTH Aachen, (in preparation).
Bogner, S.: Experimentelle Validierung und Analyse stationärer und dynamischer Zustandsüberwachung von Hubschrauber-Gasturbinen. Lehrstuhl für Flugantriebe, Technische Universität München, 1993.
Boie, C.: Simulation der Vorgänge in einer Magnetoplasmaströmung eines MPD-Triebwerkes mit hochauflösenden numerischen Verfahren, Universität Stuttgart, IRS, 2000.
Bönisch, T.: Die Berechnung von Wiedereintrittsphänomenen auf hierarchischen Supercomputern mit einem effizienten parallelen Multiblockverfahren. Universität Stuttgart, RUS, (in progress).
Bregman, E. R.: Flugoptimierung von aerodynamisch gestützten Orbittransferfahrzeugen unter besonderer Berücksichtigung der aerothermodynamischen Fluglasten, Universität Stuttgart, IRS, 1994.
Breitsamter, C.: Turbulente Strömungsstrukturen an Flugzeugkonfigurationen mit Vorderkantenwirbeln. Lehrstuhl für Fluidmechanik, Technische Universität München, 1997.
Brodbeck, C. M.: Entwicklung eines strukturiert/unstrukturierten Verfahrens zur Lösung der Navier-Stokes-Gleichungen. RWTH Aachen, (in preparation).
Burkhardt, J.: Konzeptioneller Systementwurf und Missionsanalyse für einen auftriebsgestützten Rückkehrkörper. Universität Stuttgart, IRS, 2001.
Butzek, S.: Aspekte des Zusammenhangs von Optimalsteuerproblemen und zugehörigen relaxierten Problemen. Lehrstuhl für Numerische Mathematik und Optimierungstheorie, Universität Greifswald, 1997.
Callies, R.: Optimale Flugbahnen einer Raumsonde mit Ionentriebwerken. Fakultät für Maschinenwesen, Technische Universität München, 1990.
Chudej, K.: Optimale Steuerung des Aufstiegs eines zweistufigen Hyperschall-Raumtransporters. Fakultät für Mathematik, Technische Universität München, 1994.
Cvrlje, T.: Instationäre Aerodynamik des Separationsvorgangs zwischen Träger und Orbiter. Lehrstuhl für Fluidmechanik, Technische Universität München, 2001.
Dabalà, P.: Massenspektrometrische Untersuchungen zum Erosionsverhalten von Hitzeschutzmaterialien für Wiedereintrittskörper, Universität Stuttgart, IRS, 1998.
Daiss, A.: Modellierung katalytischer Reaktionen auf Siliziumdioxidoberflächen unter Berücksichtigung von Verdünnungseffekten, Universität Stuttgart, IRS, 1997.

Decker, F.: Experimentelle und theoretische Untersuchungen zur Aerodynamik der Hyperschallkonfiguration ELAC-1 im Niedergeschwindigkeitsbereich. RWTH Aachen, 1996.
Decker, K.: Instationäre Aerodynamik zu ein- und zweistufigen Raumtransportsystemen. Lehrstuhl für Fluidmechanik, Technische Universität München, 2003.
Denk, G.: Ein neues Diskretisierungsverfahren zur effizienten numerischen Lösung rasch oszillierender Differentialgleichungen. Fakultät für Mathematik, Technische Universität München, 1992.
Deutschmann, O.: Modellierung heterogener Reaktionssysteme, Universität Stuttgart, ITV, 1995.
Dewell, L.: Extensions of Optimal Periodic Control and Regulation of Constrained Hypersonic Cruise. University of California Los Angeles, 1996.
Dietl, U.: Beeinflussung des Haftschichtoxidationswiderstandes von Wärmedämmschichtsystemen für Gasturbinen. Lehrstuhl für Werkstoffe im Maschinenbau, Technische Universität München, 1994.
Dinkelmann, M.: Reduzierung der thermischen Belastung eines Hyperschallflugzeugs durch optimale Bahnsteuerung. Lehrstuhl für Flugmechanik und Flugregelung, Technische Universität München, 1997.
Dobler, K.: Untersuchung zu räumlich integrierten Flugführungsanzeigen. Lehrstuhl für Flugmechanik und Flugregelung, Technische Universität München, 2000.
Döhner, N.: Notwendige Optimalitätsbedingungen höherer Ordnung und Abstiegsverfahren. Lehrstuhl für Numerische Mathematik und Optimierungstheorie, Universität Greifswald, 1996.
Drexler, M.: Untersuchung optimaler Aufstiegsbahnen raketengetriebener Raumtransporter-Oberstufen. Lehrstuhl für Flugmechanik und Flugregelung, Technische Universität München, 1995.
Eissler, W.: Numerische Untersuchungen zum laminar-turbulenten Strömungsumschlag in Überschallgrenzschichten, Universität Stuttgart, IAG, 1995.
Endlich, P.: Experimentelle Simulation des Eintritts eines Raumflugkörpers in die Marsatmosphäre, Universität Stuttgart, IRS, (in progress).
Engler, V.: Bewertung und Optimierung von Raumflugzeugen unter Berücksichtigung der Sensitivität gegenüber Entwurfsunsicherheiten. Shaker Verlag, RWTH Aachen, 1999.
Esch, Th.: Zur Integration des Antriebs in ein Hyperschallflugzeug unter besonderer Berücksichtigung der Kräftebilanzierung. Lehrstuhl für Flugantriebe, Technische Universität München, 1997.
Essebier, S.: Berechnung dynamischer Delaminationsvorgänge in Faserverbundplatten, Universität Stuttgart, ISD, 1999.
Fasoulas, S.: Experimentelle und theoretische Charakterisierung einer hochenthalpen Stickstoffströmung zur Wiedereintrittssimulation, Universität Stuttgart, IRS, 1995.
Feigl, M.: Laserdiagnostische Untersuchungen an lichtbogenbeheizten Plasmaströmungen, Universität Stuttgart, IRS, (in progress).
Fertig, M.: Modellierung reaktiver Prozesse auf Siliziumkarbidoberflächen in verdünnten Nichtgleichgewichts-Luftströmungen, Universität Stuttgart, IRS, (in progress).
Fezer, A.: Numerische Simulation des laminar-turbulenten Strömungsumschlages in supersonischen Platten- und Kegelgrenzschichten, Universität Stuttgart, IAG, (in progress).
Fink, A.: Ein Grenzflächenmodell zur Beschreibung des mechanischen Verhaltens faserverstärkter Keramiken, Universität Stuttgart, ISD, 1995.
Fiola, R.: Berechnung des instationären Betriebsverhaltens von Gasturbinen unter besonderer Berücksichtigung von Sekundäreffekten. Lehrstuhl für Flugantriebe, Technische Universität München, 1993.
Franke, M.: Beitrag zur Berechnung und Bewertung der Wechselwirkungen zwischen Außenströmung und Antrieb bei einem Hyperschall-Raumtransporter vom Typ ELAC-1. RWTH Aachen, 1997.
Fühling, St.: Untersuchung transitioneller, ablösender und gestörter Grenzschichten mit der Multisensor-Heißfilmtechnik, Verlag Mainz, RWTH Aachen, 1999.

Gabler, W.: Gemischbildung, Flammenstabilisierung und Verbrennung in einer gestuften Überschallbrennkammer. Lehrstuhl A für Thermodynamik, Technische Universität München, 1996
Gaisbauer, U.: Untersuchungen zur Stoß-Grenzschicht-Wechselwirkung an einer Doppelrampe unter verschiedenen Randbedingungen, Universität Stuttgart, IAG, (in progress).
Gerlinger, P.: Numerische Berechnung turbulenter Verbrennungsvorgänge mit einem impliziten LU-Verfahren, Universität Stuttgart, IVLR, 1994.
Gern, F. H.: Kapillarität und Infiltrationsverhalten bei der Flüssigsilicierung von C/C-Bauteilen, DLR Stuttgart, 1995.
Gleis, S.: Diagnostik von Verbrennungsschwingungen – Untersuchungen zur Übertragung vom Kleinen in Große. Lehrstuhl B für Thermodynamik, Technische Universität München, 1995.
Glößner, C.: Kurzzeit Kraftmessungen an auftriebsgestützten Wiedereintrittskörpern in Hochenthalpie-Hyperschallströmungen. RWTH Aachen, (in preparation).
Gogel, T. H.: Numerische Modellierung von Hochenthalpieströmungen mit Strahlungsverlusten, Universität Stuttgart, IRS, 1994.
Gräßlin, M. H.: Erprobung eines prädiktiven Lenkkonzepts für Rückkehrmissionen auftriebsgestützter Raumfahrzeuge, Universität Stuttgart, IRS, (in progress).
Grau, Th.: Numerische Untersuchungen von Plasmawindkanalströmungen zur Wiedereintrittssimulation, Universität Stuttgart, IRS, 2000.
Grigat, E.: Berechnung von Optimalflugbahnen in Fallwindgebieten mittels eines Homotopieverfahrens variabler Ordnung. Lehrstuhl für Flugmechanik und Flugregelung, Technische Universität München, 1997.
Grotowsky, I. M. G.: Ein numerischer Algorithmus zur Lösung der Navier-Stokes-Gleichungen bei Überschallmachzahlen. RWTH Aachen, 1994.
Gruhn, P.: Düsenströmung für Hyperschallantriebe. RWTH Aachen, (in preparation).
Grünig, C.: Gemischbildung und Flammenstabilisierung bei Pylon-Einblasung in Überschallbrennkammern. Lehrstuhl für Thermodynamik, Technische Universität München, 1999.
Haarmann, T.: Numerische Berechnung des gekoppelten Wärmetransports in Strömungen und Strukturen. RWTH Aachen, (in preparation).
Haber, J.: Konstruktion und Implementierung eines neuen Verfahrens zur Kompression von Bilddaten. Fakultät für Mathematik, Technische Universität München, 1999, ISBN 3-89675-576-5.
Habermann, M.: Erzeugung von Hochenthalpieströmungen im Hyperschall mit einem detonationsgetriebenen Stoßwellenkanal. Shaker Verlag, RWTH Aachen, 2001.
Habibie, J. A.: Euler-Lösungen für instationäre längsbeschleunigte Strömungen um Tragflügelprofile. Lehrstuhl für Fluidmechanik, Technische Universität München, 1994.
Habiger, H.: Elektrostatische Sonden und Fabry-Perot-Interferometrie zur Untersuchung lichtbogenbeheizter Plasmen für Triebwerksanwendungen und Wiedereintrittssimulation, Universität Stuttgart, IRS, 1994.
Haibel, M.: Gemischbildung und Struktur schneller Wasserstoff-Luft-Flammen im Nahbereich turbulenter Rezirkulationsgebiete. Lehrstuhl A für Thermodynamik, Technische Universität München, 1994.
Haidinger, F. A.: Numerische Untersuchungen turbulenter Stoß/Grenzschicht-Wechselwirkungen. Lehrstuhl für Fluidmechanik, Technische Universität München, 1993.
Haindl, H.: Einfluß der Fertigungsparameter der Haftschicht auf die Lebensdauer keramischer Wärmedämmschichtsysteme. Lehrstuhl für Werkstoffe im Maschinenbau, Technische Universität München, 1997.
Hald, H.: Faserkeramiken für heiße Strukturen von Wiedereintrittsfahrzeugen – Simulation, Test und Vergleich mit experimentellen Flugdaten, DLR Stuttgart, 2001.
Hanke, M.: Eine numerische Methode zur Bestimmung erweiterter flugdynamischer Derivativa durch aerostrukturdynamische Simulation. RWTH Aachen, 2003.
Hänsel, C.: Charakterisierung von Schäden in Faserverbundwerkstoffen mittels Schwingungsanalyse, Universität Stuttgart, ISD, 1999.

Hase, R.: Über den Einfluss der Fernfeldrandbedingungen auf die numerische Berechnung von Außenströmungen mittels Eulerverfahren. Lehrstuhl für Fluidmechanik, Technische Universität München, 1993.
Hauser, Th.: Hyperschallströmung um stumpfe Kegel bei externen wellenartigen Störungen. Lehrstuhl für Fluidmechanik, Technische Universität München, 1998.
Heiermann, J.: Numerische Verfahren zur Lösung der Erhaltungsgleichungen von Magneto-Plasmen, Universität Stuttgart, IRS, 2003.
Heim, A.: Modellierung, Simulation und optimale Bahnplanung bei Industrierobotern. Fakultät für Mathematik, Technische Universität München, 1998.
Heimann, D.: Oxidationsschutzschichten für kohlefaserverstärkte Verbundwerkstoffe durch Polymer-Pyrolyse, Universität Stuttgart, MPI, 1996.
Heinold, B.: Optimalitätsbedingungen für Steueraufgaben mit diskreten Zustandsbeschränkungen – Integral- und Differentialgleichungsprozesse. Lehrstuhl für Numerische Mathematik und Optimierungstheorie, Universität Greifswald, 1995.
Heiser, W.: Experimentelle Untersuchungen von supersonischen Stoß-Grenzschicht-Wechselwirkungen mit Hilfe der Laser-Doppler-Anemometrie. Institut für Luftfahrttechnik und Leichtbau, Universität der Bundeswehr München, 1992.
Heller, G.: Aerodynamik von Deltaflügelkonfigurationen bei Schieben und Gieren. Lehrstuhl für Fluidmechanik, Technische Universität München, 1997.
Heller, M.: Untersuchung zur Steuerung und Robusten Regelung der Seitenbewegung von Hyperschall-Flugzeugen. Lehrstuhl für Flugmechanik und Flugregelung, Technische Universität München, 2000.
Henkner, J.: Phänomene der instationären Strömungsablösung bei schiebenden Tragflügeln. Lehrstuhl für Fluidmechanik, Technische Universität München, 1999.
Henze, A.: Integration der Navier-Stokes-Gleichungen für Hyperschallströmungen mit Realgaseffekten. RWTH Aachen, 1996.
Herdrich, G.: Aufbau und Charakterisierung eines induktiv beheizten Plasmawindkanals, Universität Stuttgart, IRS, (in progress).
Herrmann, C.: Optimierung von Isolatoren bei Überschalleinläufen. RWTH Aachen, (in preparation).
Hilfer, G.: Beitrag zur Analyse der Plasmawechselwirkungen keramischer Hitzeschutzmaterialien unter Wiedereintrittsbedingungen, Universität Stuttgart, IRS, 1995.
Hillesheimer, M.: Entwicklung eines Quasi-Expertensystems zur Flugbahn- und Systemoptimierung zukünftiger Raumtransporter, Universität Stuttgart, IRS, 1994.
Hiltmann, P.: Numerische Lösung von Mehrpunkt-Randwertproblemen und Aufgaben der optimalen Steuerung über endlichdimensionalen Räumen. Fakultät für Mathematik und Informatik, Technische Universität München, 1990.
Hodapp, M.-A.: Schädigungsabhängige Modellierung funktionell gradierter Faserverbundkeramiken, Universität Stuttgart, ISD, (in progress).
Hollmeier, S.: Simulation des Betriebsverhaltens von Antrieben für Raumtransporter/Hyperschallflugzeuge. Lehrstuhl für Flugantriebe, Technische Universität München, 1997.
Hönig, R.: Konzentrations- und Temperaturbestimmung in Brennkammern luftatmender Antriebe mit Hilfe laserspektroskopischer Messverfahren. Lehrstuhl für Flugantriebe, Technische Universität München, 1995.
Hornik, A.: Beiträge zur Optimierung von Trägerraketen. Lehrstuhl für Raumfahrttechnik, Technische Universität München, 1995.
Hornung, M.: Entwurf einer luftatmenden Oberstufe und Gesamtoptimierung eines transatmosphärischen Raumtransportsystems. Institut für Luftfahrttechnik, Universität der Bundeswehr München, 2003.
Huber, Th.: Fluktuationssplitting auf hierarchischen Netzen. Lehrstuhl für Fluidmechanik, Technische Universität München, 1994.
Infed, F.: Real-, Katalyse- und Verdünnungseffekte beim Wiedereintritt von Raumflugkörpern unter Berücksichtigung von Strahlungswechselwirkung und Wärmeleitung an der Oberfläche, Universität Stuttgart, IRS, (in progress).
Jahn, G.: Theoretische und experimentelle Untersuchung des Hitzeschutzverhaltens einer Rückkehrkapsel, Universität Stuttgart, IRS, 1998.

Jahnen, W.: Untersuchung von Strömungsinstabilitäten in einem mehrstufigen Axialverdichter unter dem Einfluß von Eintrittsstörungen. Institut für Strahlantriebe, Universität der Bundeswehr München, 1998.

Janovsky, R.: Analyse und Bewertung horizontal startender Raumtransportsysteme. VDI Verlag, Reihe 12, Nr. 242. RWTH Aachen, 1995.

Jentschke, H.: Spektroskopische Untersuchung eines luftähnlichen Plasmafreistrahls, Universität Stuttgart, IPF, 1995.

Jonas, S.: Implizites Godunov-Typ-Verfahren zur voll gekoppelten Berechnung reibungsfreier Hyperschallströmungen im thermo-chemischen Nichtgleichgewicht, Universität Stuttgart, IRS, 1993.

Kanne, S.: Zur thermo-chemischen Relaxation innerer Freiheitsgrade durch Stoß- und Strahlungsprozesse beim Wiedereintritt, Universität Stuttgart, IRS, 1999.

Kasper, B.: Phasengleichgewichte im System B-C-N-Si, Universität Stuttgart, MPI, 1998.

Kiehl, M.: Vektorisierung der Mehrzielmethode zur Lösung von Mehrpunkt-Randwertproblemen und Aufgaben der optimalen Steuerung. Fakultät für Mathematik und Informatik, Technische Universität München, 1989.

Kipp, A.: Spline-Galerkin-Approximation elliptischer Randwertprobleme, Universität Stuttgart, MIA, 1996.

Klenk, W.: Spektroskopische Untersuchungen der Relaxationszone hinter Verdichtungsstößen in Luft mit einem Infrarot-Dioden-Laser, Universität Stuttgart, ITLR, 1997.

Kloker, M.: Direkte numerische Simulation des laminar-turbulenten Strömungsumschlages in einer stark verzögerten Grenzschicht, Universität Stuttgart, IAG, 1993.

Klomfaß, A.: Hyperschallströmungen im thermodynamischen Nichtgleichgewicht. RWTH Aachen, 1994.

Knab, O.: Konsistente Mehrtemperatur-Modellierung von thermochemischen Relaxationsprozessen in Hyperschallströmungen, Universität Stuttgart, IRS, 1996.

Kopp, S.: Dynamische Echtzeit-Leistungssyntheserechnung mit Sekundär- und Störeffekten für Hyperschall-Luftstrahlantriebe. Lehrstuhl für Flugantriebe, Technische Universität München, 2000

Kornmann, R.: Berührungslose Verformungsmessung bei hohen Temperaturen, Universität Stuttgart, ISD, (in progress).

Kreichgauer, O.: Quantitatives Modell zur Simulation von Systembelastungen in Luftverkehrsabläufen. Lehrstuhl für Raumfahrttechnik, Technische Universität München, 1995.

Kreins, A.F.: Wärmestromverteilung und Strömungsfelduntersuchung in gestörten Hyperschall-Plattengrenzschichten. RWTH Aachen, 1994.

Kreiselmaier, E.: Berechnung instationärer Tragflügelumströmungen auf der Basis der zeitlinearisierten Euler-Gleichungen. Lehrstuhl für Fluidmechanik, Technische Universität München, 1998.

Krenkel, W.: Entwicklung eines kostengünstigen Verfahrens für die Herstellung von Bauteilen aus keramischen Verbundwerkstoffen, DLR Stuttgart, 2000.

Kropp, M.: Reagierende Über- und Hyperschallströmung eines Raumtransportsystems mit Außenverbrennung. Shaker Verlag, RWTH Aachen, 1998.

Kuhnle, J.: Verhalten des Werkstoffs C/SiC bei hohen Temperaturen und Sauerstoffkorrosion, Universität Stuttgart, MPA, 1997.

Lachner, R.: Mathematische Modellierung und experimentelle Untersuchung der Schadstoffentstehung in Brennkammern luftatmender Triebwerke. Lehrstuhl für Flugantriebe, Technische Universität München, 1998.

Lang, N.: Sichtbarmachung und Geschwindigkeitsmessung in Überschall-Leeseitenwirbeln mit Lichtschnittverfahren. Books on Demand, Norderstedt, RWTH Aachen, 2000.

Laschütza, H.: Thermisches Nichtgleichgewicht in den Löschgrenzschichten von stationären Flammen und gekühlten Wänden, Universität Stuttgart, ITLR, 1993.

Laure, S.: Experimentelle Simulation der Staupunktsströmung eines mit hoher Geschwindigkeit wiedereintretenden Raumflugkörpers und deren Charakterisierung mittels mechanischer Sonden, Universität Stuttgart, IRS, 1995.

Laux, T.: Untersuchung von Wärmedämmschichten in hochenthalpen Strömungen, Universität Stuttgart, IRS, 1996.

Le Duc, A.: Etude de stabilité d'écoulements faiblement compressibles, de giration, puis d'impact sur paroi, par théorie linéaire et simulation numérique directe. Ecole Centrale de Lyon und FG Strömungsmechanik, TU München, 2001.

Lechner, R.: Kompressible turbulente Kanalströmungen. Fachgebiet Strömungsmechanik. TU München, 2001.

Lenzner, S.: Numerische Analyse der Plasmaströmung in einer induktiv beheizten Plasmaquelle, Universität Stuttgart, IRS, 2000.

Léonard, G.: Experimentelle Untersuchungen zum Hochtemperaturverhalten nichtoxidkeramischer Sinterwerkstoffe bei gleichzeitiger strömungs-dynamischer und chemischer Belastung. RWTH Aachen, 1992.

Link, T.: Simulation von Nichtgleichgewichtsströmungen in Expansionsdüsen mit einem Finite-Elemente-Verfahren. RWTH Aachen, 2003.

Loesener, O.: Pyrometrische Temperaturmessung an Oberflächen von Hitzeschutzmaterialien im Plasmawindkanal, Universität Stuttgart, IRS, 1993.

Lösch, M.: Betrachtungen zum Leistungsverhalten parallelgestufter Trägerraketen. Lehrstuhl für Raumfahrttechnik, Technische Universität München, 1995.

Ludäscher, M.: Auswirkungen von Triebwerksstörungen auf die Flugdynamik eines zweistufigen Hyperschallfluggerätes. Lehrstuhl für Flugmechanik und Flugregelung, Technische Universität München, 1996.

Maier, H.-P.: Untersuchungen zur Eignung von C/C-SiC und ODS PM 2000 als Strukturwerkstoffe in Hochtemperaturwärmetauschern, Universität Stuttgart, MPA, (in progress).

Mayrhofer, M.: Verbesserung der Missionssicherheit eines zukünftigen Raumtransportsystems mittels Flugbahnoptimierung. Lehrstuhl für Flugmechanik und Flugregelung, Technische Universität München, 2002.

Mehlhorn, R.: Integrierte Arbeitsumgebung zur numerischen Berechnung von Problemen der optimalen Steuerung. Lehrstuhl für Flugmechanik und Flugregelung, Technische Universität München, 1996.

Merk, J.: Hierarchische, kontinuumbasierte Schalenelemente höherer Ordnung, Universität Stuttgart, ISD, 1996.

Metzler, T.: Triebwerksleistungsrechnung mit Wärmeübertragungs-Kennfeldern für stationäre und instationäre Vorgänge basierend auf Versuchsergebnissen, Universität Stuttgart, ILA, (in progress).

Miermeister, M.: Statik und Stabilität nichtrotationssymmetrischer Schalenstrukturen – Allgemeine Grundgleichungen in analytischer Darstellung und ihre Anwendung. RWTH Aachen, 1996.

Mihatsch, O.: Früherkennung kritischer Zustände in chemischen Reaktoren mit Hilfe neuronaler Netze. Fakultät für Mathematik, Technische Universität München, 1998.

Möbus, H.: Euler und Lagrange – Monte-Carlo-PDF-Simulation turbulenter Strömungs- Mischungs- und Verbrennungsvorgänge, Universität Stuttgart, ITLR, 2001.

Möller, H.: Computergenerierte synthetische Sicht zur Verbesserung der Flugführung bei schlechter Außensicht. Lehrstuhl für Flugmechanik und Flugregelung, Technische Universität München, 1997.

Moravszki, C.: Improvement of Trajectory Control of Hypersonic Vehicles by Means of Predictive 3D Display Technique. Lehrstuhl für Flugmechanik und Flugregelung, Technische Universität München, (in progress).

Moser, L. U.: Experimentelle Untersuchungen stoßbedingter Realgaseffekte von Stickstoffmonoxid. Lehrstuhl für Strömungsmechanik, Universität der Bundeswehr München, 1990.

Müller, C. A.: Ein Berechnungsverfahren für abgelöste Unterschallströmungen um schlanke Delta-Konfigurationen mit glatter Oberfläche. RWTH Aachen, 1994.

Müller, S.: Erweiterung von ENO-Verfahren auf zwei Raumdimensionen und Anwendung auf hypersonische Staupunktprobleme. RWTH Aachen, 1993.

Müller, W.: Numerische Untersuchung räumlicher Umschlagsvorgänge in dreidimensionalen Grenzschichtströmungen, Universität Stuttgart, IAG, 1995.

Niestroj, O.: Grundlagenuntersuchung zur Anpassung von Einlaufdiffusor und Triebwerk eines zukünftigen Hyperschallflugzeuges mit Hilfe eines Zeitschrittverfahrens. RWTH Aachen, 1996.
Ohmenhäuser, F.: Zur Rissorientierung in Materialmodellen der fiktiven Rissbildung, Universität Stuttgart, ISD, 2001.
Olawsky, F.: Effizientes Newton-GMRES-Verfahren zur Berechnung stationärer Hyperschallströmungen im thermochemischen Nichtgleichgewicht, Universität Stuttgart, IRS, (in progress).
Olejak, D.: Grundlagenuntersuchungen zu supersonischen Strömungsfeldern sowie quantitative Erfassung kanalspezifischer Phänomene mittels Laser-Doppler-Anemometrie. Institut für Luftfahrttechnik und Leichtbau, Universität der Bundeswehr München, 1995.
Pagella, A.: Numerische Untersuchungen zur transitionellen Stoß-Grenzschicht Interaktion bei Ma=4,8 und Ma=6, Universität Stuttgart, IAG, (in progress).
Peiter, U.: Beeinflussung der Aerodynamik der Konfiguration ELAC durch Oberstufe, Triebwerk und Bodeneffekt. RWTH Aachen, (in preparation).
Peng, J.: Thermochemie und Konstitution von Si-(B-)C-N-Precursorkeramiken, Universität Stuttgart, MPI, 2002.
Penski, Ch.: Numerische Integration stochastischer differential-algebraischer Gleichungen in elektrischen Schaltungen. Fakultät für Mathematik, Technische Universität München, submitted 2000.
Qatanani, N.: Lösungsverfahren und Analysis der Integralgleichung für das Hohlraumstrahlungsproblem, Universität Stuttgart, MIA, 1996.
Rahn, M.: Eine numerische Methodik zur simultanen Flug- und Systemoptimierung von Raumtransportsystemen, Universität Stuttgart, IRS, 1998.
Raible, T.: Bewertung und Optimierung zweistufiger Raumflugzeuge mit Nurflügler- und Wellenreiter-Unterstufe. RWTH Aachen, (in preparation).
Rau, C.: Kristallisationsverhalten dünner SiC-Schichten aus Precursoren, Universität Stuttgart, MPI, 1998.
Reisinger, D.: Experimentelle Untersuchungen zur stoßinduzierten Ablösung an zweidimensionalen Rampen im Überschall. Institut für Luftfahrttechnik und Leichtbau, Universität der Bundeswehr München, 1997.
Reitenbacher, F. S.: Ein Verfahren zur approximativen Stabilitätsanalyse von nichtlinearen, dynamischen Systemen und seine Anwendung auf flugmechanische Problemstellungen. RWTH Aachen, 1997.
Resch, M.: Metacomputing für Simulationsanwendungen, PhD-thesis, RUS-51, Universität Stuttgart, RUS, 2001.
Riedel, R.: Erprobung eines Stoßwindkanals zur Untersuchung des laminar-turbulenten Umschlags in Überschallgrenzschichten, Universität Stuttgart, IAG, 2000.
Riedel, U.: Numerische Simulation reaktiver Hyperschallströmungen mit detaillierten Reaktionsmechanismen, Universität Stuttgart, ITV/Universität Heidelberg, 1992.
Riegler, C.: Modulares Leistungsberechnungsverfahren für Turboflugtriebwerke mit Kennfelddarstellung für Wärmeübertragungsvorgänge, Universität Stuttgart, ILA, 1997.
Rihaczek, C.: Anwendung objektorientierter Techniken zur Berechnung des gekoppelten Wärmetransports durch Leitung und Strahlung, Universität Stuttgart, ISD, 1993.
Rinderknecht, S.: Delamination in Faserverbundplatten – ein vereinfachtes Berechnungsmodell, Universität Stuttgart, ISD, 1994.
Rochholz, H.: Eulerlösungen für den Separationsvorgang von Träger-Orbiter-Systemen im Hyperschall. Lehrstuhl für Fluidmechanik, Technische Universität München, 1994.
Röck, W.: Simulation des Eintritts einer Sonde in die Atmosphäre des Saturnmondes Titan in einem Plasmawindkanal, Universität Stuttgart, IRS, 1999.
Roennecke, A. J.: Ein bordautonomes Verfahren zur Flugführung und Regelung von Rückkehrfahrzeugen, Universität Stuttgart, IFR, 2001.
Rössle, A.: Asymptotische Entwicklungen für dünne Platten im Rahmen der linearen Elastostatik, Universität Stuttgart, MIA, 1999.

Rüggeberg, T.: Experimentelle Untersuchungen des Betriebsverhaltens eines Triebwerkseinlaufs bei Überschall-Anström-Machzahlen. RWTH Aachen, 1996.
Rupp, O.: Vorhersage von Instandhaltungskosten bei der Auslegung ziviler Strahltriebwerke. Lehrstuhl für Flugantriebe, Technische Universität München, 2000.
Schinköth, D.: Laser-Streu-Diagnostik im Vergleich mit Emissionsspektroskopie an einem Freistrahlplasma, Universität Stuttgart, IPF, 2001.
Schlaich, F.: Experimentelle Untersuchungen zu Stoß-Grenzschichtwechsel-wirkungen an der kurzen Doppelrampe im Überschall, Universität Stuttgart, IAG, 1996.
Schmid, G.: Zur Entwicklung objektorientierter Finite-Elemente-Programme, Universität Stuttgart, ISD, 1998.
Schmid, N. R.: Leistungsverhalten von Kombinationstriebwerken unter dem Einfluß von Hyperschallflug-typischen Eintrittsstörungen. Institut für Strahlantriebe, Universität der Bundeswehr München, 2000.
Schmidt, D.: Modellierung laminarer reaktiver Strömungen mit Hilfe reduzierter Kinetik, Universität Stuttgart, ITV, 1995.
Schmitz, E.: Kalte Triebwerkssimulation am Raumtransportsystem ELAC 1. RWTH Aachen, 2002.
Schoder, W.: Untersuchung zur Steuerung und Regelung eines Hyperschall-Flugsystems beim Separationsmanöver. Lehrstuhl für Flugmechanik und Flugregelung, Technische Universität München, 1995.
Schöll, E.: Ein dreidimensionales Rechenverfahren für reibungsbehaftete Strömungen in Schaufelreihen von Turbomaschinen, Universität Stuttgart, IRS, 1995.
Schönemann, A.: Massenspektrometrie zur Untersuchung lichtbogenbeheizter Plasmen im Niederdruck-Plasmawindkanal, Universität Stuttgart, IRS, 1994.
Schreiber, E.: Verbesserung der Messgenauigkeit von Emissionsgradmessungen durch den Einsatz eines neuartigen Multispektralpyrometers, Universität Stuttgart, IKE, (in progress).
Schulte, D.: Beeinflussung viskoser Strömungseffekte in Hyperschall-Einläufen. ISBN 3-8265-8882-7, Shaker Verlag, RWTH Aachen, 2001.
Schulte, H.: Leistungsverhalten von Kombinationsantrieben mit Zweistrom-Turboteil, Universität Stuttgart, ILA, 2003.
Schulte-Fischedick, J.: Untersuchungen zur Entstehung des Rissmusters während der Pyrolyse von CFK-Vorkörpern zur Herstellung von C/C-SiC-Werkstoffen, DLR Stuttgart, (in progress).
Schulz, M.: A-Posteriori Fehlerschätzer für mittels Finiter Elemente modellierte elastoplastische Verformungsvorgänge, Universität Stuttgart, MIA, 1997.
Seling, F.: Dreidimensionale Simulation von Problematiken der Überschallverbrennung, DLR Stuttgart, (in progress).
Sleziona, P.C.: Numerische Analyse der Strömungsvorgänge in magnetoplasmadynamischen Raumfahrtantrieben, Universität Stuttgart, IRS, 1992.
Stantschev, G.: Langfaserverstärkte SiBCN-Keramiken, DLR Stuttgart, (in progress).
Steinbach, O.: Gebietszerlegungsmethoden mit Randintegralgleichungen und effiziente numerische Lösungsverfahren für gemischte Randwertprobleme, Universität Stuttgart, MIA, 1996.
Steinebach, D.A.: Untersuchung zur Auslegung von luftatmenden Antriebssystemen für horizontal startende Raumtransporter. RWTH Aachen, 1997.
Steppe, P.: Haftschichtoxidation ausgewählter MCr-AL-Y-Legierung. Lehrstuhl für Werkstoffe im Maschinenbau, Technische Universität München, 1991.
Steppe, P.: Modellversuche zum Hochtemperatur-Oxidationsverhalten plasmagespritzter Wärmedämmschichtsysteme auf Flug-Gasturbinen-Schaufeln in Luft. Lehrstuhl für Werkstoffe im Maschinenbau, Technische Universität München, 1991.
Stich, R.: Flugeigenschaftsuntersuchung zur Langzeitdynamik der Längsbewegung im Hyperschallflug. Lehrstuhl für Flugmechanik und Flugregelung, Technische Universität München, 2003.

Stöckle, T.: Experimentelle und numerische Untersuchung der Oberflächenkatalyzität metallischer und keramischer Werkstoffe in Hochenthalpieströmungen, Universität Stuttgart, IRS, 2000.

Stoll, P.: Entwicklung eines parallelen Mehrgitterverfahrens zur Verbrennungssimulation in kompressiblen und inkompressiblen Strömungen, Universität Stuttgart, ITLR, 2001.

Störkmann, V.: Kraftmessungen an Modellen in Hyperschallkanälen. Shaker Verlag, RWTH Aachen, 1998.

Stromberg, A.: Experimentelle Untersuchungen an der Hyperschallkonfiguration ELAC I bei Unter- und Überschallanströmung. RWTH Aachen, 1994.

Strube, G.: Struktur und Brenngeschwindigkeit turbulenter, vorgemischter Wasserstoff-Flammen. Lehrstuhl A für Thermodynamik, Technische Universität München, 1993.

Therkorn, D.: Fortschrittliches Leistungs-Berechnungsverfahren für luftatmende Turbotriebwerke, Universität Stuttgart, ILA, 1992.

Theuer, A.: Experimentelle Untersuchungen zum thermomechanischen Verhalten von Faserkeramik, Universität Stuttgart, IFB, 1996.

Thumm, A.: Numerische Untersuchungen zum laminar-turbulenten Strömungsumschlag in transsonischen Grenzschichtströmungen, Universität Stuttgart, IAG, 1991.

Ting, C.-C.: Strömungs- und Wärmeübergangsmessung für das zweistufige Raumtransportsystem ELAC. RWTH Aachen, 2003.

Tremblay, F.: Direct and large-eddy simulation of flow around a circular cylinder at subcritical Reynolds numbers. FG Strömungsmechanik, TU München, 2001.

Tscharnuter, D.: Optimale Auslegung des Antriebsstrangs von Kraftfahrzeugen. Fakultät für Mathematik, Technische Universität München, submitted 2000.

Ullmann, T.: Plasmagespritzte keramische Oxidationsschutzschichten für thermo-mechanisch belastete C/C-SiC-Strukturen rückkehrfähiger Raumfahrzeuge, DLR Stuttgart, (in progress).

van Keuk, J.: Numerische Analyse und Bewertung von Upwind-Verfahren für die Anwendung auf mehrdimensionale Über- und Hyperschallströmungen von Gasen. Shaker Verlag, RWTH Aachen, 2000.

Vetter, M.: Hyperschallumströmung von Modellen im Stoßwellenkanal. RWTH Aachen, 1993.

Volk, G.: Zeitaufgelöste Bestimmung von Plasmaparametern der stillen Entladung in Stickstoff, Universität Stuttgart, IPF, 1994.

von Stryk, O.: Numerische Lösung optimaler Steuerungsprobleme: Diskretisierung, Parameteroptimierung und Berechnung der adjungierten Variablen. Fakultät für Mathematik, Technische Universität München, 1994.

Wagner, S.: Thermodynamik der Grenzflächenreaktionen in Schutzschichten für Kohlefaserverbundwerkstoffe, Universität Stuttgart, MPI, 2003.

Walther, C.: Systemtechnische Verfahren zur Bestimmung der Zusammenhänge zwischen Eigenschaften und Funktionsstruktur technischer Systeme. Lehrstuhl für Raumfahrttechnik, Technische Universität München, 1994.

Wegmann, T.: Untersuchung der Grenzen des Betriebsbereichs von magnetoplasmadynamischen Raumfahrtantrieben, Universität Stuttgart, IRS, 1994.

Wegner, W.: Vollständige Riemannlösung der ein- und zweidimensionalen Euler-Gleichungen. Lehrstuhl für Fluidmechanik, Technische Universität München, 1992.

Weigel, N.: Ein Mechanismen-basiertes Werkstoffmodell für Faserkeramiken, Universität Stuttgart, ISD, 2000.

Weisgerber, H.: Experimentelle Untersuchungen zur Nichtgleichgewichtsexpansionsströmung nach Wasserstoff-Luft-Verbrennung in Hyperschall-Staustrahltriebwerken. RWTH Aachen, 2001.

Weiss, J.: Experimental Determination of the Free Stream Disturbance Field in the Short Duration Supersonic Wind Tunnel of Stuttgart University, Universität Stuttgart, IAG, 2002.

Werle, M.: Numerische Simulation der Fluid-Festkörper-Interaktion dünnwandiger Strukturen im Überschall unter Berücksichtigung des Wärmeaustauschs. RWTH Aachen, (in preparation).

Widdecke, L.: Relaxationsprozesse hinter einfallenden Stoßwellen in $N_2$-$O_2$-NO-Gemischen, Universität Stuttgart, ITLR, 1995.
Wimmer, G.: Kombination von direkten und indirekten Verfahren zur Optimalsteuerung. Fakultät für Mathematik, Technische Universität München, submitted Sept. 2003
Winter, M.: Emissionsspektroskopische Untersuchungen an Plasmabeschleunigern, Universität Stuttgart, IRS, (in progress).
Zeutzius, M.: Überschalldüsen mit Schubvektorsteuerung. RWTH Aachen, 1994.
Zhou, F.W.: Numerische Simulation des Strömungsfeldes des zweistufigen Raumflugzeugs ELAC. RWTH Aachen, 2003.
Zimmermann, F.: Optimierung der seilgestützten Rückkehrmission einer gelenkten Wiedereintrittskapsel, Universität Stuttgart, IRS, 2001.

## 10.3 Habilitations

Bill, J.: Keramische Si-(B-)C-N-Materialien aus molekularen Vorstufen. MPI, INAM, Universität Stuttgart, 2001.
Callies, R.: Entwurfsoptimierung und optimale Steuerung. Differential-algebraische Systeme, Mehrgitter-Mehrzielansätze und numerische Realisierung. Fakultät für Mathematik, Technische Universität München, Juli 2000.
Gerlinger, P.: Effiziente numerische Simulation turbulenter Verbrennung. Universität Stuttgart, IVLR, 2003.
Glaser, S.: Gekoppelte thermomechanische Berechnung dünnwandiger Strukturen mit der Methode der Finiten-Elemente. Universität Stuttgart, ISD, 1999.
Kiehl, M.: Simulation reaktionskinetischer Prozesse. Modellierung, Partitionierung, Sensitivitätsanalyse und Parallelisierung. Zentrum Mathematik, Technische Universität München, 1995.
Kugelmann, B.: Ein paralleles Rückkopplungsverfahren zur Lösung von Optimalsteuerungsproblemen. Zentrum Mathematik, Technische Universität München, 1995.
Manhart, M.: Direkte numerische Simulation – Ein Werkzeug zur Vorhersage und Analyse komplexer turbulenter Strömungen. Fachgebiet Strömungsmechanik, Technische Universität München, 2002.
Pomp, A.: Eine Rand-Gebiets-Integralmethode zur numerischen Lösung der Schalengleichung eines elliptischen Differentialgleichungssystems mit variablen Koeffizienten. Universität Stuttgart, MIA, 1996.
Rist, U.: Zur Instabilität und Transition in laminaren Ablöseblasen. Universität Stuttgart, IAG, 1998.
Sleziona, C.: Hochenthalpieströmungen für Raumfahrtanwendungen. Universität Stuttgart, IRS, 1998.
Steinbach, O.: Stability estimates for hybrid-coupled domain decomposition methods. Universität Stuttgart, MIA, 2001.
von Stryk, O.: Numerical Hybrid Optimal Control and Related Topics. Fakultät für Mathematik, Technische Universität München, submitted 2000.

## 10.4 Patents

Aldinger, F., Bill, J.: Herstellung einkristalliner dünner Schichten aus SiC, (195 03 976.9), Deutsche Patentanmeldung.

Aldinger, F., Bill, J.: Production of thin single-crystal SiC Layers, (PCT) (WO 96/24709), Internationale Patentanmeldung.

Henke, T., Krenkel, W.: Verfahren zum Erzeugen einer SiC enthaltenden Schutzschicht, DE0019834018C1.

Hirsch, K., Kaiser, M., Greiner, A., Bill, J., Aldinger, F.: Beschichtung zur Verminderung der Erosion an thermisch hochbelasteten Oberflächen aus faserverstärkter Keramik und Verfahren zu deren Herstellung, Patent Nr. 199 28 173.4, 19. 08. 1999, Deutsches Patentamt München, (Fraunhofer Gesellschaft).

Hirsch, K., Kaiser, M., Greiner, A., Bill, J., Aldinger, F.: Verwendung einer Beschichtung mit faserverstärkter Keramik und Verfahren zu deren Herstellung. PCT/EP 00/05555, 16.06.2000, Europäisches Patentamt München, Offenlegung: August 2000, (Fraunhofer Gesellschaft).

Hirsch, K., Roth, B., Altmann, I., Schinköth, D., Lützenburger, N., Krenkel, W., Kochendörfer, R.: Bauteil, das eine Schutzschicht gegen Erosion durch thermische Belastung aufweist und Verfahren zur Herstellung einer Schutzschicht auf einem Bauteil, 10048764.5, 29. 9. 2000, Deutsche Patentanmeldung, (DLR).

Kochendörfer, R.: Verfahren und Vorrichtung zur Übertragung von Kräften zwischen zwei Fügeteilen (195 24 708.6-32).

Krenkel, W., Kochendörfer, R.: Verfahren zum dauerhaften Verbinden von wenigstens zwei Bauteilkomponenten zu einem Formkörper, DE0019636223C2.

Messerschmid, E.W., Glocker, B.: Hochtemperatur-Pyrolyse, P 44 17 646.5-33.

## 10.5 Number of Diploma Theses

Number of diploma theses conducted at the three Collaborative Centres: 491.

## 10.6 Visiting Researchers

Dipl.-Ing N.A. Adams, DFLR-Institut für Theoretische Strömungsmechanik: 11.–12.02.1992, SFB 255.

Dr. N.A. Adams, ETH Zürich, Zürich, Schweiz: 17.–19.11.1997, SFB 255.

Prof. Agarwal, National University of Singapore, Singapur: 19.–24.11.1995, SFB 255.

Prof. I.Sh. Akhatov, Vice President, Russian Academy of Sciences, Moskau, Russland, SFB 255.

Doz. Dr. Ananev, IMM Uro RAN, Jekatarinenburg, Russland: 23.–28.09.1996, SFB 255.

Prof. M.D Ardema, Department of Mechanical Engineering, Santa Clara University: 26.05.–01.06.1991, SFB 255.

Prof. Dr. D. Arnal, ONERA, Cert, Toulouse, France: 01.04.–04.04.2001, SFB 259.

Dr. V. Avrashkov, Moscow Aviation Institute, Moskau, Russland: 03.–11. 11. 1994/10.–20. 01. 1995/01. 10.–20. 12. 1995/07. 02.–31. 03. 1996/01. 06.–15. 07. 1996, SFB 255.
Prof. S. Baranovsky, Moscow Aviation Institute, Moskau, Russland: 27.–30. 10. 1992, SFB 255.
Prof. Z. P. Bazant, Northwestern University, USA: 01.–30. 06. 1991, SFB 259.
Prof. H. Behnke, Universität Osnabrück, Osnabrück: 11.–12. 11. 1997, SFB 255.
Dipl.-Ing. F. Bergmann, DFLR-Institut für Theoretische Strömungsmechanik, Göttingen: 14.–15. 01. 1992, SFB 255.
Dr. W. Berry, Dept. of Propulsion and Aerothermodynamics, Noordwijk, Niederlande: 07.–08. 02. 1996, SFB 255.
Dr. F. P. Bertolotti, DLR Göttingen, Göttingen: 07.–08. 05. 1996, SFB 255.
Dr. J. Betts, Boeing, Seattle, USA: 26. 05.–01. 06. 1991/27.–31. 01. 1996, SFB 255.
Prof. Dr. R. Bialecki, Silesian Technical University, Gliwice, Poland: 16.–18. 02. 1998, SFB 259.
Prof. Dr. L. Bittner, Universität Greifswald, Sektion Mathematik: 19. 02.–10. 03. 1990, SFB 255.
Prof. Dr. V. Boltyansky, Scientific Institute of Systems Research, Moskau, Russland: 12. 05.–04. 06. 1991/16.–19. 02. 1993/05.–25. 07. 1993/27. 04.–15. 05. 1994/06.–13. 12. 1995/ 03.–12. 11. 1996, SFB 255.
Dr. Nicola Botta, Potsdam Institute for Climate Impact Research, Potsdam: 05. 02. 2002, SFB 255.
Dr. Michael H. Breitner, TU Clausthal, FB Mathematik und Informatik: 06.–15. 08. 1999, SFB 255.
Dr. M. Brodetsky, Institute of Theoretical and Applied Mechanics, Russian Academy of Sciences, Siberian Divison, Novosibirsk, Russia: 22.–24. 11. 2000, SFB 255.
Prof. M. Brokate, Universität Kiel, Kiel: 04.–06. 03. 1996, SFB 255.
Prof. N. Chokani, North Carolina State University, USA: 12.–16. 12. 2001/17. 04.–30. 04. 2002, SFB 259.
Prof. G. Comte-Bellot, Ecole Centrale de Lyon, France: 14.–15. 12. 2001/17.–30. 04. 2002, SFB 259.
Prof. Dr. A. Bruce, Conway, Dept. of Aeronautical-Astronautical Engineering, University of Illinois at Urbana: 26.–28. 06. 2000, SFB 255.
Dr. C. Brun, Commissariat à l'Energie Atomique, Grenoble, Frankreich: 19.–20. 03. 1998, SFB 255.
Doc.Rndr. Miroslav Brzezina, Technische Universität Liberec, Tschechien: 19.–24. 07. 1999, SFB 255.
Prof. Dr. James R. Bunch, Dept. of Mathematics, UC San Diego: 08.–15. 04. 1999, SFB 255.
Prof. J. A. Burns, Department of Aerospace and Ocean Engineering, Virginia Tech, Blacksburg: 26. 05.–01. 06. 1991, SFB 255.
Dr. C. Büskens, Lehrstuhl für Ingenieurmathematik, Universität Bayreuth: 01. 02. 2001, SFB 255.
Prof. John Butcher, Dept. of Mathematics, University of Auckland, New Zealand: 20.–30. 10. 2000, SFB 255.
Prof. A Calise, Department of Aerospace Engineering, Georgia Tech, Atlanta: 26. 05.–04. 06. 1991, SFB 255.
Dr. W. Castillo, Escuela de Matematica Universidad de Costa Rica, Costa Rica: 20.–24. 11. 1992, SFB 255.
Prof. F. L. Chernousko, Russische Akademie der Wissenschaften, Moskau, Russland: 01.–10. 09. 1992, SFB 255.
Prof. Dr. F. L. Chernousko, Akademie der Wissenschaften der UdSSR, Moskau: 25. 03.–15. 04. 1990, SFB 255.
Dr. K. Chudej, Lehrstuhl für Ingenieurmathematik, Universität Bayreuth: 31. 05. 2000, SFB 255.
Prof. E. M. Cliff, Department of Aerospace and Ocean Engineering, Virginia Tech, Blacksburg: 26. 05.–04. 06. 1991, SFB 255.

Prof. Dr. Bruce A. Conway, Dept. of Aeronautical-Astronautical Engineering, University of Illinois at Urbana: 26.–28.06.2000, SFB 255.
T. Cox, NASA Dryden Flight Research Center, Edwards, USA: 20.01.–01.02.1992/25.11.–04.12.1993, SFB 255.
Prof. C. de Boor, University of Wisconsin, Madison, USA: 01.06.–31.07.1993, SFB 255.
Dr. G. Deng, Ecole Central Nantes, Nantes, Frankreich: 28.01.–05.02.1998, SFB 255.
Prof. S.M. Deshpande, Indian Institute of Technology, Bangalore, Indien: 05.–09.07.1997, SFB 255.
Prof. J.A. Desideri, INRIA, France: 08.–10.12.1991, SFB 259.
L. Dewell, University of California Los Angeles, zur Zeit Politecnico di Milano, Mailand, Italien: 07.–12.01.1996/04.–07.03.1996/10.–13.06.1996, SFB 255.
Prof. Dr. Harijono Djodjodihardjo, Department of Aeronautics and Astronautics, Faculty of Industrial Technology, Institute of Technology Bandung, Bandung, Indonesien: 01.07.–20.10.2002, SFB 255.
Prof. Dr. A. Draux, INSA de Rouen, Cedex, France: 16.–18.02.1998, SFB 259.
Prof. Dr. M. Dudeck, CNRS, Orléans, France: 15.–21.08.1993, SFB 259.
Prof. V. Duganov, Moscow Aviation Institute, Department of the Theory of Air-Breathing Engines, Moskau, Rußland: 15.–19.06.1998, SFB 255.
Dr. H. Eckelmann, Universität Göttingen, Göttingen: 02.–03.12.1997, SFB 255.
Dr. A. Emelyanov, Moscow-Chernogolorka, Russia: 25.–30.08.1992, SFB 259.
E. Enevoldson, NASA Dryden Flight Research Center, USA: 27.04.1994/28.03.2001, SFB 255.
Dr. M. Farge, Ecole Normale Supérieure de Paris, Paris, Frankreich: 22.–23.09.1994, SFB 255.
Prof. Ferzinger, Stanford University: 26.–27.05.1991, SFB 255.
Dr. D.G. Fletcher, NASA Ames Research Center, USA: 15.–25.11.1999, SFB 259.
Dr. D. Fu, Peking Research Institute of Materials and Technology, Peking, China: 01.–31.01.1996, SFB 255.
Prof. Gabasov, Universität Minsk, Minsk, Weißrußland: 08.–12.12.1997, SFB 255.
Jörg Gablonsky, Center for Research in Scientific Computation, North Carolina State University, USA: 15.–18.06.1999, SFB 255.
Prof. Gad-el-Hak, University of Notre Dame, USA: 09.–10.07.2001, SFB 255.
Dr. Eduardo Gallestey, Dept. of Engineering, Australian National University, Canberra: 20.–26.08.1999, SFB 255.
Prof. R. Gamkrelidze, Steklov Institut Moskau, Moskau, Russland: 25.–28.05.1994, SFB 255.
J. Gera, NASA Dryden Flight Research Center, USA: 04.06.2001, SFB 255.
Prof. C. Gantes, Technical Universität, Athens, Greece: 11.–15.12.1996, SFB 259.
Prof. Dr. S.A. Gaponov, SB RAS ITAM, Novosibirsk, Russia: 15.–25.11.1999, SFB 259.
Prof. M. Gaster, Queen Mary&Westfield College, London, England: 27.–31.03.1996, SFB 259.
Prof. Ph. E. Gill, University of California, San Diego, USA: 16.–31.12.1995, SFB 259.
Dr. T. Grabowski, Warsaw University of Technology, Warschau, Polen: 22.–25.10.1997, SFB 255.
Dr. M.D. Grigorjan, Staatliche Universität Erewan: 17.–19.12.1992, SFB 255.
Prof. Takishi Goto, Tohoku University Sendai, Japan: 01.–30.11.2002, SFB 259.
Prof. Dr. R. Grundmann, TU Dresden, Germany: 21.–26.10.1993, SFB 259.
Prof. V.F. Gubarew, Ukrainian National Space Agency, Kiew, Ukraine: 22.–25.10.1997, SFB 255.
Prof. Y. Gupalo, Abt. Chem. Technik des Institutes für Probleme der Mechanik der Akademie der Wissenschaften, Moskau, Russland, SFB 255.
Prof. V. Gusev, TsAGI Moscow, Russia: 18.–19.02.2002, SFB 259.
Prof. Dr. L. Györfi, TU Budapest, Ungarn: 15.–25.11.1993, SFB 259.
Dipl.-Ing. E. Haile, Ecole Centrale Paris, Paris, Frankreich: 02.–03.05.1996, SFB 255.
Prof. J. Hansen, University of Toronto, Canada: 22.–24.06.1994, SFB 259.
Dr. C. Härtel, ETH Zürich, Zürich, Schweiz: 16.–18.02.1998, SFB 255.

Dr. St. Hechz, Delft Univ. of Techn, Delft, Niederlande: 26.–29.07.1998, SFB 255.
J. Hicks, NASA Dryden Flight Research Center, Edwards, USA: 11.–19.05.1991/20.–22.09.1991/12.–23.10.1993, SFB 255.
Prof. Th. J.R. Hughes, Stanford University, USA: 17.–19.11.1993, SFB 259.
Prof. Ivanov, Institute of Theoretical and Applied Mechanics, Novosibirsk, Russland: 10.–12.11.2002, SFB 255.
Prof. Dr. M.S. Ivanov, Academy of Science (ITAM), Novosibirsk, Russia: 22.–24.06.1994/03.–04.11.1996, SFB 259.
Prof. Dr. G. Jarkova, Academy of Science (ITAM), Novosibirsk, Russia: 30.08.–01.09.2000, SFB 259.
Dr. C. Jimenez-Sanchez, University Zaragoza, Zaragoza, Spanien: 15.–25.10.1996, SFB 255.
Ph.D. B.V. Johnson, Hartford University, Manchester, Connecticut, USA: 08.–12.06.2002, SFB 259.
Dr. H.J. Jung, Physikalisch-Technische Bundesanstalt, Berlin, Germany: 27.–28.11.1996, SFB 259.
Prof. Y.S. Kachanov, Academy of Science (ITAM), Novosibirsk, Russia: 01.–03.03.1993/10.06.–13. 07. 2001, SFB 259.
Prof. Kalimin, Universität Minsk, Minsk, Weißrußland: 22.–25.10.1997, SFB 255.
Dr.-Ing. H.-J. Kaltenbach, Technische Universität Berlin: 26.01.2002, SFB 255.
Dipl.-Ing. M. Kalter, TU Berlin, Berlin: 16.–17.12.1997, SFB 255.
Prof. R.K. Kapania, Virginia Polytechnic and State University, USA: 14.–29.11.1996/24.11.–04.12.1997, SFB 259.
Dipl.-Math. Kassmann, Universität Bonn, Bonn: 26.11.1996, SFB 255.
Prof. Dr. D. Keefer, University of Tennessee, Tullahoma, USA: 20.10.–20.11.1991, SFB 259.
Prof. A.M. Kharitonov, Dr. M.D. Brodetsky and other scientists from the Institute of Theoretical and Applied Mechanics, in Novosibirsk, Russia, SFB 253.
Prof. A.M. Kharitonov, Institute of Theoretical and Applied Mechanics, Russian Academy of Sciences, Siberian Divison, Novosibirsk, Rußland: 20.–21.11.1997/22.–24.11.2000, SFB 255.
Prof A. Kharitonov, SB RAS, ITAM, Novosibirsk, Russia: 18.–19.02.2002, SFB 259.
Prof. Dr. Belinda B. King, Interdisciplinary Center for Applied Mathematics, Virginia Tech, Blacksburg, USA: 06.–27.11.1998, SFB 255.
Prof. F.M. Kirillova, Byelorussian Academy of Sciences, Minsk, Weißrußland: 01.–10.1992/23.–28.10.1995/21.01.–04.02.1996, SFB 255.
Prof. M. Kisielewicz, TU Zielona Gora, Zielona Gora, Polen: 14.–17.06.1998, SFB 255.
Prof. Dr. P. G. Klemens, University of Connecticut, USA: 21.02.–06.03.1993, SFB 259.
Prof. R. Klötzler, Univ. Leipzig, Mathematisches Institut: 08.06.–01.07.2001, SFB 255.
Dr. A. Kolesnikov, Academy of Science (IPM), Moscow, Russia: 15.–28.11.1999/07.–12.09.2001, SFB 259.
Prof. V. Kolmanovski, Universität für Elektronik und Mathematik, Moskau, Russland: 16.01.–16.02.1994, SFB 255.
Dipl.-Ing. Konstantin Kondak, Fachgebiet Prozessdatenverarbeitung und Robotik, FB Informatik, TU Berlin: 17.–18.07.2000, SFB 255.
Dr. V. Kopchenov, Central Inst. of Aviation Motors CIAM, Moscow, Russia: 18.–19.02.2002, SFB 259.
Dr. W. Korobow, Charkow: 01.–31.12.1994, SFB 255.
Prof. Dr. A. Koshelev, University St. Petersburg, Russia: 29.11.–24.12.1995, SFB 259.
Dr. A.D. Kosinov, Academy of Science (ITAM), Novosibirsk, Russia: 05.–31.12.1997/16.11.–15.12.1998/28.11.–28.12.1999/02.11.–06.12.2000/16.11.–14.12.2001/14.04.–05.05.2002, SFB 259.
Prof. Olga Kostyukova, Belorussian Academy of Sciences, Minsk, Weißrußland: 12.–22.06.1995/01.10.–15.11.1996/21.09.–02.10.1997, SFB 255.
Prof. Dr. K. Kozel, Faculty of Mechanical Engineering, Prag, Czech Republic: 02.–06.12.1996, SFB 259.

Prof. J.-P. Kremer, Universite Libre de Bruxelles, Brüssel, Belgien: 22.–25.10.1997, SFB 255.
Dipl.-Ing M. Kulkarni, McDonnell Douglas/NASA, USA: 19.–23.12.1994, SFB 255.
Prof. Y. Kurosaki, Tokyo Institute of Technology, Faculty of Engineering, Department of Mechanical Engineering for Production, Tokyo, Japan, SFB 255.
Dr. M. Laitinen, University of Jyvaskylä, Finland: 16.–20.05.1999, SFB 259.
Dr. R. Lamour, Institut für Angewandte Mathematik, Humboldt Universität Berlin: 07.–10.12.1992/09.–13.12.1996, SFB 255.
Dr. V. Lapygin, TsNIIMASh, Korolev, Moscow, Russia: 18.–19.02.2002, SFB 259.
Prof. Dr. V. Lebiga, Academy of Science (ITAM), Novosibirsk, Russia: 10.–17.10.2001, SFB 259.
Prof. G. Leitmann, Berkeley University, Berkeley: 26.05.–01.06.1991, SFB 255.
Prof. Dr. F. Lepoutre, CNRS, Paris, France: 31.03.–03.04.1998, SFB 259.
Dr. M. Lesieur, University Grenoble, Grenoble, Frankreich: 06.–07.12.1997, SFB 255.
Dr. S. Liang, Ecole Centrale de Lyon, Lyon, Frankreich: 15.07.–20.09.1993/08.–22.11.1993, SFB 255.
Prof. Chao-Quiang Lin, Northwestern Polytechnical University Xi'an, China: 1990/91, SFB 255.
Dr. L. Lollini, Ecole Central Lyon, Lyon, Frankreich: 06.–07.12.1997, SFB 255.
Dr. W.C. Loomis, International Corporation, Fountain Valley, CA, USA: 12.–19.10.1996, SFB 259.
Prof. S.J. Losev, Moscow State University, Moscow, Russia: 19.–23.02.1996, SFB 259.
Prof. E. Maiburg, Univ. of Southern California, Los Angeles, USA: 02.–04.06.1997, SFB 255.
Prof. Marek, CTU Prag, Prag, Tschechische Republik: 20.–22.11.1997, SFB 255.
Prof. Dr. A. Maslov, Academy of Science (ITAM), Novosibirsk, Russia: 15.–18.02.1994/31.03.–11.04.2001/20.–25.09.2002/28.09.–11.10.2003, SFB 259.
Prof. J. Mathew, Indian Institute of Science, Bangalore, Indien: 15.05.–31.07.2000/23.11.–02.12.2000/18.06.–31.07.2001, SFB 255.
Prof. Dr. Helmut Maurer, Universität Münster: 18.–20.07.1991, SFB 255.
Prof. K. Mease, Department of Mechanical and Aerospace Engineering, Princeton University, Princeton: 26.05.–04.06.1991, SFB 255.
Prof. K. Mease, Princeton University, USA: 29.06.–01.07.1992, SFB 259.
E. Medvedeva, SB RAS, ITAM, Novosibirsk, Russia: 18.–19.02.2002, SFB 259.
Prof. Dr. A. Melikan, USSR Academy of Science, Moskau: 15.10.–05.11.1989, SFB 255.
Prof. A. Miele, Rice University, Houston: 26.05.–04.06.1991, SFB 255.
Prof. Dr. S. Mikhailov, Stankin University, Moscow, Russia: 16.–31.12.1995, SFB 259.
Dr. L. Mikulski, Institut für Baumechanik, TH Krakau, Polen: 19.–25.11.1992/17.–19.02.1993/23.08.–09.09.1993/06.–20.12.1994/10.–24.03.1996/22.–25.10.1997/19.–23.11.1997, SFB 255.
Prof. G.R. Miller, University of Washington, USA: 18.06.–09.07.1991, SFB 259.
Prof. I. Morozov, Russische Akademie der Wissenschaften, Semenov Institute of Chemical Physics, Moskau, Rußland: 14.–16.07.1995/01.–31.10.1996/04.08.–23.09.1997/31.03.–07.04.2000/07.–11.03.2002, SFB 255.
Prof. C.-D. Munz, Universität Stuttgart, Stuttgart: 15.12.1997, SFB 255.
Prof. R. Naslain, University Bordeaux, France: 18.–19.11.1999, SFB 259.
Prof. I.R. Nigmatulin, USSR Academy of Sciences, Siberian Branch, Tyumen Institute of Mechanics and Multiphase Systems, Tyumen GSP 625, Russland, SFB 255.
Prof. M.S. Nikolskii, Steklov Institute of Mathematics, Moskau, Russland: 07.–14.10.1992, SFB 255.
Prof. M. Nishida, Nagoya University, Japan: 02.10.1996, SFB 259.
Dr. M. Oberlack, RWTH Aachen, Aachen: 16.–18.06.1997, SFB 255.
Dr. D. Obrist, University of Washington, Seattle, USA: 01.04.–31.05.2000, SFB 255.
Prof. G. Pantelidis, Polytechnische Universität, Athen, Griechenland: 01.–07.10.1992/21.–23.11.1997, SFB 255.
Prof. Dr. C. Park, ELORET Corporation, CA, USA: 01.–31.12.1995/15.07.–10.08.1996/15.07.–15.08.1997/01.–31.07.1998/09.–15.11.2001/13.–15.11.2002, SFB 259.

Prof. Dr. A. Paull, University of Queensland, Australia: 12.–13. 12. 1995, SFB 259.
Prof. M. Peric, Univ. Hamburg, Hamburg: 09.–11. 02. 1998, SFB 255.
Prof. L. Petrosjan, Universität St. Petersburg, St. Petersburg: 19.–20. 12. 1993, SFB 255.
V. A. Petrov, Institute of High Temperature, Moscow, Russia: 22.–29. 10. 1995, SFB 259.
Prof. H.X. Phu, Institute of Mathematics Hanoi, Hanoi, Vietnam: 06.–13. 11. 1995/20.–24. 11. 1997, SFB 255.
Prof. C. Pinski, Stanford University, USA: 22. 06.–08. 07. 1995, SFB 259.
Prof. Dr. H. Poppe, Hochschule für Seefahrt, Warnemünde-Wustrow: 05.–10. 11. 1990, SFB 255.
Ir Eddy Priyono, Department of Aeronautics and Astronautics, Faculty of Industrial Technology, Institut of Technology Bandung, Bandung, Indonesien: 25. 08.–13. 10. 2002, SFB 255.
Prof. C.D. Pruett, James Madison University, USA: 16. 06.–28. 06. 2003, SFB 259.
Prof. Dr. N. Qatanani, College of Science and Technology, Jerusalem, Israel: 12.–18. 10. 2001, SFB 259.
Dipl.-Math. U. Rettig, TU Clausthal-Zellerfeld, Clausthal-Zellerfeld: 12. 01. 1998, SFB 255.
G. Riccio, University of Florence Department of Energetics, Italien: 01. 10. 2000–28. 02. 2001, SFB 255.
Prof. G. Rill, Fachhochschule Regensburg, Regensburg: 14. 11. 1994, SFB 255.
Prof. M. Rosenblatt, University of California, San Diego, USA: 31. 10.–15. 11. 1992, SFB 255.
Dr. Rosenfeldt, SCHENCK PEGASUS GmbH, Darmstadt: 18. 03. 1999, SFB 255.
Dr. T. Roubícek, Karls-Universität Prag, Prag, Tschechische Republik: 19. 02.–01. 03. 1996/03.–12. 07. 1998, SFB 255.
Dr. A. Rudakov, Central Inst. of Aviation Motors CIAM, Moscow, Russia: 18.–19. 02. 2002, SFB 259.
Prof. Rudolph, HTWK Leipzig/Leopoldina Halle: 01. 04.–31. 08. 1993, SFB 255.
Prof. V. Sabelnikow, Central Aerohydrodynamic Institute (TsAGI), Moscow, Russia: 12. 12. 1997, SFB 259.
Prof. P. Safonov, Russische Akademie der Wissenschaften, Moskau, Russland: 30. 11. 1993, SFB 255.
Dr. S. Sakar, ICASE, NASA Langley, Hampton, VA, USA: 10.–19. 07. 1992/10.–22. 10. 1993, SFB 255.
Dr. N. Sandham, DFLR-Institut für Theoretische Strömungsmechanik: 09.–11. 06. 1991, SFB 255.
Dr. K. Schadow, Naval Air Warefare Center Weapons Division, China Lake, USA: 26. 11.–02. 12. 1994, SFB 255.
Dr. J. Schellekens, TU Delft, NL: 01. 03.–30. 06. 1993/10.–22. 12. 1992, SFB 259.
Prof. Dr. K. Schittkowski, Mathematisches Institut Universität Bayreuth: 20. 03. 1990, SFB 255.
Ph. D.S. Schlamp, ETH Zürich, Institut für Fluiddynamik, Suisse: 16. 05. 2002/16.–19. 07. 2002/29.–31. 07. 2002/20.–23. 08. 2002, SFB 259.
Prof. P. Schmid, University of Washington, Seattle, USA: 28.–29. 05. 2001, SFB 255.
Prof. Dr. S.P. Schneider, University of Purdue, West Lafayette, USA: 11.–13. 05. 2000, SFB 259.
Prof. G. Schnerr, Universität Karlsruhe, Karlsruhe: 11. 06. 1996, SFB 255.
Dr. H. Schramm, Mathematisches Institut Universität Bayreuth: 20. 03. 1990, SFB 255.
Prof. Dr. C. Schwab, University of Maryland, Baltimore, USA: 13.–30. 08. 1991/01.–14. 08. 1992, SFB 259.
Dr. C.D. Scott, NASA Johnson Space Centre, USA: 20.–23. 09. 1992, SFB 259.
Prof. T. Sengupta, Indian Institute of Technology, Kanpur, Indien: 08. 05.–12. 06. 2001, SFB 255.
Dipl.-Math. R. Seppelt, Sonderforschungsbereich 179, Braunschweig: 16. 12. 1996, SFB 255.
Dr. J. Sesterhenn, ETH Zürich, Zürich, Schweiz: 13.–15. 02. 1996, SFB 255.
Prof. E. Shafranovsky, Russische Akademie der Wissenschaften, Semenov Institute of Chemical Physics, Moskau, Russland: 01. 07.–30. 09. 2000/11. 09.–31. 12. 2001/15. 01.–31. 03. 2002, SFB 255.

Dr. Ev. Shestakow, Academy of Science, Moscow, Russia: 15.–20.12.1993, SFB 259.
Prof. Shinar, Technion, Israel Institut of Technology, Haifa, Israel: 06.–08.12.1992, SFB 255.
Dr. D. Silin, Staatliche Universität Moskau, Moskau, Russland: 16.–19.07.1992, SFB 255.
Prof. Simpson, Virginia Polytechnic Inst, Bladesburg: 07.–08.05.1990, SFB 255.
Prof. A.M. Slinko, Russische Akademie der Wissenschaften, Moskau, Russland: 05.–09.08.1992, SFB 255.
Prof. Dr. I. Smirnov, Samara State Aerospace University, Russia: 15.–20.12.1997, SFB 259.
Prof. B. Smith, University of Chicago, USA: 25.10.–08.11.1992, SFB 259.
R. Smith, früher, NASA Dryden Flight Research Center, USA: 16.05.2001, SFB 255.
Prof. Dr. D.R. Smith, University of California, San Diego, Vorlesung über Theorie und Anwendung der singulären Störungen (DFG-Gastprofessur): 15.10.89–15.01.90, SFB 255.
Prof. Dr. D.R. Smith, University of California, San Diego: 05.–09.06.1991, SFB 255.
Dr. B. Smorodsky, Academy of Science (ITAM), Novosibirsk, Russia: 01.08.–31.10.1999/01.09.–31.10.2000/15.09.–15.12.2001/03.11.2002–03.01.2003, 16.11.2003–16.02.2004, SFB 259.
Prof. T. Soga, Nagoya University, Japan: 02.10.1996, SFB 259.
Prof. L. Hung Son, University of Hanoi, Vietnam: 01.08.–30.10.1991, SFB 259.
Dipl.-Ing. T. Spägele, MTU Friedrichshafen, Friedrichshafen: 17.12.1996, SFB 255.
Prof. J. Steinhoff, The University of Tennessee Space Institute, Tullahoma, SFB 253.
Prof. Sundberg, Universität Växjö, Växjö, Schweden: 18.–20.05.1997, SFB 255.
Prof. J. Suris, Universität Bremen, Bremen: 17.–19.01.1994, SFB 255.
Prof. A.J. Surkan, University of Nebraska, Lincoln, USA: 28.06.–01.07.1992, SFB 255.
PhD. PEng.Th. Szirtes, University of Toronto, Canada: 23.–27.11.1999, SFB 259.
Dr. Y. Takakura, Tokyo Noko University, Tokyo, Japan: 20.–26.08.2000, SFB 259.
Prof. Dr. K. Takayama, Tohoku University, Sendai, Japan: 14.–16.07.1998, SFB 259.
Prof. Dr. N. Tanatsugu, Institute of Space and Astronautical Science, Kanagawa, Japan: 17.–21.10.1994, SFB 259.
Dr. A.M. Tereza, SB RAS ITAM, Moscow, Russia: 07.01.–23.03.1992, SFB 259.
Dr. T. Tiihonen, University of Jyvaskylä, Finland: 16.–18.02.1998, SFB 259.
Prof. H. Tokunaga, Kyoto Institute of Technology, Kyoto, Japan: 27.–31.07.1997, SFB 255.
Prof. Turchak, Akademie der Wissenschaften, Moskau: 31.10.1990, SFB 255.
Dr. R. Tye, Consultant, London, England: 14.–15.12.1995, SFB 259.
Dr. A. Ulanov, Samara State Aerospace University, Russia: 26.10.–11.11.1997, SFB 259.
Prof. M. Usami, Mie University, Japan: 02.10.1996, SFB 259.
Dr. L. Vervisch, Université de Rouen, Frankreich: 18.–19.12.1994, SFB 255.
Prof. Vesely, CTU Prag, Prag, Tschechische Republik: 21.–22.11.1997, SFB 255.
Prof. N.X. Vinh, University of Michigan: 26.05.–01.06.1991, SFB 255.
Dr. Vitek, University Prag, Czech Republic: 21.09.–10.10.1992, SFB 259.
Prof. Vollhein, Wissenschaftsbereich Strömungsmechanik TU Dresden, Dresden: 28.–29.01.1990, SFB 255.
Prof. E. von Lavante, Univ. Essen, Essen: 11.–13.12.1995, SFB 255.
Dr. E.V. Vorozhtsov, Institut für theoretische und angewandte Mechanik, Akademie der Wissenschaften, Novosibirsk: 09.07.1991, SFB 255.
L. Walberg, Saab Military Aircraft, Linkoping, Schweden: 12.12.1996, SFB 255.
Dipl.-Ing. F. Wang, TU Peking, April–Okt. 1991, SFB 255.
Dipl.-Ing. C. Weinberger, TH Darmstadt, Darmstadt: 21.–22.02.1996/12.–13.10.1997, SFB 255.
Dr. Wille, IWR, Universität Heidelberg, Heidelberg: 23.–26.01.1994, SFB 255.
Prof. D.R. Williams, Illinois Institute of Technology, Chicago, USA: 24.–28.04.1995, SFB 259.
Prof. V.A. Yaroshevsky, Central Aerohydrodynamik Institute (TsAGI), Moscow, Russia: 05.–12.07.1995, SFB 259.
Prof. B. Younis, University of California, USA: 03.–15.09.2003, SFB 259.
Prof. Zampieri, Universidade Estadual de Campinas, z. Zt. Fachhochschule Regensburg, Brasilien: 21.–22. Juni 1993, SFB 255.
Prof. I. Zaslonko, Russische Akademie der Wissenschaften, Moskau, Russland: 06.–08.11.1997/01.03.–30.04.1999, SFB 255.

Prof. Fan Zhaolin, Direktor des Instituts für experimentelle Hochgeschwindigkeitsaerodynamik, China Aerodynamic Research and Development Center (CARDC), Mianyang, China: 21. 02. 2001, SFB 255.

Prof. G.M. Zharkova, Academy of Science (ITAM), Novosibirsk, Russia: 30.08.–01.09.2000, SFB 259.

## 10.7 Organization and Projects

### Collaborative Research Centre 253

#### 1. Chairmen

Prof. Egon Krause, Ph.D. (1989–1998)
Lehrstuhl für Strömungslehre und Aerodynamisches Institut

Prof. Dr.-Ing. Dieter Jacob (1998–2003)
Lehrstuhl und Institut für Luft- und Raumfahrt

#### 2. Institutions

Lehrstuhl für Luft- und Raumfahrt, RWTH Aachen (A1, A4)
Lehr- und Forschungsgebiet Hochtemperatur-Gasdynamik, RWTH Aachen (A2, B3)
Lehrstuhl für Strömungslehre, RWTH Aachen (A3, B1)
Lehrstuhl für Flugdynamik, RWTH Aachen (A 4.2, A7)
Lehr- und Forschungsgebiet für Mechanik, RWTH Aachen (A5, B5, C3)
Lehrstuhl für Leichtbau, RWTH Aachen (A6)
Deutsches Zentrum für Luft- und Raumfahrt, Abteilung Windkanäle, Köln (B2, C8)
Lehr- und Forschungsgebiet Betriebsverhalten der Strahlantriebe, RWTH Aachen (C2, C5)
Lehrstuhl für Strahlantriebe und Turboarbeitsmaschinen, RWTH Aachen (C1, C4)
Lehrstuhl für Industrieofenbau und Wärmetechnik im Hüttenwesen, RWTH Aachen (C7)
Lehrstuhl für Mathematik, RWTH Aachen (B5)
Lehr- und Forschungsgebiet Elektrische Antriebe in der Raumfahrt, RWTH Aachen (C6)

3. Research Projects and Projects Leaders

*Project Area A: Hyperschallkonfigurationen*

Research Project A1 (1989–2003)
Konzepte und Entwurfsanalysen von Hyperschallkonfigurationen
Prof. Dr.-Ing. D. Jacob (1994–2003)
Prof. Dr.-Ing. R. Staufenbiel (1989–1994)
Dr.-Ing. D. Coors (1989–1999)

Research Project A2 (1989–2001)
Messung aerothermodynamischer Größen an der Oberstufe
Prof. Dr.-Ing. H. Olivier (1996–2001)
Prof. Dr. rer. nat. H. Grönig (1989–1996)

Research Project A3 (1989–2003)
Berechnung von Strömungsfeldern für Komponenten von Raumflugzeugen
Prof. Dr.-Ing. W. Schröder (1998–2003)
Prof. E. Krause, Ph. D. (1989–1998)
Dr.-Ing. M. Meinke (1996–2003)

Research Project A4 (1989–2001)
Aerodynamik von Hyperschallkonfigurationen im Langsamflugbereich
Prof. Dr.-Ing. D. Jacob (1994–2001)
Prof. Dr.-Ing. R. Staufenbiel (1989–1994)
Dr.-Ing. G. Neuwerth (1989–2001)

Research Project A4.2 (1989–1998)
Aerodynamik von Hyperschallkonfigurationen im Langsamflugbereich (Flugmechanik)
Prof. Dr.-Ing. W. Alles (1997–1998)
Prof. Dr.-Ing. R. Staufenbiel (1989–1994)
Prof. Dr.-Ing. D. Jacob (1994–1997)
Dr.-Ing. W. Kasberg (1989–2001)

Research Project A5 (1989–1995)
Hypersonische Konfigurationen im Langsamflug
Prof. Dr.-Ing. J. Ballmann (1989–1995)

Research Project A6 (1989–2001)
Strukturgewicht
Prof. Dr.-Ing. H.-G. Reimerdes (1992–2001)
Prof. Dr.-Ing. Dr. h.c. (H) H. Öry (1989–1992)
Dr.-Ing. J. Albus (1996–1998)

Research Project A7 (1999–2002)
Stabilität und Steuerung von Hyperschallkonfigurationen im Langsamflugbereich
Prof. Dr.-Ing. W. Alles (1999–2002)
Dr.-Ing. W. Kasberg (1999–2001)

*Project Area B: Hyperschallanströmung hochbelasteter Flächen*

Research Project B1 (1989–2001)
Experimentelle Untersuchungen an Modellen von Raumflugzeugen
Prof. Dr.-Ing. W. Schröder (1998–2001)
Prof. E. Krause, Ph. D. (1989–1998)
Dr. rer. nat. W. Limberg (1989–2001)

Research Project B2 (1989–2003)
Experimentelle Optimierung eines Hyperschall-Einlaufs
Dr.-Ing. A. Henckels (DLR) (1996–2003)
Dr.-Ing. F. Maurer (DLR) (1989–1995)

Research Project B3 (1989–2003)
Wechselwirkung der Rumpfgrenzschicht mit Anbauten
Prof. Dr.-Ing. H. Olivier (1996–2003)
Prof. Dr. rer. nat. H. Grönig (1989–1996)

Research Project B5 (1989–1995)
Berechnung hypersonischer Staupunktströmungen bei hohen Temperaturen
unter Berücksichtigung chemischer Reaktionen
Prof. Dr. sc. math. R. Jeltsch (1989–1994)
Prof. Dr.-Ing. J. Ballmann (1992–1995)

## *Project Area C: Einlauf- und Düsenströmung*

Research Project C1 (1989–2001)
Theoretische Grundlagenstudie über Antriebskonzepte für Raumtransporter
Prof. Dr.-Ing. R. Niehuis (1998–2001)
Prof. Dr.-Ing. H.E. Gallus (1989–1998)
Dr.-Ing. W. Kühl (1989–2000)

Research Project C2 (1989–2001)
Experimentelle Untersuchung des Betriebsverhaltens
eines Triebwerkseinlaufs bei Überschall-Anström-Machzahlen
Prof. Dr.-Ing. W. Koschel (1989–2001)

Research Project C3 (1989–2003)
Berechnungsverfahren für Einlaufströmungen luftatmender Triebwerke
von Raumtransportern bei M>1
Prof. Dr.-Ing. J. Ballmann (1989–2003)

Research Project C4 (1989–1995)
Grundlagenuntersuchung zur aerothermodynamischen Anpassung
von Einlaufdiffusor und nachgeschaltetem Triebwerk (TL, SL)
unter Berücksichtigung variabler Geometrie
Prof. Dr.-Ing. H.E. Gallus (1989–1995)

Research Project C5 (1989–2001)
Berechnung der Expansionsströmung eines Hyperschall-Antriebssystems
unter Berücksichtigung der Realgaseffekte eines Wasserstoff-/Luft-Verbrennungsgases
Koschel (1989–2001)

Research Project C 6 (1989–1998)
Wechselwirkung flacher Überschalldüsen-Düsenfreistrahlen
mit Wänden und Außenströmung
Prof. Dr.-Ing. A.E. Beylich (1989–1998)

Research Project C7 (1998–1991)
Thermische Wandbelastung im Schubdüsenbereich bei Regenerations-,
Transpirations- und Filmkühlung
Prof. Dr.-Ing. H. Wilhelmi (1989–1991)

Research Project C8 (1993–2001)
Experimentelle Untersuchung einer Düsenströmung für Hyperschallantriebe
Prof. Dr.-Ing. H. Weyer (DLR) (1993–2001)
Dr.-Ing. A. Henckels (DLR) (1999–2001)

## Collaborative Research Centre 255

### 1. Chairman

Prof. Dr.-Ing. Gottfried Sachs (1989–2003)
Lehrstuhl für Flugmechanik und Flugregelung

### 2. Institutions

Lehrstuhl für Fluidmechanik, TU München (A1, A2, A3)
Institut für Strömungsmechanik, UniBw München (A4)
Institut für Luftfahrttechnik und Leichtbau, UniBw München (A5)
Lehrstuhl für Flugantriebe, TU München (B1, B4, B12)
Lehrstuhl A für Thermodynamik, TU München (B3, B10, B11)
Lehrstuhl B für Thermodynamik, TU München (B4)
Lehrstuhl für Werkstoffe im Maschinenbau, TU München (B6)
Institut für Strahlantriebe, UniBw München (B9)
Lehrstuhl für Flugmechanik und Flugregelung, TU München (C1, C4)
Institut für Dynamik der Flugsysteme, DLR, Oberpfaffenhofen (C2)
Lehrstuhl für Höhere und Numerische Mathematik, TU München (C3)
Lehrstuhl für Raumfahrttechnik, TU München (C6)
Lehrstuhl für Numerische Mathematik und Optimierungstheorie,
Ernst-Moritz-Arndt-Universität Greifswald (C7)

### 3. Research Projects and Projects Leaders

#### *Project Area A: Aerothermodynamik*

Research Project A1 (1989–2001)
Instationäre Hyperschallströmungen
– zwei- und dreidimensionale Berechnungsmethoden
Prof. Dr.-Ing. B. Laschka (1989–2001)
Akad. Oberrat Dipl.-Ing. F. Matyas (1989–1998)
Dr.-Ing. C. Breitsamter (1998–2001)

Research Project A1 (2002–2003)
Instationäre Hyperschallströmungen bei Simulation der dynamischen Stufentrennung – Zwei- und dreidimensionale Berechnungsmethoden
Prof. Dr.-Ing. B. Laschka (2002–2003)
Dr.-Ing. C. Breitsamter (2002–2003)

Research Project A2 (1989–1995)
Berechnung komplexer zweidimensionaler turbulenter
Über- und Hyperschallgrenzschichten
Prof. Dr.-Ing. habil. R. Friedrich (1989–1995)

Research Project A2 (1995–2001)
Berechnung turbulenter Verbrennungsvorgänge in kompressiblen Scherschichten
Prof. Dr.-Ing. habil. R. Friedrich (1995–2001)

Research Project A2 (2002–2003)
PDF/FDF-Methoden zur Prognose turbulenter Überschallverbrennung
Prof. Dr.-Ing. habil. R. Friedrich (2002–2003)

Research Project A4 (1989–1991)
Experimentelle Ermittlung stoßbedingter Realgaseffekte
unter besonderer Berücksichtigung der Umweltbeeinflussung
Prof. Dr. rer. nat. F.J. Hindelang (1989–1991)

Research Project A5 (1989–1991)
Experimentelle Untersuchung von Wechselwirkungen
zwischen Verdichtungsstößen, Grenzschicht und Absaugung
in supersonischen Einläufen
Prof. Dr.-Ing. S. Wagner (1989–1991)

Research Project A5 (1992–1995)
Experimentelle Untersuchung von Stoß-Grenzschicht-Wechselwirkungen
und Stoß-Stoß-Wechselwirkungen bei turbulenten Grenzschichten
an variablen Geometrien
Prof. Dr. rer. nat. F.J. Hindelang (1992–1995)
Prof. Dr.-Ing. S. Wagner (1992–1995)

Research Project A5 (1995–2001)
Experimentelle Untersuchungen zu Stoß/Grenzschicht-Wechselwirkungen
und multiplen Wechselwirkungen bei turbulenten Grenzschichten
Prof. Dr.-Ing. W. Staudacher (1995–2001)

Research Project A6 (1992–1995)
Störungsanfachung im Nasenbereich abgestumpfter Körper im Überschall
Prof. Dr.-Ing. habil. R. Friedrich (1992–1995)
Prof. Dr.-Ing. B. Laschka (1992–1995)

Research Project A6 (1995–2001)
Dreidimensionale Störungsanfachung im Nasenbereich abgerundeter Flugkörper
im Hyperschall
Prof. Dr.-Ing. habil. R. Friedrich (1995–2001)

### *Project Area B: Antriebssysteme*

Research Project B1 (1989–1995)
Leistungsberechnung und Betriebsverhaltensanalyse luftatmender Antriebssysteme
für Raumtransporter/Hyperschall-Flugzeuge
Prof. Dr.-Ing. habil. H. Rick (1989–1995)

Research Project B1 (1995–1998)
Leistungsberechnung und Betriebsverhaltensanalyse luftatmender Antriebssysteme
für Hyperschallflugzeuge im Normalbetrieb und bei Störungen
Prof. Dr.-Ing. habil. H. Rick (1995–1998)

Research Project B1 (1998–2001)
Stationäres und instationäres Betriebsverhalten der Antriebssysteme
für Hyperschallflugzeuge unter Berücksichtigung von Störungen
Prof. Dr.-Ing. habil. H. Rick (1998–2001)

Research Project B3 (1989–1991)
Gemischbildung, Reaktionskinetik und Flammenstabilisierung
in Hochgeschwindigkeitsströmungen
Prof. Dr.-Ing. Dr.-Ing. E. h. F. Mayinger (1989–1991)

Research Project B3 (1992–1995)
Gemischbildung, Reaktionskinetik und Flammenstabilisierung
in Hochgeschwindigkeits-Wasserstoff-Luft-Flammen
Prof. Dr.-Ing. Dr.-Ing. E. h. F. Mayinger (1992–1995)

Research Project B4 (1989–1995)
Organisation der Überschallverbrennung in Überschallantrieben
Prof. Dr.-Ing. G. Kappler, M. Sc. (1989–1995)
Prof. Dr. rer. nat. D. Vortmeyer (1989–1995)

Research Project B6 (1989–1995)
Auswahl und Optimierung von keramischen Hochleistungswerkstoffen
zur Isolation von thermisch hochbelasteten Bauteilen
Prof. Dr.-Ing. K. G. Schmitt-Thomas (1989–1995)

Research Project B6 (1995–1998)
Verfahrens- und werkstofftechnische Optimierung
von keramischen Wärmedämmschichtsystemen
Prof. Dr.-Ing. K. G. Schmitt-Thomas (1995–1998)

Research Project B9 (1995–1998)
Leistungsverhalten des Turboteils von Kombinationsantrieben
unter dem Einfluss von Eintrittsstörungen
Prof. Dr.-Ing. L. Fottner (1995–1998)

Research Project B9 (1998–2001)
Stationäres und instationäres Leistungsverhalten des Turboteils
von Kombinationsantrieben unter dem Einfluss von Eintrittsstörungen
Prof. Dr.-Ing. L. Fottner (1998–2001)

Research Project B10 (1995–1998)
Untersuchung der physikalischen Mechanismen zur Stabilisierung
von Kerosin/Wasserstoff-Überschallflammen – "Dual Fuel Concept"
Prof. Dr.-Ing. Dr.-Ing. E. h. F. Mayinger (1995–1998)

Research Project B11 (1998–2003)
Selbstzündung, Flammenstabilisierung und Stoß-Flammen-Wechselwirkung
in stoß-induzierten Überschallflammen
Prof. Dr.-Ing. T. Sattelmayer (1998–2003)

Research Project B12 (1998–2001)
Teillastverhalten einer Überschallbrennkammer mit Wasserstoffverbrennung
in Über- und Hyperschallantrieben
Prof. Dr.-Ing. H.-P. Kau (1998–2001)
Prof. Dr.-Ing. Rist (1998–2001)

Research Project B12 (2002–2003)
Pilotierte Überschallverbrennung
Prof. Dr.-Ing. H.-P. Kau (2002–2003)

*Project Area C: Flugmechanik und Gesamtsystem*

Research Project C1 (1989–1995)
Aufstiegs-Reichweiten-Bahnoptimierung für transatmosphärische Flugsysteme
Prof. Dr.-Ing. G. Sachs (1989–1995)
Prof. Dr.-Ing. habil. O. Wagner (1989–1995)

Research Project C1 (1995–1998)
Optimierung von Flugleistungen transatmosphärischer Flugsysteme
Prof. Dr.-Ing. G. Sachs (1995–1998)
Prof. Dr.-Ing. habil. O. Wagner (1995–1998)

Research Project C1 (1998–2001)
Flugbahnoptimierung für transatmosphärische Flugsysteme
Prof. Dr.-Ing. G. Sachs (1998–2001)
Prof. Dr.-Ing. habil. O. Wagner (1998–2001)

Research Project C2 (1989–1991)
Optimierung und Lenkgesetz-Entwurf für Wiedereintritt und Bahnebenen-Drehung
Prof. Dr.-Ing. K.H. Well (1989–1991)

Research Project C3 (1989–2001)
Berechnungsmethoden für optimale Echtzeit- und Rückkopplungssteuerungen
sowie für optimale Systemstufung
Prof. Dr. rer. nat. Dr. h.c. mult. R. Bulirsch (1989–2001)

Research Project C3 (2002–2003)
Gesamtsystembezogene Flugbahnoptimierung
Prof. Dr. rer. nat. Dr. h.c. mult. R. Bulirsch (2002–2003)
Priv.-Doz. Dr. rer. nat. habil. Dr.-Ing. R. Callies
Prof. Dr.-Ing. G. Sachs

Research Project C4 (1989–1995)
Flugdynamik- und Flugeigenschaftsprobleme im Hyperschall
Prof. Dr.-Ing. G. Sachs (1989–1995)

Research Project C4 (1995–1998)
Flugdynamik, Flugeigenschaften und Flugführung von Hyperschall-Flugsystemen
Prof. Dr.-Ing. G. Sachs (1995–1998)

Research Project C4 (1998–2001)
Flugeigenschaften, Steuerung und Flugführung von transatmosphärischen Flugsystemen
Prof. Dr.-Ing. G. Sachs (1998–2001)

Research Project C4 (2002–2003)
Flugdynamische Probleme der Stufentrennung von transatmosphärischen Flugsystemen
Prof. Dr.-Ing. G. Sachs (2002–2003)

Research Project C5 (1989–1991)
Grundlagen für ein rechnergestütztes Entwurfsverfahren
für transatmosphärische Flugsysteme
Prof. Dr.-Ing. S. Wagner (1989–1991)

Research Project C6 (1989–1995)
Raumfahrttechnische Bewertung transatmosphärischer Flugsysteme
Prof. Dr.-Ing. H. O. Ruppe (1989–1991)
Prof. Dr.-Ing. E. Igenbergs (1992–1995)

Research Project C7 (1992–1995)
Existenz von Optimalsteuerungen und hinreichenden Optimalitätsbedingungen
Prof. Dr. L. Bittner (1992–1995)
Prof. Dr. rer. nat. W. Schmidt (1992–1995)

Research Project C7 (1995–1998)
Existenz von Optimalsteuerungen und hinreichende Optimalitätsbedingungen
für transatmosphärische Steuerprobleme und Lösung zugehöriger Syntheseaufgaben
Prof. Dr. L. Bittner (1995–1998)
Prof. Dr. rer. nat. W. Schmidt (1995–1998)

Research Project C8 (1998–2003)
Integration und Auswirkung der Ergebnisse der Teilprojekte
des Sonderforschungsbereichs 255 auf den Gesamtentwurf
eines Transatmosphärischen Raumtransportsystems (HTSM)
Prof. Dr.-Ing. W. Staudacher (1998–2003)

## Collaborative Research Centre 259

### 1. Chairmen

Prof. Dr. rer. nat. Ernst Messerschmid (1989–2000)
Institut für Raumfahrtsysteme

Prof. Dr.-Ing. Siegfried Wagner (2000–2003)
Institut für Aerodynamik und Gasdynamik

### 2. Institutions

Institut für Raumfahrtsysteme, Universität Stuttgart (A1, A3, C3, C8)
Institut für Plasmaforschung, Universität Stuttgart (A2)
Institut für Nichtmetallische Anorganische Materialien, Universität Stuttgart (A5)
Institut für Werkstoffwissenschaft Pulvermetallurgisches Laboratorium,
Max-Planck-Institut für Metallforschung, Stuttgart (A5)
Institut für Bauweisen- und Konstruktionsforschung,
Deutsches Zentrum für Luft- und Raumfahrt DLR Stuttgart (B1, B8)
Institut für Flugzeugbau, Universität Stuttgart (B2)
Institut für Kernenergetik und Energiesysteme, Universität Stuttgart (B3)
Institut für Statik und Dynamik der Luft- und Raumfahrtkonstruktionen,
Universität Stuttgart (B4, B5, B7, B8)
Mathematisches Institut A, Universität Stuttgart (B6, B7)
Institut für Thermodynamik, Universität Stuttgart (C1, C7)
Institut für Technische Verbrennung, Universität Stuttgart (C2)
Institut für Aerodynamik und Gasdynamik, Universität Stuttgart (C4, C5, C11)
Institut für Luftfahrtantriebe, Universität Stuttgart (C6)

3. Research Projects and Projects Leaders

*Project Area A: Thermophysikalische Grundlagen des Wiedereintritts*

Research Project A1 (1990–2001)
Modellierung des Plasmawindkanals
Prof. Dr. rer. nat. E. Messerschmid
Prof. Dr.-Ing. S. Fasoulas

Research Project A2 * (1990–2001)
Spektroskopie und Laserdiagnostik
Prof. Dr.-Ing. K. Behringer
Prof. Dr. rer. nat. U. Schumacher
Dr.-Ing. K. Hirsch

Research Project A3 * (1990–2001)
Diagnostik im Plasmawindkanal mit Sonden, Interferometrie und Massenspektrometrie
Prof. Dr.-Ing. M. Auweter-Kurtz

Research Project A5 * (1993–2001)
Oberflächenschutz von Si-infiltriertem CFC
Prof. Dr. rer. nat. F. Aldinger
Priv.-Doz. Dr. rer. nat. J. Bill

*Project Area B: Hochtemperaturwerkstoffe, heiße tragende Strukturen, Wärmeschutzsysteme*

Research Project B1 (1990–2003)
Materialentwicklung und Herstelltechnologie für C/C-SiC Bauteile
Prof. Dipl.-Ing. R. Kochendörfer

Research Project B2 (1990–2001)
Thermomechanisches Verhalten von C/C-SiC
Prof. Dipl.-Ing. F.-J. Arendts
Priv.-Doz. Dr.-Ing. K. Maile

Research Project B3 * (1990–2001)
Thermisches Verhalten von C/C-SiC
Dr.-Ing. G. Neuer

Research Project B4 (1990–2003)
Berechnung des thermomechanischen Verhaltens,
Mikrophysikalisch begründete Stoffmodelle für faserverstärkte Keramik
Prof. Dr.-Ing. D. Dinkler
Prof. Dr.-Ing. B.-H. Kröplin

Research Project B5 (1990–1995)
Dreidimensionale Schalentheorie
Prof. Dr.-Ing. B.-H. Kröplin
Prof. Dr.-Ing. W. Wendland

* 2002–2003
Support continued by the Government of Baden-Württemberg
"Schutzschichtentwicklung für wiederverwendbare Raumtransportsysteme"
("Heat Protection Layers for Re-Usable Space Transportation Systems")

Research Project B6 (1990–1995)
Entwicklung eines Makro-Schalenelements
Prof. Dr.-Ing. W. Wendland
Prof. Dr.-Ing. D. Dinkler

Research Project B7 (1990–2001)
Berechnung instationärer Wärmeleitung und Strahlung in Hitzeschutzsystemen
Prof. Dr.-Ing. D. Dinkler
Prof. Dr.-Ing. B.-H. Kröplin
Prof. Dr.-Ing. W. Wendland

Research Project B8 (1996–2003)
Hochtemperaturbauteil – Entwurf und Prüfung
Prof. Dr.-Ing. B.-H. Kröplin
Prof. Dipl.-Ing. R. Kochendörfer

### *Project Area C: Hochtemperatur-Aerodynamik*

Research Project C1 (1990–1995)
Relaxationsverhalten realer Gase
Prof. Dr. rer. nat. A. Frohn

Research Project C2 (1990–1995)
Thermische Nichtgleichgewichtseffekte
Prof. Dr. rer. nat. J. Warnatz

Research Project C3 * (1990–2003)
Numerische Wiedereintritts-Aerothermodynamik
Dr.-Ing. H.-H. Frühauf
Prof. Dr.-Ing. M. Auweter- Kurtz
Prof. Dr.-Ing. M. Resch
Dr.-Ing. A. Geiger

Research Project C4 (1990–2003)
Grenzschichtumschlag bei Überschallströmung
Prof. Dr.-Ing. H. Fasel
Dr.-Ing. H. Bestek
Prof. Dr.-Ing. S. Wagner
Dr.-Ing. M. Kloker

Research Project C5 (1990–2003)
Experimentelle Untersuchungen an Verdichtungsflächen und Triebwerkseinläufen für Hyperschallflugkörper
Dr.-Ing. G. Schwarz
Prof. Dr.-Ing. S. Wagner
Dr.-Ing. H. Knauss

Research Project C6 (1990–2000)
Leistungsverhalten von Hyperschall-Antrieben mit luftatmenden Turbokomponenten bei hohen Eintrittstemperaturen
Prof. Dr.-Ing. W. Braig

* 2002–2003
Support continued by the Government of Baden-Württemberg
"Schutzschichtentwicklung für wiederverwendbare Raumtransportsysteme"
("Heat Protection Layers for Re-Usable Space Transportation Systems")

Research Project C7 (1990–2003)
Vergleichende theoretische und experimentelle Untersuchungen zu Mischung, Zündung und Verbrennung in Staustrahltriebwerken für den Hyperschallflug
Prof. Dr.-Ing. J. Algermissen
Prof. Dr.-Ing. D. Brüggemann
Prof. Dr.-Ing. B. Weigand
Prof. Dr.-Ing. M. Aigner

Research Project C8 (1990–1995)
Missions- und Systemanalyse wiederverwendbarer Raumfahrtträger
Dr.-Ing. U. Schöttle

Research Project C11 (1999–2003)
Direkte numerische Simulation der Stoß/Grenzschicht-Interaktion im Hyperschall
Priv.-Doz. Dr.-Ing. U. Rist
Prof. Dr.-Ing. S. Wagner